OCCUPATIONAL SAFETY AND HYGIENE V

SELECTED CONTRIBUTIONS FROM THE INTERNATIONAL SYMPOSIUM OCCUPATIONAL SAFETY AND HYGIENE (SHO 2017), GUIMARÃES, PORTUGAL, 10–11 APRIL 2017

Occupational Safety and Hygiene V

Editors

Pedro M. Arezes
University of Minho, Guimarães, Portugal

João Santos Baptista
University of Porto, Porto, Portugal

Mónica P. Barroso, Paula Carneiro, Patrício Cordeiro
& Nélson Costa
University of Minho, Guimaraes, Portugal

Rui B. Melo
Technical University of Lisbon, Cruz Quebrada—Dafundo, Portugal

A. Sérgio Miguel & Gonçalo Perestrelo
University of Minho, Guimarães, Portugal

 **CRC Press**
Taylor & Francis Group
Boca Raton   London   New York   Leiden

CRC Press is an imprint of the
Taylor & Francis Group, an **informa** business

A BALKEMA BOOK

CRC Press/Balkema is an imprint of the Taylor & Francis Group, an informa business

© 2017 Taylor & Francis Group, London, UK

Typeset by V Publishing Solutions Pvt Ltd., Chennai, India
Printed and bound in Great Britain by CPI Group (UK) Ltd, Croydon, CR0 4YY.

Published by: CRC Press/Balkema
        P.O. Box 11320, 2301 EH Leiden, The Netherlands
        e-mail: Pub.NL@taylorandfrancis.com
        www.crcpress.com – www.taylorandfrancis.com

ISBN: 978-1-138-05761-6 (Hbk)
ISBN: 978-1-315-16480-9 (eBook)

Occupational Safety and Hygiene V – Arezes et al. (Eds)
© *2017 Taylor & Francis Group, London, ISBN 978-1-138-05761-6*

Table of contents

Occupational Safety and Hygiene V – Arezes et al. (Eds)
© *2017 Taylor & Francis Group, London, ISBN 978-1-138-05761-6*

Foreword

This book is the fifth volume of the "Occupational Safety and Hygiene" series. It presents a selection or 112 articles submitted to SHO2017—International Symposium on Occupational Safety and Hygiene, the 13th edition, which is annually organised by the Portuguese Society of Occupational Safety and Hygiene (SPOSHO). These articles were written by 395 authors from 12 different countries. Each manuscript was peer reviewed by at least 2 of the 110 members of the International Scientific Committee of the Symposium. These international experts cover all scientific fields of the event.

The editors would like to take this opportunity to thank the academic partners of the organisation of SHO2017's, namely, the School of Engineering of the University of Minho, the Faculty of Engineering of the University of Porto, the Faculty of Human Kinetics of the University of Lisbon, the Polytechnic University of Catalonia and the Technical University of Delft. We also would like to thank the scientific sponsorship of more than 20 academic and professional institutions, the official support of the Portuguese Authority for Working Conditions (ACT), as well as the valuable support of several companies and institutions, including the media partners, which have contributed to the broad dissemination of the event. Finally, the editors wish also to thank all the reviewers, listed below, which were involved in the process of reviewing and editing the included papers. Without them, this book would not be possible.

To conclude, we hope that this book will be a valuable contribution to improving the results and dissemination of research by academics involved in SHO2017. It is work done in different areas, showing new research and methodologies, giving visibility to emerging issues and presenting new solutions in the field of occupational safety and hygiene.

The Editors,

Pedro M. Arezes
J. Santos Baptista
Mónica P. Barroso
Paula Carneiro
Patrício Cordeiro
Nélson Costa
Rui B. Melo
A. Sérgio Miguel
Gonçalo Perestrelo

Reviewers involved in the process of reviewing and editing the papers included in this book

A. Sérgio Miguel	Guilherme Buest	Marino Menozzi
Alfredo Soeiro	Gyula Szabó	Mário Vaz
Álvaro Cunha	Hernâni Neto	Marta Santos
Ana Ferreira	Ignacio Castellucci	Martin Lavallière
Anabela Simões	Ignacio Pavón	Martina Kelly
Angela Macedo Malcata	Isabel Loureiro	Matilde Rodrigues
Antonio Lopez Arquillos	Isabel Nunes	Miguel Diogo
Beata Mrugalska	Isabel Silva	Mohammad Shahriari
Béda Barkokébas Junior	J. Santos Baptista	Mónica Paz Barroso
Bianca Vasconcelos	Jesús Carrillo-Castrillo	Nélson Costa

Camilo Valverde

Carla Barros

Carla Viegas

Carlos Guedes Soares

Catarina Silva

Celeste Jacinto

Celina P. Leão

Cezar Benoliel

Cristina Reis

Delfina Ramos

Denis Coelho

Divo Quintela

Ema Leite

Emília Duarte

Emília Rabbani

Enda Fallon

Evaldo Valladão

Fernanda Rodrigues

Filipa Carvalho

Filomena Carnide

Florentino Serranheira

Francisco Fraga

Francisco Rebelo

João Ventura

Jorge Gaspar

Jorge Patrício

José Cabeças

Jose Cardoso Teixeira

José Carvalhais

José Keating

Jose L. Melia

José Pedro Domingues

José Torres Da Costa

Juan C. Rubio-Romero

Laura Martins

Liliana Cunha

Luis Franz

Luiz Bueno Da Silva

M.D. Martínez-Aires

Mª Carmen Rubio-Gámez

Mahmut Ekşioğlu

Manuela Vieira da Silva

Marcelo Silva

Maria Antónia Gonçalves

Maria José Abreu

Olga Mayan

Paul Swuste

Paula Carneiro

Paulo Carvalho

Paulo Flores

Paulo Noriega

Paulo Sampaio

Pedro Arezes

Pedro Ferreira

Pedro Mondelo

Pere Sanz-Gallen

Ravindra Goonetilleke

Rui Azevedo

Rui B. Melo

Rui Garganta

Salman Nazir

Sérgio Sousa

Sílvia Silva

Susana Costa

Susana Viegas

Teresa Patrone Cotrim

Walter Correia

Occupational Safety and Hygiene V – Arezes et al. (Eds)
© 2017 Taylor & Francis Group, London, ISBN 978-1-138-05761-6

Environmental risk assessment for discharge of hazardous ship waste

F. Ferreira & C. Jacinto
*UNIDEMI, Research Unit—Mechanical and Industrial Engineering, Universidade Nova de Lisboa,
Caparica, Portugal*

H. Vaz
Cleanport, S.A., Lobito, Angola

ABSTRACT: This paper reports a study of an environmental and occupational risk assessment of two processes, regarding the transfer/discharge of hazardous ship waste (sludge and oily waters). Following the methodology suggested by the Spanish Standard UNE 150008:2008, a number of accident scenarios (n = 7) were created and their respective risks evaluated, focusing on environmental matters. The results pinpointed several aspects needing priority attention, mostly within procedures and equipment. To improve the safety of those processes, the authors proposed a number of recommendations.

1 INTRODUCTION AND BACKGROUND

In the aftermath of major accidents like Seveso or Chernobyl, specialists started to look for ways to assess the risks in their industries. International Bodies provided complex methodologies like ARAMIS (Accidental Risk Assessment for Industries in the framework of Seveso II) (Salvi & Debray 2006), or FSA (Formal Safety Analysis) established by the IMO (International Maritime Organisation) (Kontovas & Psaraftis 2009). Major accidents involving dangerous substances represent a significant threat to humans and the environment. Furthermore such accidents might cause massive economic losses and disrupt sustainable growth (EU Directive Seveso-III 2012).

The application of some well-known methodologies can be difficult due to their range being either quite specific (narrow) or, too wide, requiring high level knowledge of experts of different areas. That is why industries started to look for simpler and qualitative methods (Bahr 2006, p. 2795).

The Spanish Standard UNE 150008:2008 provides a practical and easy to understand methodology. It includes instructions for hazard identification and tools for risk evaluation.

Despite a large variety of studies on risks related to maritime transportation of hydrocarbons, less is found in the literature with regard to ship's sludge and oily waters (ship waste) (eg: Ronza et al. 2006, Zuin et al 2009).

The objective of this work was to carry out a risk assessment of two processes, regarding the transfer/discharge of hazardous ship waste. The assessment covered both occupational and environmental risks, although with an emphasis on the latter case. The processes studied were:

1. the discharge of waste in the Port of Lobito, from the ship to a truck-tanker;
2. the discharge of such waste into the company storage facilities.

The case described here focuses on one particular hazard: "spill of hazardous ship waste in the soil and the sea water".

The company studied is called Cleanport (Lobito, Angola) and is specialized in industrial waste management, industrial cleaning and naval maintenance.

2 METHODOLOGY

The methodology applied in this study was selected upon three factors: it is easy to understand, to apply and provides all tools one needs to perform the risk analysis and assessment. Another plus is that the way the results are presented makes it understandable for any person who has only a notion of the subject matter.

The standard UNE 150008:2008 is composed of 5 main steps. It is recommended a multidisciplinary team to take into account the several facets. The first step is the "hazard identification". This step is essential because everything will revolve around those identified hazards. The second step is the "accident scenario design" where accident scenarios are created. The third step is called "frequency/probability estimation". For each accident scenario, it is necessary to estimate its probability

Table 1. Environmental risk assessment of accident scenario #5 (illustration example).

Accident Scenario #5	
Accidental scenario	Spill of sludge in the facilities (grounds/soil)
Causes of the accident	Collision of truck-tanker against one of the large deposits in land (3000 m^3 capacity)
Initial element (event)	Deposit rupture
Scenario description	The truck-tanker hits the lateral wall of the deposit during maneuvering. This causes a rupture and a spill
Existing control and mitigation measures	Containment basin with 50 m^3 Spill response equipment ready at the site Workers are trained in containment procedures
Recommended control measures (additional improvement)	Maintenance and enlargement of the current containment basin Create a safety perimeter around the deposit, preventing vehicle access (e.g.: concrete or iron pins) Segmentation of the deposit Replacement of the deposit by several others with lower capacity Impermeabilization of the installation floor/soil

Frequency classification		scores
Frequency	Likely—this occurrence is directly related to an human error	2

Classification of quantity, dangerousness and extension of release		
Quantity (Q)	The tear in the wall is likely located 1.5 meters from the floor. Assuming that the deposit is almost full, with a level height of 14 meters and storing around 3 millions of liters, the release will correspond to the remaining 12.5 meters of residue, which is equivalent to 2 million and 678 thousand liters (2678 m^3) Residue average density (it depends on the quantity of water) = 0.90 kg/dm^3 Q = 2.678 m$^3 \times$ 0,90 kg/dm^3 = ~2410 ton > 500 ton (very high quantity)	4
Dangerousness	Inflammable, toxic, toxic for water species	3
Extent of release	Very extensive area—given the quantity of residue spilled, it is very likely that it could occur underground water and sea water contamination, since the installation is located near the coast	4

Natural environment classification		
Acute	Losses exceeding 50% of all species and biomass that were in contact with the residue, only a long term recovery is possible	3

Human environment classification		
Very high	Impossible to determine the number of people affected by the contaminated drinking water and the contaminated species/biomass losses (likely to be > 100 people)	4

Socioeconomic environment classification		
High	Extensive damages to the company installations and equipment. Might affect local livestock and agricultural activity	3

Severity of the outcome—classification (c.f. Table 2)		Severity Score	Classification
Natural	$4 + 2 \times 3 + 4 + 3 = 17$	4	Serious
Human	$4 + 2 \times 3 + 4 + 4 = 18$	5	Critical
Socioeconomic	$4 + 2 \times 3 + 4 + 3 = 17$	4	Serious

Environmental risk classification—final score		Risk Classification (scenario #5)
Natural	$2 \times 4 = 8$	Moderate risk
Human	$2 \times 5 = 10$	Moderate risk
Socioeconomic	$2 \times 4 = 8$	Moderate risk

Table 2. Formulas for severity and environmental risk classification.

Severity on natural environment =	Quantity	$+ 2 \times$ Dangerousness	$+$ Extent	$+$ Quality of the environment
Severity on human environment =	Quantity	$+ 2 \times$ Dangerousness	$+$ Extent	$+$ Affected population
Severity on socioeconomic environment =	Quantity	$+ 2 \times$ Dangerousness	$+$ Extent	$+$ Patrimony and productive capital

Environmental risk classification = Frequency $\times$ Severity scores.

or frequency of occurrence. In this study, the estimation of frequency was based on the company's accident historical data.

The forth step is "risk assessment". This step can be divided into two phases: "severity of the outcome" (or consequences) and "environmental risk estimation". The severity of the outcome is estimated for three dimensions (or environments): human, natural environment and socioeconomic. For that, it uses four variables, three of which are common to all three dimensions. The common variables are: the *quantity* of spilled/released/burned residue, the *dangerousness* of the residue and the *extent of release* (i.e., affected area). The forth variable is specific for each environment. For the human dimension, it evaluates the affected population, whereas natural environment is assessed in terms of damage caused to natural species and alike. Finally, the socioeconomic environment is evaluated through monetary and capital losses. The standard UNE 150008:2008 provides tables with criteria and scores for all the variables used in the assessment process. Some of these variables are rated 1–4 (being 4 the worst case), while others are rated 1–5 (being 5 the worst case). At the end, after combining all variables, one will obtain the final level of risk, for which the highest level is "very high risk" (scored 21–25). In this approach the final assessment returns three risk results, one for each dimension (human, natural environment and socioeconomic). As in any other methodology, the last step (5th) refers to "recommendations" where analysts should suggest measures for improvement.

The methodology is partially demonstrated in Tables 1 and 2.

3 RISK ASSESSMENT

3.1 Description of case study

This application case-study was carried out in the company Cleanport, located in Lobito (Angola). Being a fairly new company (3 years of activity), it was created to help complying with the MARPOL (marine pollution convention) in Angola, which was neglected. Their activity started in Port of Lobito but after performing 100 services to ships,

they expanded their activity to naval maintenance, as well as industrial cleanings and industrial waste management. Their liquid waste storage capacity is over 7400 m³ of which 6000 m³ are reserved for oily waste from ships and industrial machinery (two high capacity deposits, approximately 3000 m³ each).

The first process analysed was the discharge of ship waste in the Port of Lobito. This waste is discharged to a truck-tanker parked in the dock, using the ship's pump and high pressure hoses. At each end, there is an emergency stop button for the pump. This process requires one chief of operations and 2 field workers. The workers are responsible for the preparation and connections (hoses and gaskets), as well as verifying the level of residue in the truck-tanker. The chief of operations, in addition to supervision, is responsible for the equipment inspection before and after the pump starts.

The second process takes place at the company's facility where the land deposits are located. The procedures are very similar to each other (i.e., from ship to truck-tanker and from this to land deposit). The amount of residue handled in both processes may vary from 5 m³ to 30 m³ (maximum capacity of tankers), and pumps that can have a flow over 100 m³ per hour. For instance, in one of the scenarios analysed the estimated amount of spilled residue exceeded 2000 m³, which indicates a major spill event.

3.2 Results and discussion

As mentioned before, the main hazard analysed was the spill of dangerous residues, i.e., oily ship waste. The 7 accidental scenarios created involved hoses and deposit ruptures, incorrect connections, insufficient inspections and valve malfunctions.

The first scenario (Acc Scenario #1) considers a rupture in a high pressure hose during the discharge from the ship to the truck-tanker. This task is illustrated in Figure 1.

This kind of accident has already occurred in the company and it was due to the hose degradation. Since this work is performed outdoors, next to the sea, in a tropical country (extreme humidity, radiation, high temperature), the equipment suffers a fast ageing and degradation.

3

Figure 1.   Discharge from ship. Connecting hoses.

The frequency of hose rupture due to ageing was estimated based on the company's records, which indicated that this type of occurrence happened 2 times in the first 3 years of activity. This is translated in a score of 3 (likely) by the UNE standard. Then, the variables relative to the residue (sludge) were estimated. The *quantity* of spilled residue was estimated using an average time of the potential spill and the pumping flow.

Based on past experience, an average of 15 seconds spill was considered plausible; such time, with a flow of 20 m³/h, resulted on a calculated spill of 83.3 litres, which received a score of 1 (very little quantity, <5ton).

Since hydrocarbons are inflammable and toxic, the *dangerousness* was classified with a score of 3 (dangerous). Additionally, 5 occupational risks were identified: 1) contact with hazardous substances (through nose and mouth, via inhalation), 2) contact with a hot fluid or hose, 3) struck by object (whiplash effect), 4) fall of worker against the floor, and 5) drowning in the sea. All these occupational risks were assessed as "high risk level", with the exception of "being struck" by the projected hose (#3), which was classified "very high risk".

As for the *extent of release*, in scenario #1, it was established that this accident might cause contamination of sea water, which can affect an area over 1 km radius (score 4 - very extensive area). Even though the company uses containment booms to control the spill, there is always some residue left behind and solid sediments that sink into the bottom. Those can never be scooped from the water and are usually eaten by wild life.

For the *human dimension*, the effect was hard to classify, since a large number of people with food poisoning (from contaminated fish) were likely to be affected, hence score 4 (very high, >100 people).

With regard to the *natural environment* (in Acc Scenario #1), even though this type of residue is toxic for marine species, the quantity involved was fairly low, resulting in a score of 2 (chronic). It is important to note that the sediments eaten by the species can cause mutations at medium/long term.

In relation to the *socioeconomic environment*, it was found that this accident #1 would have caused very few losses to the company, which represents a score of 1 (very low impact in economic terms).

Using the previous scores and following step-by-step the whole assessment procedure the final risk for scenario #1 was found to be:

- "moderate" (9) for human environment;
- "medium" (12) for natural environment;
- "moderate" (9) for socioeconomic.

For demonstrating purposes, Table 1 gives a better understanding of the assessment details, in this case for scenario #5, in which case all three environmental risks were classified as "moderate".

Table 2, on the other hand, gives the formulae for combining all the criteria.

Table 3 summarises the results obtained for all 7 accident scenarios, including a brief description of each scenario.

3.3   Recommendations

This risk assessment provided the necessary information as to where the main problems reside. The majority were related to the poor condition of the equipment. In the course of time, this kind of failures (blowing hoses, leaks, deficient joints and valves) are likely to occur more often. This can be avoided with an efficient and frequent maintenance, as well as by performing a good rotation of the available equipment. A record with hours of use, date of last maintenance, replaced parts and description of the malfunctions, can be useful to prevent/control future equipment failures.

As for the deposits area (storage tanks), it was found that a restricted perimeter should be created using concrete or metal pins (safety pins) to disable the circulation of vehicles in the vicinity. This would prevent any contact between the deposits and an "object" that can cause a rupture.

To avoid soil contamination in the premises, the waterproofing of the soil could be a solution. This way, even after a spill of large proportions, the contamination could be prevented and the cleaning process would be quicker, without resorting to soil removal. It should too be performed the enlargement of the containment basin, as well as its proper and regular maintenance.

Table 3. Environmental risk assessment results (summary table for the 7 scenarios).

Accident Scenario	Brief description	Severity of the outcome			Environmental risk—final risk level		
		Natural	Human	Socioeconomic	Natural	Human	Socioeconomic
#1	Sludge spill on the dock (hose rupture & whiplash effect)	3 Moderate	4 Serious	3 Moderate	9 Moderate	12 Medium	9 Moderate
#2	Oily waters spill in the dock (blowing hose)	3 Moderate	4 Serious	3 Moderate	9 Moderate	12 Medium	9 Moderate
#3	Waste spill on board of the vessel (pump premature initiation)	1 Light	1 Light	1 Light	9 Moderate	9 Moderate	9 Moderate
#4	Residue spill in the dock (Deficient joint connections)	3 Moderate	4 Serious	3 Moderate	6 Moderate	8 Moderate	6 Moderate
#5	Waste spill in the installation soil (Deposit rupture)	4 Serious	5 Critical	4 Serious	8 Moderate	10 Moderate	8 Moderate
#6	Waste spill in the installation soil (blowing hose/deficient joint)	2 Light	2 Light	2 Light	6 Moderate	6 Moderate	6 Moderate
#7	Waste spill in the installation soil (malfunction of the deposit valve)	4 Serious	4 Serious	4 Serious	4 Low	4 Low	4 Low

Even though the company has in place an Internal Emergency Plan, it should also have an External Emergency Plan established with the local authorities (fire department, hospitals, etc.).

To finish, it was recommended providing some kind of information and training to workers on the hazards. This way, they would be aware of what could go wrong and be prepared if something happens.

3.4 Comments on the standard

In the course of the study, it was felt that the risk matrix proposed by standard UNE 150008:2008, to estimate the environmental risk, does not promote a conservative attitude toward the protection of the environment, contrary to what one should expect.

Of all possible interactions between *frequency* and *severity*, 40% represent "low risk", 28% "moderate risk", 16% "medium risk", 12% "high risk" and only 4% return "very high risk". In the authors' opinion this aspect should be taken into account before suggesting "light" or no measures for an apparent "low risk" or "moderate risk", as if "nothing too serious might result from such scenario".

4 CONCLUDING REMARKS

Even though this methodology estimates an environmental risk in a way that is not conservative for the environment, it provided the authors a good insight on the risks existing in the processes scrutinised. Moreover, it also allowed identifying measures to prevent or mitigate potential consequences of the accidents mapped in the 7 scenarios considered.

This method can be used in various industries. A more thorough analysis can be achieved with more resources, e.g., time, personal, knowledge of the processes, etc.

The innovative aspects in this work are twofold:

- the methodology itself, which is not yet well disseminated in the specialty literature,
- the country where it took place (Angola), showing genuine efforts towards safer workplaces and environments in developing countries.

ACKNOWLEDGEMENTS

The authors are grateful to Cleanport SA for opening their doors and participating in this assessment. Study made under UNIDEMI (ref. PEst-OE/EME /UI0667/2014).

REFERENCES

Bahr, N. J. 2006. System safety engineering and risk assessment. *International Encyclopaedia of Ergonomics and Human Factors*. Vol. 3: 2794–2797. Boca Raton: CRC Press, Taylor & Francis Group.

EU—Directive Seveso III. Directive 2012/18/EU, of 4 July 2012, on the control of major-accident hazards involving dangerous substances. Official Journal of the European Union L 197/1.

Kontovas, C. & Psaraftis, H. 2009. Formal safety assessment: critical review. *Marine Technology*, 46(1): 45–59.

Ronza, A., Carol, S., Espejo, V., Vílchez, J. A., Arnaldos, J. 2006. A quantitative risk analysis approach to port hydrocarbon logistics. *Journal of Hazardous Materials*, 128(1): 10–24.

Salvi, O. & Debray, B. 2006. Convergence in risk assessment for SEVESO sites from ASSURANCE results to ARAMIS method. Symposium "Quantitative risk analysis: Quo vadis?", Mar 2006, Tutzing, Germany: 56–59.

UNE 150008. 2008. Análisis y evaluación del riesgo ambiental. Madrid: Asociación Española de Normalización y Certificación (AENOR) (in Spanish only).

Zuin S., Belac E., Marzi B. 2009. Life cycle assessment of ship-generated waste management of Luka Koper. *Waste Management*, 29(12): 3036–3046.

Occupational Safety and Hygiene V – Arezes et al. (Eds)
© 2017 Taylor & Francis Group, London, ISBN 978-1-138-05761-6

Failure Mode and Effect Analysis (FMEA) as a model for accessing ergonomic risk

Eduardo Ferro dos Santos
University of São Paulo, Engineering School of Lorena, EEL/USP, Lorena, Brazil

Karine Borges de Oliveira
Salesian University Center of São Paulo, UNISAL, Brazil

André Solon de Carvalho
University of São Paulo, Engineering School of Lorena, EEL/USP, Lorena, Brazil

ABSTRACT: The existence of methodologies for priority and control the risk in different areas support the realization of safety and security process practice. However, endeavors to create tools and methodology to rank the ergonomic risk activities are mostly devoted on product design. Motivated by such scarcity, the goal of this study is attempted to develop a modified FMEA (Failure Mode and Effect Analysis) as means to access the criticality of ergonomic risk in operations. An improved model to the ranking ergonomic risk by using Risk Priority Number (RPN) is proposed and applied in a case study, in the attempt to help decision makers to debate the problems occurred.

1 INTRODUCTION

The ergonomic work analysis is proposed to conduct the analysis of work activities in an organization with the presupposition that makes the worker in the production process identifies the ergonomic risks (Mhamdi et al., 2015). The primary objective of the FMEA is the prevention of problems in processes or products before they occur (Ghasemi et al, 2016).

In this paper is presented an ergonomic risk analysis by a model integrated to FMEA for appreciation of ergonomic risk to identify priorities and to control the observed conditions. This model can contribute to companies to develop the improvement in ergonomics projects and process (Santos & Nunes, 2016).

2 FMEA AS A MODEL FOR ERGONOMICS

2.1 *Focus*

During a meeting in a case study, members of the company and an ergonomist researcher got together and presented what could be grouped and defined as homogeneous to the execution of modeling of the process in analysis.

The packing area is presented to be the object of the evaluation, specifically the Packing Machine Operator (workstation). In this machine, the process takes the following steps: pack the product, check the weight of the bag; seal the bag; send them to belts for palletizing.

In order to better organize these elements, the group led a simple study of crono-analysis in the process flow. The timings were systematized through a sample of 10 workers, in which the arithmetic averages of the execution times of each element of the task were observed.

2.2 *Preliminary*

Through a review, gathering findings of the study with the appreciation of all members of the group, the following considerations were observed:

– The plastic bag was instable. The operator acted many times so that he could prevent some waste of product. In three of the observations made, it was necessary to add some more product with a specific spoon after the inspection to reach the right weight, because the way the plastic bag was handled (instability) caused some waste of product that fell out of the bag. A bucket was placed next to the scale to meet this need. What could represent a waste of time in the process, waste of product and what could make the workers to perform unnecessary movements;

– In 60% of the observation, it was necessary to hammer the output funnel of the packing line with a rubber hammer so that the product could be pushed out of the machine more easily. This problem is a result from the high relative air humidity of the place (72%) once the job is performed next to the cooking area that produces vapors going around the packaging area as it was an open shed divided in two areas. These situations caused the waste of time once they do not allow a continuous and uniform flow of product, besides demanding unnecessary movement from the workers;

– There was not any proper place to put away the hammer and the plastic bag. The space designated to the plastic bags store could only prevent them from being damaged or torn. The bags were stored in an area that was out of the reach of the operator, so that they had to bend the body to get them. This results in unnecessary movements of the worker. If the relative air humidity could be improved it would also reduce the use of the hammer;

– Equipment layout was also inappropriate, demanding the operator to turn his body, to carry the load and to move unnecessarily. This item was also considered a waste of time and biomechanical overload due to the execution of unnecessary movements (the efforts to hold the bags), that may victimize the worker with muscle fatigue and strain injuries;

– The sealing machine was placed higher in relation to the height of the scale that may overload the body and the spine because the operator needed to lift the bags to compensate the depression victimizing the workers with muscle fatigue and strain injuries. The image of the process is illustrated in Figure 1.

Figure 1. Packing machine operator.

2.3 Mapping

Given these findings, it was proceeded with the ergonomic mapping by assessing variants. The group went to the field for this activity, filmed, interviewed and confronted the initial findings with workers experiences. The problems were discussed at a formal meeting attended by all involved (group, ergonomist) and jot down into the form with the initial description of each of the problems found. This assessment demonstrates the understanding of the group about the problems involved in the manual packaging.

Based on the previous mapping and on the modeling of the activities, it is possible to look deeper and systematically into the response to the problems found in this step. Items related to the kinesiology of upper limbs were added (displacement, catching and positioning) and cognitive demands (adjustments and inspections). These items allow the elements of the process to be more precisely analyzed in terms of its needs (unnecessary movements) and possible losses in the process. The sample has represented the same made by the focus group once all of them were recorded in a video.

The findings were reviewed in its root cause and the conclusions are as follow:

– The waiting time represents 48.2% of the cycle time. It is assumed that if the "building up" did not exist, the flow could be higher and the waiting time reduced;

– The transportation represents the biggest time waste in the cycle. 19.2% of the time is wasted due to the transportation of the loads from one point to another. It is assumed that the improvement of the layout of the machines could minimize or even eliminate this waste. This item is also considered the one that demands the body turns, load handling and unevenness of the heights (scale-sealing) which caused overloads to the upper and lower limbs and spine;

– The item preparation consumes 12.5% of the cycle. The longest time of elements occurs when the packing is adjusted in the scale so that it can be placed vertically. Due to the plastic bag, the operator spends on average 4 seconds (out of the total 7) trying to place the package in a vertical position;

– Relative humidity can contribute to the increase of waiting time and unnecessary movements of hammering which can also damage the equipment;

– The displacement of materials is the highest risk of the process. Besides the waste of time, it can be responsible for human costs (medical leaves and complaints on spine and shoulders) and administrative costs (lawsuits);

- The quality of some input items of the process, such as the position of the packaging material and the package itself can contribute with the improvement of timing in the cycles of production and with the organization of the work;
- Possibilities of introducing shift rotation or programmed pauses in this job post should be considered, where there is a strong risk of physiological overload due to the long lasting standing position the activity demands.

2.4 Ergonomic Risks Analysis

The results of the evaluation were grouped in a form based on FMEA that was adapted to the Ergonomic Risks Analysis shown in Appendix 1. The following steps were taken:

- The variants were classified according to the form of field research and described in systematic observations through a text written by the member of the focus group and the ergonomist. The application had come up, then, the criteria were standardized (specific form) and will be the base for future application in the OHSAS 18001 management in the company (Fernández-Muñiz, 2012);
- Many documents mentioned in the observed items were studied (tables) throughout the form. The illustration and tools were also mentioned and attached to the reports;
- The existent means of control of each situation were mentioned and assessed in terms of efficiency by the focus group and ergonomist, against scales elaborated to the organization to control them;
- The definition of Risk Indicators (Severity x Probability × Control) was discussed in specific events (Kaizen) through brainstorming and reviews.

Then, a specific legend was created for each one of the items, as presented in Tables 1 to 3, with the results of Table 4. When the indicator showed duplicity (two items in Severity and Probability fields), the higher value was proposed.

The major risks were in the surroundings of displacement activities that take Based on the findings, a spread sheet was created and the reviews made consider that:

- Place around equipment because they are acted simultaneously with load handling. Such situations do not only causes hazards of biomechanical overload on the shoulders and spine but also causes waste of time in the process. The reorganizing of the layout stands out, so that it could be projected in a way that the equipment could be placed closer to each other and that body rotation and displacement of loads could be eliminated;

- High relative humidity contributes with the reduction of the speed of the flow of the product. The high rate of humidity is originated in the cooking area that stayed next to the packing area. Ways to isolate the area should be studied;
- A program of prevention against fatigue should also be taken into consideration, so that brings along the specific or compensatory pauses on the muscular groups that were under stress at the

Table 1. Defining FMEA—Severity.

Indicator	Human	Organization
1	Do not generate human overload	Little or no interference in the process.
2	Generate situations of discomfort and fatigue.	The isolated agent may interfere in temporary stops and minor losses of productivity.
3	Hazards that can jeopardize health, causing injuries and leaves.	Entails significant delays in the process, reduction of planned work. Items that do not meet the legislation in effect.

Table 2. Defining FMEA—Probability.

Indicator	History	Exposure
1	Any event related to the agent	Shortly, less than 10% of the time sampling (day or cycle).
2	There are complaints and events in terms of verbalization	Reasonable time of 11 to 30% of sampling time (day or cycle).
3	The complaints are frequent and specific to the agent, with indicators and demonstrative records	Shortly, less than 10% of the time sampling (day or cycle).

Table 3. Defining FMEA Control.

Indicator	Control
1	There are good control plans to handle the risk.
2	There is a plan to handle the risk, but there is a lack of formal procedures and the efficiency is uncertain.
3	There is no plan and awareness to deal with the risk. The operational practices show the exposure seems to be out of control.

Table 4. Risk priority number.

Risk Level		General Biomedical Classification	OHSAS 18001 Equivalence (BSI, 2007)
1	Minor	Normal technical Action or with no significant risk	No action is required and no documents to be filed are necessary
2–3	Tolerable	Unlikely risk of injury, they are more related to the sporadic difficulties.	The control practices should be maintained and monitored
4–9	Moderate	Situations which cause muscle fatigue if performed for a long period and/or with no control practices	Control/Prevention practices should be implemented
12–18	Substancial	Situations which cause injuries	The activities should be sistematically studied, an implementation plan should be approved by the high leadership team so that the hazard could be mitigated or eliminated within a given period.
27	Intolerable	These situations can potentially cause serious injuries, diseases and accidents that can generate medical leaves or functional disabilities. The company does not give the situations the proper attention, neglecting them.	Systematic studies should be implemented, and an immediate implementation improvement plan should be created. This should be approved by the high leadership team so that the hazard could be mitigated or eliminated. The execution of the plan should be monitored and evaluated.

Table 5. Action plan.

What to do	How to do it
Rearrange the layout project, where the filling, sealing and belt are placed side by side, eliminating the displacement of loads, body rotation and weight management.	Projecting new filling equipment. Acquiring new paper seam system, with compatible height to be placed next to the scale. Insert a semi-automated belt at the output of the filling system
Replacing the plastic packaging for packing paper (less polluting and more resilient).	Projecting new design of packing paper more resistant and test the material. Implement recycled packing paper, and inform customer about the change.
Isolate the packing area from the cooking area.	Building brick and mortar walls to separate the areas
Insert seats next to the job posts.	Placing semi-sitting seats next to the job post so they can be used as an alternative relay posture.
Inserting pauses for stretching (labor gym).	Hiring a specialized company to implement the program.
Function rotation with the pallet position at every 2 hours.	Carrying out ELSS project to make the process viable.
Provide safety shoes with bi-density soles.	Buying bi-density shoes to replace the existing model.
Place an anti-fatigue mat opposite the workplace	Acquiring the product and place them on the job posts.

performance of the activities. Once the activity takes place 100% of time, without any pause, full time;
– The package of the product should be substituted by another one less flexible.

An action plan that involved the responsible and future controls was created then. (Table 5).

The high leadership of the company assured the implementation and the extension of ELSS project all over the company through the standardization of the system and definition of an operational system. Due to the success of the method four more four-hour-events were conducted and the procedure of ergonomic management was installed.

3 CONCLUSIONS

In general, the process of generating a plan of step by step implementation, aligned to the reality of the organization, may result significant gains and may positively feedback the efforts to change on the ergonomic management.

Then, it is possible to join FMEA and Ergonomics in an integrated way on the solution of problems. This proposition was validated inside the scope of this study, once, through FMEA the improvement process go beyond the results of risk assessment and can also represent priority classifications. In this study, could be noticed that the problems were analysed in a structured way, presenting a hybrid model that could meet the current methodologies of management around the world.

REFERENCES

BSI. 2007. British Standards Institute. BS OHSAS 18001:2007: Occupational health and safety management systems—Requirements. London: BSI Global.

Fernández-Muñiz, B.; Montes-Peón, J.M.; Vázquez-Ordás, C.J. 2012. *Occupational risk management under the OHSAS 18001 standard: analysis of perceptions and attitudes of certified firms, Journal of Cleaner Production*, Volume 24, March, Pages 36–47. doi: 10.1016/j.jclepro.2011.11.008.

Ghasemi, S.; Mahmoudvand, R.; Yavari, K. 2016. Application of the FMEA in insurance of high-risk industries: a case study of Iran's gas refineries.

Stoch Environ Res Risk Assess, 30: 737. doi:10.1007/s00477–015–1104–7

Mhamdi, A.; Magroun, I.; Youssef, I.; Damak, N.; Amri, A.; Ladhari, N. 2015. *Analyse ergonomique du travail dans une entreprise de confection en Tunisie, Archives des Maladies Professionnelles et de l'Environnement*, Volume 76, Issue 5, October, Pages 449–457. doi:10.1016/j.admp.2015.01.006.

Santos, E.F.; Nunes, L.S. 2016. *Methodology of Risk Analysis to Health and Occupational Safety Integrated for the Principles of Lean Manufacturing. Advances in Social & Occupational Ergonomics*, July 27–31, 2016. doi: 10.1007/978–3-319–41688–5_32

Appendix 1.   FMEA in ergonomic risk analysis.

Issue	Aspects and hazards	Existing Administrative and Controls	Main Effect	P	G	C	RPN	Root Cause of the problem
Deambulation	Displacement with 20 kg bags on the filing, sealing and belt areas.	The employees receive a training course on how to handle heavy materials with efficiency. There are procedures to prevent individual lifting of loads heavier than 30 kg.	9.37% waste in the time of the process and Fatigue of lower limbs	2	2	3	12	The layout is distributed so that there is displacement of the employee with unnecessary load. There was no attention to this item in the design process
			Overload in efforts on upper limbs and spine	3	3	2	18	
	Displacement without any load between the belt and filling area.	Inexistent.	7.14% waste in the time of the process.	2	1	3	6	
			Fatigue of lower limbs	2	2	3	12	
Horizontal Distance	Displacement filling-sealing and sealing-belt transporting 20 kg loads.	Inexistent.	Effort overload by rotating with load the lumbar spine.	3	3	3	27	As the previous one.
Vertical Distance	Efforts on the displacement of bags through the areas with different levels	Inexistent.	Overload in flexor efforts on shoulder and spine.	3	3	3	27	The leveling of the room was not taken into consideration at the conception.
Quality	The plastic pack is difficult to handle.	Inexistent.	12.5% waste in the time of the process.	2	2	3	12	As a matter of design, the packaging used is plastic, and the quality of operational handling was not tested.

(Continued)

Appendix 1. (*Continued*).

Issue	Aspects and hazards	Existing Administrative and Controls	Main Effect	P	G	C	RPN	Root Cause of the problem
Positioning	The plastic bag is stored in an inappropriate place (suspended on the frame of the filling equipment)	There is a good positioning of the load on the side protections of the filling device.	Inadequate specific site at the input of the process.	2	1	2	4	There was a concern about the proper place, since the bar ensures the proper positioning of the pack.
	The hammer is an area that is difficult to be reached.	The worker can place it his way.	Body inclination in the area of difficult access	2	2	2	8	The hammer was not predicted to take part in the process, and then there was no conception on the plan for its location.
Movements	Repetitive tasks.	There are pauses for lunch (60 m) and two (15 m) pauses for coffee and personal needs.	Physiological overload with the risk of muscle fatigue in upper and lower limbs	2	3	2	12	
Humidity	High Relative Humidity (72%).	There are evaluations of occupational hygiene that do not consider risks since the legislation refers to the minimum met by the company (40%).	Unnecessary effort with the upper limbs to perform hammer in an attempt to decongest the product in the filling nozzle with the possibility of damaging the equipment.	2		2	8	The shed has no separation between the rooms, which makes that the humidity of the kitchen facility be transferred to the packaging area, building up the product by grouping wet particles.
Pauses	Absence of necessary pauses.	There are pauses for lunch (60 m) and two pauses (15 m) coffee and personal needs.	Physiological overload with the risk of muscle fatigue in upper and lower limbs.	2		2	8	There is not staff enough to ensure the pause at the current demand of the production procedures
Movements	Repetitive tasks.	There are pauses for lunch (60 m) and two (15 m) pauses for coffee and personal needs.	Physiological overload with the risk of muscle fatigue in upper and lower limbs	2		2	12	
Statics	Long lasting standing posture	Inexistent.	Physiological overload with the risk of muscle fatigue in lower limbs.	2		3	12	For the current layout and need to travel, it can not work with the rotating position.

Occupational Safety and Hygiene V – Arezes et al. (Eds)
© *2017 Taylor & Francis Group, London, ISBN 978-1-138-05761-6*

Risk evaluation in the transportation of dangerous goods

D.M.B. Costa
FEUP—Faculty of Engineering of the University of Porto, Porto, Portugal

M.V.T. Rabello
UFRJ—Federal University of Rio de Janeiro, Rio de Janeiro, Brazil

E.B.F. Galante
IME—Military Institute of Engineering, Rio de Janeiro, Brazil

C.V. Morgado
UFRJ—Federal University of Rio de Janeiro, Rio de Janeiro, Brazil

ABSTRACT: This work presents a risk assessment of a road transport company that delivers fuel to distribution stations and seaports by tanker trucks in Brazil. Road transportation of dangerous goods is a critical activity due to the occupational exposure of workers to these products as well as potential consequences following accidents, which may include personal injury, property damage, fuel spills, among others. Considering the need of including environmental aspects in the occupational health and safety assessment towards a more integrated evaluation, the Methodology of Integrated Evaluation of Risks (MIAR) was applied in the product transportation, identified through a risk matrix as the most critical activity developed by the company. An evaluation of legal compliance to national standards was also performed for the verification of gaps in performed activities, revealing 73 items requiring corrective actions.

1 INTRODUCTION

Final refinery operation consists on loading gases and liquid hydrocarbons into pipelines, tanker cars, tanker trucks and marine vessels and barges for transport to terminals and consumers. This process requires clear determination of product characteristics, distribution needs, shipping requirements, fire prevention, environmental protection and operating criteria to ensure safety in the activities performed (Kraus 1988). Accidents involving hazardous products may lead to spills, which in turn leads to hazards such as fire, explosion, chemical burn, or environmental damage, representing serious consequences to life and property.

Generally, these consequences cannot be fully contained or otherwise reduced, requiring risk management strategies such as preventive measures to mitigate either the probability of occurrence or the magnitude of the consequences (Laarabi et al. 2014). Considering this, a risk assessment of the activities of a company in the transport sector engaged in road transportation of dangerous goods in Brazil was conducted by the evaluation of its legal compliance and by the application of Methodology of Integrated Evaluation of Risks (MIAR) in its operational activities.

This work unfolds into five distinct sections. After the present introduction, the case study is described in Section 2. In Section 3, risk evaluation methodologies are applied and their relevance examined. Finally, results and discussion are presented in Section 4 followed by the conclusions.

2 CASE STUDY

The case study is a company that operates within the transport sector and is primarily engaged in transportation of dangerous goods in Brazil. The transported goods are mostly petroleum-based fuels (except ethanol) for fuel distribution stations, seaports, and vessels.

The company started its operations in 2010 and currently employs 23 people, 15 of whom are tanker truck drivers, and possesses 24 Tanker Trucks (TT). It delivers fuels in the modality FOB (Free on board) to gas stations and CIF (Cost, Insurance and Freight) to seaports and ships. All operational activities are performed by the TT drivers and the workflow can be summarized as follows: (i) TT enters the facility towards the loading platform, (ii) TT enters loading platform, (iii) loading inspection, (iv) hatch opening for inspection, (v) fuel pumping,

(vi) fuel loading and unloading and (vii) product transportation to customer/receiver.

Although no serious environmental incidents took place in 2015, the company has recorded occasional leaks and spills, including five episodes of marine diesel oil leakage during the unloading operation (overall 153 liters leaked), demonstrating the need of evaluation of this operational aspect. Environmental accidents can be defined as unplanned and unwanted events that can cause, directly or indirectly, damage to the environment and health. In 2014, 744 environmental accidents in Brazil were reported, widely spread in the country, but more frequently in the southeastern region (64.3%), and mostly associated with road transport (28.3%), a trend also observed in past years (IBAMA 2014).

Among all existing risk assessment methodologies, MIAR was selected since it allows the integration of quality, environment, and occupational health aspects of the activities. MIAR allows the evaluation of synergistic effects and reduces subjectivity, resulting in a comprehensive and balanced evaluation (Ferreira & Santos Baptista 2013). Furthermore, this method is a tool that supports the improvement of critical aspects in Integrated Management System (IMS) of the company.

3 MATERIALS AND METHODS

To perform the quantitative and qualitative assessment of the case study, audits were conducted in the company and its activities. Legal compliance aspects and the quantitative methodology applied are described in sections 3.1 and 3.2.

3.1 Legal compliance assessment

Legal compliance was based on the Brazilian Occupational Health and Safety (OHS) regulatory standards (which are also known as "Normas Regulamentadoras" or NR). These standards are a responsibility of the Ministry of Labor and Social Welfare[1], supported by the Law 6.514/77, among other legal documents. Following the identification of noncompliance with regulatory standards, a Gravity, Urgency and Tendency (GUT) Priority Matrix was applied to identify preventive and corrective measures.

3.2 MIAR Methodology

The choice of this method is based on the possibility of integrating environmental and occupa-

Table 1. Risk indexes. Source: Antunes et al. (2010).

RI	Value	Meaning
1	1–90	Minor
2	91–250	Medium
3	251–500	High
4	501–1.800	Very High

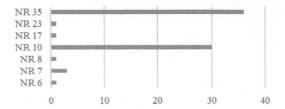

Figure 1. Distribution of non-compliances.

tional factors towards a more representative risk evaluation and identification of opportunities of improvement. The MIAR is a recently developed method in University of Porto which has been continuously validated in other case studies (Antunes et al. 2010, Ferreira et al. 2013, Bessa et al. 2015).

The method is based in the composition of a Risk Index (RI) based in the following parameters: G—gravity, E—impact extension, EF: exposure frequency, PC—prevention and control systems and C—costs and complexity of corrective measures. G is a result of the sum of P (Potential Hazards) and Q (quantification aspect). All parameters receive a score based on its consequences or intensities, as fully described in Antunes et al. (2010). The RI is built following Equation 1 and the meanings of possible results are presented in Table 1.

$$RI = G \times E \times EF \times PC \times C \qquad (1)$$

The methodology was applied only in the operational activity of the company, namely, the transport phase. This process was considered the most critical out of the operational procedures of the company, as evidenced by a previous evaluation of all existing sectors using a hazard matrix (Haddad et al. 2008).

4 RESULTS AND DISCUSSION

4.1 Legal compliance assessment

After the verification of compliance to national safety standards, a compilation of the audits was performed evidencing a total of 73 nonconformities distributed across 7 standards, as presented in Figure 1. These standards are related to: (i) the usage

[1]Free translation from Portuguese of "Ministério de Trabalho e Previdência Social"—MTPS.

Table 2. MIAR methodology applied to dangerous goods transportation.

(1)	Aspect	Aspect Characterization	Hazard	G (Q+P)	E	EF	PC	C	RI	(2)
1	TT enters the facility to the loading platform	Internal traffic—cars	Motor vehicle collision (MVC)	2	3	2	2	3	72	1
		Internal traffic—people	Run overs	5	1	2	2	3	60	1
2	TT enters loading platform	Excessive proximity to equipment	Entrapment	3	1	2	2	3	36	1
			Crushing	10	1	2	2	3	120	2
3	Loading inspection	Work in high places > 2,0 meters	Slips, trips and falls	10	1	3	1	3	90	1
4	Hatch opening for inspection	Exposure to chemical hazards and toxic substances	Dermal and respiratory exposure to chemical agents	1	4	3	2	3	72	1
		Equipment accident	Hand injury	2	1	1	4	3	24	1
5	Fuel pumping	Static electricity	Fire	5	1	2	1	3	30	1
			Explosion	10	1	2	1	3	60	1
6	Fuel loading and unloading	Spill or accidental release of fuel or other combustible material	Cutaneous conditions	1	1	2	3	3	18	1
			Environmental contamination	4 (1+3)	1	2	4	2	64	1
7	Product transportation to customer/ receiver	Traffic Accident	MVC	10	3	2	4	2	480	3
			Rollovers	10	3	2	4	2	480	3
			Run overs	10	1	2	4	2	160	2
			Fuel or another combustible material spill	8 (5+3)	3	1	4	2	192	2
		Atmospheric emissions	Atmospheric pollution	5 (2 +3)	2	3	4	2	240	2

(1) Activity / (2) Classification.

of Personal Protective Equipment (PPE) – NR 6, (ii) occupational health control program—NR 7, (iii) safety in workplace facilities—NR 8, (iv) electric systems—NR 10, (v) ergonomics—NR 17, (vi) fire prevention systems and (vii) work at heights—NR 35.

A priority sequence was established for the actions with the implementation of GUT Priority Matrix, leading to 18 recommendations to achieve their legal compliance. Among corrective actions identified, 8 are related to work at heights and the most critical aspect was the ergonomics risks involved in the activities performed by drivers.

In terms of ergonomics, it was evidenced that bottom loading system may be prioritized to avoid the need of climbing in the truck. For work at heights, definition of working procedures, risk analysis of tasks and inspection at workplaces are some of the measures necessary to fulfill the legal compliances.

4.2 MIAR Implementation

The first phase of the study consisted in the systematic and organized collection of the existing information related to OHS in the company, followed by a survey of existing hazards in the transportation process. For the evaluation of atmospheric emissions, carbon dioxide (CO_2) emis-

sion factor for road transport using diesel as fuel was considered equivalent to 74.10 kg de CO_2/GJ (IPCC 2006). Following these steps, a risk evaluation accordingly to MIAR was performed and summarized in Table 2.

All 7 stages of the workflow were evaluated and further associated to 16 hazard categories or scenarios. From these 16 categories, 62.5% were classified as minor, 25% as medium and 12.5% as high. Product transportation to customer/receiver can be considered as the most critical activity since the associated risk categories concentrated the highest classification.

Considering this result, preventive measures for minimizing traffic accidents and transportation risks were delimited focusing on causes identified in literature. The main causes of accidents involving dangerous goods have been assigned as drivers' errors (44.3%), others (23.61%), vehicle failure (21.83%) and road conditions (3.71%) (Ferreira 2003).

Considering this aspect, the main preventive measures for the process is the reduction of drivers' errors by intensification of training as well as better compliance with safety regulations, such as Law 13.103/2015, which provides specifications for the exercise of professional drivers. Furthermore, organizational factors may be improved such as the monitoring of working hours, resting hours, and establishment of psychological monitoring.

In terms of vehicle failure, the company maintains a rigid maintenance control and pursues a contract of preventive maintenance with the TT manufacturer authorized agency to ensure safety and comfort conditions within the fleet. Additionally, all TT are less than five years old, which reduces both mechanical failures and spill risks.

In terms of atmospheric emissions, all existing TT run on low Sulphur diesel, reducing emissions of particles and white smoke. The TT are also equipped with Selective Catalytic Reduction (SCR) systems and make use of a reagent to chemically reduce emissions of nitrogen oxides present in exhaust gases.

In addition to these aspects, it is worth mentioning that emergency response and information systems are important gaps in controlling the consequences of these events at national level. In the USA, e.g., accidents involving the transportation of dangerous goods and hazardous materials can be reported and registered in proper databases, such as the one maintained by the Pipeline and Hazardous Materials Safety Administration (PHMSA), an agency of the national Department of Transportation.

In Brazil, national databases are still limited. The creation of a database integrating different national institutions would allow the development of better transportation routes and accident indicators, especially in driver related incidents.

5 CONCLUSIONS

Dangerous products transportation can have very serious consequences for the population in general and particularly for workers exposed to these products, as well as property and environmental damage. The assessment conducted evidenced several noncompliance items according to the national OHS standards, which were ranked in a GUT Priority Matrix.

In the evaluated company, a previous risk assessment was conducted through a hazard matrix, which identified the most critical sector as the operational activities. Following this, the application of MIAR methodology supported the ranking of risks among its workflow, identifying transportation itself as the most critical stage in the operational process, in which more urgent interventions are required to minimize their risks.

Future works should explore the possibility of conducting other risk evaluations in the operational process, such as the identification of risks on route. Transportation of dangerous goods is an important subject to ensure the mobility of goods and should be the focus of improved enforcement to ensure the safety of people and the preservation of the environment.

REFERENCES

Antunes, F., Santos Baptista, J. & Tato Diogo, M. 2010. Methodology of integrated evaluation of environmental and occupational risks. In *SHO 2010 - International Symposium on Occupational Safety and Hygiene*. Guimarães.

Bessa, R., Santos Baptista, J. & Oliveira, M. J. 2015. Comparing three risk analysis methods on the evaluation of a trench opening in an urban site. In *Occupational Safety and Hygiene III*.

Brazil. 1997. Lei n° 6.514, de 22 de Dezembro de 1997. Brasília, Brasil: Presidency of the Republic.

Brazil. 2015. Lei n° 13.103, de 2 de Março de 2015. Brasília, Brasil: Presidency of the Republic.

Ferreira, C. & Santos Baptista, J. (2013). The risk in choosing the method of Risk Assessment. *Occupational Safety and Hygiene*, 1(1), 138–139.

Ferreira, C. E. C. (2003). Acidentes com motoristas no transporte rodoviário de produtos perigosos. *São Paulo em Perspectiva*, 17, 68–80. doi: 10.1590/S0102-88392003000200008

Haddad, A. N., Morgado, C. V. & Desouza, D. I. 2008. Health, safety and environmental management risk evaluation strategy: Hazard Matrix application case studies. Paper read at 2008 IEEE—International Conference on Industrial Engineering and Engineering Management (IEEM), at Singapore.

IBAMA, Instituto Brasileiro do Meio Ambiente e dos Recursos Naturais Renováveis. 2014. Relatório de Acidentes Ambientais. edited by IBAMA. Brazil: IBAMA.

IPCC, Intergovernmental Panel on Climate Change. 2006. Mobile Combustion. In *IPCC Guidelines for National Greenhouse Gas Inventories* edited by IPCC.

Kraus, Richard S. 1988. "Oil and Natural Gas." In *ILO Encyclopaedia of Occupational Health and Safety*, edited by Jeanne Mager Stellman. Geneva: ILO, International Labour Organization.

Laarabi, Mohamed Haitam, Boulmakoul, Azedine, SACILE, Roberto & Garbolino, Emmanuel. (2014). A scalable communication middleware for real-time data collection of dangerous goods vehicle activities. *Transportation Research Part C: Emerging Technologies*, 48, 404–417.

Mtps, Ministério do Trabalho e Previdência Social. 1978a. NR 17 – Ergonomia. Brasilia, Brazil.

Mtps, Ministério do Trabalho e Previdência Social. 1978b. NR 23 – Proteção Contra Incêndios. Brasilia, Brazil.

Mtps, Ministério do Trabalho e Previdência Social. 2001. NR 8 – Edificações. Brasilia, Brazil: MTPS.

Mtps, Ministério do Trabalho e Previdência Social. 2004. NR 10 – Segurança em instalações e serviços em eletricidade. In *Brasilia, Brazil*.

Mtps, Ministério do Trabalho e Previdência Social. 2009. NR 6 – Equipamento de Proteção Individual. Brasilia, Brazil.

Mtps, Ministério do Trabalho e Previdência Social. 2012. NR 35 – Trabalho Em Altura. Brasilia, Brazil.

Mtps, Ministério do Trabalho e Previdência Social. 2013. NR 7 – Programa de Controle de Controle Médico de Saúde Ocupacional. Brasilia, Brazil.

Occupational Safety and Hygiene V – Arezes et al. (Eds)
© 2017 Taylor & Francis Group, London, ISBN 978-1-138-05761-6

Understanding the management of occupational health and safety risks through the consultation of workers

I. Castro
Polytechnic Institute of Cávado and Ave, Barcelos, Portugal

D.G. Ramos
School of Engineering, Polytechnic Institute of Cávado and Ave, Barcelos, Portugal
Algoritmi Centre, University of Minho, Guimarães, Portugal

ABSTRACT: Understanding the perceptions of employees on the risk management system is very important, but there is still a lack of full understanding of how the workers characterize the risk. This study aimed, through the consultation of workers, understanding the perception of risk by these and thus improve the risk management system. A questionnaire addressed to 201 employees of a metalworking industry has been prepared. The results of the questionnaire showed that it is necessary a higher involvement of the workers in the process of management of occupational hazards and accidents at work and the corresponding communication. Since the organization is preparing the process of certification of the Occupational Health and Safety (OHS) management system, according to ISO/DIS 45001, it was important in this study to analyse, discuss and understand the perception of workers of OHS risk.

1 INTRODUCTION

The risk perception reflects the needs, issues, knowledge, beliefs and values of the stakeholders (ISO Guide 73, 2009). The understanding of workers' perceptions of the risk management system is very important, but there is still a lack of full understanding of how the workers characterize the risk (Alexopoulos et al., 2009). According to Wachter and Yorio (2014) in a behavioral perspective, workers bring their beliefs, culture, values and vision when performing its work, and this is important when designing and implementing an Occupational Health and Safety Management System (OHSMS). Also according to Alexopoulos et al. (2009), there are cultural differences in the perception of risk of activities that pose threats to occupational health and safety. Thus, it is important to evaluate the perception of risk by workers in order to develop an adequate safety culture.

Experience has a vital role in the perception of risk because, for example, misleading experiments may be linked to the tendency for a worker to believe that he is personally immune to many dangers. According to the study made by Wachter and Yorio (2014), employee involvement in occupational health and safety reduces the probability of human error.

Workers are more involved and aware of their tasks and the associated risks. According to Rundmo (1996) the perception of risk is significantly related to risk behaviours, providing an important vision for safety management. According to ILO (2011) training in OHS is essential and should be ongoing in order to assure the system of knowledge and also that the instructions related to changes in the organization are always updated. In this sense, communication between the different levels of the company should be effective and in both directions, once the health and safety related contributions offered by workers should be taken into consideration by the management, which demonstrates the importance of focusing the system on workers.

According to ISO Guide 73 (2009) communication and consultation are continuous and iterative processes that an organization conducts in order to provide or share information, and to engage in dialogue with stakeholders with regard to risk management. The authors Arocena and Núñez (2009) and Torp and Grøgaard (2009) also report that the legislative compliance influences the implementation and improvement of OHSMS. The consultation of workers on OHS in Portugal must be made in written form once a year, covering the content of all the paragraphs contained in section 1 of art. 18 of Law No. 102/2009 of 10 September, amended by Law No. 3/2014, of 28 January, as there is the assumption that it is the employer's obligation to take appropriate measures to carry out the activities listed in article 18.

Risk management is one of the priorities for research related to OHS in Europe over the period 2013–2020 according to the European Agency for Safety and Health at Work (EU-OSHA, 2013). An adequate risk management system leads to a reduction in the costs of accidents, incidents and diseases, increased competitiveness and productivity of organizations and contributes effectively to the sustainability of social protection systems. According to EU-OSHA (2016) and Ramos (2013), it must be observed that poor OHS has costs. Ramos et al. (2016) show that there is a direct relationship between good management of OHS in the organization and the improvement of performance and profitability. Everyone loses when OHS is neglected, from the workers at the individual level to the national health systems. However, this means that everyone can benefit from the implementation of adequate policies and practices that inevitably will prove to be the most effective. Ramos and Costa (2016) emphasized the importance of a risk management system appropriate to increase the competitiveness and productivity of the organization. Such a system depends on technical and human aspects and can lead to fewer accidents and result in increased employee motivation.

The aim of this study was, through the consultation of workers, understanding the perception of risk by these and thus improve the risk management system in an organization of the metalworking sector.

2 MATERIALS AND METHODS

2.1 *Case study*

This study was conducted in an organization linked to the area of metalworking. The organization is distinguished by different areas of intervention. The organization is certified by the quality management system according to ISO 9001 and is in process of certification of the occupational health and safety management system, according to ISO 45001 and the innovation management system according to NP 4457. The organization has 201 workers.

2.2 *Research methodology*

According to the legislation, the health and safety technician of the organization prepared a questionnaire to all the 201 employees of the company. The questionnaire was divided into 10 sections, with the possible answers "yes", "no" and "no opinion". The questionnaire was tested in a pilot survey that was conducted for a sample of 2 workers to detect any possible weak points, get feedback

on the intel-ligibility and ambiguousness of the questions. Some improvements in the instrument formatting was car-ried out to facilitate its fill. The full questionnaire, validated by the administration, is presented in Appendix. The query data to employees concerns the year 2015.

3 RESULTS AND DISCUSSION

Out of the 201 workers of the company, 166 workers both of productive and non-productive sectors have responded to the questionnaire. The response rate was thus 83%.

The perception of risk is the stakeholder view on a risk (ISO Guide 73, 2009; ISO/DIS 45001, 2016). In this study, the stakeholders are the employees. For the present study were selected only the issues of sections 2 "professional risks" and 6 "work accidents" because they are the most relevant for this study.

Table 1 shows the results for the questions of sections 2 and 6 of the consultation questionnaire to employees in OHS.

Figure 1 shows the results of the consultation questionnaire to employees concerning the section 2 "professional risks".

Given the importance of understanding the perceptions of the employees on the management

Table 1. Results of the questionnaire of consultation of workers (n = 166).

Questions	n	% Answers		
		Yes	No	No opinion
2. Occupational Risks				
2.1 Do you receive information about the risks to which you are exposed in your workplace?	166	71	23	6
2.2 Do you receive information about the preventive measures to eliminate risks in your workplace?	166	68	25	7
6. Work accidents				
6.1 Have you had any accident at work?	166	45	53	2
6.2 The results of investigations of accidents have been communicated?	166	27	22	51
6.3 If you have suffered any accident, have you been involved in the research?	166	28	38	34

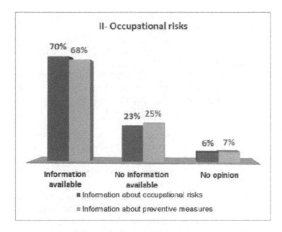

Figure 1. Results of the questionnaire of consultation of workers. Section 2 "professional risks".

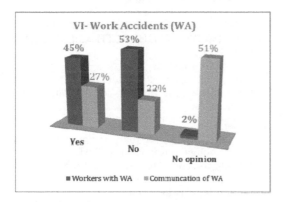

Figure 2. Results of the questionnaire of consultation of workers. Question 6 "work accidents".

system of occupational risks in the organization, the results of Question 2 shown in Figure 1 confirm that a greater involvement in the process of management of occupational risks is necessary. Given these results, it is proposed that the organization should implement the following measures: i) restructuring of the risk assessment for the workplace, ii) promote training activities held in the previously completed risk assessment, in order to provide to the workers knowledge concerning the risks associated with their professional activity, as well as preventative measures to be considered.

Figure 2 shows the results of the consultation questionnaire to employees concerning section 6 "work accidents".

45% of employees who responded to the questionnaire have already suffered accidents. However, most workers who have suffered work accidents have not been involved in the corresponding research.

These results reflect the need to prevent work accidents in a more efficient / effective way. It is also important that the company develops channels to better communicate to employees the results of investigations of accidents. Given these results, it is proposed that the organization should implement the following measures: i) make quarterly disclosure of accidents occurred in the company, ii) promote actions awareness to prevent work accidents, iii) better involve workers in the investigation of accidents.

4 CONCLUSIONS

An appropriate risk management system is vital to increase the competitiveness and productivity of the organization. The conditions of workers and their perceptions in terms of OHS should be observable. According to Alexopoulos et al. (2009), the understanding of workers' perceptions of the risk management system is very important, but there is still not a full understanding of how the workers characterize the risk.

Risk management is one of the priorities for research related to OHS in Europe over the period 2013–2020, according to the European Agency for Safety and Health at Work. The new ISO 45001 will force organizations that adopt this new framework to have a management system with greater focus on risk, in order to succeed in controlling health and safety at the level required by the organization.

Since the organization under study is preparing the certification process of the occupational health and safety management system, according to ISO 45001, the organization should pay special attention to the results of the annual consultation of employees, reflecting on them. It was central to this study to investigate, analyse, discuss and understand the perception of risk of the workers in terms of OHS.

Thus, communication between the different levels of the organization must be effective and in both directions, as in fact the health and safety related contributions given by the workers should be taken into consideration by the administration, which demonstrates the importance to focus the OHS system on workers.

From this research, important academic and managerial implications on risk management in OHS can be addressed. Indeed, it is necessary to understand the dynamics of adoption and institutionalization of risk management practices in OHS systems.

The results of the questionnaire of consultation of workers show that there are important measures that should be taken into account by the organization, especially in terms of professional risks and of work accidents.

Risk Management is a shared continuous and interactive process, for which all company employees can and must contribute at different times and for different stages of the risk management process. The contribution of different employees can occur in different ways, considering the various process steps.

This contribution may evolve as the risk management process becomes mature and institutionalized (definitely embedded in day-to-day business).

ACKNOWLEDGEMENTS

The authors would like to thank the company for supplying the data and for its collaboration in the questionnaire.

REFERENCES

Alexopoulos, E. C., Kavadi, Z., Bakoyannis, G. and Papantonopoulos, S. (2009). "Subjective Risk Assessment and Perception in the Greek and English Bakery Industries". Journal of Environmental and Public Health Volume 2009. Hindawi Publishing Corporation.

Arocena, P. & Núñez, I. (2009). The effect of occupational safety legislation in preventing accidents at work: traditional versus advanced manufacturing industries. Environment and Planning C: Government and Policy, 27(1), 159–174. Retrieved from http://ideas.repec.org/a/pio/envirc/v27y2009i1p159–174.html.

EU-OSHA (2013). European Agency for Safety and Health at Work. Priorities for occupational safety and health research in Europe: 2013–2020. ISBN 978-92-9240-068-2. Luxembourg.

EU-OSHA (2016). Agência Europeia para a Segurança e Saúde no Trabalho. Bons níveis de SST são um bom negócio. Consulted in April 2016, available at https://osha.europa.eu/pt/themes/good-osh-is-good-for-business.

ISO/DIS 45001 (2016). Occupational health and safety management systems—Requirements with guidance for use.

ISO Guide 73 (2009). Risk Management. Vocabulary.

OIT—Organização Internacional do Trabalho (2011). Sistema de Gestão da Segurança e Saúde no Trabalho: um instrumento para uma melhoria contínua. ISBN 978-92-2-224740-0. Torino.

Ramos, D. G. (2013). "Análise Custo-Benefício em Avaliação de Risco Ocupacional". PhD Thesis is Industrial and Management Systems. University of Minho.

Ramos, D., Arezes, P.M. & Afonso, P. (2015). Analysis of the Return on Preventive Measures in Musculoskeletal Disorders through the Benefit-Cost Ratio: a Case Study in a Hospital. International Journal of Industrial Ergonomics. DOI: 10.1016/j.ergon.2015.11.003.

Ramos, D. & Costa, A. (2016). Occupational Health and Safety Management System: a case study in a waste company. Arezes et al. (eds), Occupational Safety and Hygiene IV, Taylor & Francis Group, London, ISBN 978-1-138-02942-2, pp. 597–601.

Rundmo, T. (1996). Associations between risk perception and safety. Safety Science 24, 197–209. http://dx.doi.org/10.1016/S0925-7535 (97)00038-6.

Torp, S. & Grøgaard, J. B. (2009). The influence of individual and contextual work factors on workers' compliance with health and safety routines. Applied Ergonomics, 40(2), 185–193. doi:10.1016/j.apergo.2008.04.002.

Santos, G., Ramos, D., Almeida, L., Rebelo, M., Pereira, M., Barros, S. & Vale, P. (2013). Implementação de Sistemas Integrados de Gestão: Qualidade, Ambiente e Segurança., 2ª Edição, ISBN: 978-989-723-038-7. Publindústria, Edições Técnicas.

Wachter, J. K. & Yorio, P. L. (2014). A system of safety management practices and worker engagement for reducing and preventing accidents: An empirical and theoretical investigation. Accident Analysis and Prevention 68, 117–130.

APPENDIX: Questionnaire

CONSULTATION TO WORKERS IN TERMS OF OCCUPATIONAL HEALTH AND SAFETY SYSTEM (OHS)

General Information

Name, gender, age, education, number of years working in this firm, function/activity, number of work accidents.

1. General
1.1. Do you consider that the company fulfils its obligations in terms of OHS?
1.2. Do you consider that the company has adequate OHS conditions?

2. Occupational risks
2.1. Do you receive information about the risks to which you are exposed in your workplace?
2.2. Do you receive information about the preventive measures to eliminate risks in your workplace?

3. Emergency
3.1. Do you have some kind of knowledge of firefighting?
3.2. Can you properly handle a fire extinguisher?
3.3. Do you have basic knowledge in first aid?

4. Personal Protective Equipment (PPE)
4.1. Do you have available adequate PPE to prevent the risks inherent to your job function?
4.2. When you use PPE, do you know against what kind of risk you are being protected?
4.3. Do PPE present good condition?
4.4. Are PPE adjustable and comfortable?
4.5. Are they compatible with other PPE?
4.6. Are they available in sufficient number?

5. Equipment and Machinery
5.1. Equipment and machines have adequate security conditions?

5.2. The maintenance is up to date?

5.3. Do equipment and machines have electrical cords in good condition?

6. Work accidents

6.1. Have you had any accident at work?

6.2. The results of investigations of accidents have been communicated?

6.3. If you have suffered any accident, have you been involved in the research?

7. Comfort and hygiene and sanitary conditions

7.1. Are workplaces clean, maintained, organized and sanitized?

7.2. Are workplaces properly illuminated in order to perform its function?

7.3. Do workplaces ensure a suitable temperature?

7.4. Are sanitary facilities cleaned, preserved and well organized?

8. Social Spaces

8.1. Do you regularly eat food in your job?

8.2. Has the company space for the meals with means to heat food, benches/chairs and tables?

8.3. Do you consider important that the company provides a space for meals?

9. Training

9.1. Does the company provide training for workers?

9.2. o you consider it appropriate?

10. Occupational health

10.1. Does the company provide health screenings?

10.2. Do you consider the occupational health services appropriate?

Occupational Safety and Hygiene V – Arezes et al. (Eds)
© 2017 Taylor & Francis Group, London, ISBN 978-1-138-05761-6

Influence of cold thermal environment on packing workers from the frozen food processing industry

T. Zlatar
Research Laboratory on Prevention of Occupational and Environmental Risks (PROA/LABIOMEP), University of Porto, Portugal

B. Barkokébas Jr
University of Pernambuco, Pernambuco, Brazil

L. Martins
Federal University of Pernambuco, Recife, Brazil

M. Brito
Centre of Mathematics of the University of Porto CMUP, Portugal

J. Torres Costa, M. Vaz & J. Santos Baptista
Research Laboratory on Prevention of Occupational and Environmental Risks (PROA/LABIOMEP), University of Porto, Portugal

ABSTRACT: In the fresh food industry the working activities are conducted in environmental temperatures from 0°C to 10°C, while in the frozen food industry usually are at temperatures below −20°C which influences the variations in core and skin body temperatures and affects the working performance, health and safety of the employees. The aim of this work is to contribute with a study on the influence of cold thermal environment on core and skin body temperatures in packing workers from the frozen food industry. By using the core body pill sensor and 8 skin temperature sensors a study was conducted on 4 workers during 11 days. The lowest recorded temperature was for hand 14.09°C, mean skin temperature had variations of 1.10 to 3.20°C along the working period and the mean body temperature on two occasions decreased below 35°C. The core temperature was found to increase. The mean body temperature showed small changes along the time.

1 INTRODUCTION

The frozen food deliver high quality, good value, safe foods with an extended storage life, helping the dietary portion control and reducing waste, offer the possibility to preserve and use seasonal foods all year round (Young et al. 2010). In emerging markets like Latin America, South East Asia and Eastern Europe, an increase demand for richer and more varied diets will occur and, importantly, increase demand for large domestic appliances such as freezers (Kennedy 2000). Further on, shifts in global economic, social and demographic trends will continue to put pressure on food supplies, as we already witness today, more frozen foods to be sold each year and new products introduced to swell the total sales (Artley, Reid, and Neel 2008). The structure of the labour market is constantly undergoing change, away from fresh and homegrown towards chilled and frozen food, and its future trend is shown clearly through the development in recent years (Baldus, Kluth, and Strasser 2012).

Indoor working exposure to cold offer constant and predictable climate conditions, which facilitates cold risk management and workers cold adaptation. The different types of cold adaptations are related to the intensity of the cold stress and to individual factors such as body fat content, level of physical fitness and diet. The hypothermic general cold adaptation seems the most beneficial for surviving in the cold but the interest of the development of general cold adaptation in workers in the cold are questionable since occupational activities can be organized to avoid cold disturbances (shelter, clothes, heat sources, time sharing). For the workers working in cold environments, adaptations of extremities are beneficial, as are developing cold induced vasodilation, improving manual dexterity and pain limits

(Launay and Savourey 2009). Workers with less years of activity seem to be more satisfied with the cold thermal ambient than veterans with more than 10 years (Oliveira et al. 2014). Cold work involves several adverse health effects that are observed in indoor work. Many of these adverse outcomes may be further aggravated in persons having a chronic disease (Mäkinen and Hassi 2009).

Severe Cold thermal Environment (SCE) reduces skin body temperature (Tskin) and therefore physical working performance, lower muscle performance, maximal grip frequency and grip strength, hand and finger dexterity, maximal voluntary contraction, while increase muscle fatigue (Zlatar, Baptista, and Costa 2015). In cold, musculoskeletal complains and symptoms are common (Oksa, Ducharme, and Rintamäki 2002), which further on might lead to work accidents (Mäkinen and Hassi 2009). By a questionnaire conducted by Taylor, Penzkofer, Kluth and Strasser (Penzkofer, Kluth, and Strasser 2013), it was found that order-picking work in the cold leads to frequent complaints especially in the upper part of the body. Repeated, prolonged and chronic hyperpnoea with cold dry air represents a significant environmental stress to the proximal and distal airways, leading to the development of respiratory symptoms, airway hyper-responsiveness and injury, and inflammation and remodelling of the airway (Sue-chu 2012). When the human body is exposed to cold, the initial response is to preserve heat by reducing heat loss. The skin blood flow, especially in extremities is reduced by vasoconstriction (Charkoudian 2010), which leads to increased systolic and diastolic blood pressure, lowered heart rate and body temperature in extremities (Gavhed 2003). It is very well documented that there is an increase mortality related to Acute Myocardial Infarction (AMI) during the cold season (Chang et al. 2004; Kriszbacher et al. 2009). Some authors considered fluctuation in air temperatures as a major influence on AMI and stroke, especially when it comes to sudden decrease in temperature in a 24-hour period (Gill et al. 2013). A 5°C reduction in mean air temperature was associated with 7% and 12% increase in the expected hospitalization rates of stroke and AMI, respectively (Chang et al. 2004). On other hand, Kriszbacher suggested that beside temperature fluctuations, the barometric pressure and front movements greatly contributed to AMI cases (Kriszbacher et al. 2009). Nevertheless, cardiovascular diseases can be reduced by good management program of risk factors at work (Mitu and Leon 2011).

The objective of this work was to evaluate the influence of cold thermal environment on the core and skin temperature of packing workers in the frozen food processing industry.

2 METHODOLOGY

2.1 General data

The experiments were conducted in the cold packing sector from the frozen food processing industry. All the documents (informed consent, participation and trial forms, information for the volunteers) were translated into Brazilian Portuguese and reviewed and culturally adapted by a native Brazilian speaker. The experiment was approved by the Ethics Committee of the University of Porto, approval number: 06/CEUP/2015.

In Table 1 is shown the outside environmental conditions at near the industry location, gathered by the National Institute of Meteorology (INMET), station of meteorology of Macau, Rio Grande do Norte, Brazil.

All workers usually spent their time in moderate cold thermal environment, conducting moderate to heavy physical work (packing 400 grams packages into 20 kg packages, check the stored material, organize materials on pallets, separate materials from pallets, move pallets and heavy loads with forklifts, once a week breaking the ice on the floors in SCE chambers and do heavy lifting). The only male with low intensity work was the volunteer number 1 which was the leader of logistics. The logistic leader had to delegate working tasks, control, supervise, and count packages. He was a connection between offices and the stock, therefore no heavy work was conducted, but still, in order to check the packages and organize them, his exposure to SCE was with greater intensity compared with other workers. Other volunteers selected to be fully monitored were male packing operators, spending mostly their time in moderate cold thermal environment, and several times per day storing frozen packages in the SCE chambers. The thermometer in the moderate cold sector was showing environmental temperature of 18 ± 2°C, but as it was placed at 7 meters height, it was showing the highest room temperature. The workers were exposed to much lower temperature as they were working on a lower height, and they were located in front of SCE chambers with −25 to

Table 1. Outdoor air temperature, relative humidity and velocity data from the measuring days.

	09:00	12:00	13:00	17:00
Mean temp. ±SD (°C)	25.1 ±0.84	29.3 ±1.48	29.8 ±2.13	29.7 ±2.00
Mean Relative Humidity ±SD (%)	87.8 ±3.86	68.0 ±9.95	67.0 ±11.81	67.0 ±11.00
Mean Air velocity ±SD (m/s²)	1.9 ±0.34	3.6 ±1.66	4.2 ±1.29	5.6 ±1.00

–30°C, which would cool the moderate cold sector each time the SCE chamber was opened.

2.2 *Fully monitored workers*

Four workers from the cold packing sector were chosen to be fully monitored during their working activities. Three of them were screened during 3 working days, while one during 2 working days. In total, a sample of 11 measurements was achieved. The mean age ± sd of the successfully fully monitored workers was 29 ± 6.3 years old, mean body height was 167.4 ± 5.4 cm, mean weight of all 11 measuring days was 79.7 ± 17.9 kg, mean Body Mass Index (BMI) was 28.4 ± 6.3 kg/m². The medical examination was conducted by the industrial medical doctor. All subjects were informed about the goals and risks of the experiments, and signed the informed consent prior to participating. The subjects were examined and asked to drink the usual amount of coffee, tea, to avoid drinking alcohol for at least 12 h before the test; not to eat spicy food at least 12 h before the test; sleep normally before the test (about 8 h); not conduct greater physical exertion than it is usual for the volunteer at least 1 day before the test. According to volunteer's answers to the participation and trial forms, all were non-smokers, didn't drink tea, alcohol, eat spicy food, were right handed and had a usual physical exertion the day before the trial was conducted. Every day before the trial, it was recorded what they ate in their previous meal, what time they ate, if they took some medicines, the time when they went to sleep and when they woke up, as well the number of hours they slept. All subjects had the same working period of 8 hours, the morning part from 07:30 till 11:30, the pause of 1 hour, and finally the afternoon part from 12:30 till 16:30 (working hours varying depending on the process situation).

The experiments were conducted during a normal working day, with the subjects performing the usual tasks being exposed to SCE as usual and over the usual time period. Workers conducted usual industrial work of 8 hours at 16 to 18°C (measured at the height of 7 meters) and entered for several times in the frozen food chamber at air temperature of –25°C. Tskin was measured with Bioplux skin temperature sensors. The sensors were put on 8 measuring points (forehead, right scapula, left upper chest, right arm in upper location, left arm in lower location, left hand, right anterior thigh and left calf) according to ISO 9886:2004 (ISO 2004). For measuring the core temperature (Tcore) was used an Equivital ingestible pill sensor (thermometer telemetry capsule) with dimensions of 8.7 mm by diameter and 23 mm by length. It was swallowed with water for at least 5 hours before each test (usually before going to sleep); travelled along the digestive tract harmlessly, and leaving naturally within 24 to 72 hours. The sensors began to transmit one minute after the capsule activation by the external monitor, sending details every 15 seconds to the *EQ02 Life Monitor—Electronics Sensor Module* (SEM), which transmits the data via Bluetooth. The SEM is transported in a belt, recording the data from Tcore, chest skin temperature, heart rate and respiratory frequency.

2.3 *Clothing*

Clothes were given by the company as uniforms. The subjects normally wore normal cotton clothing (socks, underpants, t-shirt, trousers) and above it the cold protective clothing (jacket with a hood, trousers, boots), and sometimes gloves when entering the cold chamber.

2.4 *Data analysis*

The references were searched through databases by using the institutional IP address or University of Porto federate credentials. References were managed using the Mendeley 1.15.3. Tcore was recorded by using the Equivital Manager and EqView professional programs. Tskin was recorded by using the MonitorPlux program, later on to be processed by using the Matlab software program. The mean skin temperature was calculated using the weighting coefficients as suggested by ISO 9886:2004 (ISO 2004). Statistical analysis was done by using excel statistical toolbox.

3 RESULTS

As an example of Tskin and Tcore, results from two volunteers are illustrated in the Figures 1, 2, 3 and 4,

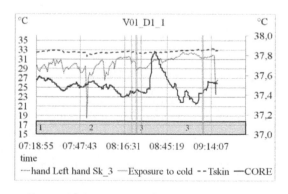

Figure 1. Results for the volunteer 1 on day 1 morning, left hand, Tcore and Tskin (1-working on a computer, 2-touching the cold package, 3-walking, counting, standing).

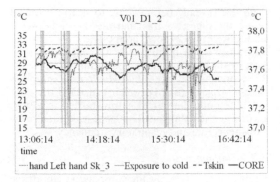

Figure 2. Results for the volunteer 1 on day 1 afternoon, left hand, Tcore and Tskin (logistic leader—delegate working tasks, control, supervise, count the packages).

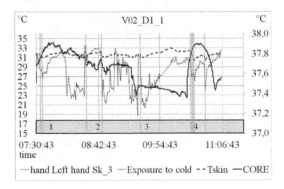

Figure 3. Results for the volunteer 2 on day 1 morning, left hand, Tcore and Tskin (1-packing packets (400 g), 2-pushing 80 kg, 3-pushing 100 kg packets, 4-hanging 20 kg packet).

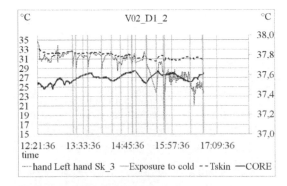

Figure 4. Results for the volunteer 2 on day 1 afternoon, left hand, Tcore and Tskin (loading packets with 20 kg of shrimps, washing out, shaking, transfer to other package).

Table 2. Minimal and maximal temperatures (°C).

	Left hand	Forehead	Core Temp.	Mean skin Temp.	Mean body Temp.
Min	14.09	18.55	36.55	29.08	34.94
Max	34.64	35.41	37.91	34.29	36.59

Temp. - temperature

where 1 and 2 represent the data of one volunteer in the morning and afternoon, while 3 and 4 represent the data of the other worker. On the left side axis are illustrated values for the hand and mean skin temperature varying from 14 to 35°C, while on the right side axis are illustrated values for the core temperature varying from 37 to 38°C. The vertical lines represent the exposure to SCE, which is in accordance with radical dropping of hand temperature.

In Table 2 are presented the mean, minimal (min) and maximal (max) values of two volunteers where the mean body temperature decreased below 35°C.

4 DISCUSSION

The workers experienced big fluctuation in air temperatures along the working days between outdoor and indoor environments, which might result with AMI and stroke (Gill et al. 2013), therefore the workers should be controlled on regular basis and the risk should be managed and reduced[21.22].

In the Figure 1, there is visible one radical dropping of hand temperature without the worker being exposed to SCE. In that case, the worker was touching the frozen food package in order to measure its temperature, but without using cold protecting gloves. The hand and forehead where found to have the highest and most frequent fluctuation in Tskin, which is reasonable as they were mostly not covered with the cold protective clothing. The forehead had a difference between the min and max recorded values, on average at least one time was recorded a drop of 5.32 ± 3.64°C, with in one case dropping even to 18.55°C.

The hand had the biggest differences between the min and max recorded values, on average at least one time was recorded a drop of 10.43 ± 4.87°C, with in one case dropping even to 14.09°C. The hand skin temperature recovery was fast, but when exposed to cold air or touching cold products or material it also dropped fast, making the changes of its temperature frequent, fast and with greater differences.

The mean Tskin show some small changes along the time, but without great and fast lowering or rising of the means Tskin, with a maximal variation between the min and max recorded value in one case of 5.14°C along all the measuring working period, and an average of between 1.10 to 3.20°C difference among the workers along all the measured working period.

The mean body temperature (Tbody) has no small change along the working period. As it is shown in the Table 2, on two occasions: for V01_D3_1 and V02_D2_1, the mean Tbody drops slightly below 35°C (on the min measured values 34.98°C and 34.94°C respectively. The current literature has described it as the start of the mild form of hypothermia (general freezing) which occurs when the body temperature drops to 32–35°C, and appears with shivering, tachycardia, tachypnea and slowness of ideation and compensated dysarthria. On appearing of mentioned clinical features, the workers should start the rewarming process[22,24].

Tcore variations were always less then 1°C difference between the min and max value, in the most of the cases between 37 and 38°C. Tcore seems to be raising with the exposure to SCE, which could be explained with vasoconstriction (Charkoudian 2010). Nevertheless, the exposure to SCE was short and should be studied with larger time of exposure to SCE before making further conclusions. The type of movement with its speed (accelerometry) and physical exertion also seem to influence the raising of Tcore, and should be therefore furthermore studied in order to explain the Tcore variations during the working period.

5 LIMITATIONS

Challenges were found in transmitting the recorded data of the Bioplux skin temperature sensors. The equipment didn't have the possibility of recording the collected data directly to the device, but only by Bluetooth with a maximal distance of 10 meters from the equipment to the computer. As the workers were constantly moving it was challenging to stay constantly on a distance of less than 10 meters from the subject, following them when they entered the chamber and staying close to them while conducting the working activities. The number of subjects was limited to the number of workers present in the cold packing sector. One of the limitations was that activities carried out under SCE exposure varied from worker to worker. In the factory sector where the activities were developed the thermometer was located at 7 meters height and not at the height where workers worked, so there is no real information on to which environmental temperature the workers were exposed in the moderate cold sector.

6 CONCLUSIONS

Highest and most frequent fluctuations were found in the hand and forehead skin temperature. The mean Tskin showed some small changes along the time, but without great or rapid changes. The mean Tbody showed small fluctuations along the working period, but in two cases dropped slightly below 35°C which in the current literature has been described as the start of the mild form of hypothermia (general freezing). Tcore variations were always less then 1°C difference between the min and max value, in most of the cases between 37 and 38°C. It was concluded that Tcore was rising when exposed to SCE (because of vasoconstriction) and with higher physical exertion. Further experiments should be conducted with a higher number of volunteers, greater exposure to SCE, having a bigger sample with same working activities, thermal environment and time of exposure.

REFERENCES

Artley, David, David Reid, and Stephen Neel. 2008. "Frozen Foods Handling & Storage." WFLO Commodity Storage Manual, 1–12.

Baldus, Sandra, Karsten Kluth, and Helmut Strasser. 2012. "Order-Picking in Deep Cold—Physiological Responses of Younger and Older Females. Part 2: Body Core Temperature and Skin Surface Temperature." Work 41: 3010–17. doi:10.3233/WOR-2012-0557-3010.

Chang, Choon Lan, Martin Shipley, Michael Marmot, and Neil Poulter. 2004. "Lower Ambient Temperature Was Associated with an Increased Risk of Hospitalization for Stroke and Acute Myocardial Infarction in Young Women." J. of Clinical Epidemiology 57 (7): 749–57. doi:10.1016/j.jclinepi.2003.10.016.

Charkoudian, Nisha. 2010. "Mechanisms and Modifiers of Reflex Induced Cutaneous Vasodilation and Vasoconstriction in Humans." Journal of Applied Physiology (Bethesda, Md. : 1985) 109 (4): 1221–28. doi:10.1152/japplphysiol.00298.2010.

Gavhed, Désirée. 2003. Human Responses to Cold and Wind. Edited by Staffan Marklund. National Institute for Working life. http://nile.lub.lu.se/arbarch/ah/2003/ah2003_04.pdf.

Gill, Randeep S, Hali L Hambridge, Eric B Schneider, Thomas Hanff, Rafael J Tamargo, and Paul Nyquist. 2013. "Falling Temperature and Colder Weather Are Associated with an Increased Risk of Aneurysmal Subarachnoid Hemorrhage." World Neurosurgery 79 (1). Elsevier Inc.: 136–42. doi:10.1016/j.wneu.2012.06.020.

Golant, Alexander, Russell M Nord, Nader Paksima, and Martin A Posner. 2008. "Cold Exposure Injuries to the Extremities." Journal of the American Academy of Orthopaedic Surgeons 16 (12): 704–15. doi:10.5435/00124635-200812000-00003.

ISO, 9886. 2004. "Ergonomics—Evaluation of Thermal Strain by Physiological Measurements." Int. Standards Organisation.

Kennedy, C. 2000. "The Future of Frozen Foods." Food Science and Technology Today.

Kriszbacher, Ildikó, József Bódis, Ildikó Csoboth, and Imre Boncz. 2009. "The Occurrence of Acute Myocardial Infarction in Relation to Weather Conditions." International Journal of Cardiology 135 (1). Elsevier Ltd.: 136–38. doi:10.1016/j.ijcard.2008.01.048.

Launay, Jean-Claude, and Gustave Savourey. 2009. "Cold Adaptations." Industrial Health 47 (3): 221–27. doi:10.2486/indhealth.47.221.

Mitu, Florin, and Maria Magdalena Leon. 2011. "Exposure to Cold Environments at Working Places and Cardiovasculare Disease." Revista de Cercetare Si Interventie Sociala 33 (1): 197–208.

Mäkinen, Tiina M, and Juhani Hassi. 2009. "Health Problems in Cold Work." Industrial Health 47 (3): 207–20.

Oksa, Juha, Michel B Ducharme, and Hannu Rintamäki. 2002. "Combined Effect of Repetitive Work and Cold on Muscle Function and Fatigue." Journal of Applied Physiology (Bethesda, Md. : 1985) 92 (1): 354–61.

Oliveira, A. Virgílio M., Adélio R. Gaspar, António M. Raimundo, and Divo A. Quintela. 2014. "Evaluation of Occupational Cold Environments: Field Measurements and Subjective Analysis." Ind. Health 52 (3): 262–74. doi:10.2486/indhealth.2012-0078.

Penzkofer, Mario, Karsten Kluth, and Helmut Strasser. 2013. "Subjectively Assessed Age-Related Stress and Strain Associated with Working in the Cold." Theoretical Issues in Ergonomics Science 14 (3): 290–310. doi:10.1080/1463922X. 2011.617114.

Sue-chu, Malcolm. 2012. "Winter Sports Athletes : Long-Term Effects of Cold Air Exposure," 397–401. doi:10.1136/bjsports-2011-090822.

Young, Brian, Warwick House, Long Bennington, Business Park, Main Road, Judith Evans, Churchill Building, Howard Street, and Charlotte Harden. 2010. "The British Frozen Food Industry—A Food Vision." British Frozen Food Federation, no. November.

Zlatar, T, J Baptista, and J Costa. 2015. "Physical Working Performance in Cold Thermal Environment: A Short Review." In Occupational Safety and Hygiene III, edited by SHO 2015 International Symposium on Safety and Hygiene, 401–4. CRC Press. doi:10.1201/b18042-81.

Occupational Safety and Hygiene V – Arezes et al. (Eds)
© 2017 Taylor & Francis Group, London, ISBN 978-1-138-05761-6

Risk mapping and prioritization—case study in a Brazilian industrial laundry

G.L. Ribeiro & A.N. Haddad
Escola Politécnica, UFRJ, Rio de Janeiro, Brazil

E.B.F. Galante
IME—Instituto Militar de Engenharia, Rio de Janeiro, Brazil

ABSTRACT: The Brazilian legislation 6.514/1977 provides guidance on how to manage health and safety aspects in the work environment. Under such legislation, this work aims to identify and map hazardous activities and substances within a case study, which is an industrial cleaning company that works on cleaning and dyeing clothes. The methodology divides the work in two main sections: the first addressing risk mapping and the second presenting and applying the Hazard Matrix (HM) risk assessment tool to this case study. The work concludes that the most critical hazards is the chemical storage room (with 29.3% of the hazards of the whole case study), followed closely by thermo-presses (25.9%), while the most severe hazards to be dealt with are associated with fire and explosion (which stands for 17.3% of the hazards) and noises (14.3%).

1 INTRODUCTION

The Brazilian legislation 6.514/1977 (Brazil 1977) provides guidance and protocols for companies to manage health and safety aspects in the work environment. Regardless of the company type, size or complexity, every single one of them must comply with this legislation and follow standards, known as "Normas Regulamentadoras" or just "NR". Among those NR, one in particular is addressed in this paper: NR15 (Brazil—MTE 2008; MTE 1978). The NR15 addresses hygiene aspects and imposes threshold limits for exposure to chemical, biological and physical agents, such as heat and noise.

The case study is a company whose activities relate to industrial cleaning, which means dealing with several chemicals and other hazardous substances.

2 OBJECTIVES

The aim of this work is to identify and map hazards (activities) related to industrial hygiene (chemical, biological and physical agents exposure), following a prioritization of such risks. Ultimately, we hope that this work could offer guidance on managing such risks in other similar companies.

3 METHODOLOGY

This paper describes a study that carried out a qualitative risk mapping, under the guidance of Brazilian legislation NR05 (Brazil—MTE 2008; MTE 1978; Brazil—MTE 2007) followed by hazards prioritization through the hazard matrix methodology (Haddad et al. 2012), which is described just before the results, for simplicity purposes.

4 CASE STUDY

The case study is an industrial cleaning company that works on cleaning and dyeing clothes. The industrial laundry has a staff of 30 people, who work a regular 44 hours/week. The company's job descriptions include financial manager, department staff, laundry aid, washer, chemical technician and driver, among others. The company has its premises inside a covered warehouse, which holds several different and isolated working environments.

The flow process starts with the arrival of the clothes into the laundry room, which are accommodated in the reception room while administrative procedure takes place. After being received and catalogued the items are machine washed, dried (if needed) and ironing. After this process, the parts are in stock "stand by" in specific rooms to be delivered to customers.

5 RESULTS AND DISCUSSION

This part of the paper is divided in two main sections: one addressing risk mapping and a second presenting and applying the Hazard Matrix (HM) risk assessment tool to this case study.

5.1 Risk mapping

Risk mapping is mandatory under Brazilian legislation (Item 5.16 of NR05) (Brazil—MTE 2007) and it should be carried out by a commission team maybe by employers and employees. It is also recommended to present the results in a graphical manner, which helps visualization and publicity of such hazards. Several guides are available to aid in mapping hazards on a workplace environment.

The Manual for Risk Mapping issued by the Government of Goiás/Brazil (Brazil—Goiais 2012) recognizes environmental risks as any related to any activity conducted by the company and suggests five hazard groups and colours code (Table 1).

Prior to applying this mapping technique to the case study, it is helpful to split the area in smaller sectors, which are classified by risk and activity similarities as much as possible. In this work, the case study has been divided in seven sectors (sectors A to G), as follows:

- Sector A: Distribution Rooms and stock stand by goods ready for delivery, reception of goods room, kitchen, dining room and bathrooms. In this sector work 13 employees.

Table 1. Hazard groups and colour codes for mapping.

Hazards type	Colour code	Examples
Physical Hazards	Green	Noise, heat, vibrations, abnormal pressures, radiation and humidity
Chemicals Hazards	Red	Dust, fumes, right-flies, gases, vapours, mists, chemicals and chemical compounds in genera
Biohazards	Brown	Viruses, bacteria, parasites, protozoa, fungi, bacilli
Ergonomic Hazards	Yellow	Physical exertion, leaks weight, poor posture, tight control of productivity, stress, work at night time, long working hours, monotony and repetitiveness and intense routine enforcement
Accident hazards	Blue	Inappropriate physical arrangement, machinery and equipment without protection, inadequate or defective tools, electricity, fire or explosion, poisonous animals

- Sector B: Dryers and effluent treatment units, compressor and pressure vessel. In this sector work 3 employees.
- Sector C: Thermo-presses, steam iron, area crafts. In this sector work 20 employees overall.
- Sector D: Chemical storage room, compressor, pressure vessel, and operational office, handling room chemicals, storage, bathroom and dressing room. In this sector work 17 employees.
- Sector E: Washing machines with different capacities, homogenization boxes, two boilers, water tank of 25.000 L. In this sector work 8 employees overall.
- Sector F: Disposal Area. In this sector work 10 employees
- Sector G: Washing Machines, ozone generator, one rest room and bathroom. In this sector work 9 employees.

In this case study the environmental risks of the above-referred 7 sectors are listed in Table 2. In sector A the risks identified as ergonomic are due to handling of laser-based machine, which pushes the operator to work under cycles of repetitive movements (thus monotonous). Accident risks are mainly originated from the presence of loose wires, hence electrical hazards. When coupling this with the combustible material (clothing), one can expect significant likelihood of both fire and explosion (there is cooking in stove). Everything is aggravated by poor layouts.

In sector B we identified risk as physical hazards those originated in gas compressor and dryers, which generate noise levels above 85 dB(A) (threshold provided by NR 15). The Effluent Treatment Plant requires handling of chemical products, thus suggesting chemical hazards. Regarding accident hazards we catalogued electricity (lack of protective conduits upon live wires), likelihood of fire or explosion (mostly because the combined electricity issues with cloths and chemicals) and unprotected ladders might cause operator falls.

The predominant hazards within sector C are heat exposure, mainly due to ironing activity, operating heat presses and irons. The safety management group working in the case study assessed heat exposures under the guidance of NR 15 (Brazil—MTE 2008) and it has been determined breach in those values. Ergonomically wise, the tasks are repetitive and monotonous, as well as demanding of wrong body posture.

During interviews, workers reported in the shoulders region (which can be consequence of operating with heads down). The operation of heat iron presses also causes the burning accident hazards, which exist along risk of falling; electricity (due to the presence of scattered wireless and electro-products); fire and explosion.

Table 2. Hazards identified and mapped within case study.

Sector	Physical Hazards	Chemicals Hazards	Biohazards	Ergonomic Hazards	Accident hazards
A	Not identified	Not identified	Not identified	Monotony and repetitiveness	Electricity, Fires and explosions, poor layout
B	Noise	Chemicals, several	Not identified	Not identified	Electricity, Fires and explosions, falls.
C	Heat	Not identified	Not identified	Monotony and repetitiveness, requirement of odd body positions	Burnings, falls, electricity, fire and explosion, hand cuts.
D	Noise	Chemicals, several	Not identified	Illumination	Chemical burns, Fires and explosions, poor layout
E	Heat and noise	Chemicals, several	Not identified	Not identified	Electricity, Fires and explosions, poor layout, unprotected machinery
F	Noise	Chemicals, several	Contaminated Air	Monotony and repetitiveness	Fire and explosions
G	Noise	Chemicals, several	Not identified	Monotony and repetitiveness, requirement of odd body positions	Fire and explosions

D sector in characterized by its significant noise levels generated by gas compressor. The chemical hazards relate to storage area and handling of chemicals. The poor illumination causes one to determine an ergonomic hazard in the area. Accident possible causes are chemical burns, poor layout (which compromises any evacuation procedure) and the possibility of fire and explosion due to activity of the gas compressor and electricity due to the presence of scattered wireless and cable ducts.

In the sector E the main physical hazards risks are noise from the operation of washing machines, and heat due to the presence of two boilers and poor ventilation in the area. The chemical hazard is characterized by handling miscellaneous chemicals. Accident risks are electricity, the inadequacy of the control cabinet, and this, easy access to all people circulating on the spot; risk of fire and explosion due to the activity of the boilers and the presence of large amounts of combustible material (clothing); poor layout allowing the presence of clothing furniture containers blocking the way as well as incorrect distance between equipment's. The last potential cause of an accident is unprotected machinery and equipment.

The physical hazards within both sectors F and G arise from the noise generated by the dyeing activity, which also causes both chemical hazards (handling chemicals) and ergonomic hazards. This last hazard is due to demand for improper posture during machine operations coupled with repeatitivity and monotony of such tasks. In the office rooms there are also ergonomic risks due to the employee's position in her job and biohazard is originated from lack of control in cleaning the air condition system. Accident hazards are mainly fire and explosion in blasting machine and electricity due to the presence of scattered wireless and cable ducts.

Usually, the next phase of risk mapping would be to plot these risks in a blueprint and make it known by everyone within the facility. This phase is not disclosed in this paper to avoid disclose the case study recognizable features.

5.2 Risk prioritization

According to Haddad et al. (2012), this methodology helps to understand and set priorities between several risks in complex systems. The technique is recommended when managing risks and is particularly efficient to highlight the criticality of the hazards and sectors within a case study. The HM consists of identifying and classifying hazards related to every activity within each sector of a company. This classification takes into account to two aspects: kind of danger (physical, chemical, biological, ergonomics and accident,

Table 3. Risk codes for HM application.

Code	Meaning
0	No hazard exposure
3	Exposure to hazard considered low. No control required
6	Exposure to hazard considered medium. Control recommended
9	Exposure to hazard considered medium. Control required, urgently.

for example) and their severity. The severity parameters are determined using the values from in Table 3.

The methodology for the calculation of the Hazard Matrix (HM) in this case study follows what has been proposed by Haddad (Haddad et al. 2012) that determines that the matrix is mapped for each sector (S_i) making a line (from 1 to y), followed by a column that stands for the number of workers on that sector (W_i). As for the other columns, they take all the hazards (H_j) identified in all the sectors, drawing column from 1 to x. The intensity level (or frequencies according to HM methodology) of hazards is calculated for each sector (f_{si}) and for each hazard (f_{H1}), using equations 1 and 2.

$$f_{si} = \sum_{j=1}^{x} W_i * R_{i,j}, \text{ Where } 1 \le i \le y \qquad (1)$$

$$f_{Hj} = \sum_{i=1}^{y} W_i * R_{i,j}, \text{ Where } 1 \le j \le x \qquad (2)$$

The final HM for this case study is presented in Table 4.

According to the HM applied to the case of study, the sector most critical is the sector D (with 29.3% of all the hazards) (Figure 1). The sector with lower criticality is B (2.6%) and less critical danger is air contamination (1.1%). The second most critical sector is C (25.9%), followed by sector E (industry, 14.7%). Sector F hazards sum 10.8%, sector G with 9.7% and sector A with 7.0%.

With regard to hazards, fire and explosion (17.3%) is the top priority, followed by noise (14.3%) and electricity (12.2%). Burnings comes next with 10.1%, followed by chemicals (8.5%) and Ergonomic factors (poor posture, 7.8%), inappropriate layout (6.8%), monotony and repetitiveness (5.6%) and the least critical hazards are heat (5.2%), hand cuts (4.3%) and unprotected machinery (2.6%). Figure 2 puts hazards levels in comparison.

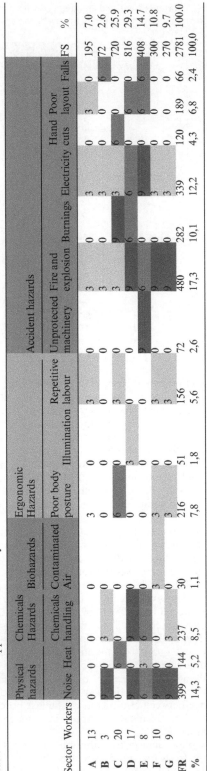

Table 4. Hazard Matrix applied to the case study.

Sector	Workers	Physical hazards Noise	Heat	Chemicals Hazards Chemicals handling	Biohazards Contaminated Air	Ergonomic Hazards Poor body posture	Illumination	Repetitive labour	Accident hazards Unprotected machinery	Fire and explosion	Burnings	Electricity cuts	Hand cuts	Poor layout	Falls	FS	%
A	13	0	0	0	0	3	0	3	0	3	0	3	3	3	0	195	7.0
B	3	9	0	3	0	0	0	0	0	3	0	3	0	0	6	72	2.6
C	20	6	6	6	0	6	3	3	0	3	0	3	0	6	0	720	25.9
D	17	6	3	6	0	0	0	3	9	6	6	6	6	6	6	816	29.3
E	8	3	0	0	3	0	0	3	0	9	0	3	0	0	0	408	14.7
F	10	9	0	3	0	3	0	3	0	0	0	0	0	6	0	300	10.8
G	9	0	0	0	0	3	0	3	0	3	0	3	0	0	0	270	9.7
FR	399		144	237	30	216	51	156	72	480	282	339	120	189	66	2781	100,0
%	14,3		5,2	8,5	1,1	7,8	1,8	5,6	2,6	17,3	10,1	12,2	4,3	6,8	2,4	100.0	

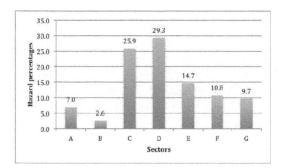

Figure 1. Hazard levels per sector.

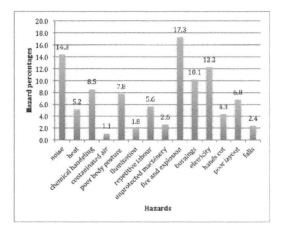

Figure 2. Hazard levels.

6 CONCLUSIONS

The results from the HM implementation indicate that the most critical area is the chemical storage room, Sector D (29.3%), followed closely by thermo-presses, Sector C (25.9%). These two sectors should be prioritized when implementing mitigation plans.

Likewise, the most severe hazards to be dealt with are associated with fire and explosion (17.3%) and noises (14.3%). Comparing these results with those from the risk mapping, it is clear that the risk perception is consistent.

REFERENCES

Brazil, 1977. *Federal Bill 6.514/1977.* Brasilia, Brazil: Ministry of Work and Employment (*available in Portuguese only*).

Brazil—Goiais, 2012. Guidelines for elaboration of Risk maps. Brazil. Available at: http://www.sgc.goias.gov.br/upload/arquivos/2012-11/manual-de-elaboracao-de-mapa-risco.pdf (*available in Portuguese only*).

Brazil—MTE, 1978. *NR 15—Annex 11—Chemical Agents with Exposure limits.* Brasilia, Brazil: Ministry of Work and Employment. Available at: http://portal.mte.gov.br/data/files/8A7C812D3F9B201201407C E4F9BC105D/Anexo n.? 11 Agentes Quimicos—Tolerância.pdf (*available in Portuguese only*).

Brazil—MTE, 2008. *NR 05—Internal Comission of Accidents Prevention.* Brasilia, Brazil: Ministry of Work and Employment (*available in Portuguese only*).

Brazil—MTE, 2008. *NR 15—Activities and unhealthy activities.* Brasilia, Brazil: Ministry of Work and Employment (*available in Portuguese only*).

Haddad, A. et al., 2012. *Hazard Matrix Application in Health,* Safety and Environmental Management Risk Evaluation. In *InTech.*

ISO, 2009. *ISO 31.000—Risk Management—Principles and Guidelines,* Geneva, Switzerland: International Organisation for Standardization.

SI, 2007. *OHSAS 18001:2007 Occupational Health And Safety Management Systems—Requirements.,* OHSAS Project Group—British Standards Institution.

USDoD (US Department of Defence), 1993. *MIL STD 882-C-SYSTEM SAFETY PROGRAM REQUIREMENTS,*

USDoD (US Department of Defence), 2000. *MIL STD 882-D—Standard Practice for System Safety,* Washington, USA: USA.

USDoD (US Department of Defence), 2012. *MIL STD 882-E—Standard Practice for System Safety,* Washington, USA: USA.

Occupational Safety and Hygiene V – Arezes et al. (Eds)
© *2017 Taylor & Francis Group, London, ISBN 978-1-138-05761-6*

Blood alcohol concentration effect on driving performance: A short review

Norberto Durães, Sara Ferreira & J. Santos Baptista
Faculty of Engineering, University of Porto, Porto, Portugal

ABSTRACT: Driving performance may be affected differently according to Blood Alcohol Concentration (BAC) levels. We conducted a short review in accordance with PRISMA statement guidelines following a search strategy which includes the combinations of keywords: *alcohol; driver errors; driving influence; Acute protracted error; Acute tolerance; Ascending and descending blood alcohol Concentrations; Cognitive performance* on 25 search engines and 34 scientific publishers. Nine studies met inclusion criteria; the studies were analyzed regarding the type of experiment and procedures including a total of 230 participants who drove under BAC influence. Results of the studies agreed to be in the descending phase of the BAC curve that more driving errors are recorded, with loss of motor skills and cognitive impairments mainly due to the effect of acute tolerance, which affects the perception of the drivers about the risks of driving by minimizing the BAC effects.

1 INTRODUCTION

Road accidents with motor vehicles and alcohol consumption have drawn attention of governments of various countries from Japan to Norway, from United States to New Zealand, from Canada to China. Although countries like Turkey, Australia, Austria, Argentina, Belgium, Italy, Netherlands, Portugal, South Africa, and Spain have a legal limit for Blood Alcohol Concentration (BAC) of 0.49 g/l, studies reveled that drivers with that value on the limit were more involved in accidents than drivers with zero BAC (Karakus *et al.* 2015). The influence of BAC on drivers, even within moderate levels (<0.5 g/l), affects the ability to drive (Desapriya *et al.* 2007).

In what can be considered as large-scale studies, in several countries significant improvements were registered in accident rates by reducing the maximum permitted BAC. In Japan the decrease of the alcohol level allowed by law from 0.5 g/l to 0.3 g/l, is associated to a significantly decrease of the number of traffic violations (Desapriya *et al.* 2007). The same occurs in USA, where reduction of the BAC to 0.8 g/l has contributed to a considerable decrease of fatalities and accidents without casualties (Kaplan and Prato 2007). In Sweden the reduction of the legal limit from 0.5 g/l to 0.2 g/l led to a decrease of 9.7% on single-accidents and a decrease of 7.5% of the total of accidents that have occurred in the six years of law enforcement (Norstrom 1997). Considering the results and conclusions pointed out by those authors, it is clear that

driving under the influence of alcohol in the blood can cause a considerable amount of traffic accidents. All those studies also observed that women and elderly comply more with the legal limits regulations and that women, even under the influence of alcohol, are more careful than men (Blomberg *et al.* 2009). Studies focusing on the consequences of BAC in drivers, which cognitive impairment and driving performance are harmed by this consumption, have been contributing to providing rules at a general and personal levels (Desapriya *et al.* 2007, Charlton and Starkey 2015).

The impairing effect of alcohol on driving performance is well established. Laboratory research indicates that moderate doses of alcohol impair a broad range of behavioral and cognitive functions, and any of these could contribute to overall driving performance (Weafer and Fillmore 2012).

Despite most of the studies focus on BAC effects early after the dose is rising, the concept of acute tolerance has been introduced on the subject referring to the diminished intensity of impairment at a given BAC during the descending compared to the ascending limb of the BAC curve. This phenomenon is of particular importance as decisions to drive are often made after a drinking episode. Acute tolerance to the effects of alcohol has been observed in several behaviors, including motor coordination, reaction time, and subjective intoxication (Fillmore *et al.* 2005, Schweizer and Vogel-Sprott 2008, Marczinski and Fillmore 2009, Cromer *et al.* 2010, Weafer and Fillmore 2012). Bearing in mind these studies it is clear that BAC seriously compromises both driving and cognitive skills.

The effects caused by the presence of BAC on drivers can be evaluated using driving simulators, naturalistic driving tests and/or other experimental tests, which allow assessing the consequences of drinking in individuals with different characteristics. Those types of laboratorial trials require specific procedures to control the measured parameters and the reliability of the results. For instance, to consider the influence of the physical differences among the participants during the tests undertaken to check the variation in time of the effect of a dose of alcohol (i.e. acute tolerance), parameters such as blood pressure, temperature, weight and height are usually measured. On the other hand, food with high fat content, sugar, or substances with caffeine can affect the absorption of alcohol. So, participants should avoid the ingestion of food or drinks that affect alcohol absorption (Liu and Fu 2007).

In this study, a short review is presented aiming to analyze the BAC effects on driving performance focusing on the comparison of that performance between ascending phase and descending phase of BAC rates in laboratorial trials.

2 MATERIALS AND METHOD

This short review was performed in accordance with **PRISMA** Statement guidelines (Liberati *et al.* 2009). The review was based on the analysis of several scientific papers reporting cases of driving under the influence of alcohol. In this sense, the following research questions were established to guide the short review:

– How alcohol ingestion affects driving performance?
– What are the differences on driving behavior for different levels of BAC?
– When driving mistakes are more pronounced, during the ascending or descending phase of alcohol curve?

Were select articles published in scientific journals describing studies involving experiments with humans, reporting the existence of an informed consent form and other relevant information regarding the experiment methodology in order to be possible to compare and reproduce the trials.

2.1 *Data sources and search strategy*

Were searched 39 electronic databases and scientific publishers: *Academic Search Complete*; CiteSeerX; Compendex; ERIC; *Inspec*; *Library,* MEDLINE; PsycArticles; PubMed; *ScienceDirect*; SCOPUS;

TRIS *Online; Web of Science; ACM Digital Library; ACS Journals; AHA Journals; AIP Journals; AMA Journals; ASME Digital Library; BioMed Central Journals; Cambridge Journals Online;* CE *Database; Directory of Open Access Journals; Emerald Fulltext; Highwire Press; Informaworld;* Ingenta; IOP *Journals; MetaPress; nature.com;* Oxford Journals; *Royal Society of Chemistry;* SAGE *Journals Online; SciELO—Scientific Electronic Library Online; Science Magazine; ScienceDirect; Scitation;* SFX A-Z; SIAM; and *Wiley Online Library*.

The search strategy was the same for all databases, using the following search key-words: (1) *alcohol and driver errors*; (2) *alcohol and driving influence*; (3) *alcohol and acute protracted error*; (4) *alcohol and acute tolerance*; (5) *driving and during ascending and descending blood alcohol concentrations*; (6) *driving and cognitive performance*.

2.2 *Procedures*

A selection procedure was carried out according to the flow diagram shown in Figure 1. Firstly, potential relevant articles were identified using the key-words aforementioned. After that, duplicate articles were removed. Secondly, articles dated

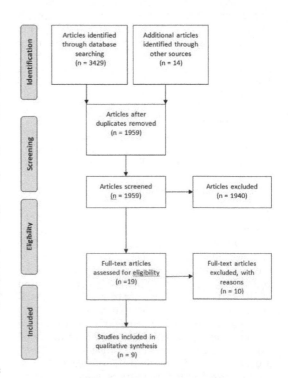

Figure 1. Flow diagram of studies selection: summary of the results in accordance with the PRISMA *Statment*.

for more than 10 years as well as articles missing relevant information were excluded. Additionally, the research questions were used to guide the screening phase. Therefore, articles that in some way do not meet those questions were excluded. Thirdly, the eligibility criteria aforementioned were applied to the final selection. To do that, data was extracted from each study to identify specific elements describing the experiment and methodology used in each study.

3 RESULTS

The search identified 3429 articles for assessment (Figure 1). Additionally, 14 more articles were included as relevant bibliography. From this first outcome, 1484 duplicate articles were excluded. Of the remaining 1959 articles, 104 articles were excluded by insufficient or incorrect identification, 846 because were published more than 10 years ago and 990 were out of the topic. At the end, 19 eligible articles were identified and evaluated to satisfy the research purpose. Data extraction and analysis was undertaken and 9 articles were selected as meeting the criterion that the experiments were supported by an informed consent form. The 10 excluded articles did not provide information about this issue and/or the used methodology was not sufficiently described to be applied and results compared.

Each article was deeply analyzed regarding the experimental procedures in order to identify and compare the following descriptors: participants (sample, gender, age, weight, height), the substance used, experiment description including type, tests, measures and procedures (Table 1).

All experiments were real-time clock and all participants have valid driver's license. All studies used a driving simulator to develop the experiment except one that applied cognitive and psychological tests supported on a visual analog scale specifically designed for the study (Cromer et al. 2010). Sample sizes of included studies ranged from 8–61 participants. Most of the selected articles (Marczinski and Fillmore 2009, Cromer et al. 2010, Starkey and Charlton 2014, Charlton and Starkey 2015) report the differences found in driving performance between the ascending and descending limb of the BAC curve. Two (Weafer and Fillmore 2012, Tremblay et al. 2015) others are focused only in the descending limb in order to analyze the effect on driving performance of acute alcohol consumption.

The 9 articles published over a period of eight years, from 2007 to 2015, are identified in Table 1.

Table 1. Selected studies: Summary of the experimental procedures.

Author (Year)	Participants	Substances
(Liu and Fu 2007)	8 S[1]: 6 M[2] and 2 F[3]; 4 S: [20–24] yo[4], 4 S: [25–30] yo	Alcohol (Vodka)
(Marczinski et al. 2008)	40 S (2 Afr-Am, 1 Asian, 36 Caucasian and 1 Nat-Am): 20 M, 20 F; [21–29] yo; 24 binge drinkers and 16 non binge drinkers.	Alcohol (Beer) and drugs.
(Marczinski and Fillmore 2009)	28 S (Afr-Am, 1 Asian American, Americano, 23 Caucasian and 1 "other"): 16 M, 12 F; [21–28] yo; At least twice a week drivers, with and without drinking habit	Alcohol (Beer) and drugs
(Cromer et al. 2010)	17 S: 9 M and 11 F at the beginning, 3 discarded after screening. Remaining total = 17; [21–25] yo; Social-drinking adults	Alcohol (vodka) and drugs (spritzer of vodka)
(Weafer and Fillmore 2012)	20 S: 10 M, 10 F; [21–31] yo; Social-drinking adults	Alcohol and drugs
(Helland et al. 2013)	20 S, M: 20 driving simulator + 10 instrumented vehicle. [25–35] yo	Alc (vodka) – drugs; (non-alc vodka); placebo pill.
(Starkey and Charlton 2014)	61 S: 33 M, 28 F; [20–50] yo; occasional drinkers	Alcohol and drugs
(Charlton and Starkey 2015)	44 participants: 21 M, 23W, [20–47]	Alcohol (vodka) or drugs
(Tremblay et al. 2015)	16 S: 10 M, 6 F; G[5]-1: (6 M, 2 F), Average age 21,6; G-2: Average age 20,9 ± 2,35 years (4 M, 4 F).	Alcohol

[1]S-subjects; [2]M-mail; [3]F-female; [4]yo-years old; [5]G-Group; Afr-Am-African-Americans; Nat-Am-Native-American; Alc-Alcohol.

4 DISCUSSION

In this short review, it becomes clear that driving under the influence of alcohol increases the risk of traffic accidents and seriously impairs driver's performance as analyzed, observed and confirmed in the nine studies. As it is difficult and dangerous to study this phenomenon with naturalistic studies, laboratory trials were developed to carry on those studies, most of them using driving simulators. In this point, the work developed by Helland et al. (2013) becomes very important. These authors compared naturalistic studies with simulator studies, and concluded that studies held in driving simulator are effective to perceive the influence of alcohol on driving performance. Nevertheless, the same authors stressed that differences were found between naturalistic and simulated experiments regarding risk perception, which is more noticeable in the naturalistic driving as expected. However, as this difference is not significant to the main objective, the driving simulator is accepted as a valuable tool for the evaluation of driving under the influence of alcohol consumption.

Regarding the influence of alcohol on driving performance, Liu and Fu (2007) found that the higher the level of alcohol in the blood, the higher the number of driving mistakes and thus, committing driving performance. It was observed that once the BAC increases, the ability of attention on the various factors that compromise the driving performance decreases. In addition, the increase of BAC is associated with a decrease of cognitive capacities and motor skills. Additionally, Marczinski, et al. (2008) state that the influence of alcohol on driving performance is correlated with the amount of alcohol ingested, whereas drivers who ingest moderate amounts of alcohol (nonbinge drinkers) have a more realistic perception of whether they are qualified to drive. In contrast, drivers who ingest large amounts of alcohol at once (binge drinkers) block the ability to realize they are not able to drive. This is due to the fact that binge drinkers are more accustomed to alcohol, which allows them to have a perception of a better response to its effect as a result of the tolerance they developed (acute tolerance) which in turn gives them a false sensation that they are capable of driving under those alcohol conditions.

From the nine included studies, six analyzed the effect of the acute tolerance on driving performance and/or driver's behavior (Marczinski and Fillmore 2009, Cromer et al. 2010, Weafer and Fillmore 2012, Starkey and Charlton 2014, Charlton and Starkey 2015, Tremblay et al. 2015), showing the BAC curve resulting from their experiments, i.e., BAC levels as a function of time after alcohol consumption. It was the case of the study carried on by Marczinski and Fillmore (2009) that have found differences between ascending and descending limb of the BAC curve regarding driver's perception. The authors observed that the perception that one is not able to drive is the same whether for drivers who consume alcohol moderately (nonbinge drinkers) or for drivers who consume large amounts of alcohol at once (binge drinkers), when asked in the ascending phase of the alcohol curve. In contrast, in the descending phase of the alcohol curve, although none of the drivers has the ability to drive, binge drinkers have the false perception that they are capable of driving. They also noticed that during the descending phase of the alcohol curve, drivers make more mistakes due to being more tired and eventually sleepy while in the ascending phase they still have some energy. Finally, it is clear that binge drinkers lose the ability to perceive their degree of intoxication, creating a certain tolerance to alcohol, i.e. acute tolerance.

Cromer et al. (2010) corroborated, to some extent, the same conclusions of the authors above, through undertaking cognitive and psychologic tests namely visual analog scale, instead of a driving simulator. Cromer et al. (2010) concluded that alcohol seriously affects cognitive abilities however, they observed that in some cases, these capabilities are more affected during the rise of BAC while other are affected during the decreasing of BAC level. Despite the descending phase of the BAC curve does not affect all the capabilities, it is at this stage that there is the illusion that one is not intoxicated by alcohol and thus, considering himself able to drive. In fact, Cromer et al. observed that a candidate in the descending phase of the alcohol curve, with a BAC of 0.8 g/l, showed the illusion of being already with zero BAC, revealing that he/she has acute tolerance.

Weafer and Fillmore (2012) also concluded that during the descending phase of the BAC curve drivers make more mistakes and the drivers with acute tolerance have a false perception of their abilities to drive safely. Curiously, the authors observed that despite drivers with acute tolerance make more mistakes not all skills are affected. In fact, acute tolerance was observed specifically to alcohol effects on motor coordination. However, and despite any acute recovery from the impairing effects of alcohol on motor coordination might allow for some degree of recovery of driving ability, the study findings showed that acute tolerance to alcohol-impaired motor coordination was not accompanied by any acute tolerance to alcohol-impaired driving performance.

In line with the previous studies, Starkey and Charlton (2014) pointed out that the driving performance is affected by the BAC level and by the acute tolerance. Also, they distinguish the ascending phase from the descending phase in which the second is associated to a decrease of driving performance.

Tremblay *et al.* (2015) studied the BAC influence on young drivers. Among others, they concluded that even with a low level of BAC, young drivers tend to increase driving speed as well as the number of driving mistakes. The authors noted that during the descending phase, after 4 hours of alcohol ingestion and with a BAC ranging between 0.5 g/l and 0.7 g/l, the number of driving mistakes is higher than in the ascending phase of the BAC curve. Additionally, Charlton and Starkey (2015) mentioned that social drinking highly affects the perception of driving capabilities as well as increases the acute tolerance. Hence, the decision of driving and drinking on a social event should be made before. In this study, more driving mistakes were reported during the descending phase of the BAC curve even for the same BAC values of the ascending phase, confirming the main conclusions of the previously mentioned studies.

5 CONCLUSIONS

An extensive review of literature was performed focusing on BAC influence on driving performance, particularly regarding the difference between ascending and descending phases of the BAC curve, and acute tolerance on drivers. Taking into account the particular condition of the participants in this type of experiments, i.e. under the influence of alcohol consumption, laboratory trials and particularly driving simulators are a safe tool in a controlled environment that provide insights into situations that are difficult to measure in a naturalistic driving study.

The experimental studies undertaken using driving simulators and cognitive and psychological tests analyzing the responses of the participants under various tasks, showed that the cognitive and motor skills are affected both on ascending and descending phases of the BAC curve, however in different ways. During the ascending phase, the mistakes regarding motor skills are more pronounced inasmuch as the response to an unexpected task is exaggerated and it is observed a variation of the speed around the desired speed as set for the experiment. In the same studies, during the descending phase, and despite in some cases are observed a reduction on the magnitude of alcohol impairment of motor coordination, cognitive mistakes are observed, affecting not only the driving performance but also the risk perception. In this point, studies showed that drivers have a wrong perception of their driving capabilities in the descending phase of the BAC curve. These findings described the acute tolerance phenomenon that may be associated with traffic conflicts and accidents when drivers are under influence of moderate BAC. Indeed, it was observed that drivers with acute tolerance feel that they are capable of driving well, minimizing the risk perception.

Overall, the studies selected in this review are in line inasmuch as they concluded that it is during the descending phase of the BAC curve that it is observed higher number of mistakes, however diverging on the type of capabilities most affected. Indeed, during the descending phase same authors found an acute tolerance to alcohol effects on motor coordination, however no acute tolerance was observed to the impairing effects of alcohol on inhibitory control as individuals remained just as disinhibited by alcohol when BAC was declining as when it was rising. Additionally, when BAC begins to decline, other factors, such as fatigue or impaired attention, might play a role in determining driving impairment.

The studies pointed to the importance of a deep analysis of BAC effect on driver's performance, especially regarding the acute tolerance. This phenomenon may be a common issue during social events, confounding the drivers and others about their driving capabilities compromising the right decision of not driving and thus, exposing them to the risk of accident. Proving this assumption, the decision of driving should be made before any alcohol consumption otherwise the accident risk increases due to poor driver performance. Moreover, the legal limits in several countries may be questioned regarding the desirable effects on accident and injury prevention.

REFERENCES

Blomberg, R.D., Peck, R.C., Moskowitz, H., Burns, M., Fiorentino, D., 2009. The long beach/fort lauderdale relative risk study. *J. of Safety Research* 40 (4), 285–292.

Charlton, S.G., Starkey, N.J., 2015. Driving while drinking: Performance impairments resulting from social drinking. *Accident Analysis & Prevention* 74, 210–217.

Cromer, J.R., Cromer, J.A., Maruff, P., Snyder, P.J., 2010. Perception of alcohol intoxication shows acute tolerance while executive functions remain impaired. *Experimental and Clinical Psychopharmacology* 18 (4), 329–339.

Desapriya, E., Pike, I., Subzwari, S., Scime, G., Shimizu, S., 2007. Impact of lowering the legal blood alcohol concentration limit to 0.03 on male, female and teenage drivers involved alcohol-related crashes in japan. *I.J. of Injury Control and Safety Promotion* 14 (3), 181–187.

Fillmore, M.T., Marczinski, C.A., Bowman, A.M., 2005. Acute tolerance to alcohol effects on inhibitory and activational mechanisms of behavioral control. *Journal of Studies on Alcohol and Drugs* 66 (5), 663–672.

Helland, A., Jenssen, G.D., Lervåg, L.-E., Westin, A.A., Moen, T., Sakshaug, K., Lydersen, S., Mørland, J., Slørdal, L., 2013. Comparison of driving simulator performance with real driving after alcohol intake: A randomised, single blind, placebo-controlled, crossover trial. *Accident Analysis & Prevention* 53, 9–16.

Kaplan, S., Prato, C.G., 2007. Impact of bac limit reduction on different population segments: A poisson fixed effect analysis. *Accid. Analysis & Prevention* 39 (6), 1146–1154.

Karakus, A., İdiz, N., Dalgiç, M., Uluçay, T., Sincar, Y., 2015. Comparison of the effects of two legal blood alcohol limits: The presence of alcohol in traffic accidents according to category of driver. *Traffic Injury Prev.* 16 (5), 440–442.

Liberati, A., Altman, D.G., Tetzlaff, J., Mulrow, C., Gøtzsche, P.C., Ioannidis, J.P.A., Clarke, M., Devereaux, P.J., Kleijnen, J., Moher, D., 2009. The prisma statement for reporting systematic reviews and meta-analyses of studies that evaluate healthcare interventions: Explanation and elaboration. *BMJ* 339.

Liu, Y.-C., Fu, S.-M., 2007. Changes in driving behavior and cognitive performance with different breath alcohol concentration levels. *Traffic Injury Prevent.* 8 (2), 153–161.

Marczinski, C.A., Fillmore, M.T., 2009. Acute alcohol tolerance on subjective intoxication and simulated driving performance in binge drinkers. *Psychology of Addictive Behaviors* 23 (2), 238–247.

Marczinski, C.A., Harrison, E.L.R., Fillmore, M.T., 2008. Effects of alcohol on simulated driving and perceived driving impairment in binge drinkers. *Alcoholism: Clinical and Experimental Research* 32 (7), 1329–1337.

Norstrom, T., 1997. Assessment of the impact of the 0.02 percent bac-limit in sweden. *Studies on Crime and Crime Prevention* 6 (2), 245–258.

Schweizer, T.A., Vogel-Sprott, M., 2008. Alcohol-impaired speed and accuracy of cognitive functions: A review of acute tolerance and recovery of cognitive performance. *Experimental & Clinical Psychopharmacology* 16 (3), 240.

Starkey, N.J., Charlton, S.G., 2014. The effects of moderate alcohol concentrations on driving and cognitive performance during ascending and descending blood alcohol concentrations. *Human Psychopharmacology: Clinical and Experimental* 29 (4), 370–383.

Tremblay, M., Gallant, F., Lavallière, M., Chiasson, M., Silvey, D., Behm, D., Albert, W.J., Johnson, M.J., 2015. Driving performance on the descending limb of blood alcohol concentration (bac) in undergraduate students: A pilot study. *PLoS ONE* 10 (2), e0118348.

Weafer, J., Fillmore, M.T., 2012. Acute tolerance to alcohol impairment of behavioral and cognitive mechanisms related to driving: Drinking and driving on the descending limb. *Psychopharmacology* 220 (4), 697–706.

Occupational Safety and Hygiene V – Arezes et al. (Eds)
© 2017 Taylor & Francis Group, London, ISBN 978-1-138-05761-6

HazOp study in a wastewater treatment unit

J.E.M. França
Departamento de Engenharia Civil, UFF, Niterói, Brazil

G.F. Reis & A.N. Haddad
Escola Politécnica, UFRJ, Rio de Janeiro, Brazil

E.B.F. Galante
IME—Instituto Militar de Engenharia, Rio de Janeiro, Brazil

D.M.B. Costa
FEUP—Faculty of Engineering of the University of Porto, Porto, Portugal

I.J.A. Luquetti dos Santos
Instituto de Energia Nuclear, IEN-CNEN, Rio de Janeiro, Brazil

ABSTRACT: This paper presents a case study that applies a Hazard and Operability Study (HazOp) into a part of an industrial wastewater treatment plant under construction. The aim of this study is to identify hazardous situations and operational risks before the occurrence of these scenarios, prior to the plant start-up. HazOp is a structured and systematic qualitative examination of a process plant operation, based on guidewords and nodes, that identifies and assess hazardous and risks. Two risk scenarios were identified and analysed by HazOp methodology, resulting in technical recommendations for the management of these risks. These recommendations target safety in plant operations and, since they are suggested prior to the beginning of plant operation, its implementation demands few resources and presents high cost-effectiveness.

1 INTRODUCTION

This work is based upon the thesis written by Reis (2016), in which the wastewater treatment unit of a siderurgy company, named "Waters" (fictitious name) is evaluated. This study presents a risk assessment of the processes of an industrial wastewater treatment plant in executive design using the Hazard and Operability (HazOp) method to analyse industrial process risks and to identify possible facility improvements, such as changes in equipment, instrumentation, automation and chemical handling before the industrial plant starts operating. Despite the existence of numerous risk analysis techniques, such as PHA, FMEA and FTA, HazOp was chosen for this project because it presents the best methodology for identifying risks during all pashes of any industrial process plants, especially during the operation.

According with Kletz (1999) apud Isimite & Rubini (2016), "*It is better to illuminate the hazards we have passed through than not illuminate them at all, as we may pass the same way again, but we should try to see them before we meet them ... unfor-* *tunately, we do not always learn from the hazards we have passed through.*" Furthermore, Dunjó et al (2011) notes that HazOp study is a risk assessment tool applied worldwide, used for studying not only hazards, but also operability problems of a system by exploring the effects of any deviations from design conditions. This method reviews equipment, instrumentation, utilities, human factors and external events that might impact the process following Piping & Instrumentation Diagrams (P&ID). This study explores this method to illuminate hazards inherent of a wastewater treatment unit that handles with dangerous substances and to identify critical process parameters.

2 METHODOLOGY

This paper carried out a quantitative HazOp study (Galante et al. 2014) of a wastewater treatment unit through a structured analysis of the system and its processes by a multi-disciplinary team. In this method, the team follows the process lines, analysing possible deviations in each stage of process design or operation (Crawley, 2000). Even being systematic and rigorous, the analysis can also be flexible. This analysis is done by a systematic usage of words in combination with system parameters to identify deviations in the process workflow.

According to Kotek & Tabas (2012) HazOp is a systemic approach best suitable to carry out risk assessment in process units. HazOp method has been used: (i) systemic approach towards assessment of safety and operability; (ii) advantage of the keywords (see Table 1) for generation of deviations from the safe situation, and (iii) principle of a brainstorming during the creative development of the considered scenarios of events, originating from the deviation from the safe situation, followed by the identification of deviation causes, safety functions and evaluation of their possible effects.

The record of the study is realized by a usual form of the discussion of the HazOp team following the scheme: deviation—causes—effects—safety functions—action/measure. Identification of significant deviations is made through guide parameter words coupled with process parameters. Table 1 lists some of the most commonly used parameters and guidewords combinations used.

In order to structure the analysis, the process is split into segments, each of them designated as a "node", which is a part of all process that usually has similar characteristics and equipment or low variation of the operating parameters. The first attempt for nodes selection is to identify main sections, which are easy to "disconnect" from each other (e.g., feeds section, reaction section, absorber section, distillation, etc.). Main sections are addressed as a wide number of equipment that is involved in achieving a sub-aim: contributing to the overall design intention of the process. This is

Table 1. Deviations and parameters.

Parameter	Guideword	Deviation
FLOW	None, Less, More, Reverse Other, Also	No flow, Less flow, More flow, Reverse flow, Other flow, Contamination
PRESSURE	More	More pressure
	Less	Less pressure
TEMPERATURE	More	Higher temperature
	Less	Lower temperature
VISCOSITY	More	More viscosity
	Less	Less viscosity
REACTION	None	No reaction
	Less	Reaction incomplete
	More	Intense reaction

Source: (Galante et al. 2014).

Table 2. Probability levels.

Cat.	Description	Aspects
A	Frequent	Likely to occur often in the life of an item.
B	Probable	Will occur several times in the life of an item.
C	Occasional	Likely to occur sometime in the life of an item.
D	Remote	Unlikely, but possible to occur in the life of an item.
E	Improbable	So unlikely, it can be assumed occurrence may not be experienced in the life of an item.
F	Eliminated	Incapable of occurrence. This level is used when potential hazards are identified and later eliminated.

Source: Mil STD 882-E (USDoD, 2012).

Table 3. Severity categories.

Cat.	Description	Mishap Result Criteria
1	Catastrophic	Could result in one or more of the following: death, permanent total disability, irreversible significant environmental impact, or monetary loss equal to or exceeding $10M.
2	Critical	Could result in one or more of the following: permanent partial disability, injuries or occupational illness that may result in hospitalization of at least three personnel, reversible significant environmental impact, or monetary loss equal to or exceeding $1M but less than $10M.
3	Marginal	Could result in one or more of the following: injury or occupational illness resulting in one or more lost work day(s), reversible moderate environmental impact, or monetary loss equal to or exceeding $100K but less than $1M.
4	Negligible	Could result in one or more of the following: injury or occupational illness not resulting in a lost workday, minimal environmental impact, or monetary loss less than $100K.

Source: Adapted from Mil STD 882-E (USDoD, 2012).

Table 4. Risk assessment matrix

RISK ASSESSMENT MATRIX				
SEVERITY / PROBABILITY	Catastrophic (1)	Critical (2)	Marginal (3)	Negligible (4)
Frequent (A)	High	High	Serious	Medium
Probable (B)	High	High	Serious	Medium
Occasional (C)	High	Serious	Medium	Low
Remote (D)	Serious	Medium	Medium	Low
Improbable (E)	Medium	Medium	Medium	Low
Eliminated (F)	Eliminated			

Source: Mil STD 882-E (USDoD, 2012).

the first chased size for breaking the process into nodes (Dunjó et al, 2011).

In a typical HazOp, for each node, the team maps possible deviation and identifies and analyses its causes and effects. This work moves beyond a typical methodology by using quantitative approach proposed by Galante et al (2014) in which causes and consequences are qualitatively assessed. Causes are converted into frequency of occurrence using a series of parameters (Table 2), while and its effects are converted into consequences (severity) using the parameters from Table 3.

Probability and severity categories determines a risk rating value, which supports decision making whereas a risk is tolerable, as well as its priority when it comes to mitigation. The combination of frequency and severity is conducted by the usage of risk matrices (Table 4).

3 CASE STUDY

The wastewater treatment plant under evaluation presents a series of hazards, which include the usage of chemicals, high and low pressures, temperatures and flow. The inventories of chemicals include basic chemicals (sodium hydroxide and calcium hydroxide) and acids (sulphuric acid, citric acid and ferric chloride).

Hazards from these chemicals include corrosively (sodium hydroxide and sulphuric acid and ferric chloride); irritant capability (calcium hydroxide and citric acid); and flammability. Flammability is mainly due to the presence of Ethanol (CH_3CH_2OH) with a flash point of 13°C, LEL (Lower Explosive Limit) of 3.3% by volume and UEL (Upper Explosive Limit) of 19% by volume (PubChem, 2016), i.e., small releases of ethanol vapours may generate explosive atmospheres in the surrounding area.

Table 5. Ethanol storage conditions.

Storage period	29 days
Required volume	7,0 m³
Number of tanks	1 (TQE-E02-005)
Tank diameter	2,0 m
Tank useful height	2,3 m
Tank useful volume	7,0 m³

Ethanol consumption is 10 L/h (equivalent of 8 kg/h) and it is storaged under condition presented in Table 5.

The ethanol tank, shown under tag TQE-E02-005 in Figure 1 is protected against spills by a reinforced concrete tank (width 4,0 m; length 4,0 m; height 1,0 m; total volume: 10,2 m³) and a level sensor (LIT-E02-011).

4 RESULTS AND DISCUSSION

The P&ID from Figure 1 corresponds to the node under evaluation in this work, equivalent to ethanol transference and storage into the tank. Although this entire HazOp was developing around 33 nodes, this work focuses on the most critical node, which links preparation, storage and dosage of ethanol operations.

For the assessment of these risk scenarios, only two parameters were considered applicable: flow and pressure. Ethanol supply to the tank is done by a dedicated truck connection coupled to a pump, which has only one direction of rotation and operates at a fixed flow rate. This assures consistency with the internal pressure generated by the fluid and supported by the hose. There are quick couplers on connections (truck and pump connections), coupling the truck-hose between these two points.

Tables 6 and 7 present the results from the application of HazOp in the selected node. Table 6 addresses the transference of ethanol from the transportation truck to the storage tank, while Table 7 stands for the analysis of the ethanol storage into the tank TQE-E02-005.

This HazOp analysis, in the operation of transference of ethanol from the transportation truck to the storage tank, for the deviation "less flow" and "no flow", has assessed one scenario as Low Risk, two as Medium Risk, and others two as Serious Risk. Once there are medium and serious risks, it is necessary to provide recommendations for risk management. In scenarios where there is the possibility of ethanol accidental liberation, causing fire or explosion due its flammability characteristics, the study has recommended to remove or eliminate explosion ignition sources during all steps

43

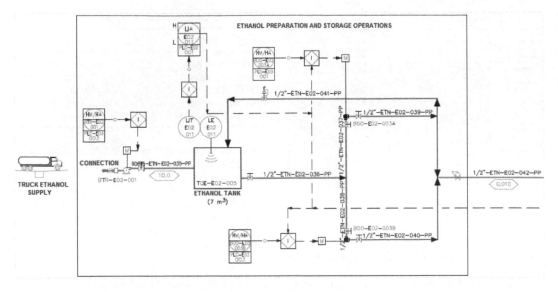

Figure 1. P&ID of ethanol preparation and storage operations.

of operation, and ensure that the hose and con-nections are correctly grounded in the same earth equipotential of the tank trunk and the transfer-ence pump. Moreover, due ethanol viscosity and its transference in pressure, static electricity build up on the walls of the hose are a concern once a spark source can potentially starts afire. Taking pump, trunk tank, hose and connections by the same earth equipotential, the risk of ignition from static electricity is greatly reduced.

The less pressure deviation had presented a seri-ous risk, having as consequence the mechanical tank truck implosion, caused by a failure in the top vents of this tank. This implosion happens due an lower pressure (vacuum) inside of the tank, break-ing the mechanical structure of the tank and caus-ing an ethanol spill that contaminates personal and environment, as well as fire or explosion, due its flammability. Recommendations for this scenario were similar to the less flow deviation, however including procedures to ensure the all top vents of the truck tank are in perfect condition and without obstruction before the operations stars, and also guarantee that all personal involved in this opera-tion has been trained to identify top vents failures and obstructions.

For ethanol storage operations in tank TQE-E02-005, two possible deviations were identified: high level and low level. As defined in the P&ID, the ethanol tank will have two LIT (Level Indica-tor Transmitter), as redundancy. Although this redundancy control, LIT can fail, and this might lead to leakage of ethanol. As recommendation for

"high level" deviation, it was suggested to enable an automation interlock between LIT of TQE-E02-005 and the transference pump, shutting down this pump when the tank level reaches HH (High High) parameters.

Furthermore, it is possible to evaluate the pos-sibility of having a backup automated system, not having LIT as a reference, which also shut down the transference pump when detects high ethanol levels. For "low level" deviations, it was identified that, during normal operation of ethanol dosing, for the wastewater treatment process, if the LIT fail during the transmission of low level signals, the metering pumps will send continuously ethanol to the process, till no more product is available inside the tank.

These metering pumps are responsible for etha-nol dosing in the process and by design, cannot run without fluids. If these pumps run dry it may cause serious cavitation damages, which can be mitigated enabling an automation interlock between LIT of TQE-E02-005 and the metering pumps, that when LIT reaches LL (Low Low) levels, all the metering pumps will be shutted down instantly.

An overall recommendation is to observe all operational procedures, check materials and ver-ify equipment integrity to ensure that all situa-tions that can cause a loss of ethanol containing are controlled and identified. Safety must be the prime concern in all actions during the operations of transference of ethanol from the transportation truck to the storage tank and ethanol storage into the tank TQE-E02-005.

Table 6. HazOp table: transference of ethanol from the transportation truck to the storage tank.

Parameter	Guide word	Deviation	Cause	Consequences	Freq.	Sev.	Risk	Recommendations
Ethanol **flow** transference between truck and tank	None	No flow	Unplanned Locking of manual valves	Delay in the operations	E	4	L	- Prevent the inadvertent closing of manual valves. - In case of loss of containment, follow procedures to avoid ethanol spreading.
Ethanol **flow** transference between truck and tank	None	No flow	Fail in the hose integrity or accidental disconnection of any quick coupler	Ethanol spill, causing contamination, fire or explosion	D	1	S	- Remove or eliminate explosion ignition sources during operation. - Ensure that the hose and connections are correctly grounded in the same earth equipotential of the tank trunk and the transference pump. - Ensure that the hose and connections are in perfect condition and operation. - Evaluate the possibility of substitution of ethanol as a carbon source for other less dangerous product.
Ethanol **flow** transference between truck and tank	None	No flow	Pump fail	Delay in the operations	B	4	M	- Ensure that the pump, its accessories and connections are in perfect condition and operation. - Evaluate the possibility of having a backup pump for this process.
Ethanol **flow** transference between truck and tank	Less	Less flow	Partial fail in the hose	Ethanol spill, causing contamination, fire or explosion	D	2	M	- Remove or eliminate explosion ignition sources during all steps of operation. - Ensure that the hose and connections are in perfect condition and operation. - Ensure that the hose and connections are correctly grounded in the same earth equipotential of the tank trunk and the transference pump.
Ethanol **pressure** transference between truck and tank	Less	Less pressure	Failure in the top vents of the truck tank, causing internal vacuum	Mechanical tank implosion, causing ethanol contamination, fire or explosion	D	1	S	- Ensure that all top vents of the truck tank are in perfect conditions and without obstructions before transference operation begins. - Ensure that all personal involved in this operation has been trained to identify top vents failures and obstructions. - Ensure that the hose and connections are correctly grounded in the same earth equipotential of the tank trunk and the transference pump.

Table 7. HazOp table: ethanol storage into the tank TQE-E02-005.

Parameter	Guide word	Deviation	Cause	Consequences	Freq.	Sev.	Risk	Recommendations
Ethanol **level** in the tank TQE-E02-005	High	High level	LIT fail during ethanol loading to the tank	Ethanol spill, causing contamination, fire or explosion	D	1	S	- Enable automation interlock between LIT (level indicator) of TQE-E02-005 and the transference pump. - Evaluate the possibility of having a backup automated system that shut down the transference pump when detected high ethanol level.
Ethanol **level** in the tank TQE-E02-005	Low	Low level	LIT fail during normal operation of the tank	Damages in the metering pumps of the ethanol dosing process	D	2	M	- Enable automation interlock between LIT of TQE-E02-005 and the metering pumps. - Evaluate the possibility of having an automated system that shut down the metering pumps when detected low ethanol level.

5 CONCLUSIONS

A HazOp analysis presents some operational recommendation to avoid economic and safety loss in the experimental works. However, besides the consideration of economic order, aspects concerning the safety of a process are becoming more important by an increasing knowledge of the environmental problems and life quality in a scenario of sustainable world growth. Safety assessment at workplaces involves operation risks as well as all other kinds of tasks and hazardous activities, which, in this paper, include the operations of a wastewater treatment unit that is under construction, but its operational risk were already identified, resulting in recommendations that will easily implemented. For instance, the recommendation of an automated system that shut down the metering pumps when detected low ethanol level, for "low level" deviations, will demand a simple installation of new two sensors, one transmitter and few connecting wire in each metering pump; a simple and not complex installation that can im-prove operational safety, and can be easily done be-cause was identified by a HazOp study during the construction phase. Although this particular HazOp attends a Wastewater Treatment Unit, this study can be performed for any industrial plant, in any phase of this life cycle: project, construction, operation, expansion and decommissioning. However, as evidenced, if HazOp is performed in the project or construction phases, the safety recommendations can be implemented quickly, economically and effectively.

REFERENCES

Crawley, F., Preston, M. & Tyler, B., 2000. *HAZOP: Guide to best practice. Guidelines to best practice for the process and chemical industries.*, Institution of Chemical Engineers (Great Britain), European Process Safety Centre. United Kingdom: The Cromwell Press.

Dunjó, J.; Fthenakis, V. M.; Darbra, R. M.; Vílchez, J. A.; Arnaldos, J. Conducting HAZOPs in continuous chemical processes: Part I. Process Safety and Environmental Protection. Elsevier, v. 89, p. 214–223, 2011.

Galante, E., Costa, D.M.B. da & Nóbrega, M. Risk Assessment Methodology: Quantitative HazOp. *Journal of Safety Engineering*, p. 8, 2014.

Ghasemzadeh, K.; Morrone, P.; Iulianelli, A.; Liguori, S.; Basile, A. H_2 production in silica membrane reactor via methanol steam reforming: Modeling and HAZOP analysis. International Journal of Hydrogen Energy. Elsevier, v. 38, p. 10315–10326, 2013.

Isimite, J. & Rubini, P. A dynamic HAZOP case study using the Texas City refinery explosion. Journal of Loss Prevention in the Process Industries. Elsevier, v. 40, p. 496–501, 2016.

Kotek, L. & Tabas, M. HAZOP study with qualitative risk analysis for prioritization of corrective and preventive actions. Procedia Engineering. Elsevier, v. 42, p. 808–815, 2012.

National Center for Biotechnology Information. PubChem Compound Database; CID = 702, https://pubchem.ncbi.nlm.nih.gov/compound/702 (accessed Nov. 6, 2016).

Reis, G. F. "Assessment of operational risk and safety of the work in a sewage treatment plant" (in Portuguese). Postgraduation, UFRJ, Escola Politécnica—Rio de Janeiro, 2016.

USDoD (US Department of Defence), 2012. *MIL STD 882-E—Standard Practice for System Safety*, Washington, USA: USA.

Occupational Safety and Hygiene V – Arezes et al. (Eds)
© *2017 Taylor & Francis Group, London, ISBN 978-1-138-05761-6*

A framework for developing safety management competence

S. Tappura & J. Kivistö-Rahnasto
Tampere University of Technology, Tampere, Finland

ABSTRACT: The safety management competence plays a crucial role when managing safety in organisations. Developing managers' safety competence is one way of achieving better safety performance. The aim of this study is to construct a Safety Management Competence Development (SMCD) framework based on previous studies and empirical results in a Finnish case organisation. The framework consists of definition of safety management competence requirements, self-assessment of the competence, definition of development needs, and implementation of competence development activities. Assessing and developing managers' safety competence provide them with knowledge of their responsibilities and expectations, company-wide safety procedures and targets, as well as tools for promoting safety. The SMCD framework provides the means for systematically improve managers' safety competence as an integral part of general competence development in organisations. Moreover, the assessment and development activities influence managers' perceptions of and attitudes toward safety and encourages their commitment to safety.

1 INTRODUCTION

1.1 Safety management competence

To remain aligned with the dynamic needs of the business environment, organisations need to ensure up-to-date competencies (Suikki et al. 2006). Competence is the ability to transform knowledge and skills into practice in a qualified way and to achieve the required level of performance (Boyatzis 1982, Dreier 2000, Königová et al. 2012). In this study, safety management competence refers to managers' ability to perform safety-related management activities and behaving appropriately for the required safety performance.

Since safety and related demands are increasingly an integrated part of business (EU-OSHA 2011, Veltri et al. 2013), managers' safety competence should be developed accordingly. Current work life presumes managers have different types of safety management and leadership competencies (Conchie et al. 2013, Hale et al. 2010, Tappura & Nenonen 2014). In organisations in which safety considerations are vital for the companies' strategy, competence requirements should be defined as a part of identifying core competencies (Prahalad & Hamel 1990, Rothwell & Lindholm 1999). However, in all organisations, safety management and leadership are part of managing other business activities, and should be closely integrated in general business management in organisations (e.g. EU-OSHA 2010, Hale 2003, Simola 2005, Veltri et al. 2013). Moreover, safety management competence

contributes to better safety performance, and, therefore, to business performance (Blair 2003, Clarke 2013, Köper et al. 2009, Wu et al. 2008).

Managers' resources, competence, and commitment are important in establishing successful organisational safety policies and procedures (Conchie et al. 2013, Fruhen at al. 2014, Hale et al. 2010, Hardison et al. 2014, Simola, 2005). Due to insufficient safety competence, managers may not be aware of their responsibilities or the organisation's safety policy, procedures, and tools. In addition, managers need leadership skills to maintain safe performance in the workplace. (Carder & Ragan 2005, Tappura & Nenonen 2014)

Occupational Health and Safety (OHS) regulations provide the foundation for safety management competence requirements. The definitions of an effective OHS management system (Gallagher et al. 2001, Frick et al. 2000) and voluntary OHS management systems (such as OHSAS 18001:2007) form another perspective on managers' safety competence needs.

Previous studies (Biggs & Biggs 2012, Hardison et al. 2014) defined safety management competencies in the construction sector. Tappura & Hämäläinen (2012) defined safety management competence requirements based on literature and two safety management training cases in Finnish organisations (Table 1). Moreover, the effective safety leadership competencies relating to safety performance were identified in Tappura & Nenonen's (2014) review.

Table 1. Safety management competence requirements (Tappura & Hämäläinen 2012).

Safety management competence requirements
OHS regulations and their mandatory requirements
Managers' role, responsibilities, and authority to intervene in violations of OHS
Motivation and justification of OHS from economic and ethical perspectives
OHS policy, goals, programs, and procedures of the organisation in question
Continuous monitoring and improvement procedures of the working environment, the work community, and work practices
Hazard identification, risk assessment, and information sharing to prevent risks from being actualised
OHS orientation and training
Occupational injuries and near-miss reporting, investigation, and subsequent learning
Work-related health problems in the working community and psychosocial work environment
Safety performance measurement and reporting
Corrective actions control
OHS communication (meetings, inspections rounds, and discussions)
Encouraging employee participation
OHS cooperation, supporting organisations, and professionals

Furthermore, Tappura et al. (2016) defined safety management tasks at different organisational levels.

1.2 Safety management competence development

Acquiring and developing competences are critical strategic factors that ensure organisations' competitiveness (Johannessen & Olsen 2003, Königová et al. 2012, Suikki et al. 2006). Thus, organisations should provide conditions to ensure sufficient competence development (Senge 1994, Viitala 2005). Much of the earlier research concentrated on management and leadership competence development (e.g. Crawford 2005, Fong & Chan 2004, Rose et al. 2007, Suikki et al. 2006, White et al. 1996). However, only a few studies have examined safety management and leadership competences (Biggs & Biggs 2012, Hale et al. 2010, Hardison et al. 2014).

Competencies can be managed with competency models (Königová et al. 2012, Rothwell & Lindholm 1999) to identify the knowledge, skills, abilities, and behaviour needed to perform effectively in an orgainsation (Lucia & Lepsinger 1999). Competency models can serve as tools for helping managers in self-reflection, identifying development needs, and developing and building collective comprehension concerning management

(Viitala 2005). In this study, the competence development framework is used as a broader concept that includes competence requirement identification, competence assessment, and development activities.

Competence development can be defined as the general development of knowledge, understanding, and cognition for a specific domain in a person (Hyland 1994), here, safety management competencies. Competence development typically involves formal and informal learning (Schoonenboom et al. 2007). Management competencies are generally learned through formal training, induction, and work experience (Suikki et al. 2006). On-the-job experience, on-the-job training, and experiential learning are considered to be the most effective methods for developing managers' competence (Fong & Chan 2004). Effective safety management training includes joint discussions with colleagues, demonstrations, and hands-on techniques to strengthen managers' commitment to safety (Tappura & Hämäläinen 2011).

Management development studies have suggested that improving self-knowledge is the basis for all true management development (Pedler et al. 1986, Viitala 2005), and managers' own interpretation of their competence development needs should be supported in organisations (Viitala 2005). A competence assessment is used to communicate goals and expectations, to give feedback on the current competence level, and to identify development needs (Bergman & Moisio 1999). Thus, the competence assessment is a basis for the competence development activities. The competence assessment concept is based on cognitive learning theories, in which the student is seen as an active participant sharing responsibility for the learning process and practices self-evaluation (Baartman et al. 2007, Birenbaum 1996). Relevant competence assessment methods include behavioural assessment, simulations, and self-assessment. However, organisational support, reflection, and mentoring must accompany the development of assessment programs (Epstein & Hundert 2002).

These observations led to the proposition that safety management competences must be taken into account in general management competence development in organisations. The main target of this study is to construct a Safety Management Competence Development (SMCD) framework to help organisations in safety promotion. The framework, based on previous studies and validated in a case organisation, provides the means for systematically improving managers' safety management competence as an integral part of general management competence development procedures.

2 MATERIALS AND METHODS

This study comprises a theory-based construction of the SMCD framework and empirical reflections. Thus, the study is located in the normative area of business studies. The study uses the qualitative approach since it is interpretative. The study is descriptive and does not state an explicit hypothesis.

The theoretical part of the study was conducted as a review of the relevant literature to chart the general competence development models and their applicability to safety management as well as the safety management competence requirements.

The empirical part of the study was conducted in a Finnish technical safety service organisation of about 200 employees. The employees work in varying and demanding environments, including office and customer sites in almost all industrial sectors. Thus, the safety management competence requirements for the managers are extensive. The motivation of the study in the case organisation arises from the need to increase line managers' safety management competence, to systematically assess it, to discover individual and organisational competence development needs, and to integrate safety issues into general competence development.

The SMCD framework was validated in the case organisation. First, a preliminary safety competence requirements list was produced based on the occupational safety regulations and the literature by the researcher. Second, the OHS and human resource professionals in the organsation (OHS committee) were interviewed (n = 7), and the list was completed based on this focus group interview. Third, a self-assessment of ten line managers (a total of 18 line managers) was conducted using the agreed 17 competence areas. The self-assessment is used as an assessment method in this study due to its reflective nature. The self-assessment communicates the safety management goals and competence requirements, as well as provides feedback on the current safety management competence level and development needs. In addition, self-assessment and reflection encourage managers to discover their individual development needs and engage them in the development process.

Based on the self-assessment, the main safety management competence development needs were identified and presented to the management team of the organisation. The safety management competence development activities and further competence evaluation were the responsibility of the case organisation and, thus, are not available for this study.

3 RESULTS

3.1 Construction of the SMCD framework

The SMCD framework was constructed to increase managers' safety management competence and to integrate safety issues into managers' general competence identification, assessment, and development. The framework is based on the previous literature and consists of the following phases: (1) definition of safety management competence requirements and needs, (2) managers' self-assessment of their safety competence, (3) definition of organisational and individual safety competence development needs and focus areas, (4) planning and implementation of related safety competence development activities, and (5) evaluation of the managers' safety competence.

Identifying safety management competence requirements was chosen as the basis for the SMCD framework. In the first phase, general and organisational safety competence requirements are defined. In the second and third phases, the relevant issues are self-assessed to define the major development needs at the organisational and individual levels. When the assessment is carried out as a self-assessment, the managers reflect on their development needs directly. Moreover, the self-assessment is a good way of improving the managers' commitment to competence development activities. In the fourth phase, safety management competence development activities are planned based on the development needs at the individual and organisational levels. When possible, the development activities are integrated in the general management competence development activities. In the fifth phase, the evaluation of the adequacy of managers' safety competence is also integrated in the general management competence evaluation.

3.2 Validation of the SMCD framework

The SMCD framework was validated in the case organisation as follows. First, a preliminary list of 13 general safety management competence areas was defined based on the employers' regulatory occupational safety requirements and previous literature (see Tappura & Hämäläinen 2012). The preliminary list was presented to the occupational safety committee of the case organisation. The competence areas were discussed and supplemented in the committee. A final version of 17 safety competence areas (Table 2) was accepted.

Second, a self-assessment was carried out with ten line managers. According to the results, the major organisational development needs are related to the OHS policy, reporting and inspection of occupational accidents, OHS plan, and support for OHS management in general. During the self-assessment, the managers reflected on and identified individual development needs and, thus, started the learning process. The self-assessment identified the main individual and organisational competence development needs. The results were presented and discussed in the management group.

Table 2. Safety management competence areas for the self-assessment in a case organisation.

Safety management competence areas
1. OHS policy
2. Reporting and inspection of occupational accidents
3. OHS plan
4. Early support and intervention guidelines
5. Support from OHS committee
6. Support from occupational health care
7. Occupational health care agreement
8. Intervention procedure in safety violations
9. Risk assessment procedure and major risks
10. Workplace survey by occupational health care
11. Alcohol abuse prevention plan
12. Personal protective equipment guidelines
13. Intervention procedure in case of harassment or other inappropriate treatment
14. Contact person (and support) of OHS committee
15. Development discussion procedure
16. Working hours, holiday and travelling guidelines
17. Recruitment procedure

Based on the competence areas, an assessment form was produced for the self-assessment.

Based on the results, appropriate individual and organisational competence development actions were planned and integrated into the general competence development activities and management training. The competence development activities, however, were not studied in this study.

4 DISCUSSION

Managers' safety competencies clearly overlap with the existing conception of "good" management behaviour and therefore should be integrated into the overall management practices (HSE 2007). Moreover, managers' leadership skills should be emphasised to improve safety performance (Tappura & Nenonen 2014). However, the OHS management perspective is generally absent in management and HR studies and literature (Zanko & Dawson 2012).

In this article, a framework for developing managers' safety management competence was constructed. The framework provides a basis for managers' safety management competence identification, assessment, and development as an integrated part of managers' general competence development. Moreover, it can be utilised in organisational-specific competence models. The framework takes the general development activities into account and complements existing

activities with relevant safety issues. For example, identifying managers' safety competence could be part of competency models (Königová et al. 2012) and identifying core competencies (Prahalad & Hamel 1990, Viitala 2005). Self-assessment could be part of the general development discussion, where safety objectives, related competence requirements, and development needs could be discussed as part of other management responsibilities. The managers' own interpretation of their safety competence needs should be taken into consideration since it reflects their motivation to develop as well as deficiencies in competence development procedures (Viitala 2005).

Safety management competence requirements for the managers are extensive, and well applicable to managerial work in many industrial sectors. However, organisational and industrial-specific competence requirements should be considered when the framework is applied. For example, in the construction sector such requirements have been defined (Biggs & Biggs 2012, Hardison et al. 2014).

After the competence development activities, safety management competence should be evaluated on a regular basis, for example, in annual development discussions or work climate surveys. Managers' safety competence needs should be evaluated and updated according to the changing regulatory and organisational OHS requirements and procedures.

The safety literature and practical experiences provide the foundation for the SMCD framework. However, such a framework requires continuous evaluation and reflection to develop competence effectively (Suikki et al. 2006) and to be utilised in competence models.

5 CONCLUSION

Safety management is an integral part of management activities in organisations and safety management competencies should be developed accordingly. This article suggests a framework for assessing and developing managers' safety management competence in different industrial sectors. The framework provides the means for systematically improve safety management competence as an integral part of managers' general competence development in organisations. The suggested framework enables better organisational support for managers when developing their safety management competencies and commitment to safety. In the future, the SMCD framework tools and applicability to other industrial sectors should be evaluated.

REFERENCES

Baartman, L.K.J., Bastiaens, T.J., Kirschner, P.A. & van der Vleuten, C.P.M. 2007. Evaluating assessment quality in competence-based education: A qualitative comparison of two frameworks. *Educ. Research Review* 2: 114–129.

Bergman, T. & Moisio, E. 1999. Käytännön kokemuksia osaamisen hallinnan kehittämisestä (in Finnish). *Työn Tuuli* 2/1999.

Biggs, H.C. & Biggs, S.E. 2012. Interlocked projects in safety competency and safety effectiveness indicators in the construction sector. *Safety Science* 52: 37–42.

Birenbaum, M. 1996. Assessment 2000: Towards a pluralistic approach to assessment. In M. Birenbaum & F.J.R.C. Dochy (eds), *Alternatives in assessment of achievement, learning processes and prior knowledge*: 3–29. Boston: Kluwer Academic Publishers.

Blair, E. 2003. Culture & leadership: Seven key points for improved safety performance. *Prof. Safety* 48(6). 18–22.

Boyatzis, A.R. 1982. *The Competent Manager: A Model for Effective Performance*. New York: J. Wiley.

Carder, B. & Ragan, P. 2005. *Measurement matters. How effective assessment drives business and safety performance*. Milwaukee, Wisconsin: Asq Quality Press.

Clarke, S. 2013. Safety leadership: A meta-analytic review of transformational and transactional leadership styles as antecedents of safety behaviours. *Journal of Occupational and Organizational Psychology* 86: 22–49.

Conchie, S.M., Moon, S. & Duncan, M. 2013. Supervisors' engagement in safety leadership: Factors that help and hinder. *Safety Science* 51: 109–117.

Crawford, L. 2005. Senior management perceptions of project management competence. International *Journal of Project Management* 23: 7–16.

Dreier, A. 2000. Organizational learning and competence development. *The Learning Organization* 7(4): 52–61.

EU-OSHA 2010. *Mainstreaming OSH into business management*. Luxembourg: Office for Official Publications of the European Communities.

EU-OSHA 2011. *Healthy workplaces. Working together for risk prevention*. Luxembourg: Publications Office of the European Union.

Epstein, R.M. & Hundert, E.M. 2002. Defining and assessing professional competence. *Journal of the American Medical Association* 287(2): 226–235.

Fong, P.S.W. & Chan, C. 2004. Learning behaviours of project managers. In *Proceeding of the IRNOP VI, Turku, Finland*: 200–219.

Frick, K., Jensen, P., Quinlan, M. & Wilthagen, T. 2000. Systematic Occupational Health and Safety Management—An Introduction to a New Strategy for Occupational Safety, Health and Well-being. In K. Frick, P. Jensen, M. Quinlan & T. Wilthagen (eds), *Systematic OHS Management: Perspectives on an International Development*: 1–14. Amsterdam: Elsevier.

Fruhen, L.S., Mearns, K.J., Flin, R. & Kirwan, B. 2014. Skills, knowledge and senior managers' demonstrations of safety commitment. *Safety Science* 69: 29–36.

Gallagher, C., Underhill, E. & Rimmer, M. 2001. *Occupational Health and Safety Management Systems: A Review of their Effectiveness in Securing Healthy and Safe Workplaces*. Sydney: National OHS Commission.

HSE 2007. *Management competencies for preventing and reducing stress at work. Identifying and developing the management behaviours necessary to implement the HSE Management Standards*. London: Health and Safety Executive.

Hale, A.R. 2003. Safety Management in Production. *Human Factors and Ergonomics in Manufacturing* 13: 185–201.

Hale, A.R., Guldenmund, R.F., van Loenhout, P.L.C.H. & Oh, J.I.H. 2010. Evaluating safety management and culture interventions to improve safety: Effective intervention strategies. *Safety Science* 48(8): 1026–1035.

Hardison, D., Behm, M., Hallowell, M.R. & Fonooni, H. 2014. Identifying construction supervisor competencies for effective site safety. *Safety Science* 65: 45–53.

Hylan, T. 1994. *Competence, Education and NVQs: Dissenting Perspectives*. London: Cassell.

Johannessen, J.-A. & Olsen, B. 2003. Knowledge management and sustainable competitive advantages: The impact of dynamic contextual training. *International Journal of Information Management* 23: 277–289.

Königová, M., Urbancová, H. & Fejfar, J. 2012. Identification of Managerial Competencies in Knowledge-based Organizations. *Journal of Competitiveness* 4(1): 129–142.

Köper, B., Möller, K. & Zwetsloot, G. 2009. The Occupational Safety and Health Scorecard—a business case example for strategic management. *Scandinavian Journal of Work, environment & Health* 35(6): 413–420.

Lucia, A.D. & Lepsinger, R. 1999. *The Art and Science of Competency Models*. San Francisco, CA: Jossey-Bass.

OHSAS 18001:2007 *Occupational health and safety management systems Requirements*. London: OHSAS Project Group BSI.

Pedler, M., Burgoyne, J. & Boydell, T. 1986. *A Manager's Guide to Self-development*. London: McGraw-Hill.

Prahalad, C.K. & Hamel, G. 1990. The Core Competence of the Corporation. *Harvard Business Review* 8(3): 79–91.

Rose, J., Pedersen, K., Hosbond, J.H. & Kræmmergaard, P. 2007. Management competences, not tools and techniques: A grounded examination of software project management at WM-data. *Information and Software Technology* 49: 605–624.

Rothwell, W.J. & Lindholm, J.E. 1999. Competency Identification Modelling and Assessment in the USA. *International Journal of Training and Development* 3(2): 90–105.

Schoonenboom, J., Tattersall, C., Miao, Y., Stefanov, K. & Aleksieva-Petrova, A. 2007. A four-stage model for lifelong competence development. In *Proc. of the 2nd TENCompetence Open Workshop, Manchester, UK, 11–12 January 2007*: 131–136.

Senge, P. 1994. *The Fifth Discipline Fieldbook—Strategies and Tools for Building a Learning Organization*. London: Nicholas Brealey Publishing.

Simola, A. 2005. *Turvallisuuden johtaminen esimiestyönä. Tapaustutkimus pitkäkestoisen kehittämishankkeen läpiviennistä teräksen jatkojalostustehtaassa* (in Finnish), Doctoral dissertation. Oulu: University of Oulu.

Suikki, R., Tromstedt, R. & Haapasalo, H. 2006. Project management competence development framework in turbulent business environment. *Technovation* 26: 723–738.

Tappura, S. & Hämäläinen, P. 2011. Promoting occupational health, safety and well-being by training line managers. In J. Lindfors et al. (eds), *Proc. nordic conf., Oulu, 18–21 September 2011*: 295–300.

Tappura, S. & Hämäläinen, P. 2012. The occupational health and safety training outline for the managers. In P. Vink (ed), *Advances in Social and Organizational Factors, Advances in Human Factors and Ergonomics Series 9*: 356–365. CRC Press.

Tappura, S. & Nenonen, N. 2014. Safety Leadership Competence and Organizational Safety Performance. In P. Arezes & P. Carvalho (eds), *Advances in Safety Management and Human Factors. Advances in Human Factors and Ergonomics Vol. 10, Section 3; proc. intern. conf., Kraków, 19-23 July 2014*: 129–138.

Tappura, S., Teperi, A.-M. & Kivistö-Rahnasto, J. 2016. Safety management tasks at different management levels. In J. Kantola et al. (eds), *Advances in Human Factors, Business Management, Training and Educa-tion; proc. intern. conf., Walt Disney World, 27–31 July 2016*: 1147–1157.

Veltri, A., Pagell, M., Johnston, D., Tompa, E., Robson, L., Amick III, B.C., Hogg-Johnson, S. & Macdonald, S. 2013. Undestanding safety in the context of business operations: An exploratory study using case studies. *Safety Science* 55: 119–134.

Viitala, R. 2005. Perceived development needs of managers compared to an integrated management competency model. *Journal of Workplace Learning* 17(7): 436–451.

White, R., Hodgson, P. & Crainer, S. 1996. *The Future of Leadership. Riding the Corporate Rapids into the 21st Century*. Lanham, Md.: Pitman.

Wu, T.-C., Chen, C.-H. & Li, C.-C. 2008. A correlation among safety leadership, safety climate and safety performance. *Journal of Loss Prevention in the Process Industries* 21: 307–318.

Zanko, M. & Dawson, P. 2012. Occupational Health and Safety Management in Organizations: A Review. *International Journal of Management Reviews* 14: 328–344.

Occupational Safety and Hygiene V – Arezes et al. (Eds)
© 2017 Taylor & Francis Group, London, ISBN 978-1-138-05761-6

Workaholism and burnout: Antecedents and effects

G. Gonçalves, F. Brito, C. Sousa & J. Santos
Faculty of Human and Social Sciences, University of Algarve, Portugal

A. Sousa
Institute of Engineering, University of Algarve, Portugal

ABSTRACT: Among the various negative indicators of well-being are workaholism and burnout. Given that health professionals have been considered the professionals with higher levels of burnout, we defined for this study to analyze some of the antecedent variables of workaholism (work passion, professional satisfaction and engagement for life) and their effect on burnout. The sample consists of 208 professionals from a Public Hospital, aged between 19 and 66 years (M = 40,14; SD = 11,43). The results of the multiple linear regression analysis allowed us to identify that some of the variables and their dimensions present significant effects of the workaholism on burnout.

1 INTRODUCTION

The technological development associated with social and economic changes has rapidly and profoundly altered the work systems, contributing to new concerns in the approach to the factors associated with the workers' health and well-being. One of the concepts that has received more attention is workaholism (Ng et al. 2007). Defined as "the compulsion or the uncontrollable need to work incessantly" (Oates, 1971, p. 1). It is still unclear whether it has positive or negatives effects on work (Burke, 2001), its relationship with physical and psychological health (Burke, 2001), as well as its antecedents and effects (Gonçalves et al. 2016). Given the emerging importance of this concept for understandding and intervening in the promotion of psychological health at work, we carried out this study with the objective of evaluating the variables work passion, engagement with life and professional satisfaction as antecedents of workaholism, and the burnout as an effect on a professional sector that presents high levels of stress and burnout – the health sector (Cooper et al. 2016). Regarding the antecedents, Work Passion (WP) is understood as a strong affective inclination toward work, in which one spends time and energy (Vallerand et al. 2003). The authors propose two types of WP: Harmonious Passion (HP) and Obsessive Passion (OP) which are distinguished by the degree of internality in personal identity and the consequent compulsion to engage in the activity (Vallerand et al. 2003). Professional Satisfaction (PS), in turn, is understood as a positive attitude towards work (War et al. 1979) arising associated with health and well-being variables (Faragher et al. 2005) and with organizational, professional and personal variables (Lu et al. 2005). WP and PS are variables that indicate pleasure and involvement with work. In turn the Engagement for Life (EL) is defined in terms of the degree to which a person engages in the activities to which he gives more value. On the other hand, a high importance and expenditure of time and energy at work can lead to a burnout situation (Innanena et al. 2014, Vallerand et al. 2010). This is understood as the result of a continuous and prolonged professional stress experience (Maslach & Schaufeli 1993). It is characterized by extreme physical and mental fatigue and emotional exhaustion and it is associated with a wide range of professions, particularly in the health sector (Doctors and Nurses) where involvement with people is central to their work (Gundersen, 2001). Thus, if an individual has signs of burnout, it is natural that his/her psychological well-being also diminishes significantly. Burnout represents costs to the organizations as well as to society (Bakusic et al. 2017). Among the various outcomes, emerge the absenteeism (Michie & Williams, 2003), turnover (Leiter & Maslach, 2009) and hospitalizations due to mental and cardiovascular disorders and poor performance at work (Bakusic et al. 2017, Maslach et al. 1996), in particular in doctors and nurses (Becker et al. 2006). In health professionals these effects are more pronounced. Research on fatigue and burnout shows negative effects on job and life satisfaction, more heart disease, depression, suicide, divorce, etc. with evident consequences in patient satisfaction, greater number of medical errors and negligence (Becker et al. 2006, Bria et al. 2014, Montgomery, 2014, Shanafelt et al. 2010). Thus, to obtain a model able to predict burnout on the basis of the independent variables, a multiple linear regression was performed. Four models were used to determine the variables studied in the dimensions of burnout: physical exhaustion, emotional exhaustion and cognitive fatigue.

2 METHOD

2.1 *Sample and procedure*

The sample of the present study was collected for accessibility in the same public health institution. It is a sample composed of 208 workers, 177 of the feminine gender and 27 of the masculine gender, with a mean age of 40.14 years (SD = 11,43). In relation to marital status, the majority of the participants are married or live together (59,1%). Regarding the educational level, 71.8% of the participants have higher education, the remaining percentage is distributed in Primary Education (2%), Basic Education (8.4%) and Secondary Education (17.8%). Regarding the type of contractual relationship, 64.6% of respondents have a permanent contract, 20% have a fixed-term contract and 14% have other types of contracts. The majority of the participants are nurses (57.5%), the rest are technical assistants (14.4%), operational (13.7%), doctors (5.2%) and still with minor presence psychologists, therapists, auxiliaries, and physiotherapists (9.2%).

2.2 *Measures*

All applied scales were evaluated on a 7-point Likert scale (1 - Strongly disagree to 7 - Totally agree) and are properly validated or have been translated and tested for the Portuguese population according to the recommendations of Hambleton, Merenda, and Spielbergers (2006): Passion Scale (Vallerand et al. 2003) - adaptation of Gonçalves, Orgambídez-Ramos, Ferrão and Parreira (2014). This scale is composed of 2 subscales of 7 items: harmonious passion and obsessive passion. Both values of internal consistency are satisfactory and similar to the original study values and the adaptation (HP = 0.92 and OP = 0.89). Engagement for Life—It is a one-dimensional measure originally constructed by Scheir et al. (2006), consisting of six items: three items framed in a positive direction (items 2, 4 and 6) and three items framed in a negative direction (items 1, 3 and 5). The intent of the scale is to provide an index of purpose in life, assessing the importance of the activities for the participant. As an example of a positively-driven item: "For me, everything I do is important." And as an example of a negative item: "I do not give much importance to the things I do". The internal consistency of the original scale presents a value of 0.77 similar to the original studies. Professional Satisfaction. It is a one-dimensional scale of 16 items originally constructed by Warr et al. (1979). The original study had an internal consistency of 0.89, and in the present study we obtained a Cronbach's alpha of 0.82. Workaholic's profile was evaluated through the two-dimensional

WorkBat scale of Spence and Robbins (1992). It is a 25-item self-report questionnaire subdivided into three subscales: the work involvement factor that refers to the generalized attitude of psychological involvement with work (items 1–8); work drive indicator questions which are related to an internal compulsion to work hard and a sense of guilt when you fail at work (items 9–15); and finally, questions of work enjoyment that are related to the pleasure derived from work (16–25). Regarding the internal consistency of the scale, it showed a good reliability in the general scale of 0.77, however, in the work involvement subscale it showed an alpha of 0.44, in the work drive subscale an alpha of 0.81, and finally in the subscale work enjoyment an alpha of 0.77. With the exception of the work involvement, the values are acceptable and similar to the original studies. Burnout—was evaluated with fourteen items of the two-dimensional measure adapted from Shirom-Melamed Burnout Measure (SMBM) by Shirom & Melamed (2006). The scale consists of three subscales, namely physical fatigue (6 items), emotional exhaustion (3 items) and cognitive fatigue (5 items). Regarding the internal consistency of the scale, it presented values between 0.79 and 0.9, similar to the values of the authors of the scale.

3 RESULTS

3.1 *Descriptive analysis*

Table 1 shows the means and standard deviations of all the variables under study. With respect to the Work Passion (WP), we can observe that the Harmonious Passion (HP) presents a mean (M = 4.98) significantly higher than the Obsessive Passion (OP) (M = 2.75) (p<0.05). The variables Engagement for Life (EL) and Professional Satisfaction (PS) present means of 6.00 and 4.14, respectively. Regarding the dimensions of workaholism (W), although the largest dimension is work involvement (M = 4.01), no significant differences were

Table 1. Mean and standard deviation of the variables.

		M	SD
Work Passion	Harmonious Passion	4,98	1,35
	Obsessive Passion	2,75	1,36
Engagement for Life		6,00	0,99
Professional Satisfaction		4,14	0,97
Workaholism	Work envolvement	4,01	1,16
	Work drive	3,65	1,32
	Work enjoyment	3,73	0,97
Burnout	Physical exhaustion	3,55	1,69
	Emotional exhaustion	2,92	1,75
	Cognitive fatigue	2,56	1,57

observed in the mean values between the three dimensions (p> 0.05). As for Burnout, the highest mean is found in the physical fatigue dimension (M = 3.55) and the lowest mean in the cognitive fatigue dimension (M = 2.56). These differences are significant (p <0.005).

3.2 *Inferential analysis*

Tables 2 to 4 summarize the results obtained by the hierarchical regressions performed in the three dimensions of burnout: physical fatigue, emotional exhaustion and cognitive fatigue. The subsequent inferential analysis focuses only on the final models (M4), i.e., on the total set of variables, after the introduction of the workaholic variable. As it can be seen in Table 2 – physical fatigue dimension of the burnout—the predictive power of the variables under study is improved ($\Delta r^2 = 10\%$, p = 0.000) by including the dimensions of workaholism (involvement, drive and enjoyment). It should be pointed out that the work drive dimension assumes a significant contribution (β = 0.353; p = 0.000) and that the global model contributes approximately 30% to the explanation of burnout—physical fatigue. In the remaining two dimensions of burnout—emotional exhaustion (Table 3) and cognitive fatigue (Table 4)—analogous situations occur, and the predictive power is improved by adding the dimensions of workaholism (involvement, drive and enjoyment) in ($\Delta r^2 = 4\%$, p = 0.000) and ($\Delta r^2 = 5\%$, p = 0.000), respectively. It is also worth noting that the work drive dimension assumes

Table2. Synthesis of hierarchical regression for burnout—physical fatigue.

M	Variables	β	T	P
1	Harmonious passion	−0.387	−5.188	0.000
	Obssessive passion	0.208	2.787	0.006
2	Harmonious passion	−0.256	−3.006	0.003
	Obssessive passion	0.121	1.534	0.127
	Engagement for life	−0.229	−2.986	0.003
3	Harmonious passion	−0.232	−2.688	0.008
	Obssessive passion	0.129	1.646	0.101
	Engagement for life	−0.203	−2.611	0.010
	Professional Satisfaction	0.117	1.671	0.096
4	Harmonious passion	−0.178	−2.099	0.037
	Obssessive passion	0.093	1.162	0.247
	Engagement for life	−0.120	−1.606	0.110
	Professional Satisfaction	0.118	1.787	0.076
	Workaholic: envolvement	−0.022	−0.355	0.723
	Workaholic: drive	0.353	5.210	0.000
	Workaholic: enjoyment	−0.163	−2.148	0.033

Table3. Synthesis of hierarchical regression for burnout—emotional exhaustion.

M	Variables	β	t	p
1	Harmonious passion	−0.430	−5.874	0.000
	Obssessive passion	0.262	3.576	0.000
2	Harmonious passion	−0.161	−2.092	0.038
	Obssessive passion	0.083	1.160	0.247
	Engagement for life	−0.471	−6.801	0.000
3	Harmonious passion	−0.161	−2.048	0.042
	Obssessive passion	0.083	1.158	0.248
	Engagement for life	−0.470	−6.641	0.000
	Professional Satisfaction	0.003	0.047	0.963
4	Harmonious passion	−0.124	−1.535	0.126
	Obssessive passion	0.056	0.738	0.461
	Engagement for life	−0.420	−5.922	0.000
	Professional Satisfaction	0.008	0.130	0.897
	Workaholic: envolvement	0.034	0.566	0.572
	Workaholic: drive	0.202	3.149	0.002
	Workaholic: enjoyment	−0.097	−1.347	0.180

Table 4. Synthesis of hierarchical regression for burnout—cognitive fatigue.

M	Variables	β	t	p
1	Harmonious passion	−0.477	−6.813	0.000
	Obssessive passion	0.427	6.104	0.000
2	Harmonious passion	−0.231	−3.107	0.002
	Obssessive passion	0.263	3.834	0.000
	Engagement for life	−0.430	−6.436	0.000
3	Harmonious passion	−0.211	−2.801	0.006
	Obssessive passion	0.270	3.944	0.000
	Engagement for life	−0.409	−6.021	0.000
	Professional Satisfaction	0.097	1.588	0.114
4	Harmonious passion	−0.186	−2.405	0.017
	Obssessive passion	0.234	3.212	0.002
	Engagement for life	−0.359	−5.281	0.000
	Professional Satisfaction	0.100	1.661	0.098
	Workaholic: envolvement	0.017	0.296	0.768
	Workaholic: drive	0.204	3.324	0.001
	Workaholic: enjoyment	−0.065	−0.944	0.346

significant contributions in the emotional exhaustion (β = 0.202, p = 0.002) and cognitive fatigue (β = 0.204, p = 0.001) dimensions, and that the corresponding models contribute, respectively, to the explanation of the burnout, associated with these dimensions, by approximately 35% and 40%.

4 DISCUSSION AND CONCLUSION

This study aimed to evaluate the effect of workaholism and its predictive variables on burnout in the health professionals. In order to obtain a model that allowed to predict burnout, as a function of the independent variables, a multiple linear regression was performed. Four models of determination were performed. The model with the greatest explanatory power was the one that included the five independent variables together. Regarding the explanatory contributions to burnout, in its various dimensions (physical fatigue, emotional exhaustion, cognitive fatigue), it can be observed that in the three hierarchical regressions performed, the significant contributions were consistent, that is, the WP (HP and OP), the EL and the workaholism work drive dimension. It should be noted that of these significant contributions, the OP had a higher explanatory power of the cognitive fatigue dimension. In turn, HP and EL presented a negative direction, that is, the more the individual experiences them, the fewer symptoms of burnout presents. In other words, it may be assumed that they may be protective factors (Vallerand et al. 2010). A harmonious positive attitude towards work allows one to take pleasure from work because the individual is free to leave and engage, equally with pleasure, in other activities of life (Wrosch et al. 2003). In this situation, when work and life are balance the individual on the one hand, is less workaholic and, on the other, develops less burnout. The work drive dimension of workaholism is the one with the greatest predictive power of burnout, corroborating other studies that show that this component of workaholism is the one that consistently presents higher predictive values with negative outcomes (see Burke, 2000; for reviews), in particular in the physical fatigue dimension (Gonçalves et al. 2016, Taris et al. 2005). However, in general the workaholism has a low predictive value compared to the other variables, in our opinion it may be negatively associated with EL and HP.

In summary, this study provides an overview of workaholism and its effects on burnout and psychological well-being in health professionals. Studies show that workaholism is equated with negative outcomes (e.g., burnout and lower levels of psychological well-being). Both entail significant losses for workers and organizations, particularly in a professional area already physical and emotionally very demanding: the health area. Although much research has been done on this construct, its antecedents and effects, many questions are still open. Perhaps because health professionals themselves are contaminated by a perspective of pathology or disease that needs to be cured (Montgomery, 2014) or prevented in terms of personal skills training (for example: resilience (Chaukos, et al. 2016)). So, the identification of the factors that potentiate both situations, will allow to delineate strategies of prevention and intervention associated to the variables of development of burnout. Interven-

tion measures are necessary in the antecedent variables of the burnout in order to promote healthier and less risky work environments for both, health professionals and patients. In practice, this study allows to identify variables to take into account in the measures of evaluation of fatigue and professional risk. Future studies should consider sources of conflict and the work-family interaction management (Sousa et al. 2016), particularly with regard to work demands and personal resources (e.g., JD-R model), the career perspectives and demands, and the value that individuals attribute to work and to family.

ACKNOWLEDGEMENTS

This paper is financed by National Funds provided by FCT—Foundation for Science and Technology through project UID/SOC/04020/2013.

REFERENCES

Bakusic, J., Schaufeli, W., Claes, S., & Godderis, L. (2017). Stress, burnout and depression: A systematic review on DNA methylation mechanisms. *Journal of Psychosomatic Research, 92*, 34–44

Becker, J., Milad, M., & Klock, S. (2006). Burnout, depression, and career satisfaction: cross-sectional study of obstetrics and gynecology residents. *American Journal of Obstetrics & Gynecology, 195*(5), 1444–14449.

Bria, M., Spânu, F., Baban, A., & Dumitrascu, D. (2014). Maslach Burnout Inventory—General Survey: Factorial validity and invariance among Romanian healthcare professionals. *Burnout Research, 1*, 103–111. http://dx.doi.org/10.1016/j.burn.2014.09.001

Burke, R. (2000). Workaholism in organizations: Psychological and physical well-being consequences. *Stress Medicine, 16, 11–16.*

Burke, R. (2001). Workaholism components, job satisfaction, and career progress. *Journal of Applied Social Psychology, 31*, 2339–2356.

Chaukos, D., Chad-Friedman, E., Mehta, D., Byerly, L., Celik, A., McCoy Jr, T., & Denninger, J. (2016). Risk and Resilience Factors Associated with Resident Burnout. *Academic Psychiatry, 1–6.* doi:10.1007/s40596–016–0628-6

Cooper, S., Carleton, H., Chamberlain, S., Cummings, G., Bambrick, W., & Estabrooks, C, (2016). Burnout in the nursing home health care aide: A systematic review. *Burnout Research, 3*, 76–87. http://dx.doi.org/10.1016/j.burn.2016.06.003

Faragher, E., Cass, M., & Cooper, C. (2005). The Relationship between Job Satisfaction and Health: A Meta-Analysis. *Occupational and Environmental Medicine, 62*(2), 105-112.

Gonçalves, G., Nené, D., Sousa, C., Santos, J., & Sousa, A. (2016). The workaholism as an obstacle to safety and well-being in the workplace. *Occupational Safety and Hygiene, IV* (pp. 81–85). London: Taylor & Francis Group. ISBN: 978-1-138-02942-2; ISBN: 978-1-315-62896-7

Gonçalves, G., Orgambídez-Ramos, A., Ferrão, M., & Parreira, T. (2014). Adaptation and Initial Validation of the Passion Scale in a Portuguese Sample. *Escritos de Psicología, 7*(2), 19–27. doi:10.5231/psy.writ.2014.2503

Gundersen, L. (2001). Physician burnout. *Annals of Internal Medicine, 135*, 145–148, 2001.

Hambleton, R., Merenda, P., & Spielberger, C. (2006). Adapting educational and psychological test for cross-cultural assessment. London: Lawrence Erlbaum Associates.

Innanena, H., Tolvanenb, A., & Salmela-Aroa, K. (2014). Burnout, work engagement and workaholism among highly educated employees: Profiles, antecedents and outcomes. *Burnout Research, 1*, 38–49.

Leiter, M., & Malasch, C. (2009). Nurse turnover: the mediating role of burnout. *Journal of Nursing Management, 17*, 331–339

Lu, H., While, A., & Barriball, L. (2005). Job satisfaction among nurses: A literature review. *International Journal of Nursing Studies, 42*, 211–227.

Maslach, C., & Schaufeli, W. (1993). Historical and conceptual development of burnout. In W. Schaufeli, C. Maslach, & T. Marek (Eds.), *Professional burnout: Recent developments in theory and research* (pp. 1–16). Washington, DC: Taylor & Francis.

Maslach, C., Jackson, S., & Leiter, M. (1996). *The Maslach Burnout Inventory* (3rd ed.). Palo Alto, CA: Consulting Psychologists Press.

Michie, S., & Williams, S. (2003). Reducing work related psychological ill health and sickness absence: a systematic literature review. *Occupational & Environmental Medicine, 60*(1), 3–9.

Montgomery, A. (2014). The inevitability of physician burnout: Implications for interventions. *Burnout Research, 1*, 50–56. http://dx.doi.org/10.1016/j.burn.2014.04.002

Ng, T., Sorensen, K., & Feldman, D. (2007). Dimensions, antecedents, and consequences of workaholism: A conceptual integration and extension. *Journal of Organizational Behavior, 28*, 111–136.

Oates, W. (1971). *Confessions of a Workaholic: The Facts about Work Addiction*. New York: World Publishing.

Scheier M., Wrosch C., Baum A., Cohen S., Martire L., Matthews K., Schulz R., & Zdaniuk B., (2006). The Life Engagement Test: Assessing purpose in life. *Journal of Behavioral Medicine, 29*, 291–298.

Shanafelt, T., Balch, C., Bechamps, G., Russell, T., Dyrbye, L., Satele, D., et al. (2010). Burnout and medical errors among American surgeons. *Annals of Surgery, 251*, 995–1000. doi: 10.1097/SLA.0b013e3181bfdab3

Shirom, A., & Melamed, S. (2006). A comparison of the construct validity of two burnout measures in two groups of professionals. *International Journal of Stress Management, 13*, 176–200.

Sousa, C., Gonçalves, G., Sousa, A., Silva, T., & Santos, J. (2016). Gestão da interface trabalho-família e efeitos na sa- tisfação com a vida e na paixão com o trabalho. *Occupational Safety and Hygiene SHO2016 - Proceedings book* (pp. 338–340). Guimarães: SPOSH. ISBN: 978-989-98203-6-4.

Spence, J., & Robbins, A. (1992). Workaholism: Definition, measurement, and preliminary results. Journal of Personality Assessment, 58, 160–178.

Taris, T., Schaufeli, W., & Verhoeven, L. (2005). Workaholism in Netherlands: Measurement and implications for job strain and work-nonwork conflict. Applied Psychology: An International Review, 54, 37–60.

Vallerand, R., Blanchard, C., Mageau, G., Koestner, R., Ratelle, C., Léonard, M., Gagne, M., & Marsolais, J. (2003). Les passions de l'a^me: On obsessive and harmonious passion. Journal of Personality and Social Psychology, 85, 756 767.

Vallerand, R., Paquet, Y., Philippe, F., & Charest, J. (2010). On the Role of Passion for Work in Burnout: A Process Model. Journal of Personality, 78, 289–312. http:// dx.doi.org/10.1111/j.1467–6494.2009.00616.x

Warr, P., Cook, J., & Wall, T. (1979). Scales for measurement of some work attitudes and aspects of psychological well-being. Journal of Occupational Psychology, 52, 129–148.

Wrosch, C., Scheier, M., Miller, G., Schulz, R., & Carver, C. (2003). Adaptive self-regulation of unattainable goals: Goal disengagement, goal reengagement and subjective well-being. Personality and Social Psychology Bulletin, 29, 1494–1508.

Occupational Safety and Hygiene V – Arezes et al. (Eds)
© 2017 Taylor & Francis Group, London, ISBN 978-1-138-05761-6

Are there health risks for teenagers using mobile phones? A study of phone use amongst 14–18 year olds

J. Fowler & J. Noyes
School of Experimental Psychology, University of Bristol, UK

ABSTRACT: Mobile phone use amongst teenagers has increased more than any other age group. However, recent studies have suggested possible health concerns from mobile phone use. A survey of 163 participants aged 14–18 years is reported. Participants completed questionnaires about their mobile phone use combined with exploratory interviews. Findings suggest users could be at risk from radiofrequency damage from lengthy phone calls and storage of the phone on the body, musculoskeletal problems, impaired performance due to multi-tasking, eyestrain, dependence, 'time wasting', anti-social behaviour and sleep disturbance. It is concluded that active promotion and application of government guidelines on mobile phone use need to be recommended.

1 INTRODUCTION

The mobile phone has become a ubiquitous gadget for daily use throughout the world. It has spread faster than any other information technology (Jonathan, 2010). Usage has increased amongst all ages but, in particular, amongst teenagers (Lenhart, 2012). In 2014, along with tablets, the mobile phone had become the most popular technology for 6–15 year olds in the UK after the television (Ofcom, 2014).

It has been suggested that the success of the mobile phone amongst young people is because of the way in which the phone facilitates teenage developmental needs (Korensis, 2013). At a stage in their life when they are defining their own identity and core values, relationships with friends and independence from parents are important concerns (Baron, 2009, Ling & Yttri, 2002). A mobile offers the teenager a tool for easy contact through texting, calling, email, and other social media. However, mobile phones now carry a UK Government health warning that highlights the scientific uncertainties about the long-term health effects of mobile use. Current government recommendations advise young users not to hold their mobile near their head if they make a voice call and recommends texting rather than calling (Government Advice, 2012). Moreover, many recent studies have identified health risks from using mobile phones. These range from musculoskeletal problems to the effects of electromagnetic radiation and also concerns regarding hearing, vision, cognition, multitasking, dependence, addiction and distracted or inappropriate use.

1.1 Musculoskeletal

Extensive usage of mobile phones, often using awkward postures, has led to a range of musculoskeletal problems, for example, pain at the base of thumb has been reported (Berolo et al., 2011, Gustafsson et al., 2010).

1.2 Hearing and vision

Hearing and vision problems include high frequency hearing loss (Velayutham et al., 2014) and damage to the inner ear (Panda et al., 2010), eyestrain (Bababekova et al., 2011), vision disturbance (Balik et al., 2005) and blurred vision (Kucer, 2008).

1.3 Radiation effects

Studies have suggested a possible link between mobile phone Radiofrequency (RF) radiation effects, the incidence of brain tumours (Hardell et al., 2008) and reduced fertility (Redmayne, 2013).

Up until the age of 18, teenagers are potentially more susceptible to the RF effects because their brains have a higher fluid content and their skulls are thinner than those of adults. This means that the penetrative effects are much greater and the specific absorption rate is considerably higher for 14–18 year olds than for adults.

A meta-analysis of 101 publications of RF radiation showed that 49 people reported a genotoxic effect while 42 did not (Hardell et al., 2008). A comprehensive review of published scientific literature found 10 human studies that had identified changes in sperm exposed to phone radiation. Men who carried their phones in a pocket or slung on a belt were more likely to have lower sperm count and/or inactive or less mobile sperm (Ruediger, 2009).

1.4 Cognition and multi-tasking

Some computerised psychometric tests on mobile phone users aged 11–14 year olds have shown impaired cognitive function (Abramson et al., 2009). However, other studies have not (Besset, 2005). The experiments of Hyman et al. (2009) showed that attention was affected when students were walking and using their phones at the same time. Recent experiments show degraded cognitive performance (Baron, 2009) Users can also experience cognitive salience. This is when the mobile phone overrides other thought processes so that when trying to focus on an activity, they keep thinking about their phone.

1.5 Dependence, time wasting and addiction

Advantages of improved social communication from the use of the mobile phone have been well documented (Baron, 2009). Teenagers can gain a sense of freedom from being able to be contacted anywhere, any time. However, there have been negative health effects reported when voice calls and texts disrupt social interaction (Humphreys, 2005) and cause sleep disturbance (Thomee et al., 2011). A UK Mobile Consumer Survey (2015) showed that almost half of 18–24 year olds check their phone in the middle of the night.

Many teenagers feel compelled to reply to texts and emails, constantly checking for the arrival of a message. Some of this seems to be an appropriate affordant to the nature of the technology but it can become problematic. Descriptions of mobile phone use from a study in Australia (Walsh et al., 2008) revealed symptoms of behavioural addiction. It is suggested that characteristics of mobile phone behaviour are contributing to the addictive behaviour rather than a propensity to addictive behaviour *per se*. In extreme cases, mental distress and disruption of social relationships (Beranuy et al., 2009) and stress (Sansone, 2013) can occur.

1.6 Inappropriate use

Teenagers may experience inappropriate contact through cyber-bullying and/or security issues (Korensis, 2014).

There is a need to find out more about how teenagers use their phones and the possible health risks (Madell & Muncer, 2007). This study attempts to address this in the light of the literature and to find out if teenagers are aware of any health risks from using mobile phones.

2 MATERIALS AND METHOD

Questionnaire booklets about mobile phone use were developed and presented to 14–18 year-olds (n = 163) to find out how they use mobile phones. Participants completed the questionnaires in small groups across four year groups in two educational establishments (one school and one further education college) in the South West of England. It took around 20 minutes to complete each questionnaire. The questions included in the questionnaires are shown in the Appendix 1.

A pilot study was carried out; some questions were changed to make sure the terminology was appropriate for the age group. The questionnaires included open-ended and multiple choice questions.

Interviews were carried out with 15 participants in groups of three. Interview questions are outlined in Appendix 2. They were designed to encourage open-ended, undirected responses.

Ethical approval was given by the University of Bristol Ethics Committee.

3 RESULTS AND DISCUSSION

Descriptive statistics show that all 14–18 year olds own a mobile phone. Latest figures from Ofcom show that 90% of 16–24 s own a Smartphone (Statistica, 2015). The 14–18 year olds in this study make 4 to 5 calls a day (M = 4.7, SD 17.2) with a range of 0 to 213 calls.

Noteworthy points that can be observed from the figure are that 12.8% say they make no calls on their mobile. The most number of calls made in a day is one (27.4%). About a third make two calls a day (20.7%), roughly a seventh of participants make three calls a day (15.2%). The number of calls made then decreases with 0.6% saying they make 213 calls a day. This is considerably higher than most users in this study and could be indicative of problematic mobile phone use. The Pew internet

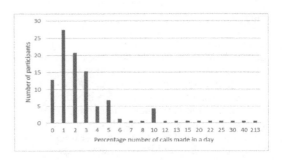

Figure 1. Shows the percentage number of calls made in a day.

study reports users making and receiving a median of 5 calls a day (Lenhart, 2012).

The participants make 71 texts a day (M = 70.9, SD = 144.7) with a range of 0 to 1000. The large standard deviation for texting could be due to them having an unrealistic assessment of how many texts they make or it could be indicative of excessive texting suggesting problematic mobile phone use. Three participants stated they sent 1000 texts a day. The 11–17 s Pew Study found that the median number of texts a day was 60 in 2011 (Lenhart, 2012).

A few participants said they did not text (2.4%). The most number of texts made in a day are 100 (12.8%) About a tenth make 10 (10.4%) or 20 texts a day (11%) and a few said they make 50 texts a day (7.3%).

Associated with usage of mobile phones by 14–18 s, health concerns are reported from the questionnaires and interviews.

3.1 Musculoskeletal

Most 14–18 year olds use only their thumbs when they are texting (64%) whilst about a third use both their fingers and thumbs (33%); however, only a few said they use their fingers (3%). Texting is the most popular use of the mobile phone for users in this study (99%) and participants show considerable skill in texting ability. For example, one skilled participant said she could "text without looking". However, the evolutionary function of the thumb has not evolved for repetitive usage, as is required for texting, and can lead to "Blackberry thumb", that is, a serious pain at the base of the thumb. Using both thumbs for texting is better than using only one thumb and it is advisable not to text at high velocity (Gustafsson, 2012).

3.2 Hearing and vision

No participants reported problems with hearing and only one participant commented on the negative impact of eye strain from using her phone at night.

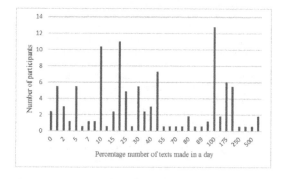

Figure 2.   Shows the percentage number of texts made in a day.

3.3 Radiation effects

Participants often used their mobiles for lengthy long distance contact, "we don't really bother texting; we just call each other because it's so much easier to catch up and everything." This could have implications for possible radiation effects of the phone. Another concern is where participants' carry their phone when on the move. The most common place was in a pocket (63.4%). Only a few kept their phone slung on a belt (4.3%). In the interviews, one participant said that her mobile phone was "quite handy, like, you can just put it in your pocket and have everything on it". Kang and Ghandi (2002) found that the specific absorption rate was increased by up to seven fold when a mobile was placed in a shirt pocket and mobile radiation could be absorbed into the surface of the heart muscle nearest to the skin (Ghandi et al., 2012). Keeping a mobile in a trouser pocket also has health implications. This study did not ask about which pocket the phone was kept but this could be ascertained in future studies.

When asked if they had any health concerns from using mobile phones, many participants were unsure. In the interviews, for example, one participant said, "I've sort of heard stories but I'm not sure how much to believe them" to which another replied, "There's one (story) about brain cancer, about putting it under your pillow, and then it's by your head when you are asleep … it's really bad for you". A third participant joined in by saying "I try to avoid putting my phone near my head and stuff. I know it's not really good because of the frequencies and stuff, like, even though you can't see them, they're still there and they do stuff to people" and "when you have the phone next to your ear, you have the radiation". Another participant said that she got "headaches during long phone calls".

3.4 Cognition and multi-tasking

Just over half of 14–18 year olds reported multi-tasking on their phones (61%), slightly more than younger users (Fowler & Noyes, 2015). Many 14–18 year olds said that they would multi-task with almost any activity that it was possible (14%). These included talking, walking, studying, watching TV, cooking, cleaning, playing games, tidying, eating, writing, using a laptop or listening to music, social networks, in fact, "any daily task or life things". Other data has shown that 61% of mobile phone users watch television and use their phones and 36% use their phones whilst eating (Facts & Figures, 2015). Texting while walking, for example, can alter walking behaviour because of the increased cognitive demands placed on working memory and executive control by performing dual tasks (Schabrun, 2014).

3.5 Dependence, time wasting and addiction

Most 14–18 year olds liked to have their phones on all the time and to be able to be contacted at all times (96%). There are many benefits to constant connectivity (Baron, 2009) but the effect of this can momentarily increase personal stress (Balding, 2013). The UK study of Balding reported that the emergence of stress was due to participants getting caught up in compulsively checking for new messages, alerts and updates. Nearly all teenagers in this study liked to be able to use a phone to keep in touch no matter wherever they were (98%).

The 'time wasting' capacity of mobile phones is commented on in the interviews. Participants enjoy the fact that they can use their phones to alleviate boredom but frustration is expressed at unintentional time wasting, "It's like, I think I'm going to bed early because I'm really tired but, then, I look at random stuff and, then, …. it's like, half past 12". Problems with cognitive salience can also be experienced by some users, as shown by one participant's comments, "It's just like, really distracting because you're like, Oh, I wanna know what it says but I'm doing my work so, you're not really concentrating (on your work)".

Most participants slept with a phone next to their bed (92%) and left their phone on when they went to sleep (88%). Many more leave their phone on when sleeping than younger users of 8–11 years (Fowler & Noyes, 2015). 53% of younger users slept with a phone next to their bed and 35% left their phone on when they went to sleep. This means that sleep disturbance from calls or texts during the night are likely.

3.6 Inappropriate use

The interviews showed that teenagers disliked antisocial behaviour connected to mobile phone use, for example, cyber-bullying and unknown callers.

One participant commented on how "People are able to get your number and text you, someone you don't know". In the interviews, eight participants voiced their concern about 'inappropriate' contact and were concerned about their safety.

3.7 Conclusions

This study has demonstrated the high involvement and investment of time in mobiles by this age group. When matched against the literature on health hazards, a number of health concerns associated with teenager's use of mobile phones are surmised.

An awareness of the health consequences of mobile phone use is important to aid our knowledge and understanding. Active promotion and application particularly of the government recommendations for mobile phone use are important to encourage educational awareness and understanding for preventative

harm. However, most teenagers, schools and families are unaware of these guidelines.

REFERENCES

Abramson, M.J., Benke, G.P., Dimitiadis, C., Inyang, I.O., Sim M.R., Wolfe, R.S. & Croft, R.J. 2009. Mobile telephone use is associated with changes in cognitive functions in young adolescents. *Bioelectromagnetics* 30: 678–686.

Bababekova, Y., Rosenfield, M., Hue, J.E. & Huang, R.R. 2011. Font size and viewing distance offhand held smartphones. *Optometry & Vision Science* 88(7): 795–797.

Balik, H.H., Turgut-Balik, D. Balicki, K. & Ozcan, I.C. 2005. Some ocular symptoms and sensations experienced by long term users of mobile phones. *Pathologie Biologie* 53(2): 88–91.

Baron, N.S. 2009. Talk about texting: Attitudes towards mobile phones. In *Proceedings of the London Workshop of Writing, University of London*, pp. 207–225.

Beranuy, M., Oberst U., Carbonell X., & Chamarro A. 2009. Problematic internet use and mobile phone use and clinical symptoms in college students: The role of emotional intelligence. *Computers in Human Behavior*, 25, 1182–1187.

Berolo, S., Wells, R.P., Amick, B.C., 2011. Musculoskeletal symptoms among mobile held device users and their relationship to device use: A preliminary study in a Canadian population. *Applied Ergonomics*. 42(2), 371–8.

Besset, A., Espa, F., Dauvilliers, Y., Billard, M., & de Seze, R. 2005. No effect on cognitive function from daily mobile phone use. *Bioelectomagnetics*, 26, 102–108.

Fowler, J., & Noyes, J. 2015. From dialling to tapping: Health considerations for young users of mobile phones, *Proceedings of the International Symposium of Occupational Safety and Hygiene, Guimaraes: University of Guimaraes* (pp. 115–117).

Ghandi, O.P., Lloyd Morgan L., de Salles, A.A., Han, Y-N., Herberman, R.B., & Davis, D.L. 2012. Exposure limits: The underestimation of absorbed cell phone radiation, especially in children. *Electromagnetic Biology and Medicine*, 31, 1, 34–51.

Government Advice, 2012. Wired Child, Protecting our Children from Wireless Technology. Retrieved September 9, 2012 from wiredchild.org/government-alias.html.

Gustafsson, E., Johnson, P.W. & Hagberg, M. 2010. Thumb postures and physical loads during mobile phone use—a comparison of young adults with and without symptoms, *J Electromy ogr Kinesiology* 20(1), 127–135.

Gustafsson, E. 2012. Ergonomic recommendations when texting on mobile phones, *Work*, 41, 5705–5706.

Hardell, L., Carlberg, M. Soderqvist, F. & Hansson Mild, K. 2008. Meta-analysis of long-term mobile phone use and the association with brain tumours. *International Journal of Oncology*, 32, pp. 1097–1103.

Humphreys, L. 2005. Cell phones in public: Social interactions in a wireless era. *New Media & Society*, 7 (6), 810–833.

Hyman, I.E., Boss, S.M., Wise, K.E., & Caggiano, M. 2009. Did you see the unicycling clown? Inattentional blindness whilst walking and talking on a cell phone. *Applied Cognitive Psychology*, 24, 5, 597–607.

Jonathan, L. 2010. Mobile phones lift poor out of poverty: U.N. study, Reuters. Retrieved September 8, 2015, http://www.reuters.com/article/2010/10/14/us-telecoms-poverty-idUSTRE69D4XA20101014.

Kang, G., & Ghandi, O.P. 2002. SARS for pocket mounted mobile telephones at 835 MHz and 1900 MHz. *Physics in Medicine & Biology,* 47, 4301–4313.

Korensis, P., & Billick, S.B. 2014. Forensic Implications: Adolescent sexting and cyberbullying. *Psychiatry Q,* 85, 97–101.

Kucer, N. 2008. Some ocular symptoms experienced by users of mobile phones. *Electromagnetic Biology Medicine*, 27:2, 205–9.

Lenhart, A. 2012. Teenager, smartphones & texting, Retrieved September 9, 2015 from http://pewinternet.org/Reports/2012/Teens-and-smartphones:aspx.

Ling, R., & Yttri, B. 2002. Hyper-coordination via mobile phones in Norway. In J. E. Katz, & M. Aakhus, (Eds.) Perpetual contact: Mobile Communication, Private Talk, Public Performance. Cambridge: Cambridge University Press.

Madell, D., & Muncer, S.J. 2007. Control over Social Interaction: An important reason for young people's use of the internet and mobile phones for communication. *Cyberpsychology & Psychology*, 10, 1, 137–140.

Mobilewise 2011. Mobile Phone Health Risks: The case for action to protect children, Retrieved March 3, 2016 from www. Mobilewise.org.

Ofcom, (2014). The communications market 2014, Retrieved September 9, 2015 from http://stakeholders.ofcom.org.uk/market-data-research/market-data/communications-market-reports/.

Ofcom, 2015, UK now a smartphone Society, Retrieved November 4, 2016 from http://www.ofcom.org/uk.../uk-now-w-smartphone-society.

Panda, N.K., Jain. R. Bakshi, J. & Munjal, S. 2010. Audiologic disturbances in long term mobile phone users. *J. Otolaryngology Head Neck Surgery*. 39 (1) 5–11.

Redmayne M. 2013. New Zealand Adolescents cell phone and cordless phone user habits: are they at increased risk of brain tumours already? A cross-sectional study. *Environmental Health*, 12 (5).

Retrieved November 3, 2014 from http://www.Biomedcentral.com/1471-2458/11/66.

Ruediger, H.W. 2009. Genotoxic effects of radiofrequency electromagnetic fields. *Pathophysiology*, 16(2–3), 67–69, doi: 10.10.16/j.pathophys.2009.02.002. PMID 19264462.

Sansone, R.A., & Sansone L. 2013. Cell phones: The psychosocial risks. *Innov Clinical Neuroscience,* 10, 1, 33–37.

Schabrun S.M., van den Hoorn, W., Moorcroft A., Greenland C., & Hodges P.W. 2014. Texting and walking: Strategies for postural control and implications for safety. *PLoS One*, 9, 2.

Smartphone multitasking: The Facts and Figures 2016. Retrieved on Jan 4, 2017 from www.techiesense.com

Statistica, 2015. Number of Smartphone users in the (UK) from 2011–2018, Retrieved on Oct 31 from www.statistica.com.

Thomee, S., Harenstam A. & Hagberg M., 2011. Mobile phone use and stress, sleep disturbances and symptoms of depression among young adults—a prospective cohort study. *BMC Public Health*, 11:66 doi.10:1186/1471 2458–11-66.

Velayutham, P., Govindasamy, G.K., Raman R., & Prepageran N. 2014. NgKH, High frequency hearing loss among mobile phone users. *Indian Journal of Otolaryngology and Head & Neck Surgery*, 66, 1, 169–172.

Walsh, S.P., White, K.M., & Young, R. M. 2008. Over connected? A qualitative exploration of the relationship between exploration of the relationship between Australian youth and their mobile phones. *Journal of Adolescence,* 31, 77–92.

Appendix 1

Questions relating to Health Issues from the Mobile Phone Use Questionnaire

What do you like least about mobile phones?

Do you get chance to use a mobile phone?

Do you own a mobile phone?

When on the move where do you keep a mobile phone? – in a bag; in a pocket; slung on a belt; in my hand; other, please say where you keep a mobile phone.

Do use your fingers or thumbs when you use a mobile phone? – fingers; thumbs; fingers and thumbs; other, please say what you use

When you use your mobile phone do you do anything else at the same time?

If yes, please say what other task you carry out?

How many calls do you make on a mobile in a day?

How many texts do you make on a mobile in a day?

Do you like to keep your phone on? All the time, to keep in touch no matter where I am, to sleep with it next to my bed, to sleep with my phone on or to turn it off when I go to bed.

What do you use a mobile phone for? Calling, texting, exchanging photos, recording a video, exchanging a video, using the internet, email, games, downloading ring tones, downloading apps, playing music, setting my alarm, Twitter, Facebook, Blogging, other Social networks.

Please tick what year group you are—Year 10, 11; 12; 13; 8; 9 (Note. Year 10 = 14–15 years; Year 11 = 15–16 years; Year 12 = 16–17 years; Year 13 = 17–18 years.)

Please tick to say if you are a girl or a boy—girl; boy.

Appendix 2

Interview questions about Mobile Phone Usage (which indicated health-related issues).

Please can you tell me what you think about mobile phones?

Why do you think the mobile phone has become such an important part of everyday life?

Do you prefer to text or call?

Do you ever think if there are any possible health risks from using a mobile phone?

Occupational Safety and Hygiene V – Arezes et al. (Eds)
© 2017 Taylor & Francis Group, London, ISBN 978-1-138-05761-6

Wearable technology usefulness for occupational risk prevention: Smartwatches for hand-arm vibration exposure assessment

I. Pavón, L. Sigcha, J.M. López & G. De Arcas
Instrumentation and Applied Acoustics Research Group (I2A2), Universidad Politécnica de Madrid, Campus Sur UPM, Madrid, Spain

ABSTRACT: This paper analyzes the opportunities offered by wearable technology for occupational risk assessment and specifically the use of smartwatches for hand-arm vibration exposure assessment. A wide range of electronic devices know like "wearables" include MEMS sensors (Micro Electro Mechanical System) that provide interesting features to the devices. The potential of a smartwatch for evaluating hand-arm vibration was assessed in laboratory by comparing a smartwatch device with a reference vibrometer. The results suggest that this kind of devices could be able to be used in specific hand-arm vibration assessment tasks. It has been found some technical limitations but also it has been identified opportunities for improvement in the near future.

1 INTRODUCTION

Wearables are devices, gadgets or sensors that are worn on the body or in clothes, often with a biometric and safety functionality. They can measure many variables like steps taken, body temperature, heart rate, sleep patterns, movement, posture and more.

The information collected by wearable sensor technology ensures risk management by enabling health and safety managers to:

- Identify of specific risk;
- Improve measures to reduce injuries;
- Develop a safety culture and education of workers;
- Ensure productivity and cost-benefit to the business. (Ronchi, 2016).

At present the wearable technology is increasingly present in society (Page, 2015). Continuously new applications appear focused on security, fitness, health, habits (Sazonov et al., 2014). The field of occupational risk assessment is no stranger to this technological innovation. There are some initiatives for risk assessment in the workplace using wearable technology (Peppoloni, 2016, Kim, 2013).

In regard to hand-arm vibration exposure, many workers are exposed to vibration during their work, for example: when using vibrating hand tools, mobile and stationary machinery or when driving heavy vehicles. Exposure to hand-arm vibration is a major cause of occupational diseases like musculoskeletal disorders, neurological disorders, vascular diseases and muscle disorders that manifest as pain, numbness, stiffness and decreased muscle strength (South, 2013). There is a specific European (Directive 2002/44/EC), as well as a catalog of technical standards about mechanical vibrations and evaluation methods. The technical procedures for the measurement of the magnitude and exposure to hand-arm vibrations are defined in ISO 5349 Standard parts 1 and 2.

The performance specifications and tolerance limits for instruments designed to measure human vibration exposure values are described in the ISO 8041: 2006 Standard. The vibration range of the signal affecting the hand-arm system is from 5.6 Hz to 1400 Hz, the amplitude range must be at least 60 dB and linearity error amplitude should not exceed 6% of the input value including the transducer and the analyzer.

The vibration exposure risk assessments are usually made from taking discrete time measurements, this evaluation method attempt to be representative of the worker exposure during an operation or a work cycle.

The risk assessments can also be performed using the manufacturer's declared vibration emission values from databases. This may be useful to help make an estimate of daily exposure and an assessment of risk (EU 2006). Although the practical advantages of using these values, it's suspected that the vibration values are underestimated. In different studies it has been detected that the measurements made following the procedures defined in ISO 5349 standards show great variability with the values declared by the manufacturers of machines and tools (Moschioni et al., 2011). The main reason for this variability could be based on the wide range of tools presents at the work place and the difference on machines used in various tasks. This imply inaccurate the risk assessment and difficulties for the comparison between measurements and the limits established by law.

Therefore, the risk assessment will be more accurate when it be performed with real measurements and take into account the factors that are influencing in the uncertainty of measurement such as the accelerometer characteristics, fixing methods and time exposure. (Ainsa et al., 2011).

In order to solve the found problems and improve the vibration risk assessing procedures is possible to use wearable technology (smartwatches) for risk assessment, exploiting the advantages of the smart devices instead the traditional vibration measurement systems.

There are some previous experiences where specific solutions have been developed for the vibration assessment in the workplace using low-cost accelerometers called MEMS that are integrated in smart devices or autonomously in a custom made device.

In (Tarabini et al., 2012) the advantages and disadvantages of the use of MEMS accelerometers for vibration measurement of hand-arm and whole body were evaluated. This approach presents a viable solution to carry out continuous and repeatable measurements for evaluation and occupational risk prevention.

In the case of vibrations transmitted to the hand-arm system, there are some experiences in the development of compact analysis devices with MEMS sensors (Aiello et al., 2014, Morello, 2010). These studies confirmed that the MEMS are a mature technology for the development of inexpensive devices for measuring vibration that can be fastened to clothing or worker equipment, even considering the limitations in the sampling rate imposed by the test devices.

In (Austad et al., 2013) a commercial wearable sensor module (IsenseU) was placed over the hand, showing that this kind of devices present adequate characteristics to perform vibration assessment and also presents useful algorithms to calculate the daily exposure. Similar results have been found in (Bo et al., 2013) where a smartphone was placed over the hand, demonstrating that this technology can be used to perform initial assessments or with educational or training purposes.

Until now, there are no evidences of the use of commercial smartwatches for the evaluation of exposure to vibrations in the workplace. However, this kind of device presents great potential to be used in occupational risk assessments for its affordable price, portability, autonomy and wireless connectivity (Wolfgang et al., 2014a, Rawassizadeh et al., 2015). The aim of this paper is to assess the possibilities offered by wearable technology for occupational risk assessment and specifically the use of smartwatches for hand-arm vibration exposure assessment by mean laboratory tests.

2 MATERIALS AND METHODS

In order to assess the usefulness if this kind of technology for vibration assessments, several tests must be done. These tests are based on frequency response, amplitude, linearity and others like background noise or over-load. Tests must be made to the device analyzers, including the accelerometers. To assess the accuracy of a smartwatch as a measuring device for hand-arm vibrations, a vibration measurement system based on a commercial smartwatch with Android platform was developed.

Although the technical specifications of MEMS accelerometers allow sampling rates up to 1 kHz and dynamic ranges of ± 2, ± 4 and ± 8 g; these are limited by the operating system and the sensor driver to reduce power consumption and increase the battery life. Actually these kinds of devices are usually set to work in a range of ± 2 g and a maximum sample rate of 200 Hz.

Similar limitations or even higher have been found in other mobile operating systems like iOS (Wolfgang et al., 2014b), therefore, Android system may be suitable for programming applications for vibration assessment.

The test device used was a Sony 3 (SWR50) smartwatch with a Bosch sensor (BMX055), with factory default configuration to perform at a maximum sampling rate of 250 Hz and a dynamic range of ± 2 g.

2.1 *Methodology followed to verify the smartwatch characteristics for vibration assessment*

– Development of an smartwatch based system for hand arm exposure assessment:
The scheme of Figure 1 was used to implement a measurement system following the guidelines of the ISO 5349-1:2001 standard. The system calculates the vector sum of the frequency weighted acceleration a_{hv} and the value of exposure to vibration A(8).

An Android application was developed to capture and save the raw accelerometer tri-axial data, and then the acceleration data were processed with a computer to get the measurements values.
– Test bench:
The evaluation of the frequency response and linearity was performed using a vibration analysis system, composed by a PULSE 7537 B & K data acquisition system connected to a LDS PA 100E power amplifier and a LDS V406 CE M4 vibration shaker.

To compare results obtained with the smartwatch a portable vibration analyzer SV106 Svantek

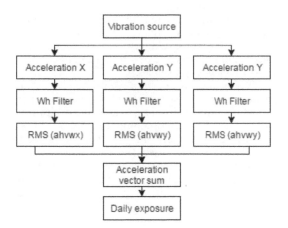

Figure 1. Measuring system according ISO 5349 Standard.

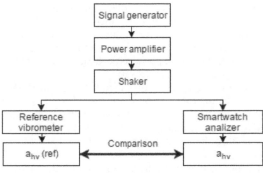

Figure 2. Block diagram of the simultaneous evaluation of vibration system.

with an accelerometer Dytran 3023M2 was used as reference.

– Adjustment and Calibration:

The smartwatch has a factory calibrated tri-axial accelerometer, however, several tests were made to verify its accuracy and variability in capturing vibrations. To perform the tests a calibration signal of 12.5 Hz with an RMS amplitude of 10 m/s² was used as reference. Then the vibration was applied to the smartwatch in order to find a sensitivity factor that was used in the developed application.

Tests performed:

– Evaluation of frequency and amplitude (Laboratory test).

The test made for evaluating the response of the system developed was performed using the shaker vibrations fed by sinusoidal signals at octave intervals. The reference accelerometer and the smartwatch were coupled simultaneously on the shaker vibrating surface (Figures 2 and 3), and then 10 a_{hv} measurements were made at each analysis point (level and frequency), the used frequencies were 6.3, 12.5, 25, 50 y 100 Hz considering the maximum sample rate of the smartwatch, the signal level was gradually increased by 200 mV$_{RMS}$ from the signal generator until get the saturation values in the smartwatch accelerometer.

– Evaluation of the analysis system using a common use machine (Outdoor test).

To verify the functionality of the device a simulated test was made using a hand drill, measuring simultaneously the weighted acceleration a_{hv} with the reference vibrometer and the smartwatch. To perform this experiment, an electric drill (Inhell Mark, model BSM 550) was used to make several drills in different materials using different drill sizes for a time at least 1 minute (Figure 4).

Figure 3. Simultaneous evaluation of system using a vibration Shaker.

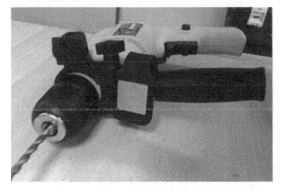

Figure 4. Simultaneous test using a common electric drill machine.

3 RESULTS

– Frequency and amplitude results (Lab. test).

The results from laboratory test (Table 1) show that the differences between a_{hv} measurements made with the two devices are small when the acceleration amplitudes are low, but they increase at high amplitude vibration levels especially in the frequencies of 50 and 100 Hz. The amplitude and frequency analysis is performed only for the first octave bands due to the limitations imposed by the operating system with respect to the sensor's sample rate, that don't allow to perform measurements at frequencies above 125 Hz.

– Common drill machine results (Outdoor test).

The results of the outdoor test show a similar behavior to the laboratory test, showing small differences for low amplitude levels that increase when reaching the signal saturation (Figure 5). It can be observed that in some points there is a higher difference between the measurements of the devices, these errors are produced mainly by the variability of the sample rate and the fitting to the hand drill.

With the current configuration the device has limitations to be used as a precision instrument, mainly due to the limitation of the sampling rate (Fs = 250 Hz) and the dynamic range (±2 g) in which the sensors of smart devices are set during manufacturing.

Table 1. Differences between the a_{hv} (m/s^2) means from the smartwatch and the reference vibrometer.

| Amplitude | Frequency | | | | |
	6.3	12.5	25	50	100
200	0.005	0.006	0.029	−0.067	0.291
400	−0.001	0.004	0.034	−0.153	2.405
600	−0.003	0.003	0.058	0.992	4.431
800	−0.007	0.000	0.066	−5.844	3.647
1000	−0.005	0.050	0.205	5.387	3.122

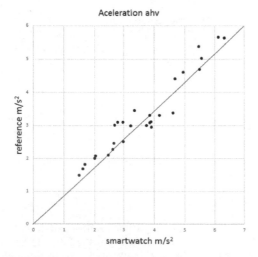

Figure 5. Comparison of the measurements made with the smartwatch and the reference vibrometer placed on the hand drill.

The results show that the developed vibration analysis system is suitable for evaluation of low-power vibrating machinery with low a_{hv}-weighted acceleration levels, the best results are obtained in low-frequency test signals.

This situation can improve in the near future when smart consumer devices get higher performance in their sensors, associated electronics or in the Programming Interfaces (APIs). Allowing developers modify these parameters to achieve the required levels of accuracy. For example, in the current market, there are high-end smartphones with sensors configured to operate in ranges of ±8 g and sampling rates from 200 to 250 Hz.

4 CONCLUSIONS

The use of the system developed in this study or similar devices even considering their technical limitations, presents considerable advantages as:

– The possibility of performing long temporal assessment, with the possibility of monitoring worker exposure on a continuous throughout the entire working day in order to reduce the uncertainties associated with the measurement (magnitude and time).

– The cost reduction by using devices and sensors with relatively low prices compared to the specific vibration measurement equipment. The costs have been characterized as an entry barrier for the evaluation of workplace vibrations.

– Identification of specific risks related with machinery, tools and vehicles with high vibration magnitude.

– Develop a safety culture and education of workers, company, managers, etc.

– And allows to comply with the principle of not increasing or causing new risks because their ease of use, portability, wireless connectivity and autonomy.

At present, this kind of devices are less precise than professional equipment, however these technologies will be more accurate in the near future. Nowadays these devices could be used to obtain better representativeness of the exposure levels be-cause they can be worn during the entire working day.

ACKNOWLEDGEMENTS

This work has been carried out within "*Beca de I+D 2016 de la Fundación Prevent*" (Prevent Foundation R & D Scholarship 2016). The authors gratefully acknowledge the Fundación Prevent their financial support.

REFERENCES

Aiello, G., La Scalia, G., Vallone, M., Catania, P., & Venticinque, M. 2012. Real time assessment of hand-arm vibration system based on capacitive MEMS accelerometers. *Computers and Electronics in Agriculture, 85*, 45–52.

Ainsa, I., Gonzalez, D., Lizaranzu, M., & Bernad, C. 2011. Experimental evaluation of uncertainty in hand–arm vibration measurements. *International Journal of Industrial Ergonomics, 41*(2), 167–179.

Austad, H. O., Røed, M. H., Liverud, A. E., Dalgard, S., & Seeberg, T. M. 2013. Hand-arm vibration exposure monitoring with wearable sensor module. *Studies in Health Technology and Informatics, 189*, 113.

De Meester, M., De Muynck, W., De Bacquer, D., De Loof, P., & Vanhoorne, M. 1998. Reproducibility and value of hand-arm vibration measurement using the ISO 5349, method and compared to a recently developed method. *Proceedings of the Eight International Conference on Hand-Arm Vibration*, 11–13.

EU Good Practice Guide HAV. Guide to good practice on Hand-Arm Vibration Non-binding guide to good practice with a view to implementation of Directive 2002/44/EC on the minimum health and safety requirements regarding the exposure of workers to the risks arising from physical agents (vibrations). 2006.

ISO 5349-1:2001 Mechanical Vibration-Measurement and evaluation of human exposure to hand-transmitted vibration-Part 1: General requirements.

ISO 5349-2:2002 Mechanical Vibration-Measurement and evaluation of human exposure to hand-transmitted vibration-Part 2: Practical guidance for measurement at the work place.

ISO 8041:2005 Human response to vibration-measuring instrumentation.

Kim, S., & Nussbaum, M. 2013. Performance evaluation of a wearable inertial motion capture system for capturing physical exposures during manual material handling tasks. *Ergonomics, 56*(2), 314–326.

Liu, B., & Koc, A. B. 2013. Hand-arm vibration measurements and analysis using smartphones. *2013 Kansas City, Missouri, July 21-July 24, 2013*, 1.

Morello, R., De Capua, C., & Meduri, A. 2010. A wireless measurement system for estimation of human exposure to vibration during the use of handheld percussion machines. *IEEE Transactions on Instrumentation and Measurement, 59*(10), 2513–2521.

Moschioni, G., Saggin, B., & Tarabini, M. 2011. Prediction of data variability in hand-arm vibration measurements. *Measurement, 44*(9), 1679–1690.

Page, T. 2015. A forecast of the adoption of wearable technology. *International Journal of Technology Diffusion, 6*(2), 12–29.

Peppoloni, L., Filippeschi, A., Ruffaldi, E., & Avizzano, C. 2016. (WMSDs issue) A novel wearable system for the online assessment of risk for biomechanical load in repetitive efforts. *International Journal of Industrial Ergonomics, 52*, 1–11.

Rawassizadeh, R., Price, B., & Petre, M. 2015. Wearables: Has the age of smartwatches finally arrived? NEW YORK: ACM.

Rimell, A. N., Notini, L., Mansfield, N. J., & Edwards, D. J. (2008). Variation between manufacturers' declared vibration emission values and those measured under simulated workplace conditions for a range of hand-held power tools typically found in the construction industry. *International Journal of Industrial Ergonomics, 38*(9), 661–675.

Ronchi, A. The Role of Wearable Technology in Risk Reduction. *Construction Executive Risk Management*. 12/7/2016. On line: https://enewsletters.constructionexec.com/riskmanagement/2016/07/the-role-of-wearable-technology-in-risk-reduction/.

Sazonov, E., Neuman, M.R., & Books 24 × 7, I. 2014. *Wearable sensors: Fundamentals, implementation and applications*. Amsterdam: AP, Academic Press is an imprint of Elsevier.

South, T. 2013. Managing noise and vibration at work Routledge.

Tarabini, M., Saggin, B., Scaccabarozzi, D., & Moschioni, G. 2012. The potential of micro-electro-mechanical accelerometers in human vibration measurements. *Journal of Sound and Vibration, 331*(2), 487–499.

Wolfgang, R., & Burgess-Limerick, R. 2014a. Using consumer electronic devices to estimate whole-body vibration exposure. *Journal of Occupational and Environmental Hygiene, 11*(6), D77-D81.

Wolfgang, R., Di Corleto, L., & Burgess-Limerick, R. 2014b. Can an iPod touch be used to assess whole-body vibration associated with mining equipment? *Annals of Occupational Hygiene, 58*(9), 1200–1204.

Occupational Safety and Hygiene V – Arezes et al. (Eds)
© 2017 Taylor & Francis Group, London, ISBN 978-1-138-05761-6

Accompanied method risk management and evaluation method of exposure risks to biological agents

R. Veiga, I. Miguel & C. Pires
ISLA—Institute of Management and Administration of Santarém, Santarém, Portugal

ABSTRACT: The presence of biological agents in the workplace may result in hazardous situations for workers. The present study was carried out in the microbiology laboratory of a Medical School and the methodology was based on the qualitative method presented in the Technical Note of Spanish Prevention (NTP 833) and the Accompanied Method of Biological Risk Management (MAGRB).

It was concluded that the MAGRB allows to distinguish and prioritize the risk associated to the different operations as a whole, taking into account safety conditions and practices, if evaluation parameters defined in the legislation of biological agents are met, something that the International method (NTP 833) does not allow doing so effectively. As the focus of worker protection, the MAGRB provides the necessary guidelines to trigger a prevention structure in all activities likely to be exposed to biological agents.

1 INTRODUCTION

Biological agents are living beings of microscopic dimensions, and all substances derived from them can have negative effects on the worker's health. The major difference between biological agents and other dangerous substances is their reproduction capacity, since under favorable conditions they can develop in a short Amount of Time (ACT, 2008). These micro-organisms can originate any type of infection, allergy or toxicity in the human body. Their presence in workplaces may result in risk situations for workers (Pinto, 2016).

The work involving exposure to biological agents can occur in a variety of situations and activities (Novás, 2008), (Perez, Mena, Watson, Prater, & McIntyre, 2015). Since microorganisms are ubiquitous in the environment, exposure to biological agents in various contexts is inevitable, implying very different risk situations and characterized in some cases by great specificity (Health, 2004) (Gershon, Pogorzelska, Qureshi, & Sherman), (Singh, 2016), (Ulutasdemir, Cirpan, Copur, & Tanir, 2015).

In Portugal the Law-Decree N°. 84/97, 16th April, regulates the protection of workers against the risks of exposure to biological agents at work by classifying the biological agents into four groups according to the level of infectious risk: Group 1—Low probability of causing illness, Group 2—Can cause disease and constitute a danger, Group 3—Can cause serious illness and constitute a serious risk, Group 4—Cause serious illness and constitute a serious risk (Law-Decree N°. 84/97). In the course of their duties, some workers are exposed to a number of living micro-organisms (viruses, bacteria, etc.) and to substances or structures that are originated from them. The activities with higher risk of exposure to biological agents are: work in food and agricultural production units; Activities in which there is contact with animals and products of animal origin; Work in health facilities, including isolation and autopsy units; Clinical and veterinary laboratories, collection, transport and waste disposal units, water and wastewater treatment facilities (Freitas, 2011).

The protection of workers is basically centered on the assessment of exposure risks, where the features of the involved agents in the activity are considered, the suitability of facilities, equipment and work practices (Nunes, 2010), (Romero, 2006), (Pinto, 2016).

Biological risk assessment is a challenge (Nácher, Alapont, Sales, & Ferrando, 2006), (Moore, et al., 2010) in first place regarding the diversity of agents and secondly because the limits of occupational exposure (VLE) for the great majority of these agents have not been defined. Pathogenic micro-organisms may be hazardous in extremely low concentrations (Larson & Aiello, 2006), and are invisible to the naked eye. "*It is imperative to be aware that a risk assessment of biological agents should be carried out on the basis of the uniqueness concerning each case*" (Teixeira, 2015). According to the World Health Organization (WHO) and the National Institute of Occupational Safety and Health—Spain (INSHT) the measurement of biological agents is not an essential element but rather its identification and assessment. The main reasons for non-measurement are: lack of confidence in the results, due to the great variability of professional activities; the high cost, time and money involved in the analysis, in particular the complete and accurate identification of biological agents in the work environment; the lack, to date, of a standardization regarding the exposure limit values for biological agents (OMS, 2004), (INSHT, 2014).

In Spain the Royal Decree N°. 664/1997 regulates the protection of workers who are exposed to biological hazards in the workplace and the NTP 833 outlines the procedures for risk assessment, by defining exposure levels and preventive measures associated with potential risk levels (Trabajo, 2009). The Accompanied Method of Biological Risk Management (MAGRB) intends to assess, in accordance with Portuguese legal requirements, the risks of exposure to biological agents. Biological agents are identified and classified according to Decree-Law 84/97 and the method allows determining the level of risk to which workers are exposed after defining the potential intrinsic and residual risk. The calculation of the intrinsic potential risk is based on two essential variables, the level of exposure (contact frequency, amount handled and production of bio aerosols) and the damage or effect (risk to workers, propagation in the community and existence or not of prophylaxis means) for a worker exposed to the biological agent. The residual risk that we can find in a facility is calculated from intrinsic potential risk, but it takes into account the already controlled risk through the different prevention and protection measures that exist. The value attributed to prevention and protection measures is obtained after the Organization has been audited, applying a checklist drawn up after the basis of Decree-Law 84/97, which quantifies percentile the level of compliance. In the end, the risk level allows to launch intervention priorities by establishing four levels of significant risks and one non-significant risk. These levels have been established as a progressive requirement and do not tolerate any deviation in the conformity procedures, facilities and containment from exposure to biological agents of groups 2, 3 and 4.

2 OBJECTIVES

Facing the difficulty of finding in Portugal, an evaluation method, which compared with the NTP 833 method complies legal requirements, the main goal in conducting this study on the management and evaluation of occupational risks in exposure to biological agents in the laboratory of microbiology was defined.

The following specific objectives have been defined:

1st Assess the risks with the simplified method (NTP 833);

2nd Assess risks using MAGRB;

3rd Verify that the methods used are in accordance with the legislation for biological agents;

4th Compare results of the evaluations obtained with both methods when evaluating safe and unsafe work practices that distinguishable according to the risks of exposure to biological agents.

3 METHODOLOGY

The study was carried out in a Medical School's microbiology laboratory, where the accomplishment of three tasks that involved the manipulation of biological agents and was developed were observed according to the in the following stages:

1st Stage—Laboratory and laboratory practices audit.

In order to carry out the laboratory and laboratory practices audit, a checklist was used on an excel spreadsheet, which allowed us to quantify the degree of conformity obtained.

This file makes it possible to analyze, for each of the tasks, the degree of conformity of the Group II, III and IV confinement measures, the individual protection measures used, the organization of SHW services, Safety at Work Handling, handling and disposal of hazardous wastes and fire safety in accordance with current legislation.

2nd Stage—Risk assessment according to the simplified method presented in NTP 833, which was done in a document prepared on an excel spreadsheet.

3rd Stage—Since the obtained result in the first two stages and considering the simplified method as insufficient, by only providing us with a qualitative evaluation, the MAGRB was developed. The method was based on NTP 833 and was developed through an excel document in order to identify and evaluate biological agents, addressing the case study to activities in which there is a deliberate intention to work with biological agents. Through the identification of hazards that may represent these agents and the possibility of exposure to them, it is intended to establish potential risk levels action priorities, magnitude and degree of requirement in the fulfillment of associated preventive actions.

The MAGRB presents as a differentiating parameter the calculation of the residual risk level. The residual risk is the risk that remains after the introduced mitigation by the control measures (prevention and protection) and is based on the value resulting the facility audit.

In the case study, the manipulation process started with receipt of a master sample confined to a maximum of four agents. In this sample there was a randomness of agents, i.e., which agents were present was unknown, only that these agents belonged to Group II was informed, and since they are the type of agents with which the laboratory in question works. For the study the data presented in Table 1 was collected.

Table 1. Data collection of the three tasks performed in the microbiology laboratory.

	Tasks		
Analyzed parameter	Reception of the sample	Work in safety chamber	Sample freezing
Group Agent	II	II	II
Production of bio aerosols	High but sporadic	Scarce	Scarce
Contact frequency	15% Working time	70% Working time	15% Working time
Handled Number	Average	Average	Average
Operators	1 Man 2 Women	1 Man 2 Women	1 Man 2 Women

Table 2. Data collection in the three tasks performed in the microbiology laboratory.

Task	Level of compliance achieved (%)
Reception of the sample	5
Work in biological safety chamber	96
Sample freezing	82

4 RESULTS

In the first phase, the level of compliance was calculated, according to the existing conditions, value of the controlled risk (risk eliminated by the preventive measures and confinement), Table 2.

In the sample receiving task, the audit determines the lowest compliance level (5%), in biological safety chamber was obtained the highest level (96%) and sample freezing (82%).

In the second phase, we performed the biological risk assessment using the simplified method presented by the Spanish Technical Note NTP 833.

The identified biological agents (bacteria) are all classified in risk group II. The amount handled is normal in all cases, while there are differences in contact frequency and bio aerosol production. The evaluation allowed to classify in level III of potential risk—high risk, all the agents in the three tasks, independently of the verified level of conformity, through the application of a Checklist, in the laboratory and laboratory practices.

In the third phase, applying the same data used in the simplified method, but taking into account the conformity value obtained in the audit, and with the method accompanied by biological risk management different results were obtained. The classification of the level of potential residual risk is different and presents in two tasks level IV - severe and imminent.

It was found that the simplified method NTP 833 did not identify the evaluation parameters provided for in Articles 6 and 7 of Decree-Law N° 84/97 of April 16th presented in the Portuguese Law, such as supplementary risk for previous illness, the recommendations of the Directorate-General for Health, technical information on related diseases and awareness of the disease in a worker. On the other hand, MIAGRB has identified all the evaluation parameters provided for in that legislation.

When comparing the results of the evaluations obtained with the two methods, in addition to the presented results being more accurate with the accompanied method, this allowed us to identify some more elements, namely:

- Description of the activity;
- Identification of the most vulnerable workers and risky or forbidden activities;
- Identification and characterization of the biological agent;
- Means and ways of contamination or transmission;
- Symptoms according to the identified biological agent;
- Specific prevention measures by contaminant agent;
- General prevention measures of the biological risk factor;
- Confinement measures according to the identified risk group;
- Hygiene measures and individual protection;
- Training and promotion of workers' health;
- Applicable legislation.

5 CONCLUSIONS

It is concluded, therefore, that the process of risk assessment is a real challenge, according to Nácher, Alapont, Sales, and Ferrando (2006) and Moore, et al. (2010). The present methodology for assessing the risk of exposure to biological agents (MAGRB) allows a distinction and hierarchize the risk associated with the different operations under study, taking into account safety conditions and practices, complying Evaluation parameters defined in the legislation of biological agents, something that the international method (NTP 833) of INSHT does not allow to do effectively.

This evaluation method, the focus is on the protection of workers, the MAGRB after identifying and characterizing the biological agent also identifies the means and forms of contamination and transmission, symptomatology, prevention, confinement and protection measures, providing the necessary guidelines to trigger a prevention structure in all biologically risky activities.

It complies with all the evaluation parameters required by Portuguese legislation for biological agents.

It allows to differentiate the level of risk between the different groups of biological agents, and to distinguish between the same four levels' group taking into account greater or less mitigation of risk by the control measures.

It is very useful to use quantitative verification lists, which allow the residual level of risk to be calculated after the degree of conformity of the procedures, installations and level of obtained containment, as a way of proving the greater or lesser degree level of mitigation of the prevention measures, in the value of the intrinsic risk.

One of the steps to be taken into account during risk assessment is to evaluate the obtained results, for which we need to sustain it with reference limits or valuation criteria. It turns out that the assessment criteria (VLE—Limit Values of Exposure) for biological agents are not yet established by standard or legislation, in part because of the huge difficulty to obtain them against the characteristics of the biological agents. Indeed:

- They are capable of reproducing in a certain environment and under suitable conditions;
- They can acquire forms of resistance (spores) that allow them to survive in adverse environments over long periods of time;
- They show differences in the degree of virulence;
- They exhibit differences in the immune system response of affected organisms.

The MAGREB does not allow, nor is its objective to be able to answer all the questions that are raised to the technicians, namely in the scope considered in the previous point.

It does not answer yet another very important question, namely how to evaluate the cumulative effects of the presence of numerous agents in the same working environment and the reaction of the organism affected to these agents, facing the multiple attack to which the subject's immune system is subjected.

REFERENCES

ACT. 2008. Agentes biológicos no trabalho. Obtido 10-4-2016, de ACT—Autoridade para as condições de trabalho: *https://www.act.gov.pt/AgentesBiologicos-Trabalho.pdf*.

Decreto de Lei nº 84/97. 1997. Prescrições mínimas de proteção da segurança e da saúde dos trabalhadores contra os riscos da exposição a agentes biológicos . Lisboa: *Diário da República—I Série.*

EU-OSHA, A. E. (s.d.). Instrumentos da avaliação dos riscos. Obtido em 9-8-2014, de h*ttps://osha.europa.eu/pt/topics/riskassessment/index_*.

Freitas, L. C. 2011. Manual de Segurança e Saúde do Trabalho (2ª Edição ed.). Lisboa. *Edições Sílabo, Lda.*

Gershon, R. R., Pogorzelska, M., Qureshi, K. A., & Sherman, M. 2008. Home health care registered nurses and the risk of percutaneous injuries: A pilot study. *AJIC-American Journal of Infection Control,* Vol.36 nº 3, 165–172.

INSHT. 2014. Guía tecnica para la evoluacion y prevencion de los riesgos relacionados con la exposicion a agentes biológicos. Madrid: Instituto Nacional de Seguridad e Higiene en el Trabajo.

Larson, E., & Aiello, A. E. 2006. Systematic risk assessment methods for the infection control professional. *AJIC practice forum,* 323–326.

Moore, D. A., Leach, D. A., Bickett-Weddle, D., Andersen, K., Castillo, A. R., Collar, C. A., et al. 2010. Evaluation of a biological risk management tool on large western United States dairies. *Journal of Dairy Science* Vol. 93 No. 9, 4096–4104.

Novás, C. M. 2008. Evaluación higiénica de riesgos biológicos del trabajo en estabulario de un centro de investigación sanitaria. *Med Segur Trab 2008; Vol. 54; Núm. 213,* 97–103.

Nunes, F. M. 2010. Segurança e Higiene do trabalho—Manual Técnico (3ª Edição ed.). Amadora. *E. G. Eiffel, Ed.*

Nácher, S. B., Alapont, M. M., Sales, I. M., & Ferrando, P. S. 2006. Evaluación de riesgo biológico en el hospital Rey Don Jaime. In M. S. 2007, & V. L. 206 (Ed.*),* Madrid. *IV jornadas nacionales de los servicios de prevención de riesgos laborales.* 9–14.

OMS. 2004. Manual de segurança biológica em laboratório (3ª Edição ed.). Genebra: *Organização Mundial de Saúde.*

Perez, V., Mena, K. D., Watson, H. N., Prater, R. B., & McIntyre, J. L. 2015. Evaluation and quantitative microbial risk assessment of a unique antimicrobial agent for hospital surface treatment. *American Journal of Infection Control,* 1201–1207.

Pinto, M. V. 2016. Perceção e Risco de Exposição Ocupacional a Agentes Biológicos em Centros de Triagem de resíduos e Aterro Sanitário. Coimbra. *Escola Superior de tecnologia e Saúde de Coimbra.*

Romero, J. C. 2006. Métodos de Evaluación de Riesgos Laborales. Madrid: *Ediciones Diaz de Santos.*

Saúde, D. G. 2004. Medidas de controlo de agentes biológicos nocivos à saúde dos trabalhadores. Obtido em 10-04-2016, de http://www.dgs.pt/saude-ocupacional/documentos-proteccao-dos-trabalhadores-contra-os-riscos-de-exposicao-aos-agentes-biologicos

Singh, K. 2016. Laboratory-Acquired Infections. Obtido em 20-4-2016. LISBOA Faculdade Motricidade: *http://cid.oxfordjournals.org*

Teixeira, M. 2015. Métodos de avaliação de riscos biológicos. Leiria. *Vertentes e Desafios da Segurança* (p. 256).

Trabajo, I. N. 2009. NTP 833. Agentes Biológicos. *Evaluación simplificada* . Espanha.

Ulutasdemir, N., Cirpan, M., Copur, E. O., & Tanir, F. 2015. Occupational Risks of Health Professionals in Turkey. Obtido em 22-10-2016, de *http://dx.doi.org/10.1016/j.aogh.*

Occupational Safety and Hygiene V – Arezes et al. (Eds)
© *2017 Taylor & Francis Group, London, ISBN 978-1-138-05761-6*

The importance of Workplace Gymnastics and of the Rehydrating Serum to the health of the rural sugarcane worker

A.R.B. Martins, B. Barkokébas Junior, B.M. Vasconcelos & A.R.M. de Moraes
University of Pernambuco, Recife, Pernambuco, Brazil

ABSTRACT: Acknowledging in the manual cut of sugarcane an exhaustive undertaking, daily demanding energy comparable to that of high-performance athletes and considering complaints and absenteeism to have a causal link with illnesses reported by workers, this paper seeks to evaluate two measures to improve health and safety conditions of rural sugarcane workers, at a sugar-alcohol unit in the countryside of the state of Pernambuco: the practice of Workplace Gymnastics and the use of Rehydrating Serum. Using as methodological bases the theoretical and scientific surveying on the subject, implementation of these measures at the sampled station, together with results reached in three crops, through monitoring the number of ambulatory occurrences and a feedback poll from workers. By checking the results, it is possible to verify the effectiveness of Workplace Gymnastics and of the Rehydrating Serum on improving the health and safety of sugarcane workers, mainly for two of the five illnesses analysed.

1 INTRODUCTION

To the Ministry of Agriculture, the branch of agribusiness is deemed as one of the activities which generate a bigger impact in the country's economic development. However, on those terms, working conditions demand a greater physical effort and, consequently, generate a greater propensity to risks of accidents (Monteiro e Gonçalves, 2010). The job undertaken by the sugarcane workers is considered to be very hard, for besides the exhausting working hours, it also exposes the worker to all sorts of ailments due to it taking place outdoors under the sun, with suits and protective equipment which result in the worker sweating much, therefore losing water and minerals essential to maintaining a good health, also leading the individual to dehydration (Alves, 2008).

Maintaining that routine for years and at intense rates, the body reaches a physical straining point with the emergence of pains and back problems (Menezes e Silva, 2010). Losses of water to the figure of 5% compared to body weight are associated with the reduction of physical capacity in 30%, and, if it goes above this mark, the risk of circulatory collapse becomes imminent, which can lead to hypothermia and death (Tirapegui, 2005). Thus, the athlete or manual labourer must ingest liquids before, during and after physical activity, in order to balance out the loss of water arising from excessive sweating.

1.1 *Justification*

Studies focus on comparing manual workers' productivity with elevated spending of energy and their nutritional condition. Researches dating from the 70's indicated that sugarcane workers' productivity was directly associated to height, quantity of lean body mass and body fat and, still, to monotonous nutritional habits. The latter has been singled out as the limiting factor in the ingestion of calories and adequate nutrients to supply the minimum needs capable of favoring its recovery and maintaining the organism's nutritional condition when working in the sugarcane cutting labour (Silva Neta, 2009.

In sugarcane workers of the sampled plant, it was possible to verify the frequent occurrence of five illnesses generated from the undertaking of the work which could be avoided or attenuated through the practice of Workplace Gymnastics and by the use of the Rehydrating Serum, as suggested by authors Polito and Bergamashi (2010) and Tirapegui (2005) about good practices while working on activities demanding intense physical exertion. Researching on CID 10 (Código Internacional de Doença\ICD—International Classification of Diseases) about the complaints highlighted by workers, it was possible to find the following descriptions: CID M54.5 – Lombalgy or Backache, which is a group of painful manifestations taking place around the lumbar region stemming from some abnormality in that region; CID J11.1 – Flu, which is an acute infectious disease which affects birds and mammals; CID R51 – Cephalea or Migraine, which are medical terms for headache; CID M79.1 – Myalgia, which is a term used to characterize muscle pains in any part of the body; CID R25.2 – Cramps, which are involuntary and painful contractions of a skeletal muscle.

By the finding of these complaints over the years and acknowledging the hardships in implanting improvements in working conditions in the sugarcane cutting industry, this paper aims to evaluate the implementation of good labour practices.

1.2 *Objectives*

The adoption of healthier habits brings consequences to health in general, not just from a physical standpoint. Investments in prevention and promotion of health in the workplace have direct results in the workers' satisfaction and wellbeing, besides bringing benefits concerning the attenuation of pains and preventing occupational illnesses. Thus, this paper seeks to evaluate the effects of Workplace Gymnastics and of the Rehydrating Serum in the sugarcane worker's health at an agricultural-industrial complex in the Mata Sul region of the state of Pernambuco.

For such, it is proposed the undertaking of a bibliographical research on the subject relating it to the legislation concerning safety and health in the workplace; implementing the practices of Workplace Gymnastics and the Rehydrating Serum at a sugarcane processing plant in the state of Pernambuco\ Brazil; to survey for complaints and occurrences of illnesses related to the activities being studied; to survey in order to find out the satisfaction rates of workers concerning Workplace Gymnastics and the Rehydrating Serum, as well as drawing conclusions on the importance of these two practices..

2 BIBLIOGRAPHICAL REVIEW

2.1 *Workplace Gymnastics*

Workplace Gymnastics highlight the importance of postural re-education and relieves the stress related to the method of work, as an instrument to promote health and prevention of diseases of the "LER" (Repetitive Effort Injuries) and "DORT" (Work-Related Musculoskeletal Disorders) types.

It is a group of practices based around the particular professional activity. The technique seeks to compensate the structures of the body which are utilized the most and activate the ones less needed, relaxing and toning them.

Said practices are comprised of sessions of educative physical exercises like stretching, breathing, postural re-education, body control, body perception, strengthening of unused structures and compensation of the muscular groups involved in the operational tasks, respecting each collaborator's physiological limits and clothing.

The ideal duration of the sessions varies between 10 to 15 minutes daily in order to reach significant results in the musculature (flexibility and strength).

2.2 *Rehydrating Serum*

Hard, energy-draining work under elevated temperatures and bad nutrition are factors which affect the health of sugarcane workers, leading them to physical exhaustion after a day's work, contributing to episodes of cramps and muscle pains which undermine their performance at work and, consequently, their monthly wage, which is directly proportional to their productivity.

Fatigue caused by intense labour is accompanied by excessive sweat and the loss of electrolytes like sodium, potassium and magnesium. Under such conditions there is still a drop in blood pressure, an increase in the heart rate and progressive hypovolemia followed by dehydration (Nadel, 1998).

The Rehydrating Serum, as it is called in "Compromisso Nacional", a program by the Brazilian government, is denominated by ANVISA, decree number 222 from March 24th of 1998, as Hydroelectrolythic Replacer and defined as being made of "products formulated through a varied concentration of electrolytes, associated to varied concentration of carbohydrates, seeking to replace the loss of water and electrolytes occurring from physical activity".

Hydroelectrolythic replacers are products formulated with the goal of replacing the loss of water and electrolytes from physical activity, being widely used both by athletes and practitioners of intense physical activities; it's a product that must contain in its formulation varied concentrations of sodium, chlorides and carbohydrates, with an optional addition of potassium, vitamins or minerals.

Beyond the nutrients allowed by ANVISA in its formula, it's common that some manufacturers, based on undertaking that demand a higher need for some micronutrients by the athletes, under allegations that such nutrients would help both the performance and the recovery of their users.

Considering that sugarcane workers are submitted to a physical activity compared to that of high-performance athletes, it is essential to ingest water and substances that guarantee the balanced replacement of liquids, salts and vitamins consumed by the intense physical activity. The increased ingestion of products aimed at practitioners of physical activity, in substitution of liquid foods such as milk, water, juice, etc., and the need or precise orientation on the nutritional supplementation of people who practice physical activities made the Ministry of Health publish specific regulation which define the criteria for "Food of Practitioners of Physical Activity", among them hydroelectrolythic replacers. And the composition is also regulated based on Resolution RDC n°269, dated

September 22nd of 2005 from ANVISA, which establishes the levels of protein, vitamins, salts and others for daily ingestion, for adults.

3 METHODOLOGY

To attest the importance of the effects of Workplace Gymnastics and the use of the Rehydrating Serum in sugarcane workers, this paper monitored the implementation of these practices (as per items 3.1 and 3.2) in a plant located at the Mata Sul region of the state of Pernambuco. For that, it was helped by professionals of the sector, and instructions acquired at other unit, situated in the countryside of the state of Alagoas, which was already using these practices. A bibliographic research was made to substantiate the subject and a survey took place to analyse the results from the two plants in the following fashion: at the unit from Alagoas, the survey of collected data, during four crops, was based on complaints of lumbar pains reported by sugarcane workers; to evaluate the benefits of Workplace Gymnastics and of the Rehydrating Serum in the activity of sugarcane cutting at the plant in Pernambuco, it was analyzed the quantity of sick notes from the five illnesses that can be avoided or attenuated by these practices during the crops of 2011\2012, 2012\2013 and 2013\2014, precisely on the months of September to March, extracted from the registries of ambulatory occurrences. It was also surveyed the opinions of 165 workers benefited with the adoption of these programs, in direct interviews, aiming to objectively ask about satisfaction, indifference and dissatisfaction concerning the practices of Workplace Gymnastics and the use of the Rehydrating Serum.

3.1 The implementation of the practice of Workplace Gymnastics in the company under study

In the crop 2011\2012, the plant adopted this practice in conformity with the following steps:

1. Hiring of a Physical Education Professional (teacher).
2. Definition of the movements to be practices by the workers based on the activities to be conducted by labour during working hours, as per chart 1.
3. Elaboration of leaflets focused on guidance, showcasing pictures of the movements to be practiced during Workplace Gymnastics.
4. Raising awareness and training rural team leaders so they can become daily monitors of Workplace Gymnastics.
5. Lectures with the rural teams, one by one, in classrooms, to introduce Workplace Gymnastics; what is it, what improvements does it

generate, operating plan, and the need of collaboration on the worker's part.
6. Establishing the schedule to begin practicing Workplace Gymnastics.
7. Inaugural class for each team, with the obligatory presence of the teacher.
8. Execution of the daily exercises under the orientation of the team leader.
9. Setting the timeline in which the teacher will visit each team weekly.
10. Elaboration of a SOP (Standard Operating Procedure) for Workplace Gymnastics.

	1-Place hands over the nape and press the head down.		2- Pull the head with one of the hands until you feel a slight pressure on the side of the neck.
	3- Make a spinning movement with the head, starting clockwise, then anti-clockwise.		4- Pull the head with one of the hands until you feel a slight pressure on the side of the neck.
	5-Take the flexed arm to the back of the head, and with the other hand, pull slightly to the other side.		6-With the knees semi-flexed and with one hand placed on the hip, raise the other hand up and bend sideways.
	7- Keep the legs well spread-out and toe tips pointing outwards, and lower the torso towards a chosen side until you feel a slight tension on the backside of the thigh.		8- Keep the hands placed on the ground and the knee muscles semi-flexed, taking the abdomen to the thighs.
	9- Stretch the arms, aligning them with the torso. Keep the abdomen slightly contracted, and the knees unlocked.		10-Keep the torso erect and the abdomen slightly contracted. Take a foot backwards until it reaches the gluteus. Lightly flex the supporting leg.

Chart 1.   Implemented Wokplace Gymnastics movements.

3.2 Implementing the use of the Rehydrating Serum in the company under study

Parallel to the adoption of workplace gymnastics, the unit also implemented the rehydrating serum, easily accepted by all, according to the following steps:

1. Explanation aimed at rural managers and team leaders about the Rehydrating Serum. What is it? What is it used for? How it must be handled and what is the best time to drink it.
2. Explanation aimed at rural sugarcane workers about the Rehydrating Serum. What is it? What is it used for? How it must be handled and what is the best time to drink it.
3. Definition of the delivering Schedule and weekly control of the boxes containing the product to team leaders.
4. Definition of the delivering Schedule and weekly control of the sachets containing the product to the workers.
5. Elaboration of a SOP (Standard Operating Procedure) for delivery and control of the Rehydrating Serum.

4 RESULTS

4.1 Results of the adoption of Workplace Gymnastics and the Rehydrating Serum at the plant in Alagoas

As a result of a visit to an agricultural-industrial unit in the state of Alagoas, it was seen that the chosen way to verify the benefits of Workplace Gymnastics was to keep tabs of the lumbar pain complaints by the workers. During the first year (2008), gymnastics were applied once a day, at the beginning of activities, evolving in 2009 to twice a day, beginning and end of activities. The benefits of such practice can be seen in chart 2, which show-

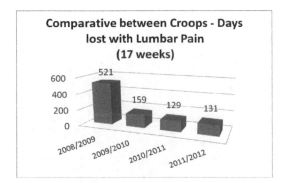

Chart 2. Complaints of lumbar pains at the plant in Alagoas.

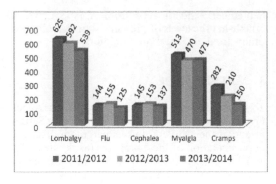

Chart 3. Occurrences of illnesses at the plant in Pernambuco.

cases a reduction of almost 75% in the number of lumbar pain complaints. This plant also adopted the Rehydrating Serum for all workers.

4.2 Results of the adoption of workplace Gymnastics and of the Rehydrating Serum at the plant under study

Implementation of these two practices took place in the crop 2011\2012 and the survey for results was based on the number of occurrences of the five diseases arising from the activity of cutting sugarcane, them being: Lombalgy, flu, cephalea, Myalgia and cramps, as per chart 3. It is noticeable that only cramps and lombalgy had an expressive and sustained fall, from 46,8% and 13,7%, respectively, from the number of occurrences between the first and last evaluated crops. Even though the other illnesses (flu, cephalea and myalgia) showing variance in reduction of occurrences throughout the crops, small drops between the first and last crops can be noticed; 13,2%, 5,5% and 8,2%, respectively, and even then a tendency of a drop in occurrences can be noticed.

4.3 Feedback poll

To keep track of the acceptance of the two practices, 165 randomly chosen workers were directly asked if they considered themselves Satisfied (likes it), Dissatisfied (doesn't like it) and Indifferent (whatever), and as result, the Rehydrating Serum received 100% of positive satisfaction, while 43,64% approved the daily practice of Workplace Gymnastics.

5 CONCLUSIONS

This study reported a significant reduction in the occurrence of lombalgia at the plant in Alagoas, which was also confirmed at the plant in Pernambuco, even though it has shown a more modest gain for that illness. The plant in Pernambuco fared better when it came to knocking down the occurrences of cramps. However, only the practice of Workplace Gymnastics and\or the use of the Rehydrating Serum are still not enough to verify expressive results in relation to the health and safety of the worker.

It can also be added the statement of the social worker for the Alagoas unit, who, by experience, affirms it is necessary to integrate long-term programs aimed at promoting health in order to attain sustained results in this area, including a re-education of the worker concerning his daily habits both at work and at home.

Faced with the reports and results, the benefits and importance of these practices become evident for the worker who performs using intense physical effort. It is also noticeable the need to maintain and improve workplace gymnastics and the use of the rehydrating serum in tandem with other ancillary practices, mainly educative ones, seeking to attain synergy between working conditions and individual habits.

REFERENCES

Alves, F. 2008. Processo de Trabalho e Danos à Saúde dos Cortadores de Cana. *Revista de Gestão Integrada em Saúde do Trabalho e Meio Ambiente* n° 2 v. 3.

FAO/OMS 2001. Human Vitamin and Mineral Requirements. In: Report 7th Joint FAO/OMS Expert Consultation. Bangkok, Thailand, 2001. xxii + 286p.

Institute of Medicine 1999–2001. Food and Nutrition Board. Dietary Reference Intakes. *National Academic Press*, Washington D.C.

Menezes, M.A. & Silva, M.S. 2010. A cana judia de nós! Impactos da Migração e da atividade de cortar cana-de-açúcar sobre a saúde dos trabalhadores-migrantes nordestinos. VIII Congresso Latino americano de Sociologia Rural, Porto de Galinhas, Recife-PE.

Monteiro, L.A.S. & Gonçalves, W.J. Aspectos metodológicos para identificar qualidade de vida no trabalho e na dignidade humana no setor do agronegócio. 48° Congresso SOBER—Sociedade Brasileira de economia, administração e sociologia rural. Campo Grande, MS, 2010.

Nadel, ER. Limitações impostas pela prática de exercícios em ambientes quentes. *Gatorade Sports Science Institute* – Revista Nutrição no Esporte, 1998, set/out n° 19.

Polito, E. & Bergamaschi, E.C. Ginástica Laboral. *Sprint*. 4ª Edição. Rio de Janeiro. 2010.

Silva Neta, Maria de Lourdes da. Perfil nutricional e de saúde de cortadores de cana nordestinos migrantes no no sudoeste do Brasil. Dissertação de mestrado em Nutrição, Programa de Pós-Graduação em Nutrição da Universidade Federal de Alagoas. Maceió, 2009.

Tirapegi, J. 2005. Nutrição, Metabolismo e Suplementação na Atividade Física. São Paulo: Atheneu, 350 p.

Occupational Safety and Hygiene V – Arezes et al. (Eds)
© 2017 Taylor & Francis Group, London, ISBN 978-1-138-05761-6

Thermal sensation assessment by young Portuguese adults in controlled settings

D.A. Coelho
Human Technology Group, Department Electromechanical Engineering and Centre for Mechanical and Aerospace Science and Technology, Universidade da Beira Interior, Covilhã, Portugal

P.D. Silva
Department Electromechanical Engineering and Centre of Materials and Building Technologies, Universidade da Beira Interior, Covilhã, Portugal

ABSTRACT: Laboratory experiments in a controlled climatic chamber enabled evaluation of fitness of the PMV equation to thermal sensation assessment by a Portuguese sample. Experiments were carried out in two occasions (May and December 2014), with two different college students cohorts, with 5 experimental conditions in each cohort. A total of 323 individual assessments of specific controlled thermal environment conditions were collected. Subjects sat down in a chair while inside the chamber and evaluated the thermal environment after a standard period of 3 minutes inside the controlled climate chamber. Each participating subject undertook the sitting in the chamber in five different days, with different climatic conditions each time. PMV and thermal sensation as assessed in the climatic chamber are compared, yielding correlation coefficients which differ between the Spring-Summer experiments and the Autumn-Winter Experiments but not between genders. The paper discusses possible factors contributing to different correlations obtained in the Autumn-Winter and the Spring-Summer experiments.

1 INTRODUCTION

1.1 Background

Thermal comfort is defined by standard ISO 7730 as the "state of mind that expresses satisfaction with the thermal environment". This is a simple, consensual and straightforward definition but one that is not straightforwardly converted into measurable physical parameters. Moreover, the thermal environment is one of a myriad of potential sources of disruption to performance, health, comfort and well-being (Parsons, 2003), e.g. noise, lighting, air quality, vibrations, and inadequately set psychosocial factors (Coelho et al., 2015).

The most important contributor to modelling thermal sensation was P.O. Fanger (1934–2006), who created a predictive model for general, or whole body, thermal comfort during the second half of the 1960s from laboratory and climate chamber research (Van Hoof, 2008). In that period, environmental techniques were improving, wealth increased and workers wanted the best indoor environment, while at the same time offices were growing larger (McIntyre, 1984). With his work, Fanger wanted to present a method for use by heating and air-conditioning engineers to predict, for any type of activity and clothing, all those combinations of the thermal factors in the environment for which the largest possible percentage of a given group of people experience thermal comfort (Fanger, 1967). It provided a solution for predicting the optimum temperature for a group in, for instance, an open plan office, which could be provided by architects and engineers (McIntyre, 1984). This Predicted Mean Vote (PMV) model has, over more than 4 decades since its creation, become the internationally accepted model for describing the predicted mean thermal perception of building occupants (Van Hoof, 2008).

1.2 Theory and aim

Calculations of thermal sensations are commonly done by energy balance models of Fanger (1972). However, previous experimental studies have shown that observed and calculated values differed considerably, leading to the development of theories explaining the differences according to factors such as "short-term thermal adaptation" and "thermal expectation" (Becker et al., 2003).

Additionally, thermal sensation of females and males in the same thermal environment, activity and clothing insulation has been reported as being consistently different (Amai et al., 2007). Moreover, other studies indicate that there is a strong interaction and influence of our experience with outdoor weather and our indoor thermal comfort (Chun et al., 2008).

The Predicted Mean Vote (PMV) model of thermal comfort, created by Fanger in the late 1960s, is currently used worldwide to assess thermal comfort. Fanger based his model on college-aged students for use in invariant environmental conditions in air-conditioned buildings in moderate thermal climate zones (Van Hoof, 2008). Growing multiple evidence shows that the PMV model is not applicable to all types of people in any kind of building in every climate zone.

Given the results of other studies and the interest in ascertaining whether the discrepancies found in other countries and climatic zones regarding the PMV model of thermal sensation, the study reported in this study focuses on a Portuguese sample of young adults. The current study hence contributes to ascertain the level of validity of the PMV model of thermal sensation, by investigating the differences (with statistical significance assessed) between self-assessed thermal sensation and PMV. These differences are evaluated across two important dimensions: climatic season (Autumn-Winter versus Spring-Summer) as well as gender of the participants (female versus male).

2 METHODS

2.1 Experimental design

Laboratory experiments in a controlled climatic chamber were the basis to evaluate fitness of the PMV equation to thermal sensation assessment by a Portuguese sample. Experiments were carried out in two occasions (May and December 2014), with two different cohorts of college students, with 5 experimental conditions in each cohort. However, only 4 of these experimental conditions are comparable across the two climatic seasons of the year, yielding six different conditions (assessed by proximity of calculated average PMV). The difficulties in conditioning the environment and compensating for different levels of clothing worn by the subjects in the two times of the year, led to the limitation of comparability reported.

A total of 323 individual assessments of specific controlled thermal environment conditions were collected. Subjects sat down in a chair while inside the chamber and evaluated the thermal environment after a standard period of 3 minutes inside the controlled climate chamber. Each participating subject in one of the two seasons undertook the sitting in the chamber in five different days, with different climatic conditions each time. Different participants

took part in the Autumn-Winter experiments in December 2014, compared to the ones that had taken part in the Spring-Summer experiments in May the same year. PMV and thermal sensation as assessed in the climatic chamber are compared, yielding correlation coefficients which differ between the Spring-Summer experiments and the Autumn-Winter Experiments. Gender based differences are also explored in the data.

A total of 37 college students (28 male and 9 female participants) took part in the May 2014 experiments. The December 2014 experiments were participated by a different group of 29 college students, with 14 female participants and 15 male participants. In both seasons, students from Arquitecture, Industrial Design and Industrial Engineering programs took part in the experiments for course credit within Human Factors and Ergonomics courses taught by the authors.

The conditions in each of the 5 trials carried out in each one of the experimental blocks were controlled. The average and standard deviation of the PMV calculations for all individuals involved in each trial are shown in Table 1 with a classification of the type of experiment according to the proximity of the mean PMV calculated value obtained, resulting in 6 conditions (A, B, C and E comparable across the two seasons; D only in the Autumn Winter block and F only in the Spring-Summer block).

2.2 Climate chamber

The climate chamber used in this work presents inner dimensions of 1.5 m, 1.2 m and 1.0 m, and allows reproducing the inside or outside climatic conditions of a building. The walls of the chamber are constituted of 15 cm of XPS, being this material limited by MDF panels with 2 cm thick. The chamber was connected to an independent air conditioning system capable to control the air temper-

Table 1. Pairing of experimental trials across seasons based on calculated PMV mean for whole cohort in each season.

Calculated PMV	Spring-Summer		Autumn-Winter	
Paired Trial	mean	st. dev.	mean	st. dev.
A	−2.4	0.46	−2.0	0.65
B	−1.4	0.40	−1.3	1.01
C	−0.5	0.19	−0.5	0.56
D	------	------	+0.2	0.82
E	+1.0	0.08	+0.6	0.51
F	+1.7	0.07	------	-----

ature, air humidity and air velocity. The imposed conditions are monitored in real time through a dedicate acquisition system.

This climatic chamber has al-ready been used in previous works and additional constructive details can be found in Pires et al. (2013). Inside the chamber was placed a chair to accommodate the occupant and simulate the sitting position. Inside the chamber, in a central position, a chair was placed so as to accommodate the occupant and simulate the sitting position. To allow access to the interior of the climatic chamber, its closing and opening was achieved through the use of a removable frame. This frame was covered with a clear plastic film in order to allow the occupant to visualize the outside environment thereby reducing the feeling of being in a confined space.

2.3 Measurements

The measurements inside the climatic chamber were performed using a thermal comfort system including a thermal comfort data logger (INNOVA 1221) and several transducers. The air temperature was measured using a radiantly shielded platinum sensor (Pt100) with an accuracy of ±1.0°C. A wet bulb temperature transducer that complies with ISO7243 was used to measure the air wet bulb temperature with an accuracy of ±0.5°C.

To take account of the thermal radiation exchange between the occupant and the climatic chamber walls, a globe temperature transducer was used. This sensor complies with the ISO7243 and consists of a Pt100 temperature sensing element situated at the centre of a globe with 150 mm in diameter and surface emissivity of 0.98. This sensor allows a global temperature measurement with an accuracy of ±0.5°C.

The indoor air velocity was measured with an omnidirectional air velocity transducer based on the constant temperature difference anemometer principle. According to the range of velocity values obtained during the experimental work the air velocity accuracy can be reported as better than ±0.06 m/s.

2.4 Predicted Mean Vote (PMV)

The Predicted Mean Vote Calculator prepared and made freely and publicly available by Håkan Nilsson (2005) was used. In order to use this calculator, yielding operative temperature, PMV and PPD (Predicted Percentage of Dissatisfied) (Fig. 1), a number of parameters should be input. The parameters, bound by pre-set intervals, are clothing (Clo), [0 to 2clo], air temperature (°C) [10 to 30°C], mean radiant temperature (°C) [10 to 40°C], activity (met) [0.8 to 4met], air speed (m/s) [0 to 1m/s] and relative humidity (%) [30 to 70%].

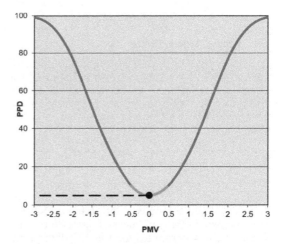

Figure 1. The relationship between Predicted Mean Vote and Predicted Percentage of Dissatisfied (Source: Nilsson, 2005; originally proposed by Fanger, 1970).

The PMV calculator was used to obtain the calculated PMV value individually, based on the values measured in real-time by the instruments and a detailed assessment of the clothing insulation level of the individual participant in each of the sitting trials in the climatic chamber and considering the level of activity of 1 Met, which corresponds to the activity of sitting.

2.5 Subjective assessment of thermal sensation

Upon arrival to the laboratory setting each participant was given a survey that assessed the participant's physical conditions and psychological state at the time of the trial. Every trial was carried out in a different day of the week. After this, an individual participant goes inside the climatic chamber, prepared and thermally stabilized, and instrumented. The transparent door of the chamber was then sealed in order to maintain the internal thermal environment. The participant is invited to sit in a relaxed manner in the chair supplied within the chamber, and stay there for three minutes. After this time is elapsed the participant registered her or his thermal sensation and then exists the climatic chamber. This process was repeated for every participant.

The scale used for thermal sensation assessment was the Portuguese version of the ASHRAE thermal sensation scale. This scale consists of: −3 (cold), −2 (cool), −1 (slightly cool), 0 (neutral), +1 (slightly warm), +2 (warm) and +3 (hot). This scale is directly equivalent to the results of the PMV calculation from Fanger's models (1972).

3 RESULTS AND ANALYSIS

3.1 Spring-Summer trials

The scatter plot of the results obtained for the trials carried out in May 2014 is shown in Figure 2. The Pearson correlation factor between the calculated PMV and the self-assessed thermal sensation was computed for the 185 data points encompassed yielded r = 0.871 (p<0.0005).

3.2 Autumn-Winter trials

The scatter plot of the results obtained for the trials carried out in December 2014 is shown in Figure 3. The Pearson correlation factor between the calculated PMV and the self-assessed thermal sensation was computed for the 137 data points encompassed yielded r = 0.667 (p < 0.0005).

3.3 Female subjects

When considering the female participants only (Fig. 4), the Pearson correlation factor between PMV and thermal sensation combining the experiments carried out in the two seasons is r = 0.762 (p < 0.0005). When looking only at the Spring-Summer data, the correlation is r = 0.686 (p < 0.0005), while it is higher, at r = 0.850 (p < 0.0005) for the Autumn-Winter data, considering only data from female subjects.

3.4 Male subjects

When focusing on the male participants only (Figure 5), the Pearson correlation factor between PMV

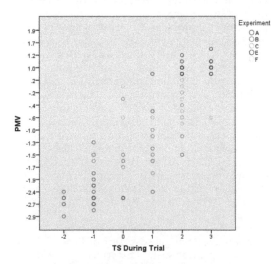

Figure 2. Overall outlook for the results of the experimental trials carried out in the Spring-Summer season (May 2014): Thermal Sensation (TS) self-assessed during trial versus calculated PMV value.

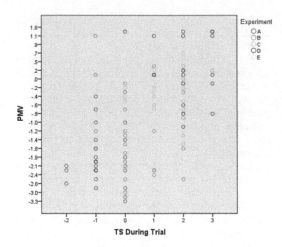

Figure 3. Overall outlook for the results of the experimental trials carried out in the Autumn-Winter season (December 2014): Thermal Sensation (TS) self-assessed during trial versus calculated PMV value.

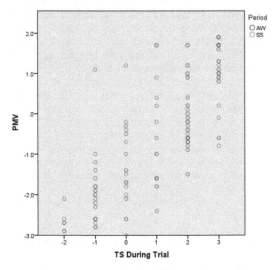

Figure 4. Overall outlook for the results of the experimental trials carried out in both seasons for female participants only: Thermal Sensation (TS) self-assessed during trial versus calculated PMV value (AW—Autumn-Winter; SS—Spring-Summer).

and thermal sensation combining the experiments carried out in the two seasons is r = 0.821 (p < 0.0005).

When looking only at the Spring-Summer data, the correlation is r = 0.656 (p < 0.0005), while it is higher, at r = 0.878 (p < 0.0005) for the Autumn-Winter data, considering only data from male subjects.

3.5 *Seasonal differences across comparable trials*

Inspection of Figure 6 reveals a V-shaped pattern in both seasons and that differences between calculated PMV and self-assessed thermal sensation display reversing behaviors, progressing towards neutral PMV values and then diminishing towards higher PMV values. Moreover, there is an interesting

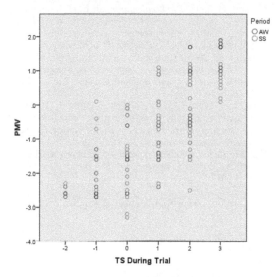

Figure 5. Overall outlook for the results of the experimental trials carried out in both seasons for male participants only: Thermal Sensation self-assessed during trial versus calculated PMV value (AW—Autumn-Winter; SS—Spring-Summer).

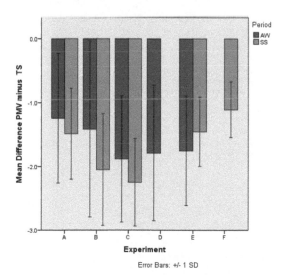

Figure 6. Seasonal comparison of differences across experimental trials A through F (see Table 1): Thermal Sensation self-assessed during trial versus calculated PMV value (AW—Autumn-Winter; SS—Spring-Summer).

reversal in that differences are greater for Spring-Summer than for Autumn-Winter in the lower PMV values region (experiments A, B and C) and vice-versa for the higher PMV values region (experiments D, E and F).

Non-parametric statistical testing (independent samples Kolmogorov-Smirnov test) yielded statistically significant differences across the two seasons for all comparable experiments: A (p = 0.048), B (p = 0.005), C (p = 0.032) and E (p = 0.003). Additionally, the independent samples Mann-Whitney U test led to significant differences for experiments B (U = 321; p = 0.009) and E (U = 678; p = 0.006).

3.6 *Discussion*

Analysis of correlation suggests improved performance of the PMV model in Autumn-Winter season as opposed to Spring-Summer season for the sample under study. Analysis by gender shows a different set of differences by season. Future studies might be beneficial to investigate further the factors that may explain the variant but statistically significant discrepancies between thermal sensation and calculated PMV, according to season, whether considering subjects from both genders, or separating subjects according to gender.

4 CONCLUSION

This paper report the results from an experimental study focuses on a Portuguese sample of young adults in order to evaluate potential discrepancies between the PMV model equation and the thermal sensation assessment. The results shows that the correlation between PMV equation and thermal sensation presents a higher determination coefficient in Spring-Summer experiments (r = 0,871) than in the Autumn-Winter experiments (r = 0,667) which indicates the need to improve the PMV model for this season. There are also some discrepancies between the thermal sensation and the PMV equation when the individual or joint analysis of genders is done with the correlation being stronger in the Autumn-Winter experiments when females or males are considered separately. The need for a revision of the PMV model taking into account regional and population differences is not dismissed by the results of this study.

REFERENCES

Amai, H., Tanabe, S. I., Akimoto, T., & Genma, T. 2007. Thermal sensation and comfort with different task conditioning systems. *Building and Environment*, 42(12), 3955–3964.

Becker, S., Potchter, O., & Yaakov, Y. 2003. Calculated and observed human thermal sensation in an extremely hot and dry climate. *Energy and Buildings*, 35(8), 747–756.

Coelho, D.A., Tavares, C.S., Lourenço, M.L., & Lima, T.M. 2015. Working conditions under multiple exposures: A cross-sectional study of private sector administrative workers. *Work*, 51(4), 781–789.

Fanger, P.O. 1967. *Calculation of thermal comfort: introduction of a basic comfort equation*, ASHRAE Trans., 73, III.4.1–III.4.20.

Fanger, P.O. 1970. *Thermal Comfort*, Copenhagen, Danish Technical Press.

Fanger, P.O. 1972. *Thermal Comfort: Analysis and Applications in Environmental Engineering*, McGraw-Hill, New York.

ISO—International Standards Organization. 1989. Standard ISO 7243: 1989. *Hot environments -- Estimation of the heat stress on working man, based on the WBGT-index (wet bulb globe temperature)*. International Standard, International Organization for Standardization (ISO), Geneva.

ISO—International Standards Organization. 2005. Standard ISO 7730: 2005. *Ergonomics of the thermal environment. Analytical determination and interpretation of thermal comfort using calculation of the PMV and PPD indices and local thermal comfort criteria*. International Standard, International Organization for Standardization (ISO), Geneva.

McIntyre, D.A. 1984. Evaluation of thermal discomfort. In: Berglund, B., Lindvall, T. and Sundell, J. (eds) *Proceedings of Indoor Air '84*, Vol. 1, Stockholm, 147–158.

Nilsson, H. 2005. *PMV calculator v.2*, English language [MS Excel Macro enriched file]. Department of Technology and Built Environment, Laboratory of Ventilation and Air Quality, University of Gävle, Sweden.

Parsons, K. 2003. *Human Thermal Environments: The effects of hot, moderate and cold environments on human health, comfort and performance*. 2nd ed, Taylor & Francis.

Pires, L., Silva, Pedro D., & Castro-Gomes, J.P. 2013. Experimental study of an innovative element for passive cooling of buildings, *Sustainable Energy Technologies and Assessments*, 4, 29–35.

Van Hoof, J. 2008. Forty years of Fanger's model of thermal comfort: comfort for all? *Indoor air*, 18(3), 182–201.

Occupational Safety and Hygiene V – Arezes et al. (Eds)
© 2017 Taylor & Francis Group, London, ISBN 978-1-138-05761-6

Measurement of physical overload in the lumbar spine of baggage handlers at a Brazilian airport

Luciano Fernandes Monteiro, José Wendel dos Santos & Veruschka Vieira Franca
Federal University of Sergipe, Sergipe, Brazil

Viviana Maura dos Santos
Federal University of Rio Grande do Norte, Rio Grande do Norte, Brazil

Odelsia Leonor Sanchez de Alsina
Institute of Technology and Research, University Tiradentes, Sergipe, Brazil

ABSTRACT: The objective of this study was to measure the physical overload in the lumbar spine of baggage handlers in a Brazilian airport and propose ergonomic solutions that could reduce the risk factors for the development of low back pain. For that, the anthropometric measurements of the agents and the instrumental apparatus used in the work activity were taken, and a systemic analysis of the biomechanics used during the tasks was performed. The data were submitted to the 3D Static Strength Prediction Program (3DSSPP) and the NIOSH Lifting Equation. The results showed that the biomechanics associated with moving the luggage and goods with a weight above recommended submitted the lumbar spine of the agents to potentially harmful compression forces. It is expected that the risk factors identified in this study will sensitize the managers so that the ergonomic proposals should be carried out systematically.

1 INTRODUCTION

Air transport has been considered the leading means of passenger and freight transport around the world by traveling long distances quickly and safely. In Brazil, the growth of the economy in the last ten years was one of the main driving forces for the popularization of this type of transport. According to data from the National Civil Aviation Agency (ANAC), in 2014 the share of the transport market reached 63.0%, compared to 34.8% in 2004.

According to Zimmermann & Oliveira (2012), in order to meet this growing demand, airlines have intensified the use of their fleet within their operating networks and expanded their installed production capacity in terms of number and size of aircraft and flight frequencies. In 2015, 1.09 million flights were made by Brazilian companies, considering the total of domestic and international operations. Regarding the number of passengers transported, the National Transportation Confederation (CNT) recorded a turnover of 102.32 million passengers in 2014. This figure represented a real increase of 210.8% over 2000, when Brazilian airlines were responsible for the handling of 32.92 million.

However, despite the apparent modernization of the Brazilian airport infrastructure, there is still a high incidence of workers involved in activities whose processes are still rudimentary, with minimal degree of technology. Unlike the baggage dispatch process, which has an automated baggage and merchandise management system, the logistics for the preparation and distribution of these are carried out by manual methods and involves a large contingent of manpower.

The physical work performed by baggage handlers is characterized by an activity that has high energy demand and requires physical force measures during the lifting and movement of baggage and merchandise of different weights, shapes and sizes. The activities of handling baggage and heavy goods, regardless of the limitations of the human body, can pose serious health risks to the baggage handlers, because lumbar spine receives potentially damaging physical overloads to the intervertebral discs, increasing the chances of the development of low back pain (NIOSH, 2015, Oxley et al., 2009, Rückert et al., 2007, Tafazzol et al., 2015).

In Brazil, although the working conditions of the baggage handlers are a subject rarely explored, several international studies have been operationalized in an attempt to identify the impacts of these activities on the health of these workers. Oxley et al. (2009) sought to identify the prevalence of symptomatic bodily manifestations between baggage handlers at Midlands Airport in the United Kingdom. In the presented case, the results indicated that most of the workers reported having pain in the lumbar spine, knees, neck and shoulder, as well as attributing the symptoms to the work performed.

Tafazzol et al. (2015), when analyzing the epidemiological and biomechanical aspects of the logistics of baggage and merchandise preparation and distribution at Mehrabad International Airport in Iran, not only found a high prevalence of pain in the lower back, knees and neck, but also identified that Manual handling of overweight baggage has promoted potentially damaging compression forces to the backbone of the workers, predisposing them to the occurrence of low back pain.

Low back pain not only affects the health of the worker himself but also social consequences, such as absenteeism, change of profession due to incapacity for work, social security expenses, among others, which should not be neglected (Merino, 1996).

In this perspective, the present study had as objective to measure the physical overload in the lumbar spine of baggage handlers in a Brazilian airport and propose ergonomic solutions that could minimize the risk of developing low back pain.

2 MATERIALS AND METHODS

2.1 Study design

This study was conducted in the baggage and goods processing sector of Aracaju Airport, located at 10° 59'12.7" S and 37° 04'20.4" W, in the state of Sergipe, Brazil. The airport was founded in 1952 and incorporated to the Brazilian Airport Infrastructure Company (INFRAERO) in 1975. This airport complex has a contingent of over 1.000 (thousand) employees to meet an average monthly demand of 115 thousand passengers and 20 scheduled daily flights by the largest airlines in the country.

The sample consisted of twenty-two baggage handlers responsible for the logistics of preparation and distribution of luggage and merchandise at the analyzed airport. All the workers are male, with mean age of 35 ± 4 years, mean height of 1.76 ± 0.05 m, mean weight of 75.18 ± 7 kg and Body Mass Index (BMI) of 24.03 ± 2.1 kg/m². The stipulated working day is 8 hours a day with an interval of 2 hours. The average time of service provided by the workers comprises 6.47 ± 3.28 years.

2.2 Collection procedures and data analysis

Data collection took place in the first half of December 2015 and was systematized in two stages. In the first stage, a pre-scheduled interview was conducted with the representatives of the company responsible for the preparation and distribution of luggage and merchandise, airlines and INFRAERO. At this stage, it was possible to know the physical facilities of the airport, the logistics of preparation and distribution of luggage, select the volunteers to compose the Homogeneous Exposure Group (HEG) and set the field research agenda.

In the second stage the anthropometric measurements of the workers and instrumental apparatus used in the processing of luggage and merchandise were taken, with the aid of the following equipments: portable vertical anthropometer, portable platform scale and anthropometric tape. In addition, the systemic observation of the biomechanics used by the workers during the execution of the task was performed. These data were documented and recorded by means of photos and filming.

The data collected were analyzed by the 3D Static Strength Prediction Program (3DSSPP), version 6.0.6, developed by the Ergonomics Center of the University of Michigan, to analyze the biomechanical demands and to measure the physical backload of the baggage handlers during the execution of the activity. For this, the database was used with records of twelve cycles of preparation of checked baggage and ten cycles of operation of loading and unloading, of such baggage as of the merchandise.

In addition, the NIOSH Lifting Equation proposed, in 1994, by the National Institute for Occupational Safety and Health (NIOSH) was used to determine the Recommended Weight Limit (RWL) and Lifting Index (LI), so that a considerable percentage of the population of the workers can perform the task over an 8-hour shift without increasing the risk of musculoskeletal disorders.

3 RESULTS AND DISCUSSION

3.1 Work and workplace characteristics

The point of induction of baggage processing dispatched at Aracaju Airport occurs at airline counters. At this point, identification tags are inserted in the luggage, which are then deposited on conveyors and directed to the security check. The baggage is then sent to the preparation zone for the flight for which it was destined, in which a specialized company acts as a provider of logistics services for the preparation and distribution of baggage to airlines and to INFRAERO.

In the logistics of preparation and distribution of luggage and merchandise, a large part of the operations is carried out manually and with an average execution time of around 14 ± 5 minutes. In the preparation stage the transfer of checked baggage to the support car, which had 120 cm in length, 60 cm in width, 53 cm in height and load capacity of 2000 kg was carried out. The average weight of baggage and goods processed during field research was 28 ± 6 kg (274.6 ± 58.8 N), with a minimum weight of 12 kg (117.68 N) and a maximum weight of 34 kg (333.43 N).

The support cars are then attached to mini tractors and routed to the storage area while awaiting the unloading of the baggage from the previous flight. In the case of merchandise, after the bureaucratic procedures carried out in the sector adjacent to the boarding hall, these are sent to the merchandise storage area, and the process of transfer to the support car is similar to that carried out with the checked baggage.

After the aircraft is parked at the aerodrome, the unloading process is initiated and requires the services of four baggage handlers and two towing tractor operators. This process lasted on average 12 ± 4 minutes and begins with the positioning of cars near the two compartments of the aircraft. Subsequently, the baggage handlers climb into the base of the support car and in a coordinated manner, remove and deposit the luggage and goods from the aircraft compartment to the base of the support car. Figures 1 and 2 show the baggage transfer operations for the support and unloading cart of the aircraft compartment.

After the aircraft is fully discharged, the mini tractors transport the baggage or goods unloaded from the aircraft to the screening area to be performed bureaucratic procedures with the Federal Police and Federal Revenue and released without any irregularities.

Concurrently, the operation of loading the checked baggage for flights of the other airlines takes place. This process is similar to the one performed in the discharge, but in the reverse flow and has an average duration of 20 ± 3 minutes.

3.2 *Measurement of physical overload*

Regarding the preparation of the luggage, it was possible to notice that in all cases, the workers flexed the trunk in front of the luggage, making an average angle of 33 degrees, intending the muscles of the neck and spine. Generally, both arms remained extended, and soon the workers flexed the right elbow and the two knees to pick up the luggage deposited on the conveyor 26 cm in height. With the use of coarse handling, the workers returned to the anterior position by extending the trunk and rotated the body at an average angle of 93 degrees. This trunk movement associated with baggage retrieval caused compressions in the L4/L5 and L5/S1 disks of the spine of the workers on the order of 3394 ± 385.1 N.

Soon after picking up the luggage, 86.36% of the agents flexed their legs, varied the movement of the arms between flexion and extension, and during the route placed the left hand in the base of the luggage. As workers positioned themselves in front of the support car, they turned the body again at an average angle of 88°, their legs stretched out and their arms flexed so that the luggage was deposited on the support car. During this task the compression on the L4/L5 and L5/S1 discs of the workers was calculated, being 2893 ± 267.7 N, which differed from the first position by the fact that most of the workers brought the luggage close to their bodies.

Figure 1.   Operation of transfer luggage to the support car.

Figure 2.   Process of unloading goods from the aircraft.

In the process of loading and unloading luggage and goods from aircraft, it was observed that in order to pick up the luggage or merchandise located at the entrance of the aircraft compartment the biomechanics common among the workers occurred in the following movements: extension of the two arms above the level of the shoulder flexion, forward trunk flexion, and variations in flexion and extension of the legs. After picking up the luggage or merchandise located in the aircraft compartment, which is 48 cm high relative to the shoulder level, the officers flexed their arms and turned the body at an average angle of 128 degrees and deposited the luggage or goods in the car which was 64 cm from the initial position of the agent. While depositing the luggage at the base of the car, trunk flexions with angle variation between 21.14° and 53.67° were observed, which depended on the height at which the luggage or goods were stacked one on the other. These flexions caused compressions in the L4/L5 and L5/S1 disks of the agents in the order of 3395.31 ± 253.67 N. Figures 3 and 4 illustrate the main biomechanical stresses used by the workers during the preparation of the checked and during the operation baggage of the goods of the aircraft.

According to Merino (1996), the multidirectional compressions in the discs L4/L5 and L5/S1 of the spine should not be greater than 3400 N, because if subjected to a compressive force above this order they cause micro traumas, that depending on the degree of evolution, could result in low back pain. Although low back pain is not characterized as a disease but as a symptom of intense pain, they tend to be potentiated, usually at the end of the working day or during production peaks, with relief during night rest and at the weekend. However, recurrences of episodes may progress to chronic pain, and in these cases, decreases the worker's chances of returning with the same physical conditions.

According to Teixeira et al. (2011), among the causes of low back pain, the activities of manual lifting of loads are the main factor. One of the main aspects considered in the investigation of the risk factors are related to the weight and position of the load. In this sense, when lifting, pulling, or pushing luggage or goods it is fundamental that the baggage handlers considers the weight and position of the load relative to the axis of the body, as these factors are narrowing related to the compressive forces generated in the spine.

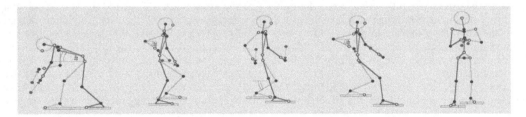

Figure 3. Main biomechanical requests during preparation of checked baggage.

Figure 4. Main biomechanical requests during the unloading of goods from the aircraft.

Table 1. Standard location variables for lifting luggage or merchandise.

Process step	23 kg	HM	VM	DM	FM	AM	CM	RWL	LI
Preparation	23 kg	1.00	0.81	0.98	0.80	0.70	0.90	9.20 kg	3.04
Loading and unloading	23 kg	1.00	0.86	0.89	0.80	0.42	0.90	7.75 kg	3.61

Note: HM—Horizontal location of the object relative to the body; VM—Vertical location of the object relative to the floor; DM—Distance the object is moved vertically; FM—Frequency and duration of lifting activity; AM—Asymmetry angle or twisting requirement; CM—Coupling or quality of the workers grip on the object.

For the analyzed activities, the result of the NIOSH Lifting Equation suggested a drastic reduction of the weight manipulated by the baggage handlers during the preparation of the luggage from 28 kg to 9.20 kg and in the process of loading and unloading of luggage and goods from 28 kg to 7.75 kg. Among the considered aspects, the asymmetric angles and vertical distance of luggage and merchandise were the factors that contributed the most to the inadequate RWL and LI values, as can be observed in Table 1.

3.3 Ergonomic solutions proposals

Considering the risk factors for low back pain identified in this study, some ergonomic recommendations were presented to the representatives of the company responsible for logistics services for preparation and distribution of luggage and merchandise, airlines and INFRAERO.

It was recommended to use devices that provide mechanical assistance in terms of reducing physical workload, manual handling and risk factors associated with handling baggage and merchandise at airports, such as Unit Load Device (ULD) or vacuum handler's devices for baggage and goods. It was pointed out that the reconfiguration of the post and working methods should be carried out in the short term. In addition, they should also be implemented in the long term if there is no possibility of establishing the propositions of automation of the process. In this sense, it was suggested that the instrumental apparatus necessary for the operation be adapted to the anthropometric measurements of most baggage handlers. It was suggested that airlines provide baggage weight marking labels so that the heavier baggage would be carried by two workers. In addition, towing tractors should position the support carriages perpendicular to the conveyor belts to reduce the angle of rotation of the trunk from 180° to 90°. Likewise, position the support cars parallel to the compartments of the aircraft.

Regarding the handling of luggage and goods by the workers, it was recommended that the knees should be flexed when picking up and unloading the luggage on the treadmill, standing upright and storing luggage or merchandise close to the longitudinal axis of the body, trunk flexion, or body rotation during the process.

4 CONCLUSIONS

This study aimed to measure the physical overload in the lumbar spine of baggage handlers in a Brazilian airport and propose ergonomic solutions that could reduce the risk factors for the development of low back pain.

In the analysis, it was evident that all the activities observed in the baggage preparation and distribution logistics of the analyzed airport demand intense muscle strength, which present risk factors that compromise the structures of the spine, especially the lumbar spine. This can be confirmed by the results of the methods applied.

The NIOSH Lifting Equation allowed the previously assumed assumption that the weight of the luggage and goods was not within the ideal limits was confirmed. The material handling was approximately 72.32% above the recommended one. That is, the maximum weight to be handled by the workers should be 9.20 kg (preparation) and 7.75 kg (loading/unloading), not 28 kg.

Biomechanical modeling reinforces these findings in that it indicates the use of too much effort for rotation, flexion and extension of the trunk, associated to the lifting of luggage and merchandise, resulting in the formation of compression forces at the lumbar spine level of the baggage handlers in the order of 3394N (preparation) and 3395.31 (loading/unloading). Values very close to the limit recommended in the literature.

In general, it is expected that the risk factors for low back pain identified in this study will sensitize managers so that the ergonomic propositions are implemented, so that the physical integrity of these professionals is preserved, and, thus, allows efficient performance in the exercise of their attributions.

REFERENCES

Agência Nacional de Aviação Civil (ANAC). 2015. Statistical data of air transportation—Brazilian companies (*published in Portuguese only*). Available at: <http://trabalho.gov.br/seguranca-e-saude-no-trabalho/normatizacao/normas-regulamentadoras>. Accessed: oct/2016.

Confederação Nacional do Transporte. 2015. Transport and economy: air passenger transport (*published in Portuguese only*). Brasília, DF: CNT. Available at: http://www.cnt.org.br/. Accessed Oct/2016.

Merino, E. A. D. 1996. Acute and chronic effects caused by the handling and movement of loads in the worker (*published in Portuguese only*). Dissertation (Master degree)—Federal University of Santa Catarina, Florianópolis: 128.

National Institute for Occupational Safety and Health (NIOSH), 1994. Applications manual for the revised NIOSH lifting equation. U.S. Department of Health and Human Services (NIOSH), Public health Service, Cincinnati, OH.

National Institute for Occupational Safety and Health (NIOSH). 2015. Reducing musculoskeletal disorders among airport baggage screeners and handlers. Department of Health and Human Services (NIOSH), Public health Service, v. 1: 4.

Oxley, L.; Riley, D.; Tapley, S. 2009. Musculoskeletal ill-health risks for airport baggage handlers—Report on stakeholder's project at East Midlands Airport. Health and Safety Executive, v. 1: 1–106.

Rückert, A.; Rohmert, W.; Pressel, G. 2007. Ergonomic research study on aircraft luggage handling. Applied Ergonomics, 35(9): 997–1012.

Tafazzol, A; Aref, S.; Mardani, M.; Haddad, O.; Mohamad, P. 2015. Epidemiological and Biomechanical Evaluation of AirlineBaggage Handling, International Journal of Occupational Safety and Ergonomics, 22(2): 218–227.

Teixeira, E. R.; Okimoto, M. L. R.; Gontijo, L. A. 2011. Index of Niosh Equation and Low Back Pain (*published in Portuguese only*). Revista Produção Online. Florianópolis, SC, 11(3): 735–756.

University of Michigan. 2012. 3 D Static strenght prediction program: version 6.0.6. University of Michigan, Center for Ergonomics. Available at: <https://c4e.engin.umich.edu/tools-services/3dsspp-software>. Accessed: fev/2016

Zimmermann, N.; Oliveira, A. V. M. 2012. Economic liberalization and universal access in air transport: it is possible to conciliate free markets with social goals and still avoid infrastructure bottlenecks (*published in Portuguese only*). Journal of Transport Literature, 6(4): 82–100.

Occupational Safety and Hygiene V – Arezes et al. (Eds)
© 2017 Taylor & Francis Group, London, ISBN 978-1-138-05761-6

Occupational exposure assessment of ramp operators of a Brazilian airport to the heat and noise

Luciano Fernandes Monteiro, José Wendel dos Santos & Veruschka Vieira Franca
Federal University of Sergipe, Sergipe, Brazil

Maria Betania Gama dos Santos
Federal University of Campina Grande, Paraíba, Brazil

Odelsia Leonor Sanchez de Alsina
Institute of Technology and Research, University Tiradentes, Sergipe, Brazil

ABSTRACT: The present study aimed to evaluate occupational exposure to heat and noise of the ramp operators of a Brazilian airport, by comparison of the levels obtained in the field research with the levels established by Brazilian law. Therefore, we collected data from occupational noise and temperature in air operational airfield at set points near each ramp operator during the execution of work activities. In the analysis of the results it was observed that the thermal and acoustic levels were off the ergonomic standards established by Brazilian law. Overall, the results showed the need to establish a planning for actualizing the exposure limits of ramp operators to thermal and acoustic overload and, above all, a more rigid supervision by the competent organizations.

1 INTRODUCTION

In recent years, Brazilian air transport has become the main modal for passenger and freight transport. According to Zimmermann and Oliveira (2012), in order to meet this growing demand, airlines have intensified the use of their fleet within their operating networks and have expanded their installed production capacity in terms of number and size of aircraft and flight frequencies.

Official data from the National Civil Aviation Agency (ANAC) show that in 2015, 1.09 million flights were made by Brazilian companies, considering the total of domestic and international operations, and that from January to August 2016, there were half a million takeoffs in the country. Regarding the number of passengers transported, the National Transportation Confederation (CNT) recorded a turnover of 102.32 million passengers in 2014. This figure represented a real increase of 210.8% over 2000, when Brazilian airlines were responsible for the handling of 32.92 million. Ratifying this trend of growth, in 2015 the number of passengers transported reached the mark of 119.8 million, and only in the first half of 2016, 73.1 million people have already used the air route (ANAC, 2015).

Not surprisingly, with the increase in the daily movement of passengers at Brazilian airports, airlines had to minimize baggage processing time, especially in connections between flights, which resulted in an increase in the workload of the professionals who provide Support services for aircraft, passengers, luggage, cargo and mail. However, despite the apparent modernization of the Brazilian airport infrastructure, it is observed that a large part of the contingent of workers involved in this process develops still rudimentary activities, with a minimum degree of technology, as is the case of ramp operators.

These professionals are responsible for loading and unloading baggage and goods on aircraft. In addition to the high-energy demand during lifting and movement of luggage and goods, they perform their tasks, most of the time, in the environment outside the airport, where are subject to the weather and occupational noise caused by aircraft turbines. These working conditions are closely related to the substantial increase in cases of absenteeism and incapacitation to work among these professionals (Gugliermetti et al., 2010).

In Brazil, although the environmental work conditions of ramp operators are a subject rarely explored, several international studies have been operationalized to identify the impacts of the risk factors inherent to the activities developed in the health of these workers. Yoopat et al. (2002) conducted a research to investigate the combined effects of physical workload and thermal stress on ramp operators at Bangkok Airport, Thailand.

From the analyzed parameters, they observed that the exposure to solar radiation associated to the workload contributed to the increase of the body temperature of the ramp operators during the execution of the task, and for this reason, they concluded that measures to combat this exposure should be adopted.

Gugliermetti et al. (2010) evaluated the exposure of ramp operators to occupational noise during the development of work activities close to the aircraft at the International Airport of Rome, Italy. In this analysis, they observed that workers were exposed to occupational noise levels above 87 dB (A), requiring preventive measures ranging from the availability of hearing aids with adequate attenuation levels to the continuous monitoring of occupational noise in the workplace and the performance of audiometric tests.

In this perspective, the present study has the objective of evaluating the occupational exposure of the ramp operators of a Brazilian airport to heat and noise, from the comparison of the levels obtained in field research with the levels established in Brazilian legislation, specifically in Regulatory Norm 15 (NR-15).

2 MATERIALS AND METHODS

2.1 Study design

This study was conducted in the baggage and goods processing sector of Aracaju Airport, located at 10° 59'12.7" S and 37° 04'20.4" W, in the state of Sergipe, Brazil. The airport was founded in 1952 and incorporated to the Brazilian Airport Infrastructure Company (INFRAERO) in 1975. This airport complex has a workforce of over 1.000 (one thousand) employees to meet an average monthly demand of 115 thousand passengers and 20 scheduled daily flights.

The sample consisted of nine ramp operators responsible for loading and unloading baggage and goods from aircraft at the analyzed airport. All the agents are male, with mean age of 32 ± 5 years, mean height of 1.72 ± 0.09 m, mean weight of 73.36 ± 9 kg, and mean Body Mass Index (BMI) of 21.94 ± 1.8 kg/m². The stipulated working day is 6 hours daily with an interval of 1 hour. The average time of service provided by the agents comprises 5.86 ± 2.91 years.

2.2 Collection procedures and data analysis

Data collection took place in the first half of December 2015 and was systematized in two stages. In the first stage a pre-scheduled interview was conducted with representatives of the company responsible for the service, airlines and INFRAERO. In this stage, it was possible to know the physical facilities of the airport, the logistics of preparation and distribution of luggage and merchandise, select the volunteers to compose the Homogeneous Exposure Group (HEG) and set the field research agenda.

In the second stage, the environmental data were collected in the operational aerodrome. The thermal levels were collected at five pre-established points, close to each worker during the execution of the work activities, considering intervals between measurements of 60 ± 10 minutes. The daily measurement had a duration of 8.05 ± 0.07 hours and the total monitoring time was 42.45 ± 3 hours occupational noise levels were taken through two continuous integrative noise meters for personal use throughout the work day of the operator.

The Portable Globe Thermometer (ITWTG/2000) was used to measure dry, wet and globe bulb temperatures. From Equation (1), the Wet Bulb Globe Temperature Index (WBGT) was calculated with exposure to solar radiation.

$$WBGT = 0.7\ T_w + 0.2\ T_g + 0.1\ T_a \qquad (1)$$

where: T_w is the natural wet bulb temperature; T_g is the globe temperature; T_a is the dry bulb temperature (air temperature).

The level of occupational exposure to continuous or intermittent noise during the workday was measured with the noise dosimeter (DOS-600), which had a noise measurement scale of 70 to 140 dB and an accuracy of ± 1.5 dB. The equipment has been set to operate in the compensation circuit (A), the slow response circuit and reference criterion of 85 dB, which corresponds to the 100% dose for an 8-hour work exposure. The overall monitoring of the sound pressure level was 61.78 ± 5 hours.

The variables collected were analyzed by comparing the levels obtained in field research with the levels established in Brazilian legislation.

3 RESULTS AND DISCUSSION

3.1 Work and workplace characteristics

After the preparation and transfer of luggage and goods to the support car, the towing tractor will route them to the area near the aircraft or to the storage area in case of the process of unloading the baggage of the previous flight. In the case of goods, after the bureaucratic procedures carried out in the sector adjacent to the boarding hall, these are sent to the warehousing sector.

After the aircraft is parked at the aerodrome, the unloading process is initiated and demands the services of four ramp operators and two towing tractor operators. This process lasted on average 12 ± 4 minutes and begins with the positioning of cars near the two compartments of the aircraft. Subsequently, the ramp operators rise in the base of the support car and in a coordinated manner, remove and deposit the luggage and goods from the aircraft compartment to the base of the support car. Figure 1 shows the cargo unloading operations of the aircraft compartment.

After the aircraft is fully discharged, the mini tractors transport the baggage or goods unloaded from the aircraft to the screening area to be performed bureaucratic procedures with the Federal Police and Federal Revenue and released without any irregularities. Concurrently, the operation of loading the checked baggage for flights of the other airlines takes place. This process is similar to the one performed in the loading, but in the reverse flow and has an average duration of 20 ± 3 minutes.

3.2 Evaluation of occupational exposure to noise and heat

The data collected in the baggage and goods processing sector were condensed in Table 1, in order to facilitate the analysis. In this way, a comparison can be made between the thresholds recommended by the Brazilian legislation (ideal condition) and the reality of the analyzed environment.

3.2.1 Evaluation of occupational exposure to noise

In view of the above in Table 1, it is observed that during the workday ramp operators are exposed to average occupational noise levels in the range of 93.63 dB (A), higher than stipulated by Brazilian law.

As shown in Figure 2, to illustrate the behavior of sound pressure levels during one workday, it was observed the occurrence of four peaks, due to landing and taking off other aircrafts. The first

peak occurred at 08h45 min, when the occupational noise level reached 117.24 dB (A), the second one at 09h57 min, reaching 125.67 dB (A), the third at 13:59, reaching 118.87 dB (A) and the fourth and final peak at 16:48, when it reached 119.38 dB (A).

Occupational noise levels between 80 and 100 dB were caused by the concurrent activities of aircraft preparation, such as positioning of tractors for loading and unloading luggage, the tanker aircraft outfitter, embarkation and disembarkation of crew.

In these activities, the ear protectors are mandatory. However, according the ear protector's manufacturer specification, the use of this equipment promotes an attenuation of 17 dB. During the workday, in the first sound pressure peak levels, the ramp operator was exposed to a pressure on the order of 100.24 dB and the use of a hearing aid promoted the percentage attenuation of only 14.50% of the noise.

According Cavalcante et al. (2013), occupational exposure to high sound pressure levels cause hearing extra trauma, physiological and Noise Induced Hearing Loss (NIHL). NIHL is an insidious disease, growing along the years, having a direct relationship with intensity, time of exposure and individual susceptibility to noise (Marques & Costa, 2006). In this sense, the NR-15 establishes that for unhealthy work effect, the time allowable daily exposure of six hours of work shall not exceed the limit of 87 dB, otherwise, offers serious and imminent risk to health of the workers.

Table 1. Assessment variables.

Variables	Levels in the workplace	Exposure limit	Legislation
Average occupational noise dB (A)	93.63	87.0	NR-15
Average WBGT index (°C)	29.78	27.5	NR-15

Source: Field results.

Figure 1.   Unloading operations of the aircraft compartment.

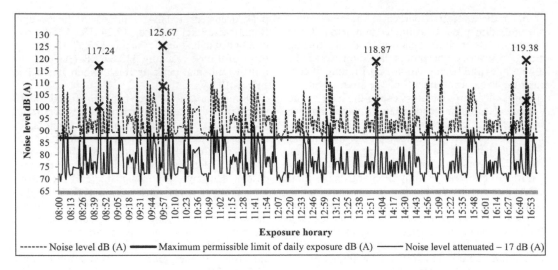

Figure 2. Continuous monitoring of operator exposure to occupational noise during the working day.

3.2.2 Evaluation of occupational exposure to heat

The average WBGT calculated for the ramp operators was 29.78°C, higher than the tolerance limit set by the NR-15. However, the intermittence of work due to the schedule arrangements and time between flights, prevents the workers to receive wage complementation and other additional rights relative to unhealthy work by the social security. In the case of ramp operators, there is a direct influence of solar radiation in the course of the working day. The average thermal overload situation occurred between 09:00 and 16:00, with a higher average thermal overload observed at 12:00, when the average temperature was 32.91°C, as can be seen in Figure 3.

For these reasons, it is observed that the activities of the ramp operators are held in unsuitable conditions. The situation may be exacerbated by the fact that besides the necessary agility coupled with intense biomechanical requests during the execution of tasks, the thermal characteristics of the asphalt paving of the aerodrome associated with internal aircraft compartment temperature causes additional load heat by convection and radiation, enhancing the sensation of thermal overload.

The human body cannot tolerate conditions exceeding 37°C. At temperatures of 27°C and a relative humidity of 40%, some healthy individuals may begin to experience heat stress with prolonged activity or exposure (Opitz-Stapleton et al., 2016). Heat stress causes fatigue, headache and muscle cramps, while heat stroke can lead to death, even among healthy people. Camargo and Furlan (2011) add that such thermal conditions affects the system of production and exchange of body heat to the environment and interferes with the thermoregula-

tory system of the body, producing physical and nervous exhaustion, decreased income, increase in errors and risk accidents at work.

Weather conditions in conjunction with health status, workload and rate, outdoor worker exposure to sunlight and wind, indoor worker exposure to radiant heat sources or without adequate ventilation, or those workers not acclimatized can lead to heat stress and stroke in the workplace (Lucas et al., 2014). However, despite this scientific and medical recognition, general business awareness of extreme heat exposure and occupational health risks remains low, and regulatory standards for heat illness prevention programs for different occupations in various countries may be lacking or inconsistent (Arbury et al., 2014; Gubernot et al., 2014; Opitz-Stapleton et al., 2016).

3.2.3 Ergonomic solutions proposals

Regarding the identified environmental problems, it was found the need of ergonomic proposals to minimize heat exposure and its negative effects ranging from progressive dehydration and cramps to more serious events, such as heat exhaustion and thermal shock. In the long term, this can cause cancerous melanoma and premature aging. Therefore, the use of sunscreens adequate to the type and sensitivity of the skin of the agents is recommended to reduce the harmful effects of sunlight on the skin. It has been suggested further that appropriate clothing be used, preferably made of light fabric and micro fibers. This kind of fabric transfers the sweat to the outer side, and favors the drying. The use of Personal Protective Equipment (PPE) such as gloves, boots and a hat with wide lateral protection for ears and neck was also

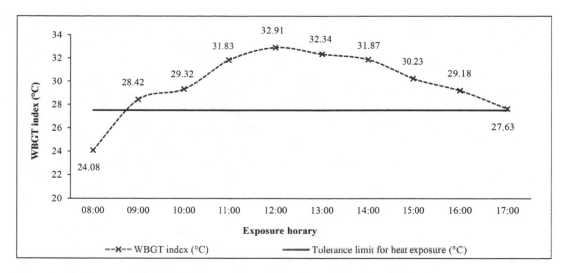

Figure 3. Continuous monitoring of exposure to occupational agent heat during working hours.

suggested. The latter was suggested because it was observed the lack of head protection by the agents during exposure to sunlight.

In relation to occupational noise exposure, although it was observed the use of ear protectors of shell type, which according to the product specifications have 17–27 dB (A) attenuation capacity, it becomes important to use physiological alternatives such as reducing the exposure time from taking regular breaks and function of rotation, even if the attenuation of hearing protection is sufficiently able to generate a safe situation without major damage to health.

4 CONCLUSIONS

This study aimed to evaluate occupational exposure of the ramp of Brazilian airport operators to heat and noise, from the comparison of the levels obtained in the field research with the levels established by Brazilian law.

In the analysis of the results it was observed that the thermal and acoustic levels were off the ergonomic standards established by Brazilian law. The average calculated for WBGT index ramp operators was 29.78, or 8.29% above the limit recommended by NR-15.

About occupational noise, it was observed that workers were exposed to high levels of sound pressure. The average occupational noise on the job of ramp operators met up 7.62% over the limit set by the NR-15. Despite the use of hearing protectors with good attenuation levels, it was shown that future studies must be conducted to develop measures to eliminate, reduce or neutralize noise in these jobs, in addition to analyzing the effectiveness of mitigation used protectors.

Overall, the results showed the need to establish a revised plan to exposure limits of ramp operators to thermal and acoustic overload and, above all, a more rigid supervision by the competent organizations.

REFERENCES

Agência Nacional de Aviação Civil (ANAC). 2015. Statistical data of air transportation—Brazilian companies (*published in Portuguese only*). Available at: <http://trabalho.gov.br/seguranca-e-saude-no-trabalho/normatizacao/normas-regulamentadoras>. Accessed: oct/2016.

Arbury, S.; Jacklitsch, B.; Farquah, O.; Hodgson, M.; Lamson, G.; Martin, H.; Profitt, A. 2014. Heat Illness and Death Among Workers—United States, 2012–2013. Morbidity and Mortality Weakly Report, 63(31): 661–665.

Brasil, MTE. 2004. Norms regulating health and safety at work: Activity and unhealthy operations (*published in Portuguese only*). Available at: <http://trabalho.gov.br/seguranca-e-saude-no-trabalho/normatizacao/normas-regulamentadoras>. Accessed: oct/2016.

Camargo, M. G.; Furlan, M. M. D. P. 2011. Physiological response of the body to elevated temperatures: exercise, temperature extremes (*published in Portuguese only*). Revista Saúde e Pesquisa, v. 4, n° 2: 278–288.

Cavalcante, F.; Ferrite, S.; Meira, T. C. 2013. Exposure to noise in the manufacturing industry in brazil. Rev. CEFAC, 15(5):1364–1370.

Confederação Nacional do Transporte. 2015. Transport and economy: air passenger transport (*published in Portuguese only*). Brasília, DF: CNT. Available at: http://www.cnt.org.br/. Accessed: oct/2016.

Gubernot, D. M.; Anderson, G. B.; Hunting, K. L. 2014. The epidemiology of occupational heat exposure in the United States: a review of the literature and assessment of research needs in a changing climate. Int. J. Biometeorol., 58(8): 1779–1788.

Gugliermetti, F.; Bisegna, F.; Violante, A. C.; Cristina Aureli, C. 2010. Noise exposure of the ramp's operators in airport apron. Proceedings of 20th International Congress on Acoustics, ICA. Available at: <www.acoustics.asn.au>. Accessed: out/2016.

Lucas, R. A.; Epstein, Y.; Kjellstrom, T. 2014. Excessive occupational heat exposure: a significant ergonomic challenge and health risk for current and future workers. Extrem Physiol Med. 3(1): 14.

Marques, F. P.; Costa, E. A. 2006. Exposure to occupational noise: otoacoustic emissions test alterations. Revista Brasileira de Otorrinolaringologia, 72(3): 362–366.

Opitz-Stapleton, S.; Sabbag, L. Kate Hawleya, Tran, P.; Lan Hoang, L.; Nguyen, P. H. 2016. Heat index trends and climate change implications for occupational heat exposure in Da Nang, Vietnam. Climate Services, 2(1): 41–51.

Yoopat, P.; Toicharoen, P.; Glinsukon, T.; Vanwonterghem, K.; Louhevaara, V. 2002. Ergonomics in practice: physical workload and heat stress in Thailand. International Journal of Occupational Safety and Ergonomics, 8(1): 83–93.

Zimmermann, N.; Oliveira, A. V. M. 2012. Economic liberalization and universal access in air transport: it is possible to reconcile free markets with social goals and still avoid infrastructure bottlenecks (*published in Portuguese only*). Journal of Transport Literature, 6(4): 82–100.

Occupational Safety and Hygiene V – Arezes et al. (Eds)
© 2017 Taylor & Francis Group, London, ISBN 978-1-138-05761-6

The health and welfare of working judges in Brazil: A quantitative and qualitative analysis

F.M. Junior
Doctoral Program in Occupational Safety and Health, University of Porto, Porto, Portugal

C. Pereira & M. Santos
Faculty of Psychology and Education Sciences of the University of Porto, Psychology Center at the University of Porto, Porto, Portugal

ABSTRACT: The purpose of this paper is to conduct an analysis of the legal environment, health and welfare of working judges in Brazil. The study employs three methods of inquiry: Effort-Reward Imbalance Questionnaire (ERI-Q), Patient Health Questionnaire 9 (PHQ-9) and the Nottingham Health Profile (NHP). It also relied on semi-structured interviews as a technique for acquiring a fuller understanding of the results obtained. These showed that unstable working conditions had a serious effect on the levels of health, welfare and inter-personal relationships of the judges. The study found that it is the restructuring of the working practices of the legal profession, by fostering a democratic and supportive system of management for the judges themselves, that can be regarded as the most important means of establishing a framework based on a binomial model of a judicial council that is just and can provide the magistracy with a good quality of life.

1 INTRODUCTION

A good deal of research has been concerned with the health disorders of workers, particularly those linked to psycho-social factors, and these studies have sought to understand their signs and effects with a view to adopting better preventive and interventionist measures and thus ensure a healthy working environment (Meurs & Perrewé, 2011). In France, there has been a systematic attempt to divide psycho-social risk factors at work into six categories. These range from questions linked to the time of work to the poor standards of social relations, and encompass issues such as emotional demands, a lack of or limited degree of autonomy, conflicting values and job insecurity (Gollac & Bodier, 2011).

In the case of the Brazilian Judiciary, the alteration of the working conditions of the judges and exacerbation of psycho-social risks, together with the deterioration of standards with regard to social relations at work and restrictions on autonomy, have taken on a new importance with the publishing of Resolution No. 70 of the National Council of Justice (18 March 2009) (Brazil 2009). By incorporating good business practices in the Brazilian Judicial System (Alves, 2014), this measure established them as pillars of strategic management and planning and operational efficiency.

The parameters of the business logic of the Brazilian Judiciary can be confirmed by the growing use of new technology in the Courts. This is especially the case with the implementation of the Electronic Judicial Process (EJP), which has triggered the physical-to-virtual conversion of court cases and also allowed a broadening of the Magistrate's Office, which is spreading out indefinitely to the whole area where there is access to the Internet.

This situation combines the mechanisms of an institutional character (with regard to the world of the judges) that addresses the problem of job insecurity, with a system that links the earnings of judges to their productivity. In other words, judges will be paid bonuses for the buildup of judicial cases and for undertaking activities that should be the responsibility of other judges, as stipulated in Resolution No. 155 of the Supreme Council of Labour Law, of 23 October 2015 (Brazil, 2015).

Thus, the purpose of this study is to analyze the possible repercussions of the judges' work on their health, while taking into account activities they undertake and the way their responsibilities are organized, as well as the mechanisms that are recognized as being necessary to create a balanced work environment.

2 METHODS

2.1 *Characteristics of the sample*

The sample used for the quantitative data analysis included 263 working judges in every region in Brazil, which represented 7.73% of the total number of 3,400 working judges in the country (Brazil, 2015b).

It was found in the sample that there was a relatively balanced gender distribution (49.6% females, 50.4% males), a large proportion in the 45–54 year old age group (39.5%), while for the 35–39 years age group, it was 17.5% and for the 40–44 years age group, it was 16%. Most of them were serving in the first stage of their judicial career as Assistant Judges (42.6%). The first stage of the judge enabled him/her to act as a circuit judge with the right to exercise judicial powers in different lower federal courts; or as a Chief Judge of a Lower Federal Court (48.3%) – the second stage in the career of the judge exercising judicial powers as a permanent judge in a lower federal court. Most of the working judges had had judicial authority for over 10 years (58.6%).

In the qualitative analysis, the sample consisted of 13 judges (4 assistant judges, 5 chief judges, 3 High Court judges—in the third probationary period of the magistracy who exercised judicial powers in the Regional Labour Courts—and 1 Minister of Justice who was in the fourth stage of the career of the magistracy and exercised judicial power in the Upper Labour Court.

2.2 *Instruments, procedures and data processing*

Three investigative methods of inquiry were employed: the Effort-Reward Imbalance Questionnaire (ERI-Q), the Patient Health Questionnaire 9 (PHQ-9) and the Nottingham Health Profile (NHP).

The ERI-Q, which had a version of 23 items, used the categories of effort, reward and overcommitment to supply features that show the occurrence of psychological disorders resulting from the mismatch between the effort of the workers and the rewards emanating from the employer (Siegrist, 2008).

The PHQ-9 consists of an instrument used for the diagnosis of common mental disturbances (Spitzer et al., 1999) based on 9 items.

The NHP consists of an instrument for assessing the quality of life, which is formed of 38 items (Hunt, 1981).

The ERI-Q, PHQ-9 and NHP inquiries take account of the transcultural adaptation into Brazilian Portuguese (Chor et al., 2008; Osório, 2009; Teixeira et al., 2004, respectively).

The inquiries took place in an anonymous and voluntary way in the period 28 October-27 November 2015, with statistical data processing by the SPSS (Statistical Package for the Social Sciences), version 22.0.

After the analysis of the data supplied by the ERI-Q, PHQ-9 and NHP inquiries, the semi-structured interview technique was employed with flexible parameters (Minayo, 2014).

The interviews, which on average, lasted for 60 min, took place in a voluntary and anonymous manner in the period February-July 2016.

The excerpts from the text were sorted out into thematic categories and subcategories and subsequently submitted to independent judges. By proceeding with the validation of the excerpts attributed to each category, they helped in producing an evolving pattern of the analytical parameters, which emerged from the data.

From this perspective, the thematic categories were arranged as follows: work of the magistracy; health of the magistracy; and magistracy and the family (Figure 1).

The thematic category "the work of the magistracy" is combined with the thematic subcategories that fall within the scope of the working activities undertaken by the judges and can be divided into the following thematic subcategories: organisation of the work by the Judiciary; complexity of judicial activities; degree of satisfaction of the magistracy with their profession; and defensive strategies adopted by the magistracy when involved in their work.

The thematic category "health of the magistracy" includes the thematic subcategories that refer to the effects of work on the welfare and

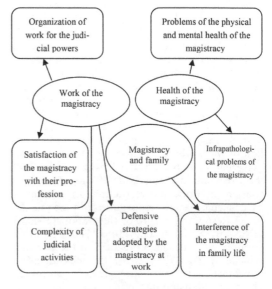

Figure 1. Thematic map with 3 categories and 7 subcategories.

health of the magistracy and are thus divided into the following: infrapathological problems of the magistracy and problems of physical and mental health.

The thematic category "the magistracy and the family" is only combined with the thematic sub-category concerned with the interference of the work of the magistracy with family life.

In the final stage of this methodological path (Braun & Clarke, 2006), a session was held with 82 working judges where there was an opportunity to provide feedback on the data (both quantitative and qualitative), validate the main results obtained and set out the measures that could assist in preventing the particular health problems that are caused or aggravated by the exercise of professional duties.

3 RESULTS

The quantitative and qualitative results are set out in a holistic way as a means of obtaining a better understanding of the complexity of judicial work, insofar as the integration of the results broadens the range of interactions and meaning of the available information.

3.1 Internal reliability of data—Cronbach's alpha

The data from the ERI-Q inquiry (Cronbach's alpha in the "effort": 0.777; Cronbach's alpha in the "reward" dimension: 0.794; and Cronbach's alpha in the "overcommitment" dimension: 0.890), PHQ-9 (Cronbach's alpha: 0.887) and PSN (Cronbach's alpha: 0.921) show a satisfactory internal degree of consistency.

3.2 ERI-Q

The parameters of the ERI-Q scale (Siegrist 2008) show that 36.7% of the working judges carry out activities with a greater psychic effort, 49.8% were remunerated with an incommensurate reward and the work of 44.9% involves overcommitment.

It was confirmed in the sample that there is a positive correlation between "psychic effort" and overcommitment (coefficient of person 0.664; $p < 0.01$).

In some items from the effort and overcommitment dimensions, the results were highly significant. In the effort dimension, the item "I constantly feel under time pressure because of my heavy workload", 25.6% chose the answer "I agree" and 67.2% the answer "I completely agree", or in other words, 92.8% of the judges admit that they feel under pressure from the responsibilities of their work. In the overcommitment dimension, the item

"When I wake up in the morning, I immediately begin to think about the problems of my work", 35.6% chose the answer "I agree" and 41.4% the answer "I completely agree", or in other words, 77% of the judges show that they are intensely preoccupied with their work.

The interpretation of the effort and overcommitment dimensions can be better understood when confirmation is given by the interviews of the complexities of their activities. This is especially with regard to the burden of decision-making, the huge number of lawsuits and the difficulties of reaching a verdict in legal proceedings and is illustrated by the following observations: "there is not enough time to settle all the lawsuits and with the increasing amount of work, I find that all the time I am sleeping fewer hours at night" (Chief Judge 5); "every Supreme Court Judge handles about 15,000 or 16,000 lawsuits a year. I write the proceedings of the Court, the plea bargaining sessions and the Full Court sessions. It's an extremely technical service" (Supreme Court Judge).

The reward dimension covers all the benefits recognized by the employer, which include the salary, respect for superiors and support when carrying out activities (Siegrist, 2008). With regard to the judges, the results suggest that the rewards are incommensurate on account of the weakness of the job structure and the lack of support it provides. This can lead to situations such as: "the pressure of the Justice Department comes from the sheer number of lawsuits. It doesn't make the least difference what the quality of the service is like and this causes a good deal of conflict between the judges" (Assistant Judge 4); "if some sentence has been delayed, the system has to immediately inform the Justice Department, to the extent that people feel they are under surveillance the whole time. This affects the standard of the service and the calm atmosphere needed to make a decision" (High Court Judge 2).

The imbalance in the effort-reward model suggests that there is a paradox between the rise in the amount of effort that has to be expended on the work and the disproportionate reward that is received. This tends to have a negative reaction on the thoughts of the worker (Siegrist, 2008), and reveals the failure of the job structure and a corrosion of interpersonal relations (Dejours, 2001). It can be best understood in the following testimony: "we have a system that gives primacy to numbers and to achieving goals. In a climate where heavy stress is laid on numbers of the sentencing, I have noticed that this area of work has swollen out of all proportions, to the detriment of the quality. This is terrible. Our system of working has even affected our relations with colleagues who are now beginning to regard themselves as competitors or adversaries." (Chief Judge 1).

The data also show that the closer ties between the dimensions of effort and overcommitment have also brought about negative feelings among the workers. This can be confirmed by the following comments: "my personal responsibilities are very heavy and related to the feeling that they must be fulfilled. I take on a great deal and in the last 5 years I have been constantly feeling ill. I had a bout of depression and a short time ago I discovered I had psoriasis"; "I am the kind of judge who likes to end the day with the task completed, but this means that I am practically incapable of doing anything else for the rest of the day" (High Court Judge).

It was found that when the Effort-Reward Imbalance Model was applied to the sample, 87.55% of them in effect suffered from effort-reward imbalance.

3.3 *PHQ-9*

By means of a judicial notice with a score of 10, based on theoretical parameters (Wittkampf et al., 2009), it was confirmed that 36.05% of the sample suffered from major depressive disorder, with some variation with regard to gender (males – 32.30% and females – 39.84%). However, according to the student's t-test, for two independent samples, the p-value is higher than 0.05 (p value = 0.179), which is evidence that the differences between the sexes is not statistically significant.

The data gathered in the interviews suggest the existence of diagnosed mental diseases ("I had psychiatric treatment for a time, after a period in which I was involved in a lengthy procedural hearing and reached the point of being diagnosed with burnout syndrome" (Assistant Judge 1)). The interviews reveal that other diseases were diagnosed that are not necessarily within the area of mental health ("I had an attack of chronic gastritis as a result of my work" (High Court Judge 3), but these are regarded by the judges as being caused by exercising their professional activities.

3.4 *NHP*

In the 38 items of the NHP, attention should be drawn to the following results: 33.1% stated that they felt tired as a result of their work; 37.3% stated that they felt downcast and depressed as a result of their work; 25.1% stated that they had ceased to take part in leisure activities as a result of their work; 39.5% stated that they lost their temper very easily on account of their work; 30.8% declared that worrying about their work kept them awake during the night; 37.6% declared that they slept badly because of their work.

The statistical data of NHP provides important information in terms of percentages about tension,

irritation and tiredness caused by work and also by the difficulty of breaking free from work. This can be better understood from the following comments of the judges: "when I get back home, I'm a mental wreck (…) I'm quite unable to do anything. My mind is vacant and this makes me annoyed. That's why I have to take some medicine to calm me down" (Chief Judge 3); "There were days when I had to take tranquillizers in the morning, afternoon and evening so that I could face my work" (Assistant Judge 4); "when I take the oath at night and go on to daybreak—this happened this week—I can't sleep any more, I feel trapped" (High Court Judge 1).

This means that apart from the problems of health that can be diagnosed medically, the judges describe pain, tiredness, and lack of energy that is the outcome of their activities, pressure and the complexity of their work.

But while recognising the existence of a commitment to health on the part of the judges, the continuation of judicial activities can only be ensured by the pleasure, which despite everything else, can be derived from carrying out their work. This is reflected in the following comment: "I think I was born to be a judge. I feel particularly satisfied with the magistracy" (Chief Judge 1); "I regard it as a vocation. I enjoy what I do" (Chief Judge 5).

4 DISCUSSION

It can be noted that the environment of judicial work is increasingly being pervaded by business management with the adoption of principles of operational efficiency, personnel management and strategic planning, which has become so widespread that it has exacerbated the problem of competitiveness and individualism (Alves, 2014). This has been recognized by the Brazilian magistracy as "a scheme for taming the magistracy by turning judicial decisions into a kind of merchandise within a productive rationalisation plan, without any involvement or debate with the judges. When we look at the business model that seeks to combine the lowest cost possible with the greatest effort within human capacity, we are working together to create a definitive state or working environment among judges and their staff that is profoundly unhealthy" (High Court Judge 1).

While working on behalf of an objective that seeks to speed up legal proceedings to the maximum extent, the judges are increasingly being pressurised by the working practices adopted by the Brazilian Judiciary, to view the legal proceedings, the hearing, the parties involved, and the sentence, as nothing more than a number (Alves, 2014), which can have a decisive effect on their salaries or even the progress of their careers. In their reaction to this situation, the judges made the following comments: "we are moving away from a model when judges were skilled craftsmen, to becoming a judge whom I don't know how to designate, except that he is someone who has to fulfil goals and hand out sentences in batches (...) The volume of work is increasing all the time and people think their colleagues at work are causing the problem. People look at everyone else as an adversary and lose any spirit of solidarity and they don't face the real problem, which is the way work is being organized" (Assistant Judge 3).

The discord in the judicial labour organisation is reflected in the results obtained from the ERI-Q tool, which indicates an effort-reward imbalance of 87.55% in the sample investigated. This is compatible with the populations who are subjected to a high degree of stress, such as in the area of educational management where the same model shows that the directors of school centers have an imbalance of 86.7% (Schreyer & Krause, 2016).

It should be noted that the deterioration of the judicial labour organisation can be accounted for by two factors, namely: a) the setting of goals by the leading judicial authorities without the participation of the judges; and b) the difficulty of establishing close ties between the judges in the light of their long days, which means they are denied emotional support and have an adverse effect on their family environment. With regard to these factors, the judges state that "the policy of setting goals goes from top to bottom and ends up by not being the best policy (...) one way out would be to give new thought to basing the labour organisation on a democratic system of management" (Assistant Judge 1); "I felt cut off from my family and found myself not taking as much care of my children as I should. My husband resented the fact and told me that I should look after my family; all I wanted to know about was my work" (Assistant Judge 3).

The defensive strategies adopted by the judges to conceal the harmful effects of the judicial organisation have not always been enough to overcome the problems caused by tension, fatigue, irritation, anxiety and even what is strictly speaking ill health.

The judges have said that in attempting to hide the problem of pressure of work, they have adopted, for example, "an arrangement with the administrative authorities whereby having reached a point in my work schedule, they will cease to pass on any more lawsuits because if they continue to do this, I'm going straight to my house to access the Internet and open an Electronic Judicial Process" (Chief Judge 2). Later on, he stated "a part of your anxiety is caused by the fact that you are worried about the sheer volume of your work (...) then symptoms began to appear of a neck pain when I began to work the Electronic Judicial Proceedings" (Chief Judge 2).

In a search for a restructuring of the Judiciary and for an environment that is conducive to healthy legal work, measures have arisen for restoring working conditions appropriate to judges through a collective, supportive and democratic management system. This is confirmed by the following comments: "it's much easier to be committed to a particular goal when I'm involved in it because there is a subjective aspect to formulating this goal or objective. Thus, I think this involvement on the part of the judge is essential both as a commitment and as a defined and feasible aim that reflects a real situation" (Assistant Judge 2); "I think that a democratic management of the goals set out by the judges involved in the work of the district would be highly beneficial because the whole network and the whole collective structure that has been planned and discussed would result in a better solution to the problem." (Chief Judge 4).

5 CONCLUSION

It can be concluded from the results of this investigation that currently judges are experiencing a severe impairment to their physical and mental health. This is aggravated by the emergence of a logic of business management that has begun to shape the way professional activities of the judiciary are carried out, and led to problems such as the exacerbation of competitiveness, individualism and the precarious nature of working conditions.

It should be underlined that during the data-retrieval session (quantitative and qualitative) and validation of the results (check Section 2.2), the judges recognized that the restructuring of the organisation of the judicial working environment is essential if one seeks to design a model that can reconcile the concept of judicial protection with the quality of life of the magistracy.

REFERENCES

Alves, G. 2014. As Condições de Produção da Justiça do Trabalho no Brasil: Uma Análise Crítica do Documentário "O Trabalho do Juiz". Projeto Editorial Praxis, Brasil.

Brasil. 2015. Resolução n. 155 do Conselho Superior da Justiça do Trabalho, de 23 de outubro de 2015. http://juslaboris.tst.jus.br/handle/1939/71289.

Brasil. 2015b. Justiça em números 2015: ano-base 2014. Conselho Nacional de Justiça, Brasília. http://www.cnj.jus.br/programas-e-acoes/pj-justica-em-numeros.

Dejours, C. 2001. Le mal-vivre ensemble. Sciences économiques et sociales en Île-de-France à Versailles,

7, 3-09. http://www.ac-versailles.fr/pedagogi/ses/vieses/hodebas/dejours16-01-01.htm

Gollac, M. & Bodier, M. 2011. Mesurer les facteurs psychosociaux de risque au travail pour les maîtriser (Relatório do Collège d'Expertise sur le Suivi des Risques Psychosociaux au Travail). Retirado do website do Collège d'Expertise sur le Suivi des Risques Psychosociaux au Travail. http://www.college-risquespsychosociauxtravail.fr/rapport-final, fr,8,59.cfm.pdf.

Hunt, S., McKenna, S., McEwen, J., Williams, J., Papp, E. 1981. The Nottingham health profile: Subjective health status and medical consultations. Social Science & Medicine. Part A: Medical Psychology & Medical Sociology 15 (3): 221–9. doi:10.1016/0271-7123(81)90005-5. PMID 6973203.

Meurs, J. & Perrewé, P. 2011. Cognitive activation theory of stress: An integrative theoretical approach to work stress. Journal of Management, 37(4), 1043–1068.

Minayo, L.C. 2014. O Desafio do Conhecimento. Pesquisa Qualitativa em Saúde. Hucitec Editora, 14ª. Edição, São Paulo, 407 p.

Osório F., Mendes A., Crippa J., Loureiro S. 2009. Study of the discriminative validity of the PHQ-9 and PHQ-2 in a sample of Brazilian women in the context of primary health care. Perspectives in Psychiatric Care. 2009;45:216–227.

Schreyer, I. & Krause, M. 2016. Pedagogical staff in children's day care centres in Germany—Links between working conditions, job satisfaction, commitment and work-related stress, Early Years, DOI: 10.1080/09575146.2015.1115390.

Siegrist, J. 2008. Effort–reward imbalance and health in a globalized economy. SJWEH Suppl 2008; (6):163–168.

Spitzer, R., Kroenke, K., Williams, J. 1999. Validation and utility of a self-report version of PRIMEMD—The PHQ primary care study. JAMA. 1999; 282:1737–1744.

Teixeira, L. et al. 2004. Adaptação do Perfil de Saúde de Nottingham: um instrumento simples de avaliação da qualidade de vida. Cad. Saúde Pública [online]. vol.20, n.4, pp.905–914. ISSN 1678–4464. http://dx.doi.org/10.1590/S0102-311X2004000400004.

Wittkampf K., et al. 2009. The accuracy of Patient Health Questionnaire-9 in detecting depression and measuring depression severity in high-risk groups in primary care. Gen Hosp Psychiatry 2009; 31: 451–9.

Occupational Safety and Hygiene V – Arezes et al. (Eds)
© *2017 Taylor & Francis Group, London, ISBN 978-1-138-05761-6*

Risk assessment of chemicals used in a brewery and safety function analysis of a large container

J. Freitas & C. Jacinto
UNIDEMI, Research Unit—Mechanical and Industrial Engineering, Universidade Nova de Lisboa, Caparica, Portugal

C. Martins & A. Correia
Safety and Environment Office, Sociedade Central de Cervejas e Bebidas SA, Vialonga, Portugal

ABSTRACT: This paper reports a study made in a manufacturer of beverages (brewery), aimed at supporting COSHH requirements (control of substances hazardous to health and the environment), under REACH regulation (for registration, evaluation, authorization and restriction of chemicals). The work was divided into two phases, the first of which consisted of a wide-range risk analysis to evaluate 291 recipients containing chemical substances. It was concluded that all recipients were within an "acceptable" risk level, but in specific cases improvement recommendations were made. The second phase was designed to strengthen the previous one, by applying the method Safety Function Analysis (SFA) to assess the relevant Safety Functions (SF) of a specific storage process. The latter covered both environmental and occupational safety issues. The results showed that the risk level of this recipient almost exceeded the "acceptability" limit. In this case 22 SF were identified and evaluated. A few shortfalls were found and, consequently, a number of recommendations were proposed to improve performance of the applicable SF.

1 INTRODUCTION

European Legislation for the control of hazardous substances, such as the EU-REACH Regulation (concerning the registration, evaluation, authorisation and restriction of chemicals) has been enforced over time, mostly because of accidents that have occurred with dangerous chemicals (Vierendeels et al 2011). This topic is particularly important for industrial companies that require large quantities of chemicals in their manufacturing processes. On the other hand, these substances can have a major impact on the environment if they are not properly controlled and dealt with.

The first stage of the current study addressed a legal requirement on a beverages' company concerning the many chemicals used in their factory. This was designed to answer a legal request made by APA (Agência Portuguesa do Ambiente), which is the governmental agency for the protection of the environment.

The objective was to assess whether this company's activity could contaminate, or not, the soil and groundwater around it. To that purpose, all the recipients containing chemical substances were systematically listed and verified. After that, a risk analyses was carried out focusing on environmental risks for water and soil. An evaluation risk matrix was created from scratch by the authors.

To expand the assessment, in a second phase, a Safety Function Analysis (SFA) was applied to a specific large storage recipient. The recipient chosen for this second assessment was a 10% ethanol solution container (tank), because this equipment had been recently installed in the company.

The above mentioned method was created by Harms-Ringdahl in 2003 to evaluate Safety Functions (SF), but it has been recently updated (Harms-Ringdahl 2013). This author defines SF as: "*a technical or organisational function, a human action, or a combination of these, that can reduce the probability and/or consequences of accidents and other unwanted events in a system*" (Harms-Ringdahl 2013). The purpose of this second analysis was to assess the performance (safety level) of the referred new storage facility.

2 METHODOLOGY

As already mentioned, the study was made in two stages: A wide-ranging analysis of all chemicals, followed by a specific study to determine the safety status of a new storage installation.

2.1 Chemical substances—general analysis

The first extensive analysis followed the steps summarised next.

Step 1 – Chemical Substances Verification—all the chemical substances and waste used, produced or released inside the company were recorded.

Step 2 – Creation of a matrix (analysis table)—a matrix was created for subsequent evaluation of each chemical. In this step, the potential to contaminate was firstly established through each substance Safety Data Sheet (SDS). Chemicals that did not represent a potential hazard to the environment were excluded from the rest of the analysis.

Step 3 – Registration in the matrix of the quantities, storage area and how the substances were used—the quantities of each chemical were registered in this step, as well as their storage places (i.e., indoors or outdoors) and how the substance was used in the premises/processes.

Step 4 – Exclusion of substances which quantities were bellow 1 tonne—all substances with accumulated quantities bellow 1 tonne were excluded from the rest of the analysis. It was considered that a substance with a storage quantity below 1 tonne would have a very low impact on the soil and groundwater around the company.

Step 5 – Exclusion of substances kept indoors, inside the building (factory)—due to the type of floor (impermeable) and the fact that all sewage grids inside the building are connected to the factory Waste Water Treatment Plant (WWTP), all substances usually kept or stored inside the building were excluded from the rest of the analysis.

Step 6 – Establishment of parameters and criteria. Initial Risk Level (IRL)—for risk evaluation purposes two parameters were established: Probability (P) and Severity (S). These parameters are obtained by addition of three characteristics, as detailed below, and which criteria are explained in Table 1 and Table 2, respectively.

Table 1. Probability criteria.

| Score | Probability (Fu + Ft + Wt) (max score = 5+5+3) | | |
	Fu	Ft	Wt*
1	Less than once a month	Less than once a month	Fed directly to the cistern/main tank; distribution by internal circuit
2	Once a month	Once a month	–
3	Twice a month	Twice a month	By forklift truck
4	Once a week	Once a week	–
5	Once a day, daily use	Once a day, daily transport	–

* Only two levels are considered for scoring Wt.

Table 2. Severity criteria.

| Score | Severity (T + Mqs + Ps) (max score = 5+5+5) | | |
	T*	Mqs (ton)	Ps**
1	No "R" sentences	1–2	Solid
2	Irritating	3–6	–
3	–	7–12	–
4	Corrosive and/or Harmful	13–20	–
5	Toxic/dangerous to the environment	>21	Liquid

* Four levels for scoring T, ** Two levels for scoring Ps.

$$P = \text{Frequency of use (Fu)} + \text{Frequency of transportation (Ft)} + \text{Way of transportation (Wt)} \quad (1)$$

$$S = \text{Toxicity (T)} + \text{Maximum quantity stored (Mqs)} + \text{Physical state (Ps)} \quad (2)$$

Applicable definitions are (c. f. Table 1):
Frequency of use (Fu): how many times the substance is used.
Frequency of transportation (Ft): how many times the substance is transported in company premises.
Way of transportation (Wt): represents the mode of transportation normally used for each substance.

Definitions for Table 2 are:
Toxicity: toxicity level, from each substance's SDS.
Maximum quantity stored: quantity (t) of substance that is stored inside the factory.
Physical state: The physical state of the substance.

Still on step 6, the Initial Risk level (IRL) is estimated by multiplying Probability and Severity, as shown below.

$$\text{Initial Risk level} = \text{Probability} \times \text{Severity} \quad (3)$$

It is noteworthy saying that, when estimating the initial risk level, any control measures already implemented by the company are not taken into account.

The two figures presented next (Figures 1–2) show all the initial risk levels.

For each range of values (risk levels), the action(s) to take were specified so that the company would know what to do.

Step 7 – Control Measures identification and recording—all control measures already implemented by the company were systematically identified and recorded, so that their impact on risk control could be taken into account. The control measures recorded were classified within one of

Risk level	Acceptability decision (action)
12 - 35 ⇨	Acceptable - the situation can stay as it is
36 -77 ⇨	Significant - the applicable counter-measures have to be planed as fast as possible and executed in less than a year
78 - 121 ⇨	Substantial - the applicable counter-measures have to be planed as fast as possible; have to be executed in less than 3 months; the possibility of stopping the activity should be considered
> 121 ⇨	Unacceptable - the activity has to be stopped until the implementation of appropriate counter measures

Figure 1.   Risk level and acceptability decision.

		Probability									
Scores		4	5	6	7	8	9	10	11	12	13
	3	12	15	18	21	24	27	30	33	36	39
	4	16	20	24	28	32	36	40	44	48	52
	5	20	25	30	35	40	45	50	55	60	65
	6	24	30	36	42	48	54	60	66	72	78
	7	28	35	42	49	56	63	70	77	84	91
Severity	8	32	40	48	56	64	72	80	88	96	104
	9	36	45	54	63	72	81	90	99	108	117
	10	40	50	60	70	80	90	100	110	120	130
	11	44	55	66	77	88	99	110	121	132	143
	12	48	60	72	84	96	108	120	132	144	156
	13	52	65	78	91	104	117	130	143	156	169
	14	56	70	84	98	112	126	140	154	168	182
	15	60	75	90	105	120	135	150	165	180	195

Figure 2.   Risk levels (depicting 4 levels of risk).

three classes: Technical (T), Organisational (O) and Emergency & Rescue (E&R).

Step 8 – Impact /effect of the control measures evaluation—for each class of control measure(s), a reduction strategy was established, based on the number of controls already in place, and their expected impact in risk reduction (Table 3).

According to Table 3, if all relevant control measures are in place and well-maintained, the company estimates that maximum possible risk reduction will be around 50% for the Technical, 20% for the Organizational, and 10% for the Emergency & Rescue classes. This means a "residual risk" of no less than 20%, since it was considered that null risk was unattainable and unrealistic.

Step 9 – Residual Risk Level (RRL) calculation— after estimating the initial risk level and deciding the impact coefficients of the control measures, the Residual Risk Level was also estimated. The residual risk level was obtained by the equation below.

Residual Risk Level = IRR × [1-(T+O+E&R)]   (4)

Step 10 – Comparing with acceptability criteria—this step involved a comparison between

Table 3.   Control measures—expected risk reduction.

	Existing control measures	Impact coefficient
Technical (T)	**(a)** The substance has a specific storage place, designed to the purpose and well-maintained (reduction of 5% on the score of initial risk level)	5%
	(b) Additionally, there is a well-maintained spill containment system (spill basin), or a WWTP connection or an automatic monitoring (sensor) of the liquid level (total reduction of 50% is now expected)	50%
Organizational (O)	**(c)** A worker has been assigned to check regularly the condition of the substance	Each measure reduces risk on 5%
	(d) Workers receive specific training to deal with dangerous substance	
	(e) Workers know where to find (and interpret) the applicable SDS	
	(f) Safety signs are displayed and kept in good condition *(cumulative effects are considered)*	
Emergency (E&R)	**(g)** Containment kits are in place (e.g.: sand bags, buoys, etc.)	Each measure reduces risk on 5%
	(h) Workers are trained on how to act in case of a leak or spill *(cumulative effects are considered)*	

the results, i.e., the estimated RRL (residual risk) and the acceptability criteria adopted (Table 3).

Step 11 – Identifying improvement opportunities— once risk assessment was concluded, the chemicals representing a significant risk (soil and groundwater contamination) were pinpointed, for which the analysts made improvement recommendations.

2.2 Safety Function Analysis

The last version of Safety Function Analysis (SFA) (Harms-Ringdahl 2013) was applied in this case for assessing the relevant Safety Functions (SF) of a particularly large recipient (tank) containing a 10% ethanol solution in water. The tank has a capacity of 160 ton, almost 19 m height, and the alcohol solution is used in the cooling systems of production equipment. There is a small control room, equipped with sensors and gauges, for monitoring the whole system (closed circuit).

Within the previous analysis, the residual risk estimated for this tank was still "acceptable" but near the "significant" level. On the other hand, this is recently installed equipment and it was considered opportune to check its safety level.

This second analysis covered both occupational and environmental aspects. Only a brief summary of the method is given here because its detailed description, as well as relevant information on safety barriers and safety functions, can be found elsewhere (e.g.: Harms-Ringdahl 2013, Beatriz 2013, Sklet 2006, Hollnagel 2004).

The first step of SFA consists of *data collection*, as depicted in Figure 3. In the current study, data collection was based on the results from the previous risk analysis, covering all chemical substances existing in the company. The technical features of the whole system, including the tank itself, were also collected for study (e.g.: manufacturer's manual, operating instructions, maintenance procedures, sensors and actuators, connections to other systems, etc.).

The second step is to *identify SF*. All the Safety Functions (SF) that are part of the tank were studied.

The third step consists on the *registration and classification of each SF*. Once again, to promote continuity between this second study and the first one, the SF were classified in the same way as the

control measures: *Technical* (T), *Organizational* (O) and *Emergency & Rescue* (E&S).

The fourth step of this method is the *evaluation of the SF*. This evaluation is made through five parameters that will be ranked. The first is Intention. This is used to evaluate whether a certain function is, or not, primarily intended for safety purposes. Intention is only an informative parameter, i.e., it does not enter the final estimation and decision. However it is useful to promote a thorough identification of all SF.

The second parameter is Importance. The meaning of this parameter is to evaluate the influence of an SF if an accident occurs; it goes from "very low influence" to a "big influence". The third one is Efficiency. This parameter is estimated by the probability of an SF working properly. It goes from "very low" (<50%) to "very high" (99.99%).

The next parameter is Monitoring. This parameter has two kinds of classification: the "need" for monitoring and the "actual" level of monitoring. The ranking code is given to the SF by a comparison between these two sub-parameters. Finally, the last parameter is Acceptability. This one considers a combination of the three previous parameters. Using a "decision tree" which shows all possible combinations, the analyst is able to derive a conclusion and decide on whether each SF needs, or not, to be improved.

All these parameters, their respective ranking, and assessment criteria have explicit tables provided by the method (Harms-Ringdahl 2013, Chapter 11).

The fifth and final step consists on proposing *recommendations for improvement* based on the results of the SFA analysis.

3 RESULTS AND DISCUSSION

3.1 *General risk analysis*

During the first phase, 291 recipients containing chemical substances were recorded. At this early stage 10 substances were excluded from the evaluation because they did not represent a significant hazard to the soil and groundwater in the vicinities. Then, all substances for which their recipient capacity was bellow one tonne were excluded too. The last set of exclusions comprised all substances that were stored inside de factory building (indoors). After this "filtering" process, the remaining 42 substances were meticulously evaluated.

For all 42 cases, Probability and Severity were scored to estimate *initial risk level* (Tables 1–2 and Figures 1–2). This resulted in 12 substances classified in "Unacceptable" level, 27 in "Substantial" level and 3 in the "Significant" level.

```
┌─────────────────────────────────────┐
│        1. Data collection            │
└─────────────────────────────────────┘
                 ⇓
┌─────────────────────────────────────┐
│ 2. Identification of Safety Functions (SF) │
└─────────────────────────────────────┘
                 ⇓
┌─────────────────────────────────────┐
│ 3. Structuring and Classification of SF │
└─────────────────────────────────────┘
                 ⇓
┌─────────────────────────────────────┐
│ 4. Evaluation of individual SF (paramet.) │
│ a- Intention                         │
│ b- Importance                        │
│ c- Efficiency                        │
│ d- Monitoring                        │
│ e- Acceptability                     │
└─────────────────────────────────────┘
                 ⇓
┌─────────────────────────────────────┐
│        5. Decide Improvements        │
└─────────────────────────────────────┘
```

Figure 3. Flowchart of SFA process (Harms-Ringdahl 2013).

However, after the control measures were also assessed (e.g.: Table 4), all the recipients showed to be in the "Acceptable" level of risk. Nonetheless, some improvement opportunities were found.

Recommendations from general analysis

Certain situations have drawn the authors' attention, and recommendations were made. Of particular interest were some tanks kept outdoors, near the cellars, whose labels were damaged and needing replacement (e.g.: caustic soda tank). These tanks have retention basins, but there is no connection with the industrial sewage, which could become a problem should their containment limit be exceeded. Connection to industrial sewage would decrease risk of spillage and contamination.

Another area of concern was the calcium chloride tank (outdoors), in the brewing area, as it is located very close to pluvial sewers. Despite the low probability, if the chemical leaks, it will drain out to the pluvial sewers, contaminating waterways. Once again, connection to the industrial swage system is recommended.

The barrels line is the weakest spot and the less protected area of the factory. The 1 tonne recipients should be stored inside the factory to have connection with the Wastewater Treatment Plant (WWTP) in the event of a spill. This situation needs revaluation on a medium-term basis.

Table 4. Example of risk evaluation (scoring results).

Substance	Initial Risk Level	Control Measures in place			Residual Risk Level
		T	O	E&S	
Ethanol 10% solution	108	50%	10%	10%	**32.4** (70% reduction in total)

Table 5. Example of an SFA evaluation process.

SF		a (intention)	Process Evaluation			
Type	Function		b	c	d	e
T (tech)	Pressure controllers—analogic and digital pressure gauge(s)	2–3 intended for safety	3 very important	4 efficiency is high	2 good monitoring	1 OK, but can improve

Finally, the last element analysed was the 10% ethanol solution storage tank. The main causes of concern seem to be its location and lack of connection to the industrial sewerage. Relocating the tank would be costly and difficult due to lack of space. This is a new project that may need adjustments and, therefore, it was selected for further analysis.

3.2 *Safety function analysis results*

This section presents the main findings from the SFA application to the 10% ethanol tank. Within this storage system (tank and associated controls), 22 SF were identified, recorded and evaluated. Table 5, below, illustrates how the SF "Pressure Control", performed by two controllers, was coded (or ranked) using the pre-defined criteria.

As the table shows, the "Pressure Control" gauges are performing well, but improvement can still be considered, such as a tight plan of calibration and maintenance. Of the 22 SF evaluated, 8 need improvement and 2 were found "not existing".

Recommendations from SFA analysis

The first SF assessed was the anti-fall protection rail. This function is very important as it prevents the fall of workers from the top of the tank (where the rails are located). Despite being well ranked, a life wire could be added to this SF to reduce risk further.

The anti-collision barrier around the tank is considered too low. If a truck bumps into it, it could overtake this barrier and hit the ethanol tank. So, increasing the barriers height is a practical recommendation.

Traffic rules for vehicles are an SF that also performs reasonably well, but it needs continual improvement. Near the tank, an audible alarm can be added to the existing speed limit warning light, to deter drivers from speeding up.

4 CONCLUSIONS

The first part of this work was designed to answer a legal duty. To carry out this risk analysis a new method was developed, which gave a significant contribution to good practice and new knowledge that will remain in the company. The methodology applied allowed to derive important and useful conclusions, namely for improving risk control measures. It also allowed establishing that, as it is now, there is a low level of environmental risk for soil and waterways in the vicinity of the company scrutinized.

The second part was focused on the analysis of safety functions in a 10% ethanol solution tank.

This more detailed study allowed confirming that this (new) tank holds a good safety level. Even so, the SFA method applied was helpful to pinpoint a few shortfalls, for which specific improvement measures were recommended.

ACKNOWLEDGMENT

The authors are grateful to the company workers who collaborated in this assessment. Study made under UNIDEMI (ref. PEst-OE/EME/UI0667/2014).

REFERENCES

Beatriz, R., Jacinto, C., Harms-Ringdahl, L. 2013. Safety Function Analysis in a Manufacturing Process of Paper Products. In: Arezes et al (Eds), *Occupational Safety and Hygiene*, OSH 2013, Balkema, Taylor & Francis Group, London, ISBN 978-1-138-00047-6: 561–566.

EU-REACH. 2006. Regulation concerning the registration, evaluation, authorisation and restriction of chemicals. Regulation (EC) No 1907/2006 of the European Parliament and of the Council.

Harms-Ringdahl, L. 2013. *Guide to safety analysis for accident prevention*. IRS Riskhantering, Stockholm, Sweden.

Hollnagel, E. 2004. *Barriers and Accident Prevention*. Ashgate Publishing Limited, England.

Sklet S. 2006. Safety barriers; definition, classification, and performance. *Journal of loss prevention in the process industries* 19: 494–506.

Vierendeels G., Reniers G.L.L. and Ale B.J.M. 2011. Modeling the major accident prevention legislation change process within Europe. *Safety Science* 49: 513–521.

Occupational Safety and Hygiene V – Arezes et al. (Eds)
© *2017 Taylor & Francis Group, London, ISBN 978-1-138-05761-6*

Levels of urinary 1-hydroxypyrene in firemen from the Northeast of Portugal

M. Oliveira
REQUIMTE-LAQV, Instituto Superior de Engenharia, Instituto Politécnico do Porto, Porto, Portugal

K. Slezakova & M.C. Pereira
LEPABE, Department de Engenharia Química, Faculdade de Engenharia, Universidade do Porto, Porto, Portugal

A. Fernandes & M.J. Alves
Escola Superior de Saúde, Instituto Politécnico de Bragança, Bragança, Portugal

C. Delerue-Matos & S. Morais
REQUIMTE-LAQV, Instituto Superior de Engenharia, Instituto Politécnico do Porto, Porto, Portugal

ABSTRACT: Polycyclic Aromatic Hydrocarbons (PAHs) are ubiquitous environmental pollutants that are mostly formed during incomplete combustion of organic materials (e.g. coal, oil, wood, and tobacco). Many PAHs and their epoxides are highly toxic, mutagenic and carcinogenic to humans. Thus, the present study assesses the total exposure to PAHs by the determination of 1-hydroxypyrene (1OHPy) in the urine of (non-smoking and smoking) firemen serving at four Portuguese fire corporations from the Northeast of Portugal. The global median concentrations of 1OHPy ranged between 8.0×10^{-3} to 6.3×10^{-2} μmol/mol creatinine in subjects. Urinary 1OHPy levels were always below the benchmark of 0.5 μmol/mol creatinine proposed by the American Conference of Governmental Industrial Hygienists for occupational exposure to PAHs. Moderate to strong correlations were found between the urinary levels of 1OHPy and the number of cigarettes consumed per individual for some fire stations, seeming to reflect the impact of tobacco consumption in firemen' total exposure to PAHs.

1   INTRODUCTION

Every year Portugal is among the most affected European countries by forest fires (JRC, 2015). In 2014, a total of 7067 fires occurred in national lands and forests, the northern and central regions of the country being the most affected areas (JRC, 2015). Surprisingly scarce studies are available in Europe concerning the occupational exposure of firemen (Barboni et al., 2010; Laitinen et al., 2010, 2012; Leyenda et al., 2010; Miranda et al., 2010, 2012; Oliveira et al., 2015, 2016a, b). The pollutants that have been under scrutiny by those reports are predominantly particulate matter, carbon monoxide, nitrogen dioxide, and some volatile organic compounds.

Polycyclic Aromatic Hydrocarbons (PAHs) are a large group of ubiquitous organic compounds that are released during the incomplete combustion of organic matter (Kim et al., 2013). PAHs are well known for their cytotoxic, mutagenic and carcinogenic properties (IARC 2002, 2010a). As a consequence, fire emissions have a significant contribution to firemen occupational exposure to PAHs (Baxter et al., 2014; Fent & Evans, 2011; Fent et al., 2013, 2014; Kirk & Logan, 2015; Oliveira et al., 2016a, b; Pleil et al., 2014; Robinson et al., 2008). Regarding PAHs, the majority of the studies available in the literature were conducted in the United States of America and Australia, being scarce the information for European firemen, particularly in the most severely affected South-eastern countries such as Portugal, Spain, Italy, Greece and France. Thus, assessment of firemen' occupational exposure (background values and after participation in fire combats) to PAHs is needed.

PAHs monitoring is complicated by the mixed aerosol/vapor composition of airborne compounds and by the absorption of these chemicals from inhalation, food ingestion and dermal contact. Human biomonitoring is the most appropriate tool to estimate the total internal dose of PAHs (Adetona et al., 2015; Leroyer et al., 2010). Urinary hydroxylated PAHs (OH-PAHs) are suitable biomarkers of exposure to PAHs (Oliveira et al., 2015). Pyrene is one of the most abundant hydrocarbons in all PAH mixtures (Cirillo et al., 2006; Gomes et al., 2013; Slezakova et al., 2010) and about 90% of the total urinary excretion of pyrene is in the form of 1-hydroxypyrene (1OHPy) (Tsai et al., 2003) making this metabolite the most widely used biological indicator of internal dose of exposure to PAHs. Bile, faeces, urine, and milk are the principal elimination routes of PAHs and/or their metabolites in human organisms (Franco et al., 2008; Jongeenelen 2001; Likhachev et al., 1992; Rey-Salgueiro et al., 2009). Among the many different elimination routes of OH-PAHs, urine is the easiest, cheapest and less invasive matrix to determine biomarkers of exposure.

The present work determines the urinary levels of 1OHPy (in non-smoking and smoking) firemen serving at four Portuguese fire corporations from the Northeast region of the country: Miranda do Douro (MDD), Sendim (SDM), Mogadouro (MGD), and Freixo de Espada à Cinta (FEC).

2 MATERIAL AND METHODS

2.1 *Firemen population and urine sampling*

The study population (Table 1) consisted in firemen serving at the fire corporations of MDD, SDM, MGD, and FEC, located in the district of Bragança, Portugal. After giving an informed consent, each fireman filled a structured questionnaire to assess relevant personal information on age, weight, number of years as fireman; time dedicated to firefighting activities in the last 48 hours; exposure to environmental tobacco smoke and tobacco smoking habits and also information on the most frequently consumed meals (boiled, roasted, and grilled) during the week before urine collection.

Table 1. Data characterizing firemen that participated in the study.

Data	MDD	SDM	MGD	FEC
	Fire corporation			
n	18	24	12	24
Age (mean ± SD; years)	35±13	33±5	41±10	35±5
Weight (mean ± SD; kg)	89±22	91±16	83±15	77±10
Years as firefighter				
≤ 10 years (%)	33	50	25	50
> 10 years (%)	67	50	75	50
Smoking habits				
Non-smoking (%)	50	25	50	25
Smoking (%)	50	75	50	75
Number of cigarettes per day (mean ± SD)	9±6	16±5	20±14	22±11

Tobacco consumption constitutes one of the most relevant sources of exposure to PAHs (IARC, 2004), reason why firemen from each fire corporation were divided into non-smoking and smoking subjects (Table 1).

Urine sampling was conducted in the same day for firemen serving at the same fire corporation. Samples were collected from each subject at the end of a regular work shift in sterilized polycarbonate containers. After collection, urine samples were frozen at –20°C until analysis.

2.2 *Chromatographic analysis*

Urinary 1OHPy was analyzed by liquid chromatography with a fluorescence detector according to Oliveira et al. (2016a). Urinary 1OHPy was detected at its optimum excitation/emission wavelengths of 242/388 nm. Concentrations were determined through calibration curves that were prepared with standards of 1OHPy in methanol. Detection (LOD) and quantification (LOQ) limits were, in these conditions, 1.5 ng/L urine and 5.0 ng/L urine. The concentration of 1OHPy was normalized with the urinary creatinine levels (μmol/mol) (Kanagasabapathy & Kumari, 2000).

3 RESULTS AND DISCUSSION

3.1 *Urinary PAH metabolites*

Globally, the mean age of the firemen that participated in this study ranged from 33 years (SDM) to 41 (MGD) years old, with a mean weight varying from 77 kg (FEC) to 91 kg (SDM) (Table 1). All the selected firemen reported not to be involved in any kind of firefighting activities during the week before the urine sampling, thus the attained values correspond to non-fire settings, i.e., firemen occupational exposure at fire stations. All subjects were healthy and reported a diet exclusive of boiled, roasted, and grilled meals within the five days before urine collection, making the contribution of food to the firemen' total exposure to PAHs negligible. Smoking subjects reported a mean consumption of 9 (MDD) to 22 (FEC) cigarettes per day (Table 1). Only firemen that were not exposed to environmental tobacco were selected to the non-smoking group.

Urinary 1OHPy was detected in 89% of the non-smoking subjects and in 82% of the samples from smoking firemen. The median concentrations of 1OHPy in the urine of Portuguese firemen (non-smoking and smoking) are presented in Table 2.

The highest levels of 1OHPy were found, by descending order, MGD > FEC > MDD > SDM for non-smoking firemen. For smoking individuals, a completely different sequence was found:

Table 2. Urinary 1OHPy levels[a] (μmol/mol creatinine; median; min-max) determined in non-smoking and smoking firemen.

	Fire corporation			
	MDD	SDM	MGD	FEC
Non-smoking				
Median	2.2×10^{-2}	1.3×10^{-2}*	6.3×10^{-2}	3.6×10^{-2}
Min	1.8×10^{-2}	1.0×10^{-3}	7.0×10^{-3}	1.9×10^{-2}
Max	8.5×10^{-2}	2.8×10^{-2}	0.13	5.2×10^{-2}
Smoking				
Median	2.1×10^{-2}	3.9×10^{-2}*	8.0×10^{-3}	2.0×10^{-2}
Min	2.0×10^{-4}	3.0×10^{-2}	3.0×10^{-4}	2.0×10^{-4}
Max	0.11	8.7×10^{-2}	1.9×10^{-2}	7.8×10^{-2}

[a] Whenever a concentration was below the LOD, the value of $LOD/\sqrt{2}$ was considered (Hornung & Reed, 1990). *Statistically significant differences ($p \leq 0.05$; non-parametric Mann-Whitney U test) between non-smoking and smoking firemen for each fire station.

SDM > MDD ~ FEC >> MGD, seeming to reflect the impact of active smoking in the levels of the urinary biomarker of PAHs. Overall, the maximum concentrations in the urine of firemen were predominantly higher in smoking than in non-smoking subjects, except for those working at the MGD fire corporation (Table 2). Urinary 1OHPy levels increased up to 75% for firemen serving the SDM fire station ($p = 0.003$).

Spearman correlation coefficients between the urinary concentrations of 1OHPy with the number of cigarettes daily consumed by each fireman were determined. Globally, positive and moderate to strong correlations were found for firemen serving at FEC ($r = 0.258$; $p > 0.05$), SDM ($r = 0.572$; $p = 0.013$), and MGD ($r = 0.949$; $p = 0.004$) fire stations. Subjects serving at the MDD fire corporation presented a weak correlation, which was attributed to the lower number of smoked cigarettes (Table 1).

There are no reference standard guidelines for urinary OH-PAHs, only few recommended limits that were proposed by some authors. In that regard, Jongeneelen (2001) established, for the first time, a no-biological effect limit of 1.4 μmol/mol creatinine (1OHPy) for workers that were actively exposed to PAHs. More recently, the American Conference of Governmental Industrial Hygienists proposed a benchmark level above 0.5 μmol/mol creatinine for the post-shift urinary levels of 1OHPy as indicative of occupational exposure to PAHs (ACGIH, 2010). The urinary concentrations of 1OHPy among non-smoking and smoking Portuguese were well below those recommendations. Still, since firemen occupational exposure is classified as a possible carcinogen to humans (IARC, 2010a; NIOSH, 2007) the monitorization of these relevant biomarkers should be mandatory.

3.2 *Comparison with other studies*

Limited studies concerning firemen' biomonitoring are available in the literature, being scarce those dedicated to PAH metabolites (Adetona et al., 2015; Caux et al., 2002; Edelman et al., 2003; Fernando et al., 2016; Laitinen et al., 2010; Oliveira et al., 2016a, b; Robinson et al., 2008) particularly those that focused on European firemen (Laitinen et al., 2010; Oliveira et al., 2016a, b). Concerning the urinary levels of 1OHPy in non-smoking subjects it was observed that the determined values were in a close range with the concentrations reported by Robinson et al. (2008) but slightly higher than the levels determined by Edelman et al. (2003) in a control group of firemen during the response to the World Trade Center collapse. The urinary concentrations of 1OHPy determined in smoking Portuguese firemen were globally lower than the levels reported in firemen that actively participated in firefighting activities (Caux et al., 2002; Laitinen et al., 2010; Oliveira et al., 2016a).

4 CONCLUSIONS

The findings achieved in this study suggest that occupational exposure of Portuguese firemen to PAHs in non-fire settings is well below the recommended limits. Still, evidences were found related with the significant contribution of regular cigarettes consumption to total firemen' occupational exposure to PAHs. Studies containing a higher number of subjects from other fire corporations and the use of other PAH metabolites, as well as the inclusion of more than one spot urine sample would help to validate these findings.

ACKNOWLEDGMENTS

Authors thank to *Escola Superior de Saúde-Instituto Politécnico de Bragança* and to all firemen involved in this work, for their collaboration. This work received financial support from European (FEDER funds through COMPETE) and National (*Fundação para a Ciência e Tecnologia* project UID/QUI/50006/2013) funds and in the scope of the P2020 Partnership Agreement Project POCI-01-0145-FEDER-006939 (Laboratory for Process Engineering, Environment, Biotechnology and Energy—LEPABE) funded by FEDER funds through COMPETE2020 - *Programa Operacional Competitividade e Internacionalização* (POCI). K. Slezakova is grateful for her fellowship (SFRH/BPD/105100/2014).

REFERENCES

ACGIH, 2010. *Documentation for a Recommended BEI of Polycyclic Aromatic Hydrocarbons.* American Conference of Governmental Industrial Hygienists. Cincinatti, Ohio, USA.

Adetona, O., Simpson, C.D., Li, Z., Sjodin, A., Calafat, A.M., Naeher, L.P., 2015. Hydroxylated polycyclic aromatic hydrocarbons as biomarkers of exposure to wood smoke in wildland firefighters. *J. Expo. Sci. Environ. Epidemiol.,* http://dx.doi.org/10.1038/jes.2015.75.

Barboni, T., Cannac, M., Pasqualini, V., Simeoni, A., Leoni, E. & Chiaramonti, N. 2010. Volatile and semivolatile organic compounds in smoke exposure of firefighters during prescribed burning in the Mediterranean region. *Int. J. Wildland Fire.* 19: 606–612.

Baxter, C.T., Hoffman, J.D., Knipp, M.J., Reponen, T. & Haynes, E.N. 2014. Exposure of firefighters to particulates and polycyclic aromatic hydrocarbons. *J. Occup. Environ. Hyg.* 11: D85-D91.

Caux, C., O'Brien, C. & Viau, C. 2002. Determination of firefighter exposure to polycyclic aromatic hydrocarbons and benzene during fire fighting using measurement of biological indicators. *Applied Occupational and Environmental Hygiene.* 17: 379–386.

Cirillo, T., Montuori, P., Mainardi, P., Russo, I., Triassi, M. & Amodio-Cocchieri, R. 2006. Multipathway polycyclic aromatic hydrocarbon and pyrene exposure among children living in Campania (Italy). *J. Environ. Sci. Heal. A.* 41: 2089–2107.

Edelman, P., Osterloh, J., Pirkle, J., Caudill, S.P., Grainger, J. & Jones, R. et al. 2003. Biomonitoring of chemical exposure among New York City firefighters responding to the World Trade Center fire and collapse. *Environ. Health Perspect.* 111: 1906–1911.

Fent, K.W. & Evans, D.E. 2011. Assessing the risk to firefighters from chemical vapors and gases during vehicle fire suppression. *J. Environ. Monit.* 13: 536–543.

Fent, K.W., Eisenberg, J., Evans, D., Sammons, D., Robertson, S. & Striley, C., et al. 2013. *Evaluation of dermal exposure to polycyclic aromatic hydrocarbons in fire fighters,* Health Hazard Evaluation Report No. 2010-0156-3196. United States Department of Health and Human Services, Centers for Disease Control and Prevention, National Institute for Occupational Safety and Health.

Fent, K.W., Eisenberg, J., Snawder, J., Sammons, D., Pleil, J.D. & Stiegel, M.A., et al. 2014. Systemic exposure to PAHs and benzene in firefighters suppressing controlled structure fires. *Ann. Occup. Hyg.* 58(7): 830–845.

Fernando, S., Shaw, L., Shaw, D., Gallea, M., Vander-Enden, L. & House, R., et al. 2016. Evaluation of firefighter exposure to wood smoke during training exercises at burn houses. *Environ. Sci. Technol.* 50: 1536–1543.

Franco, S.S., Nardocci, A.C. & Günther, W.M.R. 2008. PAH biomarkers for human health risk assessment: a review of the state-of-the-art. *Caderno Saúde Pública* 24: 569–580.

Gomes, F., Oliveira, M., Ramalhosa, M.J., Delerue-Matos, C. & Morais, S. 2013. Polycyclic aromatic hydrocarbons in commercial squids from different geographical origins: Levels and risks for human consumption. *Food Chem. Toxicol.* 59: 46–54.

Hornung, R.W. & Reed, L.D. 1990. Estimation of average concentration in the presence of nondetectable values. *Appl. Occup. Environ. Hyg.* 5: 46–51.

IARC, 2002. Monographs on the Evaluation of the Carcinogenic Risks to Humans. *Naphthalene.* World Health Organization, International Agency for Research on Cancer. 82, Lyon, France.

IARC, 2004. Monographs on the Evaluation of Carcinogenic Risks to Humans. *Tobacco smoke and involuntary smoking.* International Agency for Research on Cancer. 83, 1–1492.

IARC, 2010a. Monographs on the Evaluation of the Carcinogenic Risks to Humans. *Some non-heterocyclic polycyclic aromatic hydrocarbons and some related exposures.* International Agency for Research on Cancer. 92, 1–853.

IARC, 2010b. Monographs on the Evaluation of Carcinogenic Risks to Humans. *Painting, firefighting and shiftwork.* International Agency for Research on Cancer. 98, Lyon, France.

JRC, 2015. *Forest Fires in Europe, Middle East and North Africa 2014.* Joint Research Centre Technical Reports. Joint Report of JRC and Directorate-General Environment, Luxembourg.

Jongeneelen, F.J. 2001. Benchmark guideline for urinary 1-hydroxypyrene as biomarker of occupational exposure to polycyclic aromatic hydrocarbons. *Ann. Occup. Hyg.* 45: 3–13.

Kanagasabapathy, A.S. & Kumari, S. 2000. *Guidelines on standard operating procedures for clinical chemistry.* World Health Organization, Regional Office for South-East Asia, New Delhi: 25–28.

Kim, K.-H., Jahan, S.A., Kabir, E. & Brown, R.J.C. 2013. A review of airborne polycyclic aromatic hydrocarbons (PAHs) and their human health effects. *Environ. Int.* 60: 71–80.

Kirk, K.M. & Logan, M.B. 2015. Firefighting instructors' exposures to polycyclic aromatic hydrocarbons during live fire training scenarios. *J. Occup. Environ. Hyg.* 12: 227–234.

Laitinen, J., Mäkelä, M., Mikkola, J. & Huttu, I. 2010. Fire fighting trainers' exposure to carcinogenic agents in smoke diving simulators. *Toxicol. Lett.* 19: 61–65.

Laitinen, J., Mäkelä, M., Mikkola, J. & Huttu, I. 2012. Firefighters' multiple exposure assessments in practice. *Toxicol. Lett.* 213: 139–133.

Leroyer, A., Jeandel, F., Maitre, A., Howsam, M., Deplanque, D. & Mazzuca, M., et al. 2010. 1-hydroxypyrene and 3-hydroxybenzo[a]pyrene as biomarkers of exposure to PAH in various environmental situations. *Sci. Total Environ.* 408: 1166–1173.

Leyenda, B.C., Rodríguez-Marroyo, J.A., López-Satué, J., Ordás, C.Á., Cubillo, R.P. & Vicente, J.G.V. 2010. Exposición al monóxido de carbono del personal especialista en extinción de incendios forestales. *Rev. Esp. Salud. Pública.* 84: 799–807.

Likhachev, A.J., Beniashvili, D.Sh., Bykov, V.J., Dikun, P.P., Tyndyk & M.L. Savochkina, I.V. et al. 1992. Biomarkers for individual susceptibility to carcinogenic agents: excretion and carcinogenic risk of benzo[a]pyrene metabolites. *Environ. Health Persp.* 98: 211–214.

Miranda, A.I., Martins, V., Cascão, P., Amorim, J.H., Valente, J. & Borrego, C. et al. 2012. Wildland smoke exposure values and exhaled breath indicators in firefighters. *J. Tox. Env. Health-A.* 75: 831–843.

Miranda, A.I., Martins, V., Cascão, P., Amorim, J.H., Valente, J. & Tavares, R. et al. 2010. Monitoring of firefighters exposure to smoke during fire experiments in Portugal. *Environ. Int.* 36: 736–745.

NIOSH, 2007. *NIOSH Pocket Guide to Chemical Hazards. U.S.* Department of Health and Human Services, Public Health Service, Centers for Disease Control and Prevention. National Institute for Occupational Safety and Health, Cincinnati, Ohio.

Oliveira, M., Slezakova, K., Alves, M.J., Fernandes, A., Teixeira, J.P. & Delerue-Matos, C., et al. 2016a. Firefighters' exposure biomonitoring: impact of firefighting activities on levels of urinary monohydroxyl metabolites. *Int. J. Hyg. Environ. Health.* 219: 857–866.

Oliveira, M., Slezakova, K., Alves, M.J., Fernandes, A., Teixeira, J.P. & Delerue-Matos, C., et al. 2016b. Polycyclic aromatic hydrocarbons at fire stations: firefighters' exposure monitoring and biomonitoring, and assessment of the contribution to total internal dose. *J. Hazard. Mater.* http://dx.doi.org/10.1016/j.jhazmat.2016.03.012.

Oliveira, M., Slezakova, K., Delerue-Matos, C., Pereira, M.C. & Morais. S. 2015. Firefighters occupational exposure: Review on air pollutant levels and potential health effects. In Multi. Vol set on *Environmental Science and Engineering (in 12 vols set), Air and Noise Pollution (Vol 3)* Publishers Studium Press LLC, USA, in press.

Pleil, J.D., Stiegel, M.A. & Fent, K.W. 2014. Exploratory breath analysis for assessing toxic dermal exposures of firefighters during suppression of structural burns. *J. Breath. Res.* 8: 037107.

Rey-Salgueiro, L., Martínez-Carballo, E., García-Falcón, M.S., González-Barreiro, C., Simal-Gándara, J. 2009. Occurrence of polycyclic aromatic hydrocarbons and their hydroxylated metabolites in infant foods. *Food Chem.* 115: 814–819.

Robinson, M.S., Anthony, T.R., Littau, S.R., Herckes, P., Nelson, X. & Poplin, G.S., et al. 2008. Occupational PAH exposures during prescribed pile burns. *Ann. Occup. Hyg.* 52(6): 497–508.

Slezakova, K., Castro, D., Pereira, M.C., Morais, S., Delerue-Matos, C. & Alvim-Ferraz, M.C.M. 2010. Influence of traffic emissions on the carcinogenic polycyclic aromatic hydrocarbons in outdoor breathable particles. *J. Air Waste Manag. Assoc.* 60(4): 393–401.

Tsai, II.-T., Wu, M.-T., Hauser, R., Rodrigues, E., Ho, C.-K., Liu, C.-L. et al. 2003. Exposure to environmental tobacco smoke and urinary 1-hydroxypyrene levels in preschool children. *Kaohsiung J. Med. Sci.* 19: 97–104.

WHO, 2000. *Air Quality Guidelines*, second ed. WHO Regional Publications, European Series No. 91, Copenhagen.

Occupational Safety and Hygiene V – Arezes et al. (Eds)
© 2017 Taylor & Francis Group, London, ISBN 978-1-138-05761-6

Improving the thermophysiological behaviour of sports players with human body cooling techniques

A.M. Raimundo, M.D. Ribeiro & D.A. Quintela
ADAI-LAETA, Department of Mechanical Engineering, University of Coimbra, Coimbra, Portugal

A.V.M. Oliveira
Department of Mechanical Engineering, Coimbra Institute of Engineering, Polytechnic Institute of Coimbra, Coimbra, Portugal

ABSTRACT: The main purpose of this study is to analyse the improvement achieved on the thermophysiological behaviour of sport players by the application of four human body cooling techniques, namely: (*i*) immersion of hands and forearms in cold water at 10°C (during a period of 30 minutes before muscle warming of players and in the first 10 minutes of the interval); (*ii*) ingestion of cold fluids at 5°C (300 ml in the beginning of the game and in the interval and 200 ml in each one of the time-break periods); (*iii*) body cooling by exposure to an ambient temperature of 10°C (during a period of 30 minutes before muscle warming of players and in the first 10 minutes of the interval); and (*iv*) simultaneous application of the 3 previous techniques. Globally, the results indicate that the three cooling techniques contribute to lowering the sportsman thermal state, but with distinct efficiencies.

1 INTRODUCTION

In competitive sport the performance of athletes is crucial and body temperature, dehydration and low carbohydrate intake are among the most relevant factors affecting that performance. While dehydration and carbohydrate intake are easy to correct, as far as body temperature is concerned its control is considerably more difficult. Athletes can produce between 3 to 12 times more heat than at rest, which, when combined with high ambient temperatures, can compromise the human thermoregulatory system, leading the body to hyperthermia (Sawka *et al.* 2011). Without the action of the thermoregulatory mechanisms, 5 km of running would be sufficient for the human body temperature to reach dangerous levels. At the beginning of the physical activity the body temperature rapidly increases because heat production exceeds its dissipation. However, with the increase of the hypothalamic temperature, the human thermoregulatory system responds and the regulation of body temperature becomes more effective (Kenefick *et al.* 2007).

The practice of intense physical exercise in hot and humid environments can cause athletes' body temperature higher than 39°C (Brade *et al.* 2013, Adams *et al.* 2016). When this hyperthermic level is attained, the athlete performance substantially decreases and he is at serious risk of developing heat illness (Wendt *et al.* 2007, Wegmann *et al.* 2012). The more common exercise-related illnesses reported in the literature are muscle cramping, heat exhaustion (collapse during or after exercising), introversion (violent sweating, loss of judgment, amnesia, delusions, etc.) and heat stroke (fainting, cessation of sweating, central nervous system alteration, etc.) (Raimundo & Figueiredo 2009). Thus, several authors recommend the use of cooling strategies to mitigate the rise of body temperature, prevent athlete's performance degradation and the developing of heat related illness during exercise in hot conditions (Jones *et al.* 2012, Wegmann *et al.* 2012, Ross *et al.* 2013, Brade *et al.* 2014, Zhang *et al.* 2014, Tyler *et al.* 2015, Bongers *et al.* 2015, Racinais *et al.* 2015, Chan *et al*).

2 METHODS

In this study use was made of the *HuTheReg* software (Raimundo *et al.* 2012) to predict the human thermophysiological response. Its main module is based on the Stolwijk (1971) thermoregulation model, improved with knowledge of the literature (Fiala *et al.* 1999; Tanabe *et al.* 2002). When compared with the original Stolwijk model, the present numerical calculations show important enhancements. In this 89-node model the human body is assumed divided in 22 segments (face, scalp, neck, chest, abdomen, upper back, lower back, pelvis, left shoulder, right shoulder, left arm, right arm, left forearm, right forearm, left hand, right hand, left thigh, right thigh, left leg, right leg, left foot and right foot). Each body segment is composed by 4 layers (core, muscle, fat and skin) and the 89th node is the central blood compartment.

The loss of heat by respiration is supposed to occur across the elements of the pulmonary tract. The repartition coefficients proposed by Fiala et al. (1999) were followed and the respiratory heat loss was imputed 20% to the core of the face, 25% to the muscle layer of the face, 25% to the muscle band of the neck and 30% to the core of the chest (lungs).

The heat and water transport through clothing is accounted using the model described in Havenith et al. (2002) and ISO 9920 (2007). This model is applied to each specific human body segment. Then, individual values of clothing insulation (I_{cl}) and of vapour permeability efficiency (i_{vp}) must be specified for each of the 22 human body segments considered. Each section is either completely clothed or nude. In order to consider the reduction of insulation due to body movements, the I_{cl} value at each human body segment is adjusted according to the person activity using the relations proposed by Oliveira et al. (2011).

The convective heat transfer phenomena are described using empirical relations derived from thermal manikin experiments obtained by Quintela et al. (2004) for natural convection and by Havenith et al. (2002) for forced convection. The effective radiation areas and Gebhart absorption factors, needed for the calculation of the radiation components on each human body segment, are determined by a set of expressions established using a radiation program implemented by Raimundo et al. (2004), which are valid for a person with different postures: standing, sitting, supine and prone.

3 MATERIAL

In the present work two sports were considered: Basketball and Football (eleven). In both cases the clothing ensemble comprises the following items: t-shirt, drawers/boxers, shorts, socks and slippers (basketball) or football boots.

To set the required properties of the clothing and its distribution over the various sections of the human body, ISO 9920 (2007), Raimundo & Figueiredo (2009) and Oliveira et al. (2011) were taken into account. More specifically, vapour permeability efficiencies $i_{vp} = 1.0$ for nude segments, $i_{vp} = 0.34$ for clothed and $i_{vp} = 0.20$ for feet were considered. The values for the emissivity were $\varepsilon = 0.93$ for nude segments and $\varepsilon = 0.90$ for not-nude. In the

case of the individual intrinsic clothing insulation $I_{cl} = 0$ clo was used for nude segments, $I_{cl} = 0.1$ clo for shoulders, thighs and legs, $I_{cl} = 0.2$ clo for chest and back sections, $I_{cl} = 0.3$ clo for pelvic zone and $I_{cl} = 0.6$ clo for feet, which corresponds to a global value of $I_{cl} = 0.146$ clo.

3.1 Cooling techniques

Human beings are homoeothermic, that is, they are capable of regulating their body temperature in order to keep it within the range of "normal thermoregulation" (hypothalamic temperature between 34 and 39°C, Raimundo & Figueiredo 2009). Outside this range their thermoregulation system becomes unbalanced (Taylor 2006), being endangered their physical and mental capabilities (Wendt et al. 2007, Wegmann et al. 2012). During exercise in hot environments, the increase of core body temperature reduces both muscle contractibility and central motor drive stimulus, leading to an overall muscle performance weakening (Cheung 2008, Duffield et al. 2010, Xu et al. 2013). It is also established that the thermal body state in which the physical and mental abilities of a person reach their maximum corresponds to the midpoint of "normal thermoregulation" (Hanna & Tait 2015). Then, it can be expected that this is the thermal state in which an athlete reaches his best performance.

In order to approximate the body thermal state of athletes to its optimum physical and mental performance, four body cooling techniques were tested: (i) IHFCW—immersion of hands and forearms in cold water at 10°C (during a period of 30 minutes before player's muscle warming and in the first 10 minutes of the interval); (ii) ICF5—ingestion of cold fluids at 5°C (300 ml in the beginning of the game and in the interval and 200 ml in each one of the time-break); (iii) ECA10—exposure to a cold ambient at 10°C (during a period of 30 minutes before player's muscle warming and in the first 10 minutes of the interval); and (iv) ALL3—simultaneous application of the 3 previous techniques. To serve as a reference (Control) the software *HuTheReg* was also used to provide a prediction when no cooling technique is applied.

3.2 Sports' protocols

The human muscles are surprisingly ineffective, i.e. they tend to produce more heat than mechanical work. For a person with a 50% training level, during physical activity around 70 to 80% of the energy produced by the muscles is on the form of heat (Wendt et al. 2007, Sawka et al. 2011). With the increase of training level muscle efficiency rises substantially, improving the mechanical work produced in relation to the heat that is generated.

A basketball player with 1.90 m high, 85 kg of mass, 10% of fat and 90% of training level was assumed. Table 1 shows the protocol, which includes the game periods, those before that, the main interval and the time-breaks. The athlete's optimal thermal state was assumed at the beginning of the process.

Basketball is assumed here as a sport traditionally played indoors in sports halls. To establish the characteristics of each phase statistical information of real games available in Pereira (2005) was used. To assess the level of metabolic activity the methods proposed in ISO 8996 (2004) were taken into account. As a result of this process, the following parameters were considered for the game phases: air temperature $T_{air} = 23°C$, mean radiant temperature $T_{mr} = 23°C$, relative humidity $RH = 90\%$, athlete's average velocity $V = 4$ m/s, activity level $M = 8$ met. The muscle warming and time-break phases occurred in the same environment, but with a lower level of velocity and activity, respectively $V = 2$ m/s, $M = 2.4$ met and $V = 1$ m/s, $M = 1.2$ met. The interval and pause phases are characterized by $T_{air} = 23°C$, $T_{mr} = 23°C$, $RH = 70\%$, $V = 0.5$ m/s and $M = 1.2$ met.

It was assumed a football player with 1.75 m high, 75 kg of mass, 10% of fat and 90% of training level. Table 2 shows the protocol for this case, which includes the game periods, the ones before that and the main interval. The athlete's optimal

Table 1. Protocol of the basketball player.

Duration	Control	IAFCW	ICF5	ECA10
30 min	Pause + Neutral	Pause + IAFCW	Pause + Neutral	Pause + ECA10
5 min	Muscle warming	Muscle warming	Muscle warming	Muscle warming
5 min	Pause	Pause	Pause + ICF5	Pause
10 min	Game	Game	Game	Game
2 min	Time break	Time break	Time break + ICF5	Time break
10 min	Game	Game	Game	Game
10 min	Interval	Interval + IAFCW	Interval	Interval + ECA10
5 min	Interval	Interval	Interval + ICF5	Interval
10 min	Game	Game	Game	Game
2 min	Time break	Time break	Time break + ICF5	Time break
10 min	Game	Game	Game	Game

Table 2. Protocol of the football player.

Duration	Control	IAFCW	ICF5	ECA10
30 min	Pause + Neutral	Pause + IAFCW	Pause + Neutral	Pause + ECA10
5 min	Muscle warming	Muscle warming	Muscle warming	Muscle warming
5 min	Pause	Pause	Pause + ICF5	Pause
45 min	Game	Game	Game	Game
10 min	Interval	Interval + IAFCW	Interval	Interval + ECA10
5 min	Interval	Interval	Interval + ICF5	Interval
45 min	Game	Game	Game	Game

thermal state was once again assumed at the beginning of the process.

Football is a sport traditionally played outdoors in stadiums. In this work it was considered that the game takes place during the diurnal part of the day, so it was assumed for the outside phases a solar radiation of 500 W/m² coming from overhead, 100 W/m² from below and 200 W/m² from the horizontal directions. To establish the characteristics of each phase statistical information of real games available in Caixinha et al. (2004) was used. To obtain the level of metabolic activity the methods defined in ISO 8996 (2004) were used. As result of this process, it was considered for the game phases (outdoors) $T_{air} = 23°C$, $T_{mr} = 23°C$, $RH = 70\%$, $V = 3$ m/s and $M = 6$ met. The muscle warming phase occurred in the same environment, but with a moderate level of velocity and activity, $V = 2$ m/s and $M = 2.4$ met. The interval and pause phases (indoors) are characterized by $T_{air} = 23°C$, $T_{mr} = 23°C$, $RH = 70\%$, $V = 0.5$ m/s and $M = 1.2$ met.

4 RESULTS AND DISCUSSION

Although the software *HuTheReg* can produce a wide range of outputs, the values of the heat accumulated in the human body per square meter of skin area (Q_{stored}, Wh/m²), the rectal (T_{rectal}, °C) and the hypothalamus (T_{hyp}, °C) temperatures were selected for this work.

4.1 Basketball

Figure 1 shows the evolution of the heat stored in the body (Q_{stored}) of the basketball player for the

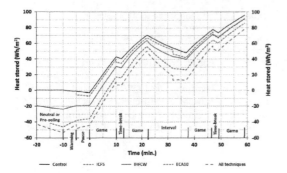

Figure 1. Evolution of heat stored of the basketball player.

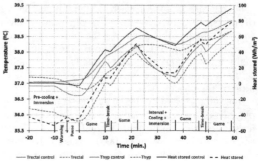

Figure 2. Thermal state of the basketball player.

reference situation (Control) and for the cooling techniques (IHFCW, ICF5, ECA10 and ALL3).

The more efficient individual body cooling technique is the ECA10. When compared to other individual techniques (IHFCW and ICF5), not only it leads to the lowest values of Q_{stored} but also assures lower human body temperatures. This is mainly due to the fact that the athletes start the game with a more reduced body temperature.

The body cooling technique by intake of cold fluid (ICF5) has a reduced efficiency. Although better than the previous, the efficiency of dipping the arms and forearms in cold water (IHFCW) is moderate.

As expected, the simultaneous application of the 3 techniques under study (ALL3) produces the best results. As shown in Figure 2, when compared to the control situation, this strategy can ensure a much more favorable body thermal state, hence enhancing the performance of basketball players.

4.2 Football

Figure 3 shows the evolution of heat stored in the body (Q_{stored}) of the football player for the reference situation (Control) and the four cooling strategies (IHFCW, ICF5, ECA10 and ALL3).

As in the case of basketball, the more effective individual cooling technique is the body cooling by exposure to cold air at 10°C (ECA10). When compared to the control situation and to the other tested techniques (IHFCW and ICF5), this method guarantees a lower level of heat stored in the human body, thus ensuring lower body temperatures.

Once again, the simultaneous application of all techniques (ALL3) promotes a better thermal state of the football player when compared with the separate application of each protective measure. According to Figure 4, when compared to the control situation, this strategy can ensure a much more favorable body thermal state, thus prone to enhance performance.

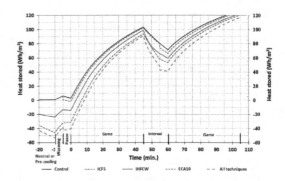

Figure 3. Evolution of heat stored of the football player.

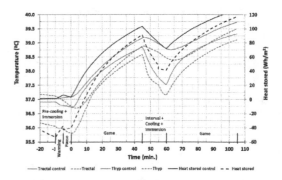

Figure 4. Thermal state of the football player.

5 CONCLUSIONS

In this study a human thermoregulation software was used to evaluate the thermophysiological response of athletes playing basketball and football. Since player's performance decreases with high body temperature, the focus of the present study was to analyze strategies to lower the sportsman thermal state.

Four human body cooling protocols were tested: (i) immersion of hands and forearms in water at 10°C (during a period of 30 minutes before player's muscle warming and in the first 10 minutes of the interval); (*ii*) ingestion of cold fluids at 5°C (300 ml in the beginning of the game and in the interval and 200 ml in each one of the time-break); (*iii*) body cooling by exposure to an ambient at 10°C (during a period of 30 minutes before player's muscle warming and in the first 10 minutes of the interval); and (*iv*) simultaneous application of all techniques.

Globally, the results indicate that the three individual cooling techniques contribute to the lowering of the sportsman thermal state, but with distinct efficiencies. In terms of human body thermal state, the intake of cold drinks is only very slightly effective. The immersion of hands and forearms in water at 10°C is more beneficial, but even so its effectiveness is moderate. The body cooling technique by exposure to an ambient of 10°C (before players warming and during the interval) presents a very good efficiency in terms of control of the athlete body temperature. Between the tested scenarios, this cooling strategy presents the more favorable body thermal state. If possible, the simultaneous application of the three cooling techniques is a good solution for lowering the player body temperature and, theoretically, to improve performance. These findings are in good agreement with other research studies (Zhang *et al.* 2014, Tyler *et al.* 2015, Bongers *et al.* 2015, Racinais *et al.* 2015, Zimmermann & Landers 2015, Onitsuka *et al.* 2015, Chan *et al.* 2016).

The effects on athlete's health of frequently application of body cooling procedures is not yet completely clarified. The only concern reported in the literature is the increase of propensity to muscle injuries, namely if vigorous active cooling methods are applied to lower limbs. Then, to avoid the appearance of undesired health effects, the athlete's body cooling active strategies should only be applied when the environment conditions can lead to hyperthermia (Tyler *et al.* 2015, Racinais *et al.* 2015).

REFERENCES

Adams WM, Hosokawa Y, Adams EL, Belval LN, Huggins RA & Casa DJ (2016). Reduction in body temperature using hand cooling versus passive rest after exercise in the heat. *Journal Science and Medicine in Sport*: 19, 936–940.

Bongers CC, Thijssen DH, Veltmeijer MT, Hopman MT & Eijsvogels TM (2015). Precooling and percooling (cooling during exercise) both improve performance in the heat: a meta-analytical review. *British Journal of Sports Medicine*: 49(6), 377–384.

Brade CJ, Dawson BT & Wallman KE (2013). Effect of pre-cooling on repeat-sprint performance in seasonally acclimatised males during an outdoor simulated team-sport protocol in warm conditions. *Journal of Sports Science and Medicine*: 12, 565–570.

Brade CJ, Dawson BT & Wallman KE (2014). Effects of different precooling techniques on repeat sprint ability in team sport athletes. *Eur J of Sport Science*: 14(S1), S84–S91.

Caixinha PF, Sampaio J & Mil-Homens PV (2004). Variação dos valores da distância percorrida e da velocidade de deslocamento em sessões de treino e em competições de futebolistas juniores, *Revista Portuguesa de Ciências do Desporto* 4(1): 7–16.

Chan Y-Y, Yim YM, Bercades D, Cheng TT, Ngo K-L & Lo K-K (2016). Comparison of different cryotherapy recovery methods in elite junior cyclists. *Asia-Pacific Journal of Sports Medicine, Arthroscopy, Rehabilitation and Technology*: 5, 17–23.

Cheung SS (2008). Neuromuscular response to exercise heat stress. Marino FE (ed): Thermoregulation and human performance—physiological and biological aspects, *Medicine and Sport Science*: 53, 39–60.

Duffield R, Green R, Castle P & Maxwell N (2010). Precooling can prevent the reduction of self-paced exercise intensity in the heat. *Med & Sc in Sports & Exercise*: 42(3), 577–584.

Fiala D, Lomas K & Stohrer M (1999). A computer model of human thermoregulation for a wide range of environmental conditions—the passive system. *J App Physiology* 87: 1957–72.

Hanna EG & Tait PW (2015). Limitations to thermoregulation and acclimatization challenge—human adaptation to global warming. *Int J Environmental Research and Public Health* 12: 8034–8074.

Havenith G, Holmér I & Parsons K (2002). Personal factors in thermal comfort assessment—clothing properties and metabolic heat production. *Energy and Buildings* 34: 581–591.

ISO 8996 (2004). Ergonomics of the thermal environment—Determination of metabolic rate. ISO, Geneva.

ISO 9920 (2007). Ergonomics of the thermal environment—Estimation of thermal insulation and evaporative resistance of a clothing ensemble. ISO, Geneva.

Jones PR, Barton C, Morrissey D, Maffulli N & Hemmings S (2012). Pre-cooling for endurance exercise performance in the heat—a systematic review. *BMC Medicine*: 10, 1–19.

Kenefick RW, Cheuvront SN & Sawka MN (2007). Thermoregulatory function during the marathon, *Sports Medicine* 37(4–5): 312–315.

Oliveira AVM, Gaspar AR & Quintela DA (2011). Dynamic clothing insulation—measurements with a thermal manikin operating under the thermal comfort regulation mode, *Applied Ergonomics* 42(6): 890–899.

Onitsuka S, Zheng X & Hasegawa H (2015). Ice slurry ingestion reduces both core and facial skin temperatures in a warm environment. *Journal Thermal Biology*: 51, 105–109.

Pereira HF (2005). Caracterização dos parâmetros de esforço em basquetebol, MSc thesis, Faculdade de Ciências do Desporto e de Educação Física, Universidade do Porto.

Quintela DA, Gaspar AR & Borges C (2004). Analysis of sensible heat exchanges from a thermal manikin. *Eur J App Physiology* 92: 663–668.

Racinais S, Alonso J, Coutts A, Flouris A, Girard O, González-Alonso J, Hausswirth C, Jay O, Lee J, Mitchell N, Nassis G, Nybo L, Pluim B, Roelands B, Sawka M, Wingo J & Périard JD (2015). Consensus recommendations on training and competing in the heat. *Sports Medicine*: 45, 925–938.

Raimundo AM & Figueiredo AR (2009). Personal protective clothing and safety of firefighters near a high intensity fire front. *Fire Safety Journal* 44: 514–521.

Raimundo AM, Gaspar AR & Quintela DA (2004). Numerical modelling of radiative exchanges between the human body and surrounding surfaces. *Climamed 2004 - 1st Mediterranean Congress of Climatization,* April, Lisbon, paper 8/1.

Raimundo AM, Quintela DA, Gaspar AR & Oliveira AVM (2012). Development and validation of a computer program for simulation of the human body thermophysiological response. *Portuguese chapter of IEEE EMBS* 1–4, Coimbra, Portugal, 23 to 25 February.

Ross M, Abbiss C, Laursen P, Martin D & Burke L (2013). Precooling methods and their effects on athletic performance—a systematic review and practical applications. *Sports Medicine*: 43, 207–225.

Sawka MN, Leon LR, Montain SJ & Sonna LA (2011). Integrated physiological mechanisms of exercise performance, adaptation, and maladaptation to heat stress. *Comprehensive Physiology* 1: 1883–1928.

Stolwijk JAJ (1971). A mathematical model of physiological temperature regulation in man. *NASA contractor report CR-1855*, NASA, Washington DC.

Tanabe S, Kobayashi K, Nakano J, Ozeki Y & Konishi M (2002). Evaluation of thermal comfort using combined multi-node thermoregulation (65MN) and radiation models and computational fluid dynamics (CFD). *Energy and Buildings* 34: 637–646.

Taylor NAS (2006). Challenges to temperature regulation when working in hot environments. *Ind. Health* 44: 331–344.

Tyler CJ, Sunderland C & Cheung SS (2015). The effect of cooling prior to and during exercise on exercise performance and capacity in the heat—a meta-analysis. *British Journal of Sports Medicine*: 49(1), 7–13.

Wegmann M, Faude O, Poppendieck W, Hecksteden A, Fröhlich M & Meyer T (2012). Pre-cooling and sports performance—a meta-analytical review. *Sports Medicine*: 42(7), 545–564.

Wendt D, van Loon LJC & Lichtenbelt WD van M (2007). Thermoregulation during exercise in the heat—strategies for maintaining health and performance. *Sports Medicine*: 37(8), 669–682.

Xu X, Karis A, Buller M & Santee W (2013). Relationship between core temperature, skin temperature and heat flux during exercise in heat. Eur J App Physiology 113(9): 2381–2389.

Zhang Y, Nepocatych S, Katica C, Collins A, Casaru C, Balilionis G, Sjökvist J & Bishop P (2014). Effect of half time cooling on thermoregulatory responses and soccer-specific performance tests. *Montenegrin Journal of Sports Science and Medicine*: 3(1), 17–22.

Zimmermann M & Landers G (2015). The effect of ice ingestion on female athletes performing intermittent exercise in hot conditions. *Eur J of Sport Science*: 15(5), 407–413.

Occupational Safety and Hygiene V – Arezes et al. (Eds)
© *2017 Taylor & Francis Group, London, ISBN 978-1-138-05761-6*

Management tools for workplace safety in building sites—implementation and evaluation

E.M.G. Lago, B. Barkokébas Junior, F.M. da Cruz & M.C.B.S. Valente
University of Pernambuco, Recife, Pernambuco, Brazil

ABSTRACT: In a scenario of economic hardships, it was understood by Brazilian civil construction businessmen that investments in the area of workplace safety minimized indemnity costs and reduces rates of absenteeism (workers being away from work). The objective was to implement five tools of workplace safety management (Daily Safety Dialogue, Investigation of Accidents, Preliminary Analysis of Risks, Permission to Work, Safety Inspections) in a medium-sized civil construction company in Pernambuco—Brasil, aiming to contribute to the reduction of accident rates and financial impact for the company. First, a diagnosis was applied with the research, and, after implantation, the research was repeated again after three years. The results reached indicate an improvement in communication within the workplace and a reduction in the number of accidents with and without worker absence, as well as an increase in training hours per worker. It is also verifiable that the tools are still being used.

1 INTRODUCTION

Brasil currently undergoes a hard time in its economy. According to the Brazilian Institute of Geography and Statistics, Instituto Brasileiro de Geografia e Estatística—IBGE (2016), between October and December of 2015, national GDP showed negative variation of 3,8% in relation to the previous year, and such a result, according to the Institute, was the greatest retraction in a historic series of similar retractions which began in 1996.

In this scenario of uncertainties, it is understandable that the entire industry took a significant hit, with civil construction in particular misfortune due to the reduction in the impetus for building in the residential sector caused by elevated stocks accumulated by the construction companies in 2012 and 2013 (IBGE, 2016).

Facing this reality, it is of utmost importance that the companies find new ways to ensure their survival in an increasingly hostile market, internally creating an environment that is suitable to increase their profits. In this scenario, the investment in actions promoting safety and health in the workplace seems like a good strategic action during a time of crisis (Lago, 2006).

Nowadays, investing in safety and quality of life of the workers became much more than an obligation or an enforced regulation. The investments in the area avoid labour-related lawsuits as well as decreasing levels of absenteeism due to accidents.

To Melo (2011), company managers have also begun to realize an ever-increasing level of requirement from clients in the real estate market. These circumstances lead those businessmen to realize the need for investments in supplementary systems like quality, professional management and new technologies of information.

Another important point to be evaluated within this context is the cost arising from workplace accidents. To measure such a cost is a hard undertaking, for it involves a series of variables, especially if it culminates in the death of the worker, because the price of a life is invaluable. So it stands to reason that businessmen understood that investing in prevention proves infinitely cheaper than dealing with the costs of each accident that occurs. It's within this context that many building companies have been searching, in other sectors of the industry, for some theories and working tools. Such experiences lived by others different branches need to be adapted to the reality of civil construction, for this sector showcases characteristics that distinguish it from the others (Todeschini, 2009).

The product of civil construction is unique to each work, not showcasing a serialised, homogeneous production found in other sectors. Other important characteristics are the lack of qualification in the general workforce and the nearly artisanal character of its processes. The incapacity of fixating itself in just one place and the outsourcing also help creating a reality that is unique to this industry.

All those specificities make the branch of civil construction peculiar and impose certain obstacles for the implementation of some management tools already broadly used in other industry sectors. However, other tools can be easily implemented and may bring a great benefit to the workers of construction companies, improving their quality of life, the relationship and communication with their colleagues and their outlook on prevention of accidents (Melo, 2011).

All those changes are very positive to the company and promote many benefits to the businessmen and their organizations.

In order to complete this paper, five tools of Safety and Workplace Health Management were studied and applied, them being: Daily Safety Dialogue (DDS), Investigation of Accidents (IA), Preliminary Analysis of Risk (APR), Permission to Work (PT), Safety Inspections (IS).

The choice for these tools was due to the low complexity in their implementation and their broad use by various industry sectors.

Over this paper, the theoretical bases of these tools were studied and, after that, they were implemented at some work sites owned by a building company acting in the metropolitan region of the city of Recife—Pernambuco—Brazil.

2 METHOD

In order to develop this paper, it was made the choice of some management tools such as Daily Safety Dialogue (DDS) aimed at awakening in the collaborator the awareness involving his daily activities, concerning his safety, the environment, health and quality. It was applied in a time period of 5 to 15 minutes, always before the beginning of the work hours, time reserved for discussions and basic instructions of the issues linked to prevention of accidents related to health and safety, Investigation of Accidents (IA) all accidents with or without absenteeism during the sampled period in the company were investigated with the objective of exploring the reasons and identifying the causes, Preliminary Analysis of Risk (APR) technique for prior assessment of the present risks involved in the undertaking of a specific work. It is based on the thorough breakdown of each step of the job, as well as the involved risks, Permission to Work (PT) authorizes, through written notice, the beginning of the service, after evaluating the risks through prior assessments, with the due proposition of applicable safety measures, Safety Inspections (IS) the main objective behind the safety inspections was detecting the possible causes which could lead to accidents, aiming to take or propose measures which eliminate or neutralize the risks of workplace

accidents. They were carried through a checklist, and instrument of control, comprised of a selection of conducts, names, items or tasks which must be remembered and\or followed.

After acquiring knowledge about these, steps were taken to implement them in three vertical building sites of the company during a six month timeframe. Afterwards, a survey was taken and elaborated for the staff working on those workplaces chosen for the case study in Recife.

The survey was taken in three distinct moments: at the beginning of the work (on September 2012) for diagnosing to what level the workplaces would be able to use the management tools to be implemented; 6 months after starting to implement the tools (on April 2013), aiming to verify the existence or lack thereof of benefits coming from the process and on September 2015, two years and five months after the last survey was taken, aiming to find out if the tools were still being used, proving functional and aggregating value to the work. In this third moment, the survey was implemented in three workplaces differing from those researched initially considering that the research on these was already concluded. The accident rates were also verified in the workplaces researched in the six months previous to the beginning of the implementation of the tools, comparing this rate with the quantity of accidents that occurred over six months of implementation and over the following two years and five months.

From the information collected in the survey and the analyses of the results of the accident figures and hours of training, it is possible to verify that implementing the tools turned out to be beneficial to the company concerning safety and health in the workplace.

3 RESULTS

As a result from the research, it was verified that the communication of issues related to Safety and Health in the workplace improved. In the first moments, the workers only knew two forms of communication, them being either leaflets or direct conversation with the Safety Officer. After implementing the Daily Safety Dialogues (DDS), one of the tools, a new channel surfaced; the meetings, or trainings taking place over the six months of implementation. In the third research we evidenced through the answers that the communication through Safety Officers and through meetings was increasing over the one conducted through the formerly broadly used leaflets.

The methods of conversation were evaluated, and it was possible to verify that the workers considered that the process improved after implementing the management tools, that is, before implementing the Daily Safety Dialogues (DDS), communication of health and safety-related issues was deficient and carried on in passive fashion (the aforementioned leaflets). The implementation of the tool made communication with the workers more direct, more active, giving the opportunity to verify their opinions and evaluate their comprehension. This change led the workers to have a more positive outlook on the evaluation of the methods. Another point worth mentioning is, after three years of the project's beginning, the tool was still being used even after the conclusion of the work.

The workers also evaluated the investigation of accidents within the work site. In the first research, only 63,13% of them evaluated the investigation method as good or great.

On the other hand, in the second research, 89,4% of the workers evaluated the method as good, great or excellent, which showcases that with the implementation of the tool to analyze accidents, the workers noted an improvement over the previous form of investigation. This can be attested taking into account that previously the investigation was based around just having a conversation with the injured individual.

After implementing the tool, "the tree of causes" and the "five whys" methods began to be utilized, and CIPA (internal commission for the prevention of accidents) began to participate in investigations and the Sheet of Analysis of Accident was implemented. These changes were assimilated by the workers. Still evaluating the methods of investigation, the research tried to find out who usually participated in such undertaking. In the first research, the vast majority (72,73%) answered that this task was carried by the Safety Officer and the injured person. In the second research, after implementing the tool, it was verified that 54,54% of the workers informed that the task was also being carried out by CIPA alongside the Safety Officer. In a third moment, it was possible to notice a constancy between this result, which grew to a number of 66,16%. This last result demonstrates, echoing what was observed in the last visit to the workplace, that the tool remains in use. The workers also voiced their opinion on how the analysis of risk inherent to the task took place before the beginning of activities at the building site. In the first research, most workers (75,3%) answers that this analysis was seldom made. In the second research, it was observed that 54,39% of the interviewees answered that it was always made. In the third research, 83,4% workers answered that it was always made, thus attesting the success of the tool's implementation, which remained in use until the last research.

This proved beneficial to the company, which began acknowledging and evaluating the risks existent in tasks carried out by workers, then proposing actions to eliminate or reduce the occurrence of such risks. The workers were also questioned about the existence of any procedure concerning the all-clear to take on a task before its initiation and what was, for them, the importance of such a procedure. In the first research, the vast majority of interviewees answered that there was no such procedure, but that they would consider it to be important. In the second research, the vast majority answered that yes, such procedure existed and 100% of workers considered the adoption of a procedure to give an all-clear to be important. This new result would be a reflection of the implementation of the tool for Permission of Work (PT) which authorized, in written form, the beginning of services of excavation and tasks undertaken in great heights, evaluating the risks inherent to these assignments in a preliminary analysis and proposing adequate safety measures for these. In the third research, it is noticeable that the percentage of workers who answered that the procedure existed was even bigger and 100% of the interviewees who considered it important remained with the same viewpoint, thus concluding the tool was still being used and unanimously deemed important.

Concerning the frequency in which safety inspections were carried out, 63,64% of the interviewees from the first research answered that most of the time the inspection was carried out and 100% of those who answered considered those to be important. From this scenario, following observations made in the building sites, we can conclude that the Safety Officers carried out inspections informally, without a definite periodicity. As a reflex of the application of the management tool, this percentage changed, with the majority of workers (66,66%) answering that the inspections were always carried out. In that phase, inspections happened in a more systematic and periodic fashion. In the third research, it was verified that 71,72% of workers who were questioned answered that the inspections always took place, which proves that such inspections became commonplace at the building site even after the conclusion of the study.

Visiting the sites, it was noticed that the initial check-list was upgraded and that CIPA was participating in a more active way when it comes to safety inspections. Concerning the performance of safety professionals, the first research shows a division in opinions which evaluated such performance as good and great. In the second research, it is noticeable that the percentage of excellent evaluations grew by 60%. In conclusion, this is due to the implementation of management tools, providing a greater degree of interaction between safety professionals and workers.

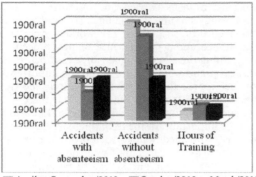

■ April to September/2012 ■ October/2012 to March/2013
■ March to August/2015

Graphic 1.   Accidents/Hours of training.

Together with the data obtained through the surveys, it was also sought out to gather information on the hours of training per worker and the number of accidents with or without absenteeism occurring in the company six months before the beginning of the implementation of the tools, in the six months during said implementation, and the six months before the third research.

Evaluating the results, it is noticeable that between the first and second moments, the number of accidents which led to a worker being absent from the site decreased 33,3% and the ones without absenteeism decreased 14,3%. The hours of training per worker increased 55,9% after the tools were implemented. This result is probably due to the increased awareness concerning workplace safety passed on to workers through the Daily Safety Dialogues. In a third moment, that is, in the six months preceding the third research (March to August 2015), we observe a discreet increase in accidents leading to absenteeism in relation to the results on the last term, something around 16% and a considerable decrease in the cases without absenteeism, 50% (Graphic 1).

4   CONCLUSIONS

The general objective of this paper was to point out that the tools chosen for this study were of easy application, susceptible to be used at building sites and helpful in decreasing the rates of accidents, thus bringing benefits to the health and safety of the worker, and consequently causing a positive financial impact to the company. The tools were applied at the sampled sites with the agreement of all those involved and without hindrances. It was verified a general acceptance coming from workers concerning the suggested changes, culminating in their reported satisfaction. According to the result of the researches, it was possible to notice that the aspects linked to safety and health in the sites went through considerable improvement. Workers became more participatory and well-informed. Safety technicians started to systematize some procedures such as safety inspections, risk analyses and the permissions to work, which generated a positive outcome concerning the prevention of accidents. These results could be felt as accidents became less frequent (both with and without absenteeism) in the months during which the tools were applied, and in the increase of hours of training, which can be attributed also to the Daily Safety Dialogue, the analyses of accidents and the other similar practices that became common at the sites. With the observation of the reality of the sites before and after the tools were implemented, alongside the results of researches conducted with the staff, it can be concluded that this paper came to contribute somehow to an improvement in the Safety Management and Occupational Health at the sampled building sites. As for the tools' longevity, the results were also satisfactory, as it was noticed that they were still being used in the company's new projects after two years and five months subsequent to their implementation, even after the study's conclusion. We believe that by the end of this paper we have attained the goals set. The tools were implemented and brought benefits to the sites, to the workers and consequently to the whole company.

REFERENCES

Associação Brasileira de Normas Técnicas. 2001. *NBR 14.280—Cadastro de Acidente do Trabalho—Procedimento e Classificação*. Rio de Janeiro.

Brasil. (2016a). *Ministério do Trabalho e da Previdência Social Anuário estatístico da previdência social*, consultado em 20/03/2016, disponível em http://www.mtps.gov.br.

Brasil. (2016b) *Ministério do Trabalho e Emprego. Normas Regulamentadoras de Segurança e Medicina do Trabalho*, consultado em 24/03/2016, disponível em: http://www.mtps.gov.br.

Instituto Brasileiro de Geografia e Estatística (IBGE). *Características da população brasileira*, consultado em 14.05.2016, disponível em http://www.ibge.gov.br/home.

Lago, E. M. G. (2006). *Proposta de sistema de gestão em segurança no trabalho para empresas de construção civil.* Dissertação Mestrado, Universidade Católica de Pernambuco—UNICAP, Recife—PE – Brasil, 194pgs.

Melo, L. E. C. (2011). *A previdência social e a luta contra os Acidentes e doenças do trabalho no Brasil*. Informe da Previdencia Social, Brasília—Brasil. v. 23, n. 7, p. 28.

Todeschini, R. (2009). *Novos caminhos da Previdência: preservação da saúde dos segurados e melhoria das condições ambientais de segurança e saúde dos trabalhadores.* Previdência social: reflexões e desafios. Coleção Previdência Social.Série Estudos, Brasília—Brasil. Cap. V, p. 139–161.

Occupational Safety and Hygiene V – Arezes et al. (Eds)
© *2017 Taylor & Francis Group, London, ISBN 978-1-138-05761-6*

Whole-body vibration exposure in forklift operators—a short review

C. Botelho
Faculty of Engineering, University of Porto, Porto, Portugal

M. Luisa Matos
Faculty of Engineering, University of Porto, Porto, Portugal
LNEG, Laboratório Nacional de Energia e Geologia, S. Mamede de Infesta, Portugal

ABSTRACT: The aim of this study was to make a literature review relative to whole-body vibration exposure when driving forklifts, in particular on: health effects, exposure constraints, assessment methodologies, reduction strategies and corrective measures. The results of this research showed a strong relationship between the Low-Back Pain (LBP) risk and driving forklifts. They showed factors that worsen exposure such as individual factors, physical factors, intrinsic factors at work. It attaches great importance to the worker's occupational and medical history when evaluating his exposure.

Keywords: Low-back pain, Assessment, Whole-body vibration, Driving forklift

1   INTRODUCTION

The Whole-Body Vibration (WBV) exposure shows consequent effects, becoming the most significant as a Low Back Pain (LBP) and spinal injuries. These effects are compounded by excessive exposure, long duration and high amplitude vibration.

The work conditions are reflected in the productivity, where in the presence of good environmental physic conditions, we have more energy to achieve a particular result, producing more with less effort. However, when the working conditions are unfavourable, the level of discomfort or irritation is quickly reached, leading to fatigue, lack of motivation, deconcentration and consequently to the drop in production. The risks associated with exposure of workers to mechanical vibration shall be eliminated or reduced. When the value of the exposure action is exceeded, we must apply organizational and technical measures to reduce exposure. The organizational measures aim to reduce the daily exposure time to vibration, taking breaks at work and rotation of jobs. Technical measures aim to reduce the vibration intensity to the human body, acting on the source of vibration or in the transmission to the body. The employer must act in order to apply measures that reduce the exposure, identify the causes of the value exceeding, and fix the protection and preventive measures.

2   RESEARCH METHODOLOGY

The literature search was based on the systematic review methodology referenced in PRISMA Statement, through SDI—Documentation and Information Service of FEUP, by an integrated search in the database in the form of Meta Search. The type of resource used was the "Database", and the survey was conducted separately for each database. Six advanced search Databases were selected, taking into account those that are relevant to this study. We selected three groups of key words each corresponding to a set of words of the same subject/area: Group I   Vibration in forklift driving; Group II—Nordic Questionnaire and associated symptoms; Group III—Storage and Transporting Industries. The search was performed in the "advanced search" mode, in which boolean terms/operators ("and"/"or") were used to connect the words in a text field and between different text fields. After the first search it was recorded the number of items found. Later exclusion criteria were applied to further refine search and direct it as much as possible for relevant content to the dissertation topic and in order to have the most current information possible. Thus the following criteria in order were applied:

– Limit only "Journal Articles" or "Article Reviews";
– Refine to: Publication year higher than 2000;
– Exclusion of repeated articles;
– Limit only the English language;
– Refine by keywords;
– Refine by title;
– Refine by content/abstract.

In the first phase of this research, 465 articles were found, of which 9 articles were considered relevant to integrate the literature review. Only in *Inspect* and *Web of Science* databases was not

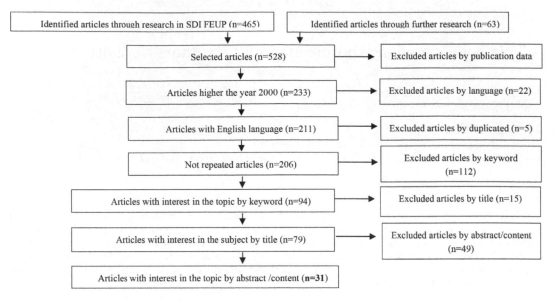

Figure 1. Articles selection diagram for the literature review.

3 RESULTS AND DISCUSSION

applied the criterion "higher than year 2000" due to the reduced number of articles found in the first phase of research. As a result of the previous research, there were only 9 articles that were not sufficiently specific and with data that did not show the conclusions made, it was necessary to make an analysis of the literature of these nine articles.

3.1 Research results

As a result of the research were found nine articles with interest for the study. Subsequently 22 articles were obtained through the bibliography of the first nine. In total, 31 articles with relevance were reached.

3.2 Health effects

The main effects associated to the exposure to vibrations are musculoskeletal or neurological injury of the spinal column. The symptoms that are more severe are chronic pain, neurological dysfunction and disturbances in movement. The most common symptom is Low Back Pain (LBP). Other effects are felt to a less extent, or not so noticeable at first glance, being relating to intestinal organs and vascular system, digestive and circulatory disorders, hearing and reproductive effects and pain in the neck and shoulders.

Massimo Bovenzi has developed several studies proving the relationship between the professional conduct in the industry and the increased risk of back pain (Bovenzi, 2010) (Bovenzi, 2006). A review by Lage Burström also shows that exposure to WBV increases the risk of developing low back pain (Burstrom, Nilsson, & Wahlstrom, 2014).

Ghuman Kuljit Singh concludes by his review that industry operators have been associated with discomfort and musculoskeletal disorders due to excessive duration of operations (Singh G. K., 2012). It was also observed that the most registered disorder was LBP, and some studies have shown possible degeneration of the spine. The development of hernias as a result of exposure to WVB when driving vehicles in industries was observed by Bovenzi, Diane E. Gregory and Jack P. Callaghan (Gregory & Callaghan, 2011). Some studies showed that the WBV had impact on sensory capacity and on the ability to maintain proper back posture (Li, Lamis, & Wilson, 2016). Stephan Milosavljevic, concluded that exposure to WBV sitting is related to the loss of balance while standing, after reviewing three studies that found this evidence, while two studies laboratory found no such relationship (Mani, Milosavljevic, & Sullivan, 2010).

3.3 Constraints

We can have present different kind of factors that influence exposure: individual factors, physical factors or factors intrinsic to the work that are the company's responsibility.

The behaviour adopted by the driver is important, such as the posture and speed adopted. Bovenzi performed a multivariate analysis, and though he may not say that factors, such as posture and vibrations, are dependent factors, he demonstrated that both increase the long-term effects (Bovenzi, 2002).

Individual characteristics, such as age and Body Mass Index (BMI), are important factors to the exposure results, having already been subject of study by Bovenzi (Bovenzi, 2006). Ivo JH Tiemessen showed that BMI does not influence the risk of LBP in professional drivers that have already been exposed to WBV (Noorloos, Tersteeg, Tiemessen, Hulshof, & Frings-Dresen, 2008).

Forklift drivers are often forced to bend the trunk sideways to be able to have a better perception of what is happening in front of him (of the forklift driver), namely loads and spaces, and also to bend the trunk back to be able to perform reverse (M Pranesh, Rakheja, & Demont, 2010). The study by Okunribido also refers the influence of sitting posture and manual handling of materials, suggesting that the combination of these factors is the cause of the visible effects (Okunribido, Magnusson, & Pope, 2008). The study by Martin Fritz states that, if the risk has to be evaluated by ISO 2631-1, the influence of the trunk inclination should be considered by multiplying a correction factor in this same exposure assessment (Fritz & Schafer, 2011).

Some physical constraints that may have an impact on the exposure intensity are: the seat type the seat suspension type and the tires type.

Regarding the seat, when the fixed and mobile models are compared, the latter has a greater attenuation when subjected only to horizontal vibrations. In other situations, the movable seat just becomes beneficial in high velocity situations (Wijaya, Jonsson, & Johansson, 2003). In general, the movable seat is characterized as more comfortable, being less discomfort in the lumbar region (Wijaya, Jonsson, & Johansson, 2003). A seat without back support can reduce the transmission of vertical vibrations. An experiment presented in the study of Anand M-Pranesh demonstrated that when the seat without backrest there is a lower magnitude when compared with the seat with backrest (M-Pranesh, Rakheja, & Demont, 2010). The effect of the presence of the support presents higher values for the lower area of the thorax and the lumbar region. It was concluded in this study that the presence of the seat back has more influence on the magnitude of the vibrations than the induced magnitude itself. Regarding the comparison between seats with air suspension and mechanical suspension is possible to confirm the presence of lower values in the z-axis when using the air suspension seats (Blood, Ploger, & Johnsson, 2010). In the analysis of the behaviour of the seats with mechanical suspension, it was shown to be dependent on the driver's weight, and offers better performance for lighter workers (Blood, Ploger, & Johnsson, 2010). In the seat with air suspension, this dependence isn't verified (Blood, Ploger, & Johnsson, 2010). Note that the pneumatic suspension seat is by rule more expensive than the seat with mechanical suspension (Blood, Ploger, & Johnsson, 2010), however, it is necessary to consider the costs associated with the injuries that arise.

Regarding the use of the belt as a preventive measure, it is not associated with reduction of the risk of LBP and should not be utilized as a control measure (Lariviere, 2008).

Vehicle tires influence the magnitude of vibration. Solid tires allow greater comfort when compared with the pneumatic tires due to the damping provided (Verschoore, Pieters, & Pollet, 2003), however, becomes more important the influence of the seat.

The hours of daily exposure, the exposure intensity, the excessive driving for many years (Bovenzi, 2002), the rest periods (Johanning, 2011) are important conditions, belonging to the company's responsibility. Relatively to the own exposure the comfort level decreases with increasing amplitude (Singh, Nigam, & Saran, 2016). The physical load inherent to other activities developed simultaneously with driving, are very heavy tasks for the worker's muscles. (Bovenzi, 2006).

3.4 Assessment methodologies

In order to evaluate, characterize and classify exposure to vibrations, different methods can be used. For general methods of risk classification, we have the example of the technique k-nearest neighbour that is a classification system used due to its simplicity and efficiency, that allows the combination of variables, independent with individual and working conditions (Sanchez, Iglesias-Rodriguez, Fernandez, & Cos Juez, 2016). This technique was applied in a study aimed to predict musculoskeletal disorders related to work.

The evaluation of vibration exposure can be done by several parameters. Bovenzi concludes that assessment parameters of the exposure to vibration derived from the duration of the exposure, as VDV, are better "predictors" of results than parameters like A(8) (Bovenzi, 2010). Some studies use the SEAT parameter that ranks seat performance and S_{ed} parameter to evaluate health in the spine. Some studies show more than one variable as exposure assessment study. The results obtained by different methods can be compared as (Verschoore, Pieters, & Pollet, 2003) tested in this study. However, the values obtained from different methods may not always be comparable.

Many studies focus on evaluation of adverse effects through transmission of vibrations by the vertical axis z, however it begins to become important to consider the effects of the three axes, x, y and z, taking into account the total effect (Johanning, 2011).

The limits set by ISO 2361-1 (ISO2631-1, 2004) may not be enough to perform an assessment, since it only takes into account the value of the acceleration, not taking into account individual and work factors such as the duration and intensity of exposure, weather conditions, body posture or additional tasks that require lifting or bending. It is felt the need to take into account the worker's history, the tasks previously performed and especially work stations with vibration exposure.

We can have another type of assessment that can also be seen as a characterization of the exposed population that is done by using a questionnaire. Bovenzi (Bovenzi, 2010) adopted a questionnaire developed with VINET (European Project Vibration Injury Network), for the evaluation of the population, that includes: personal information, occupational history, medical history and other symptoms. Also Ivo J.H. Tiemessen evaluated the back pain through the VIBRISKS WBV Questionnaire, which is divided into 5 (five) sections: personal characteristics, VINET questionnaire and a satisfaction research, complaints and musculoskeletal injuries on LBP in the last 7 (seven) and 12 (twelve) months by the Nordic Questionnaire and physical complaints. GWBQ (The Generic Work Behavior Questionnaire) was a questionnaire developed to assess general symptoms, non-specific, including emotional, cognitive (Griffiths, Cox, Karanika, & Tomás, 2006). This questionnaire allows to obtain answers about the exhaustion and strain from 12 (twelve) other symptoms. The WOAQ (The Work and Organization Assessment Questionnaire) is a questionnaire that does not evaluate the psychological part.

Laboratory tests are also performed to test different factors that can not be evaluated in the field for any reason, or because it makes it easier to use a measuring device. Some tests may use a seat with artificially induced vibrations, or a platform with sudden disruption (Lariviere, 2008). Some laboratory tests allow to evaluate the response of the muscle reflection, using electromyography to assess if the fatigue was a result of the exposure to vibrations or not (Lariviere, 2008).

3.5 Strategies for reduce and corrective measures

In order to reduce exposure to vibrations, it is possible to intervene on different fronts, such as: directly on the duration of exposure; on working conditions associated; the behaviour and attitude of the driver; in the vehicle itself or in the circulation floor.

The strategies directly linked to the vehicle fall on the suspension of the seats and the type of tires. The bibliography indicates that seats with air suspension and pneumatic tires are the best option, helping in the reduction of vibrations. With regard to the ergonomics of the seat, tests performed by M. Marksous resulted in that the backless seat demonstrated better results due to less contact between the worker and the seat and which allows lower transmission of vibration, yielding lower values in the z-axis. In the oldest seats, even when adjusted correctly it is not possible to expect attenuation (Tiemessen, Hulshof, & Frings-Dresen, 2007). Despite various strategies and corrective action, we should match the use of them. A study by Motmans R. concluded that improved floor surface was the most effective step, by following the rate parameter and finally the seat. As a combined action, the author concludes that improving the floor and reducing the speed would be the ideal situation, however the change of speed can interfere with the work management. So, the best chance will be change the floor surface and the suspension of the seat (Motmans R., 2012). The forklift characteristics and the driving performance have both influence on exposure to vibrations, therefore it would be important to consider mitigation measures that acted on both aspects. A systematic review shows that only one study presents a strategic intervention to reduce the WBV, while all others have only factors that have an effect on the exposure to vibrations (Tiemessen, Hulshof, & Frings-Dresen, 2007). These effects are divided into two categories: "design" and "skills and behaviour." The second category is less expensive but is less promising than the first. Both categories should be applied in combination. It has been found that the reduction in driving speed reduced the magnitude of the vibrations. The driving speed and the suspension seat proved to be the factors with more significance.

The company's policy towards the reality and the need for change is very important. A study conducted by Hulshof (Hulshof, Verbeek, Braam, & van Dijk, 2006) showed the influence of this condition in relation to the WBV, attitude and behaviour of truck drivers, and a trend towards greater knowledge of OSH professionals (Health and Safety at Work). One study used the ASE model (attitude, social influence and self-efficacy) which aimed to change the attitude, social influence and self-efficacy not only of drivers, as all workers (Tiemessen, Hulshof, & Frings-Dresen, 2007). This program estimates that measures such as changing the seats and the floor has a higher cost when compared to attitude and behaviour measures. The program is based on changing or trying to change the behaviour, making the effect of the intervention program permanent.

4 CONCLUSIONS

With the literature review performed, is possible to affirm the presence of the risk of LBP on forklift driving. There are factors that conjugated with vibration can aggravate the exposure of the driver: posture and speed when driving, exposure time, age and BMI of the driver, seat type, suspension type and tire type. Regarding the methods to evaluate the exposure, some studies concluded that it is better to use parameters that do not depend on the time factor. Others claim that the limits established by ISO 2361-1 are not enough, for not taking into account individual and work factors. It is observed the need to take into account the worker's history in the assessment of its exposure. Therefore, arises a form of evaluation, being this one through questionnaire, with personal, occupational, and medical questions, approaching the felt symptoms.

ACKNOWLEDGMENTS

The authors would like to acknowledge Master in Occupational Safety and Hygiene Engineering (MESHO), of the Faculty of Engineering of the University of Porto (FEUP), the support and contribution provided in the development of this work and for having allowed the publication of the study.

REFERENCES

Blood, R., Ploger, J., & Johnsson, P. (2010). Whole body vibration exposures in forklift operators comparison of a mechanical and air suspension seat. Ergonomics, 53, 1385–1394.

Bovenzi, M. (2002). Low back pain in port machinery operators. Journal of Sound and Vibration, 253 (1), 3–20.

Bovenzi, M. (2006). An epidemiological study of low back pain in professional drivers. Journal of Sound and Vibration, 298, 514–539.

Bovenzi, M. (2009). Metrics of whole body vibration and exposure-response relationship for low back pain in professional drivers: a prospective cohort study. Occupation and environmental health, 82, 893–917.

Bovenzi, M. (2010). A longitudinal study of low back pain and daily vibration exposure in professional drivers. Industrial Health, 48, 584–595.

Burstrom, L., Nilsson, T., & Wahlstrom, J. (2014). Whole body vibration and the risk of low back and sciatica:a systematic review and meta-analysis. Occupational Environmental Health, 808, 403–418.

Fritz, M., & Schafer, K. (2011). Influence of the posture of the trunk on the spine forces during whole-body vibration. Journal of low frequency noise, 30, n°4, 277–290.

Gregory, D., & Callaghan, J. (2011). Does Vibration Influence the Initiation of Intervertebral Disc Herniation. Spine Basic Science, 36, n°4, 225–231.

Griffiths, A., Cox, T., Karanika, M., & Tomás, J.-M. (2006). Work design and management in the manufacturing sector: development and validation of the Work Organisation Assessment Questionnaire. Occupational Environmental Medicine, 10, 669–675.

Hulshof, C., Verbeek, J., Braam, I., & van Dijk, F. (2006). Evaluation of an occupational health intervention programme on whole body vibration in forklift truck drivers:a controlled trial. Occupational and environmental medicine, 63, 461–468.

ISO2631-1. (2004) – Mechanical vibration and shock. Evaluation of human exposure to whole-body vibration. Part1: General requirements, 1–37.

Ivo, T.J., Hulshof, C.T., & Frings-Dresen, M.H. (2007). An overview of strategies to reduce whole body vibration exposure on drivers: a systematic review. Internation Journal of Industrial Ergonomics, 37, 245–256.

Johanning, E. (2011). Diagnosis of whole body vibration related health problems in occupational medicine. Journal of low frequency noise, 30, n°2, 207–220.

Lariviere, C. (2008). A laboratory study to quantify the biomechanical responses to whole body vibration: the influence on balance, reflex response, muscular activity and fatigue. International Journal of Industrial Ergonomics, 38, 626–639.

Li, L., Lamis, F., & Wilson, S. (2016). Wholebody vibration alters proprioception n the trunk. International Journal of Industrial Ergonomics, 38 (9–10), 792–800.

Makhsous, M., Hendrix, R., Crowther, Z., Nam, E., & Lin, F. (2005). Reducing whole body vibration and musculoskeletal injury with a new car seat design. Ergonomics, 48 (9), 1183–1199.

Mani, R., Milosavljevic, S., & Sullivan, S. (2010). The effect of whole body vibration on standing balance: a systematic review. International Journal of Industrial Ergonomics, 40, 698–709.

Motmans R. (2012). Reducing whole body vibration in forklift drivers. Work, 41, 2476–2481.

M-Pranesh, A., Rakheja, S., & Demont, r. (2010). Influence of support conditions on vertical whole body vibration of the seated human body. Industrial Health, 48, 682–697.

Murtezani, A., Ibraimi, Z., Sllamniku, S., Osmani, T., & Sherifi, S. (2011). Prevalence and risk factors for low back pain in industrial workers. Folia Medica, 53 (3), 68–74.

Noorloos, D., Tersteeg, L., Tiemessen, I., Hulshof, C., & Frings-Dresen, M. (2008). Does body mass index increase the risk of low back pain in a population exposed to whole body vibration? Applied ergonomics, 39, 779–785.

Okunribido, O., Magnusson, M., & Pope, M. (2008). The role of whole body vibration, posture and manual handling as risk factors for low back pain in occupational drivers. Ergonomics, 51 (3), 308–329.

Sanchez, A., Iglesias-Rodriguez, F., Fernandez, P., & Cos Juez, F. (2016). Applying the K-nearest neighbor technique to the classification of workers according to their risk of suffering musculoskeletal disorders. *International Journal of Indutrial Ergonomics*, 52, 92–99.

Singh, G.K. (2012). Effect of whole body vibration on vehicle operators:a review. *International Journal of Science and Research,* 34, n°4, 320–323.

Singh, I., Nigam, S., & Saran, V. (2016). Effect of backrest inclination on sitting subjects exposed to WBV. *Procedia Technology*, 23, 76–83.

Tiemessen, I., Hulshof, C., & Frings-Dresen, M. (2007). The development of an intervention programme to reduce whole body vibration exposure at work induce by a change in behaviour: a study protocol. *BMC Public Health,* 7 (329), 1–10.

Verschoore, R., Pieters, J., & Pollet, I. (2003). Measurements and simulation on the comfort of forklift. *Journal of Sound and Vibration*, 266, 585–599.

Wijaya, A., Jonsson, P., & Johansson, O. (2003). The effect of seat design on vibration comfort. *International Journal of Ocupational Safety and Ergonomics*, 9 (2), 193–210.

Occupational Safety and Hygiene V – Arezes et al. (Eds)
© 2017 Taylor & Francis Group, London, ISBN 978-1-138-05761-6

Effects of noise on the cognitive response of BAJA vehicles operators: A case study

F.M. da Cruz, B. Barkokébas Jr., E.M.G. Lago & A.V.G.M. Leal
University of Pernambuco, Pernambuco, Brazil

ABSTRACT: Cognitive faculties such as memory, attention and reaction time are fundamental for the development of labor activities in industries. The present paper intends to present the results obtained in an experiment that analyzed the influence of noise on the cognitive response of individuals after operating a prototype vehicle named BAJA. Selective attention tests were performed in engineering students before and after a 45-minute noise exposure. It was detected the existence of delay in the response time of the volunteers after the exposure. Such delay can be found in a significant way in machine operators whose exposure time are usually higher than the ones performed in the present work.

1 INTRODUCTION

Cognitive faculties such as memory, attention, reaction time, etc., are fundamental for the development of labor activities in industries. Ergonomics establish that the better the work condition, the greater the productivity and life quality of the workers involved in the process.

Among the specific segments, such as mining or construction, there is wide use of machinery whose operations rely on the workers. In this interface, between operators and machines, the occupational exposure to noise occurs, originating from the motor and the activity itself.

Noise, besides being uncomfortable, exerts influence under the human nervous system, affecting the individual's cognitive faculties. For machine operation, the degradation of these abilities, in addition to affect productivity, could unleash diseases or cause accidents while performing the task, such as collisions or run over.

This way, the investigation of the effects of physical risk factors on the aspects of human cognition could help the search for solutions that aim the elimination or neutralization of such factors.

Noise, according Tomei (2009), provoke direct effects on the auditory system, but its major impact occurs in the central nervous system, since the encoding of sound information requires energy expenditure related to brain activities. These wastages were observed by Cho (2011) while studying neurological activity in volunteers exposed to different levels of noise. The study verified that the neurons activity increases as the auditory stimuli increase, raising the energy expenditure.

Because of increased brain activity, Wong (2010) observed that the energy expenditure for sound stimuli processing alters the performance of cognitive faculties, such as attention, memory and concentration. This influence was studied by Kristiansen (2009) who demonstrated the existence of the relation between the frequency and the amplitude of the sound on the performance of mental tasks. These experiments substantiated the existence of frequency bands and sound volume that harm the attention and concentration performance.

The central nervous system is connected to the muscle fibers through the efferent nerve fibers. In this context, muscles receive constant feedback from nerve cells. According to Hotta (2014), the nervous system, when overloaded by stimuli, can cause electric micro discharges on the muscle fiber, causing an increase in its contraction and consequently higher muscular fatigue.

This effect, according to Missenard (2009) causes an increase in energy consumption, contributing to the wear of muscles and predisposition of the tissue to micro-lesions and increased stress levels.

Due to the close relationship between the nervous system and cognitive faculties, Ljungberg (2004, 2005 and 2007) conducted experiments in order to understand the impacts of noise and occupational vibrations on the cognition of university students after exposure to such factors.

In all tests, Ljungberg (2004, 2005 and 2007) used an exposure chamber to control the intern and external variables and applied the protocol for determination of the Sternberg Paradigm and the evaluation called Search and Memory Task—SAM to analyze the cognitive responses and application of the CR-10 Borg's Scale, evaluating the degree of discomfort of the task.

The tests performed by Ljunberg (2004, 2005 and 2007) relied on computers and infrastructure for their accomplishment, turning the methodology restricted to applications in simulators, since in real working conditions the method could not be applied.

In the search for alternative tools in literature, it was found the test developed by Dr. John Ridley Stroop in 1935 to study the effects on reaction time of a task when viewing colored texts. For example, when a word such as blue, green, red, etc., is printed in a different color from the color expressed by the semantic meaning (example: the red word printed in blue ink) a delay occurs in the processing of the color of the word, causing slower reaction times and increased errors. This delay is named Stroop effect, in homage to its discoverer.

With the advance of mobile technology, the Stoop test has been compatible in the form of mobile application. This facility made it possible to apply the test within the work process, allowing the collection of cognitive information in the field, which was previously only possible in laboratory conditions.

Given the above, the present work aims to study the influence of noise on the cognitive response time of individuals after operating a prototype vehicle named BAJA.

2 METHODOLOGY

2.1 Subjects

In order to perform the tests, 9 healthy volunteers between 20 and 23 years old, mechanical engineering students from University of Pernambuco were selected.

2.2 Experiment structure

The 9 students operated the BAJA vehicle (Fig. 1) for 45 minutes in a test-field of the university (Fig. 2). During the tests, the noise levels were collected through an audio dosimeter. Participants performed two selective attention tests (Stroop), the first before the operation and the second immediately after. It was analyzed the difference between the results before and after the exposure.

Figure 1.   Prototype vehicle—BAJA.

Figure 2.   Experiment area.

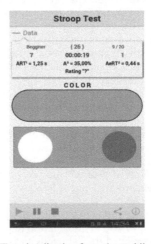

Figure 3.   Audio dosimeter (Model: The Edge5—Quest/3M) installed in the participant's auditory zone.

Figure 4.   Test visualization from the mobile application.

2.3 Noise evaluation

Noise measurement was performed by an audio dosimeter (Fig. 3), (Model: The Edge5—Quest/3M, properly calibrated) carried by an operator, installed in the auditory zone.

2.4 *Cognitive Task*

In order to analyze the cognitive response time it was applied a selective attention test, the Stroop test, that relates textual information and colors, according to Figure 4. The test was applied through a mobile application. However, pre-tests were performed, so the participants could became familiar with the tool, followed by a test before the exposure and a test immediately after the exposure.

3 RESULTS

The data collected in the test are associated to the speed response to the stimuli and the accuracy of these responses. In this way, data was organized in two ways, the first one related to the response time and the second to the exact response time.

The average response time, as presented in Table 1 shows the average response time in minutes that participants have after receiving the stimulus to respond the test.

Analyzing the data (Chart 1) it was possible to verify that 7 of 10 evaluations presented a slight increase in their response time after the exposure.

Chart 1 shows the existence of a tendency to increase the response time in the attention tests proportional to the increase of sound pressure levels of the experiment.

The average time of exact response, according to Table 2 demonstrates not only the speed of response, but also the rate of response in an assertive way.

Analyzing Chart 2 it is possible to observe that 8 out of 9 participants presented a delay in correctly responding the test.

Chart 2 data the tendency is not clear. Perhaps, the small number of participants is probably

Table 1. Average response time in minutes.

Operator	SPL dB(A)	Time Minutes	Average Response Time (ART)	
			Before (ART01 – min)	After (ART02 – min)
01	82,9	45 min	0,32	0,38
02	83	45 min	0,50	0,55
03	81,3	45 min	0,45	0,48
04	84,7	45 min	0,50	0,55
05	84	45 min	0,25	0,36
06	84,6	45 min	0,42	0,43
07	84,7	45 min	0,35	0,30
08	84,3	45 min	0,48	0,56
09	84,1	45 min	0,51	0,51

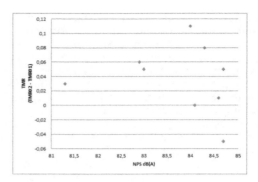

Chart 1. Relation between response time and the sound pressure level of the experiments.

Table 2. Average time of exact responses.

Operator	SPL dB(A)	Time Minutes	Average Response Time (ARTe)	
			Before (ARTe01 – min)	After (ARTe02 – min)
01	82,9	45 min	0,30	0,36
02	83	45 min	0,49	0,54
03	81,3	45 min	0,45	0,48
04	84,7	45 min	0,48	0,51
05	84	45 min	0,16	0,36
06	84,6	45 min	0,40	0,41
07	84,7	45 min	0,35	0,29
08	84,3	45 min	0,44	0,56
09	84,1	45 min	0,49	0,50

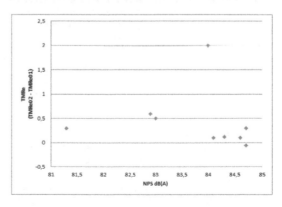

Chart 2. Relation between exact response time and sound pressure levels of the tests.

related with the difficulty to the visualization of the expected tendency.

Similar results of the present study were obtained by Ljungberg (2005) who perceived the influence of noise and occupational vibrations on memory and attention of volunteers exposed through a cognitive tool called SAM. Such influence cannot be proven statistically, but it indicated the signs of is interference over memory and attention.

In previous studies, Sherwood & Griffin (1990) used the same environmental parameters, methodological configuration and similar cognitive tools used by Ljungberg (2005), but with longer exposure time, statistically verifying the scanning memory impairment even at low energy levels.

Ljungberg & Neely, (2007) observed that the mental load of the task influences the identification of cognitive effects, which means that, activities whose mental demand are higher allow a better visualization of the influence of physical risk factors on cognitive waste.

Thereby, considering guiding a vehicle in circles a monotonous task, it may have influenced the better visualization of the effects of noise on the response time of the volunteers.

Sound pressure levels emitted by the BAJA prototype vehicle are similar to those machinery used in construction industry, mining and agriculture, however, in real working conditions, the exposure times would be 8 hours and not 45 minutes as applied in the experiment.

Thus, the delay in response times could also be identified in real working conditions, highlighting the possibility of an increased delay, given the higher exposure time directly affecting equipment and machine operators.

4 CONCLUSIONS

The innovation of the study comes from performing the experiments in external environments, given that the previous experiments found in literatures occurred in simulation chambers.

Advances in mobile technology allowed the present work to seek alternatives for cognitive data field collection with results close to those obtained by experiments performed in computers within controlled environments.

The application of Stroop cognitive test, through mobile applications, presented similar results to studies carried out by other authors who applied the SAM and the Sternberg Paradigm tests in computers.

The increased delay in the volunteer's response time presents a directly proportional relationship with the increase in sound pressure levels on the experiment.

The increased delay in the exact response times of volunteers is not clear. Perhaps, the small number of participants is probably related with the difficulty to the visualization of the expected tendency.

The delay time in correctly responding the test is higher than the delay time of the test response itself.

In the present experiment, increased noise levels influenced the volunteer's ability to make fast and accurate decisions, given the differences found between ARTe and pre exposure in relation to post exposure.

Clearer observations on the delay of cognitive responses depends on new studies with longer exposure periods, with a larger number of samples, preferably in real work environment with exposure time close to 8 hours of work.

REFERENCES

Chiang, C., Liu, C., Lin, F., Wang, W., & Chou, P. (2012). Using ryodoraku measurement to evaluate the impact of environmental noise on human physiological response.

Cho, W., Hwang, S., & Choi, H. (August de 2011). The relationship between pure-tone noise and human bio-signal response. *International Journal of Precision Engineering and Manufacturing, 12*(4), pp. 727–731.

Hotta, Y., Korakata, Y., & Kenichi, I. (2014). Verification of the muscle fatigue detection capability of a unipolar-leads system using a surface electromyogram mode. *Annual International Conference of the IEEE Engineering in Medicine and Biology—Proceedings*, pp. 110–113.

Kristiansen, J., Mathiesen, L., Nielsen, P.K., Hansen, Å.M., Shibuya, H., Petersen, H.M., . . . Sogaard, K. (2009). Stress reactions to cognitively demanding tasks and open-plan office noise. *International Archives of Occupational and Environmental Health, 82*(5), pp. 631–641.

Ljungberg, J. (2007). Cognitive degradation after exposure to combined noise and whole-body vibration in a simulated vehicle ride. *Int. J. Vehicle Noise and Vibration*, pp. 130–142.

Ljungberg, J., & Neely, G. (2005). Attention Performance after Exposure to Combined Noise and Whole-Body Vibration. *Proceedings of the Human Factors and Ergonomics Society Annual Meeting*, pp. 1549–1553.

Ljungberg, J., Neely, G., & Lundstrom, R. (2004). Cognitive performance and subjective experience during combined exposures to whole-body vibration and noise. *Int Arch Occup Environ Health*, pp. 217–221.

Ljungberg, J.K., & Neely, G. (2007). 'Cognitive aftereffects of vibration and noise exposure and the role of subjective noise sensitivity. *Journal of Occupational Health*, pp. 111–116.

Missenard, O., Mottet, D., & Perrey, S. (2009). Adaptation of motor behavior to preserve task success in the presence of muscle fatigue. *Neuroscience, 161*(3), pp. 773–786.

Sherwood, N., & Griffin, M. (1990). Effects of whole-body vibration on short-term memory. *Aviat Space Environ Med*, pp. 1092–1097.

Tomei, G., Tomaol, E., Palermo, P., Tecchio, F., Zappasodi, F., Pasquatti, P., ... Cerrarti, D. (2009). Effects on central nervous system in environmental noise exposed workers. *Giornale Italiano di Medicina del Lavoro ed Ergonomia*, pp. 358–359.

Wong, P., Ettlinger, M., Sheppard, J.P., Gunasekera, G., & Dhar, S. (2010). Neuroanatomical characteristics and speech perception in noise in older adults. *Ear and Hearing, 31*(4), pp. 471–479.

Occupational Safety and Hygiene V – Arezes et al. (Eds)
© 2017 Taylor & Francis Group, London, ISBN 978-1-138-05761-6

Health and safety guidelines in laboratories of higher education centers

F.G. dos Santos, F.M. da Cruz, B. Barkokébas Jr. & E.M.G. Lago
Polytechnic School of the University of Pernambuco, Recife, Pernambuco, Brazil

ABSTRACT: According to the Brazilian Protection Annual Report of 2015, the sectorial analysis of accidents within national range registered 5.524 typical accidents and, in Pernambuco, 485 cases of accidents occurred that were related to educational services. This paper aimed at gathering information that might contribute to the prevention of accidents and to the change of behavior of users and of those responsible for the environment, since every practical activity to be developed inside a laboratory presents risks and is accident-prone. The data were collected from the laboratories and workshop of the Higher Education Institution by means of a field research, in which the work conditions and existing risks were assessed through the preliminary analysis recommended by AHIA. The total evaluation result showed 51% of irrelevant risks, 34% of relevant risks, 11% of attention risks, 4% of critical risks and 0% of emergency risks. Based on the results of the evaluations, safety guidelines, good hygiene practices and risk mapping were structured, as well as improvement actions were proposed.

Keywords: Preliminary risk assessment, Accident prevention, Health and safety guidelines

1 INTRODUCTION

The database of the Ministry of Education (2015) showed that Brazil and the State of Pernambuco, respectively, have 391 and 31 faculties of engineering.

The professionals are exposed to a broad variety of risks in laboratories due to the presence of lethal, toxic, corrosive, irritating and flammable substances, besides the use of equipment that offers certain risks, such as temperature changes, radiation and also assignments that use biological and pathogenic agents. The main risks of greater potential are intoxication, thermal or chemical burns, electric shocks, fires, explosions, contamination by chemical agents and exposure to ionizing and non-ionizing radiation, among others.

The sectorial analysis of accidents within national range registered 5.524 typical accidents, 2.352 commuting accidents and 140 cases of diseases related to educational services. See Table 1 for the occurrence of accidents by occupation.

According to the Brazilian Protection Annual Report, in Pernambuco the latest data were registered in 2013, showing 485 accidents with workers in professional, technical and scientific activities and 186 accidents related to educational activities. The occurrence of incidents was 0,94% and 0,39%, respectively (Brazilian Protection Annual Report 2015). See Table 2 for the occurrence of accidents by risk.

Table 1. Occurrence of accidents by occupation.

Occupation	Typical accidents (%)
Polytechnic researchers and professionals	0,01
Exact sciences and engineering professionals	0,29
Teaching professionals	0,38
Mid-level technicians	1,56

Source: Brazilian Protection Annual Report (2015).

Table 2. Occurrence of accidents by risk.

Risks	Typical accidents (%)
Burns	0,63
Electric shocks	0,81
Contamination by chemical agents	0,55
Occupational exposure to risk factors	0,60
Contact with and exposure to communicable diseases	2,28

Source: Brazilian Protection Annual Report (2015).

In light of the above, this paper aims at gathering information that might contribute to the reduction of accidents in laboratories and workshops of the

Faculty of Engineering of the Higher Education Institution, as well as to contribute to the change of behavior of users and of those responsible for the environment, meeting the standards of current legislation.

2 EFFECTIVE ACCIDENT CONTROL SYSTEM

The organizational risk management refers to the systematic acknowledgement, evaluation, control and monitoring of safety risks, occupational health, environment, quality and business (Barkokébas Jr. 2005). An effective risk management is a primary responsibility of the administration and, for complex and high-risk organizations, the only viable way to control risks is through a management systems approach (Marzano 2015).

For an effective accidents control system to be designed, treating the causes is indispensable. To this end, a few questions must be asked:

- Why did the condition occur?
- What are the existing unsafe conditions?
- What are the failures of the system and of the environment?

Likewise, some concepts must be understood:

- Human failures—Failures that happen in the execution of a task, which might cause the accident. Human failures may occur by: negligence, that is the lack of attention and the use of inadequate equipment in performing a task; incompetence, which is the lack of knowledge in executing a task; malpractice, for not performing properly the job or task one is able to perform (Mendes 2014).
- Unsafe condition in the environment—a home unsafe condition is the absence of the fundamental care that one must have to properly perform a task without accidents. Therefore, there must be constant vigilance concerning what might cause accidents in the household, which must be brought to the responsible person's knowledge so that measures can be taken, avoiding possible accidents (Barkokébas Jr. 2005).
- Classification of environmental risks and risk factors—The environmental risks are classified according to their causative nature. Therefore, the types of risks may be grouped in: chemical, which are caused by chemical substances (vapors, gases, liquids, dusts, mists, fumes); physical, which are caused by energy exchange between man and environment (heat, electricity, infrared and ultraviolet radiations, ionizing radiations, pressures, impacts and others); biological, which are caused by agents that generate infections and allergies (viruses, bacteria and fungi) (Ministry of Labor 2015).

Whenever the risk factors exist in the environment, it occurs what can be named an unsafe condition. Frequently, it is believed that the danger only exists outside the perimeters of the residences, which is not quite right, for the statistics show that our homes (kitchens, bedrooms, bathrooms, gardens etc.) are responsible for various types of accidents, in which the main affected are children and the elderly (Mendes 2014).

Many are the factors that cause the accidents. On the one side, it points out those factors related to the physical and psychological conditions of people themselves, on the other side, it points out those factors connected to the conditions that are present in the physical, social and cultural environment where people live (Silva 2013).

3 PRACTICES OF SAFETY AND HYGIENE IN LABORATORIES

The practices of safety and hygiene most used by universities in teaching laboratories are safety procedures, emergency procedures and educational programs structured in safety manuals that are drawn by those responsible for the laboratories or by the responsible safety sector, that include the following subjects:

- Responsibilities of users and laboratory coordinators;
- Personal and Collective Protective Equipment;
- Facilities maintenance;
- Laboratory equipment maintenance;
- Handling and storage of reagents;
- Recommendations to avoid typical accidents;
- Educational programs: Safety Committee;
- Good Hygiene Practices, such as: to never consume or package food and beverage in the laboratory, to wash hands by the end of laboratory procedures in order to avoid contaminations, to expressly prohibit smoking inside the laboratory etc.

Furthermore, the safety manuals are available to everyone with the purpose to provide the fundamental information so that activities in the laboratories can be performed with safety, and the coordinators are available to clarify questions.

4 METHODOLOGY

The research was conducted in the laboratories and workshop of the Higher Education Institution located in the city of Recife, State of Pernambuco, Brazil. It was an *in loco* study, where 28 laboratories were visited, being 3 laboratories of Construction Industry, 1 of Fuels, 1 of Energy Efficiency, 1 of Electronics, 1 of Systems Engineering, 1 of

Mechanical Testing and Materials, 3 of Physics, 7 of Informatics, 3 of Mechanics, 1 of Mechatronics, 1 of Metrology, 2 of Workshop Practices, 1 of Chemistry, 1 of Topography and 1 of Labor Safety. To that end, a search form and a checklist were structured based on the current legislation, allowing the survey of the tasks performed by teachers and students, as well as the existing control measures.

The data analysis and the identification, evaluation and classification of risks of the laboratories' environmental conditions assumed the methodology recommended by the Association of Healthcare Internal Auditors (AIHA) that assigns numbers to characterize the exposure, the effects and the severity in order to classify the risks according to Tables 3 and 4.

According to Tables 3 and 4, the risks of each task were related and the probabilities of occurrence and the severity of the consequences were classified, generating the risk classification seen in Table 5.

Upon classification, safety measures and digital risk mapping were structured, as well as improvement actions were proposed, based on the technical and legal content applicable to the researched environment.

Table 3. Qualitative valuation of agents' exposure.

Category	Description
0 No exposure	No contact with agent or irrelevant contact.
1 Low levels	Non-frequent contact with agents.
2 Moderate exposure	Frequent contact with agent in low concentrations or non-frequent contact in high concentrations.
3 High exposure	Frequent contact with agent in high concentrations.
4 Very high exposure	Frequent contact with agent in very high concentrations.

Source: AHIA.

Table 4. Qualitative valuation of the agents' effects.

Category	Description
0	Reversible effects of little importance or unknown or simply suspected.
1	Worrying reversible effects.
2	Severe worrying reversible effects.
3	Worrying irreversible effects.
4	Life or disease threatening/ incapacitating injury.

Source: AHIA.

Table 5. Gradation of severity in the qualitative evaluation of agents (exposure/effect).

Exposure + Effect = Control	Description
0–1	Irrelevant.
2–3	Relevant.
4–5	Attention.
6–7	Critical.
8	Emergency.

Source: AHIA.

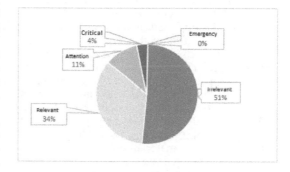

Figure 1. Classification of risks in laboratories. Source: field research.

5 RESULTS OF THE EVALUATIONS OF ENVIRONMENTAL WORK CONDITIONS

The results of the laboratories' evaluations are seen in Figure 1.

Figure 1 shows the survey of the percentage of occupational risks presented in the researched group of laboratories. As seen, 51% were considered irrelevant, with higher rates in the laboratories of Energy Efficiency, Mechanical Testing and Materials, Physics, Informatics, Mechatronics, Metrology, Computational Mechanics and Topography, in which the risks surveyed in the labor activities performed in the facilities did not present potential risks to users' health or physical integrity.

The category of relevant represents 34% of the risks related to activities that present low health hazards, which does not cause acute effects to users with technical control and the supervision of the responsible teachers during the performance of activities in the laboratories.

The relevant risks are related in the organization and structure of laboratories that show environments with packaging of flasks of products and materials disorganized on the floor, besides of having tools and equipment leaning on chairs. They also exist in environments where accidents have already occurred and where the users must wear the personal protective equipment properly and others where the facility has an improvised and limited area in which the activities must be performed. The mentioned risks expose users to the danger of tripping and/or falling on top of materials and equipment, the occurrence of new accidents in the environment and incidents with spilling of products or residue.

The attention risks represent 11% of risks in laboratories where labor and teaching activities are executed with moderate health hazard to users and that do not cause acute effects, in which the exposure is under technical control. The laboratories that presented moderate risks were the Laboratories of Workshop Practices and Informatics.

The existing risks in the Informatics laboratories are electric shocks due to old installations and exposed wires. The same occurs in the Laboratories of Workshop Practices, with machines that operate in average tension and lack grounding, susceptible to electric shock. Besides that, there were identified risks of fire outbreak due to the physical arrangement of the improvised area for welding activities, machine location, materials collection, pieces and room furnishing, leaving users exposed to many risks, such as stumbling or falling on top of a machine, tool or even a colleague, among other risks related to the activities, and what hinders an effective daily sanitation of the place.

In the category of the 4% critical risks is the laboratory of Fuels, due to malfunction and lack of maintenance of the fume hoods, which leaves the users exposed to reagents and chemical products, gases and vapors present in the daily routine of users when performing activities. The occurrence of accidents and incidents was reported by the people responsible for the laboratories, in interviews during the field research, when they mentioned cases of burns, hand cuts and allergic reactions.

6 SAFETY GUIDELINES FOR LABORATORIES

The safety guidelines structured were based on the recommendations made for the laboratories and on the need to deploy procedures of good practices with the following purposes:

1. To present teachers, students and technicians of the Higher Education Institution with the basic rules institutionally defined for access to laboratories;
2. To inform the academic community, particularly the Faculty of Engineering, about the posture and main proceedings to be adopted in laboratories, aiming to protect students, technicians and teachers from risks and accidents;

3. To facilitate the work of coordinators, teachers and technical personnel of the laboratories.

The recommendations and procedures of good practices for the laboratories and workshop of the Higher Education Institution include the following topics:

- Standards for access and permanence in the Laboratory;
- Posture and Procedures inside the Laboratory;
- Personal Protective Equipment—General;
- Specific guidelines for PPEs;
- Maintenance of facilities;
- Laboratory Equipment Maintenance;
- Safety Procedures for Fume Hoods;
- Emergency Guidelines.

7 CONCLUSIONS

As set out above, it was possible to verify through risk survey during field research that the laboratories of the Faculty of Engineering of the Higher Education Institution present low to moderate relevance risks, considering the environmental work conditions of the activities performed in the facilities.

The electrical installations need repair with a certain urgency, as seen by the reports of accidents with electric shocks in the Informatics laboratories mentioned in this paper. As for the repairs of the physical structure, they were recommended so that there are no future damages of greater proportion for the Higher Education Institution.

It is important to highlight the matter of personal and collective protective equipment, used in laboratorial activities as needed. However, it was noted in some laboratories that the PPEs were not used properly. In the Chemistry laboratory case, there are no automatic showers, eyewash units and fume hoods, which are fundamental collective safety equipment for the executed activities. The same happens in the Fuel laboratory, where the existing fume hoods are not working, leaving users exposed to chemical agents that may cause health damages due to the long periods of daily exposure in the workplace.

Another subject that calls for attention is the situation regarding an eventual need to firefighting. The lack of fire extinguishers in some laboratories increase their vulnerability in emergency cases. Still, the mere existence of a fire extinguisher in a given place is no guarantee of safety in firefighting. First, one must act on the potential fire outbreaks in order to reduce this risk to a maximum extent. In parallel, the Institution must hold a fire brigade trained to perform an initial firefighting, so that everyone's life and the material assets are protected.

It is also important to stress the matter of the Safety Committee. As noted above, some cases of accidents and incidents have already occurred. It is valid that the Safety Committee proceed to investigate these cases, both accidents and incidents, with the purpose that these occurrences do not repeat themselves, enabling a better flux of activities.

The results allowed the assessment of risks of labor conditions in the laboratories and sectors of the Institution, as well as the structure of safety guidelines in laboratories, the proposition of improvement actions for the Higher Education Institution and the manufacturing of the digital risk map so that the actions may be deployed by the coordinators responsible for the laboratories and/or the safety sector of the institution.

REFERENCES

Barkokébas, Jr. B. 2005. *Manual do Sistema de Gestão em Segurança e Saúde no Trabalho–SGSST*. Recife: Queiroz Galvão Construction Company S.A.—CQG.

Brazilian Protection Annual Report. Available from: <http://www.protecao.com.br/materias/anuario_brasileiro_de_p_r_o_t_e_c_a_o_2015/brasil/AJyAAA>. Access on 15 May 2015.

Catholic Pontifical University of Rio Grande do Sul. 2015. *Laboratórios de Ensino e Pesquisa*. Available from: <http://www3.pucrs.br/portal/page/portal/fenguni/fenguniCapa/fenguniLaboratorios>. Access on 16 June 2015.

Marzano, L. C. 2015. *A Construção da Competência Frente aos Riscos*. Available from: <https://www.dnvgl.com.br/assurance/services/aconstrucaodacompetencia.html>. Access on 29 February 2016.

Mendes, D. 2014. *Fatores humanos: Gerenciando falhas humanas*. Available from: <http://temseguranca.com/fatores-humanos-gerenciando-falhas-humanas/> Access on 28 April 2014.

Ministry of Education. 2015. *Instituições de Educação Superior e Cursos Cadastrados*. Available from: <http://mec.mec.gov.br/>. Access on 15 May 2015.

Ministry of Labor. *Decree nº 25*, December 29, 1994. Annex IV—Classification of Main Occupational Risks in Groups. Available from: <http://portal.mte.gov.br/data/files/FF8080812BE914E6012BEA44A24704C6/p_19941229_25.pdf>. Access on 10 July 2015.

Polytechnic School of the University of Pernambuco. 2015. *Laboratórios*. Available from: <http://www.poli.br/index.php?opion=com_content&view=artile&id=594&Itemid=270>. Access on 16 June 2015.

Silva, A. *Prevenção de Acidentes*. Available from: <http://pt.slideshare.net/Artursilva09/preveno-de-acidentes-25921169>. Access on 25 March 2016.

Occupational Safety and Hygiene V – Arezes et al. (Eds)
© *2017 Taylor & Francis Group, London, ISBN 978-1-138-05761-6*

Safety in the maintenance of elevators in residential buildings

B.M. Vasconcelos, B. Barkokébas Junior, E.M.G. Lago & E.N. Carvalho
University of Pernambuco—UPE, Recife, Brazil

ABSTRACT: The use of the elevator became indispensable in great metropolises due to the intense and ever-growing verticalization of buildings, and for such an equipment to perform safely, its periodic maintenance is paramount. The three places in which elevator maintenance takes place are: the machine room, the hoistway, and shaft. Aiming to minimize the most frequent accidents occurring in this kind of labour, this paper had as an objective to propose guidelines for management tools aimed at making environments more suitable for the maintenance of residential building elevators. In order to do such, data was gathered from an elevator manufacturing company, study visits were made to residential buildings in the city of Recife and employees of the sampled company were interviewed. After analyzing the data, it was possible suggest guidelines destined for tools aimed at guaranteeing a safe technical conduct for professionnals working with elevator maintenance.

1 INTRODUCTION

The so called "safety" elevator emerged in New York in 1853, when Elisha Graves Otis introduced the safety brake, and from that point on, the users became more confident in using this vertical transport, for there was a safeguard in the case of the platform falling (Santos, 2007). At the end of the nineteenth century, the mastery of elevator technology, which made them increasingly safer and faster, stimulated the construction of higher buildings. Nowadays, elevator manufacturing companies are in constant search for new technologies, for example, the evolution of microelectronics and the development of special carbon steel alloys which allow the commercialization of ropes with smaller diameter and greater resistance.

In Brazil, the use of this type of transport is intense, mainly in metropolises like the city of São Paulo, with approximately 55.000 functioning elevators making 26 million trips every 24 hours (SECIESP, 2015). These numbers highlight the importance of this equipment for modern society, and for it to work safely and perfectly, there must be a periodic maintenance service of the equipment made by a qualified, licensed company.

However, the training of elevator maintenance technicians is restricted to the training centers of the manufacturers, which occurs in the vast majority of Brazilian capitals.

During the execution of elevator maintenance, the licensed professionals are subjected to risk of accidents such as cuts, torsions and crushing. The consequences of such accidents generate both economic and human costs, due to absenteeism, damage compensations, loss of productivity, social effects reflecting on the injured people and their families, negative effects to the company's image, possible legal outcomes according to the extension and severity of the accident, among other factors.

Facing that context, this paper aimed to propose guidelines for management tools aimed at making environments in which services of maintenance of residential building elevators (with speeds up to 105 m/min) are rendered more suitable in order to minimize the most occurring accidents of this kind of labour.

2 METHODOLOGY

Initially, data was collected on statistics concerning accidents occurring with technicians of preventive maintenance in elevators with speeds up to 105 m-min for a certain elevator manufacturing company. Afterwards, technical visits were made to 10 buildings located in the city of Recife, with the application of a checklist, based on the ABNT NM 207 (1999) guidelines, with the objective to develop a diagnosis of the situation of the environments in which elevator maintenance services are being carried out, being:

- Machine Room: Environment dedicated to the traction machine, control panel, console and other components. Items like natural and artificial lighting, the presence of extinguishers and the existence of a grid for the trapdoor were also noted;
- Hoistway: Place in which the elevator moves. The presence of a safety guard rail, lighting and electrical off switch above the cabin was noted;

• Shaft: Spaces located below the level of the lowest floor. The conditions of the access stairway and lighting were checked.

Afterwards, the data collected in the application of the checklist were cross-referenced with the accident statistics of the sampled company. Concomitantly, interviews were made with 21 employees of the same company which performed activities associated with the machine room, traction machine and shaft. Then, using the qualitative and quantitative results reached, it was possible to propose guidelines for management tools which guarantee a safe access for elevator maintenance professionals.

3 RESULTS

3.1 Data from the researched company

The information concerning the workplace accidents which occur in this particular labour was obtained with the workplace safety branch of the sampled company, with data ranging from 2010 to 2014. According to the surveyed documents, the types of registered accidents were: cut, torsion, crushing, contusion, dislocation, perforation, chock, excoriation and fracture. In the case of activities performed by the preventive maintenance technician for elevators with speeds up until 105 m/min, those which happened more often were: cuts, torsion and crushing, representing more than 60% of the total of 101 accidents.

3.2 Application of the checklist

10 buildings were visited, representing a total of 24 elevators, with distinct ages, in order to emphasize that norm NBR 7192/1985 – ABNT (valid until the year of 1999), or norm NBR NM207/1999, which revoked NBR 7192:1985 (with validity start and application on December 30th of 1999), were not or are not being followed exactly in their guidelines.

The results obtained from the visits showed that all buildings had items which favoured the appearance of risky situations for the workers who undertook elevator maintenance. The machine room was either hard to reach or with obstacles which generate risk of fall, and, as a consequence, torsions and (or) cuts, as seen on Photograph 1, in which is also possible to see obstructions in the way to access the extinguishers.

In the hoistways, one of the risks observed was the presence of folded ironwork from the cabin's door (Photograph 2), with sharp extremities and burrs. Besides, according to the data obtained from the safety branch, this situation causes the majority of cut-related accidents in that place.

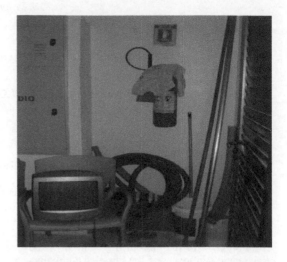

Photograph 1. Access to the machine room being obstructed by objects.

Photograph 2. Folded ironwork from the elevator's cabin door.

Photograph 3. Electric wiring present in the elevator shaft.

In the shafts, there was a frequent presence of power and signal cords, and of equipment that may cause slips, allowing for an accident to occur, and, consequently, lesions (cut, torsion, fracture, bruises), according to Photograph 3.

Interview with the employees

The interview with employees from the sampled company covered a total of 18 maintenance technicians, two safety technicians and a safety supervisor. According to the results, the greater occurrence of accidents, in the opinion of the interviewed technicians, happens in the pavement door, where the percentage compared to other places (machine room, cabin and shaft) reached 52,4%.

Concerning the typology of accidents, the cut was indicated as the most frequent for this labour activity, and "little attention" as deemed by 66,7% of collected opinions as the main causal factor for accidents. In that way, employees consider that being "attentive" while working is fundamental to diminish risks, and, subsequently, the possibility of accidents and their related lesions.

While being questioned if they receive, possess and correctly use the necessary Personal Protective Equipment (PPE), all said yes. Concerning the history of accidents, from a total of 21 employees, six have already suffered some form of accident, in the period of time between 2010 and 2014, suffering the consequences of cuts, torsions and crushing.

4 GUIDELINES FOR MANAGEMENT TOOLS

To follow a procedure for the tasks that comprise the maintenance of residential elevators is the first step to mitigate the probability of accidents happening.

Therefore, operational procedures to the services of elevator maintenance collaborate to establish criteria and subsequent definitions about how to enter these places, thus allowing a more appropriate and safe conduct by the professional.

Substantiated in the treatment and analysis of the collected data, Figure 1 details the components and conditions that offer risky conditions and present danger in the machine room, in the hoistway and in the elevator shaft.

Afterwards, guidelines were proposed for the management tools destined to the machine rooms, to the hoistways and shafts, among which:

– Ensure that the elevator won't move in the automatic mode the moment the maintenance technician accesses the top of cabin as a platform and means of locomotion, aiming to do his job. Otherwise, there is a possibility that serious accidents might occur, like crushing, cuts and torsions;

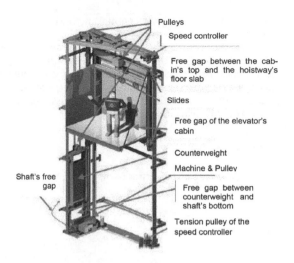

Figure 1. Red zone perspective.
Source: ThyssenKrupp elevators safety manual.

– Ensure that the elevator does not move during entry to the shaft, thus avoiding a serious accident.
– Utilize a sailor-type stair in order to enter and exit the shaft, being that the use of springs, shock absorbers or buffers as a support is not recommended due to the probability of slips which lead to accidents that may cause cuts, torsions and crushing.
– Signal objects like: steps, pipes, pipelines, floor and roof hooks, inverse beams, ceiling height smaller than two meters, ramps, curls and holes in the floor slab.
– Verify the electrical grounding and the power box before the work begins.

5 CONCLUSION

After gathering, treating and analyzing the data coming from statistic data, and from the safety conditions in workplaces and from interviews made with employees of the sampled company, it was possible to reach enough substance to propose guidelines for management tools, destined to the environments of the machine room, hoistway and shaft of elevators situated inside residential buildings.

It is worth noticing that operational procedures were elaborated with basis on those guidelines and have been tested and implemented in the maintenance services of the sampled company.

That being so, both management and tech team are already reaping the benefits of the practice and allowing it to be diffused to other professionals of the area, as well as to professionals who work with technical labour forensics.

With the use of those operational procedures, it is expected that the professional maintenance technician fosters a safe attitude while performing his work, the number of accidents falls, and consequently, the human and economic costs associated with workplace accidents.

REFERENCES

Associação Brasileira de Normas Técnicas. 1998. *Elevadores de passageiros, elevadores de carga, monta-cargas e elevadores de maca—Projeto, fabricação e instalação.* Rio de Janeiro.

Associação Brasileira de Normas Técnicas. 1999. *NBR NM 207: Elevadores elétricos—Requisitos de segurança para construção e instalação.* Rio de Janeiro.

Santos, A.B. A interface do elevador na arquitetura. Aspectos projetuais, éticos e sociais. São Paulo. Dissertação. USP. 2007. Dissertação de mestrado.

SECIESP—Sindicato das Empresas de Fabricação Instalação Modernização Conservação e Manutenção de Elevadores do Estado de São Paulo. 2015. Disponível em: <http://www.seciesp.com.br>.

Thyssenkrupp Elevadores. 2015. *Manual do usuário.* Disponível em:<http://www.thyssenkruppelevadores.com.br>.

Occupational Safety and Hygiene V – Arezes et al. (Eds)
© *2017 Taylor & Francis Group, London, ISBN 978-1-138-05761-6*

Carcinogenic polycyclic aromatic hydrocarbons in classrooms of schools: Risk assessment for primary school teachers

K. Slezakova & M.C. Pereira

LEPABE, Departamento de Engenharia Química, Faculdade de Engenharia, Universidade do Porto, Porto, Portugal

J. Madureira & E. de Oliveira Fernandes

Institute of Science and Innovation on Mechanical Engineering and Industrial Management, Porto, Portugal

M. Oliveira, C. Delerue-Matos & S. Morais

REQUIMTE–LAQV, Instituto Superior de Engenharia do Porto, Instituto Politécnico do Porto, Porto, Portugal

ABSTRACT: This study evaluated risks associated with inhalation exposure to carcinogenic PAHs for populations of primary schools teachers. Ten carcinogenic (probable, possible) $PM_{2.5}$–bound PAHs were collected in indoor air of 9 classrooms at four Portuguese primary schools (PS_C–PS_R) during winter–spring 2014. The highest total levels of ten carcinogenic ($\Sigma_{10}PAH = 21.3$–25.6 ng m^{-3}) were found in classrooms at a school with busy traffic (in the surrounding streets) whereas the lowest indoor concentrations were obtained in classrooms of a school situated in residential zone (0.47–0.72 ng m^{-3}). Dibenz[a,h]anthracene (21–29% of $\Sigma_{10}PAH$), benzo[b]fluoranthene (13–28%) and naphthalene (5–40%) were the most abundant congeners in indoor air of classrooms, whereas benz[a]anthracene was among the least abundant PAHs (1–3%). Values of total incremental lifetime cancer risks (i.e. sum of risks of ten carcinogenic PAH congeners) were below the USEPA guideline level of 10^{-6} for teachers at all studied schools (independently of the duration of the employment) and thus the risks can be considered as negligible.

1 INTRODUCTION

Teacher's well-being is an important parameter for successful education of students and their future economic success. The abilities to engage in the teaching processes are partly dependent on the well-being of teachers. Factors, such as poor work conditions, workplace relationships and intrapersonal factors can cause high levels of distress among these professionals (Zinsser et al. 2016; Friedman-Krauss et al. 2014a, b). A poor indoor environment and the quality of respective air may also lead to various health problems and adversely affect performance of teachers (Kielb et al. 2014; Muscatiello et al. 2014). Understanding the associations between school environment and the adverse health effects is important because of the amount of time teachers and students spent there. In regard to indoor air pollution, World Health Organization (WHO) designated a list of priority health–relevant air pollutants (WHO 2010) that includes Polycyclic Aromatic Hydrocarbons (PAHs).

PAHs are a class of organic ubiquitous pollutants with at least two aromatic rings. PAHs result from an incomplete combustion of organic material

(coal, wood, oil). In an ambient of densely populated urban areas, PAHs mostly result from vehicle exhausts (Ravindra et al. 2008) but emissions from cooking, various combustions (fuels, tobacco, candle burning) are the relevant sources indoors (Wu et al. 2015; Qi et al. 2014; Slezakova et al. 2014a; Shen et al. 2012). Once emitted to the atmosphere, molecular weight influences the fate of PAH mixtures. Lighter PAHs mostly stay in gaseous form until being removed (via precipitation) whereas heavier PAHs mostly bound to atmospheric particles, being predominantly associated with fine fraction (i.e. $PM_{2.5}$; aerodynamic diameter below 2.5 µm particles; Slezakova et al., 2013a, b). Due to their adverse properties (mutagenicity, genotoxicity), PAHs have gained the scientific attention. They are designated as endocrine disrupting chemicals (WHO 2013). Cancer is the most significant endpoint of PAH toxicity; occupational (i.e. industrial) exposure to PAHs has been linked with increased incidences of various cancers (lung, skin, bladder cancers; IARC 2010). In addition, selected individual PAHs are considered as carcinogenic (probable and/or possible) by several key health organizations such as International Agency for

Research on Cancer (IARC), US Environmental Protection Agency (USEPA) or US Department of Health and Human Services (USHHS).

Because of the importance of this topic, studies dedicated to assessment of PAHs in educational environments (nurseries, day care centers, preschools or primary/elementary schools) have been emerging (Oliveira et al. 2016a, 2015; Alves et al. 2014; Jovanović et al. 2014; Krugly et al. 2014; Romagnoli et al. 2014; Carpente & Bushkin-Bedient 2013; Eiguren-Fernandez et al. 2007; Cirillo et al. 2006). It should be though pointed out that children are the major focus of those works, up to this date very little is known about the health risks among the school staff or the respective teachers (Oliveira et al. 2015; Annesi-Maesano et al. 2013).

This work evaluated risks associated with inhalation exposure to carcinogenic PAHs for populations of primary schools teachers. Levels of ten carcinogenic (probable, possible) particulate PAHs were considered in indoor air of classrooms at four Portuguese primary schools and the incremental lifetime cancer risks were assessed according to the USEPA methodology USEPA 2016a).

2 MATERIALS AND METHODS

2.1 Selected educational institutes

Levels of particulate (i.e. $PM_{2.5}$) PAHs were assessed in indoor air of four primary schools (i.e. providing an education for 6–10 yrs old students) during winter/spring 2014. All schools were public institutions (with 256–446 students) situated in city of Oporto, Portugal. Primary school PS_C was situated in a city center with high density of road traffic in the neighboring streets, whereas PS_{U1} and PS_{U2} were considered as urban sites with some influences from local industries; PS_R was in a residential area and surrounded with greenery. The school buildings were built after 1940, with renovations and repairs done in all of them within the last ten years.

In each school, daily samples of PAHs were collected simultaneously in indoor air of (at least) two classrooms during three consecutive days (Tuesday–Thursday). Layout, materials and equipment existent in each room were registered. Activities, daily patterns and schedules of teachers were recorded; on daily basis they spent approximately 6 h in the classrooms. Finally, precautions were taken in order to ensure undisturbed and safe working environment for teachers and their students during the sampling campaign.

PAHs sampling was done according to USEPA IP–10 A method (USEPA 1990) by a personal environmental monitor (PEM™; SKC Ltd., Dorset,

United Kingdom) coupled with a personal air sampling pump (AirChek® 2000; SKC Ltd., Dorset, United Kingdom) using a flow rate of 2.0 L min⁻¹. Particulate samples were collected on polytetrafluoroethylene membranes (SKC Ltd., Dorset, United Kingdom; 2 μm porosity, Ø37 mm); after gravimetric mass determination (Oliveira et al. 2016a; Slezakova et al. 2014b) disks were stored in containers in freezer (–18°C) for subsequent chemical analysis.

2.2 PAHs extraction and quantification

Determination of particulate PAHs was based on liquid chromatography with fluorescence detection and described elsewhere (Castro et al. 2011, 2009). Specifically, ten carcinogenic PAHs were considered, namely benzo[a]pyrene (class 1 human carcinogen), two probable (class 2 A; dibenz[a,h]anthracene dibenzo[a,l]pyrene) and seven possible (2B; naphthalene, chrysene, benz[a]anthracene, benzo[b+j] fluoranthene, benzo[k]fluoranthene, indeno[1,2,3-cd] pyrene) carcinogens (IARC 2010).

2.3 Risk calculations

Incremental Lifetime Cancer Risks (ILCR) for school teachers were calculated according to the USEPA approach (Region III Risk-based Concentration Table; USEPA 2016a). ILCR values <10⁻⁶ are denoted as safe ones, values >10⁻⁶ and <10⁻⁴ indicate potential risks, whilst ICLR >10⁻⁴ designate potentially high risks (USEPA 1989). The details of the complete methodology and the ILCR calculations are fully given in Oliveira et al. (2016a, b, 2015).

Two periods of employments (15 and 25 yrs for younger and older staff, respectively) were considered; based on the USEPA recommendations the exposure frequency for this occupational exposure (i.e. indoor workers) was considered as 250 days per year (USEPA 2016b).

3 RESULTS AND DISCUSSION

3.1 Carcinogenic PAH levels

Concentrations of carcinogenic $PM_{2.5}$–PAHs in indoor air of four primary schools are shown in Figure 1. The concentrations of dibenzo[a,l] pyrene were below its limits of detection and quantification in all the collected samples; hence the compound is not presented. Considering all analyzed classrooms, total levels of carcinogenic PAHs (Σ_{10}PAHs) ranged between 0.47 ng m⁻³ (at PS_R) and 25.6 ng m⁻³ (PS_C) whilst the mean concentrations estimated for each school were between 0.60 and 23.6 ng m⁻³. The information regarding levels of PAHs in educational settings is in general rather limited. Furthermore, different study designs, varying number of PAH congeners included in the assessment, and the different phases/matrices considered (between PM_X–PM_{10} or even non-specified; inclusion/omission of the gas phase) confound the possible comparison of the existent studies. In addition, geographical, meteorological and cultural differences between the conducted studies further complicate assessments of the reported findings. Regarding the $PM_{2.5}$-bound PAHs, some authors (Eiguren-Fernandez et al. 2007) observed levels of PAHs (15 congeners) ranging between 0.4 and 1.8 ng m^{-3} in schools in USA. Much higher levels (20.1–131 ng m^{-3}) were though reported for schools in Lithuania (Krugly et al. 2014); Cirillo et al. (2006) observed lower levels (1.6–16 ng m^{-3}) in Italian schools that were somewhat similar to this work. In agreement, $PM_{2.5}$-bound PAHs in a range of 1.8–8.3 ng m^{-3} were reported in indoor air of schools in Rome (Romagnoli et al. 2014).

The highest levels of PAHs were found at PS_C (Figure 1). This school was situated in a city center. In a view that no indoor source of PAHs was identified at this school, it is assumed that indoor concentrations mostly resulted from penetrations of outdoor emissions indoors (by ventilations, due to inferior building isolation, *etc.*); only at PS_C classrooms were very intensively ventilated. These findings were in agreement with other studies (Rivas et al. 2015; Amato et al. 2014; Moreno et al. 2014). Specifically at PS_C, the indoor levels of Σ_{10}PAHs were about 5 to 40 times higher than in schools that were situated in urban zones (with some industrial influences) and in a residential area, respectively.

Benzo[b]fluoranthene and dibenz[a,h]anthracene were the most abundant congeners at PS_C, accounting for 28 and 21% of Σ_{10}PAHs, respectively. High abundance of these compounds

indicate traffic emissions (Ravindra 2008), in agreement with the previous findings. These two compounds were though abundant in other schools; at PS_{U1}–PS_R benzo[b]fluoranthene contributed 13–19% whereas it was 26–29% of Σ_{10}PAHs for dibenz[a,h]anthracene. Apart from these two PAHs, naphthalene was an abundant congener, with contributions ranging between 5% of Σ_{10}PAHs (at PS_C) up to 40% (at PS_{U1}) of Σ_{10}PAHs. In view of the potential health implication naphthalene is considered as priority indoor pollutants which levels in indoor air should be monitored (WHO 2010). It should be also pointed out that the current health recommendation; (given as annual guideline value of 10 mg m^{-3}; WHO 2010) is the only existing guideline for PAHs in indoor air. The maximal concentrations of naphthalene (range of 0.100 3.48 ng m^{-3}) were at all primary schools much lower than the recommended guideline. However, the levels of naphthalene in the gaseous form (which was not included in this study) typically largely surpasses its particulate content (due to the high volatility of the compound; Oliveira et al. 2016; Castro et al. 2011) so the actual concentrations in indoor air could be much higher than here estimated. Therefore, in future studies the content of this congener in the gas phases should be considered when correctly assessing its atmospheric levels. Finally, benz[a]anthracene was the least abundant PAHs (1–3%).

3.2 *Risk assessment*

Incremental lifetime cancer risks for each PAH compound as well as the total (i.e. Σ_{10}ILCR) associated with inhalation exposure to particulate PAHs for teachers employed for 15 and 25 yrs at four primary schools are summarized in Table 1.

Σ_{10}ILCR values for teachers employed both 15 and 25 yrs were below (1.4–106 times) the recommended guideline of 10^{-6} at all schools and the respective risks can be considered as negligible at PS_C–PS_R for the populations of school teachers but also for the students. The highest risks (5–44 times) were found for teachers who worked at PS_C most likely due to the overall higher PAHs pollution indoors; in agreement the lowest risks were identified for teachers working at the PS_R that was situated in residential (and relatively clean) area. The risks of teachers employed 25 yrs were approximately twice higher than of those with 15 yrs employment due to the longer occupation duration. Among the individual PAHs, the highest ICLR were observed for dibenz[a,h]anthracene (53% at PS_C – 78% of Σ_{10}ILCR at PS_{U1}) and benzo[a]pyrene (12% at PS_{U1} – 38% of Σ_{10}ILCR at PS_C).

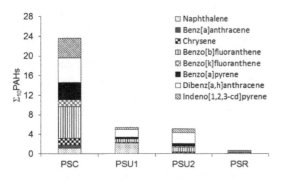

Figure 1. Levels of carcinogenic PAHs in indoor air of classrooms at four primary schools (PS_C–PS_R).

Table 1. Incremental lifetime cancer risks (ILCR) due to indoor exposure to carcinogenic $PM_{2.5}$–bound PAHs at four primary schools (PS_C–PS_R) for populations of teachers employed for 15 and 25 yrs.

15 yrs employed	PS_C	PS_{U1}	PS_{U2}	PS_R
Nap	1.45×10^{-9}	2.66×10^{-9}	1.38×10^{-10}	2.17×10^{-10}
B[a]A	2.69×10^{-9}	1.87×10^{-10}	2.98×10^{-10}	5.51×10^{-11}
Chry	5.36×10^{-10}	5.27×10^{-11}	6.69×10^{-11}	1.44×10^{-11}
B[b+j]F	2.65×10^{-8}	2.81×10^{-9}	3.90×10^{-9}	4.50×10^{-10}
B[k]F	5.33×10^{-9}	4.88×10^{-10}	7.75×10^{-10}	7.92×10^{-11}
B[a]P	1.42×10^{-7}	1.07×10^{-8}	2.28×10^{-8}	1.51×10^{-9}
D[a,h]A	2.20×10^{-7}	6.80×10^{-8}	9.91×10^{-8}	6.92×10^{-9}
InP	1.62×10^{-8}	1.78×10^{-9}	3.02×10^{-9}	2.28×10^{-10}
Σ_{10}ILCR	4.15×10^{-7}	8.67×10^{-8}	1.30×10^{-7}	9.47×10^{-9}

25 yrs employed	PS_C	PS_{U1}	PS_{U2}	PS_R
Nap	2.42×10^{-9}	4.44×10^{-9}	2.29×10^{-10}	3.62×10^{-10}
B[a]A	4.49×10^{-9}	3.12×10^{-10}	4.97×10^{-10}	8.68×10^{-11}
Chry	8.94×10^{-10}	8.78×10^{-11}	1.11×10^{-10}	2.39×10^{-11}
B[b+j]F	4.42×10^{-8}	4.68×10^{-9}	6.50×10^{-9}	7.49×10^{-10}
B[k]F	8.88×10^{-9}	8.13×10^{-10}	1.29×10^{-9}	1.32×10^{-10}
B[a]P	2.37×10^{-7}	1.78×10^{-8}	3.80×10^{-8}	2.51×10^{-9}
D[a,h]A	3.67×10^{-7}	1.13×10^{-7}	1.65×10^{-7}	1.15×10^{-8}
InP	2.69×10^{-8}	2.97×10^{-9}	5.04×10^{-9}	3.81×10^{-10}
Σ_{10}ILCR	6.92×10^{-7}	1.44×10^{-7}	2.17×10^{-7}	1.58×10^{-8}

Naphthalene (Nap); Benz[a]anthracene (B[a]A); Chrysene (Chry); Benzo[b+j]fluoranthene (B[b+j]F); Benzo[k]fluoranthene (B[k]F); Benzo[a]pyrene (B[a]P); Dibenz[a,h]anthracene (D[a,h]A); Indeno[1,2,3-cd]pyrene (InP).

4 CONCLUSIONS

This study analyzed risks associated with inhalation exposure to ten carcinogenic (possible, probable) $PM_{2.5}$–bound PAHs for populations of primary teachers. Overall, a school situated in city center (PS_C) with high density of traffic was the most polluted one and exhibited the highest levels (5–40 times) of indoor PAHs, whereas the lowest concentrations were obtained in classrooms of a school situated in residential zone (PS_R). Out of ten analyzed congeners, dibenz[a,h]anthracene (21–29% of Σ_{10}PAH) and benzo[b]fluoranthene (13–28% of Σ_{10}PAHs) were the most abundant PAHs in indoor air, whereas benz[a]anthracene was among the least abundant ones (1–3%).

Incremental lifetime cancer risks were below USEPA threshold of 10^{-6} for teachers of all primary schools independently of the duration of employment. However, it is necessary to highlight that the risk estimations were conducted with PAHs being monitored in the classrooms; evaluating the daily schedules and activity patterns teachers spent approximately 70% of their time there. Thus, other indoor environments polluted to some degree with PAHs (canteens and cafeterias, teacher lounges and offices, corridors, school passages, and *etc.*) should be also included as the respective exposures in these specific indoor environments might be relevant for the risk assessment.

ACKNOWLEDGMENTS

This work was supported by European Union (FEDER funds through COMPETE) and National Funds (Fundação para a Ciência e Tecnologia) through projects UID/QUI/50006/2013, POCI-01–0145-FEDER-007265 and POCI-01–0145-FEDER-006939, by the FCT/MEC with national funds and co-funded by FEDER in the scope of the P2020 Partnership Agreement. Additional financial support was provided by Fundação para Ciência e Tecnologia through project PTDC/DTP-SAP/1522/2012) and fellowships SFRH/BPD/105100/2014 (Klara Slezakova) and SFRH/BD/80113/2011 (Marta Oliveira).

REFERENCES

Alves, C.A., Urban, R.C., Pegas, P.N. & Nunes, T. 2014. Indoor/Outdoor relationships between PM10 and associated organic compounds in a primary school. *Aerosol and Air Quality Research* 14: 86–98.

Amato, F., Rivas, I., Viana, M., Moreno, T., Bouso, L., Reche, C., Àlvarez-Pedrerol, M., Alastuey, A.,

Sunyer, J. & Querol, X. 2014. Sources of indoor and outdoor PM2.5 concentrations in primary schools. *Science of the Total Environment* 490: 757–765.

Annesi-Maesano, I., Baiz, N., Banerjee, S., Rudnai, P. & Rive, S. on behalf of the SINPHONIE Group. 2013. Indoor air quality and sources in schools and related health effects. *Journal of Toxicology and Environmental Health - Part B: Critical Reviews* 16(8): 491–550.

Carpente, D.O. & Bushkin-Bedient, S. 2013. Exposure to chemicals and radiation during childhood and risk for cancer later in life. *Journal of Adolescent Health*, 52(5 Suppl): S21–29.

Castro, D., Slezakova, K., Delerue–Matos, C., Alvim–Ferraz, M.C., Morais, S. & Pereira, M.C. 2011. Polycyclic aromatic hydrocarbons in gas and particulate phases of indoor environment influenced by tobacco smoke: levels, phase distribution and health risks. *Atmospheric Environment* 45: 1799–1808.

Castro, D., Slezakova, K., Oliva–Teles, M.T., Delerue–Matos, C., Alvim–Ferraz, M.C., Morais, S. & Pereira, M.C. 2009. Analysis of polycyclic aromatic hydrocarbons in atmospheric particulate samples by microwave–assisted extraction and liquid chromatography. *Journal of Separation Science* 32: 501–510.

Cirillo, T., Montuori, P., Mainardi, P., Russo, I., Triassi, M. & Amodio-Cocchieri, R. 2006. Multipathway polycyclic aromatic hydrocarbon and pyrene exposure among children living in Campania (Italy). *Journal of Environmental Science and Health - Part A Toxic/Hazardous Substances and Environmental Engineering* 41: 2089–2107.

Eiguren-Fernandez, A., Avol, E.L., Thurairatnam, S., Hakami, M., Froines, J.R. & Miguel, A.H. 2007. Seasonal influence on vapor-and particle-phase polycyclic aromatic hydrocarbon concentrations in school communities located in southern California. *Aerosol Science and Technology* 41: 438–446.

Friedman-Krauss, A.H., Raver, C.C., Morris, P.A. & Jones, S.M. 2014a. The role of classroom-level child behavior problems in predicting preschool teacher stress and classroom emotional climate. *Early Education and Development* 25(4): 530–552.

Friedman-Krauss, A.H., Raver, C.C., Neuspiel, J.M. & Kinsel, J. 2014b. Child behavior problems, teacher executive functions, and teacher stress in head start classrooms. *Early Education and Development* 25(5): 681–702.

IARC Working Group on the Evaluation of Carcinogenic Risks to Humans (IARC). 2010. Some non-heterocyclic polycyclic aromatic hydrocarbons and some related exposures. *IARC Monographs on the Evaluation of Carcinogenic Risks to Humans* 92: 1–853.

Jovanović, M., Vučićević, B., Turanjanin, V., Živković, M. & Spasojević, V. 2014. Investigation of indoor and outdoor air quality of the classrooms at a school in Serbia. *Energy* 77: 42–48.

Kielb, C., Lin, S., Muscatiello, N., Hord, W., Rogers-Harrington, J. & Healy, J. 2015. Building-related health symptoms and classroom indoor air quality: A survey of school teachers in New York State. *Indoor Air* 25(4): 371–380.

Krugly, E., Martuzevicius, D., Sidaraviciute, R., Ciuzas, D., Prasauskas, T., Kauneliene, V., Stasiulaitiene, I. & Kliucininkas, L. 2014. Characterization of particulate

and vapor phase polycyclic aromatic hydrocarbons in indoor and outdoor air of primary schools. *Atmospheric Environment* 82: 298–306.

Moreno T., Rivas, I., Bouso L., Viana, M., Jones, T., Àlvarez-Pedrerol, M., Andrés Alastuey, A., Sunyer, J.& Querol, X. 2014. Variations in school playground and classroom atmospheric particulate chemistry. *Atmospheric Environment* 91: 162–171.

Muscatiello, N., Mccarthy, A., Kielb, C., Hsu, W.-H., Hwang, S.-A. & Lin, S. 2015. Classroom conditions and CO2 concentrations and teacher health symptom reporting in 10 New York State Schools. *Indoor Air* 25(2): 157–167.

Oliveira, M., Slezakova, K., Delerue-Matos, C., Pereira, M.C. & Morais, S. 2015. Exposure to polycyclic aromatic hydrocarbons and assessment of potential risks in preschool children. *Environmental Science and Pollution Research* 22 (18): 13892–13902.

Oliveira, M., Slezakova, K., Delerue-Matos, C., Pereira, M.C. & Morais, S. 2016a. Assessment of polycyclic aromatic hydrocarbons in indoor and outdoor air of preschool environments (3–5 years old children). *Environmental Pollution* 208: 382–394.

Oliveira, M., Slezakova, K., Delerue-Matos, C., Pereira, M.C. & Morais, S. 2016b. Assessment of air quality in preschool environments (3–5 years old children) with emphasis on elemental composition of PM10 and PM2.5. *Environmental Pollution* 214: 430–439.

Qi, H., Li, W.L., Zhu, N.Z., Ma, W.L., Liu, L.Y., Zhang, F. & Li, Y.F. 2014. Concentrations and sources of polycyclic aromatic hydrocarbons in indoor dust in China. *Science of The Total Environment* 491–492: 100–107.

Ravindra, K., Sokhi, R. & Grieken, R.V. 2008. Atmospheric polycyclic aromatic hydrocarbons: source attribution, emission factors and regulation. *Atmospheric Environment* 42: 2895–2921.

Rivas, I., Viana, M., Moreno, T., Bouso, L., Pandolfi, M., Àlvarez-Pedrerol, M., Forns, J., Alastuey, A., Sunyer, J. & Querol, X. 2015. Outdoor infiltration and indoor contribution of UFP and BC, OC, secondary inorganic ions and metals in PM2.5 in schools. *Atmospheric Environment* 106: 129–138.

Romagnoli, P., Balducci, C., Perilli, M., Gherardi, M., Gordiani, A., Gariazzo, C., Gatto, M.P. & Cecinato, A. 2014. Indoor PAHs at schools, homes and offices in Rome, Italy. *Atmospheric Environment* 92: 51–59.

Shen, G., Wei, S., Zhang, Y., Wang, R., Wang, B., Li, W., Shen, H., Huang, Y., Chen, Y., Chen, H., Wei, W. & Tao, S. 2012. Emission of oxygenated polycyclic aromatic hydrocarbons from biomass pellet burning in a modern burner for cooking in China. *Atmospheric Environment* 60: 234–237.

Slezakova, K., Castro, D., Delerue-Matos, C., Alvim-Ferraz, M.C.M., Morais, S. & Pereira, M.C. 2013a. Impact of vehicular traffic emissions on particulate-bound PAHs: Levels and associated health risks. *Atmospheric Research* 127: 141–147.

Slezakova, K., Castro, D., Delerue-Matos, C., Morais, S. & Pereira, M.C. 2014. Levels and risks of particulate-bound PAHs in indoor air influenced by tobacco smoke: A field measurement. *Environmental Science and Pollution Research* 21: 4492–4501.

Slezakova, K., Morais, S. & Pereira, M.C. 2014b. Trace metals in size-fractionated particulate matter in a

151

Portuguese hospital: exposure risks assessment and comparisons with other countries. *Environmental Science and Pollution Research* 21(5): 3604–3620.

Slezakova, K., Pires, J.C.M., Castro, D., Alvim-Ferraz, M.C.M., Delerue-Matos, C., Morais, S. & Pereira, M.C., 2013b. PAH air pollution at a Portuguese urban area: carcinogenic risks and sources identification. *Environmental Science and Pollution Research* 20(6): 3932–3945.

US Environmental Protection Agency (USEPA). 1989. *Risk assessment guidance for superfund, Vol. I: Human health evaluation manual.* EPA/540/1–89/002. Washington, DC: Office of Emergency and Remedial Response.

US Environmental Protection Agency (USEPA). 1990. *Compendium of methods for the determination of air pollutants in indoor air.* Research Triangle Park, NC: Atmospheric Research and Exposure Assessment Laboratory.

US Environmental Protection Agency (USEPA). 2016a. *Risk–based concentration table*, retrieved from <http://www.epa.gov/reg3hwmd/risk/human/rb-concentration_table/usersguide.htm>. Accessed in October 2016.

US Environmental Protection Agency (USEPA). 2016b. *Regional screening table user's guide*, retrieved from <http://www.epa.gov/reg3hwmd/risk/human/rb-concentration_table/usersguide.htm>. Accessed in October

World Health Organization (WHO) 2013. *State of the science of endocrine disrupting chemicals 2012.* Geneva: United Nations Environment Programme and the World Health Organization.

World Health Organization (WHO). 2010. *WHO Guidelines for indoor air quality: selected pollutants.* Copenhagen, Denmark: Regional Office for Europe of the World Health Organization.

Wu, F., Liu, X., Wang, W., Man, Y.B., Chan, C.Y., L Liu, W., Tao, S. & Wong, M.H. 2015. Characterization of particulate-bound PAHs in rural households using different types of domestic energy in Henan Province, China. *Science of the Total Environment* 536: 840–846.

Zinsser, K.M., Christensen, C.G. & Torres L. 2016. She's supporting them; who's supporting her? Preschool center-level social-emotional supports and teacher well-being. *Journal of School Psychology* 59: 55–66.

Occupational Safety and Hygiene V – Arezes et al. (Eds)
© 2017 Taylor & Francis Group, London, ISBN 978-1-138-05761-6

The process of knowledge transmission and regulation of work and its relation to the aging process

Larissa Silva
Faculty of Engineering, Univeristy of Porto (FEUP), Porto, Portugal

Liliana M. Cunha
Faculty of Psychology and Education Sciences, University of Porto (FPCE), Porto, Portugal

ABSTRACT: Population aging brings serious challenges to contemporary society, burdening social security systems and National Health Services. One of the many challenges brought by this phenomenon, which we have targeted in this study, refers to the sustainable exercise of work activity and the necessary preservation of the benefits derived from the knowledge acquired by senior workers. Figuring out how to benefit from what senior workers have to offer, their years of experience and their ability to deal with unexpected situations, is critical to the performance of certain functions. With this in mind, this paper presents a study carried out in a metal industry, aiming at understanding the process of knowledge-transfer from experienced workers that are about to retire due to their advanced age, to the younger workers that will replace them.

Keywords: aging, learning strategies, know-how, knowledge transmission

1   INTRODUCTION

The number of mature-aged workers is growing quickly. Increasing employment levels and extending the active life of people has been a priority of Portuguese and European policies since the late 1990s. The EU-28 employment rate for citizens aged 55–64 years increased from 39.9% in 2003 to 50.1% in 2013. These figures are still lower than the employment rate for the 22–64 age group, but they are progressively getting closer. The average age of people exiting the labor market increased from 59.9 in 2001 to 61.5 years of age in 2010 (OSHAS, 2016).

The present study is based on the recognition of the need to ensure the transfer of knowledge and strategies of work regulation developed by the most experienced workers to younger workers performing the same duties.

In the context of a Portuguese company confronted with this difficulty, the main concern is how to make new workers acquire the expertise acquired by senior workers, a concern directly related to productivity, particularly since we are dealing with a company that works almost exclusively for one client.

2   OBJECTIVES, MATERIALS AND METHODS

2.1   *Objectives*

The objective of the study was to study situations of knowledge transfer in the daily life of these workers, while taking into consideration their surrounding environment, that is, the aspects that determine and condition their performance in a certain way, since a worker's performance always responds to the variable constraints that he or she faces.

2.2   *Materials and methods*

Considering the defined objective, observations were made in the real work context of the people involved in the study. Semi-structured interviews were also carried out, comprising four workers and the person in charge of the company.

By observing work activities in a real-world context and through direct contact with workers, the intention was to grasp the reality of the labor activity, its demands and the strategies of regulation pursued by the workers, as well as the way this knowledge is shared between them. For seven months, the activity of the workers was observed in a daily basis, during working hours.

Based on these preliminary observations, and in the bibliographic research, data was gathered that substantiated the issues included in the interview scripts. After the interviews were held, data analysis and treatment was carried out, and parts from interviews that showed the worker's points of view on the problematic under analysis were selected.

2.3   *Sample selection*

For this study, participants were intentionally selected, taking into account the purpose of the study and the conditions of the company. In this case, two predefined criteria were met: (1) workers who were close to leaving the company for retirement; (2) workers who were to replace the previous ones in their position and in the functions that they performed.

Participants ranged from 21 to 66 years and the seniority variation in the company ranged from 1 to more than 20 years. Daily contacts were established with these workers and with the company manager throughout the research process.

3 RESULTS

The strategies of work regulation are materialized primarily through the design of equipment, which is also a form of collective knowledge transfer—building a knowledge legacy within the company—and secondly through the adaptation of the younger workers to the studied areas.

3.1 Collective transmission processes: strategies of work regulation through the (re)design of equipment

The company under study stands out for its adaptability in the face of adversity. As it works for only one customer, a multinational company in the automotive sector, over the years it had to develop mechanisms to meet the demands of that costumer, from increasing production capacity to the creation of new machines and tools to perform the work.

In an interview with the company manager, he explained how the opportunity to work with the current client emerged. It was the client that approached him with some pieces that needed to be produced, asking if he could do it, and if so, if production could start immediately. Being only a few meters away from this client was one of the company's main strengths, allowing it to be supplied anytime and almost instantaneously. Hence, in 2005 the company began to manufacture almost exclusively for the multinational.

The beginning of the production was performed with machines that the customer borrowed for this purpose, but a significant part of the production process was carried out manually, such as flange fitting.

Over time, production increased and the company could not keep up with the growth rate of its client, so it had to set up machines that could help increase its own production.

The first machine created was for the *bridas* section (pieces formed by joining a rubber component with a metal part, whose ultimate goal is to protect car pipes), as it was the most time-consuming task and involved a high number of workers in order

to meet the customer's needs. Then, together with one of his machine suppliers, the manager of the company decided to reproduce the process of wine-racking in order to fit a rubber piece into a bracket (which forms the part called "brida"). In the interview he reports: "I thought of the wine-racking machines, my supplier made a prototype, I ran some tests at home, and since we were not satisfied, we improved the machine." From there, he optimized the machines and ordered four more, the only difference being the fit, because each bracket has a specific shape and different tools are needed. Currently the company has seven flange-mounting machines, and one more is under test to assemble two more types of flanges.

In addition to these machines, the most experienced worker in the flange section felt the need to create a tool to help him open the carton boxes where the materials that he needed were stored. During the interview, he reported to have experienced difficulties and even cut himself when opening boxes, which led him to develop a cutting tool. He took a plate that came with the packaging of another raw material item, removed a part, folded it, and sharpened it until he obtained a tool similar to a knife. Currently, there are several copies of this tool around the company, used by every worker. Workers developed other techniques for regulating their work, such as the use of a thermo-ventilator in the flange section to warm the components before using them. According to one worker, the heating process facilitates the fitting of the part when assembling it, thus optimizing production time and facilitating the worker's task, since the effort in the fitting of the components was significantly less.

3.2 Transfer of knowledge: analysis of the integration process of a new worker in the flange and cutting sections

The integration of a worker may be due to: (1) admittance of a new worker in the company; or (2) changing of position. Both demand a set of actions for their adaptation, safety and health, which are directly related to work methods developed by more experienced workers and to the risks associated with their activity. At present, the company does not have a "welcome plan" for new employees. Instead, the manager personally follows them on their first day at work, presenting them to their colleagues and to the machine/workstation on which they will work.

Through this process, which depends on numerous variables, the company manager explains how to work with the machine. In some cases, the manager is replaced in this task by a more experienced worker.

Particular attention was paid to the adaptation of new workers in two tasks, in the flange section and in the cutting section. All the dynamics that are necessary in the exercise of the activity, as well as its vicissitudes (e.g. the creation of alternatives depending on what is requested at each moment) were considered.

Throughout the observation period, the worker in the *bridas* section, who was about to retire, was concerned with his succession, explaining to the novice worker how to proceed in certain situations and the attention he should pay when assembling parts, among other situations.

4 DISCUSSION OF RESULTS

We have observed several strategies of work regulation developed and adopted by the workers and the manager during our study. As Barros-Duarte (2006) points out, regulatory strategies reveal a way for the worker to guide his or her work. This can be shown by the examples given in Results (point 3). According to Onstenk (1995), these strategies can be perpetuated within a company when training a new worker, provided that certain conditions are ensured. Regarding the knowledge transfer between experienced and younger workers, the analysis of our results shows that it occurs while workers are in the exercise of their functions, which forces them to conciliate knowledge transfer and production. This is in line with the conclusions of Thébault (2013), which states that the process of knowledge transfer takes place during production, which on the one hand is an advantage—the process takes place in a real-work situation and while performing the activity it refers to. On the other hand, not being granted a specific amount of time for this activity may translate into an overload of work (having to carry out ordinary work duties while simultaneously identifying what is relevant in terms of knowledge transfer).

As such, it is important to implement explicit forms of transfer, eventually endowed with a greater formality, to promote awareness within the company for safeguarding conditions for this process. Otherwise, the know-how acquired over several years may be wasted, including the strategies developed for so long and referred to throughout this work. There is some awareness of this threat, as a few workers even mentioned that they have benefited the productivity of the company and that, with the renewal of the workforce, this advantage could be in danger.

Therefore, it may be appropriate to conciliate informal training periods with more formal and intentional forms of training, because at present the fact that the company has not implemented a plan to welcome workers, together with a training plan,

ends up putting the burden of a timely and proficient knowledge transfer on workers alone. It is true that the person in charge of the company mentions in the interview the existence of a plan for workers' succession. However, at this time, as our observation confirms, the fact is that new workers are simply placed together with more experienced workers, and the knowledge process begins only in the final month of the senior worker's activity, which clearly points out that this succession plan is ineffective or even non-existent, despite the manager's intention.

A final question that arises in this discussion concerns the criterion of age at the time of hiring new workers. The company manager mentioned his preference for older workers, due to the experience and maturity they have. Yet, he also pointed out that the benefit of hiring older workers may be contradictory, since he does not know if in a few years they will still be able to meet the daily demands of this work. As such, the question is, how can we preserve the experience gained by older workers by ensuring them sustainable working conditions (Gollac, Guyot and Volkoff, 2008)? It should be taken into consideration that, to remain in the workplace up to the age of retirement, the worker must be granted the conditions that make his work sustainable, that is, those that do not deteriorate his health and lead to an early exit (EUROFOUND, 2012).

This balance between different age groups and experience levels, ensuring adequate work conditions for all parties involved and facilitating the process of informal and possibly formal knowledge transfer, is urgently needed. This is the only way for the renewal of the labor force (which is inevitable due to the natural process of aging) to happen without major obstacles, without wasting knowledge and without harming both the workers and the company.

5 CONCLUSIONS

From the present study, it was concluded that the workers' years of experience contributed to the development of work regulation strategies that benefit the exercise of their activity, facilitating some tasks directly related to it. From this point of view, workers are responsible for minimizing potential harm to themselves and their colleagues, as was observed in the case of the tool created by one of the workers, which prevents employees from cutting themselves when opening carton boxes. The same can be said about the heating of parts to facilitate their fitting; the placing of a carton at the machine exit so that no blades are projected; or the way found to produce tubes faster while decreasing the ergonomic impact on the worker.

These and many other strategies were identified and witnessed on the spot.

It has become clear that the knowledge transfer process happens while the workers develop their regular work activities, forcing them, through the informality of the method, to coordinate their daily tasks with the training process, either as "trainers" or "trainees".

This method has both advantages and disadvantages: on the one hand, the younger workers who are learning have the opportunity to follow the activity in its real context, with all the variables inherent in it. On the other hand, along with this process, both have to keep paying attention to their work without reducing their productivity.

We can conclude with this study, as it was shared with the company manager in the form of an intervention proposal, that developing a training plan, together with an effective succession plan that allows the long-term management of the expertise of its employees, is of the utmost importance. There is an evident need to carry out a process of gradual training and assessing that same training, always taking into account the point of view of the workers involved in the transfer process and the nature and particularities of their activity, in order to preserve the irreplaceable heritage of their collective knowledge.

REFERENCES

Barros-Duarte, C. (2006). Entre o local e o global: processos de regulação para a preservação da saúde no trabalho (resumo). Laboreal, 2, (1), 48–51. http://laboreal.up.pt/revista/artigo.php?id=48u56oTV658223376276694722

Eurofound (2012), Sustainable work and the ageing workforce, Publications Office of the European Union, Luxembourg. Disponível em: http://www.eurofound.europa.eu/sites/default/files/ef_files/pubdocs/2012/66/en/1/EF1266EN.pd f (consultado em: 11/09/2016).

Gollac M., Guyot S., Volkoff S. (2008), À propos du «travail soutenable», Rapport de recherche n° 48, Centre d'études de l'emploi, Noisy-le-Grand, juin 2008.

Onstenk, J. (1995). A aprendizagem no local de trabalho no âmbito da reforma organizativa na indústria transformadora. Revista Europeia de Formação Profissional, 5, 34–42.

Oshas (2016), Agência Europeia para a Segurança e Saúde no Trabalho. Disponível em: https://osha.europa.eu/pt/themes/osh-management-context-ageing-workforce (consultado em: 07–06–2016).

Thébault, J. (2013). La transmission professionnelle: processus d'élaboration d'interactions formatives en situation de travail. Une recherche auprès de personnels soignants dans un Centre Hospitalier Universitaire. Thèse de doctorat en Ergonomie. Paris: CNAM/CEE-CREAPT.

Occupational Safety and Hygiene V – Arezes et al. (Eds)
© *2017 Taylor & Francis Group, London, ISBN 978-1-138-05761-6*

Performance evaluation of PC mice

M.L. Lourenço
Escola Superior de Tecnologia e Gestão, Instituto Politécnico da Guarda, Guarda, Portugal

D.A. Coelho
Human Technology Group, Department of Electromechanical Engineering and Centre for Mechanical and Aerospace Science and Technology, Universidade da Beira Interior, Covilhã, Portugal

ABSTRACT: This paper reports on a novel empirically derived performance indicator. Three commercially available computer pointing devices were compared: a standard computer mouse, a vertical device (supporting neutral pronation of the forearm) and a slanted device. Pointing, dragging and steering standardized tasks were implemented by software and performed by 20 experienced users. Effectiveness and efficiency (usability parameters) were calculated based on data automatically recorded from a purpose built software that also created user tasks. Additionally, structured observations of 10 subjects' activity were carried out to estimate the proportion of computer mouse operations during CAD modelling with a 3D parametric software. The value of the mean of the performance indicator was lower for the vertical device. Results suggest that the indicator proposed offers a valid means of ranking performance of alternative pointing devices regarding operation efficiency in the performance of CAD activities.

1 INTRODUCTION

Computer usage can be associated with the development of neck and upper extremity pain, especially hand and forearm musculoskeletal pain induced by intensive mouse use (Conlon et al., 2009). A decade ago, approximately 30% to 80% of computer work involved the mouse (Dennerlein & Johnson, 2006), depending on the type of work. The PC mouse has become an essential part of computer work, even today. Extended use of computer pointing devices is bound to endure in present and future days, because in computer tasks such as pointing, dragging and steering, continuously needed, touch screens have so far not been able to replace the PC mouse, e.g. in 3D computer aided design (Lourenço et al., 2015). Hand size of the subjects seems to make a difference during computer mouse usage, affecting grasp position and the level of muscle activity, suggesting that a computer mouse must be chosen according to the size of the hand of the subject (Agarabi et al., 2004). Moreover, previous tests performed on a standard PC mouse (model A in the present study) revealed statistically significant association between hand width and effectiveness of dragging with the middle button of the mouse (Lourenço et al., 2016). The activity of Computer Aided Design (CAD) typically involves the intensive use of a handheld computer pointing device, which is dominant over the computer keyboard (Cail & Aptel, 2003), raising ergonomic concerns. In this regard, in addition

to postural and biomechanical aspects related to computer handheld pointing devices' usage, it is also important to perform usability assessment. Hence, selection of a user friendly pointing device involves consideration of many aspects, including usability, which is comprised of the efficiency, effectiveness and satisfaction of the person in the activity of task completion (ISO 9241-11:1998). In this paper, the efficiency aspect of usability is focused, reporting on the proposal of an empirically derived performance indicator based on the results of two related experimental studies. In the first study objective measures of usability were collected while comparing three commercially available PC mice, having a major difference between them in what concerns the orientation of the devices and their shape, although with additional differences in size and weight. In a second study, observations were carried out to estimate the proportion of simple computer mouse operation during CAD activity using a parametric 3D CAD software. Twenty subjects participated in the first study, from which average times and efficiency per operation were obtained, while ten other subjects took part in the second study. In both studies, participants were volunteering students recruited from product design programmes of study and young engineers and architects.

The paper reports on the initial (pilot) development of the proposed indicator, which may be used in selecting between alternative devices, in an approach akin to the work of Nunes *et al.* (2014) and

McCauley Bush *et al.* (2012) on alternative device selection. Hence, the example followed through in the paper yields a tentative composite performance indicator value for each one of the three types of pointing devices evaluated, within the context of use of the selected CAD software. The devices selected are quite different from each other. Each of the devices included in the evaluations represents its own archetype geometry, which is expected to result in different levels of usability, and, in particular, of efficiency per device. In the realm of pointing devices, standardized approaches to usability evaluation could benefit from greater specificity, which is aimed with the performance indicator proposed herewith.

2 METHODS

2.1 *Elemental operations time length (study 1)*

Figure 1 shows the devices used in study 1; model A is a Microsoft® standard horizontal PC mouse (reference Optical 200), while model B is an Evoluent® vertical PC mouse (supporting the adoption of a neutral forearm pronation posture; reference VM4R) and model C is an Anker® slanted PC mouse (with mouse buttons plane angled at approximately 60 degrees with horizontal; reference TM137U). Standard PC mouse model A (Figure 1) has a mass of 57 grams (taken from weighing the device on a precision scale with the cable horizontally supported; the total weight including cable and USB plug is 78 grams). Analogously, vertical PC mouse (model B) has a mass of 137 grams and the total weight including cable and USB plug is 170 grams, while the slanted PC mouse (model C) has a mass of 119 grams and the total weight including cable and USB plug is 145 grams.

A set of tasks representative of a CAD operator's activity were standardized and recreated by a tailor made computer software application to support the experimental studies undertaken (Lourenço *et al.*, 2016). The standardized tasks included pointing, dragging and steering targets. This set of task were collected and adapted from previous studies (Odell & Johnson, 2007; Houwink *et al.*, 2009). All 20 subjects (10 female and 10 male, all right handed and with normal or corrected to normal vision) used each one of the devices performing the standard tasks in the following order: pointing at large targets (pointing large), pointing at medium targets (pointing medium) and pointing at small targets (pointing small) at first. Then, dragging targets with the left button (dragging left), dragging with the middle button (dragging middle), and steering targets inside a representation of a tunnel. The devices were randomly sorted and the participants performed the tests using the same device across the tasks in the sequence described above, and they then repeated the same sequence of tasks with another device after a resting period. The pointing tasks consisted of alternately clicking on 12 targets, randomly sorted, from 18 equally distributed locations round targets arranged in an imaginary circle. Participants clicked on the center circle to start the task and then would move the cursor and click on the first active circle target (black-highlighted), if the click hit the target it would disappear, enabling the target on the diametrically opposite side of the circle, which when hit, would lead to the next target to randomly go active, and so on. The pointing task ran in pairs, one target was randomized and the next target stood opposite to it. The dragging tasks consisted of alternately dragging 8 equally distributed round targets arranged concentrically and participants would click and drag the circle to the diametrically opposite side matching the targets with another click. The steering task partially resembled the dragging task, it was necessary to hit the black-highlighted circle, release the mouse button, and then drive the circle to the diametrically opposite side matching the targets and trying not to get outside of the tunnel. The purpose-built software collected several parameters of the trials including time to complete tasks and errors undergone, enabling calculation of

Table 1. Mean time [s] for single operation completion (per device focused and per task); standard deviation shown within parentheses (n = 20).

Operation/ Device	Pointing at large targets	Pointing at medium targets	Pointing at small targets	Dragging with left button	Dragging with middle button	Steering
Model A times	0.500 (0.086)	0.519 (0.096)	0.688 (0.106)	1.403 (0.346)	1.344 (0.266)	2.856 (0.699)
Model B times	0.586 (0.084)	0.642 (0.102)	0.982 (0.102)	1.651 (0.316)	1.744 (0.346)	3.102 (0.530)
Model C times	0.577 (0.100)	0.602 (0.119)	0.797 (0.128)	1.770 (0.568)	1.709 (0.448)	3.052 (0.711)
Average of mean times	0.554	0.588	0.822	1.608	1.599	3.004

Figure 1. PC mice studied (model A, model B and model C).

effectiveness and efficiency usability parameters (Lourenço *et al.*, 2016).

Mean times to perform single operations within each task were also obtained from dividing the total task completion time by the number of targets in the task and then averaging over all twenty subjects participating in study 1. For the purpose of calculating the mean time per operation, results obtained concern model A and the non-conventional models B and C, considered separately at first, and then average times are obtained across the three models. Table 1 depicts mean time [s] for single operation completion (per device focused and per task). Objective evaluation parameters are compared across the sample between the three devices under focus. Statistical analysis was carried out using IBM SPSS version 23. Each session in study 1 lasted between 10 and 12 minutes per device. An additional set of several non-commercial alternative pointing devices was evaluated in the same experiment, and the order of evaluation was randomized for each subject across the several devices evaluated. This paper focuses only on the commercially available devices tested.

2.2 Mix of elemental operations in example CAD software operation (study 2)

A naturalistic observation study of the activity of 3D CAD operators was carried out. Ten participants' individual activity was recorded simultaneously in an advanced 3D modelling course lab for 60 minutes. All participants were right handed with normal or corrected to normal vision. Each participant had a similar standalone 3D CAD workstation running the same software and using the same computer peripherals (Figure 2). Recording ensued through screen recording software running in each standalone workstation. During the recording time, participants executed solid modelling of distinct 3D designs including assembly and 2D technical drawing generation. All the participants had at least two years of experience in the use of the 3D CAD software.

Observation was made indirectly from the on-screen recordings of the activity, by a highly experienced user and tutor (with over 10 years of experience and continued practice). Observation was made for every 5th minute, counting the number of operations carried out in a total of 11 minutes. observations were categorized according to six of the operations involving the PC mouse (pointing at large targets, pointing at medium targets, pointing at small targets, dragging with the left button, dragging with the middle button and steering; dragging with the right button was not an active function in the CAD software used. Table 2 depicts some types of the actions within the software environment that were considered in identifying the device operations.

The number of operations of each of the categories observed within the 11 minutes actually analysed was averaged over the ten participants. Considering the mean single operation times obtained from study 1, coefficients representing the percentage of time each operation category was undertaken were then calculated for each of the six operations. Considering the efficiency values obtained from study 1 for each of the tasks and computer pointing devices, combined with the aforementioned coefficients, enabled computing a

Figure 2. Overview of the 3D modelling laboratory where study 2 took place.

Table 2. Typical examples of correspondence considered between device operations and actions in the software environment exemplified (Autodesk Inventor® v.2016).

Pointing device operations	Pointing at large targets (poi_l)	Pointing at medium targets (poi_m)	Pointing at small targets (poi_s)	Dragging with left button ($drag_l$)	Dragging with middle button ($drag_m$)	Steering (st)
Software Environment Actions (examples)	Tool selection	Selecting tree elements	Selecting points in drawn entities	Free rotate, move plane	Panning	Creating splines, dimensions

weighted efficiency efi_w (obtained from equation (1)) for each of the pointing devices, specific to the particular software and context of the studies.

$$efi_w = \frac{a.efi_{poi_l} + b.efi_{poi_m} + c.efi_{poi_s} + d.efi_{drag_l} + e.efi_{drag_m} + f.efi_{st}}{a+b+c+d+e+f}$$

$$(1)$$

3   RESULTS

3.1   *Comparative evaluation of usability (study 1)*

Participants ranged in age from 20 to 38 years old (mean = 25 years, SD = 4.8 years) and all of them were right handed. Hand width (hand breath) and hand length were measured using a retractable steel tape measure, resulting, respectively on female hand width with a mean of 79.9 mm (SD = 4.06 mm), female hand length with a mean of 177.3 mm (SD = 5.73 mm), male hand width with a mean of 88.8 mm (SD = 4.02 mm) and male hand length with a mean of 191.7 mm (SD = 4.67 mm).

Mean efficiency of task completion is given in Table 3. This shows that the mean efficiency of

tasks completion is comparably lower in model B. Model C achieves the highest efficiencies across the board, with model B scoring the lowest efficiencies across the board, while model A stands in between the other two models.

Mean time per single operation is shown in Table 1, for the six tasks included in study 1. Results shown derive from software automatically recorded time data and are specific to each pointing device. Moreover, the results are obtained by averaging across the participants individual time for task completion divided by the number of targets in the task. All mean times concerning both models B and C are higher than the ones for model A, hinting at the participants familiarity with model A, and possibly, the increased difficulty to control model B (as the operating fingers are bound to become out of view of the operator obstructed by the vertically positioned body of the device). The mean times for model C are positioned in between the times for models A and B in all but one case (dragging with the left button, where mean times for model C are higher than the times for the other two models).

3.2   *Observation of 3D CAD actions and performance indicator development (study 2)*

Participants' (6 males and 4 females) mean age was 23.9 years (standard deviation of 3.6 years). Observed operation counts, averaged over the 10 participants in study 2 are presented in Table 4. In addition, coefficients a, b, c, d, e and f, from equation (1) are also included. The operation of dragging with the right button is not included in calculating the coefficients, as this action is not used in the CAD software focused in this study.

Table 3.   Mean efficiency (per device focused and per task); standard deviation shown within parentheses (n = 20).

Operation/ Device	Pointing at large targets	Pointing at medium targets	Pointing at small targets	Dragging with left button	Dragging with middle button	Steering
Model A effici- ency	0.734 (0.091)	0.751 (0.092)	0.502 (0.091)	0.621 (0.109)	0.559 (0.158)	0.416 (0.113)
Model B effici- ency	0.672 (0.090)	0.597 (0.095)	0.349 (0.084)	0.535 (0.121)	0.499 (0.131)	0.400 (0.065)
Model C effici- ency	0.797 (0.112)	0.767 (0.124)	0.535 (0.095)	0.634 (0.160)	0.664 (0.149)	0.456 (0.108)

Table 5.   Performance indicator mean and standard deviation per device, for the focused CAD software (n = 20).

Model	Performance indicator mean (SD)
A	0.600 (0.063)
B	0.503 (0.053)
C	0.637 (0.082)

Table 4.   Mean observed operation counts (standard deviation shown within parentheses) and computed coefficients.

Pointing device operations	Pointing at large targets	Pointing at medium targets	Pointing at small targets	Dragging with left button	Dragging with middle button	Steering
Mean operation count (n = 10)	30.9 (10.6)	63.1 (24.0)	37.4 (15.3)	24.4 (8.7)	11.9 (5.1)	8.9 (5.8)
Performance indicator coefficients	a = 0.101	b = 0.218	c = 0.181	d = 0.231	e = 0.112	f = 0.157

Finally, combining the results depicted in Tables 1, 3 and 4 enables computing equation (1) for the three devices focused in the study. Results are shown in Table 5. The statistical T-test for paired comparisons yielded significance ($p < 0.01$) for the differences in the performance indicator across two pairs of devices (models A and B, models B and C).

4 DISCUSSION

The differences reached in efficiency among the three models, for the tasks under interest, is statistically supported, in spite of the small sample size and short session time that may have benefited the classic device (model A), showing clearly better performance results for model C. The tasks pointing at medium size targets, pointing at small size targets, dragging with the middle button of the PC mouse and steering targets play a key role in several computer aided design software tools, hence the present study may help users to better choose their PC mice. The results of the comparison reported in the paper are in line with those of Hedge et al. (2010) suggesting the use of slanted configurations of handheld pointing devices, in order to enable achieving a compromise between the expected long term effects on health and the objective and subjective task completion usability parameters.

REFERENCES

Agarabi, M. Bonato P., and De Luca C.J. 2004. A sEMG-based method for assessing the design of computer mice. In: 26th Annual International Conference of the IEEE Engineering in Medicine and Biology Society. *IEEE Engineering in Medicine and Biology Society*, vol. 1, 2450–2453, San Francisco. http://dx.doi.org/10.1109/IEMBS.2004.1403708

Cail, F., & Aptel, M. 2003. Biomechanical stresses in computer-aided design and in data entry. *International Journal of Occupational Safety and Ergonomics*, 9(3), 235–255.

Conlon C.F., N. Krause, and David M. Rempel (2009). A Randomized Controlled Trial Evaluating an Alternative Mouse or Forearm Support on Change in Median and Ulnar Nerve Motor Latency at the Wrist". *American Journal of Industrial Medicine*, 52 (4): 304–310.

Dennerlein, J.T., & Johnson, P.W. 2006. Different computer tasks affect the exposure of the upper extremity to biomechanical risk factors. *Ergonomics*, 49, 45–61.

Hedge, Alan; David Feathers, and Kimberly Rollings. 2010. "Ergonomic comparison of slanted and vertical computer mouse designs." *Proceedings of the Human Factors and Ergonomics Society Annual Meeting*. 54 (6) Sage Publications, 561–565. http://dx.doi.org/10.1177/154193121005400604.

Houwink, A., Hengel, K.M.O., Odell, D., & Dennerlein, J.T. 2009. Providing training enhances the biomechanical improvements of an alternative computer mouse design. *Human Factors: The Journal of the Human Factors and Ergonomics Society*, 51 (1): 46–55.

ISO 9241-11. (1998). Ergonomic requirements for office work with visual display terminals (VDTs): Guidance on usability. International Standards Organisation.

Lourenço, L.M., Pitarma, R.A., and Coelho, D.A. 2016. Association of Hand Size with Usability Assessment Parameters of a Standard Handheld Computer Pointing Device. *Proceedings of International Symposium on Occupational Safety and Hygiene—SHO 2016*, Guimarães, Balkena and CRC Press.

Lourenço, L.M., Pitarma, R.A., and Coelho, D.A. 2015. Ergonomic Development of a Computer Pointing Device—a Departure from the Conventional PC Mouse towards CAD Users. In *Proceedings of the 19th Triennial Congress of the International Ergonomics Association*, Melbourne. http://ergonomics.uq.edu.au/iea/proceedings/Indicator_files/papers/1200.pdf

McCauley Bush, P., S. Gaines, A. Watlington, M. Jeelani, L. Curling & P. Armbrister. 2012. *The Development of a Device Selection Model for Wireless Computing Devices in High Consequence Emergency Management*. Advances in Usability Evaluation Part I, Edited by Francisco Rebelo, CRC Press 2012, Chapter 52, pp. 486–499.

Nunes, I., Patriarca, D., Matos, A. 2014. Usability-based mobile phone selection for communications in emergency situations. Proceedings of the *5th International Conference on Applied Human Factors and Ergonomics AHFE 2014*, 19–23 July 2014, Kraków, Poland. Edited by T. Ahram, W. Karwowski and T. Marek, pp. 498–509.

Odell, D.L., & Johnson, P.W. 2007. Evaluation of a Mouse Designed to Improve Posture and Comfort. In *Proceedings of the 2007 Work with Computing Systems Conference*-International Ergonomics Association.

Occupational Safety and Hygiene V – Arezes et al. (Eds)
© *2017 Taylor & Francis Group, London, ISBN 978-1-138-05761-6*

Nanomaterials in textiles and its implications in terms of health and safety

D.G. Ramos
Polytechnic Institute of Cávado and Ave, Barcelos, Portugal
Algoritmi Centre, School of Engineering, University of Minho, Guimarães, Portugal

L. Almeida
Department of Textile Engineering, School of Engineering, University of Minho, Guimarães, Portugal

ABSTRACT: Nanotechnologies are already present in many consumer products, including textiles. The effects of nanomaterials in safety and health are not yet fully known. It is expectable that "nanotextiles" may release nanoparticles, which can interfere with workers in the textile and clothing industries, consumers and natural ecosystems. Skin contact in particularly relevant for textiles, although inhalation and ingestion should also be taken into account. It is important to understand the type of integration of the nanoparticles in textiles and the mechanism of release. In this paper, a brief overview of the effects obtained with nanotechnology in textiles is presented, as well as the possible release mechanisms of nanoparticles. In terms of skin contact, the development of test methods to evaluate skin exposure is presented. The method can be easily applied for instance in the case of nano-silver and nano-titanium dioxide, which are most frequently used in textiles.

1 INTRODUCTION

In the present decade there has been a rapid emergence of nanotechnology into several consumer products, which has led to concerns as regards the potential risk for human health following consumer exposure. But there is also a concern in terms of occupational safety and health, related to the exposure of workers involved in manufacturing, processing and handling of consumer goods containing nanomaterials. Nanosafety is in fact a growing concern Exposure to engineered nanomaterials has been associated with a number of health effects including pulmonary inflammation, genotoxicity, carcinogenicity and circulatory effects (Savoilainen et al., 2010).

Textiles are one of the most heavily traded commodities in the world. The industry is very diverse and its products are used by virtually everybody from private households to large businesses. The textile industry is already an important user of nanotechnologies and there are a significant number of "nanotextiles" in the market, including many consumer goods, with the incorporation of nanoparticles. These include many textiles used in direct contact with the skin, such as underwear, shirts and socks but also interior textiles like cushions, blankets or mattress covers.

There is a knowledge gap between the technological progress in nanotechnology and nanosafety research, which is estimated to be 20 years, and it is likely to expand. The European Agency for Safety and Health at Work has established as priority for research related to the safety and health in Europe, during the period 2013–2020, the increase of knowledge on nanomaterials in occupational settings, including new generation nanomaterials and understand their characteristics in relation to toxicity in biological systems (EU-OSHA, 2013).

The risk for the workers and for the consumers is linked to the characteristic properties of certain nanomaterials that make them different from their macroscale counterparts and will be determined by the chemical composition of the nanomaterial, its physicochemical properties, the interactions with the textile materials and the potential exposure levels. Ingestion exposure via the gut, airborne exposure via the lungs and dermal exposure are the most important exposure routes to be considered in a risk analysis. In addition, the increasing use of nanomaterials, including for industrial purposes, raises specific concerns regarding their disposal at the end of their life cycle with the unavoidable release to the environment that may lead to indirect human exposure.

The European Commission, in its action plan for Europe 2005–2009, defined a series of actions for the immediate implementation of a safe, integrated and responsible approach for nanosciences and nanotechnologies. In line with this commitments, it is important to determine the applicability of the existing regulations to the potential risks of nanomaterials. It was then decided to publish an official definition of nanomaterial, which is a basis for the adoption and implementation of legislation, policy and research programmes concerning products of nanotechnologies. According to this definition, "nanomaterial" means a natural, incidental or manufactured material containing particles, in an unbound state or as an aggregate or as an agglomerate and where, for 50% or more of the particles in the number size distribution, one or more external dimensions is in the size range 1 nm–100 nm. But in specific cases and where warranted by concerns for the environment, health, safety or competitiveness, the number size distribution threshold of 50% may be replaced by a threshold between 1 and 50% (European Commission, 2011).

At present, some EU Regulations already include a specific mention to nanomaterials. This is the case of food (including additives and packaging), biocides and cosmetic products. But this is still not the case of textiles.

The General Products Safety Directive 2001/95/EC (GPSD) is intended to ensure a high level of product safety throughout the EU for consumer products that are not covered by specific sector legislation, which is the case of textiles. It is foreseen to modify GPSD into a Regulation, including possibly a specific mention to nanomaterials.

The growing concern about the possible negative effects of nanomaterials on humans and on the environment can lead to restrictions to nanotextiles. In fact, for instance, the 2014 version of the ecological label GOTS (Global Organic Textile Standard) fully bans the presence of nanofinishes in textiles. Also in the recent discussion of the new version of the EU ecolabel for textiles, there were several voices to exclude nanomaterials.

In the present paper, after making an overview of the use of nanotechnology in textiles, with special emphasis on textiles for major consumer applications, the safety and health concerns related to nanotextiles are presented. The paper includes then a case study concerning the development of test methods to evaluate the skin exposure to nanoparticles, mainly directed to the transfer of the nanoparticles from the textile to the skin.

This paper does not deal with the penetration of the nanoparticles into the skin. There are many studies about this topic, related for instance to sunscreens and cosmetics, which are often based on nanomaterials. In fact, only the smaller nanoparticles seem to be able to penetrate in the undamaged skin, although in the skin is injured, larger nanoparticles can penetrate (Labouta et al, 2011). In a recent study, Larese Filon et al. (2015) made a literature survey involving 129 relevant publications and concluded that the smaller nanoparticles,

with dimensions below 4 nm, can easily penetrate the skin, while those with dimensions from 4 nm to 20 nm can potentially penetrate intact skin. Nanoparticles with size between 21 and 45 nm can also penetrate and permeate in damaged skin.

The present paper also does not concern the toxicological aspects related to nanoparticles. This topic has been and is being extensively studied in many research works. The safety and health concerns related to nanoparticles is very important for instance for the registration of chemical substances in nanoform in the REACH system. Nanotoxicology is a very important science that will serve as a basis for future regulations, including possibly textile products.

2 USE OF NANOTECHNOLOGIES IN TEXTILES

When we speak about nanotextiles, we refer normally to traditional textile products in which engineered nanomaterials (normally nanoparticles) are incorporated or on which a nanostructured surface has been applied. In fact, nanotechnology is already very often used in textile products and involves the incorporation of nanoparticles in textile materials or the nanostructuration of the surface, in order to obtain specific functionalities (Wong et al., 2006; Gowri et al., 2010). Table 1 presents the nanomaterials that are more commonly used in the functionalization of textiles.

Table 1. Nanomaterials used in the functionalization of textiles (adapted from Som et al., 2010).

Nanomaterial	Function
Silver (Ag)	Antibacterial (odour) electrically conductive
Titanium dioxide (TiO_2)	UV protection self-cleaning water and dirt repellent
Zinc oxide	UV protection antibacterial self-cleaning abrasion resistance, stiffness
Silicon dioxide (SiO_2)	Water and dirt repellent abrasion-resistant, reinforcement improved dyeability
Aluminium oxide (Al_2O_3)	Abrasion resistance flame retardant
Nanoclays (e.g. montmorillonite)	Abrasion resistance flame retardant support of active ingredients

Rivero et al. (2015) also report other nanomaterials that can be used namely for flame retardancy, antibacterial or superhydrophobic surface effects.

Nanomaterials can be incorporated in textiles in two different ways: during fibre production or during textile finishing.

In fibre manufacturing, the nanoparticles are introduced by mixing in the polymer, before fibre spinning. The incorporation is made either in the melted polymer, in the case of melt spinning, or in the polymer solution, in the case of wet or dry fibre spinning processes. In both cases, the nanoparticles are evenly distributed inside the fibre volume. We can speak in this case of a "nanocomposite" material. The content of nanoparticles in the fibre can be as low as 0.1% to obtain a sufficient functionalization (Som et al., 2009). When the incorporation is done by this process, the nanoparticles are firmly incorporated in the textile fibre and the effect is highly durable. The nanoparticles are inside the fibre and normally can only be released by means of abrasion. This fact poses low risk in terms of safety and health both for the workers involved in the subsequent textile processing and for the consumers.

Although the use of fibres functionalized by means of the incorporation of nanoparticles before fibre spinning has the advantage of having a very durable effect, it has two major inconvenients: the process cannot be applied to natural fibres and there is a dependence on the fibre manufacturer. In fact, the textile manufacturing chain is very long, from yarn spinning to the final product, and it is normally much easier and cheaper to work with traditional textiles and apply the functionalization at the end of the textile process, during fabric finishing.

The incorporation of nanoparticles during fabric finishing can be obtained by traditional processes like dipping (exhaustion process), padding, printing or coating. To assure a good adhesion of the nanoparticles to the textile surface, organic polymers are normally used.

In this case, there are further concerns for the safety and health of the workers involved in the fabric finishing processing, due to the manipulation of the chemicals containing nanoparticles and of the textile itself. There must also be a concern for the workers in the garment manufacturing processes, which will be manipulating the textiles containing nanoparticles in all the cutting, sewing, pressing and packaging operations.

Most of the nanotextiles present in the consumer products are based on the application of nanoparticles by means of the fabric finishing process. The release of nanoparticles during the daily use is a safety and health concern, which will be dealt in the next section.

3 SAFETY AND HEALTH CONCERNS RELATED TO NANOTEXTILES

As there are different manufacturing processes by which nanoparticles can be integrated in fibres or textiles, there can be a variation in how tightly bound the nanoparticles are into the textile material (fibre or coating). These factors and the use to which the textile is subjected determine whether and to what extent nanoparticles can be released from it.

Depending on the location of the integration of nanoparticles in the textile, they can be more or less heavily exposed to external influences. In addition, there can be different degrees of binding of the nanoparticles to the textile material (covalent bonding, ionic bonding, hydrogen bonding, Van der Waals bonding).

The stability (durability) of the nanoparticles present in the textile is not only dependent on their binding to the fabric, but also on the impacts on the fabric during its life cycle (production, use, recycling/disposal), which can damage the textile material or the bond between the nanoparticle and the fibres: abrasion, mechanical stress (such as strains, pressure), ultraviolet radiation, body fluids (sweat, saliva, urine), water (rain, washing), solvents (during textile processing or dry cleaning), detergents (either in textile processing or during laundry) and temperature changes; high temperatures (up to 225°C in textile finishing).

It is known that textiles can lose between 5% (in the case of continuous filaments) and up to 20% (in the case of staple fibre loose materials) of their mass during use as a result of abrasion, mechanical influence, irradiation, water, sweat, washing detergents or temperature variations. Even if the nanoparticles remain attached to the fibres, they will be released from the textile material to the environment (either to the human body, air, water, soil) together with the fibres. In fact, nanotextiles may release individual nanoparticles, agglomerates of nanoparticles or small particles of textile with or without nanoparticles.

For instance, in the case of textiles containing nano-silver, there are several studies concerning the release of silver nanoparticles during washing (e.g. Benn & Westerhoff, 2008, Geranio et al., 2009, Farkas et al., 2011, Mitrano et al., 2014). Some investigations that have been made show that some products can lose up to 35% of the silver in the washing water after only one wash; nevertheless, some suppliers of silver-based antibacterial finishes claim that there is practical no release during washing and that the finish remains effective after more than 50 washes.

Safety and health concerns related to nanotextiles should consider all the life cycle of the materials:

safety for the workers during all the manufacturing stages, especially in the case of the application of nanostructured materials in the fabrics, as mentioned in the previous section, safety for all the people involved in the trade phases (distribution and retail), safety for the consumers. Nanoparticles released from textiles to the air, to the water and to landfill must also be taken into consideration, as they can directly or indirectly affect humans.

The nanoparticles can interact with the human body via three different ways: inhalation, ingestion and skin contact. Although all the three pathways can be related to textiles, skin contact is of course the most relevant. This topic will be discussed in the next section.

4 CASE STUDY: DEVELOPMENT OF TEST METHODS FOR SKIN EXPOSURE TO NANOPARTICLES

The European Commission has recently issued a mandate to the three European Standardization bodies (CEN, CENELEC and ETSI) specifically asking to develop standards for testing methods and tools for the characterization, behaviour and exposure assessment of nanomaterials. The exposure takes into account aspects of health and safety of workers, as well as of consumers and the environment itself.

This mandate is being handled by the Technical Committee CEN/TC352 – Nanotechnologies. A roadmap identified 45 standardization projects in the field of "characterization of and exposure to nanomaterials" and "health, safety and environment".

Within this mandate, an extensive standardization programme has been prepared and is being developed, with conclusion foreseen for 2018.

One of the possible exposures of humans to nanomaterials is skin exposure. As this topic can be very relevant for textiles, the Technical Committee CEN/TC248 (Textile and Textile Products) is at present developing a technical report devoted to a guidance on measurement techniques to simulate nanoparticle release – skin exposure. Although the methods are specifically being developed for textiles, they can in principle be applied also to other products containing nanomaterials that can enter into direct contact with the skin.

In order to evaluate the migration of nanoparticles from textiles to the skin, a possible test can be based on the use of an artificial perspiration solution under physical stress. Artificial perspiration solutions (acid and alkaline) are described in standard EN ISO 105-E04 (Textiles – Tests for colour fastness – Part E04: Colour fastness to perspiration). The artificial perspiration acid solution

is already used in the following document: EN 16711–2:2015 (Textiles – Determination of metal content – Part 2: Determination of metals extracted by acidic artificial perspiration solution).

In the standard EN ISO 105-E04, there is a simulation of contact with the skin of the textile to be tested, together with white standard adjacent fabrics, during 4 h at 37°C, but there is no mechanical stress. A possible alternative is to use a test similar to the test EN ISO 105-C06 (Textiles – Tests for colour fastness – Part C06: Colour fastness to domestic and commercial laundering), by immersing the textile to be tested in the artificial sweat solutions, under mechanical agitation in a thermostatic bad at 40°C. Test is made both with the acid (pH 5.5) and with the alkaline (pH 8.0) artificial sweat solutions, with separate test specimens. The treatment is carried out during 30 minutes and involves the use of acrylic plastic balls to simulate physical stress. Nanoparticles, as well as their aggregates and agglomerates that can release nanoparticles, are then analysed in the extract solution.

This method has been developed by von Goetz et al. (2013). These authors have tested commercially available textile products intended for sports or outdoor activities (shirts, pants trousers and socks, for adults and also for children) which included finishes based on nano-silver (antimicrobial effect) and nano-titanium dioxide (UV protection). The content of nanomaterials in the tested textiles was up to 183 mg Ag/kg of textile or 8543 mg Ti/kg of textile. The tests show a significant release of nano-silver, which reached up to 14%. These results are much higher than other reported in the literature and contradict the producers that the claim that there is practically no release of nano-silver during the use of textiles and that the finish is permanent. In the case of titanium dioxide, no relevant release has been detected. It must be emphasized that sun protective lotions, which are put on the skin, are often based on nanoparticles of titanium dioxide or zinc dioxide.

Another possibility, which is also foreseen for inclusion in the standard test method to be developed, is the release from the textile by mechanical action, through a "linting" mechanism. This test could possibly be based in the method mentioned in the standard EN ISO 9073-10:2004 (Textiles – Test methods for nonwovens – Part 10: Lint and other particles generation in the dry state).

It should be mentioned that the technical committee CEN/TC 137 "Assessment of workplace exposure to chemical and biological agents" is also developing a test method relevant for occupational exposure to nanomaterials: "Workplace exposure - Guidance for the assessment of dermal exposure to nano-objects and their aggregates and agglomerates", which will describe a systematic approach to assess potential occupational risks to nano-objects, and their agglomerates and aggregate arising from the production and use of nanomaterials and/or nano-enabled products. This approach provides guidance to identify exposure routes, exposed body parts and potential consequences of exposure with respect to skin uptake, local effects and inadvertent ingestion. This document also considers occupational use of nano-enabled personal care products, cosmetics and pharmaceuticals and is aimed to occupational hygienists, health and safety professionals, and researchers to assist recognition of potential risks, and their control.

5 CONCLUSIONS

Nanomaterials have the potential to improve the quality of life and to contribute to industrial competitiveness in Europe. However, the new materials may also pose risks to the environment and raise health and safety concerns. The Scientific Committee on Emerging and Newly Identified Health Risks has concluded that, even though nanomaterials are not per se dangerous, there is still scientific uncertainty about the safety of nanomaterials in many aspects; therefore, the safety assessment of the substances must be done on a case-by-case basis. In the case of textiles, there are still very few studies on the possible health risks involved with nanotextiles.

The release of nanoparticles from textiles is particularly relevant when the incorporation is made by fabric finishing. It can occur by different mechanisms. In this paper, the release resulting from skin contact, involving abrasion and sweat, has been analysed in more detail, involving a possible standard text method. The studies made up to now involve silver and titanium dioxide nanoparticles, which are present in the most common nanotextiles in the consumer market. Nevertheless, there are still a lot of discussions on if these nanoparticles can really penetrate into the different skin layers and on the negative effects on human health.

Nanosilver is used in deodorants, deliberately put on the skin, food packaging or even in toothpastes. The relevance of the silver nanoparticles from the textiles to the human body can be questioned.

In the case of titanium dioxide, it is very commonly used in sunscreens, the nanoparticles being deliberately spread over a large surface of the skin. Comparatively, the dermal exposure coming from textiles is much lower, so the relative relevance of skin exposure coming from textiles can also be questioned. Although it is allowed in cosmetics, an EU Regulation has recently been published, forbidding the use of titanium dioxide in nanoform

in applications that may lead to exposure of the end-user's lungs by inhalation (European Commission, 2016).

Further research work involving experts in the area of textiles, of safety and health and of toxicology is needed before the emergence of any regulation concerning nanotextiles.

REFERENCES

Benn, T. & Westerhoff, P. (2008). Nanoparticle Silver Released into Water from Commercially Available Sock Fabrics, Environmental Science & Technology, 42 (11), 4133–4139.

European Commission (2011). Commission Recommendation of 18 October 2011 on the definition of nanomaterial (2011/696/EU), Official Journal of the European Union of 20.10.2011, L275/38–40.

European Commission (2016). Commission Regulation (EU) 2016/1143 of 13 July 2016 amending Annex VI to Regulation (EC) No 1223/2009 of the European Parliament and of the Council on cosmetic products.

EU-Osha (2013). European Agency for Safety and Health at Work. Priorities for occupational safety and health research in Europe: 2013–2020. ISBN 978-92-9240-068-2. Luxembourg.

Farkas J., Peter H., Christian P., Gallego Urrea J.A., Hassellöv M., Tuoriniemi J., Gustafsson S., Olsson E., Hylland K. & Thomas K.V. (2011) .Characterization of the effluent from a nanosilver producing washing machine. Environment International 2011 Aug; 37(6):1057–62.

Geranio, L., Heuberger, M. & Nowack, B. (2009). The Behavior of Silver Nanotextiles during Washing. Environmental Science & Technology, 43 (21), 8113–8118.

Gowri, S., Almeida, L., Amorim, T., Carneiro, N., Souto, A. & Esteves, M. (2010). Polymer Nano Composites for Multifunctional Finishing of Textiles—A Review. Textile Research Journal, 80 (13), 1290–1306.

Labouta, H. El-Khordagui, L., Krause, T. & Schneider, M. (2011). Mechanism and determinants of nanoparticle penetration through human skin. Nanoscale, 3, 4989–4999.

Larese Filon, F., Mauro, M., Adami, G., Bovenzi, M. & Crosera, M. (2015). Nanoparticles skin absorption: New aspects for a safety profile evaluation. Regulatory Toxicology and Pharmacology. 72 (2015) 310–322.

Mitrano, D., Rimmele, E., Wichser, A., Erni, R., Height, M. & Nowack, B. (2014). Presence of Nanoparticles in Wash Water from Conventional Silver and Nanosilver Textiles. ACS Nano, 8 (7), 7208–7219.

Rivero, P-J., Urrutia, A., Goicoechea, J. & Arregui, F.J., Nanomaterials for Functional Textiles and Fibers. Nanoscale Research Letters (2015) 10:501.

Savolainen, K. Pylkkänen, L., Norppa, H., Falck, G., Lindberg, H., Tuomi, T., Vippola, M., Alenius, H., Hämeri, K., Koivisto, J., Brouwer, D., Mark, D., Bard, D., Berges, M., Jankowska, E., Posniak, M., Farmer, P., Singh, R., Krombach, F., Bihari, P., Kasper, G. & Seipenbusch, M. (2010). Nanotechnologies, engineered nanomaterials and occupational health and safety—A review. Safety Science 48 (2010) 957–963.

Som, C., Halbeisen, M & Köhler, A. (2009). Integration von Nanopartikeln in Textilien Abschätzungen zur Stabilität entlang des textilen Lebenszyklus. EMPA, Swiss Federal Laboratory for Materials Testing and Research, St. Gallen.

Som, C., Nowack, B., Wick, P. & Krug, H. (2010). Nanomaterialien in Textilien: Umwelt-, Gesundheits- und Sicherheits-Aspekte. EMPA, Swiss Federal Laboratory for Materials Testing and Research, St. Gallen.

von Goetz, N,. Lorenz, C., Windler, L., Nowack, B., Heuberger, M. & Hungerbühler, K. Migration of Ag- and TiO_2-(Nano)particles from Textiles into Artificial Sweat under Physical Stress: Experiments and Exposure Modeling (2013). Environment Science & Technology, 47 (17), 9979–9987.

Wong, Y., Yuen, C., Leung, M., Ku, S. & Lam, H. (2006). Selected Applications of Nanotechnology in Textiles. Autex Research Journal, 6(1), 1–10.

Occupational Safety and Hygiene V – Arezes et al. (Eds)
© 2017 Taylor & Francis Group, London, ISBN 978-1-138-05761-6

Evaluation of environmental risks and security of work in a metallurgical industry

Maria Betania Gama Santos, Stella Amorim Colaço, Taíse Caroline Fernandes Da Silva & Yuri Igor Alves Nóbrega
Universidade Federal de Campina Grande, Campina Grande, Brazil

Luciano Fernandes Monteiro
Universidade Federal de Sergipe, Brazil

ABSTRACT: In industrial environments, workers end up exposed to a series of environmental risks, whether physical, chemical, biological, ergonomic or accident risks. The application of security measures that contribute to the elimination or minimization of the risks found in the work environment guarantees the preservation of the health and the environmental comfort of these individuals. Therefore, this study aimed to identify and analyze the environmental risks of a metallurgical industry located in the city of Campina Grande/PB, Brazil. The methodology used includes bibliographic research, a qualitative analysis, carried out through on-site visits and using check sheets to identify the non-conformities with the current legislation, the most critical environmental risk points in the company. In addition, quantitative measurements of the environmental agents, noisiness, illumination and ambient temperature were carried out. Environmental risk control measures were proposed in accordance with the Brazilian Regulatory Standards, in the electrical installations, in some machines, in the climatization system, in the ergonomic aspects, in order to protect the health and physical integrity of all those who participate in the operation of the company.

1 INTRODUCTION

According to the Statistical Yearbook of the Metallurgical Sector (2012), the metallurgical sector presents significant importance in the Brazilian economic scenario, with a large productive chain in the sectors related to metallurgy, machining and production of metallic manufactures, being the basis of other important activities for the country, such as Automobile industry, construction and capital goods.

Occupational hygiene is dedicated to the recognition, evaluation and control of environmental agents that arise at work and which are liable to cause harm to workers' health.

Thus, this study aimed to evaluate the aspects of occupational hygiene and work security in a small metallurgical industry installed in the city of Campina Grande, Brazil. A survey about the noisiness, light and temperature conditions was carried out and the magnitude of the potential risks was verified, covering all the functions performed by all the employees in the company's premises.

This way, to differentiate acceptable exposures from unacceptable ones, as well as to propose the adoption of security measures that contribute to elimination, minimization or neutralization of the risks found in the work environment.

2 ACCIDENTS AT WORK AND OCCUPATIONAL DISEASES IN THE METALLURGICAL SECTOR

The metallurgical sector is one of the most frequent occurrences of work accidents in Brazil. Hoeltgebaum et al. (2014) sought to identify the risks to which workers are exposed in the metallurgical industries and evaluated the need for educational actions or changes in the work methods of these professionals. It was verified that although the workers of these sectors are better informed and cautious about the prevention of accidents, it is still an environment considered as high occupational risk.

The old practice of employers to always delegate guilt to employees for the accident has been increasingly abolished by laws—which consider that not only does a human agent interfere with the system, but also physical, chemical, and biological agents can negatively influence the system. This was dealt with by Gonçalves (2007), when he did a study in a metallurgical industry over a period of three years in which there were work accidents. The agents that most contributed to the accident were noisiness, heat and non-ionizing radiation; Chemical fumes; As well as ergonomic and mechanical agents.

Regarding the damages from the metallurgical industry, Régis et al (2013) elaborated a work about the incidence of hearing loss of workers of a metallurgical industry. The research was elaborated with the participation of 1499 workers undergoing audiometric tests. The statistical study applied in a sample of 763 workers indicated that noisiness-induced hearing loss was higher in workers over 45 years old and with more than 21 years of work, which led to the conclusion that the prevalence and incidence of hearing loss increased with Age and length of service.

Matheus and Daher (2009) investigated the chemical risks caused by welding fumes and metallic dust. These elements can cause several health damages by being abundant, diverse and releasing toxic gases. The worker will experience problems according to their exposure to risk, which can be classified as short or long term. The results indicated that, although they appear harmless, inhalation of dust and metallic smoke may lead to lung diseases. Ideally, according to the authors, it is to provide a clean, risk-free environment for occupational diseases in order to improve conditions in the workplace.

Other types of accidents common to the metallurgical environment were studied by the authors Kaschuk and Lopes (2013), where they sought to understand the main causes of accidents in a steel profile industry located in Sinop/MT. What could be noticed was that the most common accidents were superficial cutting, finger crushing and finger cutting in a guillotine-type machine.

Faccin et al (2008) carried out an interdisciplinary study, combining knowledge based on the areas of phonoaudiology, production engineering and architecture. The methodology used was quite diverse, contemplating actions such as the study of the auditory profile of the workers; Individual questionnaire; Evaluation of the ear protector; Sound pressure survey, etc.—which have contributed to a good analysis of the problems in the factory environment. For acoustic comfort and environmental improvement, Faccin et al (2008) understood that there should be a constant evaluation of the problems that arise, since each one alone can contribute to the emergence of other problems. Finally, a technical solution was proposed in three instances: the worker's, the machinery and the environment.

The research of Graziela et al (2014) was elaborated with the purpose of verifying the main occupational accidents (as well as occupational diseases) that occurred and that were registered in the respective responsible body. The study was developed for the metallurgical sector of Piracicaba—SP from the period of 2009 to 2011. It was observed that men tend to suffer more accidents than women and that there are areas of the body that were classified as having a higher incidence of accidents. Among these areas, it was found that the regions that were most affected by the accidents are, in the order: hands, eyes, lower and upper limbs.

In view of the above, occupational hazards may be the most varied in the metallurgical sector. This requires a more in-depth reflection on the maintenance of workers' health and well-being, since, as in other sectors of industry, metallic material factories use tools and equipment that produce high levels of noisiness - caused by the very mechanics of their operation. Hence the need for companies to invest, not only in accident prevention, but also in the renovation of their machines and tools by increasingly silent ones and in compliance with the legislation in force to the Security.

3 METHODOLOGY

This research has an exploratory nature, since it involves a bibliographical survey, interviews with people show practical experiences with the researched problem. With the purpose of knowing and characterizing the environmental aspects and conditions associated with the production process, a bibliographical research was carried out in articles, books and periodicals.

Regarding technical procedures, it is a descriptive case study that will seek to diagnose issues related to the company, as well as to delimit the variables that act in the study. As for the method of approach, it can be described as qualitative-quantitative, whose company environment is the direct source for data collection, which will be analyzed. Exploratory visits to the project were carried out with the purpose of conducting a qualitative analysis, using verification sheets to evaluate the conditions of the installations related to related to the Brazilian laws that deal with the subject being studied, they are the NR 06 – Personal Protective Equipament PPE, NR-8 – Buildings, NR-10 – Facilities NR-12 – Boilers and Pressure Vessels, NR-17 – Ergonomics, NR-23 Fire Protection, NR-26 Security Signs.

Therefore, the check sheet was adapted according to the NR's quoted for the study company, in addition, informal interviews were conducted with the workers, investigating the possible impacts that could be caused. We used the method of evaluation by the questioning of YES and NO, and score by the affirmative markings as index of quality in the research, that is, the higher the score of YES responses, the better the working conditions of the evaluated company. When an item is not applicable, it will not count towards the score, and the result will be adapted to the number of applicable questions as a percentage. The result had

5 classifications: terrible (0 to 20%), bad (20.1 to 40%), regular (40.1 to 60%), good (60.1 to 80%) and optimum (80.1 to 100%).

As for the measurements, a quantitative in loco and in situ quantification of the risk agents observed was performed on November 26, 2015, with thermal aspects (AT—Ambient Temperature (°C), Illumination (LUX) and acoustic (Noisiness) in the company, which ran from 08:00 to 17:30, due to the need to cover complete work cycles, a study was carried out on the jobs and activities related and the ones exposed to environment agents indicated, according to a previous analysis of risks, the need for quantification with equipment technologically appropriate to risk assessments.

The measurements of noisiness, illumination and temperature were made every hour from 8:00 a.m. to 12:00 p.m., and after 1:30 p.m. to 5:30 p.m. Five jobs were selected to be analyzed, due to the greater environmental bottlenecks, such as Torno 01 (point 1), Fresa (point 2), Torno 02 (point 3), Injector (point 4) and office (Point 5). All measurements of temperature and illumination were made using thermo-hygrometer type equipment and noisiness with a dosimeter—manufacturer INSTRUTHERM.

Noisiness measurements were performed in places where the need for quantification of said agent was observed, where representativity was made covering a representative part of the workday. Measurements were performed every hour, four measurements between 15-second intervals, resulting in an average for each hour, such as the apparatus operating with the readout close to the worker's ear.

The procedures followed by measurement of analysis are based on the current legal system, Regulatory Norm NR-15, through Annexes n ° 1 – Tolerance Limits for Noisiness or Intermittent, Annex n ° 3 – Tolerance Limits for Heat Exposure, As well as based on norms of procedures recommended by research institutions that deal with the subject, such as: Jorge Duprat Figueiredo Foundation for Occupational Security and Medicine—FUNDACENTRO; NHO – 01 – Evaluation of Occupational Exposure to Noisiness, NHO – 06 – Standard for the evaluation of Occupational Exposure to Heat.

4 COMPANY CHARACTERIZATION

The company, object of this study, has a national code of economic activity, CNAE, of the main activity, being the manufacture of tools (25.43-8-00) and for the secondary activities, these being the services of machining, turning and welding (25.39-0.01); The production of non-ferrous metal forgings and their alloys (25.31-4-02) and the manufacture of standardized metal drawing products (25.92-6-01). According to the IBGE (2016), it presents risk degree 03, through NR 04 (Specialized Service in Security Engineering and Occupational Medicine—SESMT) for main activity and degree of risk 04 for secondary activities.

The industry develops activities of mechanical workshop, more precisely of making of molds and machining of pieces, injection of collectors for biological examinations and components for the area of security in locks. The company employs 7 employees, working in the morning and afternoon shifts, are distributed in four different sectors: Machining of parts, Welding and Painting, Injectors and Mills, in addition to the support sectors as administrative and warehouse. For the accomplishment of their tasks, the employees make use of the equipment of type air compressor, machines of type, milling machine, injectors, lathe, milling machine, planer, band saw, and drill of column, hydraulic press, furnace for tempere, emery, benches, and tools in general, etc.

5 RESULTS AND DISCUSSION

From the check sheets applied, it could be concluded that the company is in a regular state. The company offers the appropriate PPE to the employees, however, it was recommended the study to implement collective control measures, in order to minimize the possible impacts to the worker and the company. On the floors, it was recommended to install ramps of non-slip materials where there is danger of slipping.

As for the electrical installations, all the machines must have grounding, therefore, a periodic revision was proposed throughout the electrical network, checking the points that are in need of correction. The company does not adopt any kind of preventive maintenance, it has been recommended that it be done and must always be performed with the machines stopped, in accordance with the new NR 12 (Machinery and Equipment). The company has a pressure vessel in its premises (compressed air compressor). It was recommended to perform periodic security inspection of pressure vessels, according to NR-13, by a legally qualified professional, and their respective Technical Report should be issued.

As for inadequate posture, if for long standing or sitting, it was recommended that workers try to remain in the anatomical position, ie, straight spine in order to avoid possible problems in the spine; That there is alternation between sitting/standing position; The workstations should be designed in such a way as to offer the best possible ergonomic conditions to employees, especially in activities with a predominance of seated work where they must have ergonomic chairs, that is, swivel with an adjustable accent and back. It was also proposed that a security signage be adopted with the purpose of preventing accidents, identifying security equipment, delimiting areas and warning against risks.

After analysis and measurements in the company, the heat was verified as not being a risk factor, subject to external climatic changes, meets the conditions of comfort, security and health, ensuring adequate ventilation through ceiling fans, in addition to openings in the Site, meeting NR 8 (Buildings).

5.1 Noisiness evaluation

Table 1 shows the values obtained for the average noisiness levels measured at each workstation. Based on Regulatory Norm 15 (Unhealthy activities and operations), the data of noisiness doses obtained in each work station were related to what governs the norm.

After analyzing the equivalent noisiness levels of the sectors, it was found that only one working environment—which contains the injection machine—is in disagreement with the NR-15. According to this standard, in order to be healthy in the environment, the LEQ value (daily exposure dose to noisiness) should not exceed 1.0 (one). The environment in which the injector is has a noisiness level in which the LEQ calculation resulted in 1.33

Table 1. Noisiness measurements at NR-15 based workstations.

| Work-station | Average Noisiness Measurement Performed | | | |
	LAVG (dB A)	Dose exposure (8 hours)	Legal base	Conclusion
Lathe 1	69,9	–	NR-15	Below the tolerance limit
Milling cutter	63,9	–	NR-15	Below the tolerance limit
Lathe 2	72,4	–	NR-15	Below the tolerance limit
Injector	86,8	1,33 (Leq)	NR-15	Above the tolerance limit
Office	51,5	–	NR-15	Below the tolerance limit

Table 2. Evaluation of illumination in the workstations based on NR-17.

Workstation	Measured Value (Lux)	Min. Permissible (Lux)	Conclusion
Office	320	300	Meets
Lathe 1	215	300	Does not meet
Milling cutter	145	300	Does not meet
Lathe 2	230	300	Does not meet
Injector	450	300	Meets

(surplus of 33% of the tolerance limit), therefore, it is unhealthy. It is necessary the intervention of control measures, for example the use of auricular protector.

5.2 Illumination evaluation

Measurement of illumination was carried out in the workstations and near to equipment in which artificial lighting is used. The methodology used in Technical Standard ABNT NBR ISO/CIE 8995-1/2013 was used.

The illumination values must conform to Technical Standard ABNT NBR ISO/CIE 8995-1/2013, since they are below the recommended values. Artificial illumination sources need to be replaced by sources with higher luminous flux.

6 FINAL CONSIDERATIONS

From the measurements and analyzes, it was observed that noisiness was the most alarming risk, since it presented high values that exceeded the limits of tolerance established by NR 15. Therefore, it was recommended, in addition to the adoption of the PPE's already used, that these are authorized by the Ministry of Labor, and it is still the responsibility of the company to advise its employees to make the correct use of these so that there is no exposure to doses of noise considered unhealthy. Also, the adoption of collective measures of protection, as for example, the isolation of the machine. In addition to noisiness, ergonomic risks of posture in the workplace can become a long-term problem for workers' health.

As for illumination, this does not cause discomfort to employees, however in places where the level of illumination does not meet, it was recommended to carry out periodic cleaning to remove dust accumulation in the lamps, improve the positioning of luminaires, replace burned or defective lamps; Increase the number of lamps and/or install additional spot lighting in the workstations in accordance with ISO/IEC 8995-1: 2013.

Still, as for temperature, it was seen that even the company guarantees thermal comfort, at an average of 25.7°C, and this does not become a risk factor. Therefore, it is not characterized as an unhealthy factor, according to NR 15, in its Annex 3.

REFERENCES

BRASIL, Ministry of Labor. Regulatory Standards (NRs). Available at: <http://trabalho.gov.br/index.php/seguranca-e-saude-no-trabalho/normatizacao/normas-regulamentadoras>.

CNAE (National Classification of Economic Activities)—IBGE. Available at: <http://www.cnae.ibge.gov.br/>.

Faccin, Renata; Gonçalves, Cláudia Giglio de Oliveira; Vilela, Rodolfo Andrade de Gouveia; Bolognesi, Tatiani de Moraes. Industrial acoustics and worker health: proposals for improvements. UNAR. Araras, SP, v.2, n.2, p.1–12, 2008.

Fundacentro. Security and Health at Work: Prevention of Repetitive Strain Injury—RSI. N° 4. São Paulo, 1999. (Folhetim).

Graziano, Graziela Oste; Oswaldo, Yeda Cirera; SPERS, Valéria Rueda Elias; CASTRO, Dagmar Pinto de. Occupational Health: Survey and Analysis of Accidents and Occupational Diseases of Companies of the Industrial Sector of Piracicaba/SP in the Period of 2009/2011. Available at: <http://www.faccamp.br/ojs/index.php/RMPE/article/view/657/pdf>.

Hoeltgebaum, Danielle; Munhoz, Jonas Ricardo; Lini, Renata Sano; Menotti, Vinícius Stela; Madia, Mariana Aparecida Oliveira; Nishiyama, Paula; Mossini, Simone Aparecida Galerani. Occupational exposure in metallurgical industries. Available at: <http://www.dex.uem.br/forum/images/exposi%C3%A7ao%20ocupacional%20em%20industrias%20metalurgicas.pdf>.

Kaschuk, Odirlei Rodrigo; Lopes, Vinicius José Santos. Risks of Work Accidents in the Operation of Steel Cutting and Folding Machines in an Industry in Sinop/MT Municipality. Available at: http: // www.webtartigos.com/artigos/riscos-de-acidentes-de-trabalho-na-operacao-de-maquinas-de-corte-e-dobra-dea-aco-em-an-industrial-in-the-city-of-sinop-Mt/109006/>.

Matheus, Bruna; Daher, Maria José E. Available at: <http://publicacoes.unigranrio.br/index.php/rcs/article/viewFile/509/568>.

Régis, Ana Cristina Furtado de Carvalho; Crispim, Karla Geovanna Moraes; Ferreira, Aldo Pacheco. Incidence and prevalence of noisiness—induced hearing loss in workers of a metallurgical industry, Manaus—AM. Rev. CEFAC, Sep / Oct, 2014.

Saliba, Tuffi Messias. Basic Course of Occupational Health and Security—São Paulo, SP: Ed. LTr, 2004. 453 p. ISBN 85-36.

Occupational Safety and Hygiene V – Arezes et al. (Eds)
© 2017 Taylor & Francis Group, London, ISBN 978-1-138-05761-6

Occupational risks in the Disassembly, Transportation and Reassembly (DTR) operations of drilling probes

A.R. Thais, M.B.G. Santos & O.O. Ronildo
Federal University of Campina Grande, PB, Brazil

ABSTRACT: Regarding the land wells drilling operations, the process of DTR is characterized by the need to carry the probe of the location in which was concluded the drilling of a well for the location in which it plans to drill the next well. In this activity, there are occupational hazards that can lead to threats to the security and health of the worker. This research aimed to analyze the occupational hazards present in these activities and suggest anticipatory measures, recognition, evaluation and control of risks to perform a safe process. Among the methodological aspects used in this work, there is the use of the risk management tool consisting of the GUT matrix (Gravity, Urgency, Tendency) for prioritization of the specific risks on the steps of the DTR activities and the respective mitigation measures of these risks, such as the use Personal Protective Equipment (PPE), training of workers and inspection of machines. The results indicate that the risks of accidents are prevalent during both assembly and disassembly as in transportation of equipment and probes. Thus, electric shock, falling of suspended loads, rupture of hoses and lines, machinery and equipment without protection are among the risk agents that require prioritization of implementation of mitigation measures and risk control agents.

1 INTRODUCTION

In land drilling operations, soon after the completion of a well, the probe must be dismantled in several parts (loads) using cranes. The loads are transported by carts to the next location, in the proper order, where they are reassembled. This operation is called DTR. The risks that these workers are subject to are originated from the process itself and from contact with chemical products, besides, of course, non-specific risks of the petroleum sector, such as noise, heat, etc. Within this context, it is important to note the importance of implementing effective mitigation measures, identifying which procedures can contribute to the elimination of the hazard and consequently the risk reduction in these operations. In order to protect these workers, 36 regulatory norms have been created in Brazil by the Ministry of Labor and Social Security, NR's which regulate and make feasible guidelines on mandatory measures to be taken by companies related to safety and occupational health. However, there is still no current regulatory standard addressing the specificities of the operations involved in the petroleum sector, mainly existing in relation to specialized services in safety engineering and occupational medicine such as NR-4, personal protective equipment and safety in machinery and Equipment such as NR-06 – Personal Protective Equipment (PPE), the NR-7

which deals with medical occupational health control program as well as NR-12 – Safety at work in machinery and equipment. In view of the above, this research is justified by presenting a study of the risks present during these operations, proposing improvements to the DTR process aiming at the worker's safety and the operation itself, by means of the high investments necessary to carry out the activities, with the purpose of analyzing the present occupational risks. In the DTR activity as well as propose measures of anticipation, recognition, evaluation and control of existing risks that contribute to the accomplishment of DTR activities in a safe and effective way.

1.1 *DTR activities and accidents in the sector*

After receiving the well program and requesting all materials and equipment for intervention in the same, the transportation and assembly of the probe and peripheral equipment begins, in these activities there are several risks that if not analyzed and administered, contribute significantly. To increase the rate of work accidents [Rocha; Silva; Araújo Filho, 2015].

According to data from the SINDIPETRO-BA (tankers' union) of 2011, a worker died from a work accident when the DTR activity of the rig was about to be completed, only missing the mud tank transport of the rig. When performing the operation of coupling the tank to the mechanical horse, the worker was sandwiched between the tank and the mechanical horse, and died on the way to the hospital.

2 METHODOLOGY

This research is characterized in a theoretical and exploratory way seeking, initially bibliographical through consultations to books, magazines, technical and scientific articles and websites addressing the theme portrayed. The data were obtained through secondary sources, these come from observations of videos and photos provided by the work team of the study site and videos via the Internet, as well as the collection of data through observations acquired through semi-structured interviews with a work safety engineer Active in the activities of DTR, petroleum engineer and other workers. In this way, it was possible to deepen and apply the concepts of Safety related to the conduction of the disassembly, transportation and Reassembly of probes and equipment.

From the risk assessment in the three stages of the process (D, T and R) were observed the risk agents, damages and consequent impacts of each risk found and elaborated possible solutions to the problem. Risks were classified as physical, chemical, ergonomic and accident risks, each risk agent was related to its operating segment (D, R and/or T) and, finally, the priorities were chosen according to the risk assessment tool. Risk management, urgency and trend (GUT).

2.1 Application of risk management technique GUT (Gravity, Urgency and Tendency)

A survey and risk study was carried out in the three segments of the DTR activity, namely disassembly, transportation and reassembly of probes and equipment. These risks were prioritized according to the GUT (Gravity, Urgercy and Tendency) risk management technique.

According to Klassmann [2011], the GUT technique should be used to determine priorities in eliminating problems, in particular if they are several and related to each other. This matrix acts by selecting and staggering the problems, taking into consideration the positive and negative aspects that may happen during its correction. Risks are appreciated by three classes:

Gravity: Represents the impact of the analyzed problem on tasks, people, results, processes and organizations should it happen.

Urgency: Represents the deadline, the time available or needed to solve a specific problem that has been analyzed.

Table 1. Score attributed to the GUT matrix.

Points	Gravity	Urgency	Tendency
5	The damage or difficulties are extremely serious	Immediate action is required	If nothing is done, the aggravation will be immediate.
4	Very serious	Some urgency	It will get worse in the short term.
3	Serious	As soon as possible	It will get worse
2	Not very serious	You can Wait a little	It will get worse
1	No Gravity	There is no hurry.	It Will not get worse or may even improve

Tendency: Represents the growth potential of the problem, the probability of the problem becoming larger over time. It is the evaluation of the tendency of growth, reduction or disappearance of the problem.

Then a score of 1 to 5 is assigned to each problem listed within the aspects cited above. In this way, the matrix seeks to classify in descending order of points the problems to be attacked in the process improvement, according to Table 1.

At the end of the assignment of notes to the problems, following the GUT aspects, a number is assigned to the result of the whole analysis, it will define the degree of priority of the identified problem. The scale is idealized in a configuration that proposes that the higher the analyzed result, the higher the priority, $G \times U \times T = $ PRIORITIZATION.

3 RESULTS AND DISCUSSION

In the GUT Matrix (Table 2), it is demonstrated the identification, classification and prioritization of the risks found in the process of disassembling, transporting and reassembling probe and equipment.

Table 2 shows the classification of the occupational risks to which the workers involved with the DTR activity are submitted. This table was based on the observation of the activity and consequently the risk exposure to which 11 workers are submitted to a DTR activity in the state of Rio Grande do Norte, Brazil. In column 01 are listed the risk agents, in columns 02, 03 and 04 are listed respectively the gravity, urgency and tendency adopted, in column 05 is the product of the values adopted for each risk, ranging from 1 (no gravity)

Table 2. Matrix G.U.T. For disassembly, transport and reassembly according to environmental risks.

Matrix GUT	G	U	T	Total	Priority
Physical Hazards					
Noise exposure (D, T and R)	5	5	5	125	1st
Exposure to Mechanical Vibration (T)	2	2	1	4	11th
Direct exposure to solar radiation (D, T and R)	5	4	2	40	7th
Chemical Risks					
Exposure of the worker to a chemical agent (D, T and R)	3	3	1	9	10th
Ergonomic Risks					
Inadequate posture (D and R)	4	4	4	64	4th
Accident Risk					
Contact with heated parts (D)	4	5	5	100	2nd
Leaks of flammable and chemical products (D, T and R)	4	5	5	100	2nd
Contact with sharp parts (D and R)	2	4	3	24	9th
Contact moving parts (coils, cables, etc.) (D, R)	4	4	5	80	3rd
Electric shock (D, R)	3	4	5	60	5th
Mechanical shock (equipment, people, cargo) (D, R)	5	4	4	80	3rd
Slides, stumbles and falls (D, R)	4	5	5	100	2nd
Brake or clutch failure on the winch (D, R)	5	5	5	125	1st
Danger of fall and height (D, R)	5	5	5	125	1st
Danger of falling load/ equipment (D, R)	5	5	5	125	1st
Solid or liquid particle projection (D, T, R)	4	3	4	48	6th
Breaking of steel or chain cables (D,T,R)	5	4	4	80	3rd
Breaking of hoses and lines (D, R)	5	5	5	125	1st
Holes of anchors, pit and crane struts (D, R)	5	4	4	80	3rd
Wells, oil or steam lines, power lines in the location and vicinity	5	5	5	125	1st
Venomous animals	3	3	3	27	8th
Danger of passing under low mains	5	5	5	125	1st

to 5 (extremely serious) and finally in column 06 is the prioritization of risks.

In DTR, it is possible to find accidents, chemical, physical and ergonomic risks, being the accident risk which demonstrates the first priorities among the risks.

Physical Risks—Priority 1st, 7th, 11th: The time the worker is exposed to this agent of risk is considerable and can cause harmful effects not only on hearing but also stress throughout the circulatory, respiratory and digestive system.

Measures should be taken to mitigate these risks, such as the use of ear protectors in environments with high noise levels and/or where signaled, and worker turnover, and pause in activities, if possible, as this will not result in excessive exposure of workers to this Risk agent Meet the standard NR 15 – Unhealthy activities and operations.

However, there are risks due to mechanical vibration (11th place of prioritization) and indirect exposure to solar radiation (7th place of prioritization). Constant sun exposure, without proper use of PPE, can cause serious damage to workers' health, such as skin diseases.

Alternative work methods that reduce exposure to mechanical vibration; Adequate maintenance programs for equipment, premises and existing facilities in the workplace; Limitation of duration and intensity of exposure are some methods that should be deployed in the workplace.

Chemical Hazards—Priority 10th Solutions to this problem could be obtained from the use of PPE's such as rubber gloves, masks and other protective equipment that act as a barrier to contact with this product; Always have available in the lease safety data sheets of the chemicals and their medicines for the treatment of injuries; Take care to check the toxicological data of the products to be handled and in case of contact with the skin, immediately wash with potable water; Change immediately if contamination occurs. Ergonomic risks—Priority 4th: These risks come from the great efforts of the worker, since the work of disassembling, transporting and reassembling is carried out in uncomfortable environments, with heavy machinery and long working hours, that is, excessive physical effort is very peculiar to the working methods in the petroleum sector, an aggravating factor for the presence of this risk that occupies the fourth position of prioritization among the risks found in the activity. In order to reduce damage that improper posture during the execution of the DTR may cause the worker, measures should be taken, such as: to keep the back straight, using the muscles of the legs and arms to perform the work; Seek help when needed; Use suitable tools/ancillary equipment; Intercalary periods of rest; Biological hazards—Not found in the activity.

Accident Risks—Priorities 1st, 2nd, 3rd, 4th, 5th, 6th and 8th: This risk is present in the three DTR segments. The conditions and infrastructure of the workplace contribute significantly to the occurrence of accident hazards. It was found that most of the accident risk agents encountered were classified as priority 1st, because this operation is surrounded by situations that favor human error, such as improper handling of machines and tools.

A. Contact with heated parts: Contact with the heated parts is placed second in the risk prioritization where immediate action is required to protect the worker against burns. Mitigating measures for these rich people are in the use of helmets, safety goggles, boots with non-slip soles and PVC or leather gloves, it is also extremely important to provide adequate insulation/protection of the heated parts and to observe the working environment carefully, Exposing these parts.

B. Moving parts contact: It occupies the third place in the prioritization of actions. Possible solutions to this risk are: Wear helmets, safety glasses, safety boots with non-slip soles and gloves PVC or leather; Protect the moving parts (winches, pulleys, etc.) and observe the working environment carefully, and do not position yourself in the region of moving parts.

C. Danger of electric shock: This risk agent occupies the fifth place of prioritization. Mitigating measures for this risk include actions such as: the use of helmets, goggles and safety boots, PVC gloves (gloves of insulating material if given with energized equipment); Checking electrical voltage before touching parts of equipment or installations; Certification that the equipment is properly grounded or disconnected from the voltage source; Signaling to prevent third party reclosing; Do not work with energized equipment while the body or clothes are wet; Requesting help in dangerous conditions; Use of suitable tools and in perfect conditions of use (e.g: do not use tools with poor insulation).

D. Mechanical shock hazard (equipment/persons/load). These risks come from the collision of equipment with the workers and represent the third place in the priority ranking. In order that these actions do not happen, it is necessary to constantly signal to the crane operator during the movement of the loads, especially in areas difficult to visualize the load; Perform environmental inspection and cargo handling equipment before starting the service; Isolate and signal the area; Do not stand under suspended load; isolate, where possible, the area underneath where service is being performed at a height; Always work at high altitudes with the tools tied and wear safety helmets, gloves, boots and glasses.

E. Slides, stumbles and falls: During the operation, some liquid substance may fall on the floor and the worker may slip off unattended through the area. Possible solutions for these risk agents are to visualize general recommendations; Realize the environment before starting the service; Keep the environment clean; Use safety boots with the soles in good condition and non-slip.

F. Brake failure or winch clutch: It is necessary to wear a helmet, safety goggles, safety boots with sound and non-slip soles; Do not position yourself in the area of descent of the mast or under the region of movement of the traveling block and its loads and comply with the plan of maintenance of the brake and clutch of the winch.

G. Danger of falling from height: For this agent of risk it is necessary to realize the recognition of the environment before starting the service; To work on the mast of the probe use parachute type belt, crash-locks and counterbalance with dynamic brake. In the case of someone else's climb use a parachute-type belt, fall arrester and retractable lifts; The floor must be non-slip and clean without oil or grease; Use tailor made clothing.

H. Danger of fall of suspended load/equipment: It is due to the lifting of loads during the operation. It is necessary to use PPE's; Do not stand under suspended load; Safe distance of cargo to be transported; Inspect load lifting steel cables; Guiding load with rope (another element that does not lead to electricity), do not guide directly with hand or foot.

I. Breaking of steel cable and/or chains: These cables and chains in use need to be visually inspected for corrosion or kneading, replacing them if necessary; The worker must protect himself and keep away from intended cables and chains and make sure that the cables or chains are correctly lashed before they are intended.

J. Breaking of hoses and lines: Possible solutions for this risk agent are wearing helmets, safety goggles, safety boots with non-slip soles and PVC gloves; Depressurize the equipment before disconnecting it; Know the parameters of operation and state of the equipment (eg, be certain of the contained fluid, internal pressure, mechanical resistance of equipment and etc); Keep away from pressurized equipment and out of the way of buffers, valves and the like; Anchor the high pressure lines; During line testing expect to reduce pressure before approaching to get leak site.

K. Leaks of flammable and chemical products: This agent is considered important, as any leakage of these substances can cause explosions.

This requires containment at the location and storage area of products; Always have reagent products available that can neutralize the aggressive agents; If necessary, trigger the emergency plan; Personnel should be trained in emergency evacuation and prior testing should be performed on valve lines with higher than expected pressures in the process to ensure the safety of the operation.

4 CONCLUSION

Through the concepts of occupational hygiene and work safety in the stages of disassembly, transportation and reassembly of probes and equipment, it was possible to discern the main agents of occupational risks related to DTR operations. In this sense, the risk management tool GUT (severity, urgency and tendency) used enabled the establishment of priorities in the suggestions for implementing measures to be taken to control or mitigate occupational and environmental risks.

Workers in the DTR sector are exposed to physical, chemical, ergonomic and accident risks, so mitigation and risk control measures must be implemented continuously. Using the GUT matrix, it was observed that the risks of accidents and the physical risks as the noise were the predominant ones in the three segments of the activity. In this way, it is these risk agents that require prioritization of the implementation of mitigation and correction measures. This way, there would be fewer accidents, fewer production stops and consequently less direct and indirect costs to a company and society.

REFERENCES

Amorim, E.L.C. de. Risk Analysis Tools. Apostila of the Environmental Engineering course of the Federal University of Alagoas, CTEC, Alagoas, 2010. Available at: <http://pt.scribd.com/doc/71505557/Apostila-de-tools-de-analysis-de-risco>. Accessed on: 08 May 2016

ASSOCIAÇÃO BRASILEIRA DE NORMAS TÉCNICAS—ANBT. ISO31000: Risk management—Principles and guidelines. Rio de Janeiro, 2009.

Klassmann, A. B.; Brehm, F. A.; Moraes, C.A. Employees' perception of risks and hazards in operations in the foundry sector. Rev. Est. Tecnológicos. V. 7, n. 2, p. 142–162, May/December 2011.

Rocha, Sandra Patricia; Silva, Andre Vieira da.; Araújo Filho, Alcides Anastacio de. Analysis of environmental risks and work accidents in an oil and gas company. In: XXXV National Meeting of Production Engineering. Global Perspectives for Production Engineering. Fortaleza, CE, Brazil, October 13 to 16, 2015. Available at: http://www.abepro.org.br/biblioteca/TN_STO_209_238_28006.pdf. Accessed on: April 09, 2016.

SINDICADO DOS PETROLEIROS DA BAHIA—SINDIPETRO-BAHIA (Nazaré, BA). Available at: http://www.sindipetroba.org.br/novo/. Accessed on: April 09, 2016.

Occupational Safety and Hygiene V – Arezes et al. (Eds)
© *2017 Taylor & Francis Group, London, ISBN 978-1-138-05761-6*

Recognition of hazards in worker exposure to salt dust: Practical study on saltworks of Rio Grande do Norte, Brazil

P.C. Neto
Federal Institute of Education, Science and Technology of Rio Grande do Norte, Brazil

T.C.N.O. Pinto, J.A. da Silva, A.M.T. Bon & C.C. Gronchi
Jorge Duprat Figueiredo Occupational Safety and Health Foundation, São Paulo, Brazil

ABSTRACT: To identify the main sources and activities generating of sea salt dust in the beneficiation process in four saltworks located in Mossoró and Grossos, Rio Grande do Norte, Brazil. The Recognition of Hazards technique was used to identify the sources of salt dust generation and existing control measures. The grinding operations, refining and bagging of sea salt were identified are those that generate more dust into the work environment. The sources and activities identified are the mechanical vibrating screens, manual bag filling, the unclogging of the local exhaust system machines, the use of Big-Bag without clamping system, the defect on valve of the plastic bag, damage in the tissue interconnection joints, and dry floor cleaning. According to the International Labour Organization, the Recognition of Hazards is sufficient for recommending of control measures. For the surveyed saltworks the implementation of preventive control measures should be considered to preserve the health and physical integrity of workers.

1 INTRODUCTION

The International Labour Organization (ILO) defines Occupational Hygiene (OH) as the science which treats anticipation, recognizing, evaluation and control of hazards arising in or from the workplace, and which could impair the health and well-being of workers (Goelzer, 2011).

For the ILO, when the health hazards are obvious, the control should be recommended, even before the quantitative evaluations are carried out. Thus, the classical concept of "recognition- evaluation-control" can shift to "recognition-control-evaluation" or even "recognition-control" (Goelzer, 2011).

One of the classic steps of OH practice is the recognition of possible health hazards in the workplace (Goelzer, 2011). The goal of Recognition of Hazard is to identify the existing risk factors and the conditions of occupational exposure (Goelzer, 2013). In a simplified way, the Canadian Centre for Occupational Health and Safety (CCOHS), states that in this phase all possible risk factors in the workplace are to be found and registered (CCOHS, 2009). The Health and Safety Executive (HSE) remarks the importance of a systematic and structured approach in which it is assured that the identified risks depict the process, the system and the operations currently developed in that work environment (HSE, 2003).

The step of Recognition of Hazards for any professional activity involves the characterization of the workplace, identification of risks and of the group of workers who are potentially exposed (Lillienberg, 2011). To Goelzer (2013), during this step, it is necessary to investigate the possibilities of use, formation or dispersion of agents or factors which are potentially harmful and are related to different labor processes as well as understand how they affect health and recognize the real exposure conditions.

To HSE (2003), the present risks in a work environment may be the result of a combination of the following factors: substances, machinery, processes, work organization, tasks, procedures as well as the circumstances to which the activities are carried on, including the physical aspects of the work facilities. Also according to HSE (2003), the Recognition of Hazards is a crucial step, for an unknown risk is an uncontrolled risk.

Even though Brazil is the 9° biggest salt producer in the world (DNPM, 2014), there are few studies regarding the risks salt workers are exposed. Studies are restricted to the ones from Hatem et al. (1987) and from Ali (1998).

One of the poorly assessed risks approached by Hatem et al. (1987) and Ali (1998) is the worker exposure to marine salt dust. These studies carried out in the Rio Grande do Norte saltworks are only concerned with occupational hazards related to skin and eyes. These studies do not correlate the possible effects of inhaling salt dust. They are limited to mentioning "salt dust in suspension state" as one of the "risks to the environment" which is present during salt processing step.

In 2013, Brazil produced 7.2 Mt (millions of tons), of those, 5.9 Mt (82%) were marine salt and 1.3 Mt of rock salt (18%). Rio Grande do Norte holds is ranked high in the production of marine salt, for in 2013 the state was responsible for 5.6 Mt (78%) of the national salt production. In 2013, the Rio Grande do Norte salt-producing municipalities were: Mossoró (1.8 Mt), Macau (1.7 Mt), Porto do Mangue (0.60 Mt), Areia Branca (0.59 Mt), Grossos (0.45 Mt), Galinhos (0.39 Mt) and Guamaré (0.06 Mt) (DNPM, 2014).

Data collected from Brazilian Government indicate that a total of 289 companies from the salt sector which employ a total of 5.030 workers. In Rio Grande do Norte has 143 companies totaling 3.704 employees (RAIS, 2012).

Sodium Chloride (NaCl) or salt is regarded an essential substance to human and animal life (Paiva; Penna, 2002). Salt is a product broadly used as a food complement, in the chemical industry for the production of chlorine and derivants as well as in livestock farming as part of animal diet (DNPM, 2014).

The process of marine salt production is divided in 4 stages: sea water concentration, the crystallization of sodium chlorine, harvest and wash-down and processing (Norsal, 2014). The dust production takes place during the grinding and refinement of the salt. During the grinding stage, the salt is placed into a mechanical grinder as to obtain smaller grains. In refinement, the salt is sieved through mechanical vibration in which salt crystals are sorted by size.

According to the World Health Organization (WHO), the dust is usually originated through a mechanical rupture of great masses of the same material. And "if clouds of dust are seen in suspension, it is almost certain that dust of potentially harmful size is present" (WHO/SDE/OEH/99.14, 1999, p. xi).

As far as the possible effects to health due to salt inhalation, the few references found in scientific database report perforation of nasal airways of workers who were exposed to salt dust (Mackenzie, 1905; Adamek, Witkowska, Zysnarska, 2010; IFA, 2011), and the increase of arterial blood pressure (Haldiya et al., 2005; CCOHS, 2009; IFA, 2011).

The Recognition of Hazards step in Occupational Hygiene is important technical tools which enables a broader awareness as well as describe in more detail the worker exposure to marine salt dust during the processing stage in the salt industry.

Within this perspective, this risk assessment study of exploratory nature was performed in four saltworks in Rio Grande do Norte and had as main objective to identify the main sources of dust and dust generating activities during the processing of marine salt.

2 METHODOLOGY

This study was of exploratory and descriptive nature and based on the Recognition of Hazards. The present research is part of the project "Worker Health and Safety in the marine salt extraction, transport and refinement industry" from the Fundação Jorge Duprat Figueiredo de Segurança e Medicina do Trabalho (Fundacentro).

Due to the fact that this study only encompassed an observational analysis of the work environments, there was no need to have it undergo a Research Ethics Committee.

After having the authorizations from the saltworks owners and counting on the logistics support from Fundacentro from November 2013 to January 2014, visitations were made in all nine previously contacted saltworks with the intent of becoming familiar with the production process as well as the end product (ground and refined salt). Thus, being able to select the saltworks where the risk and salt dust exposure assessment would take place.

The first criteria of choice used for saline selection was the granulometry of the salt produced, being ground or refined. Priority was given to companies that produce refined salt because of their smaller particle size, as they are more likely to be suspended in the environment.

The second criteria adopted were the quantity of workers in salt refining or grinding activities. Priority was given to companies with a larger number of employees.

After visitation in nine saltworks distributed in five Rio Grande do Norte municipalities (Areia Branca, Grossos, Mossoró and Porto do Mangue), four saltworks were selected suitable for studies. The companies were alphabetically identified in the present study as "A", "B", "C" and "D" as a non-disclosure measure the identity of the companies. The selected companies are presented in Table 1.

In its assessment stage, the study identified operations during the salt processing stage in the salt-

Table 1. Saltworks approved in the selection criterias.

| Company | Municipalitie | Quantity of exposed workers | |
		Grounded salt	Refined salt
A	Mossoró	1	18
B	Grossos	0	8
C	Grossos	8	0
D	Grossos	0	8

works, which generate salt dust in the workplace, the characteristics of product refinement, the structure of the work environment, the sources of salt dust as well as the existing control measures.

The following criteria were used as to identify the sources of salt dust generation: a) presence of salt dust clouds visible to the naked eye (WHO/SDE/OEH/99.14, 1999); b) location of dust-pilled spots (walls, ceiling support structures, floor and equipment as well as workers clothes) (SANTOS, 2001).

All information raised during observations such as data, beginning and end daily schedule, Collective Protective Equipment (CPE), Personal Protective Equipment (PPE) and further data, were registered in a specific formulary.

The workers evaluated in this study were those considered as Maximum Risk Employees (MRE). In order to identify these workers, the following criteria were used: proximity to the source of particulate material, moving of the worker, time of exposure and differences in operational habits (Leidel; Bush; Lynch, 1977).

3 RESULTS

The following operational steps were identified as salt dust generators in the workplace: marine salt grinding, refinement and packing for the industry of human and animal use.

Ground salt, produced by factory C for the livestock farming and chemical industry undergoes a poorly controlled grinding process; the refined salt produced by factories A, B and D which is produced for the food industry is ground and sieved as to obtain greater crystal uniformity. As for the micronized salt, used in food, paper and clothing industry, goes into suspension during the sieving process of the refinement. This salt is sucked and transported through the piping for further packing.

Regardless of the salt industrializing process being just grinding or refinement, there may be 1, 10, 20, 25 and 50 kg packing. There can also be flexible containers called "big bags" which are 500, 1.200 or 1.500 kg plastic bags. After packing, the product may be stored or taken to its clients' means of transport.

In regards to where the grinding and refinement take place, all four saltworks are similar in terms of facilities and building structure. They were brick built warehouses with a concrete industrial flooring, lit by both artificial and natural light and for the vast majority ventilation is natural.

3.1 Sources of salt dust

At the Recognition of Hazards step, the sources or salt dust generating activities were acknowledged in all four salt companies during their processing stage. Those risk sources are presented in Table 2.

After assessing the risks, it was concluded that the vibrating sieves used to obtain greater salt crystal uniformity did not have any sort of dust contention system as to minimize the dispersion of salt dust into the work environment. According to WHO (1999), the contention or isolation of a sieve is regarded as an control measure of easy implementation once a mere cover to the sieve may drastically reduce the exposure of workers to dust.

As for the cleaning of the machinery for sack dispense, packing and sealing, the operator is periodically obliged to disconnect the exhaust pipes (responsible for the salt dust suction), and unclog them by using compressed air, practice that produces a great amount of dust into the work environment. The same technique using the compressed air could also be observed in the sealing machine as to avoid the interference of salt in the sealing of plastic bags. Such practice disperses once more a great amount of dust in the air (WHO/SDE/OEH/99.14, 1999).

Table 2. Sources/salt dust generating activities identified in work environments.

Source/activity	Description
Mechanical vibrating sieve (Companies A, B and D).	The sieve movement puts the salt particles in suspension.
Dosing, packing and filling machines (Companies A and D).	Machinery unclogging and cleaning are performed by using compressed air. This puts salt particles in suspension.
Manual filling of raffia fabric sacks (Company C).	Wrong sack positioning in the filling system causes salt spillage.
Filling of flexible containers (Big Bag) (Companies A, B, C and D).	The absence of the upper valve of the Big Bag for connection with the salt filling system may put salt particles in suspension due to free falling.
Sealing valve of the polyethylene bags (Companies A, B, C and D).	A malfunction in the valve causes leakage of salt around the stocking pallet.
Fabric sealing gaskets (Companies A and D).	There are several damaged gaskets as well as saturated with salt, which may cause dust dispersion in the environment.
Work environment and machinery hygiene (Companies A, B, C and D).	Using the broom without humidification and with compressed air.

At the operation of manual filling of 25 kg raffia sacks (woven polypropylene), a demand for worker quickness at the execution of the task could be observed. Fact that leads to incorrect positioning of the sack in the filling system. Thus, leading to spillage of salt on the floor. Movement of people associated to possible air flow present at the workplace, may contribute to the generation and dispersal of particulate material there (WHO/SDE/OEH/99.14, 1999).

Another source of salt dust generation is the filling of flexible containers (Big Bags), which is performed by salt being poured into those bags. The dispersion of dust is due to the pouring as well as the impact of the dust upon falling allied to the lack of an upper valve for connection with the Big Bag. According to WHO (1999), "[...] the higher the impact, greater the dust dispersion. Beyond that, the higher the fall, greater the air flow, greater the dust dissemination [...]" (WHO/SDE/OEH/99.14, 1999, p. 9).

It was also observed that 25 kg polyethylene bags are piled in pallets and that a considerable amount of spillage was found on the floor, under and near the pallets due to problems in the sealing valve. According to WHO (1999), leakage or spillage should be addressed immediately, for that may contribute to air contamination and even putting in suspension the already deposited dust.

Worn out fabric gaskets and salt saturated gaskets were found in companies A and D. Such gaskets connect various types of equipment for the transportation of salt throughout the production plant. Such situation may also contribute to the generation of dust to the work environment, for the friction of larger particles may generate dust during transportation, then being released through the cracks of the damaged gaskets (WHO/SDE/OEH/99.14, 1999).

In all companies studied, cleaning is carried out in the dry with the assistance of brooms, thus dispersing or putting back in suspension a great amount of dust in the workplace. Company A also employs compressed air for the cleaning of machinery. However, the use of both compressed air and broom sweeping is not recommended for the removal of sitting dust, for they disperse back in the air a great amount of dust (WHO/SDE/OEH/99.14, 1999). For WHO (1999), wet sweeping is frequently recommended with the intention of minimizing dust suspension and further air contamination.

3.2 Salt dust control measures

The present study also verified the existing individual and collective prevention measures for dust control and exposition of workers to dust in salt producing companies. Those measures were summarized in Table 3.

In the hierarchy of risk agent control in Occupational Hygiene, the priority must be the control of the source of the agent (the emission), followed by its trajectory (spread). Receptor control measures (worker) are regarded as last resource for control measure (Goelzer, 2011; WHO/SDE/OEH/99.14, 1999; Santos, 2001; Goelzer, 2013).

For the WHO (1999) and for the ILO (GOELZER, 2011), the control of the risk agent in its source is accepted is the most efficient one. For, once the source is controlled no worker will probably be exposed.

Thus, as far as general ventilation in company B's shed is concerned, it was observed that it has six Wind-powered exhausts. However, only four of those worked partially due to a salt layer stuck throughout the inner walls of the exhausts, partially or totally limiting their rotation.

General ventilation can be used as dust control in its trajectory, for it reduces the contamination of skin and clothes, as well as deposits of dust in surfaces. However, general ventilation is not recommended to control contaminants in high concentrations for it

Table 3. Existing individual and collective prevention measures for salt dust in the studied companies.

Com-pany	Control Measures			
	Collective			Individual
	General Ventilation	Local Exhaust Ventilation	Conten-tion	Respiratory Protection
A	Natural (skylight)	Packaging Machines and batchers	There is not	PFF1 *(CA 14.104)
B	Forced (Wind-powered exhausts)	There is not	There is not	PFF1 *(CA 14.104)
C	Natural (skylight) Forced (fans)	There is not	Grinding area only	PFF1 *(CA 14.104)
D	Natural (hollowed out walls)	There is not	There is not	PFF1 *(CA 18.682) White fabric reusable mask

*CA—Certificate of Approval of the Brazilian Government.

may increase the exposure of people who are distant to the source of dust or even outside the workplace (WHO/SDE/OEH/99.14, 1999).

In company C, the existence of leakage spots were observed (isolation of source) in a contention built for the grinding operation. The WHO (1999) defines contention as the inclusion of a physical barrier between the contaminant and the worker. It also states that a contention system must have such ventilation that maintains the isolated area under negative pressure. Thus, naming one example, not allowing dust leakage through cracks.

It was observed that all assessed companies presented similar facility building features, which did not include clear boundaries between sectors, which makes it difficult to contain dust and contributes to its spread throughout the workplace. Beyond that, general ventilation of natural and forced sorts have shown to be insufficient, once salt dust could be seen in every studied work environment.

To WHO (1999), the control of dust cannot be considered as an isolated factor. It is also necessary the inclusion of studies on emission (source), transmission (means) and exposition (receptor) in the dust control strategy.

It was also observed in this study, that control measures of individual protection are the most used ones. Some workers from company D used a reusable mask made of white fabric without the Certificate of Approval (CA). According to Santos (2001), collective measures for control of dust should benefit the greatest possible number of workers. Such "[...] measures should be applied in both the source and the trajectory of dust propagation, thus sparing the worker from the use of individual protection gear [...]" (Santos, 2001, p. 46).

4 CONCLUSION

The present study included one of the preliminary stages of Occupational Hygiene named Recognition of Hazards due to the nature of the already built facilities, including machinery and activities which were already being conducted. Recognition of Hazards is sufficient for the recommendation of control measures of risk agents, even without submitting a qualitative and quantitative evaluation (Goelzer, 2011; Goelzer, 2013).

The sources of salt dust in work environment, as well as the measures of control of preventive nature, were studied and identified through means of Recognition of Hazards. Such assessment may be adopted as a way to prevent the effects marine salt dust on the health of salt workers.

The limitations of the present study encompass the existence of few bibliographical references on the occupational exposure to marine salt and the absence of quantitative evaluations which indicate the amount of dust inhaled by workers.

However, it is possible to state for the studied saltworks, that the implementation of preventive control measures such as an effective general ventilation of the work environment, exhaust local ventilation in the dosing and filing machines, confinement of mechanical vibrating sieves, vacuum aspiration of salt spillage on the floor and the implementation of a Respiratory Protection Program (RPP); should be considered as to preserve health and the physical integrity of the workers.

REFERENCES

Adamek, R.; Witkowska, A.; Zysnarska, M. 2010. Zagrożenia chorobami zawodowymi w Kopalni Soli "Kłodawa". Hygeia Public Health, n. 45, p. 202–205.

Ali, S.A. Condições de trabalho nas salinas do Rio Grande do Norte. 1988. FUNDACENTRO. p. 3–6.

CANADIAN CENTRE FOR OCCUPATIONAL HEALTH AND SAFETY. CCOHS. 2009. OSH Answers Fact Sheets. Risk Assessment. Feb. 2009.

CANADIAN CENTRE FOR OCCUPATIONAL HEALTH AND SAFETY. CCOHS. 2009. CHEMINFO. CCOHS Chemical Name: Sodium chloride. 30 jun. 2009.

DEPARTAMENTO NACIONAL DE PRODUÇÃO MINERAL. DNPM. 2014. Sumário Mineral 2014. Brasília.

Goelzer, B.I.F. 2013. A Importância da Higiene Ocupacional para a Melhoria das Condições e Ambientes de Trabalho. In: MENDES, R. (Org). Patologia do Trabalho. São Paulo: Ed. Atheneu. 3 ed. cap. 52, p.1655–1686.

Goelzer, B.I.F. 2011. Occupational Hygiene: Goals, Definitions and General Information. In: International Labour Organization. ILO. Encyclopedia of Occupational Health and Safety. 3. ed. Geneva: ILO.

Haldiya, K.R.; Mathur, M.L.; Sachdev, R.; Saiyed, H.N. 2005. Risk of high blood pressure in salt workers working near salt milling plants: a cross-sectional and interventional study. Environmental health: a global access science source;13(4):1–7, Jul.

Hatem, E.J.B.; Cavalcanti, F.M.T.B.; Filho, P.C.A.; Batista, J.H.L.; Moreira, M.C.M.O.; Beltrão, S.J.A.P.C. 1987. Levantamento de Riscos Profissionais na Indústria de Extração, Beneficiamento e Transporte de Sal Marinho. Revista Brasileira de Saúde Ocupacional, São Paulo, v. 15, n. 57, p. 14–30.

HEALTH AND SAFETY EXECUTIVE. HSE. 2003. Good practice and pitfalls in risk assessment. Research Report 151.

INSTITUT FÜR ARBEITSSCHUTZ DER DEUTSCHEN GESETZLICHEN UNFALLVERSICHERUNG. IFA. 2011. GESTIS Substance database. Sodium chloride. 10 nov.

Leidel, N.A.; Bush, K.A.; Lynch, J.A. 1977. Occupational Exposure Sampling Strategy Manual. Cincinnati: Ohio; National Institute for Occupational Safety and Health. NIOSH.

LillienberG, L. 2011. Occupational Hygiene: Recognition of Hazards. In: International Labour Organization. ILO. Encyclopedia of Occupational Health and Safety. 3. ed. Geneva: ILO.

Mckenzie, D. Perforation of Nasal Septum from Salt (NaCl) Dust. 1905. Proceedings of the Laryngol Society London, xii, p. 3–5.

Norsal. Norte S/A Indústria Comércio. 2014. Como o Sal é Produzido na Salina Miramar. Areia Branca: Salina Miramar.

Paiva, U; Penna, M. 2002. Império do Sal. Super Interessante. v. 180, set.

RELAÇÃO ANUAL DE INFORMAÇÕES SOCIAS. RAIS. 2012. Ministério do Trabalho e Emprego.

Santos, A.M.A. 2001. O tamanho das partículas de poeira suspensas no ar dos ambientes de trabalho. 1/1. ed. São Paulo: FUNDACENTRO, v. 1000. 96p.

WORLD HEALTH ORGANIZATION. WHO. 1999. Prevention and Control Exchange. PACE. Hazard Prevention and Control in the Work Environment: Airborne Dust. WHO/SDE/OEH/99.14.

Occupational Safety and Hygiene V – Arezes et al. (Eds)
© 2017 Taylor & Francis Group, London, ISBN 978-1-138-05761-6

Contributions of participatory ergonomics in research involving people with disabilities

G.S. Ribeiro
Federal Institute of Education, Science and Technology of Rio de Janeiro, Belford Roxo, Rio de Janeiro, Brazil

L.B. Martins
Federal University of Pernambuco, Recife, Pernambuco, Brazil

ABSTRACT: This paper aims to analyze the contribution of participatory ergonomics in studies involving the participation of people with disabilities. For that matter, a field study which intended to evaluate physical accessibility conditions at the historic centre of the town of Olinda, Pernambuco, Brazil, was used as an object of study. Therefore, participatory ergonomics proved to be an efficient and important instrument of empowerment to people with disabilities in the fight for equality of rights and opportunities.

1 INTRODUCTION

We are constantly dealing with the dispute between the power forces that act in society. Such power decides society's courses in its most diverse aspects and allows decision-making, resources allocating, as well as starting or terminating activities. This said, power can be exercised either from top to bottom, as done by the state and the capital holders, who tend to set their worldview to the population, or horizontally, as common people, who in their everyday organization seek the care and respect for their demands.

Considering that society is composed by a multiplicity and a diversity of social actors, research using methods and techniques with participatory focus can contribute to structure the power disputes held between those actors. Hence, it is possible to make the said disputes more transparent, with higher possibilities to find a solution to the researched problems, optimizing a more equitable distribution of power.

Participation is not only an instrument for problem solving, it also fulfills human beings' need of self-affirmation, social interaction, creation, accomplishment, contribution, or in other words, the need to feel useful. It is a very effective instrument to increase people's motivation and enthusiasm, contributing to the expression of an organization's full potential (Cordioli, 2001). In this way, participation is an additional incentive to the integration of people with disabilities, making it possible to eliminate both physical—in the spaces used by them—and methodological, attitudinal and programmatic barriers, since the participation of these people is encouraged, as opinion maker citizens, favoring their social inclusion.

Thus, this paper seeks to problematize the role of participatory ergonomics in researches that aim to contribute to the improvement of people with disabilities' quality of life.

This problematic is justified by the importance of discussing people with disabilities' power of action in society, allowing their real social integration and the fulfillment of their demands, considering that the premises of universal design, integral accessibility and participatory ergonomics are important instruments to achieve this integration.

The object of this analysis is the experience undertaken by Ribeiro (2008) in her research with people with disabilities, to evaluate accessibility conditions in the historic centre of the city of Olinda, Pernambuco, Brazil.

2 THE IMPORTANCE OF PARTICIPATORY ERGONOMICS TO PEOPLE WITH DISABILITIES

Iida (2005) and Thiollent (1998) agree that participatory researches use methods and techniques in which the researchers and the participants of the situation are involved in a sense of cooperation and participation in the search for solutions to a phenomenon. The researcher outgrows his or her position as a simple observer and becomes an active part of the problem's solution, as well as the subject of the research gains a role as an active agent in pointing out and solving the encountered problems.

To Eerd et al (2010), participatory ergonomic based researches may vary according to the conditions in which they are implemented. For increased chances of positive results, it is necessary to invest in financial, human, organizational and physical resources, so that there can be both ergonomic and organizational training, as well as training concerning the increase of communications to support programs to be implemented.

They may be more laborious and time-consuming in relation to other types of research, because it is not just about data collecting and information processing, there must be interaction between the researches' parties. In contrast, for that same reason, they are more likely to be implemented. Changes tend to suffer less resistance and difficulties compared to proposals made by projects that do not give opportunity to all possibly involved social actors to express their needs and expectations. Since per Nagamachi (1995), when the users themselves are part of evaluation of the items to be restructured, there is a greater sense of responsibility, as they were part of that process, the results are greater satisfaction and motivation.

Tompa et al (2013) and Pinto (2015) made an economic evaluation of a participatory ergonomics process from the company perspective and concluded that the achieved benefits are greater than the costs. Such benefits were notice on both financial and personal aspects, concerning the workers' health.

The premises of participatory ergonomics seek to involve users of various organizational levels in identifying, analyzing and solving problems, especially ergonomic problems. It is a strategy to stimulate participation, since the involvement of users in solving ergonomic problems can cause greater confidence, interest and experience, leading them to see and solve problems related to their own environment (Guimarães, 2004).

Brown (1995) believes that user's participation, both in the phases of identification of ergonomic constraints and of design and implementation of projetual proposals, is justified by granting great involvement and, consequently, achieving higher levels of success in the modifications. According to Pinto (2015), the participatory ergonomics helps to reach a preventive culture regarding ergonomic problems in companies, forming people who are more committed to the safety and well-being of all.

However, it is believed that its greatest contribution is the transfer of knowledge. According to Iida (2005), effective participation involves increasing levels of knowledge acquisition, behavioral changes and feedback controls, which should occur in continuous and cumulative ways. They begin with external regulation and evolve into self-sufficient internal regulation. Which means that at the beginning of the process, with practically no internal involvement, only the external consultant/researcher dominates ergonomics knowledge. As the process unfolds, knowledge is transferred and assimilated by its other participants, as they are incorporated into the culture of the place under study and start to have their own everyday existence without relying on external stimuli, thus allowing dismissing of the external consultant/researcher.

Participatory ergonomics is capable of promoting improvements in different sectors of society. Per Dias et al (2006), participatory ergonomics is social and the problem solution proposed by it involves the interaction of three basic elements: universe, agent and external representations.

When considering brazilian reality, where a large part of the population does not have real instruments to exercise their citizenship and social participation, participatory ergonomics is an effective instrument, especially for people with disabilities, who already suffer daily both from social barriers and the lack of their social participation.

Although this country has extensive legislation and regulations regarding the minimum aspects to be contemplated to assure that environmental, communicational and attitudinal aspects are not impediments to these people participation in society, we can still notice design professionals carrying a speech demanding "common sense" in their projects to this audience. Considering that the understanding of "common sense" is flexible, the collaborative and participatory work among users and researchers is fundamental to achieve total accessibility, making disability nullified as it encounters environments that fulfill their needs.

It is understood that integral accessibility is possible when, in a satisfactory and equitable way, the six types of accessibility described by Sassaki (2003) are incorporated into society. They are:

- physical or architectural accessibility: it refers to the elimination of physical barriers;
- communicational accessibility: it allows the elimination of barriers in interpersonal communication (face-to-face, sign language, body language, gesture language etc.), in written communication (newspaper, magazine, book, course pack, etc., including texts in Braille, extended letter texts for those with low vision and other communicational assistive technologies) and in visual communication;
- methodological accessibility: aims to eliminate barriers in methods and techniques of study (curricular adaptations, lessons based on

multiple intelligences, use of all learning styles, full participation of each student, new learning evaluation concept, new education concept, new concept of didactics logistic, etc.), of work (methods an techniques of training and development of human resources, ergonomics, new concepts of flowchart, empowering, etc.), of community action (social, cultural, artistic methodology, etc. based on participatory action), of children education (new methods and techniques in family relations, etc.) and other areas of activity;

- instrumental accessibility: aims the elimination of barriers in the instruments and tools of study (pencil, pen, computer, pedagogical materials, etc.), of work (tools, machines, equipment), of daily living activities (assistive technology related to communication, personal hygiene, dressing, eating, etc.), of leisure, sport and recreation (devices that meet the physical, sensorial and mental limitations, etc.) and other areas of activity;
- programmatic accessibility: aims to eliminate invisible barriers embedded in public policies, regulations and general use norms. Programmatic barriers are in the decrees, laws, regulations, norms, public policies and other written documents. Barriers are not explicit, but in practice obstruct or raise difficulties in the full participation of people in various sectors of society;
- attitudinal accessibility: aims a society without prejudice, stigmas, stereotypes and discrimination as a result of programs and practices to raise awareness of people in general and the coexistence in human diversity.

When a society manages to reach all spheres of accessibility, it provides better quality of life for its citizens. According to Sansiviero (2004), as a consequence of lack of accessibility, there is an increase in the risk of accidents (e.g.: falls, slippage due slippery and uneven floor surfaces), increased costs related to health and productivity loss, reduced autonomy of people with disabilities and difficulty in access, permanence, perception and communication with the environment.

The importance of participatory ergonomics is evident when giving active voice to users, justified by the quality that it is expected to be incorporated into the results, both in short term – in the search for accessibility solutions that reflect the real needs of users – and in long term, regarding the maintenance of these solutions. Therefore, its greater contribution consists of the philosophy to incorporate both the profile and the voice of the users in all the stages of the project. Thus, participatory ergonomics enters as a foundation, a basic premise for achieving the remaining stages of any project or any research in a satisfactory way for the largest part of the population.

3 METHODOLOGY

The methodological procedures developed can be divided in two moments. Firstly, were carried out bibliographical researches regarding both matters related to participatory research, as well as those related to the universe of people with disabilities, followed by a critical analysis of the researched subjects.

Once the critical analyses of the bibliographic researches were performed, they were followed by a practical activity to evaluate the premises of participatory ergonomics with a group of people with disabilities.

The group was composed of 18 persons, aged between 28 and 62 years. Of these, the majority were male, with a female representation of only 22.2% of the total. All participants presented some kind of disability, 6 of them were wheelchair users, 6 used crutches as an assistive device, 3 used prostheses/orthoses and 3 presented reduced mobility, but did not use any assistance device.

The group was selected from an association of people with disabilities. This association was selected due its strong action power in the state of Pernambuco with people with physical and/or motor limitations. Selecting the participants from this association facilitated the selection process and allowed higher comfort while performing the sessions with focus on participatory ergonomics, since the association opened its doors so that the encounters could happen there, providing more comfort to the participants in a more favorable and accessible environment.

The sessions, in a total of two, were given in the form of collaborative class followed by a discussion. Each lasted about two hours and fifteen minutes. Power Point® slides were presented to the participants, both to raise awareness of their role as citizens in society and to detail the experience to which they would participate in the historic site of the city of Olinda, in Pernambuco, Brazil.

4 RESULTS AND DISCUSSIONS

During the research (Ribeiro, 2008), which this paper uses as object of study to analyze the contributions of participatory ergonomics, it was necessary to develop a methodological proposal to evaluate the physical accessibility in urban historic sites, to contemplate users' participation, mainly people with disabilities'. At that point, methods and techniques that focused on user participation were researched and, after critical analysis, participatory ergonomics was listed as an important instrument to the study.

Its potentiality was already known from some previous studies (Ribeiro, 2007; Leite; Ribeiro,

2014) based on the Macro Ergonomic Analysis of Work (Guimarães, 2004), where participatory ergonomic was contemplated in its initial stage, named Phase 0, the release of the project (by the explanation of the project's objectives to the workers, its phases, methods and techniques that would be used). Its main objective is to address all doubts about the project to be performed and make the workers aware of the importance of their participation in pointing out ergonomic problems, so that they could be solved, improving the quality of life in the work environment of these people.

Throughout the research under analysis, the effectiveness of participatory ergonomics was confirmed.

The research participants, after having the opportunity to discuss constructively and critically about their power to act in society, were invited to experience the physical accessibility conditions at the historic centre of the city of Olinda, Pernambuco. The majority (83,3%) reported that they had already been in that place, but that, even when they suffered from the difficulties related to the lack of physical accessibility, they didn't realize that that situation wasn't simply "normal" and acceptable.

The perception of 100% of the participants in relation to the covered sidewalks is an example of how participatory ergonomics contributes to the empowerment of these people. All of them perceived the sidewalks as one of the greatest accessibility problems of the historical site, since they do not allow independent movement and locomotion of people with disabilities. They need help of other people to walk the paths and cross the barriers. Among the users' reports, regarding the accessibility conditions, it was widely noted statements demonstrating their indignation at the environment. Adjectives such as "absurd", "terrible", "horrible", "very difficult", among others were constant in the users' reports.

Considering it is a rough terrain, with several slopes, the sidewalks present unevenness that often prevent their use by people using wheelchairs. This perception was also felt by users of crutches and even people with disabilities who did not use assistive devices for locomotion, such as people with shortened limbs or paralysis in some limbs of the body. 77,8% of the respondents reported that they would simply use the street for their walk, as a way of overcoming the barriers, but they considered that after the conversation they had with the researchers and other participants, it became clear that they should not expose themselves to the risk of collision with vehicles. Instead, the sidewalks must be suitable for the use of all people, independent of their anthropometric characteristics.

This situation can be illustrated by Figures 1 and 2.

Figure 1. Excessive height between streets and sidewalk and steps along sidewalks compelling users to move along the roads. Source: author's collection (2008).

Figure 2. Narrow sidewalk making circulation difficult. Source: author's collection (2008).

The following statements, made by some of the participants of the research, reinforce their perception regarding the studied space:

"Here you see, it's not a sidewalk, it's a wall for the wheelchair, it's very high … this is absurd …" (wheelchair user).

"Most of the time, one prefers to go down the street, right? The space is disputed with the cars, because it is no use seeking a sidewalk. You can already see that there is not the minimum condition to walk using a wheelchair on the sidewalk" (crutch user).

"The sidewalks between one street and another have no access ramp. (…) the sidewalks do not have access and are damaged. Paving should be better. The height of the curb (sidewalk) is almost 50 cm, which wheelchair or crutch user will be able to climb there without a support? Look at the car there! There is no handrail, at least on one side I think it should have a handrail, but there is none. The widths of the sidewalk are also very short, it is not possible for the chair to pass in some points, in others it is possible to width up to 3 side by side, but it is not in every corner that you can see that, there are others that not even one chair can pass." (user that does not use assistive device for locomotion).

From the last two reports, in addition to understanding the difficulties faced by people with disabilities through the historic site studied, it was perceived another important contribution of participatory ergonomics: it provides the possibility of people to put themselves in other people's position, to acknowledge difficulties that other people may encounter. It provides tools so that people can think beyond their own problems and put themselves in others' place, so that the problems pointed out aim at solutions that can improve the quality of life of an even greater number of people.

This possibility of people analyzing and seeking solutions to the problems encountered, to meet a greater anthropometric range, contributes to the understanding of the environment's need to adjust to people, not the opposite. That corroborates with the defended idea about universal accessibility for people with disabilities.

According to Medeiros and Diniz (2004), disability is an experience resulting from the interaction of individual's physical characteristics and the conditions of the society in which he lives, that is, from the combination of limitations imposed by the body with some type of loss or reduction of functionality to a social organization that is not sensitive to bodily diversity.

Such concept of disability does not exclude the physical, mental or sensorial limitations that people may present, but consider that the environment is responsible for increasing or decreasing these characteristics. When a locality presents accessibility conditions to its population, the disability will not constrain the individual, since it will not prevent him from exercising his activities. The common action is that, when a place gives conditions of a "normal" life for these people, by easing their limitations, they will live like any other person. Likewise, the more precarious the environment, more people will be seen only by their disabilities. Therefore, it is believed that is possible and fundamental that people with disabilities have the right to live in equal terms with any other citizen in society.

5 CONCLUSIONS

Participatory ergonomics has proved to be an important tool for empowerment, as a citizen, of people with disabilities in the fight for their rights. More than just offering solutions to the problems encountered, it contributes to the self-esteem of these people by raising awareness that their rights should be respected, as well as those of any person in society.

Research that promotes the integration of people directly involved in their study objects is more likely to succeed, since the voice of the actual user in the study is a priority. Regarding people with disabilities, in addition to many professionals who do not know their particularities, believing its only necessary "common sense" to design solutions to the lack of accessibility that these people are forced to face, the diverse of realities is vast. The more people involved in research, pointing out the problems and discussing solutions, more likely these solutions will meet their needs and expectations. Participatory ergonomics, as well as the premises of universal design and integral accessibility are fundamental to this process, promoting improvements throughout the whole society, not only to people with disabilities.

REFERENCES

Brown, O. Jr. 1995. The development and domain of participatory ergonomics. In: International Ergonomics Association World Conferencie/Brazilian Ergonomics Congress, 7, 1995, Rio de Janeiro. *Proceedings*. Rio de Janeiro: ABERGO, p. 28–31.

Cordioli, S. 2001. Enfoque participativo no trabalho com grupos. In: BROSE, M. (Org.). *Metodologia Participativa*: uma introdução a 29 instrumentos. Porto Alegre: Tomo Editorial.

Dias, S.I.S.; Mukai, H.; Feiber, F.N.; Merino, E. 2006. Ergonomia e urbanismo: a visão sociotécnica, e a atuação do ergonomista urbano. In: Congresso Brasileiro de Ergonomia, 14, 2006, Curitiba. *Anais*. Curitiba: ABERGO.

Eerd, D.V.; Cole, D.; Irvin, E.; Mahood, Q.: Kcown, K.; Theberge, N.; Village, J.; St. Vincent, M.; Cullen, K. 2010. Process and implementation of participatory ergonomic interventions: a systematic review. *Ergonomics*. Vol. 53, Iss. 10.

Guimarães, L.B.M. 2004. *Ergonomia de Processo* - volume 2. Série monográfica Ergonomia. 4 ed. Porto Alegre: FEENG/UFRGS.

Iida, I. 2005. *Ergonomia*: projeto e produção. 2. ed. rev. e ampl. São Paulo: Edgard Blücher.

Leite, C.; Ribeiro, G.S. 2014. Ergonomic evaluation as a holistic evaluation in company assembly hydrometers. In: Marcelo Soares; Francisco Rebelo. (Org.). *Advances in Ergonomics In Design, Usability & Special Populations* - Part I. 1ed. Devens, Massachusetts, USA: Applied Human Factors and Ergonomics International.

Medeiros, M.; Diniz, D. 2004. Envelhecimento e deficiência. SérieAnis 36, *LetrasLivres*, p. 1–8.

Nagamachi, M. 1995. Requisites and practices of participatory ergonomics. *International Journal of Industrial Ergonomics*, 15, p. 371–377.

Pinto R. 2015. Programa de Ergonomía Participativa para la Prevención de Trastornos Musculoesqueléticos. Aplicación en una Empresa del Sector Industrial. *Ciencia & Trabajo*. v. 17, n. 53, p. 128–136.

Ribeiro, G.S. 2007. *Intervenção numa indústria de montagem de hidrômetros: do aprofundamento dos problemas mapeados aos testes de predições*. Monography (Specialization in Ergonomics). Recife: UFPE/PPGDesign.

Ribeiro, G.S. 2008. *Proposta de procedimentos metodológicos para avaliação da acessibilidade física em sítios históricos urbanos*. Dissertation (Master's Degree in Design). Recife: UFPE/PPGDesign.

Sansiviero, S. 2004. *Acessibilidade na Hotelaria* – uma questão de hospitalidade. Dissertation (Master's Degree in Tourism). São Paulo: UAM.

Sassaki, K.R. 2003. *Inclusão no lazer e no turismo*: em busca da qualidade de vida. São Paulo: Áurea.

Thiollent, M. 1998. *Metodologia da Pesquisa-ação*. 8 ed. São Paulo: Cortez.

Tompa, E.; Dolinschia, R.; Natale, J. 2013. Economic evaluation of a participatory ergonomics intervention in a textile plant. *Applied Ergonomics*. V. 44, Iss. 3, p. 480–487.

Occupational Safety and Hygiene V – Arezes et al. (Eds)
© 2017 Taylor & Francis Group, London, ISBN 978-1-138-05761-6

Neck and upper limb musculoskeletal symptoms in assembly line workers of an automotive industry in Portugal

M. Guerreiro & F. Serranheira
Escola Nacional de Saúde Pública, Universidade Nova de Lisboa, Lisboa, Portugal

E.B. Cruz
Escola Superior de Saúde, Instituto Politécnico de Setúbal, Setúbal, Portugal

A. Sousa-Uva
Escola Nacional de Saúde Pública, Universidade Nova de Lisboa, Lisboa, Portugal

ABSTRACT: Assembly lines are naturally related to health risks and Work-Related Musculoskeletal Disorders (WRMSD), particularly of the neck and upper limbs (WRULMSD). The assessment of perceived musculoskeletal symptoms is essential to WRULMSD prevention, but studies in this field are lacking. A descriptive cross sectional survey on assembly line workers (n = 270) was performed. The objective of this study was to analyze the frequency and distribution of upper limb musculoskeletal symptoms in assembly line workers. Most reported symptoms were neck and upper limbs (35.9%) and baseline participants were predominantly men, with ages mainly between 30 and 40 years of age and high body mass index values. This survey highlights the high number of workers with neck and upper limb musculoskeletal symptoms and that are currently working; in the future, longitudinal studies should be carried out, as standardized assessment methods should be used.

1 INTRODUCTION

Although the introduction of new processes and automation decreased considerably some physical demands for the worker, assembly lines are associated to dynamic and high physical work demands, repetitive movements, awkward positions, particularly in extreme joint positions and with force application, as to poor recovery time (Buckley, 2016; Landau et al., 2008; Sancini et al., 2013; Sluiter, 2006; Sundstrup et al., 2013, 2016). In consequence, assembly lines are naturally related to Work-Related Musculoskeletal Disorders (WRMSD), particularly of the neck and upper limbs (WRULMSD) and absenteeism (Hagberg et al., 2012; Sancini et al., 2013).

Pain and disability, reduced quality of life, and loss of mental wellbeing are some of the WRULMSD consequences, with impact either in workers as in organizations (Andersen et al., 2010; Cole et al., 2009). These are prevalent and costly health conditions (Eerd et al., 2016)—in european countries, an average of 14.1 lost days was counted in 2015-16 (Buckley, 2016).

To understand the burden of upper limbs musculoskeletal disorders and develop prevention measures, it is essential to determine the consequences, as musculoskeletal pain or disability, and then its relation to the causes (Woolf, Vos, & March, 2010).

This highlights an early identification of workers at risk to develop WRULMSD or at risk of worsening their symptoms. The assessment of perceived musculoskeletal symptoms is for the most importance. Pain and discomfort should be assessed, once they are the first ones reported by workers and the main causes to search clinical help (P. Buckle & Devereux, 1999; Zebis et al., 2011), being related to higher prevalence of musculoskeletal disorders (P. W. Buckle & Devereux, 2002; Schneider, Irastorza, & Copsey, 2010). Discomfort is a perception phenomenon related to pain, fatigue and perceived effort, and has been used as a subjective outcome for short term effects (Corlett & Bishop, 1976). Its evolution to chronic musculoskeletal pain suggests discomfort as a WRMSD predictor (Hamberg-van Reenen et al., 2008).

Nevertheless, upper limb musculoskeletal symptoms and disorders incidence rates and prevalence are difficult to find and compare, probably because there is not an universal standard for classification and diagnosis (da Costa & Vieira, 2008; Huisstede, Bierma-Zeinstra, Koes, & Verhaar, 2006). In Portugal, the PROUD Study was the first attempt identifying WRMSD complaints and reported, for 11% of the working population, WRULMSD as more common in the Automotive Industry and Electronic and Electrical Assembly Lines (Miranda, Carnide, & Lopes, 2011). At the present, studies analyzing the distribution and frequency of upper limb musculoskeletal symptoms overtime are still missing (Kennedy et al., 2010; Zoer, Frings-Dresen, & Sluiter, 2014).

The objective of this study is to analyze the frequency and distribution of upper limb musculoskeletal symptoms in assembly line workers, according to individual factors, symptoms existence and job characteristics.

2 METHODS

2.1 Study design and population

A descriptive cross-sectional survey was performed between the 17th September 2014 to 1st October 2014 in an automotive industry of Portugal.

After the authorization of the company and ethics considerations (CNPD—Portuguese Data Protection Authority), a 20 minutes briefing to invite to participation was prepared to both shifts (n = 1100). The workers that wanted to participate signed an informed consent (n = 400) and filled a form.

2.2 Online questionnaire (Work4Health® platform)

The *Work4Health*® platform is an online questionnaire that collected information using Survey-Monkey.inc. Data was organized in demographic information, health data, health status perception, work-related symptoms and job related information. It's construction had as basis the Nordic Musculoskeletal Questionnaire (Kuorinka et al., 1987), the SALTSA criteria document for WRULMSD (Sluiter, Rest, & HW Frings-Dresen, 2001) and a baseline survey conducted by Bohr (Bohr, 2000). To determine the intensity of musculoskeletal symptoms, it was used the Numeric Pain Scale (NPS) and a body-chart to mark the body area affected. Regarding the health status, we used the first question of the Portuguese version of the 12-item short form Health Survey (Sf-12:v2) (Cunha-Miranda L., Vaz-Patto J, Micaelo M., Teixeira A, Silva C, Saraiva-Ribeiro J, 2010). Job related information gathered the subjective information concerning characteristics of the tasks and work activities (Serranheira, Uva, & Lopes, 2008), musculoskeletal symptoms and its association to work.

The questionnaire was available during 15 days.

Information regarding job characteristics, as time of service, was extracted from the company.

2.3 Data analysis

First, we will describe frequencies distribution of the baseline and job characteristics, health data and musculoskeletal symptoms. Then we will explore the possible existing associations between the variables in study. Descriptive statistics will be carried out with SPSS software (IBM SPSS Statistics for Windows, Version 22.0. Armonk, NY: IBM Corp.).

3 RESULTS

3.1 Baseline characteristics

From the informed consent, the 400 workers received an email with a link to the questionnaire. 272 answered the questionnaire and 270 completed it.

3.1.1 Demographic variables

Assembly line participants were mainly men (87%), right handed (88%), with an average age of 36 years and the majority (65.2%) had secondary school education. Regarding the time working in the company, 15 years was the average period.

3.1.2 Health data

Health data shows that 46.7% had a normal Body Mass Index (BMI)—according to World Health Organization (WHO) guide. 21% of the participants reported having a health disease and 24% were using medication. From these, 41 (72%) reported diseases.

56.7% of the workers referred a musculoskeletal injury in the past; from these 56.7% reported perceived general health status as good—13.7% had diseases and took medication. 18% answered fair for health status and of these 38.7% had diseases and were using medication.

At that moment, 21% of the workers were receiving health care support; 21% of these workers reported the use of medication, having a disease and a musculoskeletal injury in the past.

49% referred regular exercise; 36.7% of the sample smoked and 17.8% had alcohol consumption.

3.1.3 Work-related musculoskeletal symptoms

Reporting musculoskeletal symptoms, 68.1% had discomfort or pain in the last month and 68.5% at that moment. The most reported locations for both periods were neck/cervical (7.4 last month and at that moment), lumbar spine (15.9 and 13%), shoulder (9.2 and 9.3%), wrist (5.9 and 5.5%) and the hand (5.5 and 5.6%). Upper limbs were the most reported in both periods (26.6% last month; 28.1% at the moment)—Table 2. Concerning gender, women reported mainly neck/cervical, shoul-

Table 1. Demographic and health data characteristics.

Variable	N (%)	Mean	SD
Age (y)	—	36.2	5.03
Gender			
Female	33 (12.2)	—	—
Male	237 (87.8)		
Work Admission (y)	—	15.07	6.25
BMI			
Underweight	1 (0.4)	—	—
Normal	126 (46.7)		
Overweight	121 (44.8)		
Obese	22 (8.1)		
Education			
Lower school	4 (1.5)	—	—
Middle School	73 (27)		
Secondary School	176 (65.2)		
Higher education	17 (6.3)		
Diseases			
Yes	57 (21.1)	—	
No	213 (78.9)		
Medication			
Yes	67 (24.8)	—	—
No	203 (75.2)		
Musculoskeletal injury (past)			
Yes	153 (56.7)	—	—
No	117 (43.3)		
Current Health Care Support			
Yes	57 (21.1)	—	—
No	213 (78.9)		
Exercise		—	
Yes	133 (49.3)		—
No	137 (50.7)		
General Health Status (1)			
Excellent	18 (6.7)	—	—
Very good	47 (17.4)		
Good	153 (56.7)		
Fair	49 (18.1)		
Poor	3 (1.1)		

Table 2. Musculoskeletal symptoms distribution.

Discomfort/Pain	Last month – n (%)	At the moment – n (%)
Yes	184 (68.1)	185 (68.5)
No	84 (31.9)	85 (31.5)
Body area		
Cervical/neck	20 (7.4)	20 (7.4)
Dorsal spine	6 (2.2)	8 (3)
Lumbar spine	43 (15.9)	35 (13)
Lower limbs	43 (15.9)	46 (17)
Upper limbs	72 (26.6)	76 (28.1)
Shoulder	25 (9.2)	25 (9.3)
Arm	6 (2.3)	7 (2.6)
Elbow	3 (1.1)	7 (2.6)
Forearm	7 (2.6)	7 (2.6)
Wrist	16 (5.9)	15 (5.5)
Hand	15 (5.5)	15 (5.6)

Table 3. Musculoskeletal symptoms intensity.

Discomfort/Pain intensity	Mean (SD)
Last month	4.02 (3.23)
At the moment	3.66 (3.06)

Table 4. Intensity by category*.

Discomfort and pain intensity	n (%)
Mild discomfort/pain (1–3)	18 (18.8)
Moderate discomfort/pain (4–6)	39 (40.6)
Severe discomfort/pain (7–10)	39 (40.6)

*(Salaffi, Ciapetti, & Carotti, 2012).

der and hand discomfort/pain; the majority of men reported symptoms in neck/cervical, shoulder and wrist.

The average intensity of those complaints was 4.02, for the last month and in that moment, respectively (Table 3). The majority of workers (81.25%) with neck and upper limbs musculoskeletal symptoms at the moment of the survey (n = 96), reported intensity above 4 (Table 4).

3.1.4 *Job related information*

Workers were mainly line operators (46.7%), although inspectors and break-down mechanics also participated (14.4 and 12.6%, respectively). Considering the work characteristics, Manual Material Handling (MMH) was present in 59.3% of the work (46.88% of these participants mainly reported in 25% of the working day), repetitive movements in 91.9% of the work (100% of a working day for the majority of participants), tasks with force demands were reported by 83.3% of the workers (62.22% referred 25 to 50% of a working day), static positions had 76.3% (25% of the day for 41.26% of the participants) and vibration was mentioned by 52.6% of the participants (in 33% of the participants was present in 25% of their working day).

54.4% of the participants answered positively to musculoskeletal complaints and its association to work demands.

3.2 *Association between variables*

We chose the "at the moment" data to analyze the possible associations between musculoskeletal symptoms with work characteristics and with individual variables. Although most of associations were reduced or nonexistent, an association between complaints with injury in the past and perceived health status was found (negative association). When analyzing symptoms and work characteristics no significant associations were found.

Despite musculoskeletal symptoms intensity in those who reported neck and upper limb symptoms was associated with other variables, as perceived health status or admission time, the significance is reduced. The period of admission of 4 to 7 years and the group over 14 years are more associated to higher intensity of symptoms (moderate or severe) as lower perception of health (good, fair).

4 DISCUSSION

A baseline of workers from a Portuguese assembly line was created. Most reported symptoms were neck and upper limbs, as other studies with similar job characteristics. Being men, with age between 30 and 40 years, high BMI and secondary education were the most common characteristics. The use of medication, the number of workers reporting musculoskeletal injuries in the past and discomfort/ pain at the survey moment, highlights one particular issue: *presenteeism*. Further analysis should determine if those findings could be related to low levels of work ability and quality of life (Sjøgaard, Justesen, Murray, Dalager, & Søgaard, 2014).

Still concerning individual and health data, having neck and upper limb musculoskeletal symptoms is independent to individual factors as age, gender and education (Loeppke, Edington, Bender, & Reynolds, 2013; Santos & Moreira, 2013).

Regarding work factors, repetitive movements and force application were the most mentioned.

Our results pointed reduced associations among the group of variables and the existence of neck and upper limbs symptoms. The association to musculoskeletal injury in the past could be explained as a musculoskeletal system more fragile and could be associated to longer admission periods. This study shows an average of 15 years and our analysis associates it to higher intensity of symptoms, but the existence of high intensity also in the first 4 to 7 years of work enhances prevention measures in younger workers (MacDonald, Cairns, Angus, & Stead, 2012).

An association of symptoms and its intensity with perceived health status was also found. Some studies reported the relation of lower perceived health status to chronic musculoskeletal conditions—this could emphasize the psychosocial factors for WRMSD prevention studies and interventions (Koolhaas, van der Klink, de Boer, Groothoff, & Brouwer, 2013).

Although BMI is high in this sample, there was not an association to musculoskeletal discomfort/ pain; past clinical history and perceived health data were more relevant.

This survey has not found any associations between neck and upper limbs musculoskeletal symptoms and job characteristics—the number of

participants could be one of the explanations, as the missing data of work risk assessment to compare.

There are some limitations to point out on this survey. Firstly, this study had low rate of participation and for that is limited to compare with similar populations. On the other hand, WRULMSD have a multifactorial etiology and some risk factors (concerning job satisfaction, motivation, for example) were not addressed in this study. Additionally, self-reported data was the main source, lacking other factors to analyze together with participants responses.

5 CONCLUSION

This survey highlights the frequency of workers with neck and upper limb musculoskeletal symptoms. Future studies should consider standardized assessment methods and combine work ability analysis. Sick leave rates, shifts and rotation plans should also be included.

Longitudinal studies on musculoskeletal disorders, as studies on the effectiveness of WRULMSD prevention measures should be made. The need to identify the first symptoms highlights the importance of occupational health surveillance.

ACKNOWLEDGEMENTS

This study was funded by the FCT—Portuguese National Funding Agency for Science, Research and Technology.

REFERENCES

Andersen, L.L., Christensen, K.B., Holtermann, A., Poulsen, O.M., Sjøgaard, G., Pedersen, M.T., & Hansen, E. a. (2010). Effect of physical exercise interventions on musculoskeletal pain in all body regions among office workers: a one-year randomized controlled trial. *Manual Therapy, 15*(1), 100–4.

Bohr, P.C. (2000). Efficacy of Office Ergonomics Education. *Journal of Occupational Rehabilitation, 10*(4), 243–255.

Buckle, P., & Devereux, J. (1999). *Work-related neck and upper limb musculoskeletal disorders.* Luxembourg.

Buckle, P.W., & Devereux, J.J. (2002). The nature of work-related neck and upper limb musculoskeletal disorders. *Applied Ergonomics, 33*(3), 207–17.

Buckley, P. (2016). Work-related Musculoskeletal Disorder (WRMSD) Statistics, Great Britain. *ww. Hse. Gov. Uk/ Statistics,* 1–20.

Cole, D.C., Theberge, N., Dixon, S.M., Rivilis, I., Neumann, W.P., & Wells, R. (2009). Reflecting on a program of participatory ergonomics interventions: A multiple case study. *Work, 34*(2), 161–178.

Corlett, E.N., & Bishop, R.P. (1976). A technique for measuring postural discomfort. *Ergonomics, 9*, 175–182.

Cunha-Miranda L., Vaz-Patto J, Micaelo M., Teixeira A, Silva C, Saraiva-Ribeiro J, F.S.G. (2010). The 12-Item

Short Form Health Survey (SF-12:v2) – Validação da escala para uso em Portugal Resultados do FRAIL Study. *Acta Reumatologia Port.*

da Costa, B.R., & Vieira, E.R. (2008). Stretching to reduce work-related musculoskeletal disorders: a systematic review. *Journal of Rehabilitation Medicine: Official Journal of the UEMS European Board of Physical and Rehabilitation Medicine, 40*(5), 321–8.

Eerd, D. Van, Munhall, C., Irvin, E., Rempel, D., Brewer, S., Van Der Beek, A.J., … Amick, B. (2016). Effectiveness of workplace interventions in the prevention of upper extremity musculoskeletal disorders and symptoms: an update of the evidence. *Occup Environ Med, 73*, 62–70.

Hagberg, M., Violante, F.S., Bonfiglioli, R., Descatha, A., Gold, J., Evanoff, B., & Sluiter, J.K. (2012). Prevention of musculoskeletal disorders in workers: classification and health surveillance—statements of the Scientific Committee on Musculoskeletal Disorders of the International Commission on Occupational Health. *BMC Musculoskeletal Disorders, 13*(1), 109.

Hamberg-van Reenen, H.H., van der Beek, A.J., Blatter, B.M., van der Grinten, M.P., van Mechelen, W., & Bongers, P.M. (2008). Does musculoskeletal discomfort at work predict future musculoskeletal pain? *Ergonomics, 51*(5), 637–48.

Huisstede, B.M. a, Bierma-Zeinstra, S.M. a, Koes, B.W., & Verhaar, J. a N. (2006). Incidence and prevalence of upper-extremity musculoskeletal disorders. A systematic appraisal of the literature. *BMC Musculoskeletal Disorders, 7*, 7.

Kennedy, C. a, Amick, B.C., Dennerlein, J.T., Brewer, S., Catli, S., Williams, R., … Rempel, D. (2010). Systematic review of the role of occupational health and safety interventions in the prevention of upper extremity musculoskeletal symptoms, signs, disorders, injuries, claims and lost time. *Journal of Occupational Rehabilitation, 20*(2), 127–62.

Koolhaas, W., van der Klink, J.J.L., de Boer, M.R., Groothoff, J.W., & Brouwer, S. (2013). Chronic health conditions and work ability in the ageing workforce: the impact of work conditions, psychosocial factors and perceived health. *International Archives of Occupational and Environmental Health.*

Kuorinka, I., Jonsson, B., Kilbom, a, Vinterberg, H., Biering-Sørensen, F., Andersson, G., & Jørgensen, K. (1987). Standardised Nordic questionnaires for the analysis of musculoskeletal symptoms. *Applied Ergonomics, 18*(3), 233–7.

Landau, K., Rademacher, H., Meschke, H., Winter, G., Schaub, K., Grasmueck, M., … Schulze, J. (2008). Musculoskeletal disorders in assembly jobs in the automotive industry with special reference to age management aspects. *International Journal of Industrial Ergonomics, 38*(7–8), 561–576.

Loeppke, R., Edington, D., Bender, J., & Reynolds, A. (2013). The association of technology in a workplace wellness program with health risk factor reduction. *Journal of Occupational and Environmental Medicine/American College of Occupational and Environmental Medicine, 55*(3), 259–64.

MacDonald, L., Cairns, G., Angus, K., & Stead, M. (2012). *Evidence review: social marketing for the prevention and control of communicable disease.* Stockholm.

Miranda, L.C., Carnide, F., & Lopes, F. (2011). Estudo PROUD. Retrieved from http://shst.cgtp.pt/index.php?option = com_content&view = article&id = 135:estudo-proud&catid = 53:noticias&Itemid = 147

Sadi, J., MacDermid, J.C., Chesworth, B., & Birmingham, T. (2007). A 13-year cohort study of musculoskeletal disorders treated in an autoplant, on-site physiotherapy clinic. *Journal of Occupational Rehabilitation, 17*(4), 610–622.

Salaffi, F., Ciapetti, a, & Carotti, M. (2012). Pain assessment strategies in patients with musculoskeletal conditions. *Reumatismo, 64*(4), 216–29.

Sancini, A., Capozzella, A., Caciar, T., Tomei, F., Nardone, N., Scala, B., … Ciarrocca, M. (2013). Risk of upper extremity biomechanical overload in automotive facility. *Biomedical and Environmental Sciences : BES, 26*(1), 70–5.

Santos, C.S., & Moreira, S. (2013). PNSOC - Programa Nacional de Saúde OcupacionaL: 2° ciclo - 2013/2017. Lisboa.

Schneider, E., Irastorza, X., & Copsey, S. (2010). *OSH in figures: Work-related musculoskeletal disorders in the EU — Facts and figures.* Luxembourg.

Serranheira, F., Uva, A.S., & Lopes, F. (2008). *Cadernos/Avulso #05 Lesões Músculo-Esqueléticas e Trabalho: alguns métodos de avaliação do risco.* Lisboa: Sociedade Portuguesa de Medicina do Trabalho.

Sjøgaard, G., Justesen, J.B., Murray, M., Dalager, T., & Søgaard, K. (2014). A conceptual model for worksite intelligent physical exercise training—IPET—intervention for decreasing life style health risk indicators among employees: a randomized controlled trial. *BMC Public Health, 14*(1), 652.

Sluiter, J.K. (2006). High-demand jobs: age-related diversity in work ability? *Applied Ergonomics, 37*(4), 429–40.

Sluiter, J.K., Rest, K.M., & HW Frings-Dresen, M.H. (2001). Criteria document for evaluating the work-relatedness of upper extremity musculoskeletal disorders. *Scandinavian Journal of Work, Environment & Health, 27*(c), 1–102.

Sundstrup, E., Jakobsen, M.D., Andersen, C.H., Jay, K., Persson, R., Aagaard, P., & Andersen, L.L. (2013). Participatory ergonomic intervention versus strength training on chronic pain and work disability in slaughterhouse workers: study protocol for a single-blind, randomized controlled trial. *BMC Musculoskeletal Disorders, 14*(1), 67.

Sundstrup, E., Jakobsen, M.D., Brandt, M., Jay, K., Aagaard, P., & Andersen, L.L. (2016). Associations between biopsychosocial factors and chronic upper limb pain among slaughterhouse workers: cross sectional study. *BMC Musculoskeletal Disorders, 17*(1), 104.

Woolf, A.D., Vos, T., & March, L. (2010). How to measure the impact of musculoskeletal conditions. *Best Practice & Research. Clinical Rheumatology, 24*(6), 723–32.

Zebis, M.K., Andersen, L.L., Pedersen, M.T., Mortensen, P., Andersen, C.H., Pedersen, M.M., … Sjøgaard, G. (2011). Implementation of neck/shoulder exercises for pain relief among industrial workers: a randomized controlled trial. *BMC Musculoskeletal Disorders, 12*(1), 205.

Zoer, I., Frings-Dresen, M.H.W., & Sluiter, J.K. (2014). Are musculoskeletal complaints, related work impairment and desirable adjustments in work age-specific? *International Archives of Occupational and Environmental Health, 87*(6), 647–654.

Occupational Safety and Hygiene V – Arezes et al. (Eds)
© 2017 Taylor & Francis Group, London, ISBN 978-1-138-05761-6

The prevalence of Work-related Musculoskeletal Disorders (WMSDs) in professional bus drivers—a systematic review

A. Cardoso
Faculty of Engineering, University of Porto, Porto, Portugal

M. Luisa Matos
Faculty of Engineering, University of Porto, Porto, Portugal
LNEG, Laboratório Nacional de Energia e Geologia, S. Mamede de Infesta, Portugal

ABSTRACT: The aim of the present study is review systematically relevant literature and synthesize the scientific knowledge in order to explore the prevalence of Work-related Musculoskeletal Disorders (WMSDs) in professional bus drivers. This search was performed using the methodology of PRISMA statement. Eleven studies were included in the review after using key words, inclusion and exclusion criteria. This studies investigated the prevalence, risk factors and symptoms of musculoskeletal disorders in bus drivers in many countries. Numerous studies have shown that bus drivers have high prevalence rates of musculoskeletal problems. Low back pain is the more evident symptom among bus drivers, besides that, the drivers suffered from stress, excessive work pressure, extreme posture conditions and monotonous job.

Keywords: Musculoskeletal disorders, low back pain/neck pain, bus drivers, Nordic questionnaire, exposure, whole-body vibration

1 INTRODUCTION

Work-related musculoskeletal disorders are severe disorders of muscles, bones, nerves, tendons and another soft tissues that appear during the professional activity (Mozafari, Vahedian, Mohebi, & Najafi, 2015) (Yasobant, Chandran, & Reddy, 2015). WMSDs are considered to be multifactorial that are caused due to the interactions between various risk factors, which result in conditions that vary across different occupation (Yasobant & Rajkumar, 2014).

This disorders, especially low back pain, cause substantial losses to individuals as well as to the community (Widanarko, et al., 2011). Professional drivers have been found to have high prevalence rates of musculoskeletal problems (Gangopadhyay, Dev, Das, Ghoshal, & Ara, 2012), especially low back pain (Gangopadhyay & Dev, 2012), due to prolonged sitting and vibration (Alperovitch-Najenson, Katz-Leurer, Santo, Golman, & Kalichman, 2010) (Thamsuwan, Blood, Ching, Boyle, & Johnson, 2013).

Low back pain among professional drivers is prevalent worldwide (Thamsuwan, Blood, Lewis, Rynell, & Johnson, 2012). These disorders lead to financial losses associated with workers compensation insurance or similar forms of social security claims (Tamrin S. B., Yokoyama, Aziz, & Maeda, 2014). The purpose of this study was to investigate the literature and the scientific knowledge of the prevalence and characteristics of WMSDs among urban bus drivers. This article has the pretension to join all the latest and relevant information related to these issues, as well as the latest methodologies for measuring different parameters and others adverse effects of these injuries in the lives of urban bus drivers.

2 METHODOLOGY

The literature review was based on PRISMA (Preferred Reporting Items for Systematic Reviews and Meta-Analysis).

For the preparation of the PRISMA Statement were made researches in different databases. The resources used were exclusively databases (*Compendex, Medline, PubMed, Scopus, Inspec, Web of Science*) and scientific journals (*BioMedCentral, Cambridge, Emerald, Diary of O. Journals, Informaworld, Metapress, Oxford Journals, Science Direct., Science Magazine, While Online Library*). Were also consulted others articles using a different research sources (such as bibliographies of selected articles to the PRISMA).

The keywords used were: Musculoskeletal disorders, low back pain/neck pain, bus drivers, Nordic questionnaire, exposure, whole-body vibration. After introduced the different combinations of research, were implemented certain inclusion and exclusion criteria to achieve the most loyal articles.

The exclusion criteria were:

a. Publication date: were excluded articles published before 2010;
b. Language: were excluded articles not published in Portuguese or English;

Relevance: were excluded the articles that were not on the topic (not addressing musculoskeletal disorders among urban bus drivers, repeated article and articles with specific items of medical nature).

The inclusion criteria were:

On selected articles is applied the inclusion criteria and those articles that were eligible complied with the requirements (include in the abstract the keywords, address the subject at issue and they have not been excluded by the exclusion criteria).

3 RESULTS AND DISCUSSION

3.1 Research results

A total of 1217 articles were retrieved from literature search for its potential on the subject. After

following the above scheme previously, eleven articles were included in review. The procedure of the scheme is shown in Figure 1.

Many of the articles were exclude for not meeting the stipulated requirements and many because they duplicate.

3.2 Musculoskeletal disorders in urban bus drivers

In recent years, Musculoskeletal Disorders (MSDSs) have been increasing (Lee & Gak, 2014). According to the (EU-OSHA (European Agency for Safety and Health at Work), 2010), musculoskeletal disorders in the (lower) back, neck and upper limbs are the most frequent causes of occupational diseases, sickness and absenteeism among European workers (Bovenzi, 2015).

There are countless studies related injuries in various branches of transport companies.

There are several similarities between studies, but only those relating to bus drivers were considered. Exist several studies that found positive results regarding the prevalence of musculoskeletal disorders in bus drivers.

Work-related Musculoskeletal Disorders (WMSDs) are painful disorders of muscles, bones, nerves, tendons and other soft tissues due to workplace activity (Yasobant, Chandran, & Reddy, 2015). For professional drivers, the most common cause of lower back pain is the combination of a

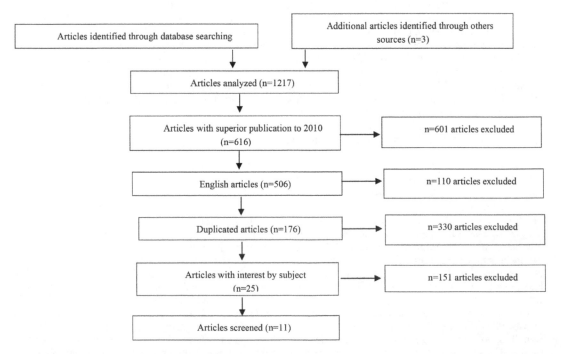

Figure 1. Flow diagram of the studies selected for review.

forced posture and extensive (Mozafari, Vahedian, Mohebi, & Najafi, 2015) sedentary work (Kresal, Roblek, Jerman, & Mesko, 2015).

High prevalence of low back pain among professional drivers has been reported in many parts of the world (Rufa'i, et al., 2013). There are many physical factors that may contribute to increase physical loading in the bus conductors' musculoskeletal system, resulting in discomfort and pain (Gangopadhyay, Dev, Das, Ghoshal, & Ara, 2012). Physical and psychological health of the bus driver is a critical factor in driving performance (Murtezani, Ibraimi, Sllamniku, Osmani, & Sherifi, 2011) (Dev & Gangopaddhyay, 2012).

3.3 Main effects of musculoskeletal disorders and others considerations

Musculoskeletal disorders may cause substantial losses to individuals as well as to the community. Professional drivers have been found to be at high risk for developing low back pain due to prolonged sitting and vehicle vibration (Alperovitch-Najenson, et al., 2010). Bus drivers are exposed to physical and chemical risk factors such as vibration (mentioned above) and exhaust emissions (Lee & Gak, 2014). In terms of individuals factors, age, gender, weight and height, the Body Mass Index (BMI) as well as the general health status of the drivers are also important risk factors associated with WMSDs (Gangopadhyay, Dev, Das, Ghoshal, & Ara, 2012).

The symptoms associated with musculoskeletal disorders vary from individual to individual. The symptoms are related to constant fatigue during and after a day's work (Tamrin S. B., Yokoyama, Naing, & Guan, 2014). Work-related musculoskeletal disorders can cause symptoms such as pain, numbness, tingling, as well as reduced worked productivity, lost time from work, and temporary or permanent disability (Tamrin S. B., Yokoyama, Naing, & Guan, 2014).

3.4 Methodologies applied to assessment the prevalence of musculoskeletal disorders in urban bus drivers

For the study of prevalence of musculoskeletal disorders, many studies speak in randomly recruiting from bus corporation (Lee & Gak, 2014), others studies are made to all drivers of a city in order to select randomly drivers to represent the central, northern, southern and eastern regions (Tamrin S. B., Yokoyama, Aziz, & Maeda, 2014). One of the worldwide tools used to evaluate the prevalence of musculoskeletal disorders is the Modified Nordic Questionnaire. This questionnaire (series of objective questions with multiple-choice responses) has information about musculoskeletal disorders (current-pain in immediate past 7 days and previous-pain in the last 12 months). This type of questionnaire contains even basic demographic information, data of the prevalence of musculoskeletal disorders and its impact (Rufa'i, et al., 2013) and others relevant information (for instance, years of experience). A few studies recruited only male drivers (Gangopadhyay, Dev, Das, Ghoshal, & Ara, 2012) but orders are performed included male and female drivers (Kresal, Roblek, Jerman, & Mesko, 2015).

Although the methodology used in this study was the Nordic Questionnaire, some authors use other methods to supplement their study as the assessment of the postures of drivers were analysed with REBA (Rapid Entire Body Assessment) (Gangopadhyay, Dev, Das, Ghoshal, & Ara, 2012), the RULA (Rapid Upper Limb Assessment) to assessed the exposure and risks of developing WMSDs (Yasobant, Chandran, & Reddy, 2015).

4 CONCLUSIONS

Our research concluded that there are still few studies in relation to musculoskeletal disorders in urban bus drivers. The studies from the literature review indicated that there is a high prevalence of musculoskeletal disorders among bus drivers, especially the prevalence of low back pain and neck pain. This problem affected the drivers' performance with negative economic implications. A review of the literature of the etiology of low back pain and injury among professional drivers suggest that ergonomics problems with the driver's seat plan, whole-body vibration and job stress may contribute to LBP among bus drivers. The bibliography presents similar methodologies, which is important for the comparison, because it shows us better results. To bypass the high rates of musculoskeletal disorders in bus drivers is necessary do more studies, organized awareness campaign program on the importance of proper ergonomics to reduce the risk of low back pain. These associations should be further confirmed in prospective studies. Several studies also indicate that there are some limitations and some leaves open questions for future research.

ACKNOWLEDGMENTS

This research was support by Master in Occupational Safety and Hygiene Engineering (MESHO), of the Faculty of Engineering, of the University of Porto.

REFERENCES

Alperovitch-Najenson, D., Katz-Leurer, M., Santo, Y., Golman, D., & Kalichman, L. (2010). Upper Body Quadrant Pain in Bus Drivers. *Archives of Environmental & Occupational Health*, 218–223.

Alperovitch-Najenson, D., Santo, Y., Masharawi, Y., Katz-Leurer, M., Ushvaev, D., & Kalichman, L. (2010). Low Back Pain among Professional Bus Drivers: Ergonomic and Occupational-Psychosocial Risk Factors. *The Israel Medical Association Journal*, 26–31.

Bovenzi, M. (2015). A prospective cohort study of neck and shoulder pain in professional drivers. *Ergonomics*, 1103–1116.

Dev, S., & Gangopaddhyay, S. (2012). Upper Body Musculoskeletal Disoders among Professional Non-Government City Bus Drivers of Kolkata. *Southeast Asian Network of Ergonomics Societies Conference (SEANES)*.

EU-OSHA (European Agency for Safety and Health at Work). (2010). *Work-related Musculoskeletal Disorders in the EU-Facts and Figures.* Luxembourg: Publications Office of the European Union.

Gangopadhyay, S., & Dev, S. (2012). Effect of Low Back Pain on Social and Professional Life of Drivers of Kolkata. *Work 41*, 2426–2433.

Gangopadhyay, S., Dev, S., Das, T., Ghoshal, G., & Ara, T. (2012). An Ergonomics Study on the prevalence of Musculoskeletal Disorders Among Indian Bus Conductors. *International Journal of Occupational Safety of Ergomonics (JOSE)*, 521–530.

Kresal, F., Roblek, V., Jerman, A., & Mesko, M. (2015). Lower back pain and absenteeism among professional public transport drivers. *International Journal of Occupational Safety and Ergonomics*, 166–172.

Lee, J.-H., & Gak, H.B. (2014). Effects of Self Stretching on Pain and Musculoskeletal Symptom of Bus Drivers. *Journal of Physical Therapy Science*.

Mozafari, A., Vahedian, M., Mohebi, S., & Najafi, M. (2015). Work-Related Musculoskeletal Disorders in Truck Drivers and Official Workers. *Acta Medica Iranica*, 432–438.

Murtezani, A., Ibraimi, Z., Sllamniku, S., Osmani, T., & Sherifi, S. (2011). Prevalence and Risk Factors for Low Back Pain in Industrial Workers. *Folia Medica*, 68–74.

Rufa'i, A.A., Saidu, I.A., Ahmad, R.Y., Elmi, O.S., Aliyu, S.U., Jajere, A.M., & Digil, A.A. (2013). Prevalence and Risk Factors for Low Back Pain Among Professional Drivers in Kano, Nigeria. *Archives of Environmental & Occupational Health*, 251–255.

Tamrin, S.B., Yokoyama, K., Aziz, N., & Maeda, S. (2014). Association of Risk Factors with Musculoskeletal Disorders among Male Commercial Bus Drivers in Malaysia. *Human Factors and Ergonomics in Manufacturing & Service Industries*, 369–385.

Tamrin, S.B., Yokoyama, K., Naing, L., & Guan, Y.N. (2014). The effectiveness of simplified intervention program for preventing and reducing low back pain among malaysian bus drivers. *American Journal of Applied Sciences*, 818–832.

Thamsuwan, O., Blood, R.P., Ching, R.P., Boyle, L., & Johnson, P.W. (2013). Whole body vibration exposures in bus drivers: A comparison between a high-floor coach and a low-floor city bus. *International Journal of Industrial Ergonomics 43*, 9–17.

Thamsuwan, O., Blood, R.P., Lewis, C., Rynell, P.W., & Johnson, P.W. (2012). Whole Body Vibration Exposure and Seat Effective Amplitude Transmissibility of Air Suspension of Different Bus Designs. *Human Factors and Ergonomics Society*, 1218–1222.

Widanarko, B., Legg, S., Stevenson, M., Devereux, J., Eng, A., Mannetje, A., ... Pearce, N. (2011). Prevalence of musculoskeletal symptoms in relation to gender, age and occupational/industrial group. *Internacional Journal of Industrial Ergonomics*, 561–572.

Yasobant, S., & Rajkumar, P. (2014). Work-Related Musculoskeletal disorders among care professionals: A cross-sectorial assessment of risk factors in a tertiary hospital, India. *Indian Journal of Occupational & Environmental Medicine*, 75–81.

Yasobant, S., Chandran, M., & Reddy, E.M. (2015). Are Bus Drivers at an Increased Risk for Developing Musculoskeletal Disorders? An Ergonomic Risk Assessment Study. *J Ergonomics*.

Occupational Safety and Hygiene V – Arezes et al. (Eds)
© 2017 Taylor & Francis Group, London, ISBN 978-1-138-05761-6

Determinants of workplace occupational safety and health: A case study

A. Kawecka-Endler & B. Mrugalska
Faculty of Engineering Management, Poznan University of Technology, Poznan, Poland

ABSTRACT: The paper outlines the problems of occupational health and safety in industrial environment. In order to define their determinants the investigation was conducted among 71 respondents in Poland. Its results showed the discrepancy in the perception of this concept in theory and practice. The compliance with health and safety rules and regulations was indicated as the most important factor in both approaches. However, the need of the application of collective and individual protections was perceived as less crucial in reference to the definition. Furthermore, such determinants as safety culture and fear of accidents were evaluated almost on the same level. All these investigations enabled to specify actions for improvements.

1 INTRODUCTION

The main objective of the multiannual national program "Improvement of safety and working conditions", which has been implemented in Poland since 2008, is to significantly reduce occupational risk related to exposure to harmful, dangerous and onerous factors at workplaces what will decrease the number of accidents at work (CIOP 2016). Therefore, in recent years, the causes and circumstances of the accidents at work have become the subject of many investigations and analyses. It results from the fact that such studies allow not only to determine the causes of accidents but also introduce prevention and control strategies (Michie 2002, Rivara et al. 2009, Harms-Ringdahl 2013).

In the EU countries prevention and control strategies are treated as a supervisory principle for Occupational Safety and Health (OSH) legislation. Their fundamental concepts are risk assessment and risk management, and include assessment of all relevant risks, checking the efficiency of the adopted measures, documentation of the outcomes and regular reviewing of the assessment. However, the main focus should be given to prevention of risk by elimination of hazards management, substitution, engineering and administrative (organizational) controls and personal protective equipment (OHSAS 18001:2007). Thus, working safe means "working in an environment where the probability of accident occurrence is small" (RIVM 2016).

The risk of occupational accidents is highly correlated with human activity (Butlewski & Sławińska 2014, Kawecka-Endler & Mrugalska 2010). For example, in 2015 in Poland 59,2% of

accidents at work were the result of improper worker behavior, whereas 8,6% of them related to material defects, 7,5% were the consequences of lack or improper operation of material factor, 5,3%—improper workstation organization and 4,6%—improper work organization (GUS 2016).

The paper deals with the problems of OSH in the terms of humanization of work environment. For this purpose, the determinants related to creating safety and health environment at work were characterized. The results of the investigation concerning understanding of the OSH concept of referring to the theory and practice were presented and discussed. It allowed to indicate the areas for improvement in industrial setting in Poland.

2 HEALTH AND WORK SAFETY ASSURANCE

The contemporary companies operate in dynamic changing environment, and their success in the market is enviably determined by innovation, flexibility (quick adaptation to changes) and entrepreneurship (Doganova & Eyquem-Renault 2009). However, the foundation and the most important factor is human factor and the humanization of the work environment in any business. The employee, who is the executor of all work processes, must be ensured health and safe working conditions (Kawecka-Endler 2014, Kawecka-Endler & Mrugalska 2012, 2014).

Despite of the introduction of many modern safety guards, systematic inspections and preventive measures the number of accidents is

Table 1. Accidents at work (GUS 2016).

Accidents	2011	2012	2013	2014	2015
Total	96 136	91 000	88 267	88 642	87 622
Fatal	404	350	277	262	303
Serious	683	627	538	520	495
Minor	95 049	90 023	87 452	87 860	86 824

slightly changing (EUROSTAT 2016). The similar situation can be noticed in Poland (Table 1).

Furthermore, the results of the study carried out by GUS (2016) also show that almost 90% of all accidents appears during performing five types of activities. The most of them (ca. 35%) are slips, trips and falls accident while walking or occur while manipulate objects work (17%), manual handling (15%), operating hand tools (13%) and machinery (9%).

The accidents are complex processes involving the entire socio-technical system (Qureshi 2007, Belmonte et al. 2011). They are a combination of events encompassing diverse technical, ergonomic and organizational aspects, which influence human directly and indirectly (Kawecka-Endler & Skulska 2011). Their essence and mutual interdependence is shown in Figure 1.

However, it should be underlined that for many years, the impact of work organization on the economic and social effects has not been noticed. Nowadays, the effect of this approach can be clearly visible (i.e. poor work and production organization, working conditions inconsistent with the standards and ergonomic requirements (Bullinger et al. 2003). There are also other organizational factors having impact on risk such as failures by front line personnel influenced by training strategies, poor maintenance priorities, inadequate supervision, a failure to undertake effective hazard identification or inadequate auditing (Anderson 2005). Thus, undertaking activities related to the improvements of working conditions is necessary not only to fulfill the legal requirements. It is evident that they can directly contribute to reducing the number of accidents at work. Furthermore, the analysis of accidents at work allows to distinguish their most common causes. However, in industrial practice working conditions consists of a large number of different factors which can influence (to a different extent and range) accidental hazard. Thus, the likelihood of near-miss incident can be the effect of appearance of one specific factor of particular importance in one situation (Kawecka-Endler & Skulska 2011).

The research studies show that it still happens that the near-miss incidents are ignored by both employees and employers (McKinnon

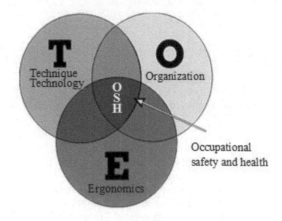

Figure 1. Model of analysis of occupational safety and health.

2012, Van der Schaaf et al. 2013). Furthermore, under-reporting is higher in case of work environments where organizational safety climate is poorer or where supervisor safety implementation is unpredictable (Probst & Estrada 2010). On the other hand, it is worth to emphasize that the level of culture in reference to occupational health and safety is still lower in Eastern European countries in comparison to Western EU countries (Gabryelewicz et al. 2015, Mościcka-Teske et al. 2016, Znajmiecka-Sikora & Boczkowska 2016) and it should be systematically increased.

In many companies health and safety rules are still not respected, occupational risk is evaluated incorrectly, or not carried out at all, and the employees are not informed either of the occupational risks associated with work, nor the rules of protection against hazards (Przenniak 2006, Cremers 2011). It should be remembered that in the case of ignoring the possibility of occurrence of accident risks or situations of underestimating the risk of accidents there is a lack of any motivation to detect hazards or to take prevention actions (Lis & Nowacki 2005). Many employers are not aware how difficult it can be to predict the consequences of dangerous, arduous and harmful working conditions. They do not recognize that the direct losses (damaged or destroyed machines, downtime, stops of the operation of the plant or its parts), the costs of treatment and rehabilitation of the victims and other costs of benefits should be added (Zakrzewska-Szepańska 2008). Thus, a systematic awareness of both employees and employers about the negative consequences of non-compliance with safety standards is required (Gołaś et al. 2016).

In reference to these facts, in order to improve occupational health and safety it is required to know the determinants which influence the present situation. In reference to particular the EU countries it is possible distinguish five broad categories (European Agency for Safety... 2013):

– influences (political and policy influence, requirements of accession and economic crisis),
– national governance and regulation and the OSH system,
– labour relations, trade unions and employers' organizations and processes,
– economic restructuring,
– related systems (e.g. social welfare, health).

3 MATERIALS AND METHODS

Knowledge is the condition of correct solutions in the field of safety and working conditions. It becomes a key determinant of the development of potential of enterprises and complex systems built by them (Grzybowska & Lupicka, 2016). In this case, the engineering knowledge - accumulated during the studies, is necessary to design workstations and their organization and then use in practice. It allows identification of abnormalities that can be diagnosed by making observation and evaluation of workstations in the company.

In order to obtain information how the notion of workplace occupational health and safety is understood and can be described in industrial practice, a research study was carried out in Poznan University of Technology, in Poland. The research participants were 71 safety engineering students (last semester of master studies).

The questionnaire consisted of two parts. The first one encompassed theoretical approach to work safety, whereas the second one was in reference to the company where they did their internships and/or worked.

In the study the following companies were described:

– 6 micro enterprises (up to 10 employees),
– 11 small enterprises (10 to 49 employees),
– 29 medium enterprises (50 to 249 employees),
– 19 large enterprises (more than 250 employees).

Overall, 65 Polish enterprises were investigated in the study.

4 RESULTS AND DISCUSSION

The results of the questionnaire investigation concerning workplace occupational health and safety are shown in Figures 2 and 3.

The first question concerned the theoretical considerations. Almost half of the respondents (48%) claimed that it is a compliance with health and safety rules and regulations at the workplace

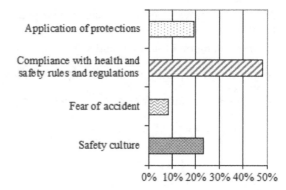

Figure 2. Theoretical understanding of the notion "workplace occupational health and safety".

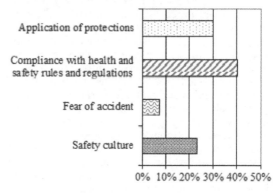

Figure 3. Perception of workplace occupational health and safety in practice.

and plant. One-fourth of responses (23%) referred to safety culture, and 19% of respondents indicated the need of the application of collective and individual protections. However, only 8% of the students indicated that this notion can be associated with the fear of an accident.

In the second part of the questionnaire the respondents were asked to discuss the health and safety issues in reference to industrial practice. Its results are presented in Figure 3.

As it can be noticed the respondents less often indicated the significance of compliance with the rules and regulations at the workplace and plant (40%). On the other hand, the use of protective equipment was pointed out by 30% of the students, so this value increased by 11%. The perception of safety culture remained on the same level (23%) and fear of accident decreased by 1%.

In addition, in both sections of the study, it was possible to develop the answers by describing its own observations. Finally, it allowed to achieve the extended point of view on the discussed problems. The respondents indicated such activities as follows to improve the current situation in the analyzed enterprises:

– implementation of a health and safety management system,
– application of protective measures at work and outside,
– planned premiums for health and safety units' workers,
– direct connection of the existing gaps with the financial consequences (penalties).

5 CONCLUSIONS

As the importance of knowledge and information about innovative processes is growing, what allows for the development and improvement of products and processes, more and more often the question arises about costs, both productive activities, as well as labor costs. In these considerations a man—an employee, who is a direct performer of processes, is not always taken into account. And it is the knowledge of how to develop a work system to ensure a good and safe working conditions, good health, the basis for the protection of man in the work environment. The costs of accidents at work, occupational diseases and any kinds of compensation for the loss of health, are social costs. Therefore, the importance and influence of socio-cultural adjustments related to changes in work processes, improving quality of life, the acceptance and understanding of the concept of entrepreneurship, has been increasing as they influence psychological distress, role ambiguity, job satisfaction, organizational commitment and work performance.

REFERENCES

Anderson, M. 2005. Behavioural Safety and Major Accident Hazards: Magic Bullet or Shot in The Dark? *Process Safety and Environmental Protection* 83(2): 109–116.

Belmonte, F., Schön, W., Heurley, L., & Capel, R. 2011. Interdisciplinary safety analysis of complex socio-technological systems based on the functional resonance accident model: An application to railway traffic supervision. *Reliability Engineering & System Safety*, 96(2): 237–249.

Bullinger, H.J., Warnecke, H.J. & Westkämper, E. 2003. *Neue Organisationsformen im Unternehmen.* (2 ed.). Berlin-Heidelberg: Springer-Verlag. ISBN 978-3-662-08934–7, s. 421–461.

Butlewski, M., & Sławińska, M. 2014. Ergonomic method for the implementation of occupational safety systems. In Occupational Safety and Hygiene II-Selected Extended and Revised Contributions from the International Symposium Occupational Safety and Hygiene, SHO (pp. 621–626).

CIOP 2016. National programme "Improvement of safety and working conditions" – phase III (2014–2016). Retrieved on 11 November 2016 from: http://www.ciop.pl/CIOPPortalWAR/appmanager/ciop/en?_nfpb = true&_pageLabel = P264 00 121511406886174136

Cremers, J. 2011. In search of cheap labour in Europe. *Working and living conditions of posted workers.* International Books. 978(90), 5727.

Doganova, L., & Eyquem-Renault, M. 2009. What do business models do?: Innovation devices in technology entrepreneurship. *Research Policy*, *38*(10): 1559–1570.

European Agency for Safety and Health at Work 2013. Analysis of the determinants of workplace occupational safety and health practice in a selection of EU Member States. European Risk Observatory. Retrieved on 10 November 2016 from: https://osha.europa.eu/en/tools-and-publications/publications/reports/analysis-determinants-workplace-OSH-in-EU.

EUROSTAT 2016. *Accidents at work statistics.* Retrieved on 11 November 2016 from: http://ec.europa.eu/eurostat/statistics-explained/index.php/Accidents_at_work_statistics #Number_of_accidents.

Gabryelewicz I., Sadłowska-Wrzesińska J., Kowal E., *Evaluation of safety climate level in a production facility.* Procedia Manufacturing: 6th International Conference on Applied Human Factors and Ergonomics (AHFE 2015) and the Affiliated Conferences, AHFE 2015, pp. 6211–6218.

Gołaś H., Mazur A. & Mrugalska B. 2016. Evaluation of the Quality of Work of the Health and Safety Service Using the Servqual Method—a Case Study in Polish Company. In: P. Arezes et al. (eds.) *12th International Symposium on Occupational Safety and Hygiene of Portuguese-Society-of-Occupational-Safety-and-Hygiene (SHO),* Guimaraes, Portugal, MAR 23–24, 2016, pp. 96–98.

Grzybowska K. & Łupicka A., 2016, Knowledge Acquisition in Complex Systems. In: Yue X.-G., Duarte N.J.R. (eds.), *Proceedings of the 2016 International Conference on Economics and Management Innovations, Advances in Computer Science Research,* 57: 262–266.

GUS 2016. Wypadki przy pracy. Warszawa: Główny Urząd Statystyczny.

Harms-Ringdahl, L. 2013. *Guide to safety analysis for accident prevention.* IRS Riskhantering.

Kawecka-Endler, A. 2014. Humanizacja a nowe formy pracy. *Zeszyty Naukowe Organizacja i Zarządzanie.* 63: 115–129.

Kawecka-Endler, A. & Mrugalska, B. 2010. Contemporary aspects in design of work. In: Karwowski, G. Salvendy (eds.). *Advances in Human Factors, Ergonomics and Safety in Manufacturing and Service Industries* (pp. 401–411). Boca Raton: CRC Press.

Kawecka-Endler, A. & Mrugalska, B. 2012. Analysis of changes in work processes. In: P. Vink (ed.). *Advances in Social and Organizational Factors* (pp. 672–681). Boca Raton: CRC Press.

Kawecka-Endler, A. & Mrugalska, B. 2014. Humanization of Work and Environmental Protection in Activity of Enterprise. W: C. Stephanidis, M. Antona (red.). *Universal Access in Human-Computer Interaction. HCII 2014, Part III, LNCS 8515,* 700–709.

Kawecka-Endler A. & Skulska M., 2011. Wpływ projektowania ergonomicznego stanowisk pracy na zmniejszenie liczby wypadków przy pracy. In: A. Kawecka-Endler, B. Mrugalska (eds.), *Praktyczne aspekty projektow-*

ania ergonomicznego w budowie maszyn, Wydawnictwo Politechniki Poznańskiej, Poznań 2011, 71–90.

Lis T. & Nowacki K. 2005. *Zarządzanie bezpieczeństwem i higieną pracy w zakładzie przemysłowym*, Wyd. Politechniki Śląskiej, Gliwice.

McKinnon R.C. 2012. *Safety Management: Near Miss Identification, Recognition, and Investigation*. Boca Raton: CRC Press.

Michie, S. 2002. Causes and management of stress at work. *Occupational and environmental medicine*, 59(1): 67–72.

Mościcka-Teske A., Sadłowska-Wrzesińska J., Butlewski M., Misztal A., Jacukowicz A. 2016. Stressful work characteristics, health indicators and work behavior. The case of machine operators. *JOSE, International Journal of Occupational Safety and Ergonomics*.

OHSAS 18001:2007. Occupational health and safety management systems.

Probst, T. M., & Estrada, A. X. 2010. Accident under-reporting among employees: Testing the moderating influence of psychological safety climate and supervisor enforcement of safety practices. *Accident Analysis & Prevention*, 42(5): 1438–1444.

Przenniak W. 2006. Uświadamiać konsekwencje wypadków. *Atest—Ochrona Pracy*, 12; 25.

Qureshi, Z. H. 2007. A review of accident modelling approaches for complex socio-technical systems. In: *Proceedings of the twelfth Australian workshop on Safety critical systems and software and safety-related programmable systems-Volume 86* (pp. 47–59). Australian Computer Society, Inc.

Rivara, F. P., Cummings, P., & Koepsell, T. D. 2009. *Injury control: a guide to research and program evaluation*. Cambridge University Press.

RIVM 2016. Occupational accidents. National Institute for Public Health and the Environment. Retrieved on: 10 November 2016 from: http://www.rivm.nl/en/Topics/O/Occupational_Safety/Occupational_accidents.

Van der Schaaf, T. W., Lucas, D. A., & Hale, A. R. (Eds.). 2013. *Near miss reporting as a safety tool*. Butterworth-Heinemann.

Zakrzewska-Szepańska K. 2008. BHP w Polsce 2007: nie jest dobrze, *Służba Pracownicza*, 9: 9.

Znajmiecka-Sikora M., Boczkowska K., 2016. Analysis of Safety Culture on the Example of Selected Polish Production Enterprises. In: P. Arezes et al. (eds.) *Occupational Safety and Hygiene SHO 2016*, Guimaraes Portugal 2016, Portuguese Society of Occupational Safety and Hygiene (SPOSHO), 349–352.

Occupational Safety and Hygiene V – Arezes et al. (Eds)
© *2017 Taylor & Francis Group, London, ISBN 978-1-138-05761-6*

Joinpoint regression analysis applied to occupational health indicators: An example of application to Spain

S. Martorell & V. Gallego
Department of Chemical and Nuclear Engineering, Universitat Politècnica de València, Valencia, Spain

A.I. Sánchez
Department of Applied Statistics and Operational Research, and Quality, Universitat Politècnica de València, Valencia, Spain

ABSTRACT: Occupational injuries impose significant burdens on individuals, employers and society. Thus, the identification of trend in time series related to health occupational indicators, is important to detect whether economic or structural changes influence the behaviour of these indicators. In this paper a joinpoint regression model is used to identify trends in the occupational health indicadors. Joinpoint regression is a statistical modeling technique that explain the relationship between two variables by means of straight lines which are connected by change points. The study is focused on the trend analysis of occupational injuries in Spain from 1995 to 2015. The analysis covers an exceptional instable period in the spanish labour market including a period of economic growth and of deep recession and different legislatives changes. The results obtained show that the economic and structural changes affect differently the trend of the analyzed indicators.

1 INTRODUCTION

Occupational injuries impose significant burdens on individuals, employers and society. Thus, the identification of trend in time series related to health occupational indicators, e.g. Incidence Rate, is important to detect whether economic or structural changes influence the behaviour of these indicators (Carnero & Pedregal 2013 & Gallego et al. 2016).

Trend analysis is defined as a process of estimating gradual change in a series of observations over time. In the literature, different nonparametric and parametric techniques are used to estimate trends (Sharma et al. 2016). In this paper a jointpoint regression model is used to identify trends in the occupational health indicators. Jointpoint regression is a statistical modeling technique that explain the relationship between two variables by means of straight lines which are connected by change points.

The study is focused on the trend analysis of occupational injuries in Spain from 1995 to 2015. The analysis covers an exceptional instable period in the spanish labour market including a period of economic growth and of deep recession. So, the reference period considered 2000–2014 shows several years of growth (e.g., 2000–2007) and recession (e.g., 2008–2012) of the Spanish GDP (Gross Domestic Product).

In addition, a number of changes affecting the coverage of occupational accidents have been introduced during the reference period considered in this paper (1995–2015). New regulations have been released in the frame of the Spanish Occupational Health and Safety Strategy (Gobierno de España, 2007), which was launched in 2007 with the aim of achieving a continuous and significant reduction in the level of accidents at work and this way approaching the mean figures for the European Union in work accidents and occupational illness. Thus, Law 20/2007 (BOE 166, 2007) established mandatory coverage of occupational accidents for self-employed workers, but only for those workers economically dependent. Next, Law 27/2011 (BOE 184, 2011), which entry into force has been delayed, will extend mandatory coverage of occupational accidents for new self-employed workers with more than 30 years. Moreover, the special sector of self-employed agricultural workers moved to the self-employed regime with mandatory coverage in 2008 and the rest of agricultural workers moved to a new special regime of general social security with mandatory coverage in 2012 (BOE 229, 2011 & BOE 277, 2011). These factors are relevant for the risk of injury in the workplace.

The main indicators used in the statistics of work accidents to analyze the evolution of the occupational health are the following: Fatal Incidence Rate, Incidence Rate, Severity Rate, Fatal Accident

Rate and median number of days lost per case of occupational injury. These indicators are analyzed with the objective to detect trend changes. The detection of these changes is important as it allows provide relevant information in order to detect periods where greater control and additional preventive actions should be taken. The results obtained show that the economic and structural changes affect differently the trend of the analyzed indicators.

The paper is organized as follow. In section 2 occupational health and safety indicators and risk metrics are presented. Section 3 is focused on joinpoint regression which has been used to detect changes in the trend of occupational health indicators. Section 4 presents the results obtained in the trend analysis of different indicators and section 5 presents the conclusions.

2 OCCUPATIONAL HEALTH AND SAFETY INDICATORS AND RISK METRICS

The main indicators used in Occupational Health and Safety (OHS) can be divided into two types: leading indicators and lagging indicators. Leading indicators are associated with active, positive steps that organizations can take to avoid an OHS incident, e.g. percentage of workers with adequate OSH training, number of workplace inspections,... Lagging indicators on the other hand are failure focused and measure OHS incidents that have already happened, e.g. incidence rate, fatal incidence rate,... (Sheehan et al 2016 & Reiman & Pietikäinen 2012). This paper is focused on the last ones, that is on lagging indicators.

The main lagging indicators used in the statistics of work accidents are showed in Table 1 (EURO-STAT 2001) where NA represents the number of accidents resulting in occupational injury, NW is the average number of workers, NH the number of person-hours worked and DL the days lost.

On the other hand the risk of an occupational accident can be formulated mathematically in terms of the two components of the risk, i.e. probability or frequency and damage, adopting either eqn. (1) or (2):

$$R = P \cdot D \tag{1}$$
$$R = F \cdot D \tag{2}$$

where P is the probability of occurrence of the occupational accident and F is the frequency of occupational accidents. The first component of risk, P, can be calculated by the Incidence Rate (IR). Alternatively, F can be calculated by the Fatal Rate (FR). The second component of risk, D, represents the severity of the occupational injury resulting

Table 1. Main indexes used in occupational health and safety.

Index	Description	Expression
IR	Incidence Rate	$IR = \dfrac{NA}{NW} \cdot 10^5$
FR	Fatal Rate	$FR = \dfrac{NA}{NH} \cdot 10^6$
DLA	Median number of days lost per case of occupational injury	$DLA = \dfrac{DL}{NA}$
SR	Severity Rate	$SR = \dfrac{1}{10^3} \cdot FR \cdot DLA$
FIR	Fatal Incidence Rate	$FIR = \dfrac{NFA}{NW} \cdot 10^5$
FAR	Fatal Accident Rate	$FAR = \dfrac{NFA}{NH} \cdot 10^3$

from the occupational accident. It can be expressed qualitatively, e.g. death in case of fatal occupational injury, or quantitatively, e.g. days lost per injury.

The risk of an occupational accident linked to an occupational hazard can be defined as the severity of the consequences (damage) of the occupational injury resulting from the occupational accident and its associated likelihood. Thus, risk of an occupational accident consists of two components: likelihood and damage.

It is worthy to be aware of the relationship between risk, risk components and accident indicators (Martorell et al. 2016). First risk component, likelihood, can be formulated mathematically in terms of either a probability or a frequency as shown in eqns. (1) and (2). Thus, P represents the ratio of NA resulting in occupational injury to the average number of workers NW, i.e. the probability of a worker having an occupational accident resulting in an occupational injury per year, while F represents the ratio of NA to the number of person-hours worked NH, i.e. frequency of occupational injuries per hour in a year.

The second component, damage, can be evaluated quantitatively. So, DLA represents a measure of damage, i.e. the median number of days lost per case of occupational accident causing injury.

A social risk of an occupational accident can be derived from DLA and FR, thus is:

$$RS = \frac{SR}{10^3} \tag{3}$$

The FIR represents a measure of individual/personal risk of death, RI:

$$RI = \frac{FIR}{10^5} \qquad (4)$$

By last, the *FAR* represents a measure of personal risk of death of workers, *RC*:

$$RC = \frac{FAR}{10^8} \qquad (5)$$

3 JOINTPOINT REGRESSION

Trend in occupational health indicators were modelled using jointpoint regression (Kim et al. 2000 & Sharma et al. 2016) which is a technique for identifying when a series of annual rates change trend. Joinpoint regression model is a piecewise linear regression model that characterizes the trend behaviour in the data by identifying the significant points where changes occur. This will be carried out by detecting the points and their locations with in the data range.

Suppose that we have *n*-pairs of observations $(t_1, y_1), (t_2, y_2), ..., (t_n, y_n)$ where y_i denote the mortality or incidence outcome process that describe the behavior as a function of time t_i. The jointpoint regression is given by:

$$y_i = \beta_0 + \beta_1 \cdot t + \delta_1 (t - \tau_1)^+ + ... + \delta_\kappa (t - \tau_\kappa)^+ + \varepsilon_i \qquad (6)$$

where κ's is the unknown number of jointpoints, τ_i's are the unknown jointpoints, β_i's and δ_i's are the regression parameters, $w^+ = w$ for $w > 0$ and 0 otherwise and ε_i's are random errors.

To determine the optimal number of jointpoints, the model selection method used sequential permutation tests (Kim, 2000). Each one of the permutation tests performs a test of the null hypothesis

$$H_0 : \kappa = \kappa_a \qquad vs \qquad H_1 : \kappa = \kappa_b \qquad (7)$$

where $\kappa_a < \kappa^b$. The procedure begins with $\kappa_a = 0$ or minimum number of joinpoints and $\kappa_b = M$ or maximum number of joinpoints. Monte Carlo simulations is used to calculate the permutation *p*-value for each hypothesis test. The *p*-value is compared against the significance level adjusted for overall over-fitting error probabilities, where the adjusted

$$\alpha(\kappa_a; \kappa_b) = \alpha/(M - \kappa_a) \qquad (8)$$

If the null is rejected, then κ_a is increased by one, otherwise, κ_b is decreased by one. The procedure continues until $\kappa_a = \kappa_b$, and the final value of $\hat{\kappa} = \kappa_a = \kappa_b$ is the selected number of joinpoints. In the final model, each joinpoint indicates a statistically significant change in trend.

Once the joinpoints are established, the estimated Annual Percent Change (*APC*) can be used to describe and test the statistical significance of the trends in the model. The estimated *APC* is the percentage change (increase or decrease) in the estimated rates per year in the time trend. *APC* is computed for each of those trends by means of generalized linear models assuming a Poisson distribution. Significant changes include changes in direction or in the rate of increase or decrease. The *APC* form year *t* yo year (*t*+1) can be obtained using the following formula:

$$APC(\%) = \frac{R_{(t+1)} - R_t}{R_t} \cdot 100 = (e^\alpha - 1) \cdot 100 \qquad (9)$$

where R_t is the rate in year *t* and α is the slope coefficient in the equation

$$\ln(R_t) = \alpha \cdot t + \beta \qquad (10)$$

When describing trends over a fixed pre-specified interval a p-value < 0.05 was considered statistically significant.

In this paper, Joinpoint analyses were performed using the 'Joinpoint' software from the Surveillance Research Program of the US National Cancer Institute (NCI, 2016).

4 APPLICATION CASE

The study is focused on Spain and it considered the period between 1995 and 2015. The data used in this paper on work-related accidents were obtained via Subdirectorate General for Statistics of the Ministry of Employment and Social Security.

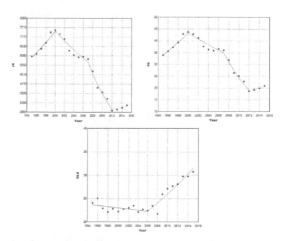

Figure 1. Evolution of the selected indexes and joinpoints detected.

Table 2. Estimates from joinpoint analyses of *FIR*, *FR* and *DLA* in Spain, 1995–2015.

Index	Period 1	APC (95% CI)	Period 2	APC (95% CI)	Period 3	APC (95% CI)	Period 4	APC (95% CI)
IR	1995–2000	4.6* (2.3, 7)	2000–2007	−3.8* (−5.4, 2.1)	2007–2012	−12* (−14.8, −9.2)	2012–2015	1.9* (−3.1, 7.1)
FR	1995–2000	5* (2.7 7.3)	2000–2007	−3.1* (−4.7, −1.4)	2007–2012	−11.2* (−13.9, −8.4)	2012–2015	2.4 (−2.5, 7.6)
DLA	1995–2006	−0.6 (−1.4, 0.3)	2006–2015	4* (2.8, 5.1)				

*Significant at *p* < 0.05 CI Confidence Interval.

Among different occupational health indicators presented in section 2 Incidence Rate (*IR*), Fatal Rate (*FR*) and Median number of days lost per case of occupational injury (*DLA*) have been selected to develop the application case. *IR* and *FR* represent the first risk component, likelihood, and *DLA* denotes a measure of damage.

Figure 1 shows the evolution of the indexes selected in the period analysed and the model obtained using jointpoint regression.

In Figure 1 is showed as different behavior is observed to the indexes selected and the model final adjusted. So, for example, if the evolution of the Incidence Rate (*IR*) and Fatal Rate (*FR*) are analyzed three change points are detected corresponding to the years 2000, 2007, and 2012. In the case of *DLA* only a change point is observed corresponding to the year 2006.

Table 2 shows the estimates obtained from joinpoint regression. From Table 2 the following conclusions can be obtained for the different indicators.

4.1 Incidence Rate

Based on multiple permutation tests that kept the overall level of type I error to less than 0.05, the final model selected for *IR* detected 3 joinpoints in 2000, 2007 and 2012. After a significant annual increase between 1995 and 2000 (4.6%), the *IR* declined significantly by 3.8% per year between 2000 and 2007. Subsequently, starting in 2007 and continuing to 2012, the *IR* continued to decline significantly but at a faster rate of 12% per year. Between 2012 and 2015, the rate increased nonsignificantly.

4.2 Fatal Rate

The final model selected for *FR* detected 3 joinpoints in 2000, 2007 and 2012. Between 1995 and 2000 the rate increased significantly by 5% per year. After between 2000 and 2007 the *FR* declined significantly a rate of 3.1% per year. Starting in 2007 the rate declined significantly by 11.2% per year until 2012. After 2012 the *FR* increased nonsignificantly.

4.3 Median number of days lost per case of occupational injury

The final model selected for *DLA* detected 1 joinpoint in 2006. A nonsignificant annual decline is observed between 1995–2006 followed by a significant annual increased (4% per year) until 2015.

5 CONCLUDING REMARKS

The joinpoint regression method can be a useful and feasible method to understand trends in occupational health indicators. The method does not allow to prevent occupational incidents but can be useful to analyze how economic and structural changes have influenced the evolution of occupational health indicators. During the period considered (1995–2015) it is showed a decrease in Incidence Rate. This trend was also observed for Fatal Rate. In both cases three joinpoints have been detected in the years 2000, 2007 and 2012. In the period 2007–2012 an important decrease of both indicators is observed. The beginning of this period coincides with the onset of the economic crisis in Spain and the launch of the Spanish Occupational Health and Safety Strategy. From the year 2012, there is no statistically significant trend in the behavior of indicators coinciding with the economic recovery in Spain. The trend observed in the median number of days lost per case of occupational injury is different. There was no statistically significant trend in the period 2000–2006. However, from 2006, an increase in *DLA* is detected.

REFERENCES

BOE 166 (2007). Gobierno de España. Ley 20/2007, de 11 de julio, del Estatuto del trabajo autónomo, pp. 29964–29978 (in spanish). Retrieved from https://www.boe.es/boe/dias/2007/07/12/pdfs/A29964-29978.pdf.

BOE 184 (2011). Gobierno de España. Ley 27/2011, de 1 de agosto, sobre actualización, adecuación y modernización del sistema de Seguridad Social, pp. 87495–

87544 (in spanish). Retrieved from https://www.boe.es/boe/dias/2011/08/02/pdfs/BOE-A-2011-13242.pdf.

BOE 229 (2011). Gobierno de España. Ley 28/2011, de 22 de septiembre, por la que se procede a la integración del Régimen Especial Agrario de la Seguridad Social en el Régimen General de la Seguridad Social, pp. 100547–100565 (in spanish). Retrieved from http://www.empleo.gob.es/es/Guia/pdfs/pdfsnuev/L2811.pdf.

BOE 277 (2011) Gobierno de España. Real Decreto 1620/2011, de 14 de noviembre, por el que se regula la relación laboral de carácter especial del servicio del hogar familiar 119046–119057 (in spanish). Retrieved from https://www.boe.es/boe/dias/2011/11/17/pdfs/BOE-A-2011-17975.pdf.

Carnero, M.C. & Pedregal. D.G. (2013). Ex-ante assessment of the Spanish Occupational Health and Safety Strategy (2007–2012) using a State Space framework. Reliability Engineering and System Safety, 110, 14–21.

Gallego, V. Martorell S. Sánchez A.I. Analysis of economic and structural factor son occupational accidents using a generalized linear model: Example of application in Spain. ESREL 2016. Glasgow September 2016.

EUROSTAT 2001. European Statistics on Accidents at Work (ESAW) – Methodology – 2001 Edition. Directorate-General for Employment and Social Affairs (DG EMPL), Statistical Office of the European Union (EUROSTAT), Luxembourg.

Gobierno de España (2007). Estrategia Española de Seguridad y Salud en el Trabajo 2007–2012 (in spanish) http://www.insht.es/InshtWeb/Contenidos/Instituto/Estrategia_Seguridad_Salud/Doc.Estrategia%20actualizado%202011%20ultima%20modificacion.pdf.

Kim, H.J., Fay, M., Feuer, E. J., and Midthune, D. N. (2000). Permutation tests for joinpoint regression with applications to cancer rates. Statistics in Medicine, 19, 335–351.

Martorell, S., Gallego, V. and Sánchez A.I. (2016). On the use of accident indicators in risk based management of occupational safety and health: Example of application to Spain. Occupational Safety and Hygiene IV. pp. 327–331.

NCI (2016) Joinpoint Regression Program, Version 4.3.1 - April 2016; Statistical Methodology and Applications Branch, Surveillance Research Program, National Cancer Institute.

Reiman, T. & Pietikäinen, E. (2012). Leading indicators of system safety—monitoring and driving the organizational safety potential. Safety Science, 50, 1993–2000.

Sharma, S., Swayne D. & Obimbo C. (2016). Trend analysis and change point techniques: a survey. Energy, Ecology and Environment. Volume 1, Issue 3, pp 123–130.

Sheehan, C., Donohue, R., Shea, T., Cooper B. & De Cieri H. (2016). Leading and lagging indicators of occupational health and safety: The moderating role of safety leadership. Accident Analysis and Prevention, 92, 130 138.

Occupational Safety and Hygiene V – Arezes et al. (Eds)
© *2017 Taylor & Francis Group, London, ISBN 978-1-138-05761-6*

Psychosocial risk management in the service industry: Linking motivation and performance

J. Guadix, J.A. Carrillo-Castrillo & D. Lucena
Universidad de Sevilla, Spain

M.C. Pardo-Ferreira
Universidad de Málaga, Spain

ABSTRACT: The main purpose of this study is to research the explanatory mechanisms that link managers' motivation regarding risk management and how these organizational issues explain the performance of enterprises' control of psychosocial risks. The paper is based on the European Survey of Enterprises on New and Emerging Risks. The analysis is performed with structural equation modelling. The main finding of this study is that improvement in the general safety management improves psychosocial performance of enterprises. These results provide an important clue as to the future of public policymaking in this area and highlight the importance of safety management for psychosocial risk management. These results can be used to design more effective programs for management motivation in this area. Activities related to better psychosocial outcomes should be promoted and given higher priority. This research provide evidence that safety management and psychosocial safety management has a positive influence on lowering psychosocial risks, while at the same time potentially reducing negative health outcomes, and improving the overall well-being of individuals within an organization. Policy makers need to take into account these results in order to better promote psychosocial risk management.

1 INTRODUCTION

1.1 Psychosocial risks

A psychosocial hazard is any occupational hazard that affects the psychological well-being of workers, including their ability to participate in a work environment among other people. Psychosocial hazards are related to the way work is designed, organized and managed, as well as the economic and social contexts of work and are associated with psychiatric, psychological and/or physical injury or illness (Leka et al. 2008).

A psychosocial hazard is any occupational hazard that affects the psychological well-being of workers, including their ability to participate in a work environment among other people. Psychosocial hazards are related to the way work is designed, organized and managed, as well as the economic and social contexts of work and are associated with psychiatric, psychological and/or physical injury or illness (Kïrsten, 2012).

Psychosocial risk is currently an issue of great importance for employers across Europe, although it is considered an emerging form of risk by the European Risk Observatory (European Agency for Safety and Health at Work, 2007). The top ten emerging psychosocial risks, according to experts'

forecasts, fall within the following five main topics: (i) new forms of employment contracts and job insecurity, (ii) the ageing workforce, (iii) work intensification, (iv) high emotional demands at work, and (v) poor work-life balance.

The service industry is a unique sector with specific concerns from the point of view of working conditions, according to the Fourth European Working Conditions Survey, which is previous to ESENER 1 (European Foundation for the Improvement of Living and Working Conditions, 2007). Some specific threats, such as lack of support from colleagues, increases in threatened and actual violence from customers, and long working hours, are prevalent in various service activities.

Psychological safety influences individual contributions in customer groups where multiple customers co-create a service experience (Kuppelwieser & Finsterwalder, 2011). Psychosocial outcomes also have important effects on the effective implementation of marketing strategies. Employee reward systems are used for this purpose because of their capacity to modify or sustain desirable psychosocial outcomes (Chimhanzi & Morgan, 2005).

Social support is the perception that one is cared for. For a long time, the construct of social support has been used as a way of exploring what it means for customers to have positive interpersonal contact with a service provider (Abelman & Ahuvia, 1995). Management factors affect the performance of a service enterprise, enhancing the ability of its employees to give social support to customers.

1.2 Safety and health management and prevention activities

Certain mandatory elements of health and safety management in the European Union are included in the Framework Directive 89/391/EEC. Employers have to evaluate all risks. Management of psychosocial risks, however, is an underdeveloped field in comparison with other risks such as safety, hygiene, or ergonomic issues.

A management system is a network of interrelated elements. These elements include responsibilities, authorities, relationships, functions, activities, processes, practices, procedures, and resources. A management system uses these elements to establish policies, plans, programs, and objectives, and to develop ways of implementing and achieving such actions (Cagno et al. 2011).

In the decision as to which health and safety management model to adopt, there are important factors such as motivation, strategy and perceived risks to consider (Guadix et al. 2015). Although injury can ensue from psychosocial risks, and certain health outcomes may stem from these antecedents, there are very few cases of such injury or illness reported as a consequence of psychological risks. Therefore, for the analysis of the performance of psychosocial risk management, it is more useful to use intermediate outcomes such as job satisfaction. Moreover, safety management is more easily evaluated through the analysis of specific activities to control and prevent risks.

1.3 Evidence on psychosocial risks

Most of the published research on psychosocial risk focuses on the influence of psychosocial risk on health outcomes. Psychosocial factors can affect mental health and life balance, and are also a risk factor for occupational accidents and musculoskeletal disorders. High psychological and emotional demands, and conflicts with supervisors and colleagues at work are factors known to contribute to the risk of incurring an injury in an occupational accident (Swaen et al. 2004). Psychological factors play a significant role not only in chronic pain but also in the aetiology of acute pain, particularly in the transition to chronic problems (Linton, 2000; Ariëns et al. 2001; Malchaire et al. 2001; Bongers et al. 2002).

In business strategy, psychosocial risks are a serious concern for organizations in terms of the health outcomes of workers. The etiological fraction attributable (i.e. the proportion of outcomes directly related) to differences in work environment exposures has been estimated at 40% (Labriola et al. 2006). The demand–control–support model emerged during the 1980s and remains the state of the art for explaining which organizational variables affect the psychosocial state of the worker (Bakker & Demorouti, 2007). The model includes three main dimensions: job demands, latitude in job decisions, and social support at work. According to reviews of published studies, workers with jobs characterized by high demands, low decision latitude, and low social support have a higher risk of poor psychological well-being and a higher incidence of cardiovascular disease (Kristensen, 1995; Rusli et al. 2008).

From a management perspective, psychosocial risks therefore constitute an important area for improvement within any organization. The control of psychosocial risks is a powerful tool for preventing accidents and absenteeism. Involvement of managers is one of the most influential variables in the improvement of the overall health of employees at the company level (Oliver et al. 2002). As a consequence, organizations that manage psychosocial risks effectively should be more competitive.

Zohar (1980) uses the term safety climate to describe a construct that captures employees' perceptions of the role of safety within the organization. The psychosocial safety climate is thus an emerging construct that refers to shared perceptions regarding policies, practices, and procedures for the protection of the psychological health and safety of workers. The psychosocial safety climate in organizations captures the way job demands are managed. In addition, the psychosocial safety climate is negatively correlated with psychological health problems (Idris et al. 2012). The development of several tools such as PSC-12 (Hall et al. 2010) helps analyze the psychosocial safety climate.

Therefore, management activities should look to improve the psychosocial safety climate as an intermediate step towards improving the overall well-being of employees. One possible intervention is to increase job support from both managers and colleagues (Dollard et al. 2000). Other actors play a role in psychosocial risks, however. Worker representation may also contribute to bringing down barriers and broadening opportunities, to achieve a better preventive scenario in terms of psychosocial risks (Walters, 2011).

The main outcomes at an organizational level are health related, but can also be seen in job satisfaction, absenteeism, and productivity (Leka & Cox, 2008).

For a long time, psychosocial hazards have been a marginal area for the activity of official inspection bodies (Johnstone et al. 2011). As a result of the inactivity of the regulators on this issue, managers have historically paid little attention to psychosocial risks in comparison with physical and hygiene hazards. As an indicator of the increasing importance of these risks, however, the attention of regulators towards psychosocial risks is increasing.

2 MATERIALS AND METHODS

This paper is based on the first European Survey of Enterprises on New and Emerging Risks (ESENER), performed in 2009. Data analysis started on 2013 (a new ESENER data gathering started in 2014). The individual worker (micro) level perspective is excluded from ESENER. On the other hand, at the organization (meso) level, which is the scope of this study, information about manager motivation and risk management is more important.

2.1 ESENER: A European survey

ESENER's purpose is to fill an important information gap in the world of health and safety at work. Data on work-related accidents and illnesses is available, but little is known about the way in which health and safety risks are managed in practice, especially regarding psychosocial risks (European Agency for Safety and Health at Work, 2010). Although there are other sources of information (Dollard et al. 2007), ESENER is the most detailed in terms of enterprise information. The survey explores the management of health and safety, and the preventive measures in place, such as the extent of risk assessment, the management's commitment to health and safety, and the sources of expertise, advice, or information.

The survey also collects data about the main reasons for addressing health and safety, and psychosocial risks, the main difficulties in dealing with these issues, and the role of worker participation. The statistical population comprises all establishments with ten or more employees in 31 participating countries, including all European Member States, covering most sectors of economic activity. Around 36,000 interviews were conducted.

In each establishment surveyed, the highest-ranking manager responsible for health and safety at work was interviewed. Additionally, an interview with the workers' health and safety representative

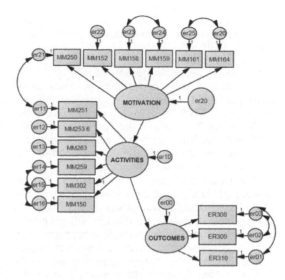

Figure 1. Model analyzed.

was carried out in those establishments where a formally designated representative takes specific responsibility for the health and safety of employees, and permission for the interview was granted by the management. According to the scope of the research, only answers from the service sector are considered.

Stratification of the sample is based on a matrix of two groups of sectors (Producing Industries and Service Sector) and five size classes (10–19 employees, 20–49, 50–199, 200–499, and 500+ employees). In the view of respondents, the psychosocial risks are much more pertinent in the service sector than in producing industries.

2.2 Model analyzed

The model analyzed is presented in Figure 1. Model was constructed using Structural Equations Modelling in AMOS v18.0.

In this model, there are three constructs, two of them related to the organization activities and motivations for psychosocial risk management and the other related to the psychosocial performance of the organization measuring the outcomes.

3 RESULTS

Latent variables in the model are related to the main issues in the process of risk prevention. The first latent variable, Motivation, is related to the motives that lead managers to address psychosocial risks. The second latent variable, Activities, is related to the activities undertaken by an enterprise to control and prevent those risks. Finally, the third latent variable, Outcomes, identifies the effectiveness of the activities performed.

For the initial identification of observed variables loading the latent variables, we used previously published reports of the European Agency for Safety and Health at Work based on ESENER. Variables for Motivation and Activities are gathered from the interviews with managers, whereas variables for Outcomes are gathered from the interviews with worker representatives.

The variables loading the construct *Motivation* are MM152 (if sickness absence is analysed), MM158 (if health and safety is addressed in management meetings), MM159 (if managers are involved in health and safety management), MM161 (if risk assessment is regularly performed), and MM164 (how many areas are assessed). The variables loading the construct *Activities* are MM250 (existence of procedures for stress), MM251 (existence of procedures for bullying), MM253 (existence of procedures for violence), MM259 (information for workers), MM302 (use of external sources of information), and MM150 (if health and safety management is outsourced). The variables loading the construct *Outcomes* are: ER308 (request from employees related to stress), ER309 (request from employees related to bullying), and ER310 (request from employees related to violence).

The structural equation model includes paths between the three latent variables. The observed variables loading each latent variable and the estimated loading factors are included in Table 1. The standardised estimated path coefficients are presented in Table 2, and the indices of the fit of the model are presented in Table 3.

Table 1. Standardized coefficients.

Latent	Observed	Estimate	Significance
Motivation	MM152	0.45	<0.01
	MM158	0.25	<0.01
	MM159	0.30	<0.01
	MM161	0.21	<0.01
	MM164	0.08	<0.01
	MM250	0.50	<0.01
Activities	MM251	0.43	<0.01
	MM253	0.41	<0.01
	MM259	0.49	<0.01
	MM302	0.52	<0.01
	MM150	0.45	<0.01
Outcomes	ER308	0.63	<0.01
	ER309	0.58	<0.01
	ER310	0.51	<0.01

Table 2. Path coefficients.

Path	Type	Estimate
Motivation = > Activities	Direct	0.97
Activities = > Outcomes	Direct	0.35
Motivation = > Outcomes	Indirect	0.34

Table 3. Model fit indices.

Index	Value
CFI	0.90
GFI	0.99
RMR	0.01
PNFI	0.68
Chi-square	373.7

4 DISCUSSION AND CONCLUSIONS

The global fit of the model developed is considered acceptable, according to the thresholds that appear in the majority of the literature (Kline, 2010). Most structural equation models dealing with health and safety issues are difficult to fit due to unobserved variables that are outside the scope of ESENER, such as safety or psychosocial climate. According to the path coefficients, an improvement in psychosocial outcomes, such as a reduction in complaints of stress, bullying, or violence at work, is achieved through the mediating concept of activities, in the form of real interventions to improve the psychosocial situation of the enterprise.

At the enterprise level, the model implies that improvements in the information for workers and implementing procedures for psychosocial risks are associated with better outcomes. In addition, enterprises that implement effective activities to control psychosocial risks are more likely to have a well-motivated and heavily involved management.

With respect to previous studies (Oliver et al. 2002), these results incorporate a mediating variable, Activities. Activities are the main outcomes of risk management. Therefore, motivation in the service sector leads enterprises to psychosocial risk management activities that will allow a better performance. Specifically, this result suggests that a good policy is to intervene to improve the motivation of managers. If so, certain activities such as developing procedures for stress, violence and bullying or providing information to workers show significant lower levels of request from employees regarding stress, violence and bullying.

The fit model shows that if managers' motivation is oriented towards an increase in preventive activities, the overall psychosocial outcomes will improve. There are limitations in the results owing to the data set. One limitation is that only cases where worker representatives were interviewed are included, so there is a bias towards the existence of such representatives in an organisation and their influence on the management of psychosocial risks (Walters, 2011). Another limitation is the heterogeneity of activities in the service sector and the lack of information about other psychosocial issues such as task organisation and organisational environment. In addition, potential differences between countries, and enforcement and promotion programmes at the country (macro) level were ignored.

ESENER is a useful tool for researching the management of psychosocial risks, at least at the enterprise level. The inclusion, however, of complementary questions regarding psychosocial climate and the decision process would facilitate a better understanding of this topic. Neither the use of management systems nor the development of innovation practices in service firms increases workers' psychosocial risks.

The main finding of this study is that improvement in the management of psychosocial risk must be achieved through the motivation of managers in this area. The use of collaborative technologies may be useful in motivating managers in service firms in the near future. These results provide an important clue as to the future of public policymaking in this area. New analysis need to be done on the new data from ESENER 2, already available.

REFERENCES

Abelman, M.B. & Ahuvia, A.C. (1995). Social support in the service sector. The antecedents, processes, and outcomes of social support in an introductory service. *Journal of Business Research* 32(3): 273–282.

Ariëns, G.A. van Mechelen, W. Bongers, P.M. Bouter, L.M. & van der Wal, G. (2001). Psychosocial risk factors for neck pain: a systematic review. *American Journal of Industrial Medicine* 39(2): 180–193.

Bakker, A.B. & Demerouti, E. (2007). The Job Demands-Resources model: state of the art. *Journal of Managerial Psychology* 22(3): 309–328.

Bongers, P.M. Kremer, A.M. & ter Laak, J. (2002). Are psychosocial factors, risk factors for symptoms and signs of the shoulder, elbow, or hand/wrist?: A review of the epidemiological literature. *American Journal of Industrial Medicine* 41(5): 315–342.

Cagno, E. Micheli, G.J. & Perotti, S. (2011). Identification of OHS-related factors and interactions among those and OHS performance in SMEs. *Safety Science* 49(2): 216–225.

Chimhanzi, J. & Morgan, R.E. (2005). Explanations from the marketing/human resources dyad for marketing strategy implementation effectiveness in service firms. *Journal of Business Research* 58(6): 787–796.

Dollard, M.F. Winefield, H.R. Winefield, A.H. & de Jonge, J. (2000). Psychosocial job strain and productivity in human service workers: A test of the demand-control-support model. *Journal of Occupational and Organizational Psychology* 73(4): 501–510.

Dollard, M. Skinner, N. Tuckey, M.R. & Bailey, T. (2007). National surveillance of psychosocial risk factors in the workplace: An international overview. *Work & Stress* 21(1): 1–29.

European Agency for Safety and Health at Work (2007). *Expert forecast on emerging psychosocial risks related to occupational safety and health*. Luxembourg: Office for Official Publications of the European Communities.

European Agency for Safety and Health at Work (2010). *European Survey of Enterprises on New and Emerging Risks Managing safety and health at work*. Luxembourg: Publications Office of the European Union.

European Foundation for the Improvement of Living and Working Conditions (2007). *Fourth European Working Conditions Survey*. Luxembourg: Office for Official Publications of the European Communities.

Guadix, J., Carrillo-Castrillo, J.A., Onieva, L. & Lucena, D. (2015). *Journal of Business Research* 68(7): 1475–1480.

Hall, G.B. Dollard, M.F. & Coward, J. (2010). Psychosocial safety climate: Development of the PSC-12. *International Journal of Stress Management* 17(4): 353–383.

Idris, M.A. Dollard, M.F. Coward, J. & Dormann, C. (2012). Psychosocial safety climate: Conceptual distinctiveness and effect on job demands and worker psychological health. *Safety Science* 50(1): 19–28.

Johnstone, R. Quinlan, M. & McNamara, M. (2011). OHS inspectors and psychosocial risk factors: Evidence from Australia. *Safety Science* 49(4): 547–557.

Kïrsten A. Way (2012). *Psychosocial Hazards and Occupational Stress*. In: OHS Body of Knowledge. Australia: Safety Institute of Australia.

Kline, R.B. (2010). Principles and practice of structural equation modelling. New York: The Guilford Press.

Kristensen, T.S. (1995). The demand-control-support model: Methodological challenges for future research. *Stress Medicine* 11(1) 17–26.

Kuppelwieser, V.G. & Finsterwalder, J. (2011). Psychological safety, contributions and service satisfaction of customers in group service experiences. *Managing Service Quality* 21(6): 617–635.

Labriola, M. Lund, T. & Burr, H. (2006). Prospective study of physical and psychosocial risk factors for sickness absence. *Occupational Medicine* 56(7): 469–474.

Leka, S. & Cox, T. (2008). *The European Framework for Psychosocial Risk Management: PRIMA-EF*. UK: Institute of Work, Health and Organizations.

Leka, S. Griffiths, A. & Cox. T. (2008). *Work Organization and Stress: Systematic problem approaches for employers, managers and trade union representative*. In: Protecting Worker Health Series: Volume 3. Geneva: World Health Organization.

Linton, S. (2000). A review of psychological risk factors in back and neck pain. *Spine* 25(9): 1148–1156.

Malchaire, J. Cock, N. & Vergracht, S. (2001). Review of the factors associated with musculoskeletal problems in epidemiological studies. *International Archives of Occupational and Environmental Health* 74(2): 79–90.

Oliver, A. Cheyne, A. Tomas, J.M. & Cox, S. (2002). The effects of organizational and individual factors on occupational accidents. *Journal of Occupational and Organizational Psychology* 75(4): 473–488.

Rusli, B.N. Edimansyah, B.A. & Naing, L. (2008). Working conditions, self-perceived stress, anxiety, depression and quality of life: A structural equation modelling approach. *BMC Public Health* 48(8).

Swaen, G.M.H. van Amelsvoort, L.P.G.M. Bültmann, U. Slangen, J.J.M. & Kant, I.J. (2004). Psychosocial work characteristics as risk factors for being injured in an occupational accident. *Journal of Occupational & Environmental Medicine* 46(6): 521–527.

Walters, W. (2011). Worker representation and psychosocial risks: A problematic relationship? *Safety Science* 49(4): 599–606.

Zohar, D. (1980). Safety climate in industrial organizations: Theoretical and applied implications. *Journal of Applied Psychology* 65(1): 96–102.

Occupational Safety and Hygiene V – Arezes et al. (Eds)
© *2017 Taylor & Francis Group, London, ISBN 978-1-138-05761-6*

Safety culture in Andalusian construction sector

J. Guadix, J.A. Carrillo-Castrillo & V. Pérez-Mira
Universidad de Sevilla, Spain

M.C. Pardo-Ferreira
Universidad de Málaga, Spain

ABSTRACT: The main purpose of this study is to analyze the safety culture of Andalusian workers in the construction sector. NOSACQ-50 is a—state of the art—tool for safety culture assessment. Questionnaires analyzed were gathered by Fundación Laboral de la Construcción in 2015 from workers attending safety courses (2,133 questionnaires). Results show that safety culture scores in our sample is lower than the grand mean for all international databases gathered by October 2016 in NOSACQ-50 but similar to other studies in construction sector. Model fit is good enough. Further research should look for associations between dimensions of safety culture and specific management practices of enterprises to look for means to improve safety culture and possible association between safety culture dimensions and safety outcomes such as accidents.

1 INTRODUCTION

1.1 *Safety culture*

Safety climate may be defined as shared perceptions among the members of a social unit, of policies, procedures and practices related to safety in the organization (Zohar, 1980).

Safety culture and climate are key concepts in modern safety research. They are close and difficult to distinguish concepts. In this paper they will be differentiated according to a general framework based on work by Schein (1992) on organisational culture.

Schein (1992) defines organizational culture as a pattern of shared basic assumptions that a group has learned as it solved issues of external adaptation and internal integration. These basic assumptions are not readily observable or measurable as they are unconscious, taken for-granted beliefs that are the ultimate source of values and actions.

This framework distinguishes three levels at which organisational culture can be studied: basis assumptions, espoused values and artefacts. At the level of espoused values we find attitudes, which are equated with safety climate. The basic assumptions, however, form the core of the culture (Guldenmund, 2000).

Safety culture can be understood as the aspects or parts of the organizational culture that influence attitudes and behaviours, which have an impact on the level of safety in the organization (Hale, 2000).

Although is a growing field of research, not many studies have been published in Spain so far. This is the first evaluation of safety culture in the construction sector of Spain using NOSACQ-50 published so far.

1.2 *NOSACQ-50*

The Nordic Safety Climate Questionnaire (NOSACQ-50) was developed by a team of Nordic occupational safety researchers based on organizational and safety climate theory, psychological theory, previous empirical research, empirical results acquired through international studies, and a continuous development process (Kines et al., 2011).

Safety climate is defined in Kines et al. (2011) as workgroup members' shared perceptions of management and workgroup safety related policies, procedures and practices. NOSACQ-50 consists of 50 items across seven dimensions presented in Table 1.

Initial versions of the instrument were tested for validity and reliability in four separate Nordic studies using native language versions in each respective Nordic country. NOSACQ-50 was found to be a reliable instrument for measuring safety climate, and valid for predicting safety motivation, perceived safety level, and self-rated safety

Table 1.   Dimensions in NOSACQ-50 (latent variables).

Dimensions (Latent variables)
D1. – Management safety priority, commitment and competence
D2. – Management safety empowerment
D3. – Management safety justice
D4. – Workers' safety commitment
D5. – Workers' safety priority and risk non-acceptance
D6. – Safety communication, learning, and trust in co-workers
D7. – Workers' trust in the efficacy of safety systems

behaviour. The validity of NOSACQ-50 was further confirmed by its ability to distinguish between organizational units through detecting significant differences in safety climate (Kines et al., 2011).

Construction sector was the first sector (Study 1) analyzed to check validity of the NOSAC-50 questionnaire (Kines et al. 2011). Our study is compared to those first results and with the mean results gathered so far since the project NOSACQ-50 begun. Another reference in the construction sector is Kjestveit et al. (2011) where 456 workers were surveyed in Norway with NOSACQ-50.

2 MATERIALS AND METHODS

2.1 NOSACQ survey

This paper is based on questionnaires submitted by the workers participating in safety courses organized by Fundación Laboral de la Construcción. Fundación Laboral de la Construcción is a non-profit organization. Both representatives of employers and workers unions are members.

Each year a number of workers are trained by Fundación Laboral de la Construcción, mostly in health and safety courses for the TPC—Tarjeta Profesional de la Construcción—accreditation (TPC is the Construction Professional ID in Spain). All participants in 2015 were invited to participate in the NOSACQ-50 survey. Questionnaires were administered by instructors at the end of safety courses. There are 2,133 questionnaires available for analysis, 3% of them correspond to supervisors, and the rest are blue-collar workers.

Spanish version of the survey was used is available at http://www.nrcwe.dk/NOSACQ

2.2 Statistical analysis

Data is analyzed using descriptive statistics and CFA (Confirmatory Factor Analysis). SPSS and AMOS were used.

Descriptive statistics are presented using the recommended scores included in the NOSACQ-50 guide available at http://www.nrcwe.dk/NOSACQ. Scores are compared with the overall grand mean at October 2016 with 37,634 'worker' respondents from 279 different work sites or studies in 37 industrial sectors on 6 continents, using 26 different language versions gathered by NOSACQ-50 international database and with the results in Kjestveit et al (2011).

CFA results are compared with the ones published by Kines et al (2011) for study one in which NOSACQ-50 questionnaire was administered in the construction industry in all five Nordic countries in October 2005 to February 2006.

The CFA support the construct validity of the seven safety climate scales. The Cronbach Alpha coefficient measures the internal consistency of the latent variables (Bach, 1951) and loading factors identify items with significant contribution to each of the latent variables. Criterion for Cronbach acceptance is higher than 0.60 as proposed by Al-Refaie (2013). Model fit tests are explained elsewhere (Guadix et al., 2015).

3 RESULTS

Mean scores for each dimension in this study and for the overall grand mean at October 2016 of NOSACQ-50 is presented in Table 2.

Standard deviation of scores for each dimension in this study and for the overall grand mean by October

Table 2. Mean scores (see dimensions in Table 1).

	Score		
Dimension	This study	Mean NOSACQ	Kjestveit et al. (2011)
D1	2.89	3.62	3.06
D2	2.78	3.57	2.93
D3	2.77	3.70	3.04
D4	3.06	3.67	3.11
D5	2.87	3.23	2.89
D6	3.01	3.79	3.06
D7	3.19	3.88	3.15

Table 3. Standard deviation.

	Score		
Dimension	This study	Mean NOSACQ	Kjestveit et al. (2011)
D1	0.32	0.24	0.45
D2	0.33	0.23	0.41
D3	0.28	0.24	0.42
D4	0.24	0.22	0.40
D5	0.31	0.25	0.49
D6	0.20	0.17	0.40
D7	0.25	0.20	0.43

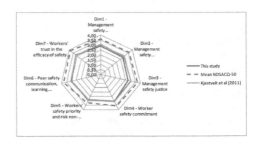

Figure 1. Radar comparison between this study, overall mean in the NOSACQ-50 project until October 2016 and Kjestveit et al. (2011).

2016 of NOSACQ-50 is presented in Table 3. A radar representation is presented in Figure 1.

Table 4. Cronbach coefficient (after depuration, 38 items) for dimensions defined in Table 1.

Dimension	Cronbach coefficient	
	This study	Kines et al. (2011)
D1	0.63	0.87
D2	0.60	0.73
D3	0.70	0.71
D4	0.76	0.77
D5	0.71	0.80
D6	0.67	0.79
D7	0.75	0.82

Table 5. Factor loading. Dimensions defined in Table 1.

Dimension	Item	Factor loading
D1	a1	1.00
	a2	0.98
	a4	1.14
	a6	0.99
	a7	1.15
	a9	−0.67
D2	a10	1.00
	a11	0.86
	a12	0.91
	a14	0.99
	a15	−0.64
	a16	0.79
D3	a17	1.00
	a19	1,01
	a20	0,91
	a22	0.99
D4	a23	1.00
	a24	0.92
	a27	0.87
D5	a29	1.00
	a30	1.29
	a31	1.41
	a32	1.77
	a33	−0.49
	a34	1.20
	a35	1.54
D6	a36	0.84
	a37	1.00
	a38	0.99
	a39	0.89
	a40	1.05
	a41	−0.61
	a42	0.58
	a43	0.99
D7	a44	1.00
	a46	1.08
	a48	1.05
	a50	0.95

Table 6. Correlations among dimensions. Dimensions defined in Table 1.

Model fit indicator	Value
Nº items	38
Alfa de Cronbach	0.82
Probability level	0.00
Chi-square	1,382.41
Degrees of freedom	603
X^2/df	2.29
CFI	0.97
NFI	0.95
TLI	0.97
RMSEA	0.02

Note: Model fit test explained elsewhere (Guadix et al., 2015).

Table 7. Correlations among dimensions. Dimensions defined in Table 1.

	D1	D2	D3	D4	D5	D6	D7
D1	1.00						
D2	0.74	1.00					
D3	0.71	0.72	1.00				
D4	0.50	0.47	0.46	1.00			
D5	0.54	0.47	0.50	0.58	1.00		
D6	0.49	0.50	0.48	0.52	0.43	1.00	
D7	0.35	0.31	0.33	0.49	0.45	0.52	1.00

Note: All correlations are significant with p > 0.01.

The Cronbach coefficient for the 50 items is 0.75. The items were depurate using the loading factors as criterion and twelve items were discarded improving overall Cronbach coefficient to 0.82 (see Table 4).

Factor loadings of each item in de latent variables are presented in Table 5.

Model fit are presented in Table 6. Indicators are considered acceptable according to (Hsu et al, 2012; Schermelleh-Engelet al, 2003).

Correlations (at individual level) among dimensions in this study are presented in Table 7.

4 DISCUSSION AND CONCLUSIONS

The measure of safety culture in the construction sector in Spain from this study show lower levels than the overall grand mean so far in NOSACQ-50 project but similar to the measure published in Kjestveit et al. (2011).

Cronbach coefficients in this study are smaller than the presented in Kines et al (2011). This can be explained because we did not discard any questionnaire as Kines et al. (2011) did. Nevertheless latent variables have enough internal reliability to be used in further research.

Model fit is good enough. Further research should look for associations between dimensions of safety culture and specific management practices of enterprises to look for means to improve safety culture and possible association between safety culture dimensions and safety outcomes such as accidents.

REFERENCES

Al-Refaie. A. (2013). Factors affect companies' safety performance in Jordan using structural equation modelling. *Safety Science* 57: 169–178.

Guadix, J., Carrillo-Castrillo, J., Onieva, L. & Lucena, D. (2015). Strategies for psychosocial risk management in manufacturing. *Journal of Business Research 68*: 1475–1480.

Cronbach. L. J. (1951). Coefficient alpha and the internal structure of tests. *Psychometrika* 16: 297–334.

Guldenmund. F.W. (2000). The nature of safety culture: a review of theory and research. *Safety Science* 34: 215–257.

Hale, A.R. (2000). Culture's confusions. *Safety Science* 34: 1–14.

Hsu, I.-Y., Su, T.-S., Kao, Ch.-S., Shu, Y.-L., Lin, P.-R. & Tseng, J.-M. (2012). Analysis of business safety performance by structural equation models. *Safety Science* 50(1): 1–11.

Kjestveit, K., Tharaldsen, J. & Holte, K. A. (2011). Young and strong: What influences injury rates within building and construction? *Safety Science Monitor* 15(2) article no. 5: 1–15.

Khanzode. V.V. Maiti. J. & Ray. P. 2012. Occupational injury and accident research: A comprehensive review. *Safety Science* 50(5): 1355–1367.

Schein. E.H. (1992). Organizational Culture and Leadership. 2nd Edition. San Francisco: Jossey-Bass.

Schermelleh-Engel, K., Moosbrugger, H. & Muller, H. (2003). Evaluating the fit of structural equation models: Tests of significance and descriptive goodness-of-fit measures. *Methods of Psychological Research*, 8: 23–74.

Zohar. D. (1980). Safety climate in industrial organizations: Theoretical and applied implications. *Journal of Applied Psychology* 65(1): 96–102.

Occupational Safety and Hygiene V – Arezes et al. (Eds)
© 2017 Taylor & Francis Group, London, ISBN 978-1-138-05761-6

Historical and scientific review—whole body vibration exposure in urban bus drivers

S. Barbosa
Faculty of Engineering, University of Porto, Porto, Portugal

M. Luisa Matos
Faculty of Engineering, University of Porto, Porto, Portugal
LNEG, Laboratório Nacional de Energia e Geologia, S. Mamede de Infesta, Portugal

ABSTRACT: Many professional drivers suffer from low back pain which is thought to be associated with exposure to whole-body vibration. The aim of this study is to review the historical and scientific studies, to explore the evidences and effects of the whole-body vibration exposure in urban bus drivers. The PRISMA methodology was used to do the search and it allowed to find 21studies. The analysed studies applied the Standards (ISO 2631-1 and ISO 2631-5) to guide researchers in measuring and analysing WBV levels and take impulsive exposures into account. It was found that a relationship exists between the road type and the vibration magnitude. In some cases, workers are exposed to higher vibration values than the regulated exposure limits and action values but the risks can be reduced using preventive strategies.

Keywords: Whole Body Vibration, Professional Drivers, Transmissibility, Daily Exposure; Exposure limits

1 INTRODUCTION

One of the foremost risk factors for the development of low back disorders in professional drivers is constant exposure to whole-body vibration. (Tamrin et al., 2007).

Many activities related to daily life and occupation may result in episodes of back or neck pain. However, none is more consistently implicated than driving a motor vehicle for extended periods of time, (Tamrin et al., 2007). Many studies have shown that urban bus drivers are part of the group of workers who are potentially exposed to WBV (Bovenzi & Hulshof, 1999)

The human response to exposure to WBV has essentially five effects: reduced comfort, health impact, interference with daily activities and nausea (Griffin, 1990; Miguel, 2010).

The consequences from exposure to WBV is not easy to analyse, since they can be felt either from the physiological point of view or from a psychological point of view and vary with the frequency range to which the worker is exposed, (Griffin, 1990; Miguel, 2010).

The feeling of discomfort due to vibration exposure is frequently felt by professional drivers, and it is described to grow along with increasing magnitude and duration of exposure to vibration, (Sekulic, Dedovic, Rusov, & Obradovic, 2016).

The long exposure of drivers to whole body vibration has been recognized as a stress mechanic factor for the accelerated development of degenerative diseases of the spine, back pain and prolapsed intervertebral discs (Johanning, 1991; Bovenzi & Hulshof, 1999; Okunribido, Magnusson, & Pope, 2006; Yasobant, Chandran, & Reddy, 2015).

The health condition of public transport drivers is important, as it is one of the factors that plays a role in assuring the safety of passengers using this service, (Tamrin et al., 2007).

Biodynamic experiments have shown that whole body vibration exposure, combined with a constrained sitting posture, can put the lumbar intervertebral disc at risk of failure. The Musculoskeletal injuries related to work, are painful injuries in the muscles, bones, tendons and other soft tissues that arise due to activity in the workplace. These lesions are responsible for morbidity in many people who work and are known to cause significant occupational problems with increasing compensation and health costs, reduced productivity and lower quality of life, (Yasobant, Chandran, & Reddy, 2015).

Over the years several studies have identified different relationships between the intensity of exposure to whole body vibration in professional drivers and other factors. In fact, the vibrations are transmitted to the bodies of drivers and passengers via their seats. It was found that intensity of vibrations that bus users are exposed to depends on the bus seat position. However, there are other factors, like the road type, the velocity, the bus typology, and others, that can amplify or attenuate the intensity of transmitted vibration for the drivers.

The goal of this study was to review the historical and scientific studies in order to explore the evidences and effects of the whole-body vibration exposure in urban bus drivers.

2 METHODOLOGY

The literature search was based on PRISMA (Preferred Reporting Items for Systematic Reviews and Meta-Analyses).

The resources used for researching scientific articles was the database of reference Scopus and the bibliography of the articles was selected from the database.

Four pairs of two keywords were selected to do the search on the database. The groups of keywords selected were: 1.º – Vibration" + "Bus driver"; 2º – "Whole body Vibration" + "Bus"; 3º – "Bus driver"; + "Musculoskeletal Disorders"; 4.º – "Vibration" + "articulated bus".

Then, an exclusion criteria was applied to select the articles: 1) Language: articles not published in Portuguese or English were excluded; 2) Publication Date: articles published before 2000 were excluded; 3) Methodology: there was only the inclusion of studies that applied the Standards ISO 2631-1 and ISO 2631-5. 4) Relevance: articles not addressing occupational vibration among public bus drivers were excluded. Very specific items of medical nature, the diseases caused by vibrations and everything that was not directly related to occupational exposure in urban bus drivers were also excluded.

3 RESULTS AND DISCUSSION

3.1 Literature search

A total of 206 articles were retrieved from literature search. Forty-eight duplicates were identified, others twenty-three were refused by the language exclusion criterion and eighty-eight studies were excluded by the methodology and the relevance exclusion criterion. Thus, 21studies were included in this review. After applying the methodology on literature search, it was possible to systematize the relevant information that contributes to the purpose of this literature review.

3.2 Historical review

The first study identified in the literature, was developed by Thomas Clarkson in the year 1913 in Lon-don. This was the first published study that made a brief reference to the importance of vibration felt by the passengers and drivers in buses. The study was based on a comparison between the steam engine bus and the internal combustion engine bus, wherein the wear caused by the vibration of wheels and vibration felt inside the bus are factors considered in the best type of bus selection. (Clarkson, 1913)

In 1954, Ross A. McFarland conducted the first study on safe driving on the roads. McFarland intended to study the conditions of the work-place in truck drivers, including the study of vibrations and other physical factors. This study aimed to study and prevent car accidents occurrences and ensure the safety of the driver. (McFarland, 1954)

In 1974, the National Institute for Occupational Safe-ty and Health (NIOSH) in the United States, produced a major study of vibrations in jobs, corresponding to the launch of this great theme worldwide. The study was designed to identify the number and types of workers exposed to vibrations and for more scientific data on the health effects due to this type of exposure. In this study, it was determined that transport (including truck drivers and drivers, heavy equipment operators and farmers driving tractors and other agricultural vehicles) is the professional category most exposed to WBV (Whole Body Vibrations) (Lehmann, 1974).

In 1992, Bovenzi & Zadini developed the first study of the lumbar symptoms reported by drivers of city buses exposed to whole body vibration through surveys. The study showed that there was an increase of lumbar symptoms occurrence with the increasing exposure of professional drivers to whole-body vibration, (Bovenzi & Zadini, 1992).

After the nineties, there was a considerable increase in publications until 2014. The increase registered until 2014 was largely due to great technological development in recent years, related to the publication of many studies of computational models of automotive technology and advanced mechanical engineering.

3.3 Methodologies applied to the measurement and assessment of whole body vibration

3.3.1 ISO standard

The WBV exposures measurement method is in accordance with the ISO 2631-1:1997, (Melo & Miguel, 2000; Bovenzi, 2009; Lewis & Johnson, 2012; Thamsuwan, Blood, Ching, Boyle, & Johnson, 2013), and ISO 2631-5:2004, (Lewis & Johnson, 2012; Thamsuwan, Blood, Lewis, Rynell, & Johnson, 2012; Thamsuwan, Blood, Ching, Boyle, & Johnson, 2013; Jonsson, Rynell, Hagberg, & Johnson, 2014; Barreira, Matos, & Baptista, 2015) respectively, and provide guidance on the assessment of health effects.

Both ISO 2631-1 and ISO 2631-5 standards should be used for better understanding of the exposure.

3.3.2 Accelerometer

In general, data collection was performed using tri-axial accelerometer ICP (model 356B40; PCB Piezometrics, Depew, NY, USA) (Bovenzi, 2009; Blood, Ploger, Yost, Ching, & Johnson, 2010; Blood & Johnson, 2012; Lewis & Johnson, 2012; Thamsuwan, et al., 2013). The measurements were collected from the seat pan (Blood & Johnson, 2012; Bovenzi, 2009; Lewis & Johnson, 2012; Melo & Miguel, 2000; Okunribido, et al., 2008; Thamsuwan, et al., 2013) and from the seat backrest (Melo & Miguel, 2000) and the floor of the bus (Lewis & Johnson, 2012; Thamsuwan, et al., 2013).

In most cases, the evaluation on the backrest is justified by the fact that the transmission of vibrations by this means is significant, however the measurements obtained on the floor of the vehicle allow to quantify the parameter SEAT (Seat Effective Amplitude Transmissibility). In addition, (Blood & Johnson, 2012; Thamsuwan, et al., 2013) use the Global Positioning System (GPS) to collect data to record the location and velocity of the bus, and type of road associated with the WBV exposures.

3.3.3 Statistical analysis

For most part of the studies, a statistical software was used (JMP, SAS Institute, Cary, SC) to perform the statistical analysis, (Blood, Ploger, Yost, Ching, & Johnson, 2010; Thamsuwan, Blood, Lewis, Rynell, & Johnson, 2012; Lewis & Johnson, 2012; Thamsuwan, Blood, Ching, Boyle, & Johnson, 2013; Jonsson, Rynell, Hagberg, & Johnson, 2014; Blood, Yost, E., & Ching, 2015)

3.4 Evaluation of Whole-body vibration exposures

To reduce the whole-body vibration exposure, previous studies have investigated various factors, including: road types, bus types and seat design.

3.4.1 Road type

The literature search included different road types: city streets, freeway and speed humps (Blood, Ploger, Yost, Ching, & Johnson, 2010; Blood & Johnson, 2012; Lewis & Johnson, 2012; Thamsuwan, et al., 2012; Thamsuwan, et al., 2013; Jonsson et al., 2014).

According (Melo & Miguel, 2000 and Barreira, Matos, & Baptista, 2015), the itinerary chosen included asphalt and brick-paved roads.

3.4.2 Seat design and bus type

The Table 1 presents some of the factors evaluated in some studies.

3.4.3 Discussion of literature main results

Melo & Miguel, 2000, calculated the parameter RMS for different road types. In this study, the vertically transmitted vibrations were more intense on the asphalt surface than along the brick-paved road.

O. Okunribido, Shimbles, Magnusson, & Pope, 2007, analysed the RMS values for different bus types, and the lowest values were seen during idling and the highest values generally occurred in the z axis during travel on cobble, particularly for the single-decker bus. While the idling accelerations were highest for the mini-bus, the asphalt and cobble accelerations were lowest for this vehicle. During the travel, the double-decker bus generated highest x-axis average acceleration and the single-decker bus highest z-axis average acceleration.

Table 1.

Author	Bus type	Seat design
(Melo & Miguel, 2000)	–	–
(O. Okunribido, Shimbles, Magnusson, & Pope, 2007)	Minibus; double-decker bus; Standard bus	–
(Bovenzi, 2009)	Standard bus	–
(Blood, Ploger, Yost, Ching, & Johnson, 2010)	Standard bus	standard foam; silicone foam. different manufactures;
(Lewis & Johnson, 2012)	High floor bus	–
(Thamsuwan, Blood, Lewis, Rynell, & Johnson, 2012)	articulated bus; Standard bus; High floor bus	standard foam
(Thamsuwan, Blood, Ching, Boyle, & Johnson, 2013)	High floor bus Standard bus	–
(Jonsson, Rynell, Hagberg, & Johnson, 2014)		Seat with and out air suspension
(Barreira, Matos, & Baptista, 2015)	Standard bus	–
(Blood, Yost, E., & Ching, 2015)	–	Air ride suspension Seat; Active suspension seat;

In the study by Lewis, 2012 and Thamsuwan, 2013, the road type had a significant effect on propagation of vibration, however the motorway provides a slightly larger exposure on the z axis compared to other types of roads. City streets and speed curves also had higher z-axis impulsive exposures (Lewis & Johnson, 2012), (Thamsuwan, Blood, Ching, Boyle, & Johnson, 2013). The study of Thamsuwan, et al., 2012, the predominant axes of WBV exposures on three buses were different. The greater $A(8)x$ on the high-floor bus seemed to be related to the movement of the bus as the bus driver sat much higher over the bus suspension and road relative to the standard buses. When the high-floor bus went over the speed humps, due to the greater height, the driver could have undergone higher angular and linear x and y translations. The $A(8)y$ on the articulated bus were greater than on the other two non-articulating buses (Thamsuwan, Blood, Lewis, Rynell, & Johnson, 2012).

In the study of Jonsson, et al. 2014, the air-suspension seat attenuated the vibration on the low-floor bus but amplified the vibration on the high-floor bus may indicate that there is a seat and bus suspension mismatch on the high-floor bus. The amplification of the WBV exposures by the air-suspension seat in the high floor bus may indicate that a different type of seat may be better suited for this type of bus.

In the study elaborated by Blood, Yost, E., & Ching, 2015, there were significant differences in seat-pan vibration exposures between the air-ride bus seat and the EM-active seat in the city-street, freeway, and rough-road segments. The air-ride bus seat universally transmitted the highest seat pan vibration in the z-axis, with the largest vibration occurring in the rough-road segment. Unlike the truck signal, the highest seat-pan exposures for the EM-active seat on the bus signal were in the z-axis, with the largest vibration also being in the rough-road segment, although at levels significantly lower than those from the air-ride bus seat.

3.4.4 *Exposure limits*

Comparing the results of whole body vibration exposure $(A(8))$ with the stipulated values on the Directive 2002/44/CE, it was visible that the most part of the studies did not have higher values than the exposure limit values. Although, sometimes the exposure values were greater than the exposure action values, (Thamsuwan, Blood, Lewis, Rynell, & Johnson, 2012; Thamsuwan, Blood, Ching, Boyle, & Johnson, 2013).

3.5 *Prevention of whole body vibration*

The creation of preventive strategies to reduce exposure to WBV can contribute significantly to reduce the risk of musculoskeletal injuries (Tiemessen, Hulshof, & Frings-Dresen, 2007).

The concept of prevention and adoption of vibration control measures is based on the use of technical, organizational and management engineering to reduce the exposure related to vibrations legally acceptable values (Thamsuwan et. al., 2013).

As the road surface's improvement is not the Bus Companies' responsibility, the most reasonable actions capable of reducing the effects of vibration transmission to these bus drivers are: 1) use of a seat equipped with vertical suspension and lumbar support; 2) driving speed adjustment to the driving conditions (namely, ground roughness, traffic flow and distance from pedestrian crossings and bus stops); 3) maintenance of correct pressure and routine maintenance of both the seating and the engine mounts (Melo & Miguel, 2000).

From an engineering design aspect, improving the seat or vehicle suspension system could help reduce the vibration exposures as well as the transmission of vibration to the drivers. Active suspension systems have been shown to reduce vibration-related discomfort under several driving conditions developed an optimization procedure that included all the vehicle suspension components and the interaction between road and tire to minimize vibration to improve the ride and handling behaviour. In addition, seat selection plays important role in reducing WBV exposures. As well, long term effectiveness and health outcomes of such ergonomic risk reduction programs need to be further assessed.

4 CONCLUSION

The studies indicate a causal relationship between whole-body vibration and the effects to health in different typologies of urban bus, included the articulated bus. This depends on frequency and magnitude of vibration. The factors that seem to determine the magnitudes of vibration are: design considerations of vehicle, skills and behaviour of the driver, seat characteristics, velocity and road conditions, as a major contribute. The dynamic response of seat can perform differently in their ability to attenuate vibration exposure for the driver and this can be a factor used to reduce human exposure vibration. Exposure to whole-body vibration is now recognized as associated with discomfort, reduced work efficiency and health problems. In addition, low back pain has been reported more extensively in research.

To control whole body vibration exposure and risks associated can be adopted engineering, administrative, and work organizational controls.

ACKNOWLEDGMENTS

The authors would like to acknowledge Master in Occupational Safety and Hygiene Engineering (MESHO), of the Faculty of Engineering of the University of Porto (FEUP), the support and contribution provided in the development of this work and for having allowed the publication of the study.

REFERENCES

Barreira, S., Matos, M. L., & Baptista, J. S. (2015). Exposure of urban bus drivers to Whole-Body Vibration. Occupational Safety and Hygiene III—Arezes et al. (Eds), 321–324.

Blood, R. P., Yost, M. G., E., C. J., & Ching, R. P. (2015). Whole-body Vibration Exposure Intervention among Professional Bus and Truck Drivers. A Laboratory Evaluation of Seat-suspension Designs. Journal of Occupational and Environmental Hygiene, 351–362.

Blood, R., Ploger, J., Yost, M., Ching, R., & Johnson, P. (2010). Whole body vibration exposures in metropolitan bus drivers: Journal of Sound and Vibration, 109–120.

Bovenzi, M. (2009). Metrics of whole-body vibration and exposure–response relationship for low back pain in professional drivers: a prospective cohort study. International archives of occupational and environmental health, 82(7), 893–917.

Bovenzi, M., & Hulshof, C. T. (1999). An updated review of epidemiologic studies on the relationship between exposure to whole-body vibration and low back pain (1986–1997). Int Arch Occup Environ Health, 351–365.

Bovenzi, M., & Zadini, A. (1992). Self-reported low back symptoms in urban bus drivers exposed to whole-body vibration. Spine, 1048–1059.

Clarkson, T. (1913). Steam omnibuses. SAE Technical Papers. Elsevier B.V.

Griffin, M. J. (1990). Handbook of Human Vibration. London: Elsevier Academic Press.

Johanning, E. (1991). Back disorders and health problems among subway train operators exposed to whole-body vibration. Scand J Work Environ Health, 414–419.

Jonsson, P. M., Rynell, P. W., Hagberg, M., & Johnson, P. W. (2014). Comparison of whole-body vibration exposures. Taylor & Francis, 58, 1133–1142.

Lehmann, P. (1974). Vibration. Job Safety Health, 5–12.

Lewis, C.A., & Johnson, P. (2012). Whole-body vibration exposure in metropolitan bus drivers. Occupational medicine, 62(7), 519–524.

McFarland, R. A. (1954). Human engineering: A new approach to driver efficiency and transport safety. Boston: Elsevier B.V.

Melo, R. B., & Miguel, A. S. (2000). Occupational Exposure to Whole-body Vibration Among Bus Drivers., (pp. 177–180). Lisboa.

Miguel, A. S. (2010). Manual de Higiene e Segurança do Trabalho. Porto: Porto Editora.

O. Okunribido, O., Shimbles, S. J., Magnusson, M., & Pope, M. (2007). City bus driving and low back pain: A study of the exposures to posture demands, manual materials handling and whole-body vibration. Applied Ergonomics, 38, 29–38.

Okunribido, O. O., Magnusson, M., & Pope, M. (2006). Delivery drivers and low-back pain: A study of the exposures to posture demands, manual materials handling and whole body vibration. Industrial Ergonomics, 265–273.

Sekulic, D., Dedovic, V., Rusov, S., & Obradovic, A. (2016). Definition and determination of the bus oscillatory comfort zones. International Journal of Industrial Ergonomics, 328 e 339.

Tamrin, S. B., Yokoyama, K., Jalaludin, J., Aziz, N. A., Jemoin, N., Nordin, R., Abdullah, M. (2007). The Association between Risk Factores and Low Back Pain among Commercial Vehicle Drivers in Peninsular Malaysia: A Preliminary Result. Industrial Health, 268–278.

Thamsuwan, O., Blood, R. P., Ching, R. P., Boyle, L., & Johnson, P. W. (2013). Whole body vibration exposures in bus drivers: A comparison between a high-floor coach and a low-floor city bus. International Journal of Industrial Ergonomics, 9–17.

Thamsuwan, O., Blood, R. P., Lewis, C., Rynell, P. W., & Johnson, P. W. (2012). Whole Body Vibration Exposure and Seat Effective Amplitude Transmissibility of Air Suspension Seat in Different Bus Designs. Human Factors and Ergonomics Society.

Tiemessen, I. J., Hulshof, C. T., & Frings-Dresen, M. H. (2007). An overview of strategies to reduce whole-body vibration exposure on drivers: A systematic review. International Journal of Industrial Ergonomics, 245–256.

Yasobant, S., Chandran, M., & Reddy, E. M. (2015). Are Bus Drivers at an Increased Risk for Developing Musculoskeletal Disorders. J Ergonomics.

Occupational Safety and Hygiene V – Arezes et al. (Eds)
© *2017 Taylor & Francis Group, London, ISBN 978-1-138-05761-6*

Exposure to chemical mixtures in occupational settings: A reality in oncology day services?

S. Viegas

Environment and Health RG—Escola Superior de Tecnologia da Saúde de Lisboa, Instituto Politécnico de Lisboa, Portugal
Centro de Investigação e Estudos em Saúde Pública, Escola Nacional de Saúde Pública, ENSP, Universidade Nova de Lisboa, Lisbon, Portugal

A.C. Oliveira & M. Pádua

Environment and Health RG—Escola Superior de Tecnologia da Saúde de Lisboa, Instituto Politécnico de Lisboa, Portugal

ABSTRACT: Antineoplastic drugs are essential to cancer treatment. Its widespread use in hospitals leads to an increased probability of occupational exposure of the healthcare professionals that manipulate the drugs. The fact that chemotherapy involves administration of multiple drugs only amplifies the problem resulting in the possible occupational exposure to a mixture of these chemicals. The aim of this study was to determine the surfaces contamination by antineoplastic drugs in one Oncology Day Service and understand if it promotes exposure to chemical mixtures. Selected workplaces surfaces were wipe-sampled. 5-Fluorouracil, Paclitaxel™ and Cyclophosphamide in the samples were simultaneously analyzed by HPLC-DAD. Of the 45 collected samples, 4.4% were not contaminated, 40% presented contamination by only one drug and 55.6% showed contamination by more than one drug. The study allows the recognition that workers are exposed to mixtures of drugs. Interactions between drugs can be a reality and this justifies further research to support a cumulative risk assessment.

1 INTRODUCTION

Current chemicals legislation is based predominantly on assessments carried out on individual substances. However, humans are exposed commonly to a wide variety of substances simultaneously found in food, air and drinking water, in household and consumer products, in cosmetics and, of course, in the workplaces. Therefore, awareness is growing about the potential adverse effects of the interactions between substances when present simultaneously in a mixture (Scher, 2012).

In this case, it should be considered developing a cumulative exposure assessment. This assessment should consider the simultaneous, overlapping, and/or sequential exposure to multiple environmental stressors that may contribute to harmful outcomes (Sexton & Hattis, 2007).

Assessment of cumulative risk to human health should be done, however it is a challenge to estimate cumulative exposure. In the absence of adequate epidemiological data, the effects of chemical mixtures on human health are based on experimental studies. However, their qualitative and quantitative transposition from the laboratory animal data to human populations can only be addressed using science-based safety factors (Evans et al., 2016).

Since the 1970s antineoplastic drugs have been recognized as hazardous to healthcare professionals such as oncology nurses, pharmacists and technicians (Falck et al., 1979).

In Hospitals, the workers potentially exposed include shipping and receiving personnel, pharmacists and pharmacy technicians, nursing personnel and cleaning and maintenance services personnel (Sessink & Bos, 1999).

Exposure to these drugs may occur mainly in three ways: aerosol inhalation, dermal absorption caused by direct contact with the skin, and, although less frequent, ingestion due to contact with contaminated hands. At present, due to the evolution of technical protection resources available, such as biological safety cabinets and chemical fume hoods, the most common route of exposure is dermal rather than inhalation or ingestion (Viegas et al., 2014).

Considering the above, contamination measurement on workplaces surfaces is the best approach to estimate workers exposure.

Several authors (Connor et al., 1999; Schmaus et al., 2002; Hedmer et al., 2005; Fransman et al., 2007; Sottani et al., 2008; Viegas et al., 2014; Fleury-Souverain et al., 2015) consider that despite recommended safe handling practices for these compounds and the improvement in the technical protection measures, exposure is not decreasing.

Moreover, the most common approach to chemotherapy involves administration of multiple agents to target as many types of cancer cells as possible. Usually, the antineoplastic treatment is made by a combination of two or more drugs. Although the toxicological profile and mechanism of action of each individual drug is well characterized and the toxicological interactions between drugs are likely, the link with an occupational exposure context is poorly established (Ladeira et al., 2016).

The aim of this study was to determine the contamination of workplace surfaces from one Oncology Day Service (with preparation and administration of drugs) of a Portuguese Hospital and understand if the characteristics of the contamination found contributes for workers exposure to a mixture of antineoplastic drugs.

2 MATERIALS AND METHODS

2.1 *Materials*

5-Fluorouracil $\geq$ 99% (HPLC) and Paclitaxel™ from *Taxus brevifolia* $\geq$ 95% (HPLC) were purchased from Sigma-Aldrich, Acetonitrile gradient grade for HPLC from Merk, Ethyl Acetate for HPLC and Water HPLC grade from VWR, Methanol HPLC grade from Chem-Lab and Cyclophosphamide, 97% from Acrós Organics.

Kimwipes were used to collect the surface samples and were purchased from Kimberly-Clark Kemtech.

HPLC System was composed by a quaternary LC pump, an autosampler and a Diode Array Detector (DAD) from Thermo Finnigan and data was processed with Excalibur 2.0.

2.2 *Selection of sampling surfaces*

Direct and passive observations at each site were conducted by members of the research team to visually identify which surfaces/objects may be contaminated with antineoplastic drugs and judge as the most potentially contaminated and more frequently handled or touched by healthcare workers in their routine tasks.

2.3 *Wipe sampling*

Surface samples were collected using the wipe sampling method. In a 10×10 cm stainless steel grid, surfaces were wiped using a Kimwipe impregnated with Ethyl Acetate. Samples were accommodated in petri dishes sealed with Parafilm™ and stored until analysis at –20°C.

All wipe samples were extracted as described elsewhere (Schmaus et al. 2002). The wipes were extracted with 15 mL of acetonitrile/methanol/water (10:25:65) for 20 min at room temperature in a bottleroller homogenizer (50rpm). Extracts were filtered through a 0.22 μm filter prior to injection.

2.4 *HPLC quantification*

Separation and quantification was performed according to Larson and colleagues (2002) with 100-μL sample loop; column Hypersil-GOLD $15 \times 5 \times 4.6$ with a guard column; mobile phase of acetonitrile/methanol/water (19:13:68) at a flow of 0.8 mL min^{-1}. A sample was considered positive for a particular drug if the value was above the Limit of Detection (LOD).

3 RESULTS AND DISCUSSION

In total, forty five samples were collected in the Oncology Day Service. Cyclophosphamide (CP), 5-fluorouracil (5FU), and Paclitaxel™ (PTX) were used as surrogate markers for surfaces contamination by antineoplastic drugs. Most of the studies reported used the same strategy since measuring all drugs use is a challenge concerning analytic methods and cost associate (Castiglia et al. 2008; Hedmer et al. 2005, 2008; Touzin et al. 2009; Kopp et al. 2013; Viegas et al., 2014).

In this study, 55% of the samples showed contamination and 5FU presented the higher number of positive samples and also the higher concentration found (Table 1). These drugs are among the most frequently employed in the hospital enrolled in the study and are commonly used in several chemotherapy protocols.

Of the forty five samples collected, 2 (4.4%) samples do not presented contamination, 18 (40%) presented contamination by only one drug and 25 (55.6%) showed contamination by more than one drug (Figure 1).

Regarding the proportion of positive samples, our results showed a widespread contamination that can be due to several factors. The fact that this service only has a Biological Safety Cabinet (BSC) instead of a Closed-System Drug Transfer (CSDT)

Table 1. Results overview.

Drugs	CP (μg/cm^2)	5FU (ng/cm^2)	PTX (ng/cm^2)
LOD	0.41	0.55	0.61
Positive samples	23	38	10
Mean Range	16.1	15.4	—
	<LOD–38.9	<LOD–75.24	<LOD–1.97

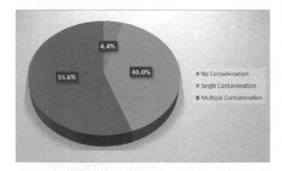

Figure 1. Description of samples contamination.

can explain the level of contamination found (Roland et al., 2016). Previous publications (Vyas et al., 2014; Simon et al., 2016) showed that CSDT can be effective in reducing contamination since ensures no escape of hazardous drugs or vapours, no transfer of environmental contaminants and, also, prevention of microbial ingress (European Parliament, 2016). However, in order to guarantee that CSDT offer the highest level of safety this equipment should be tested by a certified independent body (European Parliament, 2016).

Other factor that can be of influence is related with workers practices since probably workers are not removing the gloves immediately after drug handling, spreading the contamination in several other surfaces. This aspect has been mentioned in previous publications as responsible for the contamination found (Castiglia et al. 2008; Viegas et al., 2014).

The cleaning process is also critical in reducing the spreading of contamination (European Parliament, 2016). However, an optimized cleaning procedure should guarantee the use of adequate products, the correct identification of the surfaces to clean and the cleaning products should be in contact with the surfaces enough time to ensure decontamination (Raghavan et al., 2000; Touzin et al., 2010; European Parliament, 2016). These and other aspects need probably improvement to guarantee proper surfaces decontamination.

The multiple contamination found in most of the samples confirmed that workers exposure is probably to a mixture of antineoplastic drugs in small doses and most likely over a long period of time (Roussel & Connor, 2013). In a previous study developed by Viegas et al. (2014) was also found samples with contamination by more than one drug but in a smaller number (8%).

Chemicals with common modes of action, like antineoplastic drugs, will possibly act jointly to produce combination effects that are larger than the effects of each mixture component applied singly (SCHER, 2012). Recently, Gajski et al., (2016) verified a possible additive and/or synergistic action of the combination of 5-FU and etoposide, also an antineoplastic drug. This suggested that the risk of a multiple exposure is higher than the risk that could be assumed from evaluating the effect of a single substance.

Considering the risk assessment of chemical mixtures, the most common knowledge gap is the lack of exposure information and the rather limited number of chemicals for which there is sufficient information on their mode of action (SCHER, 2012). However, in the case of occupational exposure to antineoplastic drugs, there is data available concerning exposure and drugs mode of action, allowing to proceed with research that can contribute to unveil in more detail what can be expected in workers' health if preventive measures are not applied to reduce workplaces contamination.

Moreover, the severity of effects that can be related with the exposure to this mixture suggests risks that are likely to be unacceptable. In this case this mixture should be considered with high-priority mixtures deserving particular attention from researchers, risk assessors and regulators (Sexton & Hattis, 2007).

In a recent document made by European Parliament (European Parliament, 2016) is stated the following *"EU must commit to a new era of safe hazardous medicine handling by specifically addressing in its legislation the issue of healthcare workers' exposure to hazardous drugs and chemicals, in particular during the preparation and administration of cytotoxic drugs. This must be addressed by harmonizing national provisions relating to the protection of workers by measures to prevent exposure or to keep exposure at a level as low as possible."*

The priority and importance seems already identified by regulators but still missing actions at national level that can ensure contamination reduction. No acceptable or recommended limits exist for total quantities of antineoplastic drugs on workplace surfaces, but due to their mode of action (genotoxic most of them) the goal should be the absence of contamination or the lowest amount possible.

4 CONCLUSION

In conclusion, the study allows to recognize that workers from this Oncology Day Service are probably exposed to a mixture of antineoplastic drugs.

Data obtained, draw attention to the importance of employing adequate safety measures and proper training of personnel in order to avoid surfaces contamination and the health effects that can be caused by exposure to antineoplastic drugs.

These results emphasize the need to estimate cumulative risks as an integral part of risk management decisions. To allow this is crucial to develop further research into chemicals interactions and in vitro studies can be an approach to be employed to elucidate about the mechanism of toxicity.

REFERENCES

Castiglia L, Miraglia N, Pieri M, Simonelli A, Basilicata P, Genovese G, et al. (2008). Evaluation of occupational exposure to antiblastic drugs in an Italian hospital oncological department. *Journal of Occupational Health* 50: 48–56.

Connor TH, Anderson RW, Sessink PJM, Broadfield L and Power LA (1999). Surface contamination with antineoplastic agents in six cancer treatment centers in the United States and Canada. *Am J HealthSyst Pharm* 56: 1427–1432.

European Parliament. (2016). Preventing occupational exposure to cytotoxic and other hazardous drugs: European policy recommendations. Strasbourg: Author. Retrieved from: http://www.europeanbiosafetynetwork.eu/Exposure%20to%20Cytotoxic%20Drugs_Recommendation_DINA4_10–03-16.pdf.

Evans RM, Martina OV, Faust M, Kortenkamp A. (2016). Should the scope of human mixture risk assessment span legislative/regulatory silos for chemicals? *Science of the Total Environment* 543:757–764.

Falck K, Gröhn P, Sorsa M, Vainio H, Heinonen E, Holsti LR (1979) Mutagenicity in urine of nurses handling cytostatic drugs. *Lancet*. 9;1(8128):1250–1.

Fleury-Souverain S, Mattiuzzo M, Mahl F, Nussbaumer S, Bouchoud L, Falaschi L, Gex-Fabry M, Rudaz S, Sodeghipour F, Bonnabry P. (2015). Evaluation of chemical contamination of surfaces during the preparation of chemotherapies in 24 hospital pharmacies. *Eur J Hosp Pharm* 22:333–341

Fransman W, Roeleveld N, Peelen S, de Kort W, Kromhout H, Heederik D. (2007). Nurses with dermal exposure to antineoplastic drugs: reproductive outcomes. *Epidemiology*. 18(1):112–9.

Gajski G, Geri M, Domijan A, Garaj-Vrhovac V. (2016). Combined cyto/genotoxic activity of a selected antineoplastic drug mixture in human circulating blood cells. *Chemosphere* 165: 529–538.

Hedmer M, Tinnerberg H, Axmon, A., & Jönsson, B. A. (2008). Environmental and biological monitoring of antineoplastic drugs in four workplaces in a Swedish hospital. *International Archives of Occupational and Environmental Health*, 81(7), 899–911.

Hedmer, M, Georgiadi, A, Bremberg, ER, Jönsson, BAG, & Eksborg, S. (2005). Surface contamination of cyclophosphamide packaging and surface contamination with antineoplastic drugs in a hospital pharmacy in Sweden. *Annals of Occupational Hygiene*, 49(7), 629–637.

Kopp B, Schierl R,. Nowak D. (2013). Evaluation of working practices and surface contamination with antineoplastic drugs in outpatient oncology health care settings. *International Archives of Occupational and Environmental Health*, 86(1), 47–55.

Ladeira C, Viegas S, Costa-Veiga A. (2016). How to deal with uncertainties regarding the occupational exposure to antineoplastic mixtures: additive effect should always be considered? In: Topical Scientific Workshop—New Approach Methodologies in Regulatory Science, Helsinki (Finland). Available from: http://hdl.handle.net/10400.21/6263 e http://echa.europa.eu/documents/10162/22322016/13_viegas_en.pdf.

Larson RR, Khazaeli MB, Dillon HK. (2002). Monitoring method for surface contamination caused by selected antineoplastic agents. *American Journal of Health System Pharmacy*, 59(3), 270–277.

Raghavan R, Burchett M, Loffredo D, et al. (2000). Low-level (PPB) determination of cisplatin in cleaning validation (rinse water) samples. II. A high-performance liquid chromatographic method. *Drug Dev Ind Pharm* 26: 429–40.

Roland C, Ouellette-Frève JF, Plante C, Bussières JF. (2016). Surface Contamination in a Teaching Hospital: A 6 Year Perspective. *Pharm. Technol. Hosp. Pharm.* DOI 10.1515/pthp-2016–0016.

Roussel C, Connor TH. (2013). Chemotherapy and Pharmacy: A toxic mix?. *The Oncology Pharmacist*. 6(2):1, 32–33. Available at: http://issuu.com/theoncologynurse/docs/top_may2013_issue_web.

Schmaus G, Schierl R, Funck S. (2002). Monitoring surface contamination by antineoplastic drugs using gas chromatography-mass spectrometry and voltammetry. *Am J Health Syst Pharm*. 15;59(10):956–61.

Scientific Committee on Health and Environmental Risks, Scientific Committee on Emerging and Newly Identified Health Risks, Scientific Committee on Consumer Safety. Toxicity and assessment of chemical mixtures. Brussels: European Commission; 2012. ISBN 9789279307003. doi: 10.2772/21444

Sessink PJ, Bos RP. (1999). Drugs hazardous to healthcare workers. Evaluation of methods for monitoring occupational exposure to cytostatic drugs. *Drug Saf.* 20(4):347–59.

Sexton K, Hattis D. (2007). Assessing Cumulative Health Risks from Exposure to Environmental Mixtures—Three Fundamental Questions. *Environmental Health Perspectives* 115: 825–832.

Simon N, Vasseur M, Pinturaud M, Soichot M, Richeval C, Humbert L, et al. (2016). Effectiveness of a Closed-System Transfer Device in Reducing Surface Contamination in a New Antineoplastic Drug—Compounding Unit: A Prospective, Controlled, Parallel Study. *PLoS ONE 11(7): e0159052*. doi:10.1371/journal.pone.015905.

Sottani C, Rinaldi P, Leoni E, Poggi G, Teragni C, Delmonte A, Minoia C. (2008). Simultaneous determination of cyclophosphamide, ifosfamide, doxorubicin, epirubicin and daunorubicin in human urine using high-performance liquid chromatography/electrospray ionization tandem mass spectrometry: bioanalytical method validation. *Rapid Commun Mass Spectrom* 22:2645e59.

Touzin K, Bussières JF, Langlois E, Lefebvre M, Métra A, (2010). Pilot study comparing the efficacy of two cleaning techniques in reducing environmental contamination with cyclophosphamide. *Ann. Occup. Hyg.* 54 (3): 351–359.

Touzin, K., Bussières, J.F., Langlois, É., Lefebvre, M. (2009). Evaluation of surface contamination in a hospital hematology–oncology pharmacy. *J. Oncol. Pharm. Pract*.15, 53–61.

Viegas A, Pádua M, Veiga AC, Carolino E, Gomes M (2014). Antineoplastic drugs contamination of workplace surfaces in two Portuguese hospitals. *Environ Monit Assess* 186:7807–7818.

Vyas N, Turner A, Clark JM, Sewell GJ. (2014). Evaluation of a closed-system cytotoxic transfer device in a pharmaceutical isolator. *JNOncol Pharm Pract. Available at: http://opp.sagepub.com/content/early/2014/07/29/1078155214544993.*

Occupational Safety and Hygiene V – Arezes et al. (Eds)
© *2017 Taylor & Francis Group, London, ISBN 978-1-138-05761-6*

Management system maturity assessment based on the IMS-MM: Case study in two companies

J.P.T. Domingues, P. Sampaio & P.M. Arezes
Department of Production and Systems, University of Minho, Braga-Guimarães, Portugal

I. Inácio
LIPOR, Serviço Intermunicipalizado de Gestão de Resíduos do Grande Porto, Porto, Portugal

C. Reis
Yazaki-Saltano de Ovar Produtos Eléctricos, Ltd., Ovar, Aveiro, Portugal

ABSTRACT: This paper intends to report the assessment of integrated management systems maturity in two companies adopting the IMS-MM. The companies targeted in this study develop their businesses in different activity sectors and are certified, at least, according to the ISO 9001, ISO 14001 and OHSAS 18001 standards. Results show that the selected companies are at different stages concerning the maturity of their integrated management systems. This seems to relate with the evolution and the ultimate purpose of their organizational structures, to the stakeholders to whom the companies align their management subsystems and to the fact that the IMS-MM is a generic tool (current version) that does not take into account the specific guidelines of each activity sector.

1 INTRODUCTION

1.1 Integrated Management Systems (IMSs)

Standardized Management Systems (MSs) are adopted by companies in order to address more efficiently and systematically the requirements from the stakeholders (Almeida *et al.*, 2014; Domingues *et al.*, 2012). In the early 80 s of the last century, companies were more prone to address the issues raised by the customers and the shareholders. Due to an increasing awareness of the context where companies operate, the notion of stakeholder encompasses today not just the customers and the shareholders but also the employees, the suppliers and the involving society amidst others (Sampaio *et al.*, 2012). To incorporate the requirements from these stakeholders, companies often implement and certify their MSs: according to the ISO 9001 standard to address the requirements of customers, according to the ISO 14001 standard to address the environmental requirements of the society and according the OHSAS 18001 to attend the health and safety requirements from the employees. Currently several companies are managed through an Integrated Management System (IMS) that usually encompasses (but is not restricted to) the quality, environmental and occupational health and safety issues that should be taken into account. Although the existence of some guidelines and frameworks the integration of these MSs into a single system of systems is often developed within each company constrained solely by the resources available. So, it is difficult for the

responsible to assess and benchmark the resulting IMS against other IMSs from other companies. To fulfill this scientific gap, a maturity model was developed and proposed by the end of 2013- the IMS-MM (Domingues *et al.*, 2016). The current paper intends to report the adoption of this tool in two different companies each one with unique business and organizational peculiarities. This paper follows with an up to date revision of literature concerning the IMSs and the description of the IMS-MM. The following section describes the research methods adopted and in the "Results" section are compiled and listed the main achievements by the adoption of the tool. The last section summarizes the most relevant conclusions from the research conducted.

1.2 Cases studies

LIPOR activity is focused on the waste management (Landfill, sorting plant, energy recovery plant and composting plant) of several municipalities that comprise the greater Porto area (Northern Portugal), notably, Póvoa do Varzim, Vila do Conde, Matosinhos, Maia, Porto, Valongo, Gondomar and Espinho. Yazaki-Saltano produces electrical goods mainly for the automotive industry. The two companies adopted different strategies to proceed with the integration of their MSs into a single IMS. LIPOR has proceeded with the successive implementation of the quality MS (QMS—ISO 9001), the environmental MS (EMS—ISO 14001) and the occupational health and safety MS (OHSMS—OHSAS 18001). The ISO/TS 16949 was the primordial subsystem from where the organizational structure of Yazaki-Saltano evolved to an IMS encompassing additionally the EMS (ISO 14001) and the OHSMS (OHSAS 18001). In this latter case, the audit function acted as an integrator concept due to the need to optimize their number and frequency thus avoiding excessive disruptions to the normal production process. A third company from the healthcare sector was contacted but a deeper research on the evolution of the IMS was not possible due to the information access constrains raised by top management.

2 LITERATURE REVIEW

2.1 Integrated Management Systems (IMSs)

The scientific topic of IMSs has been increasingly addressed by scholars and researchers throughout the last years. If one considers some of the contributions published solely throughout the year of 2016 (and late 2015) it is possible to collect a snapshot of the issues commonly targeted in this domain. It is possible to stress that this scientific domain is mature enough to enable some papers focusing literature review such as those published by Domingues et al. (2015) and Nunhes et al. (2016a). In this latter paper the authors identify the main scientific gaps in the literature and point out the open research paths to where future research should be directed. In a following paper, these same authors listed the functions suitable to integration in an IMS (Nunhes et al., 2016b) and the conclusions converge with those of mainstream researchers that not all the MSs functions should be integrated. Meanwhile, Abad et al. (2016) listed the main difficulties arising during the implementation of IMSs and Bernardo et al. (2016) dissected the main peculiarities of IMSs in a non-leading country in certifications such as Greece. The work of Kafel & Casadesus (2016) and Domingues et al. (2016) should be stressed out. These latter authors proposed a maturity model to assess IMSs while the former described the changes during the time of the order and level of MSs standards implementation. The work authored by Tompa et al. (2016), highlighting the benefits of joint management practices on safety and operational outcomes, should be referenced similarly to the work authored by Hernández-Vivanco et al. (2016) that promotes the linkage between the concepts of innovation and integration of MSs. It should be noted that some papers, from other scientific domains, incorporate some references from the IMSs domain. Anholon et al. (2016), based in a case study at Embraer, and Bernal-Conesa et al. (2016) point out that the concept of corporate social responsibility relates with a proper

integration of MSs and Aquilani et al. (2016) stress that a successful sustainability may be attained (or encompasses) a successful IMS. The work authored by Lindo et al. (2016) and Cook et al. (2016) address the audit function as a research topic and the current trend towards integrated audits. Several papers from the EMS and OHS backgrounds, such as those authored by Jilcha & Kitaw (2016), Kontogiannis et al. (2016) and Mustapha et al. (2016), implicitly address the issue of MSs integration.

2.2 The IMS-MM

The IMS-MM front-office component is presented in Fig. 1. This maturity model has a three-dimensional nature considering the following axes: the Key Process Agents (KPAs), externalities and the quality management principles. Integration excellence may be achieved throughout an itinerary encompassing six (5+1 base level) maturity levels. A more thorough, detailed and complete description of this model is available at Domingues et al. (2016).

To assess an IMS each question related to each KPA should be evaluated by selecting one of the multiple option answers (five options Likert scale) or, for some cases and in alternative, by categorical type answers. In the first case (five options Likert scale) to score the Weighing (W) ascribed to the KPA the answer should be "Agree" or "Totally Agree". In the second case the answer should be "Yes". The final score is attained by multiplication of the weighing of the KPAs that comply with the above-mentioned. Moreover, the critical KPAs (KPA*) are *must be* characteristics, i.e., both the score and the critical KPAs should be accomplished.

Finally, the ascribed externality to the level should be assessed as "Agree" or "Totally Agree" based on a five option agreement Likert scale. Table 1 displays the requirements to be fulfilled in order to an IMS evolve to an upper maturity level.

Table 2 presents the KPAs to be assessed and those that are critical (KPA*).

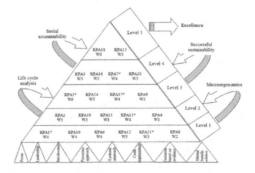

Figure 1. The front-office component of the IMS-MM.

Table 1. Assessment framework.

Level	Score	Requirements	Actions
5	...	KPA; KPA1518	Excellence
4	≥ 60	KPA7*; "Social responsibility" externality assessed, at least, as "Agree"	^ Level 5
3	≥ 72	KPA17*; "Successful sustainability" externality assessed, at least, as "Agree"	^ Level 4
2	≥ 60	KPA13*; "Life cycle analysis and management" externality assessed, at least, as "Agree"	^ Level 3
1	≥ 160	KPA21*; KPA1* "Macroergonomics" externality assessed, at least, as "Agree"	^ Level 2
Base		Assessment of all principles, at least, as "Agree"	^ Level 1

Table 2. KPAs and critical KPAs*.

KPA	Obs.
Policies integration.	KPA1*
Top management integrated vision.	KPA2
Implementation process supported on a guide or framework.	KPA3
Top management training concerning systems integration.	KPA4
Organizational tools, methods and objectives alignment.	KPA5*
Perception that the IMS originates organizational interactions.	KPA6
Non-residual authority by environmental and/or OHS managers.	KPA7*
At least, one integrating concept was considered during the integration process.	KPA8
System bureaucratization.	KPA9
Management procedures integration.	KPA10
Documental integration.	KPA11
Integrated objectives adoption.	KPA12
At the organizational structure there is an IMS manager.	KPA13*
Processes monitoring by KPIs, OPIs and MPIs.	KPA14
Integrated indicators adoption.	KPA15
Good correlation between the integrated organizational structure and the integrated level perception.	KPA16
Integrated audit typology.	KPA17*
Identification of organizational features not susceptible of integration.	KPA18
Integration strategy.	KPA19
MSs performance perceived better in an integrated context.	KPA20
The IMS perceived as an add value.	KPA21*

3 RESEARCH METHOD

The integration of MSs is being developed by an increasing number of companies in order to improve and optimize their organizational issues. A large stream of the available literature concerning this topic relies on quantitative methodologies, such as surveys, to identify and describe some of the issues that impact on the phenomenon. However, a deeper understanding of it asks for qualitative methodologies such as case studies in order to perceive the underlying relationships between these issues. Several research techniques (interviews, documental analysis and direct observation) were adopted to assure a proper data collection. The results reported were carefully scrutinized based on the convergence of the information collected from different sources. The assessment of maturity based on the IMS-MM took solely into account the axes "KPAs" and "Excellence Management Principles" (Level 0). Hence, the results presented later on do not consider the dimension "Externalities" (Macroergonomics; Life Cycle Analysis; Successful Sustainability; Social Accountability) due to the absence of suitable information.

4 RESULTS

The assessment of maturity was carried out according the assessment framework (Table 1) with the exception of the axis "Externalities" as described in the "Research Method" section. Concerning the base level (Level 0) it was possible to verify that both companies comply with the excellence management principles. In addition, both companies address the requirements of both EMS and OHSMS standards. The main purpose of the adoption of the IMS-MM was to determine at which extent integration was carried out.

4.1 LIPOR

Related to the Level 1 LIPOR rates with "Agree" or "Totally Agree" all the KPAs including the two critical KPAs: "Policies integration" and "The IMS perceived as an add value". LIPOR achieves a total score of 800 units which allow the access to Level 2 according to the assessment framework (160 units needed). Concerning Level 2, LIPOR achieves a total score of 60 units, *i.e.*, attains the minimum score to access to the following level. The critical KPA 13 is rated with the highest score but KPA 11 (Documental integration) is not rated at least as "Agree" (the minimum required). Although the access to level 3 is granted, a deeper documental integration seems to be necessary to prevent any future inconveniences related to this maturity level of the IMS. With respect to Level 3 all the KPAs (based on the agreement Likert scale) were rated at least with "Agree". Concerning critical KPA 17 it was possible to check that "Integrated Audits" was the audit typology commonly adopted by LIPOR. LIPOR attains an overall score of 15 units in Level 4 which does not allow the access to Level 5 (60 units needed). KPA 3 and KPA 7 are rated "Not Agree" and "Totally Disagree", *i.e.*, the IMS would be more mature if an implementation guideline or framework was adopted and if an effective authority was ascribed to the MSs responsible.

4.2   *Yazaki-Saltano*

Concerning Level 1, Yazaki-Saltano achieves a total score of 25 units which does not allow the access to Level 2 according to the assessment framework (160 units needed). Moreover, critical KPA 1 is not assessed with "Agree" or "Totally Agree" although the other critical KPA (KPA 21) is indeed evaluated with the highest score. In addition, KPA 8 (Integration concept) and KPA 10 (Integration of management procedures) are not evaluated with the minimum requirements. It is possible to define a path or, at least, a set of activities in order to the IMS of Yazaki-Saltano reach highest maturity levels. Yazaki-Saltano should promote an effective integration of policies and, at least, comply with one of the following issues: to define an integration concept, *i.e.*, a common concept that may be included throughout all MSs that encompass the IMS and/or proceed with a deeper integration of management procedures.

4.3   *Overall analysis of the results*

It is possible to posit that the differences observed between the two companies are related with the evolution, the ultimate purpose of their organizational structures and to the stakeholders to whom the companies align their MSs. Figs. 2 and 3 present

both the organizational structure of LIPOR and the organizational structure of the IMS.

Based on Figs. 3 and 4 it is possible to point out that LIPOR addresses simultaneously different stakeholders. From its conception, LIPOR took into account these stakeholders at the same level of the customers. In addition, the activity sector where LIPOR operates (with obvious linkages to the EMS) seems to facilitate the integration of the different MSs leading to higher maturity levels of the resulting IMS.

Fig. 4 presents the evolution of the multiple certifications achieved by LIPOR always supported and led by a strong commitment from top management.

Figure 5 presents the organizational structure of Yazaki-Saltano.

Figure 3.   Organizational structure of the MSs that encompass the IMS of LIPOR.

Operating Unit	Year								
	2002	2003	2004	2005	2006	2008	2009	2012	2014
Sorting Plant	◐◑				✳				
Energy Recovery Plant		◯	◑		✳				
Landfill of Maia				◯◑	✳				
Composting Plant							◯◑		
All the organization						●	✳	⊜	◯

◯	Certification of the QMS (ISO 9001)
◑	Certification of the EMS (ISO 14001)
✳	Certification of the OHSMS (OHSAS 18001)
●	Certification of the SRMS (SA 8000)
⊜	Certification of the IDIMS (NP 4457)

Figure 4.   Evolution of the IMS of LIPOR.

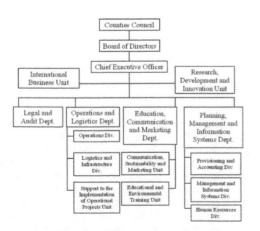

Figure 2.   Organizational structure of LIPOR.

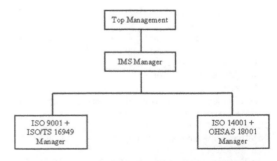

Figure 5.   Organizational structure of Yazaki-Saltano.

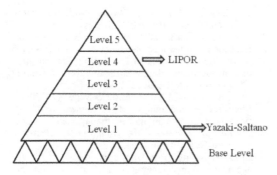

Figure 6. Position of LIPOR and Yazaki-Saltano according to the IMS-MM.

From its initial conception the main purpose of the MS of Yazaki-Saltano was to achieve the quality requirements from the customers and to address more prescriptive requirements related to environmental and OHS issues than those that comprise ISO 14001 and OHSAS 18001 standards. The audit function was paramount in the evolution of the IMS of Yazaki-Saltano since it was the need to optimize the number of audits (within the QMS) that "unveiled the path" to further optimization between the different MSs.

5 CONCLUSIONS

According to the IMS-MM, currently, LIPOR has a level 4 maturity and Yazaki-Saltano a level 1 (Fig. 6).

The evolution of the IMSs and its ultimate purpose considering the alignment to the requirements of the significant stakeholders seems to condition the maturity level that a company may attain. Yazaki-Saltano operates in an activity sector that is strongly impacted by the quality (customer) requirements. Yazaki-Saltano addresses environmental and OHS issues very specific from the automotive industry such as ROHS and ELV directives. This fact seems to be embedded in the culture of the company and efforts are done in order to assure that other requirements from less prescriptive documents (such as ISO 14001 and OHSAS 18001) be integrated through a systemic approach. On the other hand, the results achieved by LIPOR seem to be aligned with the notion that an IMS attains higher maturity when, from its original inception, all the MSs implemented are regarded as equal and supported by a top management with vision and effective leadership. Although both companies comply and proceeded with the integration of MSs adopting different approaches, the IMS-MM (due to its nature) seems to differentiate between those companies that operate in activity sectors with specific EMS and OHS requirements and those that

adopted the less prescriptive approach pointed out by ISO standards. This fact may be interpreted as a shortcoming of the IMS-MM since this model is based solely on the integration of the ISO 9001, ISO 14001 and OHSAS 18001 standards.

ACKNOWLEDGEMENTS

The authors would like to thank to LIPOR— *Serviço Intermunicipalizado de Gestão de Resíduos do Grande Porto* and Yazaki-Saltano *de Ovar Produtos Eléctricos, Ltd.* This study had the financial support of FCT (*Fundação para a Ciência e Tecnologia of Portugal*) under the project ID/ CEC/00319/2013. P. Domingues is supported by FCT Grant Ref. SFRH/BPD/103322/2014.

REFERENCES

Abad, J., Cabrera, H.R. & Medina, A. 2016. An analysis of the perceived difficulties arising during the process of integrating management systems. *Journal of Industrial Engineering & Management* 9(3): 860–878.

Almeida, J., Domingues, J.P.T. & Sampaio, P. 2014. Different perspectives on management systems integration. *Total Quality Management & Business Excellence* 25(3–4): 338–351.

Anholon, R., Quelhas, O.L.G., Filho, W.L., Pinto, J.S. & Feher, A. 2016. Assessing corporate social responsibility concepts used by a Brazilian manufacturer of airplanes: A case study at Embraer. *Journal of Cleaner Production* 135: 740–749.

Aquilani, B., Silvestri, C. & Ruggieri, A. 2016. Sustainability, TQM and value co-creation processes: The role of critical success factors. *Sustainability* 8(10).

Bernal-Conesa, J.A., Briones-Peñalver, A. & Nieves-Nieto, C. 2016. The integration of CSR management systems and their influence on the performance of technology companies. *European Journal of Management & Business Economics*, DOI: 10.1016/j.redeen.2016.07.002, in press.

Bernardo, M., Gotzamani, K., Vouzas, F. & Casadesus, M. 2016. A qualitative study on integrated management systems in a nonleading country in certifications. *Total Quality Management & Business Excellence*, DOI: 10.1080/14783363.2016.1212652, in press.

Cook, W., van Bommel, S. & Turnhout, E. 2016. Inside environmental auditing: effectiveness, objectivity, and transparency. *Current Opinion in Environmental Sustainability* 18:33–39.

Domingues, J.P.T., Sampaio, P. & Arezes, P.M. 2012. New organisational issues and macroergonomics: integrating management systems. *International Journal of Human Factors & Ergonomics* 1(4): 351–375.

Domingues, J.P.T., Sampaio, P. & Arezes, P.M. 2015. Analysis of integrated management systems from various perspectives. *Total Quality Management & Business Excellence* 26(11–12): 1311–1334.

Domingues, J.P.T., Sampaio, P. & Arezes, P.M. 2016. Integrated management systems assessment: a maturity model proposal. *Journal of Cleaner Production* 124: 164–174.

Hernandez-Vivanco, A., Bernardo, M. & Cruz-Cázares, C. 2016. Relating open innovation, innovation and management systems integration. *Industrial Management & Data Systems* 116(8): 1540–1556.

Jilcha, K. & Kitaw, D. 2016. A literature review on global occupational safety and health practice and accidents severity. *International Journal for Quality Research* 10(2): 279–310.

Kafel, P. & Casadesus, M. 2016. The order and level of management standards implementation: Changes during the time. *The TQM Journal* 28(4): 636–647.

Kontogiannis, T., Leva, M.C. & Balfe, N. 2016. Total Safety Management: Principles, processes and methods. *Safety Science*, DOI:10.1016/j.ssci.2016.09.015, in press.

Lindo, J., Stennett, R., Stephenson-Wilson, K., Barrett, K..A., Bunnaman, D., Anderson-Johnson, P., Waugh-Brown, V. & Wint, Y. 2016. An audit of nursing documentation at three public hospitals in Jamaica. *Journal of Nursing Scholarship* 48(5): 508–516.

Mustapha, M.A., Manan, Z.A. & Alwi, S.R.W. 2016. Sustainable Green Management System (SGMS) —An integrated approach towards organisational sustainability. *Journal of Cleaner Production*, DOI: 10.1016/j.jclepro.2016.06.033, in press.

Nunhes, T., Barbosa, L.C.F. & de Oliveira, O.J. 2016a. Evolution of integrated management systems research on the Journal of Cleaner Production: Identification of contributions and gaps in the literature. *Journal of Cleaner Production* 139: 1234–1244.

Nunhes, T., Barbosa, L.C.F. & de Oliveira, O.J. 2016b. Identification and analysis of the elements and functions integrable in integrated management systems. *Journal of Cleaner Production*, DOI: 10.1016/j. jclepro.2016.10.147, in press.

Sampaio, P., Saraiva, P. & Domingues, J.P.T. 2012. Management systems: integration or addition?. *International Journal of Quality & Reliability Management* 29(4): 402–424.

Tompa, E., Robson, L., Sarnocinska-Hart, A., Klassen, R., Shevchenko, A., Sharma, S., Hogg-Johnson, S., Amick, B.C., Johnston, D.A., Veltri, A. & Pagell, M. 2016. Managing safety and operations: The effect of joint management system practices on safety and operational outcomes. *Journal of Occupational & Environmental Medicine* 58(3): e80–e89.

Occupational Safety and Hygiene V – Arezes et al. (Eds)
© *2017 Taylor & Francis Group, London, ISBN 978-1-138-05761-6*

Occupational noise in urban buses—a short review

M. Cvetković & D. Cvetković
Faculty of Occupational Safety and Health of the University of Niš, Niš, Serbia

J. Santos Baptista
PROA/LABIOMEP, Faculty of Engineering, University of Porto, Porto, Portugal

ABSTRACT: The problem of noise in the public transportation bus and its effect to the driver must be considered systematically and examined in detail. The main aim of this short review is to approach the problem as detailed as possible, compare results and determine the effect(s) of the noise to the driver. For the purpose of this work the PRISMA Statement Methodology was applied. Were found 1119 articles and 14 included in this short review. Most of the results have been obtained using ANOVA Analysis and measurements that comply with standards (ISO 9612:2009, ISO 5128:1980, ISO 1999:1990). This paper illustrates the noise level variations depending on the city bus model, its age, engine position, and other factors that disturb the comfort of working space. Regarding the analysed results from the different papers, the noise depends mostly on the engine position in the bus.

1 INTRODUCTION

It is well known fact that drivers in the public transportation are not only affected by the traffic outside the vehicle, i.e. imperils caused by serious injuries, but also internally. The bus drivers, conductors and passengers are permanently threatened by noise, heat, exhaust gases of diesel engines, vibrations of the whole body, etc. (Mohammadi, 2014; Nassiri, Ebrahimi, Monazzam, Rahimi, & Shalkouhi, 2014). The noise producing does not depend only on the kind, type, and position of engine (Mohammadi, 2014; Mondal, Dey, & Datta, 2014; Mukherjee et al., 2003; Nadri, Monazzam, Khanjani, & Reza, 2012; Portela & Zannin, 2010; Zannin, Diniz, Giovanini, & Ferreira, 2003; Zannin, 2008), but also on other factors producing impulse noise or other forms of noise at different frequencies (Frost & Ison, 2007) such as vehicle' speed, stop, starting, breaking, type of tires, opening and closing of bus doors, sirens, the presence of passengers, and the like (Mukherjee et al., 2003).

All these factors influences discomfort of driver's working environment, in other words the noise increase while driving or when stop. Passing through various traffic zones daily, the city bus faces different traffic density and the quality of road where the noise produces aggravation of discomfort in the bus (Damijan, Skrzyniarz, & Kwasniewski, 2011; Mukherjee et al., 2003).

Various surveys illustrate that engines of certain models of city buses provide different levels of daily exposure to noise (Lex$_{8h}$) (Bruno, Marcos, Amanda, & Paulo, 2013; Damijan et al., 2011; Mohammadi, 2014; Portela & Zannin, 2010; Zannin, 2008). The scope of daily exposure to the noise (depending on the vehicle model) is 67.9 dB≤Lex$_{8h}$≤85.5 dB (A) (Bruno et al., 2013; Zannin, 2008). It is also already pointed that the noise level depends on the type of city bus route, so the equivalent continual noise level can go from 84dB to 91dB (Mukherjee et al., 2003).

There are no big changes in the maximum noise level during year seasons, though the minimum equivalent noise level varies during the winter/summer as Mukherjee (2003) showed. Further increase in noise in the driver's cabin is produced by the level of the vehicle speed, siren/external effect and the brakes (Mukherjee et al., 2003). However, an important role in noise increase plays the engine position (Portela & Zannin, 2010; Zannin et al., 2003; Zannin, 2008). Some authors (Portela & Zannin, 2010; Zannin et al., 2003; Zannin, 2008) stat that the level of daily exposure with front engine location is 78.6 dB (A) ≤Lex$_{8h}$≤ 85.5 dB (A), with middle engine location is 67.9 dB (a) ≤Lex$_{8h}$≤ 76.9 dB (A), and with rear engine position this factor is 74.7 dB (A) ≤Lex$_{8h}$≤ 76.9 dB (A). According to some researchers (Bruno et al., 2013; Mohammadi, 2014) and WHO (World Health Organization) it is necessary to keep the noise level within the allowed limits in order to avoid negative impacts on our health and to contribute to less discomfort of work environment.

The main objective of this paper is a short systematic review with respect to determine, first the sources of the noise of urban buses, and second the methodologies applied in measure, impact, prevention and control of the noise. Moreover, it was decided to examine the city bus not only from the point of view of internal noise, but from all aspects to achieve understanding of all possible noise sources. Creating discomfort of the work environment, and that is in our case bus driver's place, can cause various negative effects such as nausea, partial hearing loss, amongst others. There are extreme cases where, as a result of prolonged exposure to the noise, certain trauma and extra-auditory psychological changes occur (Bruno et al., 2013). Furthermore very important is preventive measures application, the noise level control regarding usage of the buses with rare engine position, regular maintenance of engine, optimal routes, and proper tires for the buses; all aiming to reduce the noise level, save the health of drivers and improve conditions of the working environment. So, the main aim of this short review is to compare results and determine the effect(s) of the noise to the driver.

2 MATERIALS AND METHODS

This study included papers between January 2007 and October 2016 when 1.119 works were discovered (1.107 of which were identified by database searching and 12 papers were found by other works' literature review); afterwards, from the article's references, some other older papers were also included. Of all that material 283 works were duplicated and excluded, while 805 were excluded as do not comply with the theme. Later 31 works have been selected as full text articles (which were oriented to noise sources of the city bus) of which 17 were excluded as not-focused to the noise sources but to the noise impact on the health of driver. The research includes 14 works, but only 12 of which are represented in the tables and were analysed.

This search included 27 Scientific Journal editors: ACM Digital Library, ACS Journals, AHA Journals, AIP Journals, AMA Journals, Annual Reviews, ASME Digital Library, BioMed Central Journals, Cambridge Journals Online, CE Database (ASCE), Directory of Open Access Journals, Emerald Fulltext, Geological Society of America, Highwire Press, IEEE Xplore, Informaworld, Ingenta, IOP Journals, MetaPress, Oxford Journals, Royal Society of Chemistry, SAGE Journals Online, SciELO, Science Magazine, ScienceDirect, Scitation, SFX A-Z, SIAM, Sociology: A SAGE Full-Text Collection, SpringerLink, The Chronicle of Higher Education, Wiley Online Library and 25 Index – Database: Academic Search Complete, Beilstein, Business Source Complete, CiteSeerX, Compendex, Criminal Justice Periodicals, Current Contents, Energy Citations, ERIC, Inspec, Library, MEDLINE, PubMed, ScienceDirect, SCOPUS, Social Sciences & Humanities Proceedings, SourceOECD, TRIS Online, Web of Science, Zentralblatt MATH.

The keywords combination used were: "occupational noise" + "urban bus"; "occupational noise" + "bus driver"; "occupational noise" + "urban bus" + "bus driver"; "noise exposure" + "urban bus"; "noise exposure" + "bus driver"; "noise exposure" + "urban bus" + "bus driver"; "noise impact" + "urban bus"; "noise impact" + "bus driver"; "noise measurements" + "urban bus"; "noise measurements" + "bus driver"; "public transport" + "noise measurements" + "urban bus"; "public Transport" + "noise measurements" + "bus driver"; "source of noise" + "urban bus"; "source of noise" + "noise impact" + "urban bus". This approach clearly shows that the systematic approach and detailed analysing applied in this paper led to best articles. Further review of literature included works that offered additional information on the noise state, as well as on other sources of noise in the city bus.

3 RESULTS

This work includes 14 papers from 5 different countries (Figure 1). When defining measuring strategies, i.e. work methodology, it is important to define standards and measuring devices for use.

In the Table 1 is illustrated which standards and devices were applied. The most of measurements were carried out at 0.10 ± 0.01 m from the driver's ear (Bruno et al., 2013; Damijan et al., 2011; Nadri et al., 2012; Portela & Zannin, 2010; Zannin, 2008).

Some papers have imparities (Mohammadi, 2014; Mukherjee et al., 2003; Nassiri et al., 2014) or measuring instruments were located outside the vehicle (to measure environmental noise of the city bus), (Frost & Ison, 2007) or data for measuring

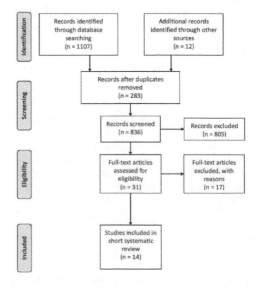

Figure 1. Prisma Flow Diagram.

Table 1. Standards, equipment and procedures.

Author	Country	Standard	Device	Position of device	Data analysis	Calibrator
Mohammadi G. (2014)	Iran	N.A.	BK 2238; B&K investigator 2260	20 cm from the external ear of the bus driver	ANOVA with Tukey test	N.A.
Nassiri P. (2014)	Iran	ISO 5128–1980	B&K 2250	0.7 ± 0.05 m height from the passengers' seats	N.A.	B&K calibrator 4231
Bruno P. (2013)	Brazil	ISO 9612:2009	B&K Mediator 2238	0.10 ± 0.01 m from the external ear of bus driver	Kruskal-Wallis; ANOVA with Tukey	N.A.
Nadri F. (2011)	Iran	ISO 5128:1980	Sound level meter CEL450	0.10 ± 0.01 m distance of the drivers right ear	ANOVA	model CEL450
Damijan Z. (2011)	Poland	ISO 9612; EN 60651: 2002 (U);	SVAN 948, I, SVANTEK, 6549.	0.10 ± 0.01 m of the driver's head	ANOVA test; Bonferroni test	N.A.
Portela, B. S (2010)	Brazil	ISO 1990: 1999	B&K 2238, class 1	0.10 ± 0.01 m away from driver's right hand	ANOVA	N.A.
Frost M. (2007)	UK	N.A.	Bruel & Kjaer integrating noise meter	1 to 1.5 m above ground level	ANOVA	N.A.
Zannin P. H. T. (2008)	Brazil	ISO 1990: 1999; BrS NHO-01, (2001); BrS NR-17 (1997).	B&K Mediator 2238; B&K Investigator 2260	0.10 ± 0.01 m of the driver's head	N.A.	B&K 4231 B&K, 1998a,b
Lacerda A. (2004)	Brazil	N.A.	N.A.	N.A.	N.A.	N.A.
Mukherjee A.K. (2003)	India	Bruel & Kjaer Co., 1986	N.A.	1 m approximately from the driver's ear	N.A.	N.A.
Zannin P. H. T. (2003)	Brazil	BrS NHO-01, 1999, BrS NR-17 (1997)	N.A.	N.A.	ANOVA; SPSS 16.0 software	N.A.
Patwardhan M.S. (1991)	India	N.A.	"IRD Mechanalysis, INC., U.K., Vibration/ Sound meter, model 308M, Approval 2G-2556" dB scale	N.A.	N.A.	1964 ISO Values, Model 10.

BrS – Brazilian Standard; B&K – Bruel & Kjaer.

instrument installation are not known (Lacerda, Ribas, Mendes, & Andrade, 2004; Patwardhan, Kolate, & More, 1991; Zannin et al., 2003).

Table 2 illustrates general characteristics of city buses and its average speed. (Lacerda et al., 2004) executed measuring with his team at three positions in the bus (when engine was in the front) and found that the noise level at the front and the middle bus sections is almost the same 80 dB ±1, while at the rear is drastically different from these two sections and is 67 dB ±2.

Table 3 shows the research results of each work. The ratio of gained results directly depends on data in the Tables 1 and 2, as well as in Table 3 results.

Some authors (Portela & Zannin, 2010; Zannin et al., 2003; Zannin, 2008) searched relations between the engine position, age and model with produced noise level, and the conclusion is that the level of daily exposure with the front engine location is of 76.6 dB (A) ≤Lex8h≤ 85.5 dB (A),

middle engine 67.9 dB (A) ≤Lex8h≤ 76.9 dB (A) and rear engine position 74.7 dB (A) ≤Lex8h≤ 76.9 dB (A). Some results should be noted (Frost & Ison, 2007; Mohammadi, 2014; Patwardhan et al., 1991) approximately engine performance in the full speed (max SPL 101 dB(A)), in low speed (max SPL 96 dB(A)), breaks (max SPL 95 dB(A)), neutral work (max SPL 92 dB(A)) and engine starting (max SPL 92 dB(A)) (Patwardhan et al., 1991). Moreover, in the same works can be found data telling that the minimum equivalent constant of noise exposure L_{eq} dB (A) is different in winter and summer (±2dB(A)) for 3 different bus models, while the maximum have approximately the same values. Patwardhan (1991) and his team illustrates possible conclusion that the noise level depends directly on the speed level of engine. He presents data in which is possible to see that the noise is the highest (100 dB ±2) in third speed level for all types of city buses.

Table 2. Bus data.

Author	N.°	Tip of bus	Manufacturing year	Engine position	Fuel	Speed average
Mohammadi G. (2014)	50	(1) Mega trance, (2) Benz 457, (3) Benz 457, (4) Benz 355, (5) Renault	(1) 2004; (2) 2002; (3) 2002; (4) 2003; (5) 2007.	(1) front, (2) rear, (3) rear, (4) front, (5) rear	N.A.	N.A.
Nassiri P. (2014)	30	Type 1, 2, 3	N.A.	(1) front; (2) rear; (3) rear	(1) Gas oil, (2) Gas, (3) Gas oil	(1) 22; (2) 34; (3) 17 km/h
Bruno P. (2013)	80	(1) Conventional, (2) light, (3) micro? bus, (4) articulated	N.A.	(1) front; (2) rear; (3) front; (4) front/rear	N.A.	N.A.
Nadri F. (2011)	80	(1) 0355 Benz; (2) 0457 Benz; (3) Renault; (4) Megatrans	N.A.	(1) rer; (2) rear; (3) rear; (4) rear.	Diesel	N.A.
Damijan Z. (2011)	1	Solaris Urbino 18	N.A.	N.A.	N.A.	<80 km/h
Portela, B. S (2010)	80	(1) Conventional, (2) Speedy, (3) Micro, (4) Articulated	(1) 1992–2001, (2) 1991–2006, (3) 1994–2003, (4) 1986–2006	(1) front, (2) rear, (3) middle, (4) rear	N.A.	N.A.
Frost M. (2007)	N.A	N.A.	N.A.	N.A.	Diesel	40 km/h.
Zannin P.H.T. (2008)	60	(1) Bi-articulated, (2) Speedy, (3) Feeder	(1) 1995–2000; (2) 1996–2002; (3) 1991–1999	(1) <middle; (2) <rear/middle; (3) <rear	N.A.	N.A.
Lacerda A. (2004)	N.A	N.A.	N.A.	front	N.A.	N.A.
Mukherjee A.K. (2003)	3	(1) S-18; (2) S-23; (3) S–22.	N.A.	N.A.	N.A.	30 mph
Zannin P.H.T. (2003)	25	(1) Feeder, (2) Rapid, (3) Bi-articulated	(1) 1995–2000; (2) 1997–2001; (3) 1991–1998	(1) middle, (2) (most) rear/middle, (3) (most) front/rear	N.A.	N.A.
Patwardhan M.S. (1991)	N.A	Type 1, 2, 3, 4, 5, 6, 7	N.A.	N.A.	N.A.	N.A.

Table 3. Measurement procedures.

Author	Time of expression	Measurement Period	Position	Noise level [dB] Max	Noise level [dB] Min	Frequency [Hz]	Passengers
Mohammadi G. (2014)	8 h	5 min	Inside	(1) 83, (2) 99.5, (3) 90, (4) 91, (5) 85.4	(1) 76, (2) 98.2, (3) 78.6, (4) 87.6, (5) 79.2	N.A.	With
Nassiri P. (2014)	30 min	5 min	Inside	(1) 88.1, (2) 83.5, (3) 81.7	(1) 79.4, (2) 74.8, (3) 73.9	N.A.	With
Bruno P. (2013)	8 h	N.A.	Inside	(1) 83.3, (2) 77.2, (3) 82.2, (4) 82.1	(1) 75, (2) 69.6, (3) 72.2, (4) 70.5	N.A.	N.A.
Nadri F. (2011)	8 h	10 min	Inside	(1) 79, (2) 76.7, (3) 76.9, (4) 73.6	(1) 69, (2) 69.2, (3) 65.9, (4) 70	1/3 octave	With
Damijan Z. (2011)	3 min (i.r.) + 3 h (d.r)	N.A.	Inside	(1) 89.5	(1) 60.9	1/3 octave	With
Portela, B.S (2010)	8 h	5 min	Inside	(1) 90.6, (2) 87.3, (3) 90, (4) 89.8	(1) 68.5, (2) 66.3, (3) 66.6, (4) 63.4	N.A.	Without
Frost M. (2007)	N.A.	N.A.	Outside	(1) 85,	(1) 79,	31.5, 63, 125, 250, 500, 1000, 2000, 4000 and 8000	N.A.
Zannin P.H.T. (2008)	8 h	N.A.	Inside	(1) 79, (2) 78.2, (3) 85.5	(1) 73.8, (2) 67.9, (3) 74.7	N.A.	N.A.
Lacerda A. (2004)	8 h	N.A.	Front, middle and back seats	(1) 82, (2) 80, (3) 81, (4) 80	(1) 58, (2) 60, (3) 60, (4) 61	N.A.	N.A.
Mukherjee A.K. (2003)	8 h	30 s	Inside	(1) 91.1, (2) 90.5, (3) 91	(1) 79.5, (2) 81.7, (3) 84.3	N.A.	N.A.
Zannin P.H.T. (2003)	8 h	N.A.	Inside	(1) 76, (2) 75.8, (3) 83.7	(1) 73.9, (2) 67.9, (3) 75.1	N.A.	With
Patwardhan M.S. (1991)	8 h	N.A.	Inside	(1) 102.5; (2) 102; (3) 100; (4) 102.5; (5) 101.5; (6) 100; (7) 100.5	(1) 91.5, (2) 90, (3) 90.5, (4) 91.5, (5) 89, (6) 90, (7) 91.5	31 (dominant)	N.A.

4 DISCUSSION

The problem of internal noise in the majority of analysed works is presented as the total equivalent noise level (Damijan et al., 2011; Lacerda et al., 2004; Mohammadi, 2014; Nadri et al., 2012; Portela & Zannin, 2010; Zannin et al., 2003; Zannin, 2008), where individual noise sources are not emphasized as contributing to the noise increase.

The noise is one of the greatest cities problems so, any contribution to reduce it is good. When waiting at the bus terminal the city bus produces SEL (Sound Equivalence Level) 72 dB (A), and when leave it in full throttle contributes $SEL_{(max)}$ 87 dB(A) (Frost & Ison, 2007). The most important influence to the noise level increase is the wheels interaction with tarmac and the engine position. It also directly depends on the bus route (Mukherjee et al., 2003).

The average age of the city buses, for each model in (Zannin, 2008) paper is about 6.5 years, and the equivalent noise level for 8 hours of work is 82 dB meaning that only maintenance of the buses is recommended due to the satisfactory noise level.

Damijan (2011) also illustrates 80 dB as acceptable noise level in 8 hours of work time and it does not exceed the driving level of 60.9 dB in driver's cabin. However, according to the Brazilian national standard (NR-17), the exposure to the noise level higher than 65 dB during 8 hours' work is considered as discomfort at working place (Zannin, 2008). Furthermore, according to the WHO permanent hearing loss may happen at $Lex_{8h} \leq 75$ dB (A) in 40 years period of time (Mohammadi, 2014; Tse, Flin, & Mearns, 2006). The study (Mohammadi, 2014) quotes that even 84% of bus drivers reported increased noise level and 74% of them a certain headache caused by the noise affecting them when drive. Consequently, this can lead to the attention disorder producing faults in driving process. Until present, only one research paper (Bruno et al., 2013) provides the ratio of passengers distance from a bus driver and the noise which passengers produce affecting directly noise level. Calculations are done on the basis of SIL method (Speech Interference Levels) relating to the already existing equations presented in the paper and it is 60 dB ±1 (the shortest distance is 0.36 m and the longest is 0.59 m) depending on the bus model.

5 CONCLUSIONS

The result of this short systematic review mirrors in attempt to find out, compare and further analyse sources. Huge contribution to ambient noise level of buses is in usual city noise, presence of passengers, (Mohammadi, 2014) wheels interaction with tarmac, buses' route, doors opening/closing, sirens, braking, movement in full throttle (Zannin, 2008), and many other factors. Only some papers use assessment of bus model, age and engine position with gained noise level data after measuring (Mohammadi, 2014; Mukherjee et al., 2003; Nadri et al., 2012; Portela & Zannin, 2010; Zannin et al., 2003; Zannin, 2008).

Due to making a full picture of the noise sources in city buses is necessary to add certain factors such as the passenger's number, the type of fuel, the average speed, as well as data of noise level measuring outside of the city buses aiming to grasp the problem in all its aspects. Different motor positions on the bus led to different noise levels: in front motors the noise level range is between 72.2 ± 2dB and 85 ± 2dB, in the middle, the noise ranges between 70 ± 3dB and 80 ± 3dB and motors in the rear produce noise levels between 70 ± 2dB and 86 ± 2dB. Most of the measures were taken 0.10 ± 0.01 m from the driver's ear.

ACKNOWLEDGMENTS

This publication has been funded with support from the European Commission under the Erasmus Mundus project Green-Tech-WB: Smart and Green technologies for innovative and sustainable societies in Western Balkans (551984-EM-1-2014-1-ES-ERA MUNDUS-EMA2).

REFERENCES

Bruno, P. S., Marcos, Q. R., Amanda, C., & Paulo, Z. H. T. (2013). Annoyance evaluation and the effect of noise on the health of bus drivers. *Noise & Health*, *15*(66), 301–6. doi 10.4103/1463–1741.116561

Damijan, Z., Skrzyniarz, S., & Kwasniewski, J. (2011). Investigation of the vibroacoustic climate inside the buses solaris urbino 18 used in public transport systems. *Acta Physica Polonica A*, *119*(6 A), 1068–1072.

Frost, M., & Ison, S. (2007). Comparison of noise impacts from urban transport. *Proc. Institution of Civil Engineers, Transport*, *160*(4), 165–172. doi 10.1680/tran.2007.160.4.165

Lacerda, A., Ribas, A., Mendes, J., & Andrade, P. (2004). Noise level and its perception in urban buses of Curitiba. *Canadian Acoustics/Acoustique Canadienne*, *32*(53), 4.

Mohammadi, G. (2014). Noise exposure inside of the Kerman urban buses : measurements, drivers and passengers attitudes. *Iranian J- of Health, Safety & Env., Vol.2, No.1, pp.224-228 Noise*, *2*(1), 224–228.

Mondal, N. K., Dey, M., & Datta, J. K. (2014). Vulnerability of bus and truck drivers affected from vehicle engine noise. *International Journal of Sustainable Built Environment*, *3*(2), 199–206. doi 10.1016/j.ijsbe.2014.10.001

Mukherjee, A. K., Bhattacharya, S. K., Ahmed, S., Roy, S. K., Roychowdhury, A., & Sen, S. (2003). Exposure of drivers and conductors to noise, heat, dust and volatile organic compounds in the state transport special buses of Kolkata city. *Transportation Research Part D: Transport and Environment*, *8*(1), 11–19. doi 10.1016/S1361–9209(02)00015–9

Nadri, F., Monazzam, M. R., Khanjani, N., & Reza, M. (2012). An Investigation on Occupational Noise Exposure in Kerman Metropolitan Bus Drivers. *Int. J. of Occupational Hygiene*, *4*(1), 1–5.

Nassiri, P., Ebrahimi, H., Monazzam, M. R., Rahimi, A., & Shalkouhi, P. J. (2014). Passenger Noise and Whole-Body Vibration Exposure-A Comparative Field Study of Commercial Buses. *Journal of Low Frequency Noise, Vibration and Active Control*, *33*(2), 207–220. doi 10.1260/0263–0923.33.2.207

Patwardhan, M. S., Kolate, M. M., & More, T. A. (1991). To assess effect of noise on hearing ability of bus drivers by audiometry. *Indian Journal of Physiology and Pharmacology*, *35*(1), 35–38.

Portela, B. S., & Zannin, P. H. T. (2010). Analysis of factors that influence noise levels inside urban buses. *Journal of Scientific and Industrial Research*, *69*(9), 684–687.

Tse, J. L. M., Flin, R., & Mearns, K. (2006). Bus driver well-being review: 50 years of research. *Transportation Research Part F: Traffic Psychology and Behaviour*, *9*(2), 89–114. doi 10.1016/j.trf.2005.10.002

Zannin, P. H. T. (2008). Occupational noise in urban buses. *Int. Journal of Industrial Ergonomics*, *38*(2), 232–237. doi 10.1016/j.ergon.2006.06.014

Zannin, P. H. T., Diniz, F. B., Giovanini, C., & Ferreira, J. A. C. (2003). Interior noise profiles of buses in Curitiba. *Transportation Research Part D: Transport and Environment*, *8*(3), 243–247. doi 10.1016/S1361–9209(03)00019–1

Occupational Safety and Hygiene V – Arezes et al. (Eds)
© *2017 Taylor & Francis Group, London, ISBN 978-1-138-05761-6*

Integrated management systems—short review

Catarina Correia, Maria João Mendes & J. Santos Baptista
LAETA, Faculty of Engineering, University of Porto (FEUP), Porto, Portugal

ABSTRACT: This study provides an overview on Integrated Management Systems (IMS). It aims to know how organizations see it, such as gains, profits, drawback and relevant points in implementation. In order to reach some results, it was used a research methodology PRISMA—Preferred Reporting Items for Systematic Reviews and Meta-Analyses. The results included basic information about every paper/article studied that was considered relevant and they are presented as table. The discussion exhibits the results of the review of the articles that were considered in the study. The conclusion is that IMS despite showing some difficulties at the beginning of the implementation, have several advantages to companies, so, from an evolutionary perspective, it could be the next step to take.

1 INTRODUCTION

Nowadays, one of the main contributes to competitiveness among companies is implementation of Integrated Management Systems (IMS), with posterior certification. Long-term success depends directly on improving the capacity of companies' operations through its self-reorganization, to respond to environmental, social and economic needs in succession (Vinodkumar & Bhasi, 2011).

The certification of one or more management systems evidences good management practices in the respective area, behaves as business card for potential customers and for the market in general (Footwear Technology Centre of Portugal—CTCP, n.d.).

The management system is introduced to satisfy the certification, which is dedicated to a specific area (environment, quality, health and safety, etc.), always with the goal of reaching the pre-set objectives and overcome obstacles (that may or may not appear). Any management system aims to improve its results, by applying the principle of continual improvement, based on the PDCA (Plan—Do – Check—Act) known as the Deming cycle.

Thus, continual improvement approach (PDCA cycle) cannot be based only on identified problems or casuistic situations, resulting opportunities or corrective actions for non-conformities identified. The principle of continual improvement comes from the established political commitment and, resulted from this commitment, the organization must define its priorities and objectives in line with the strategic direction (APCER, 2015).

Thereby, as this is the basic principle of different standards, it is also often that an organization applies more than one standard, which leads to the creation of an integrated system.

ISO 9001 (Quality), ISO 14001 (Environment) and OHSAS 18001 (Health and Safety) have all similar principles and management requirements. All three require the formulation of policies, allocation and assignment of responsibilities. (Zeng, Shi, & Lou, 2007).

When analyzing other management systems in terms of structure and content, it appears that ISO 9001:2007 and ISO 14001:2004 are somehow included on OHSAS 18001:2007 due to their related content. It ascertains that, when placed side by side, some standard requirements are established in all three:

- The involvement of management;
- The need to establish written requirements;
- The amount resulting from the audits;
- The importance given to training;
- The need for review of the system as a privileged moment for the analysis of their effectiveness (Gestão, Saúde, & Trabalho, n.d.)

The purpose of this study is to evaluate and identify the benefits, disadvantages, key points and crucial factors on implementation of IMS.

To do that, first is showed the basic information of the articles that were studied and then the result of the research.

2 METHOD

Towards to improve knowledge concerning the implementation of integrated management systems it was used a method called Preferred Reporting Items for Systematic Reviews and Meta-Analyses (Liberati et al., 2009). PRISMA is an evidence-based minimum set of items for reporting in systematic reviews and meta-analyses. This methodology system focuses on the reporting of reviews evaluating randomized trials, but can also be used as a base for reporting systematic reviews of other types of research, particularly evaluations of interventions (PRISMA).

This methodology allows to simplify literature review from distinctive scientific articles or other literary documents, focusing always on its objectivity and adequacy regarding the subject in study. PRISMA analysis method contain several phases: identification, screening, eligibility and, at last, included (PRISMA).

At first—IDENTIFICATION—there was an integrated search in which the keywords were used: *Integrated Management Systems, Safety, Occupational Safety and Health, OHSAS 18001*, this allowed the achievement of some results. Various scientific databases and articles scientific platforms (articles/journals) were used, such as *IOPJournals, Oxford Journals, SciELO, Science Magazine, Science Direct, Biomed, Compendex, Inspec* and *Web of Science*. In order to filter the search, there were only considered published articles since 2004. The selection criteria was based on the article title. Eighty was the number of articles collected, considered relevant to the study.

As the method allows, there were added, after the INCLUSION phase, 3 more articles because they were considered as relevant to the study. However, it is important to notice that they were added after. This articles appear to complete all the information already studied. To search for this articles the keywords were: *ISO 14001 + ISO 9001 + OHSAS 18001 (+ ISO 50001); ISO 45001 VS OHSAS 18001 and the Management System; Management System; Environmental, Quality, Safety and Energy Management.*

Thereafter, the second step—SCREENING— adopts an exclusion criterion items, allowing selection of those who would be most suitable to the topic in discussion. For that reason, the first exclusion criteria was the structure of the article itself, which serves as an indicator of the article credibility. During this assessment, it was possible to detect the existence of duplicates, which have been excluded, since the research was compiled in various databases using repeatedly the same keywords. By the end this stage, the number of articles remaining was 41. For phase ELIGIBILITY as inclusion criteria, were used the analysis of the abstract, objectives and method; these topics were assessed to determine whether they were in accordance with what was intended to apply, as well as to include necessary information either quantitatively or qualitatively. At the end of this stage, 30 scientific articles were accounted.

The data was organized in a table for a quick and easy access to information.

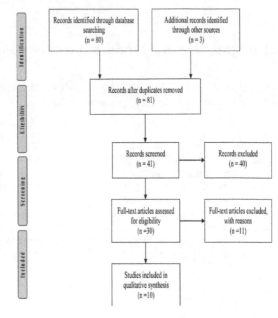

Figure 1. Schematization of the Prisma Methodology Steps with the appropriate number of articles analyzed.

All items were selected and stored in the *Mendeley Software*, which allows them to be always available for consultation. Of the 30 articles, after glimpsing the entire article, emphasizing the method and evaluating its usefulness concerning the objective of the current management integrated system, only 10 articles were left. All this process is illustrated in Fig. 1. The basic information of these articles, that were analyzed in detail for the study, as shown on Table 1.

3 DISCUSSION

It is important to note that, despite the selection of articles, not all of them contain the amount of information considered relevant to analysis.

After the selection of articles, the important is to analyze in detail, in order to concentrate the maximum information of each article.

After analyzing the selected articles, indicated in the Table 1, it was draw to and end the pros and cons regarding the Implementation of Integrated Systems.

For the implementation of integrated systems to be carried out, it is necessary to understand that a prerequisite for integration is an understanding of generic processes and tasks in the management cycle PDCA, and the potential benefits of such an integration related to internal coordination. Even more ambitious level of integration is concerned

Table 1. Presentation of the articles, taking account of their information, after the systematic review methodology PRISMA.

Author	Title	Activity/Standards/Company/Country	Type of study
Badri et al., 2012	Occupational health and safety risks: Towards the integration into project management	1) Industrial 2) --------- 3) --------- 4) Canada	Integration analysis OHS regulations in industrial enterprises
Chen et al., 2009	A comparative analysis of the factors affecting the implementation of occupational health and safety management systems in the printed circuit board industry in Taiwan	1) Industrial (chemistry) 2) OHSAS 18001 3) --------- 4) Thailand	Survey (using a scale of 5 degrees of satisfaction)
Khodabocus et al., 2010	Implementing OHSAS 18001:2007: A Case Study of Hazard Analysis from the Printing Industry	1) Industrial (printing) 2) ISO 9000:2008; OHSAS 18001:2007 3) ---; 4) Switzerland	Survey, study case (documental review)
Santos et al., 2011	Certification and integration of management systems: The experience of Portuguese small and medium enterprises	1) Services, industrial, electronics/telecommunications, construction 2) ISO 9001; ISO 14001; OHSAS 18001 3) Some companies—small and medium enterprises 4) Portugal	Surveys
Ulloa, 2012	Riesgos del Trabajo en el Sistema de Gestión de Calidad	1) Hazard identification, risk assessment, determination of necessary controls 2) ISO 9001; OHSAS 18001 3) ---; 4) Equator	------------------
Vinodkumar et al., 2011	A study on the impact of management system certification on safety management	1) Industrial (chemistry) 2) ISO 9001; OHSAS 18001 3) 8 Companies 4) India	Survey (empirical study)
Zink, 2005	From industrial safety to corporate health management.	1) ---; 2) ---; 3) ---; 4) Germany	Empirical study
Holdsworth, 2008	Practical Applications Approach to Design, Development and Implementation of an Integrated Management System	2) Petrochemical and chemical manufacturers industries 3) ISO 9001; ISO 14001; OHSAS 18001 4) One company 5) USA	Discusses a practical applications approach
Jørgensen et al., 2006	Integrated Management Systems—Three Different Levels of Integration	1) ---; 2) ISO 9001; ISO 14001; OHSAS 18001 and SA 8000 3) ---; 4) Denmark and Spain	Discuss three ambition levels of integration: from increased compatibility of system elements over coordination of generic processes to embeddedness of integration management systems
Maier et al., 2015	Innovation as a Part of an Existing Integrated Management System	1) ---; 2) ISO 9001; ISO 14001; OHSAS 18001 3) ---; 4) Romania	The bases for developing a model of integration management systems that include innovation as a main part of it.
Ferreira et al., 2015	Critical Factors in the Implementation Process of Integrated Management Systems	1) ---; 2) ---; 3) 12 Companies 4) Brazil	Structured interviews with the managers in 576charge of the implementation of the system

249

with creating a culture of learning, involvement of all stakeholders and performance in order to contribute to a sustainable development. To realize this, management needs to be focused on the synergy between customer-based quality, product-oriented environment and corporate social responsibility (Tine et al., 2006).

The combination of two or more standards, having as an objective the creation/implementation of a single system—IMS—shows benefits such as:

− Resource optimization (financial, economic and social), cost reduction;
− The unification of internal audits, improvement of employee performance;
− Better definition of responsibilities and authority;
− Reduction of bureaucracies, facilitated analysis of legislation, risk reduction;
− Increased performance and efficiency;
− A higher level of organization, a better external image of the organization, easier communication and, finally;
− Simplification in management which, consequently, results in less confusion, redundancy, and avoids conflicts in documentation.

As negative aspects regarding the implementation of an integrated management system are point it out the following:

− The initial costs are higher, the initial organizational problems;
− The increase of nonconformities, the continuous updating of documentation, which has a negative impact on the management of the activity itself, and;
− The complexity of the organizational system.

Concerning the difficulties that may occur during the implementation process, the following stand out:

− Insufficient regulations for integration, high difficulty and cost of simultaneous implementation of systems when compared to the integration of a single management system, major changes in the management of the system;
− Considerable alterations in the system due to operational changes;
− High difficulty in associating behaviors with changing organizational methods and culture, and the time spent (high) for integration (Santos, Mendes, & Barbosa, 2011).

To achieve success in the implementation of an integrated management system there are key factors such as:

− Top Management Support;
− Participation of workers involved;

− Good learning;
− Good internal auditing system, selection of good organization;
− Good teamwork, that need to be present in all stages of implementation.

It is relevant to emphasize what defines a successful implementation process of integrated management systems: in technical terms, it's characterized as the one introduced to the company that can operate all its modules, generating and providing reliable and updated information to the company managers, also easy to be accessed by the company's authorized users at management and operational levels (Kuniyoshi, 2015).

As mentioned in this analysis, certification is a tool that adds credibility, because it intersects the product or service that companies provide with the expectation of the client. Besides, in some industries certification is a legal or contractual requirement.

In a later phase, it's necessary to keep the quality of the integrated management system that was implemented, to do that, the easiest answer is through innovation (Maier et al., 2015).

4 CONCLUSIONS

The **PRISMA** method was a good approach to achieve the objectives of the work.

The Integrated Management System is a merge of various ISO standards and/or OHSAS, to implement them together in one only system.

Even though, at first, companies experience some difficulties when implementing, the integrated management systems are the natural next step to follow, by the several advantages concerning operational, economic and/or social level. Besides, in some companies there are legal or contractual requirement.

REFERENCES

APCER. (2015). Guia do Utilizador—ISO 14001:2015. http://doi.org/10.1007/s13398–014–01737.2.
Badri, A., Gbodossou, A., & Nadeau, S. (2012). Occupational health and safety risks: Towards the integration into project management. *Safety Science*, 50(2), 190–198. http://doi.org/10.1016/j.ssci.2011.08.008.
Chen, C.-Y., Wu, G.-S., Chuang, K.-J., & Ma, C.-M. (2009). A comparative analysis of the factors affecting the implementation of occupational health and safety management systems in the printed circuit board industry in Taiwan. *JNadeau*, Sylvie, 22(2), 210–215. DOI: 10.1016/j.jlp.2009.01.004.
Ferreira, A. A., & Kuniyoshi, M. S. (2015). Critical factors in the implementation process of integrated management systems. *Journal of Information Systems and Technology Management*, 12(1), 145–164. https://doi.org/10.4301/S1807–17752015000100008

Footwear Technology Centre of Portugal—CTCP (n.d.). Gestão da Qualidade, Ambiente, Segurança e Saúde no Trabalho. Guia do Empresário, #15 CTCP.

Gestão, S., Saúde, S. E., & Trabalho, N. O. (n.d.). Sistema de Gestão de Segurança e Saúde no Trabalho, 1–28.

Jørgensen, T. H., Remmen, A., & Mellado, M. D. (2006). Integrated management systems e three different levels of integration, *14*, 713–722. https://doi.org/10.1016/j.jclepro.2005.04.005.

Holdsworth, R. (2008). Practical applications approach to design, development and implementation of an integrated management system, *104*(2003), 193–205. https://doi.org/10.1016/j.jhazmat.2003.08.001.

Karapetrovic, S., & Casadesús, M. (2009). Implementing environmental with other standardized management systems: Scope, sequence, time and integration. *J. of Cleaner Production*, 17(5), 533–540. doi 10.1016/j.jclepro.2008. 09.006.

Khodabocus, B. F., & Constant, K. C. (2010). Implementing OHSAS 18001:2007: A Case Study of Hazard Analysis from the Printing Industry. *I. J. of Eng. Research in Africa*, 1, 17–27. Doi 10.4028/www.scientific.net/JERA.1.17

Kuniyoshi, M. (2015) Critical Factors in the Implementation Process of Integrated Management Systems. JISTEM - *Journal of Information Systems and Technology Management* Vol. 12, No. 1, Jan/Apr., 2015 pp. 145–164.

Liberati, A., Altman, D.G., Tetzlaff, J., Mulrow, C., Gøtzsche, P.C., Ioannidis, J.P.A., Clarke, M., Devereaux, P.J., Kleijnen, J., Moher, D., 2009. The prisma statement for reporting systematic reviews and meta-analyses of studies that evaluate healthcare interventions: Explanation and elaboration. *BMJ* 339.

Maier, D., Mariana, A., Keppler, T., & Eidenmuller, T. (2015). Innovation as a part of an existing integrated management system. *Procedia Economics and Finance, 26*(15), 1060–1067. https://doi.org/10.1016/S2212-5671(15)00930-2.

Santos, G., Barros, S., Mendes, F., & Lopes, N. (2013). The main benefits associated with health and safety management systems certification in Portuguese small and medium enterprises post quality management system certification. *Safety Science*, 51(1), 29–36. doi 10.1016/j.ssci.2012.06.014.

Santos, G., Mendes, F., & Barbosa, J. (2011). Certification and integration of management systems: The experience of Portuguese small and medium enterprises. *J. Cleaner Production*, 19(17–18), 1965–1974. doi 10.1016/j.jclepro.2011.06.017.

Ulloa-enríquez, M. Á. (2012). Riesgos del Trabajo en el Sistema de Gestión de Calidad. *Ingeniería Industrial, XXXIII* (2), 100–111. Retrieved from http://rii.cujae.edu.cu/ index.php/revistaind/article/view/443/448.

Vinodkumar, M. N., & Bhasi, M. (2011). A study on the impact of management system certification on safety management. *Safety Science*, 49(3), 498–507. Doi 10.1016/j.ssci.2010.11.009.

Zeng, S. X., Shi, J. J., & Lou, G. X. (2007). A synergetic model for implementing an integrated management system: an empirical study in China. J. Cleaner Production, 15(18), 1760–1767. Doi 10.1016/j.jclepro.2006.03.007.

Zink, KJ (2007). From industrial safety to corporate health management. ISSN: 0014–0139 (Print) 1366–5847 (Online) Journal homepage: http://www.tandfoline.com/loi/ter20.

Occupational Safety and Hygiene V – Arezes et al. (Eds)
© 2017 Taylor & Francis Group, London, ISBN 978-1-138-05761-6

Typification of the most common accidents at work and occupational diseases in tunnelling in Portugal

M.L. Tender
C-TAC-University of Minho, Guimarães, Portugal

J.P. Couto
Department of Civil Engineering, University of Minho, Guimarães, Portugal

ABSTRACT: This paper aims to fill a gap in scientific knowledge, through a statistical analysis of the most typical accidents at work and occupational diseases in tunnelling in Portugal, using Eurostat variables. It will compare the statistics in tunnelling with other types of works, considering the same indicators. First, it presents an overview of statistical data of accidents at work and occupational diseases in the Portuguese construction sector, and then move on to the tunnelling statistical data. To obtain the data, the official Portuguese Statistical Authorities' databases were searched. With the information from 84 accidents at work (year 2013) and 42 cases of occupational diseases (2000–2015) in tunnelling, the paper compares the main indicators in construction to those of tunnelling, establishing the differences that set tunnelling apart from other construction works.

1 INTRODUCTION

There are several studies analysing Accidents at Work (AW) and Occupational Diseases (OD) in the field of Civil Construction (CC) as a whole, but none specific to Tunnelling (T). The present study wants to fill this important gap. While measuring a negative indicator, knowing the conditions in which an AW occurred or a OD appeared has several advantages: 1) it gives us an important foundation to monitor and prioritize preventive measures (European Comission—Eurostat, 2000); 2) it decreases the probability of making wrong decisions (Araújo, 2011); 3) it helps companies to fulfil their legal requirements when it comes to risk assessment, and to make better decisions that minimize their costs (Hale et al., 2007); 4) it allows for learning to emerge from past mistakes (Reis, 2007). The two most important excavation methods (Tender & Couto, 2016b) (CEM-Conventional Excavation Method and TBM-Tunnelling Boring Method) will be analysed.

2 METHODOLOGY

The methodology to be used is intended to characterize as objectively as possible the circumstances in which an AW or OD occurred. In order to be able to compare between CC and T, a statistical mapping was done for each of them.

The methodology chosen for the analysis of AW was the European statistics on accidents at work (European Comission—Eurostat, 2001), promoted by Eurostat. The variables analysed will be: "occupation"; "age"; "time"; "specific physical activity"; "deviation"; "material agent-mode of injury"; "contact-mode of injury"; "type of injury"; "part of body injured"; "days lost (severity)". In order to adapt to the reality of T, some changes were made to the answer options to the variable "Occupation". The data considered were supplied by "*Gabinete de Estratégia e Planeamento*", because they are taken from the insurance companies' data and they meet the European guidelines, which makes them more reliable (Reis, 2007). The data for 2013 regarding CC were retrieved through a direct consultation of the GEP's website. The ones regarding T were obtained through a specific request the authors addressed to the GEP, and the statistical data concerning 84 AW were retrieved.

As for OD, the methodology was also the one Eurostat proposes, based on the European Occupational Diseases Statistics. For this study, the variables chosen were "Occupation", "Age" and "Diagnosis", because they are deemed to be the ones that enable a minimum characterization of the OD. The data analysed were supplied by the Social Security Institute at the author's request: CC – 1615 certified OD between 2000 and 2015; T – 42 certified OD between 2001 and 2015.

3 RESULTS AND DISCUSSION

3.1 Accidents at work

3.1.1 Occupation of the victim
CC—no available data;

T—"Operators/drivers" (19,8%), "Excavation face worker (13,5%).

The high percentage of "Operators/drivers" that suffer accidents can be explained by the specificity of T, which makes (in a narrow space) (Tender et al., 2015b) massive use of operators and drivers—for drilling, removal of debris or concreting, in the case of CEM, or for the equipment to support the works, in the case of TBM. As for the workers "Excavation face worker", the high number of accidents can be explained with the high degree of exposure to risks such as falling blocks or run-overs in those occupations.

3.1.2 Age of the victim
CC—"35 to 44 years old" (31,8%) and "45 to 54 years old" (28,0%), with an average age of 42,0.

T—"35 to 44 years old" (33,3%) and "45 to 54 years old" (26,2%), with an average age of 39,7.

The average age of the victims is lower in T. There are several possible explanations for this: they are less experienced (Reis, 2007) and therefore less able to identify risks (Ling et al., 2009); absent-mindedness or lack of family responsibilities (Jeong, 1998); carelessness (Chi et al., 2005) and lack of physical and mental maturity. The age group "55 to 64 years old" has lower accident percentages in T than in CC. That can be due to the fact that older groups are usually vastly experienced in this type of work, both in tunnels and in mines.

3.1.3 Time
CC—"2pm to 5pm" and "10am to 12pm" (Reis, 2007)

T—"10am to 12pm" (18,9%) and "2pm to 5pm" (16,2%). If we join the time periods "5pm to 8pm" and "8pm to 8am", then percentage obtained become relevant (28,8%).

In CC, the time period with a higher number of AW is mid-morning and mid-afternoon, while, in T, most AW happen in the period usually considered supplemental or nocturnal (from 5pm to 8pm and from 8pm to 8am). This can be due e.g. to tiredness or to the implications shift work has on the human body.

3.1.4 Specific physical activity
CC—"Working with hand-held tools" (31,7%), "Carrying by hand" (24,6%).

T—"Working with hand-held tools" (39,3%) and "Movement" (20,2%).

Both in CC and in T, the activity where most accidents occur is "Working with hand-held tools".

The higher percentage of AW in T with "Working with hand-held tools" can be explained by the use of tools for the installation of stabilization devices (in CEM) and prefabricated parts (in TBM), as well as for the equipment maintenance, which means that workers usually spend a lot of time working with hand-held tools. In CC, the second cause is "Carrying by hand", which does not have a high percentage in T. This can be explained by the almost non-existent manual transport of parts or materials in T (usually, they are carried my multipurpose loaders or other machinery). The second cause of AW in T is "Movement". The AW occurred in "Movement" can be attributed to run-overs by mobile equipment (vehicles) or falls from height or at the same level.

3.1.5 Deviation leading to the AW
CC—"Body movement under or with physical stress (generally leading to an internal injury)" (26,6%), "Loss of control (total or partial) of machine, means of transport or handling equipment, handheld tool, object, animal" (20,8%), "Slipping or stumbling with fall of person to a lower level"/"Slipping or stumbling on the same level" (20,7%)

T—"Slipping or stumbling with fall of person to a lower level or on the same level" (33,3%) and "Body movement under or with physical stress (generally leading to an internal injury)" (27,4%). When taken jointly, they account for a significant percentage of the whole (60,7%).

In this variable, there is a difference in the relevance of causes between CC and T.

The high percentage of "Slipping or stumbling with fall of person to a lower level or on the same level" in T can be explained by the irregularity and humidity of the ground, and by the amount of works to assemble structures (eg for final lining) or working machinery, such as the TBM. The high figure of "Body movement under or with physical stress (generally leading to an internal injury)" can be explained by the need to handle/come into contact with tools or objects, and the body is subjected to physical stress during those activities.

3.1.6 Material agent of contact—mode of injury
CC—"Objects, machine components, splinters, dust, incandescent particles" (38,0%)

T—"Objects, machine components, splinters, dust, incandescent particles" (14,3%)

In this variable, there is a similarity in the responses between CC and T.

"Objects, machine components, splinters, dust, incandescent particles" encompass a wide variety of elements, many of which very present in T, as is the case of rock blocks, sprayed concrete, formwork, among others. However, it is hard to make an individual assessment of each, since they are noted in a group.

3.1.7 *Contact—mode of injury*
CC—"Horizontal or vertical impact with or against a stationary object (the victim is in motion)" (30,9%), "Struck by object in motion" (21,4%)

T—"Horizontal or vertical impact with or against a stationary object (the victim is in motion)" (34,5%) e "Physical or mental stress" (27,4%).

"Horizontal or vertical impact with or against a stationary object (the victim is in motion)" is the biggest cause of contact-mode of injury, both in CC and in T.

This may be due to the high number of falls, meaning that the person collides against something stationary, such as the ground. This type of contact has a connection to the highest cause of deviation found, which is fall of person to a lower level or on the same level.

3.1.8 *Type of injury*
CC—"Wounds and superficial injuries" (50,6%), "Dislocations, sprains and strains" (22,5%), "Concussion and internal injuries" (9,3%)

T – "Wounds and superficial injuries" (59,5%) and "Dislocations, sprains and strains" (19,0%)

T values are similar to CC values.

The high value for "Wounds and superficial injuries" may be related to the amount of parts and objects handled manually this type of works requires. The "Dislocations, sprains and strains" can be related to fall of person on the same level, which has been identified as one of the main deviations, or with material handling.

3.1.9 *Part of body injured*
CC—"Upper Extremities" (31,7%) and "Lower Extremities" (23,9%)

T—"Lower Extremities" (33,3%) and "Upper Extremities" (31,0%).

In CC, the "Upper Extremities" are the part of body most injured, while in T it is "Lower Extremities" that are mostly hurt.

It is worthy of note the high percentage reached by both main parts of the body injured. The fact that the part of body most injured in T is "Lower Extremities" can be connected to objects falling or to strains, namely in CEM, since there are events of fall of blocks and the ground is usually not even. As for "Upper Extremities", these figures can be explained by the high amount of work with hand-held tools, which can boost this type of AW. Also, the high amount of loads handled, both in CEM (eg, when it comes to stabilizing devices), and in TBM (eg., handling parts of the TBM during its assembly and disassembly, or prefabricated segments during their positioning and assembly)

means there is a high tendency to have contact with hands, arms, legs and feet, whether due to falls or to other types of contact.

3.1.10 *Number of lost days*
CC—"7 to 13 days" (17,5%) and "30 to 90 days" (16,9%) with an average of 26,1 days.

T—"30 to 90" (38,5%) and "90 to 180 days" (19,2%), with an average of 68,4 days.

T presents an average of days lost higher than construction. This fact can be justified by a greater severity of "Wounds and superficial injuries" (haematoma, lacerations or open wounds) that results in a high number of days lost.

3.2 *Occupational diseases*

3.2.1 *Occupation of the victim*
CC—"No available data"

T—"Excavation face worker" (47,6%), followed by "Water-proofing operator" (21,4%) and "Formwork carpenter" (11,9%).

The two main occupations have a high combined percentage of 69.0% of the total. As for excavation face workers, they are often close to the excavation face, which is often rock with a high level of silica. They are also exposed to fumes from the use of explosives, particles of dust from sprayed concrete, oil mists (for example, for protecting the surface of concrete spray robots), and exhaust gases, which are all present in the confined space of the excavation face. As for the water-proofing workers, they are likely to be exposed to chemicals, namely in the form of vapours, e.g. from products to heat the water-proofing lining. On the case of "Formwork carpenters", there is a high exposition to cement and formwork release agents.

3.2.2 *Age of the patient*
CC—"50 to 59 years old" (43,9%)

T—"50 to 59 years old" (42,9%)

"50 to 59 years old" is the main age where OD appear, both in CC and in T. This can be explained by the increase of patients with chronic conditions which require long treatments and have longer recovery periods (Marica et al., 2015).

3.2.3 *Diagnosis*
CC—"Hearing disorders" (34,1%), "Musculoskeletal problems" (28,0%) and "Respiratory/pulmonary disorders" (25,9%)

T—"Respiratory/pulmonary disorders" (45,2%) and "Hearing disorders" (26,2%)

While in CC "Hearing disorders" and "Musculoskeletal problems" are at the top, in T the top place goes to "Respiratory/pulmonary disorders", followed by "Hearing disorders".

Both diagnosis, taken together, account for 71.4% of OD in T, a striking percentage. The prevalence of pulmonary diseases can be explained by the high exposure (Tender & Couto, 2016a) to agents harmful to the respiratory tract: dust from the rock mass (usually with a high content of silica, in which case it is generally associated with silicosis); dust from the sprayed concrete; aerosols from the oils used to protect the machinery against concrete splatters and build-up (Bakke et al., 2001); fumes from explosives and smoke from fires; combustion particles and gases from machinery (Tender et al., 2015a) (this last case is usually connected to cases of asthma and chronic bronchitis) (Oliver & Miracle-Mcmahill, 2006). For example, it has been established that dust from the cement used in the sprayed concrete can contribute to worsening asthma conditions, which translates into a reduction of the lung function of operators of concrete spray robots (Bakke et al., 2001).

4 CONCLUSIONS

From what has been said, the following conclusions can be drawn:

– The typical accident at work in tunnelling involves mobile equipment operators, with an average age of 39, 7 years, between 5pm and 8am, while working with hand-held tools, by falling to a lower level or on the same level against a hard surface, by contact with objects or machine parts, which cause wounds and superficial injuries in the legs or feet, leading to around 43 lost days.
– The typical occupational disease in tunnelling are respiratory/pulmonary disorders, mainly affecting excavation face workers, aged 50 to 59 years old.

ACKNOWLEDGEMENTS

The authors would like to thank:
– Infraestruturas de Portugal and Teixeira Duarte/ EPOS Joint venture, for their support in this study; the companies taking part in the R&D Project "SegOS-Segurança e Saúde em Obras Subterrâneas" associated to the Doctoral Program: MOTA-ENGIL, ORICA, SIKA, DST; Alexandra Valle Fernandes, for the translation.

REFERENCES

Araújo, J. 2011. *Analysis of fall from height accidents in the construction industry.* Msc Thesis in Human Engineering, University of Minho—School of Engineering.

Bakke, B., Stewart, P., Ulvestad, B., & Eduard, W. 2001. Dust and gas exposure in tunnel construcion work. *American Industrial Hygiene Association Journal,* 62(4): 457–465.

Chi, C.-F., Chang, T.-C., & Ting, H.-I. 2005. Accident patterns and prevention measures for fatal occupational falls in the construction industry. *Applied Ergonomics,* 36: 391–400.

European Comission—Eurostat. 2000. European Occupational Diseases Statistics—Phase 1 Methodology

European Comission—Eurostat. 2001. European statistics on accidents at work—Methodology. In Eurostat.

Hale, A., Ale, B., Goossens, L., Heijer, T., Bellamy, L., Mud, M., Oh, J. 2007. Modeling accidents for prioritizing prevention. *Reliability Engineering and System Safety,* 92: 1701–1715.

Jeong, B. 1998. Occupational deaths and injuries in the construction industry. *Applied Ergonomics,* 29(5): 355–360.

Ling, F., Liu, M., & Woo, Y. 2009. Construction fatalities in Singapore. *International Journal of Project Management,* 27: 717–726.

Marica, L., Irimie, S., & Baleanu, V. 2015. Aspects of occupational morbidity in the mining sector. *Procedia Economics and Finance,* 23: 146–151.

Oliver, L., & Miracle-McMahill, H. 2006. Airway disease in highway and tunnel construction workers exposed to silica. *American Journal of Industrial Medicine,* 49: 983–996.

Reis, C. 2007. *Improving the effectiveness of safety and health plans in reducing construction accidents.* PhD Thesis, University of Porto—Faculty of Engineering.

Tender, M., & Couto, J. 2016a. Analysis of health and safety risks in underground excavations—identification and evaluation by experts *International Journal of Control Theory and Applications,* 9(6): 2957–2964.

Tender, M., & Couto, J. 2016b. "Safety and Health" as a criterion in the choice of tunneling method. In Arezes et al (Ed.), *Occupational Safety and Hygiene IV*: 153–157. London: Taylor & Francis.

Tender, M., Couto, J., & Ferreira, T. 2015a. Prevention in underground construction with Sequential Excavation Method. In Arezes et al (Ed.), *Occupation Safety and Hygiene III*: 421–424. London: Taylor & Francis.

Tender, M., Couto, J., & Gomes, A. 2015b. Portuguese strengths and fragilities on Safety and Health practices. *Promoting Tunneling in SEE Region—International Tunneling Assocation World Tunneling Congress Dubrovnik. May 22–28, 2015.* Dubrovnik: Hubitg

Occupational Safety and Hygiene V – Arezes et al. (Eds)
© 2017 Taylor & Francis Group, London, ISBN 978-1-138-05761-6

Topics for the prevention of accidents in tunnels—the Marão Tunnel experience

M.L. Tender
C-TAC-University of Minho, Guimarães, Portugal

J.P. Couto
Department of Civil Engineering, University of Minho, Guimarães, Portugal

J. Baptista
Infraestruturas de Portugal, SA, Lisboa, Portugal

A. Garcia
EPOS—Empresa Portuguesa de Obras Subterrâneas, Lisboa, Portugal

ABSTRACT: This paper aims to share the main health and safety practical lessons retrieved from the underground excavation and final lining stages of the 2nd phase of the Marão Tunnel ("MT") construction. This is the longest tunnel built in the Iberian Peninsula in recent years and its excavation has been interrupted for three years, only with primary support applied in the excavated areas. First, the case study is presented. Afterwards, the lessons learnt when dealing with the usual main risks in underground works (run-overs, falling blocks, burying, untimely blasts, fall from heights and gases and dust inhalation, among others) are described. Adding to these were the specific conditions of this particular work, some of which stemmed directly from the suspension of the works. This paper pretends to be a legacy for the health and safety in the construction industry since it presents a successful example of risk management using new approaches.

1 INTRODUCTION

1.1 Introduction

Tunnelling has been growing more and more. Besides presenting the same risks as traditional civil construction, tunnelling presents a high a level of uncertainty in terms of geotechnical, geological and hydrological conditions, adding to work in confined spaces and work under pressure. All of these factors increase its risk level (Tender & Couto, 2016).

1.2 Overview of the work

The MT allows the underground crossing of the Marão Mountain between Amarante and Vila Real and it consists of two parallel galleries, 5667 m in length. The construction of the tunnel was awarded by the Project Owner, "IP—Infraestruturas de Portugal, S.A." to the joint venture "Teixeira Duarte—Engenharia e Construções, S.A." and "EPOS—Empresa Portuguesa de Obras Subterrâneas, S.A.", for a budget of €88 099 873,47. EPOS was responsible for the excavation and final lining and Teixeira Duarte was responsible for the installation of active and passive safety systems, and for the drainage, the paving and the technical buildings.

1.3 Construction process

The method used for excavation was the Conventional Excavation Method (CEM). This method involves the following stages (Tender et al., 2015a): Geological assessment of the excavation face; sequential excavation of the rock mass, by mechanical means or with explosives; application of stabilizing devices; ventilation of the gallery, to expel contaminated air from the tunnel after blasting; cleaning and removing the debris to landfill; scaling of the excavation site; and application of final lining.

2 METHODOLOGY

For this study, to assess the result of implementing preventive measures, the methodology chosen was expositive, based on information gathered in situ by three of the authors, while the works were under way (October 2014–May 2016). The preventive measures, chosen from known state of art, were defined in a framework of close cooperation between all of the stakeholders (namely, Site Management), considering the risk analysis carried out by the Contractor and validated by the Safety and Health Coordinator in the construction stage and approved by the Project Owner. The selection of this work was due to two main reasons: 1) this is the most relevant road tunnel in terms of size in the Iberian Peninsula; 2) this work became known for adopting exemplary practices in the field of health and safety at work.

3 RESULTS AND DISCUSSION

3.1 Training the workers is the first step for success

In the MT, two types of training were implemented: an 8 hour-long basic training (to overcome challenges due to hiring local man-power and the relative lack of time to recruit them); and the specialization training, which will be further detailed ahead. This further training was specifically given to mobile equipment operators. In order to do so, the training sessions were designed to be less theoretical and more focused on production activities. Also, there was a continuous close follow-up of the training activities, for which an in-house trainer was procured. One of the training options was the action "Hands-On in Work Context—Driver Operator", in a real work context, to improve the trainee's technical skills in operating, servicing, and safety of mobile equipment. The length of training depended on the difficulties identified by the trainer (initial learning needs assessment). Another training path was aimed at the technical improvement of the drivers and operators, giving them knowledge that will lead to a reduction in loss of productivity, namely at the level of prevention of accident risks and of 1st level maintenance (training length: 2 to 5 days). In some cases, the equipment suppliers' cooperation was sought. There was also a process of practical assessment/ certification in real work context. If the worker was considered below par, he would be prevented from operating the equipment pending specific training and reassessment. There were 87 workers assessed with highly satisfactory results, and only one worker had a negative assessment. The in-house trainer was fundamental for the success of these initiatives, for his knowledge of the equipment used, his experience and understanding of the company's culture, and his flexibility to adjust to the working schedules were crucial.

It is also worthy of mention the 14 training sessions delivered to 150 explosives operators, aiming at increasing their technical skills and preparing them to sit the exams to access the certificate of explosives operator by the PSP (Portuguese police force and the certification body in this case), with an approval rate of 93%. Also important were the training sessions with the Fire Department of Amarante, concerning the correct handling and use of hand-held fire extinguishers.

The number of training hours given to the workers allowed them to be more open to welcome safety rules and to have a proactive attitude when it comes to risk situations.

3.2 The workers' well-being is paramount

Prevention should start at the surface, by creating conditions for the well-being of the workers. The first step was the creation of an area to serve as dormitory, allowing around 100 workers to be lodged in individual bedrooms, dully equipped and air-conditioned.

To guarantee that the workers would have proper hygienic conditions, portable chemical toilets with lighting were installed in the work places, and they were moved as needed as the work progressed. Since there was no drinking water in the tunnel, a system to supply drinking water with plastic cups was installed, thus decreasing the amount of travel to the outside of the tunnel while guaranteeing the good quality of the water supplied.

3.3 Electricity is an added risk factor in the presence of water

Electricity is one of the great risks in tunnelling, namely when water is present, as was the case. All of the electric installations in the tunnel were periodically assessed by the technician in charge of electric infrastructures. Another measure to avoid contact with unprotected electrical elements was to use suspended bundled conductors along the sidewalls of the tunnels, sometimes reinforcing their support with removable clamps, and insulating the branches.

3.4 It's necessary to delimit areas to protect workers

The vicinities of the lateral undercuts dug to install the shoes for the final support of the tunnel are always areas of risk: these openings must follow the excavation as it advances, but there is a high risk of tipping for vehicles. Since there were almost always other activities being carried out in this area, it was necessary to guarantee suitable safety conditions for all the drivers/operators who were removing muck from the excavation and had to drive through this area. Thus, all of the excavation perimeter was delimited, signalled and illuminated, with signs at the ends to draw attention to the presence of risks and help improve the visibility conditions in the area.

3.5 Minimizing traffic and transportation risks to drivers/operators and pedestrians is key to succeed

Most accidents in tunnelling are connected to traffic and transportation (Vogel & Kunz-Vondracek, 2013) since the visibility in the tunnel is weak and there are also blind spots from within the vehicles (Tender & Couto, 2016). The excavation method chosen requires extensive use of heavy machinery for loading and removing debris to the landfill, and the risk of run-overs is significant. Installing camera systems is an alternative to rear view mirrors or signallers. The video camera system, in use in the MT for dumpers and loader shovels consists of a camera installed in the rear of the machinery and a screen near the driver's seat (Figure 1), which allows the driver to have better visibility, reducing the risk of run-overs.

Another preventive measure adopted was the physical separation of the pedestrian pathways and the vehicles pathways: workers were only allowed to walk along one of the sidewalls of the tunnel, the one with artificial light. Afterwards, this measure was changed, since there were no workers walking along the tunnels anymore and there was a high concentration of workers close to the working fronts. Therefore, the restriction on the pathway of the pedestrians was applied only on the affected areas, and there was the reinforcement of warning signs. Also, to decrease the risk of run-overs along the tunnels, it was forbidden to walk to the work places.

The combination of video systems and tutorial training resulted in a better approach to risk by the operators and in a decrease of the risk of run-overs. The physical separation between the areas has proven essential to avoid proximity between vehicles and workers.

3.6 The stability of excavated areas demands a daily joint effort of all the stakeholders

The excavation works (1st phase) started in July 2009 but were suspended, due to financial constraints, in July 2011, with 3961 m left to be

Figure 1. Screen and camera systems (source: EPOS).

excavated. Since that date and until the resumption of the project, there were only activities of maintenance of the part of the work already carried out and monitoring of the instruments installed. In 2014, for the preparation of tenders for the second phase of construction, since this time gap could have led to weaknesses in the primary support, thus increasing the risk of blocks falling over workers and equipment, and since the structural safety issues interfered directly and permanently with work safety issues, the Safety and Health Design Coordinator asked for the tunnels to be inspected to assess their overall condition. This inspection showed that the rock mass was globally stable and no alarming convergence situations were identified. Most of the pathologies identified were related to slabbing and cracking of shotcrete and, at places, falling of small blocks. The second stage of excavation works started in October 2014. In order to minimize the risk of further damage to the primary support, the works started by reinforcing the areas which has been excavated in the first phase. Following the indications of the Safety and Health Design Coordination, the following was done:

- Regular inspection of the condition of the sprayed concrete, conducted by the representative of the Designer, Geology, Site Management and Health and Safety Coordination in the construction stage;
- Mechanical and manual scaling of the excavation site in areas where the inspection revealed instability;
- Reinforcing the primary support in the areas already built as soon as possible;
- Definition/Signalling of "risk area", with circulation ban in critical areas;
- Observation of installed supports as part of the daily routine of those involved;
- Establishment of a strict and easy to interpret instrumentation plan;
- Ban on pedestrian circulation in the galleries, minimizing the exposure to the risk of falling blocks;

3.7 The simultaneous work on excavation and primary support/final lining demands the use of physical barriers

The set completion time, which was tight, implied that the excavation took place at the same time as the final lining. This option created air circulation and air quality constraints: as air was inflated in the excavation face, the entire length of the galleries (including the place where final lining operations were being carried out and the portal) would be continuously crossed by a cloud of air which could contain mineral dust (originating from the excavation face and with respirable crystalline silica, which can be fatal (Chapman et al., 2010)) fumes created after blasting operations. Also, the air pollution increases the lack of visibility, with all the risks associated with it (Velasco et al., 2010).

The initial decision to separate spaces and minimize risks to the workers in charge of the final coating was to install curtains to contain the contaminated air cloud (and direct it, through crosscutting galleries, to the parallel gallery). However, this system failed to achieve the desired effect, for two reasons: first, the curtain was fragile and easily damaged by repeated opening and closing, and secondly, sometimes the curtain was opened before the cloud of fumes from the fire blast crossed that section. To solve this problem, a gate was installed, which had a warning indicating the need to keep it closed, shown in Figure 2. The resulting air quality readings after that were well below the Exposure Limit Values set.

The risk of inhaling hazardous fumes has thus been drastically reduced, allowing excavation and primary support activities to occur somewhat close to areas where final lining activities were taking place. The use of explosives with less toxic fumes may, in the future, help reduce this risk.

3.8 The risk of fresh concrete fragments falling down restricts access to the excavation place

Applying a layer of sprayed concrete as initial support is one of the characteristics of the CEM. After the spraying of concrete for primary lining, and before the shotcrete gained sufficient strength, there was a risk of fresh concrete fractions falling from the crown arch or the side walls. For this reason, it was important to obtain a suitable resistance in a timely manner, which is a key element affecting productivity and also safety. The creation of exclusion zones is currently the measure internationally accepted as most suitable. Exclusion zones are based on the average time needed for the concrete to set, obtained in tests. The exclusion zone corresponds to the area next to the spraying area (where

Figure 2. Delimitation of final lining area (Source: authors).

the concrete spray robot operator is), and all of the area up to the excavation face. This measure made it possible, even in cases with greater influx of water and where the sprayed concrete application was harder, to reduce the workers' exposure to the risk of fresh concrete fragments falling down.

3.9 The movement of temporary structures can affect their integrity

In order to carry out the waterproofing and assembly of reinforcement elements, mobile platforms were used, with wheelsets to move along the tunnels, and their configuration followed the shape of the galleries. This solution resulted in obvious gains in productivity and contributed to ensure the necessary safety conditions that otherwise would be difficult to achieve due to the height of the galleries, the difficult access angles and the simultaneity of tasks with circulation of support vehicles in the various places where the activity was being carried out. To verify the conditions of such structures, a minimum quarterly periodicity was defined for their documented inspection, in accordance with the requirements established in the Decree-Law 50/2005. Since all the elements of these structures had to be secured and tightened, in order to ensure their stability after each movement, it was necessary to establish the documented inspection of the structures, by the person in charge of their use, as a means of responding to eventual reconversions of the structure, deformations and twisting of the elements resulting, for example, from small clashes with the excavated terrain.

3.10 The risk of falling to a lower level is paramount in the execution of the definitive lining

The risk of persons falling to a lower level was assessed, especially when assembling, using and disassembling temporary structures for the application of waterproofing, reinforcement and concreting of final support. To minimize risks, preventive measures were taken for assembly, use and disassembly: use of lifting platforms to support assembly/disassembly; use of guardrails and/or lifelines, in places where the span between the sidewalls of the gallery and the platform would allow a worker to fall down; installation of interior stairs and footbridges to access structures; access platforms and work platforms made of non-slip material.

3.11 External relief entities are an active part of the rescue process

The emergency planning began with a series of meetings with the National Civil Protection

Figure 3. Place reserved for the heliport (source: authors).

Authority, studying the arrival times of emergency services at the site. These meetings led to the conclusion that one of the ways to evacuate the injured would be the helicopter (Tender et al., 2015b), a usual option in construction sites difficult to reach by road (Arup, 2012). A space for a heliport was reserved for this purpose (Figure 3). This sharing of information also led to the creation of a space to install a field hospital, if needed.

As a way of testing the implementation of the Internal Emergency Plan, an accident simulacrum was performed, simulating an untimely blast near the excavation front. The simulacrum counted with the participation of the Firefighting Corporation of Vila Real.

It was found that the various stakeholders were able to coordinate, with a strong involvement and complementarity between the various responsible persons, with a well internalized action process, which they have dully followed.

4 CONCLUSIONS

In the present case study, the preventive measures mentioned in the scientific literature were applied, and authors confirmed their suitability and applicability to the field. However, new approaches to risk have also been tried out, leading to preventive measures with a positive impact in terms of production and safety. There were no fatal accidents at work during the excavation and final lining stages of the tunnel. The last 216 days of works had no reportable accidents at work. These numbers shows that the joint work between all those involved was fruitful and that the solutions applied were fit for their purpose. The proven success of the preventive measures presented could also help to define a much needed legal framework in Portugal for this field.

ACKNOWLEDGEMENTS

The authors would like to thank: Infraestruturas de Portugal and Teixeira Duarte/EPOS Joint venture, for their support in this study; the companies taking part in the R&D Project "SegOS-Segurança e Saúde em Obras Subterrâneas" associated to the Doctoral Program: MOTA-ENGIL, ORICA, SIKA, DST; Alexandra Valle Fernandes, for the translation.

REFERENCES

ARUP. (2012). Southern Nevada Water Authority Contract 070F 01 C1 *Lake Mead intake shafts and tunnel Project.*

Chapman, D., Metje, N., & Stark, A. 2010. *Introduction to Tunnel Construction.* London: Spons Architecture Price Book.

Tender, M., & Couto, J. 2016. Analysis of health and safety risks in underground excavations—identification and evaluation by experts *International Journal of Control Theory and Applications,* 9(6): 2957–2964.

Tender, M., Couto, J., & Ferreira, T. 2015a. Prevention in underground construction with Sequential Excavation Method. In Arezes et al (Ed.), *Occupation Safety and Hygiene III:* 421–424. London: Taylor & Francis.

Tender, M., Couto, J., & Gomes, A. 2015b. Portuguese strengths and fragilities on Safety and Health practices. *Promoting Tunneling in SEE Region—International Tunneling Assocation World Tunneling Congress Dubrovnik. May 22–28, 2015.* Dubrovnik: Hubitg

Velasco, J., Herrero, T., & Prieto, J. 2010. Metodología de diseño, observación y cálculo de redes geodésicas exteriores para túneles de gran longitud. *Informes de la Construcción,* 66(533): 1–10

Vogel, M., & Kunz-Vondracek, I. 2013. Safety and health in long deep tunneling-lessons learned in Swiss transalpine tunnel projects. *World Tunnel Congress 2013, Genebra.*

Occupational Safety and Hygiene V – Arezes et al. (Eds)
© 2017 Taylor & Francis Group, London, ISBN 978-1-138-05761-6

Livestock-associated MRSA colonization of occupational exposed workers and households in Europe: A review

E. Ribeiro
Research Group Environment and Health, Lisbon School of Health Technology/Polytechnic Institute of Lisbon, Lisbon, Portugal
Research Center LEAF—Linking Landscape, Environment, Agriculture and Food—Instituto Superior de Agronomia, Lisbon University, Lisbon, Portugal

A.S. Zeferino
Hospital Curry Cabral—Centro Hospitalar de Lisboa Central, Lisboa, Portugal

ABSTRACT: Human Methicillin-Resistant *Staphylococcus Aureus* (MRSA) carriers have increased risk for subsequent clinical associated diseases and become a bacterial reservoir with associated high risk to transfer the infection to others, including household members. Animals such as pigs are important reservoir of livestock-associated clones, which can be transferred to humans. The colonization of animals with MRSA is a professional hazard for workers that spend several daily hours in direct contact with MRSA-positive animals and consequently have a high risk of nasal colonization. Although MRSA human infections are one of the leading causes of morbidity and mortality in industrialized countries, exposure assessment procedures in occupational environments are not adapted to animal production settings at least in European countries. This work raise awareness to the relevance of monitoring MRSA strains associated with animal carriers and the requirement to create occupational standards and take effective preventive measures to protect exposed workers from MRSA colonization/infection.

1 INTRODUCTION

Human Methicillin-Resistant *Staphylococcus Aureus* (MRSA) infections are well-known worldwide as a cause of numerous hospitalizations and deaths associated with extremely high mortality rates for invasive infections (Klevens et al. 2007). Relevantly, humans can also become bacterial reservoirs of MRSA in which colonization predisposes to staphylococcal acquisition in clinical settings. (Ghasemzadeh-Moghaddam et al. 2015). The World Health Organization (WHO) describes antimicrobial resistance to human pathogens as a global health challenge (WHO, 2015). For the past years, with the worldwide escalation in antibiotic resistant microorganisms (Morris & Masterton 2002) social and scientific concerns have emerged regarding the over prescription of human antibiotics and it extensive use in agriculture (Smith et al. 2002) and animal food industry. Since the 90 s the amount of large Animal-Feeding Operations (AFOs) in swine, poultry, and cattle has increased expressively (USEPA, 2001) and a large variety of feed additives and drugs are approved for use in food-animal agriculture (Bloom, 2004). Some of the most extensively administered drugs are antibiotics, utilized in the management of animal health,

and more recently to growth enhancement and feed efficiency in healthy livestock. This widespread use can result in an antibiotic selection pressure that is driving the emergence of resistant strains such as MRSA.

Although, it is largely assumed that MRSA strains originated in humans, the emergence of the first pig-associated strain (ST398), which in very few years spread worldwide into diverse livestock species, reflects the prompt emergence of new pathogens. Even though most of these MRSA strains, now referred to as Livestock Associated MRSA (LA-MRSA), were isolated form healthy animals (colonized) some strains were also isolated from pathological lesions in pigs (Pomba et al. 2009). In addition to the animal to animal spread of these strains one of the early features of LA-MRSA was its ability to transfer from pigs to humans, reviewed in Barton (2014). Remarkably LA-MRSA colonization can be transient, which is suggestive of repeated contamination (Bangerter et al. 2016), and is a potential professional hazard for individuals that work in the meat production chain since these workers must spend several hours per day in direct contact with MRSA-positive animals and thus are indisputably exposed to a high risk of nasal colonization (Moodley et al.

2008; Denis et al. 2009). Considering that human MRSA carriers become bacterial reservoirs with associated high risk for subsequent occurrence of clinical disease and to transfer the infection to others (Hatcher et al. 2016), or contaminate foods and food surfaces during handling (Jordan et al. 2011), the assessment and biomonitoring of occupational exposed individuals is fundamental.

In the European context, although the main focus of current research regarding MRSA carriers is still associated with nosocomial strains (Espadinha et al. 2013), colonization by LA-MRSA have also become extremely relevant particularly in the context of occupational exposure. Goerge et coworkers recently published a review focused on the types of occupation-related infections caused by LA-MRSA CC398 and potential preventive strategies, however although since that publication new studies have emerged, the authors clearly reported that the available data is insufficient to effectively evaluate the described occupational risk (Goerge et al. 2015).

The major goal of this review was to evaluate the current knowledge regarding the prevalence of occupational exposed carriers of LA-MRSA in a European context. Furthermore we also aim to evaluate the prevalence of LA- MRSA colonization associated with direct contact with livestock farmers. Our drive is to raise awareness to the necessity of creating occupational biomonitoring standards and take effective preventive measures to protect exposed workers.

2 MATERIALS AND METHODS

An exhaustive search was made for papers available in scientific databases reporting occupational exposure to LA-MRSA in European countries. Additionally, assessment of worldwide studies that evaluated the colonization of occupational exposed workers family members was also performed. The articles considered were obtained using different scientific databases such as PubMed, Google Scholar and Scielo using the keywords: occupational exposure, LA-MRSA, colonization and household members. The articles that, besides written in English, also presented findings regarding the prevalence, persistence and effects of livestock-associated methicillin-resistant *Staphylococcus aureus* were chosen for further analyses. At the end of the selection process, 26 articles were considered important to this work including a 2015 review.

3 RESULTS

In 23 from the 26 articles studied, prevalence of MRSA colonization among occupational livestock exposed individuals in Europe, namely slaughterhouse workers, farmers and veterinaries was analyzed and summarized in Table 1.

The prevalence of LA-MRSA occupational colonization is exceedingly higher in farm workers, reaching 85% of the analyzed individuals followed by attending veterinaries which reach 45% and finally slaughterhouse workers 8%. Presently, higher levels of LA-MRSA colonization were reported in Germany. However the limited data obtained from other countries may not reflect the reality of the communities, whereas the prevalence may be higher than those currently reported. Netherlands and Germany appear to endorse greater concerns regarding the problematics of LA-MRSA colonization, since most of the analyzed studies

Table 1. Prevalence of MRSA carriers amongst livestock occupational exposed individuals.

County	Exposure setting	MRSA Prevalence	References
Netherlands	Slaughterhouse workers	3.2%–5.6%	Gilbert et al. 2012; Van Cleef et al. 2010a; Mulders et al. 2010
Netherlands	Farmer	8.9%–38.2%	Van Cleef et al. 2010b; Van Den Broek et al. 2009; Van Cleef et al. 2014; Graveland et al. 2011; Graveland et al. 2010; Geenen et al. 2013.
Netherlands	Veterinarian	3.9%–43.8%	Wulf et al. 2006; Paterson et al. 2013;
Spain	Slaughterhouse workers	8%	Morcillo et al. 2011
Germany	Farmer	0%–85.8%	Cuny et al. 2009; Köck et al. 2012; Dahms et al. 2014;
Germany	Veterinarian	45%	Cuny et al. 2009
Belgium	Farmer	26.1%	Vandendriessche et al. 2013
Belgium	Veterinarian	5.5%	Garcia-Graells et al. 2012
Italy	Farmer	8.2%–35.4%	Caggiano et al. 2016; Antoci et al. 2013
UK	Veterinarian	2.6%	Paterson et al. 2013
Romania	Farmer	6.8%	Huang et al. 2014
Switzerland	Veterinarian	3.8%	Wettstein et al. 2014
Portugal	Farmer	40%	Pomba et al. 2009

Table 2. Prevalence of MRSA carriers in household members of livestock farmers.

County	MRSA Prevalence	References
Germany	4.3%	Cuny et al. 2009
Netherlands	16%	Graveland et al. 2011

were performed in those countries. Nevertheless, countries such as Italy, Romania and Switzerland have also recently started to address this concern.

In Portugal, only a small study published in 2009, reported LA-MRSA colonization prevalence in healthy pigs 16.7%, farm workers 40%, the attending veterinarian and isolated strains from lesions of exudative epidermitis (EE) in three pig farms (Pomba et al. 2009).

We have also identified 2 studies that assessed the prevalence of MRSA colonization in family members of farmers occupationally exposed to LA-MRSA. According to the studies performed by Cuny et al. (2009), Graveland et al. (2011) and Hatcher et al. (2016), the prevalence of MRSA carriers in household members of livestock farmers, varies from 4 to 16% (Table 2).

4 DISCUSSION

Here we performed an extensive review regarding the prevalence of LA-MRSA colonization in both occupational exposed individuals and their households in a European context. LA-MRSA prevalence has been assessed in occupationally exposed workers from pig and poultry slaughterhouses; pig, cattle, veal calves, poultry and turkeys farms as well as attending veterinarians. The higher LA-MRSA occupational colonization in farm workers and the lower in slaughterhouses Table 1, indicates that the direct contact with live animal carriers is probably the main route of exposure to these bacterial strains. This hypothesis is also supported by a study recently published by Leibler JH et al. (2016) that demonstrated that the overall prevalence of *S. aureus* nasal carriage in beefpacking workers was 27.0% with no livestock-associated MRSA and Grøntvedt et al. (2016) demonstrated that transmission between pig farms mainly occurs through animal trade and to a minor level via humans or livestock transportation.

Moreover, it is also imperative to acknowledge that nasal MRSA colonization rates in pig farmers is maintained at 59% after time periods with no occupational exposure (holidays) (Köck et al. 2012) indicating that persistent MRSA colonization is expected to be more probable in occupationally exposed individuals reviewed in Goerge et al. (2015). Additionally, reports also demonstrate that prevalence of human MRSA carriers is increased in intensive antibiotic-using piggeries whereas in antibiotic-free it

is not, which indicate that antibiotic use is a driver for worker colonization (Rinsky et al. 2013).

Besides the detection of resistant bacteria isolated in piggery environments and related areas linked with antibiotics use in pigs, the contamination associated affords a reservoir of resistance genes for animals and humans in contact with the contaminated environment, reviewed in Barton (2014). Relevantly, besides MRSA nasal carriage, Hatcher et al. (2016) also revealed that for worker-child household pairs, multidrug-resistant *S. aureus* is significantly increased (23%). Studies scrutinized in this work clearly validate that occupational exposure to LA-MRSA not only constitutes an effective professional hazard but also constitute a significant risk to individuals that came direct in contact with exposed workers, particularly children that have high colonization prevalence when the farmer was a carrier (Graveland et al. 2011).

5 CONCLUSIONS

There is an urgent need to monitor MRSA strains associated with animal carriers, occupational exposed individuals and potential sources of environmental contamination and valuable efforts must be made to determine and regulate the antibiotic selection pressure that is driving their emergence.

ACKNOWLEDGEMENTS

The authors acknowledge the institutional support given by Lisbon School of Health Technology.

REFERENCES

Antoci, E., Pinzone, M.R., Nunnari, G., Stefani, S., & Cacopardo, B. 2013. Characteristics of methicillin-aureus (MRSA) among subjects working on bovine dairy farms Prevalenza e caratteristiche molecolari di Staphylococcus. *Le Infezioni in Medicina*. 2: 125–129.

Bangerter, P.D., Sidler, X., Perreten, V., & Overesch, G. 2016. Longitudinal study on the colonisation and transmission of methicillin-resistant Staphylococcus aureus in pig farms. *Veterinary Microbiology*. 183: 125–134. https://doi.org/10.1016/j.vetmic.2015.12.007.

Barton, M.D. 2014. Impact of antibiotic use in the swine industry. *Current Opinion in Microbiology*. 19(1): 9–15. https://doi.org/10.1016/j.mib.2014.05.017.

Bloom, R.A. 2004. Use of veterinary pharmaceuticals in the United States. In *Pharmaceuticals in the Environment*. 149–154. Springer. https://doi.org/doi: 10.1007/978-3-662-09259-0_12.

Caggiano, G., Dambrosio, A., Ioanna, F., Balbino, S., Barbuti, G., De Giglio, O., et al. 2016. Prevalence and characterization of methicillin- resistant Staphylococcus aureus isolates in food industry workers. *Ann Ig*. 28: 8–14. https://doi.org/10.7416/ai.2016.2080.

Cuny, C., Nathaus, R., Layer, F., Strommenger, B., Altmann, D., & Witte, W. 2009. Nasal colonization of humans with methicillin-resistant Staphylococcus aureus (MRSA) CC398 with and without exposure to pigs. *PLoS ONE. 4*(8): 1–6. https://doi.org/10.1371/journal.pone.0006800.

Dahms, C., Hübner, N.O., Cuny, C., & Kramer, A. 2014. Occurrence of methicillin-resistant Staphylococcus aureus in farm workers and the livestock environment in Mecklenburg-Western Pomerania, Germany. *Acta Veterinaria Scandinavica. 56*: 53. https://doi.org/10.1186/s13028-014-0053-3.

Denis, O., Suetens, C., Hallin, M., Catry, B., Ramboer, I., Dispas, M., et al. 2009. Methicillin-resistant Staphylococcus aureus ST398 in swine farm personnel, Belgium. *Emerging Infectious Diseases. 15*(7):1098–1101. https://doi.org/10.3201/eid1507.080652.

Espadinha, D., Faria, N.A., Miragaia, M., Lito, L.M., Melo-Cristino, J., de Lencastre, H., & Network, M.S. 2013. Extensive Dissemination of Methicillin-Resistant Staphylococcus aureus (MRSA) between the Hospital and the Community in a Country with a High Prevalence of Nosocomial MRSA. *PLoS ONE. 8*(4): 1–8. https://doi.org/10.1371/journal.pone.0059960.

Garcia-Graells, C., Antoine, J., Larsen, J., Catry, B., Skov, R., & Dennis, O. 2012. Livestock veterinarians at high risk of acquiring methicillin-resistant Staphylococcus aureus ST398. *Epidemiol Infect. 140*(3): 383–9. https://doi.org/doi: 10.1017/S0950268811002263.

Geenen, P.L., Graat, E.A., Haenen, A., Hengeveld, P.D., Van Hoek, A.H., Huijsdens, X.W., et al. 2013. Prevalence of livestock-associated MRSA on Dutch broiler farms and in people living and/or working on these farms. *Epidemiol Infect. 141*(5): 1099–108. https://doi.org/doi: 10.1017/S0950268812001616.

Ghasemzadeh-Moghaddam, H., Neela, V., van Wamel, W., Hamat, R.A., nor Shamsudin, M., Suhaila Che Hussin, N., et al. 2015. Nasal carriers are more likely to acquire exogenous Staphylococcus aureus strains than non-carriers. *Clinical Microbiology and Infection. 21*(11): 998.e1–998.e7. https://doi.org/10.1016/j.cmi.2015.07.006.

Gilbert, M.J., Bos, M.E., Duim, B., Urlings, B.A., Heres, L., Wagenaar, J.A., et al. 2012. Livestock-associated MRSA ST398 carriage in pig slaughterhouse workers related to quantitative environmental exposure. *Occup Environ Med. 69*: 472–478. https://doi.org/doi:10.1136/oemed-2011-100069.

Goerge, T., Lorenz, M.B., van Alen, S., Hübner, N.O., Becker, K., & Köck, R. 2015. MRSA colonization and infection among persons with occupational livestock exposure in Europe: Prevalence, preventive options and evidence. *Veterinary Microbiology.* S0378–1135(15): 30066-3. https://doi.org/10.1016/j.vetmic.2015.10.027.

Graveland, H., Wagenaar, J.A., Bergs, K., Heesterbeek, H., & Heederik, D. 2011. Persistence of livestock associated MRSA CC398 in humans is dependent on intensity of animal contact. *PLoS ONE. 6*(2): 1–7. https://doi.org/10.1371/journal.pone.0016830.

Graveland, H., Wagenaar, J.A., Heesterbeek, H., Mevius, D., van Duijkeren, E., & Heederik, D. 2010. Methicillin resistant staphylococcus aureus ST398 in veal calf farming: Human MRSA carriage related with animal

antimicrobial usage and farm hygiene. *PLoS ONE 5*(6): 4–9. https://doi.org/10.1371/journal.pone.0010990.

Grøntvedt, C.A., Elstrøm, P., Stegger, M., Skov, R.L., Skytt Andersen, P., Larssen, K.W., et al. 2016. Methicillin-Resistant Staphylococcus aureus CC398 in Humans and Pigs in Norway: A "One Health" Perspective on Introduction and Transmission. *Clinical Infectious Diseases : An Official Publication of the Infectious Diseases Society of America. 63*: ciw552. https://doi.org/10.1093/cid/ciw552.

Hatcher, S.M., Rhodes, S.M., Stewart, J.R., Silbergeld, E., Pisanic, N., Larsen, J., et al. 2016. The Prevalence of Antibiotic-Resistant Staphylococcus aureus Nasal Carriage among Industrial Hog Operation Workers, Community Residents, and Children Living in Their Households: North Carolina, USA. *Environmental Health Perspectives.* https://doi.org/10.1289/EHP35.

Huang, E., Gurzau, A.E., Hanson, B.M., Kates, A.E., Smith, T.C., Pettigrew, M.M., et al. 2014. Detection of livestock-associated methicillin-resistant Staphylococcus aureus among swine workers in Romania. *Journal of Infection and Public Health. 7*(4): 323–332. https://doi.org/10.1016/j.jiph.2014.03.008.

Jordan, D., Simon, J., Fury, S., Moss, S., Giffard, P., Maiwald, M., et al. 2011. Carriage of methicillin-resistant Staphylococcus aureus by veterinarians in Australia. *Australian Veterinary Journal. 89*(5): 152–159. https://doi.org/10.1111/j.1751-0813.2011.00710.x

Klevens, R.M., Morrison, M.A., Nadle, J., Petit, S., Gershman, K., Ray, S., et al. 2007. Invasive methicillin-resistant Staphylococcus aureus infections in the United States. *JAMA. 298*(15): 1763–71. https://doi.org/DOI: 10.1001/jama.298.15.1763.

Köck, R., Loth, B., Köksal, M., Schulte-Wülwer, J., Harlizius, J., & Friedrich, A.W. 2012. Persistence of nasal colonization with livestock-associated methicillin-resistant staphylococcus aureus in pig farmers after holidays from pig exposure. *Applied and Environmental Microbiology. 78*(11): 4046–4047. https://doi.org/10.1128/AEM.00212-12.

Leibler, J.H., Jordan, J.A., Brownstein, K., Lander, L., Price, L.B., & Perry, M.J. 2016. Staphylococcus aureus nasal carriage among beefpacking workers in a Midwestern United States slaughterhouse. *PLoS ONE. 11*(2): 1–11. https://doi.org/10.1371/journal.pone.0148789.

Moodley, A., Nightingale, E.C., Stegger, M., Nielsen, S.S., Skov, R.L., & Guardabassi, L. 2008. High risk for nasal carriage of methicillin-resistant Staphylococcus aureus among Danish veterinary practitioners. *Scandinavian Journal of Work, Environment and Health. 34*(2): 151–157. https://doi.org/10.5271/sjweh.1219.

Morcillo, A., Castro, B., González, J., Rodriguez-Alvarez, C., Novo, M., Sierra, A., et al. 2011. High prevalence of methicillin-resistant Staphylococcus aureus in pigs and slaughterhouse workers (P900). *In: 21st European Congress of Clinical Microbiology and Infectious Diseases (ECCMID), Milano, 07.05.-10.*

Morris, A.K., & Masterton, R.G. 2002. Antibiotic resistance surveillance: action for international studies. *The Journal of Antimicrobial Chemotherapy. 49*(1): 7–10. https://doi.org/10.1093/JAC/49.1.7.

Mulders, M.N., Haenen, A.P., Geenen, P.L., Vesseur, P.C., Poldervaart, E.S., Bosch, T., et al. 2010. Prevalence of livestock-associated MRSA in broiler flocks and risk factors for slaughterhouse personnel in The Netherlands. *Epidemiol Infect.* *138*(5): 743–55. https://doi.org/doi:10.1017/S0950268810000075.

Paterson, G.K., Harrison, E.M., Craven, E.F., Petersen, A., Larsen, A.R., Ellington, M.J., et al. 2013. Incidence and Characterisation of Methicillin-Resistant Staphylococcus Aureus (MRSA) from Nasal Colonisation in Participants Attending a Cattle Veterinary Conference in the UK. *PLoS ONE.* *8*(7): e68463. https://doi.org/10.1371/journal.pone.0068463.

Pomba, C., Hasman, H., Cavaco, L.M., da Fonseca, J.D., & Aarestrup, F.M. 2009. First description of Meticillin-Resistant Staphylococcus Aureus (MRSA) CC30 and CC398 from swine in Portugal. *International Journal of Antimicrobial Agents.* *34*(2): 193–194. https://doi.org/10.1016/j.ijantimicag.2009.02.019.

Rinsky, J.L., Nadimpalli, M., Wing, S., Hall, D., Baron, D., Price, L.B., et al. 2013. Livestock-Associated Methicillin and Multidrug Resistant Staphylococcus aureus Is Present among Industrial, Not Antibiotic-Free Livestock Operation Workers in North Carolina. *PLoS ONE.* *8*(7): 1–11. https://doi.org/10.1371/journal.pone.0067641.

Smith, D.L., Harris, A.D., Johnson, J., Silbergeld, E.K., & Morris, J.G. 2002. Animal antibiotic use has an early but important impact on the emergence of antibiotic resistance in human commensal bacteria. *Proceedings of the National Academy of Sciences of the United States of America.* *99*: 6434–6439. https://doi.org/10.1073/pnas.082188899.

USEPA. 2001. Development document for the proposed revisions to the national pollutant discharge elimination sys-tem regulation and the effluent guidelines for concentrated animal feeding operations. Retrieved from https://www3.epa.gov/npdes/pubs/cafo_proposed_dev_doc_ch1–4.pdf.

Van Cleef, B.A., Broens, E.M., Voss, A., Huijsdens, X.W., Züchner, L., Van Benthem BH, et al. 2010. High prevalence of nasal MRSA carriage in slaughterhouse workers in contact with live pigs in The Netherlands. *Epidemiol Infect.* *138*(5): 756–63. https://doi.org/doi:10.1017/S0950268810000245.

van Cleef, B.A.G.L., van Benthem, B.H.B., Verkade, E.J.M., van Rijen, M., Kluytmans-van den Bergh, M.F.Q., Schouls, L.M., et al. 2014. Dynamics of methicillin-resistant Staphylococcus aureus and methicillin-susceptible Staphylococcus aureus carriage in pig farmers: A prospective cohort study. *Clinical Microbiology and Infection.* *20*(10): O764–O771. https://doi.org/10.1111/1469–0691.12582.

Van Cleef, B.A., Verkade, E.J.M., Wulf, M.W., Buiting, A.G., Voss, A., Huijsdens, X.W., et al. 2010. Prevalence of livestock-associated MRSA in communities with high pig-densities in the Netherlands. *PLoS ONE.* *5*(2): 8–12. https://doi.org/10.1371/journal.pone.0009385.

Van Den Broek, I.V., Van Cleef, B.A., Haenen, A., Broens, E.M., Van Der Wolf, P.J., Van Den Broek, M.J., et al. 2009. Methicillin-resistant Staphylococcus aureus in people living and working in pig farms. *Epidemiol Infect.* *137*(5): 700–8. https://doi.org/doi:10.1017/S0950268808001507.

Vandendriessche, S., Vanderhaeghen, W., Soares, F.V., Hallin, M., Catry, B., Hermans, K., et al. 2013. Prevalence, risk factors and genetic diversity of methicillin-resistant Staphylococcus aureus carried by humans and animals across livestock production sectors. *Journal of Antimicrobial Chemotherapy.* *68*(7): 1510–1516. https://doi.org/10.1093/jac/dkt047.

Wettstein, R.K., Rothenanger, E., Brodard, I., Collaud, A., Overesch, G., Bigler, B., et al. 2014. Nasal carriage of Methicillin-Resistant Staphylococcus Aureus (MRSA) among Swiss veterinary health care providers: Detection of livestock- and healthcare-associated clones. *Schweiz Arch Tierheilkd.* *156*(7): 317–25. https://doi.org/doi:10.1024/0036-7281/a000601.

World Health Organization (WHO). 2016. Antimicrobial resistance: fact sheet no 194. Genebra: WHO. Retrieved June 20, 2012, from http://www.who.int/mediacentre/factsheets/fs194/en/.

Wulf, M., Van Nes, A., Eikelenboom-Boskamp, A., De Vries, J., Melchers, W., Klaassen, C., & Voss, A. 2006. Methicillin-resistant Staphylococcus aureus in veterinary doctors and students, the Netherlands. *Emerging Infectious Diseases.* *12*(12): 1939–1941. https://doi.org/10.3201/eid1212.060355.

Occupational Safety and Hygiene V – Arezes et al. (Eds)
© 2017 Taylor & Francis Group, London, ISBN 978-1-138-05761-6

Using BIM for risk management on a construction site

J. Fernandes
University of Minho, Guimarães, Portugal

M.L. Tender
C-TAC-University of Minho, Guimarães, Portugal

J.P. Couto
Department of Civil Engineering, University of Minho, Guimarães, Portugal

ABSTRACT: One of the problems in construction sites is the risk of people being run-over or hit by vehicles in the access ways and roadways. Thus, it becomes important to study methods that counteract this trend. In order to minimize the said risk, BIM-based *(Building Information Modelling)* technologies can be used. A case study was carried out on the (already built) East side construction site of the 2nd phase of construction of the Marão Tunnel. The main goal was to model the site, to obtain a 3D model and thus better understand the space, in order to optimize its use and propose a solution to manage in more detail the risks associated with the movement of equipment and people. With this ability to manipulate the site, by observing in 3D the best way of working, it is possible to minimize potential risk situations that might otherwise arise.

1 INTRODUCTION

1.1 Introduction

This study aims to describe a means to minimize risks related to the construction site of the Marão tunnel (built using the Conventional Excavation Method—CEM) by using BIM technologies. The use of this methodology intends to minimize the risk of run-overs/hits associated with the usual tasks in the sites during the whole construction phase of the Work, thus contributing to increase the safety of all those involved.

Prevention-wise, most of the ongoing studies focus on the risk of people falling to a lower level, a major cause of fatal accidents in construction. Since the risk of people falling to a lower level is not one of the most common risks in tunnelling, it is important to assess the feasibility and usefulness of BIM *(Building Information Modelling)* for the production and safety planning in tunnelling.

1.2 Risks in tunnelling

The risks in tunnelling depend on the excavation method chosen: Conventional Excavation Method (CEM) or Tunnel Boring Machine Excavation Method (TBM) (Tender & Couto, 2016b). CEM is characterized by risks which are quite specific, associated to its non-automation (Tender et al., 2015a) and use of a large number of mobile equipment

(vehicles) and workers (Tender & Couto, 2016a). This means that there is high risk of people being run-over/hit by vehicles (Tender et al., 2015b).

1.3 BIM for health and safety planning

BIM is a strong methodology that took almost 20 years of development to reach its current state (Bargstädt, 2015). This tool is a methodology for sharing information and for communication between all the actors in a construction, and in all of its phases. It is based on a digital model of the work, including 3D views, accessible by software which allows the virtual manipulation of said work. It is thus possible to follow and develop the construction in a tri-dimensional way, which, in turn, enables a timely planning. The use of BIM tools has been driven by architects, engineers and consultants, to improve job performance. Below is a list of opinions (Sulankivi et al., 2010; Kumar & Cheng, 2015; Sulankivi et al., 2013; Azhar & Behringer, 2013) on the advantages of using BIM:

- It works as a common database platform for those involved in the Work, supporting information sharing and avoiding loss of information;
- It allows you to go beyond 2D visualization, enabling 3D visualization in a simple way;
- The time required to obtain detailed results is less than in the manual case, and results can be obtained in seconds;

- It is less prone to human errors in graphic modelling;
- It makes it possible to get a computerized working base, minimizing the need for manual changes, which are always harder to accomplish;
- It enables on-site use of models prepared beforehand, in the design phase;
- It minimises the need for large drawings;
- It allows the production of complex views/details;
- It allows compatibility between elements;
- It enables the user to overcome the limitations of his/her own mental capacity to predict/interpret scenarios;
- It is possible to clarify doubts on-site, without having to use a lot of paper documents;
- The sharing of information allows you to warn workers about the risks of the work in a simpler way.

Safety-wise, using BIM has many advantages (Sulankivi et al., 2010):

- 3D tools are more effective for safety planning than 2D static drawings, given that they properly simulate real working conditions, allowing you to identify risks and optimize ways to communicate safety measures to workers;
- Production planning and safety issues can be integrated;
- Planning changes can be translated, in a short period of time, into changes in terms of safety;
- It becomes simpler to detect failures/lack of collective protection equipment.

Several authors have been exploring the potential of BIM in construction sites management. One particular study, regarding dynamic construction site models for building sites, (Kumar & Cheng, 2015) concluded that use of BIM tools saves time and resources since the entire process is done through automatic calculations. Since the use of space is optimized and the location of the equipment and materials is organized, the risk of accidents at work decreases. Through the models and simulations it is possible to take effective measures in the planning phase of the project in order to eliminate (or minimize) the hazards in construction sites.

In order to analyse the usefulness of BIM as a prevention tool, this paper will focus on its use to handle one of the main risks identified in tunnelling. Thus, it tries to answer the following research question: "Can BIM be useful to minimize the risk of people being hit?"

2 METHODOLOGY

In order to try and answer the research question, the construction site of the Marão Tunnel will be used as an example. This construction site has been chosen because of its complexity and size, since it is the largest road tunnel in the Iberian Peninsula and, therefore, the largest work to have been conducted in Portugal lately.

First, an extensive bibliographical research on the safety and health in construction sites, as well as on the introduction of BIM in them, was carried out. Next, the site was modelled: the file was first imported in *dwg (Autodesk Autocad)* format into *Autodesk Revit*, which gave us the topographic terrain in 3D. *Autodesk Revit* was chosen because the authors had previously worked with it. Based on the 2D data made available (blueprints/plans, photos and temporary signalling schemes), all the facilities present in the site were modelled in 3D. In order to be able to view people, machinery and vehicles, object families were loaded, using platforms such as BIMobject. This study was carried out in the final part of this project.

3 RESULTS AND DISCUSSION

3.1 *Sites*

The type of site to be used can differ greatly depending on the method of excavation used (Tender & Couto, 2016b). The locations where the construction sites of the Marão tunnel were set are considered sensitive, both socially and environmentally. The space available for the sites was quite small, as shown in Figures 1 and 2.

As shown in Figure 1, the site on the East side is very close to dwellings. It can also be seen that there was a single access way to the site and the portal available. This access way, in particular between the entrance gate and the portal, was very limited in terms of space, which made the movement of vehicles complex.

Figure 1.   East site (Source: Google Maps).

Figure 2.   West site (Source: Google Maps).

270

The deadline for the completion of the works was limited. For this reason, the site had to accommodate several tasks simultaneously, including underground excavation, primary and final support, and buried infrastructures—it was pioneer in Portugal in that respect also. All these activities involved a large amount of equipment (concrete mixers, trucks for the transport of muck, waterproofing screens, and reinforcement elements), all of them using the same access way.

This physical feature greatly increased the risk of a work accident due to run-overs/hits.

3.2 Site model

In order to assess the effectiveness of the proposed management method, the site was modelled. BIM assigns to each object a set of characteristics, such as geometry, spatial relations, geographic information, quantities and constructive properties of components (for example, manufacturers data), which enables the establishment of links between objects. The overview of the site is shown in Figure 3.

One of the advantages of using this methodology is a more effective space and terrain recognition before the machinery, equipment and facilities needed for the works to proceed are brought in. Figure 4 shows an example of the view of the site already in place.

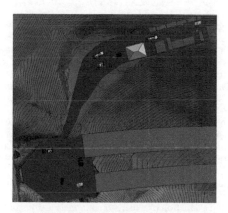

Figure 3. Overview of the East site.

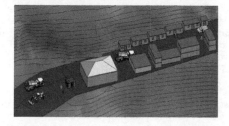

Figure 4. Distribution of equipment and facilities.

Figure 5. Movement of vehicles and people.

As mentioned, it can be seen that the space of the site was quite limited, which increased the risk of people being run-over/hit by vehicles, particularly in the narrower area between the gate and the repair shop. As a result, it became important to understand beforehand how vehicles and workers could cross in a way that would not disturb work or create dangerous conditions. Figure 5 shows a simulation in which workers are also present.

With this kind of simulations, it is possible to see beforehand the compatibility between tasks, namely how vehicles have to cross other vehicles or people in the sites, enabling a better assessment of the need to establish a physical delimitation between road and pedestrian traffic areas, for instance. In addition, it is also possible to automatically update 3D views reflecting changes to working circumstances.

The result is an improvement in planning capacity, making it possible to study beforehand the preventive measures to be implemented, and to follow the changing work conditions, creating a sort of "planning spiral".

Each construction task requires specific equipment. BIM is also helpful here: with a three-dimensional model and a correct scheduling of tasks it is possible to put in the site only the resources needed to complete the tasks at hand, and the distribution of resources can be viewed in a timely way. This way, it is possible to avoid congestion in the sites due to too much equipment being brought in without real need. Conversely, it also decreases the probability of not having all of the equipment needed on-site to complete the tasks. This is also important from an economic point of view since, as a consequence of the reduced space available, productivity can be affected if a too large number of resources are using the same space at the same time.

Figure 6. Signalling and separation between access ways.

Figure 7. Access stairs.

3.3 *Preventive measures implemented in the work*

By using BIM during planning, it was possible to identify the places where the risk of people being run-over/hit by vehicles was greater and, thus, during the construction phase, barriers were set in place to separate pedestrian pathways and vehicle pathways (using signalling as appropriate), as shown in Figure 6.

Where the width of the place did not allow for this preventive measure to be implemented, stairs were installed (Figure 7), so that the workers could access the portal without having to cross vehicles.

These and other preventive measures made it possible to achieve the desired goal of no fatal accidents at work and also made it possible to have the last 216 days of the work with no accidents at work reportable to insurance companies.

4 CONCLUSIONS

This small test of using BIM as a tool for prevention planning proved it is useful to identify risks. From this case study, it can be concluded that:

- BIM seems to be a valid instrument in risk management;
- It allows one to anticipate risk situations;
- It has the potential to become a fast means of transmitting information, namely to managers, who tend to be less able to interpret drawings;

- It helps to optimize the planning of resources to allocate to the construction site, thus potentially reducing costs and waste of time.

Finally, it is also safe to answer the research question by saying that BIM is indeed useful as a tool to reduce the risk of people being run-over/hit by vehicles.

This paper describes a first approach to integrating BIM as a tool to support prevention planning. It would be important to have future studies on the opinions of safety managers and coordinators regarding the usefulness of these kinds of tools in prevention planning.

ACKNOWLEDGEMENTS

The authors would like to thank:

- Infraestruturas de Portugal and Teixeira Duarte/ EPOS Joint venture, for their support in this study;
- Ms. Alexandra Valle Fernandes, for the translation.

REFERENCES

Azhar, S., & Behringer, A. 2013. A BIM-based Approach for Communicating and Implementing a Construction Site Safety Plan. *49th ASC Annual International Conference Proceedings, San Luis Obispo. 10–13 April 2013.* California: Associated Schools of Construction.

Bargstädt, H. 2015. Challenges of BIM for Construction Site Operations. *Procedia Engineering,* 117: 52–59.

Kumar, S., & Cheng, J. 2015. A BIM-based automated site layout planning framework for congested construction sites. *Automation in Construction,* 59: 24–37.

Sulankivi, K., Kähkönen, K., Mäkelä, T., & Kiviniemi, M. 2010. *4D-BIM for Construction Safety Planning.* Manchester, UK: CIB.

Sulankivi, K., Zhang, S., Teizer, J., Eastman, C., Kiviniemi, M., Romo, I., & Granholm, L. 2013. Utilization of BIM-based automated safety checking in construction planning *CIB World Congress, Brisbane, Australia. 05–09 May 2013.* Australia: CIB.

Tender, M., & Couto, J. 2016a. Analysis of health and safety risks in underground excavations—identification and evaluation by experts *International Journal of Control Theory and Applications,* 9(6): 2957–2964.

Tender, M., & Couto, J. 2016b. "Safety and Health" as a criterion in the choice of tunneling method. In Arezes et al (Ed.), *Occupational Safety and Hygiene IV:* 153–157. London: Taylor & Francis.

Tender, M., Couto, J., & Ferreira, T. 2015a. Prevention in underground construction with Sequential Excavation Method. In Arezes et al (Ed.), *Occupation Safety and Hygiene III:* 421–424. London: Taylor & Francis.

Tender, M., Couto, J., & Gomes, A. 2015b. Portuguese strengths and fragilities on Safety and Health practices. *Promoting Tunneling in SEE Region—International Tunneling Assocation World Tunneling Congress Dubrovnik. May 22–28, 2015.* Dubrovnik: Hubitg.

Occupational Safety and Hygiene V – Arezes et al. (Eds)
© 2017 Taylor & Francis Group, London, ISBN 978-1-138-05761-6

Factors influencing quality of accident investigations—a case study

M. Shahriari & M.E. Aydin
Konya Necmettin Erbakan University, Konya, Turkey

ABSTRACT: The main aim of the this study is to identify which factors could affect investigation and reporting of accidents. The study was performed by analyzing three different accidents, which all have had big impact on modern society. All of these accident investigations have been affected by different factors. After comparison of these accidents, it was concluded that the number of stakeholders and their economic interest has a big impact on how the investigation is performed. It is also observed that it is crucial that the group responsible for the investigation has to have the right knowledge to be able to perform a proper investigation. Political factors in the society also can have large influence on the investigation and reporting since it can control the constellation of the investigation group or what reports will be public or confidential.

Keywords: Accident investigation, investigation quality, investigation affecting factors

1   INTRODUCTION

Accidents are highly undesirable, but they cannot be fully prevented. By investigating the causes of an accident, further knowledge can be obtained and the occurrence of similar accidents can hopefully be avoided in the future. Therefore, accidents can be an important source of knowledge and improvement opportunities. But to be able to benefit from accident reporting, it is crucial that the investigation has been performed objectively and in a proper manner. Unfortunately, there are various factors that can affect the quality of an accident investigation, depending among other things on the type of accident and the circumstances.

2   MAIN PURPOSE OF THE STUDY

The main aim of this paper is to analyze three catastrophic accidents that have played a large role in the society in modern times to see how the respective investigations and reporting were performed and to identify the main factors influencing those investigations and reporting. This paper is prepared on the basis of a study carried out at Chalmers University of Technology (Svensson et al., 2010) and a literature survey as well.

3   ACCIDENT INVESTIGATION

3.1   *The purpose*

There are three main purposes for performing accident investigations, i.e. to determine the sequence of events leading to an accident, to identify the direct causes of the accident and to find risk reducing measures to prevent an accident in the future (Lappalainen et al., 2016). Accident investigation and reporting are important sources of information to prevent or minimize the risk for similar accidents to reoccur.

3.2   *Proper conduction of investigation—affecting factors*

In his article, Reason (2000) discusses causes of accidents and how investigation of accidents should be done. There he emphasizes the importance of not only analyzing active failures, the unsafe acts committed by people who are in direct contact with the system, but also, and even more importantly, latent conditions which arise from decisions. Accidents are often a combination of active failures and latent conditions. The active failures often act as trigger to the latent conditions, and the whole chain of resulting failures can lead to a severe accident. In an investigation it is therefore very important to consider the latent conditions and not only the active failures, since the accidents would not in many cases have the magnitude as it had without the latent conditions (Reason, 2000). In this study the analysis conducted based on the theory of human error developed by Reason (2000) and the report prepared by Svensson et al. (2010), Wikipedia (2016a & 2016b) and World Nuclear Association (1986).

3.3   *Common factors affecting accident investigations*

Making a proper, objective investigation can be very cumbersome, partly because there are often many parties affected by the accident and therefore many stakeholders. Poor investigation can result in false causes or only part of causes being identified, which does not support improvement opportunities. The objectivity of investigators is an important factor. It is crucial that the investigators are independent and not affected by any of the stakeholders since the stakeholders do often have own interests to protect, e.g. politically or financially. Such as when the investigation committee is deciding what stakeholder is economically responsible for an accident. Nonetheless, the fact is that in some cases, some team members are not totally objective, such as when a member of the investigation team is a representative for an insurance company. Insurance companies do often have a clear economic incentive to reach a certain result. Their representative in an investigation team can easily be influenced by this, unconsciously or not, and therefore be lacking subjectivity. A lack of subjectivity is a fact in reality, but to minimize the negative impacts it is crucial that all the other members are aware of this. It is a fact that in different countries, there is different culture and social structure. The investigators' origin can greatly affect the way he/she interprets information and circumstances. Also, the investigators' origin can affect their relationships and communication with stakeholders, e.g. whether they are dependent or not, and the surrounding view on "freedom of talking". Another important factor affecting the quality of the investigation is the knowledge, education and experiences of the investigator team. In some cases, the investigators are lacking required competence to fully investigate an accident. So from this it can be seen that how the investigation group is put together is, as expected, crucial. Apart from having the right knowledge, be politically and economically neutral, investigators should have some ethical characteristics such as sselflessness, integrity, objectivity, accountability, honesty and Leadership (Björkman, 2009).

Tenerife Accident: On the 27th of March 1977 a fatal runway collision between two Boeings 747 (KLM and Pan Am) occurred at Spanish Tenerife airport in which 583 people were killed. Investigators from both airlines as well as the Spanish authorities investigated the disaster and they all did, more or less, reach different conclusions. (Wikipedia, 2016 a). In other word, the investigation results were very contradictory depending on the investigation team involved. All the three investigation groups agreed on the cause of the accident but wanted to put as little blame as possible on its own country's citizens and its own organizations' employees. The Spanish authorities did not want to show that the airport equipment was insufficient for big international flights in bad weather and that the ground staffs couldn't

manage the traffic load and conditions present. They were perhaps afraid of putting light on issues that could affect the big and important tourist industry in Spain and the Canaries in a negative way. The other two countries involved in the accident, the U.S. and the Netherlands did all they could to minimize the effects of the accident one each respective airline and national aviation industry. The competition between airlines was during this time intense and bad publicity could have serious effects on the company's ability to prosper. This is most obvious for the Dutch investigators since all investigators teams concluded that the Dutch pilot did do a very big mistake.

Chernobyl Disaster: On 26th of April 1986, a catastrophic accident occurred in Chernobyl nuclear power plant of former Soviet Union. It was the worst historical nuclear power plant accident in terms of cost and casualties. According to UNSCEAR apart from increased thyroid cancers, "there is no evidence of a major public health impact attributable to radiation exposure 20 years after the accident." (in World Nuclear Association, 2016). The disaster resulted in the evacuation of a zone of about 30 km. Investigations were made by many different parties, the first to mention the Soviet Union, Sweden and Belarus. It should be noted that even if the cause is clear the consequences of the accident differ a lot between different reports from different organizations. The number of dead varies from 4000 who died of radiation exposure (WHO, 2005) to nearly 100000 fatal cancer (Greenpeace, 2006). When the consequences differ like this it may be worth to challenge the organizations' special interests. It is for example not unexpected that Green Peace, which is strongly opposed against nuclear power, reports the biggest consequences. In the same way, there's no surprise when an organization that is for nuclear power present reports with small consequences. The large number of consequence reports, with different conclusions, makes it hard to know what to believe. This does also make it difficult for the general public to know what report to trust and make future decisions on. Currently, both people that are strongly against and for nuclear power can find consequences that make their point.

MS Estonia: On 28th of September 1994 MS ESTONIA sank between 0:55 and 01:50 as the ship was crossing the Baltic Sea on the route from Tallinn, Estonia, to Stockholm, Sweden (Wikipedia, 2016b). The ship was carrying 989 people; 803 passengers and 186 crew members. The sinking of MS ESTONIA was the most fatal catastrophes in times of peace in Northern countries, with 852 fatalities, and only 138 being saved. As is normally done when catastrophes take place an investigation committee was recruited, consisting of a number of members from each of the three affected countries, Finland, Estonia and Sweden. In the committee's attempt to understand the course of action, the commission gathered a lot of information. In the commission's final report on the accident, the general course of events was described based on; observations on the wreck, analysis of witness' statements, analysis of the damage and evaluation of the strength of the visor and ramp attachment as well as some calculations and model tests of the vessel's behavior in waves.

3.4 Comparing factors affecting the investigations of the accidents—Investigation group composition

The members of the accident investigations group are of great importance for the validity of the report. It can be problematic to find unbiased people with the right, required knowledge. Here, it is important to note that it can be difficult to define the required competence and that in an investigation team there is need for a broad spectrum of knowledge. For example, almost all employees at an aviation organization have a positive point of view when it comes to aviation in general and they may therefore not assess the risks associated with aviation.

The Tenerife case shows that different investigation committees can reach different conclusions. The reason for this may be that the groups do not want to blame personnel from their own country or company. The MS ESTONIA case shows that the first group, even though combined as wisely as possible, met great resistance and doubts which finally resulted in the Swedish Government funding two research consortia to analyze the cause of the accident. In the case of the Chernobyl accident, it is obvious how different investigation groups can reach different results.

3.5 Stakeholders and economical interest

Different stakeholders play different roles and have different special interests in an investigation, see Table 1. The countries that are concerned have economic interest of the outcome of the investigation. The companies concerned with the accident do also have economic interests since they probably will be questioned if it is their fault and that can easily harm their competitive situation. There might also be politically instances that are of interest in the outcome of the investigation. If the form of government is of a centralized type the flow of information will be affected subjectively which will shape the investigation. These stakeholders are of interest in the outcome of the investigation and must be considered when analyzing the investigations.

Table 1. Main stakeholders for each of the three cases.

The Tenerife accident	The Chernobyl accident	MS ESTONIA accident
Spain, Netherlands, USA	Soviet, Belarus, Sweden, Ukraine	Sweden, Finland, Estonia
KLM, Pan Am	State of Soviet	MV ESTONIA operator
Relatives of the victims	Countries using nuclear power	Relatives of the victims
	Organizations and people affected	Yard where vessel was rebuilt
	Organizations against nuclear power	Inspection parties

The main difference between the three cases is that Chernobyl was owned by the state; thereby the investigation faced political pressure from a state that was strictly controlling the citizens which caused problems. Also, in the case of MS ESTONIA, some governmental institutions were being investigated which can have raised pressure from government. The Chernobyl accident also had a totally different effect afterward on the surroundings. Both Tenerife and MS ESTONIA had an immediate effect in time in the perspective of people affected.

3.6 Investigation boundaries

A reason for not reaching the same conclusions may be different scope of focus. An accident that has been investigated numerous times and many different conclusions have been drawn probably because investigation groups have had different focus. Observation document should be supported by witness statements. However, witness statement quality is very important to reach to a good judgment. This was the case in the first Estonia investigation where only staff were interviewed.

4 CONCLUSION

The analysis shows that the persons attending an accident investigation have a great influence on the investigation outcome. They are always, more or less, influenced by different factors such as economical, political and social aspects. The validity of the investigation is dependent on their ability to put these aspects aside and conduct an objective investigation. Also, awareness of possible subjectivity is important.

It is also of importance to set up teams with different professions to be able to grasp different aspects of the investigation and accident. If teams are set up from different organizations with different subjectivity the likelihood to conduct an objective investigation increase. It is however hard to define the competence needed to get the appropriate knowledge to conduct a proper investigation.

REFERENCES

Björkman, A. (2009). *Disaster Investigation.* November 29, 2010. Retrived from: www.heiwaco.tripod.com/ekatastrofkurs.htm.

Greenpeace (2006). Greenpeace new study reveals death toll of Chernobyl enormously underestimated. Available on April 18, 2006 at: http://www.greenpeace.org/international/en/press/releases/greenpeace-new-study-reveals-d/.

Lappalainen J. & Perttula P. 2016. Accident investigation techniques. OSHWIKI Network Knowledge. Available at: https://oshwiki.eu/wiki/Accident_investigation_techniques.

Reason, J. (2000). Human error: models and management. *British Medical Journal, 320,* 768–770.

Svensson, A., Börjesson, A., Danielsson, DJóhannsdóttir, F. S., & Finlöf, M. (2010). *Accident Investigation - Factors affecting quality of investigations* (unpublished project work in Risk Management and Safety course). Chalmers University of Technology, Gothenburg, Sweden, 2010.

WHO (World Health Organization), 2005. Chernobyl: the true scale of the accident. Retrieved November 19, 2010, from http://www.who.int/mediacentre/news/releases/2005/pr38/.

Wikipedia (2016a). Tenerife Airport Disaster. Available October 9, 2016 at: http://en.wikipedia.org/wiki/Tenerife_airport_disaster.

Wikipedia (2016b). MS Estonia. Available October 7, 2016 at: http://en.wikipedia.org/wiki/MV_Estonia

World Nuclear Association (1986). Chernobyl Accident. Available at: http://www.world-nuclear.org/information-library/safety-and-security/safety-of-plants/chernobyl-accident.aspx.

Occupational Safety and Hygiene V – Arezes et al. (Eds)
© *2017 Taylor & Francis Group, London, ISBN 978-1-138-05761-6*

Worker's nasal swab: A tool for occupational exposure assessment to bioburden?

C. Viegas
Environment and Health Research Group (GIAS) Escola Superior de Tecnologia da Saúde de Lisboa, ESTeSL, Instituto Politécnico de Lisboa, Lisboa, Portugal
Centro de Investigação e Estudos em Saúde Pública, Escola Nacional de Saúde Pública, ENSP, Universidade Nova de Lisboa, Lisbon, Portugal

V. Santos & R. Moreira
Undergraduated Students from Escola Superior de Tecnologia da Saúde de Lisboa, ESTeSL, Instituto Politécnico de Lisboa, Lisboa, Portugal

T. Faria
Environment and Health Research Group (GIAS) Escola Superior de Tecnologia da Saúde de Lisboa, ESTeSL, Instituto Politécnico de Lisboa, Lisboa, Portugal

E. Ribeiro
Environment and Health Research Group (GIAS) Escola Superior de Tecnologia da Saúde de Lisboa, ESTeSL, Instituto Politécnico de Lisboa, Lisboa, Portugal
Research Center LEAF—Linking Landscape, Environment, Agriculture and Food—Instituto Superior de Agronomia, Lisbon University, Portugal

L. Aranha Caetano
Environment and Health Research Group (GIAS) Escola Superior de Tecnologia da Saúde de Lisboa, ESTeSL, Instituto Politécnico de Lisboa, Lisboa, Portugal
Research Institute for Medicines (iMed.ULisboa), Faculty of Pharmacy, University of Lisbon, Lisbon, Portugal

S. Viegas
Environment and Health Research Group (GIAS) Escola Superior de Tecnologia da Saúde de Lisboa, ESTeSL, Instituto Politécnico de Lisboa, Lisboa, Portugal
Centro de Investigação e Estudos em Saúde Pública, Escola Nacional de Saúde Pública, ENSP, Universidade Nova de Lisboa, Lisbon, Portugal

ABSTRACT: The nose cavity is the primary portal of entry for inspired air and the first region of the respiratory tract to be in contact with bioaerosols. Nasal swab allows measurement of bioburden presence in the nose cavity and the collection is easy and painless. We intended to identify scientific papers reporting this tool as a surrogate to access exposure to microbiologic agents in occupational environments. Literature research was performed using scientific and academic databases. In 5 from the 11 articles studied only one parameter was analysed, being the most common Methicillin-resistant *S. aureus*. Seven studies applied culture based-methods coupled with molecular tools assay. Findings from two studies corroborate the use of nasal swab as a tool to complement the occupational exposure assessments, since was found association between the nasal swabs results and the occupational microbiota also assessed. Nasal swabs analyses should comprehend culture based-methods and molecular tools assay.

1 INTRODUCTION

Different forms of fungal diseases affecting the nose cavity and paranasal sinuses are well reported (Ebbens and Fokkens, 2008). In addition, also bacteria are implicated as pathogens in chronic rhinosinusitis (Ebbens and Fokkens, 2008), and indoor airborne bacteria can be associated with blocked nose (Haverinen-Shaughnessy et al., 2007). Several variables can affect the risk of developing different health outcomes, namely: density of nasal hair (Ozturk et al., 2011), gender differences

(Jaakkola and Jaakkola, 2004), endotoxin exposure (Kline et al., 1999) and nose size (Hall, 2005).

The nose cavity is the primary portal of entry for inspired air, and therefore, the first region of the respiratory tract in contact with airborne fungi, bacterial and viral specimens (Madsen et al., 2013; Heikkinen et al., 2002; McIntosh, et al., 1993; Ruuskanen, & Ogra 1993; Laitinen, Kontro & Kirsi, 2015). The inhalation and sedimentation of fungi and bacteria in the respiratory tract depend on the aerodynamic diameter (d_{ac}) of the bioaerossol particle. Particle deposition in the airways is directly connected to damage of the pulmonary system. Therefore, health-related particle sampling should reflect how particles penetrate and deposit in the various regions of the human respiratory system (Madsen et al., 2013). Attempts to simulate this approach in workplaces exposure assessments were already performed (Viegas et al., 2016a, 2016b), The fraction able to enter the airways through the nose or mouth is called the inhalable fraction $(d_{ac}50 = 100 \ \mu m)$ (Madsen et al., 2013).

The nasal swab procedure allows measurement of presence of fungi and bacteria in the nose cavity. In addition, the collection of a nasal swab is easy and painless, and it can be done everywhere without any additional devices (Heikkinen et al., 2002). The optimal sampling methods must be balanced with the feasibility, costs, and time required collecting the specimens (Heikinen et al., 2002; McIntosh, et al., 1993; Ruuskanen, & Ogra 1993).

Considering the above, we intended to identify in scientific papers the use of nasal swab to assess occupational exposure to bioburden and to conclude about if it is suitable to complement the exposure assessments done in the workplace environment.

2 MATERIALS AND METHODS

Extensive search was performed to identify scientific papers reporting the use of nasal swab to access occupational exposure to bioburden. Scientific papers, available on online databases, were analyzed reporting this tool as a surrogate to access exposure to microbiologic agents in occupational environments. Literature research was performed using the databases PubMed, Google Scholar and Scielo, from January 2010.

The following keywords were used: "nasal swab", "occupational exposure", "assessment", "fungi", "bacteria", "virus" and, thus, eleven scientific papers were found.

3 RESULTS

Previous studies on nasal bioburden findings coming from occupational environments are scare.

In 5 from the 11 articles studied only one parameter was analysed, namely Methicillin-resistant *S. aureus* (MRSA), *S. aureus* (Hatcher et al. 2016; Leibler et al. 2016; Diawara et al. 2014; Dorado-García et al. 2013; Marshall et al 2011) or viruses (Marshall et al. 2011). In one study also fungi were target besides bacteria (Laitinen, Kontro & Kirsi, 2015) and in one study only viruses were assessed (Marshall et al. 2011). In 7 studies culture based-methods were coupled with molecular tools assay and in 5 studies Methicillin-resistant *S. aureus* (MRSA) was target in worker's nose (Table 1).

4 DISCUSSION

The study conducted by Madsen and colleagues (2013), where a similar tool was applied (nasal lavage instead of nasal swab) corroborate the use of this biological sample to complement the occupational exposure assessments, since it was found association between the nasal swabs results and the occupational microbiota also assessed. The same association trend was reported in a previous study developed in processing biodegradable waste facilities where *Aspergillus* section *Fumigati* in the nasal mucous membranes of workers was greater in those jobs and operations, where it was also in the air (Laitinen, Kontro & Kirsi, 2015).

The detection of bioaerosols by the molecular-based PCR assay in parallel with culture based-methods is well supported (Viegas et al., 2012–2016c; Laitinen, Kontro & Kirsi, 2015). Using nasal swab samples combined with the PCR assay is an even better choice for studying workers exposure than collecting air samples only, because we breathe in ambient air all the time through our noses (Laitinen, Kontro & Kirsi, 2015, Madsen et al., 2013). The decreasing cost of molecular tools can justify their frequent use (in 9 out of 11 papers), and also the expected increase in the use of nasal swab as a tool for occupational exposure assessment to bioburden. However, we should consider gender differences due to nasal anatomy and the higher hair density from men (Hall, 2005; Ozturk et al., 2011; Madsen et al., 2013) that can constitute a larger area for retention and deposition of microorganisms (Madsen et al., 2013).

The World Health Organization (WHO) describes antimicrobial resistance in human pathogens as a global health challenge (WHO, 2015). For the past years, with the worldwide escalation in antibiotic resistant microorganisms social and scientific concerns have emerged regarding the over prescription of human antibiotics and it extensive use in agriculture (Smith et al., 2002) and animal production (Bloom, 2004). This concern is well reported for MRSA (Hatcher et al. 2016;

Table 1. Scientific papers reporting nasal swabs as a tool to access occupational exposure to bioburden.

Reference	Studied population	Analysis	Methods	Results
Hatcher et al. 2016	Swine workers and their children (n = 198) Community (n = 202)	Methicillin-resistant *S. aureus (MRSA)*	Culture, biochemical tests. PCR, Antibiotic susceptibility tests.	*S. aureus:* swine workers and children 53%, community 31%; 2–3% MRSA.
Ye et al. 2016	Swine workers	MRSA and multidrug-resistant *S. aureus (MDRSA)*	Culture. Gram staining, morphology, biochemical tests and PCR.	Higher prevalence of MDRSA and MRSA carriage.
Leibler et al. 2016	Beefpacking workers in slaughterhouses	*Staphylococcus aureus*	Culture, biochemical tests, PCR.	*S. aureus* nasal carriage 27.0%
Laitinen, Kontro & Kirsi, 2015	Biodegradable waste workers	*Actinobacteria, Campylobacter* spp., *Yersinia* spp., *Salmonella* spp. and fungi	Culture and RT-PCR	Long-term respiratory diseases.
Nadimpalli et al. 2015	Swine workers	Methicillin-resistant *S. aureus (MRSA)* and multidrug-resistant *S. aureus (MDRSA)*	Culture, antibiotic susceptibility, PCR.	Nasal carriage persistence of LA-MDRSA and MRSA.
Leahy 2015	Poultry, slaughter and processing plant workers (n = 90)	*Staphylococcus aureus* and gram-negative microorganisms	Culture, antimicrobial susceptibility	*Staphylococcus aureus* 15.6%, MRSA 1.1% and 40% gram-negative prevalence
AlWakeel 2015	Fuel workers (n = 29)	Bacterial colonization	Culture, antimicrobial susceptibility, PCR.	19 bacterial species isolated. Most prevalent *Streptococcus thoraltensis*
Diawara et al. 2014	Hemodialysis patients (n = 143) and medical staff (n = 32)	*Staphylococcus aureus*	Culture; Colony morphology, Gram staining, biochemical tests. Antibiotic susceptibility.	*Staphylococcus aureus* prevalence: 39.2% in patients; 18.8% in medical staff. 0.7% MRSA prevalence in patients.
Rosenberg Goldstein et al. 2014	Spray irrigation workers and office workers	MRSA, methicillin-susceptible *S. aureus* (MSSA), vancomycin-resistant *enterococci* (VRE), vancomycin-susceptible *enterococci* (VSE)	Culture, biochemical tests, PCR, antimicrobial susceptibility	MSSA prevalence of 26% and 29%; VSE 11% and 0% in spray irrigators and controls, respectively.
Dorado-García et al. 2013	Veal calf farms (n = 52); Farm workers and families (n = 211)	LA-MRSA	Culture, and PCR	MRSA in farmers 29.7% and 13% family members.
Marshall et al. 2011	Health care workers (n = 231) and controls (n = 215)	H1 N1 virus	Immunologic detection for antibodies in serum and RT-PCR in samples from nasal swabs	15.2% prevalence of H1 N1 virus

Ye et al. 2016; Nadimpalli et al. 2015; Rosenberg Goldstein et al. 2014; Dorado-García et al. 2013) although no information was found about antifungal resistance in the analysed studies. Nevertheless, the prevalence of triazole resistance in *Aspergillus* section *Fumigati* has been increasingly reported worldwide (Morio et al., 2012; Mortensen et al., 2010; Pham et al., 2014; Rajendran et al., 2016; Snelders et al., 2012; Buil et al., 2016; Chowdhary et al., 2015; Howard and Arendrup, 2011; Jensen et al., 2016; Lestrade et al., 2016). Therefore, we believe that the characterization of antifungal resistance in occupational settings is critical as part of microbial surveillance strategies, as information on *Aspergillus* disease and/or risk groups should be leading for decisions regarding empirical antifungal therapy in specific units and for resistance management.

Though only few occupational studies link the occupational exposure with the nasal microbiota found in workers (Madsen et al., 2013; Laitinen, Kontro & Kirsi, 2015) both assessments should always be consider to achieve association scenarios that can help bioburden exposure assessment and, consequently, risk characterization (Viegas et al., submitted). In this context, the use of nasal swab appears as a promising strategy to access occupational exposure to both susceptible and resistant bioburden.

5 CONCLUSIONS

Findings from two studies corroborate the use of nasal swab as a tool to complement the occupational exposure assessments. Culture based-methods and molecular tools should be coupled as an assay to analyse in a more detail manner the nasal swab.

ACKNOWLEDGEMENTS

This study would not have been possible to develop without the financial support given by Portuguese Authority of Working Conditions (Project reference: 005DBB/12).

REFERENCES

AlWakeel, SS. 2015. Microbiological and molecular identification of bacterial species isolated from nasal and oropharyngeal mucosa of fuel workers in Riyadh, Saudi Arabia. *Saudi J Biol Sci.* In press.

Buil, JB., Meis, JF., Melchers, WJG., Verweij, PE., 2016. Are the TR 46 /Y121F/T289 A Mutations in Azole-Resistant Aspergillosis Patient Acquired or Environmental? *Antimicrob. Agents Chemother.* 60(5): 3259–3260.

Byrnes, MC., Adegboyega, T., Riggle, A. 2010. Nasal swabs collected routinely to screen for colonization by methicillin-resistant Staphylococcus aureus in intensive care units are a sensitive screening test for the organism in clinical cultures. *Surg Infect (Larchmt).* 11(6): 511–515.

Chowdhary, A., Sharma, C., Kathuria, S., Hagen, F., Meis, J.F., 2015. Prevalence and mechanism of triazole resistance in Aspergillus fumigatus in a referral chest hospital in Delhi, India and an update of the situation in Asia. *Front. Microbiol.* 6.

Diawara, I., Bekhti, K., Elhabchi, D., et al. 2014. Staphylococcus aureus nasal carriage in hemodialysis centers of Fez, Morocco. *Iran J Microbiol.* 6(3): 175–183.

Dorado-García, A., Bos, ME., Graveland, H., et al. 2013 Risk factors for persistence of livestock-associated MRSA and environmental exposure in veal calf farmers and their family members: an observational longitudinal study. *BMJ Open.* 3(9): e003272.

Ebbens, FA., Fokkens, WJ. 2008. The mold conundrum in chronic rhinosinusitis: where do we stand today? *Curr Allergy Asthma Rep.* 8: 93–101.

Hall, RL. 2005 Energetics of nose and mouth breathing, body size, body composition, and nose volume in young adult males and females. *Am J Hum Biol.* 17: 321–30.

Hansen, V. 2013. Fungi, β-Glucan, and Bacteria in Nasal Lavage of Greenhouse Workers and Their Relation to Occupational Exposure. *Ann. Occup. Hyg.* 57(8): 1030–1040.

Hatcher, SM., Rhodes, SM, Stewart, JR., et al. 2016. The Prevalence of Antibiotic-Resistant Staphylococcus aureus Nasal Carriage among Industrial Hog Operation Workers, Community Residents, and Children Living in Their Households: North Carolina, USA. *Environ Health Perspect.* Epub.

Haverinen-Shaughnessy. U-, Toivola. M-, Alm. S, et al. 2007. Personal and microenvironmental concentrations of particles and microbial aerosol in relation to health symptoms among teachers. *J Expo Sci Environ Epidemiol.* 17: 182–90.

Heikkinen, T., Marttila, J., Salmi, AA., Ruuskanen, O. 2002. Nasal Swab versus Nasopharyngeal Aspirate for Isolation of Respiratory Viruses. *Journal Of Clinical Microbiology.* 40(11): 4337–4339.

Howard, S.J. & Arendrup, M.C. 2011. Acquired antifungal drug resistance in Aspergillus fumigatus: epidemiology and detection. *Med. Mycol.* 49(1): S90–S95.

Jaakkola, MS. & Jaakkola, JJ. 2004. Indoor molds and asthma in adults. *Adv Appl Microbiol.* 55: 309–3.

Jensen, RH., Hagen, F., Astvad, KMT., Tyron, A., Meis, JF., Arendrup, MC. 2016. Azole resistant Aspergillus fumigatus is a persistent threat in Denmark: results from a laboratory based study and focus on genotyping data. *Clin. Microbiol. Infect.* 22(6): 570.

Kline, JN., Cowden, JD., Hunninghake, GW., et al. 1999. Variable airway responsiveness to inhaled lipopolysacch ride. *Am J Respir Crit Care Med.* 160: 297–303.

Laitinen, S., Kontro, M. & Kirsi, M. 2015. Exposure to Bacterial and Fungal Bioaerosols in Facilities Processing Biodegradable Waste. Annals of Occupational Hygiene.

Leahy, K. 2015. *An Assessment Of Occupational Exposure To Gram- Negative Organisms In An Urban Poultry Slaughter And Processing Plant In Columbia,* Sc, Usa.

Leibler, J.H., Jordan, J.A., Brownstein, K., Lander, L., Price, LB., Perry, MJ. 2016. Staphylococcus aureus nasal carriage among beefpacking workers in a Midwestern United States slaughterhouse. *PLoS One.* 11(2): 1–11.

Lestrade, PPA., Meis, JF., Arends, JP., et al. 2016. Diagnosis and management of aspergillosis in the Netherlands: A national survey. *Mycoses* 59: 101–107.

Madsen, AM., Tendal, k., Thilsing, T., Bloom, R.A. 2004. *Use of veterinary pharmaceuticals in the United States. In: Pharmaceuticals in the Environment.* Berlin: Springer, pp. 149–154.

McIntosh, K., Halonen, P., Ruuskanen, O. 1993. Report of a workshop on respiratory viral infections: epidemiology, diagnosis, treatment, and prevention. *Clin. Infect. Dis.* 16: 151–164.

Morio, F., Aubin, GG., Danner-Boucher, I. 2012. High prevalence of triazole resistance in Aspergillus fumigatus, especially mediated by TR/L98H, in a French cohort of patients with cystic fibrosis. *J. Antimicrob. Chemother.* 67: 1870–1873.

Mortensen, KL., Mellado, E., Lass-Flrl, C., Rodriguez-Tudela, JL, Johansen, HK., Arendrup, MC. 2010. Environmental study of azole-resistant Aspergillus fumigatus and other aspergilli in Austria, Denmark, and Spain. *Antimicrob. Agents Chemother.* 54: 4545–4549.

Nadimpalli, M., Rinsky, JL., Wing, S., et al. 2015 Persistence of livestock-associated antibiotic-resistant Staphylococcus aureus among industrial hog operation workers in North Carolina over 14 days. *Occup Environ Med.* 72(2):90–99.

Ozturk, A.B., Damadoglu, E., Karakaya, G., et al. 2011. Does nasal hair (vibrissae) density affect the risk of developing asthma in patients with seasonal rhinitis? *Int Arch Allergy Immunol.* 156: 75–80.

Pham, CD., Reiss, E., Hagen, F., Meis, JF., Lockhart, SR. 2014. Passive surveillance for azole-resistant Aspergillus fumigatus, United States, 2011–2013. *Emerg. Infect. Dis.* 20: 1498–1503.

Rajendran, M., Mohd, T., Khaithir, N., Santhanam, J., 2016. Determination of Azole Antifungal Drug Resistance Mechanisms Involving Cyp51a Gene in Clinical Isolates of Aspergillus fumigatus and Aspergillus niger. *Med. Mycol.* 2: 1–6.

Rosenberg Goldstein, RE., Micallef, SA., Gibbs, SG., et al. 2014. Occupational exposure to Staphylococcus aureus and Enterococcus spp. among spray irrigation workers using reclaimed water. *Int J Environ Res Public Health.* 11(4): 4340–4355.

Ruuskanen, O. & Ogra, PL. 1993. Respiratory syncytial virus. *Curr. Probl. Pediatr.* 23: 50–79.

Smith, DL., Harris, AD., Johnson, JA., Silbergeld, EK., Morris, JG. 2002. Animal antibiotic use has an early but important impact on the emergence of antibiotic resistance in human commensal bactéria. *Proceedings of the National Academy of Sciences*, 99(9): 6434–6439.

Snelders, E., Camps, SMT., Karawajczyk, A., et al. 2012. Triazole fungicides can induce cross-resistance to medical triazoles in Aspergillus fumigatus. *PLoS One.* 7(3): e31801.

Viegas C, Faria T, Sabino R, Viegas S. 2016a. Fungi spores dimension matters in health effects: a methodology for more detail fungi exposure assessment. *In: Annual Meeting of International Society of Exposure Science, Utrecht (Nederland),* 9–13 October 2016.

Viegas, C., Carolino, E., Sabino, R., Viegas, S., Veríssimo, C. 2013. Fungal Contamination in Swine: A Potential Occupational Health Threat. *Journal of Toxicology and Environmental Health, Part A,* 76(4–5): 272–280.

Viegas, C., Faria, T., Caetano, LC., Carolino, E., Quintal Gomes, A., Viegas. 2016. *Aspergillus* spp. prevalence in different Portuguese occupational environments: what is the real scenario in high load settings? *Submitted.*

Viegas, C., Faria, T., Carolino, E., Sabino, R., Quintal Gomes, A. Viegas, S. 2016b. Occupational Exposure to Fungi and Particles in Animal Feed Industry. *Medycyna Pracy* 67(2).

Viegas, C., Malta-Vacas, J. Sabino, R. 2012. Molecular biology versus conventional methods—Complementary methodologies to understand occupational exposure to fungi. *International Symposium on Occupational Safety and Hygiene*, P. 478–479.

Viegas, C., Malta-Vacas, J., Sabino, R., Viegas, S., Veríssimo, C. 2014. Accessing indoor fungal contamination using conventional and molecular methods in Portuguese poultries. *Environmental Monitoring and Assessment.* 186(3): 1951–1959. ISSN 0167-6369. DOI 10.1007/s10661-013-3509-4.

Viegas, C., Neves, O., Sabino, R., Viegas, S. 2016c. Fungi occupational exposure assessment: a methodology to be followed for a more sound health effects discussion. *International Symposium on Occupational Safety and Hygiene SHO2016, Arezes, P. et al. Portuguese Society of Occupational Safety and Hygiene:* 374–376.

Viegas, C., Quintal Gomes, A., Faria, T., Sabino, R. 2015. Prevalence of Aspergillus fumigatus complex in waste sorting and incineration plants: an occupational threat. *Int. J. Enviro ment and Waste Management.* 16(4): 353–369.

World Health Organization (2014). Antimicrobial resistance: fact sheet n° 194. Genebra: WHO. Available at: http://www.who.int/mediacentre/factsheets/fs194/en/

Ye, X., Fan, Y., Wang, X., et al. 2016. Livestock-associated methicillin and multidrug resistant S. aureus in humans is associated with occupational pig contact, not pet contact. Sci Rep. 6: 19184.

Occupational Safety and Hygiene V – Arezes et al. (Eds)
© *2017 Taylor & Francis Group, London, ISBN 978-1-138-05761-6*

Pilot study regarding vehicles cabinets and elevator: Neglected workstations in occupational exposure assessment?

C. Viegas

Environment and Health Research Group (GIAS) Escola Superior de Tecnologia da Saúde de Lisboa, ESTeSL, Instituto Politécnico de Lisboa, Lisboa, Portugal
Centro de Investigação e Estudos em Saúde Pública, Escola Nacional de Saúde Pública, ENSP, Universidade Nova de Lisboa, Lisbon, Portugal

T. Faria

Environment and Health Research Group (GIAS) Escola Superior de Tecnologia da Saúde de Lisboa, ESTeSL, Instituto Politécnico de Lisboa, Lisboa, Portugal

L. Aranha Caetano

Environment and Health Research Group (GIAS) Escola Superior de Tecnologia da Saúde de Lisboa, ESTeSL, Instituto Politécnico de Lisboa, Lisboa, Portugal
Research Institute for Medicines (iMed.ULisboa), Faculty of Pharmacy, University of Lisbon, Lisbon, Portugal

E. Carolino

Environment and Health Research Group (GIAS) Escola Superior de Tecnologia da Saúde de Lisboa, ESTeSL, Instituto Politécnico de Lisboa, Lisboa, Portugal

S. Viegas

Environment and Health Research Group (GIAS) Escola Superior de Tecnologia da Saúde de Lisboa, ESTeSL, Instituto Politécnico de Lisboa, Lisboa, Portugal
Centro de Investigação e Estudos em Saúde Pública, Escola Nacional de Saúde Pública, ENSP, Universidade Nova de Lisboa, Lisbon, Portugal

ABSTRACT: In several occupational settings have been reported organic dust occupational exposure and in many is common to use vehicles to transport the raw materials and final products or even elevators to guarantee workers and materials transportation within the facilities. Measurements of particles and samples collection to assess fungal burden were done during the use of one docker's crane, one fork lift and one elevator in waste industry. All reference air samples presented much less air fungal contamination than inside the cabins. Statistically significant differences were detected in the mass concentration for almost all particles sizes between the cabins and the background where the vehicle was circulating. This pilot study suggests that workers carry fungi from their outdoor work into the vehicles cabinets and elevator and exposure to particles still occurs in the cabins. To reduce exposure focus should be on interventions related with ventilation and good hygiene practices.

1 INTRODUCTION

A large number of workers are exposed to organic dust in different occupational settings. Organic dust is usually defined an airborne mixture of viable and non-viable microorganisms (bacteria, fungi, viruses, protozoa), their metabolites (endotoxins, $(1–3)$-β-D51 glucans, mycotoxins, peptidoglycans, enzymes etc.) and solid particles of vegetable and animal origin (allergens, including pollens, vegetal fibers, epidermis, etc.) (Douwes et al., 2003; Wouters et al., 2006; Oppliger et al., 2014).

The occupational settings where organic dust have been reported are the ones related with animal handling, feed production and also farming (Seedorf et al., 1998; Hawley et al., 2015; Viegas et al., 2016). Bakeries, waste and water management, greenhouses and slaughterhouses are other types of settings also mentioned in the literature (Bunger et al., 2007; Baatjies et al., 2014; Thilsing et al., 2014, Viegas et al., 2015; Viegas et al., 2016). In all these settings exposure to particles and microorganisms is considered an occupational health problem and organic dust and related microbial exposures are the main promoters of several respiratory symptoms, such as decline in lung function, asthma, chronic bronchitis, bronchial hyper-responsiveness, wheeze, and cough and all of these have been described in several publications (Schenker et al., 1998; Douwes et al., 2003; Sigsgaard and Schlünssen, 2004; Basinas et al., 2015). To better estimate the possible health effects related with occupational exposure to organic dust, it is important to acquire detailed information on exposure sources, and also about the variables that can influence exposure (Viegas et al., 2014; Kirkeskov et al., 2016).

However, in the already mentioned occupational settings sources of exposure to microorganisms and dust are very much related with the handled material and process that is being developed (transport, chemical or physical transformation, others) (Wouters et al., 2006). Therefore, in the same unit the organic dust composition can have a significant change if the handle material is changed. Additionally, we know that in many of these occupational settings is common to use vehicles to transport the raw materials and final products or even elevators to guarantee workers and materials transportation within the facilities.

Considering the above, the aim of this pilot study was to examine exposure to fungi and dust while workers are seated in the vehicle cabinets, and to acquire knowledge if being inside the vehicles can contribute to reduce workers exposure to fungi and dust.

2 MATERIALS AND METHODS

2.1 Selected transportation vehicles

Measurements of particles and samples collection to assess fungal burden were done during the use of the three following transport equipment: docker's crane in a port, fork lift in a composting waste industry and elevator in an incineration waste industry. Besides this equipment it was possible to measured particles in additional two: inside a forklift during the movement of sugar cane and inside a forklift during the movement of waste in a waste sorting site.

2.2 Fungal contamination assessment

Air samples on the elevator from a incineration waste industry and in the docker´s crane in a port (50–100 L respectively) were collected through an impaction method with a flow rate of 140 L/min onto Malt Extract Agar (MEA) supplemented with chloramphenicol (0.05%), using the Millipore air Tester (Millipore). The sampling time was selected to be representative from the studied period (one work day) that we intend to characterize. In addition one sample from the reference indoor/outdoor site was also collected (Table 1).

Table 1. Samples performed for fungal assessment.

Vehicles	Air samples	Surface samples
Elevator	x	x
Crane	x	–
Fork lift	–	x

x Done.
– Not done.

Surface samples were collected on the fork lift (worker seat) and in the elevator (floor) by swabbing the surfaces with a 10 by 10 cm square stencil disinfected with a 70% alcohol solution between samples, according to the International Standard ISO 18593 (2004). The obtained swabs were then plated onto MEA (Table 1).

All the collected samples were incubated at 27°C for 5 to 7 days. After laboratory processing and incubation of the collected samples, quantitative (colony-forming units—CFU/m^3 and CFU/m^2) and qualitative results were obtained, with identification of the isolated fungal species or genera. For species identification, microscopic mounts were performed using tease mount or Scotch tape mount and lactophenol cotton blue mount procedures. Morphological Identification was achieved through macro and microscopic characteristics as noted by De Hoog et al. (2000).

2.3 Particulate matter assessment

Measurements were performed using a portable direct-reading equipment (Lighthouse, model 3016 IAQ) that gives information regarding mass concentration (mg $\times$ m^{-3}) in 5 different sizes (PM0.5, PM1, PM2.5, PM5, PM10). The measurements were con-ducted near each worker's nose and during the use of the equipment. One measurement with the duration of 5 min was done in each. Regarding the statistical analysis performed, the results are considered significant at a 5% significance level. To verify the normality of the data, the Shapiro-Wilk test was used.

To compare the mass particles concentration (PM0.5, PM1.0, PM2.5, PM5.0 and PM10.0) between cabin and other setting, the Mann-Whitney U test was used.

3 RESULTS

3.1 Fungal contamination assessment

The air results ranged from 800 CFU.m^{-3} from the elevator to uncountable colonies in the crane. Both surfaces from elevator and fork lift present 11×10^4 CFU.m^{-2}.

Four different species/genera of filamentous fungi were identified in air samples from elevator and 3 from the crane. In the elevator air *Penicillium* sp. was the one with higher prevalence (82.5%) but other fungi were also identified, such as: *Chrysosporium inops* and *Aspergillus* sections *Fumigati* and *Nigri*. Regarding crane air besides uncountable colonies of *Aspergillus* section *Fumigati,* also *Penicillium* sp. and *Aspergillus* section *Aspergilli* were found.

Regarding fork lift surfaces contamination 4 different species/genera of filamentous fungi were isolated being *Aspergillus* section *Nigri* the most isolated (36.4%). However, 2 more *Aspergillus* sections were identified, namely *Fumigati* (27.3%) and *Candidi* (18.2%). *Lichtheimia* sp. (18.2%) was also identified. Elevator floor presented only two fungal species, being *Penicillium* sp. the most isolated (87.1%), but *Aspergillus* section *Nigri* was also identified.

Both reference air samples (outdoor) from elevator and crane presented much less air fungal contamination, namely 180 and 12 CFU/m³, respectively. However, Maturation Park from the composting waste industry (reference floor sample from fork lift) presented higher fungal contamination (51×10^4 CFU/m²). Crane air sample present 2 different fungal species from the outdoor sample and the same happens with the surface samples from the fork lift regarding the reference surface sample from the same indoor site.

3.2 *Particulate matter assessment*

Statistically significant differences were detected in the mass concentration for all particles sizes between the cabin and the background where the vehicle was circulating, except for the PM1.0 particles. Verifying that for smaller particles (PM0.5) the cabin has a lower concentration, while for larger particles, the cabin has higher concentrations (Figure 1).

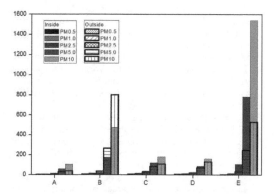

Figure 1. PMC results inside and outside of vehicles cabinet (A—Crane, B—Stacker 1, C—Stacker 2, D—Elevator, E—circular cargo transport machine).

4 DISCUSSION

Results showed that being inside a cabin in these occupational settings do not imply being protected from exposure to the fungal burden and also particles. In case of controlling organic dust exposure, containment or isolation are common measures applied in the workplaces and consists basically of placing a barrier between the dust source and the workers (WHO, 1999). Isolation of the workers by using a cabin can be a way of avoid or control exposure to particles if used closed (windows and doors) and with adequate filters in the equipment ventilation system. However, improperly maintained ventilation filters may be a source of particulate matter and microorganisms (Miaśkiewicz-Peska, 2011).

Transport of bioaerosols from one occupational environment to another have been reported in canteens from waste water treatment plants (Viegas et al., 2014), social rooms at composting plants (Liebers et al., 2012), and in offices at biofuel plants (Madsen, 2006). This pilot study suggests that workers carry fungi from their outdoor work into the vehicles cabinets and elevator as in a previous study performed by Madsen et al. (2016).

Penicillium was already reported to be dominant in waste sorting in Portugal (Malta-Vacas et al., 2012). However, *Penicillium* species are difficult to identify not only by microscopy but also by PCR-based methods (Madsen et al., 2015), and information is still lacking concerning occupational exposure at species/strains level (Viegas et al., 2015). The *Penicillium* burden found in elevator and in the crane should be also highlighted in relation to workers' health as a review study concluded that exposure to this genus can be associated with several respiratory health effects (Knutsen et al., 2012).

In all sampling sites *Aspergillus* isolates were identified being *Fumigati* and/or *Nigri* sections present in all the sampling sites. Both sections have toxigenic potential (Nielsen, 2003) and we should also consider their ability to cause invasive fungal infections. *Aspergillosis* is an invasive fungal infection that has gained more attention in the last two decades. The aspergillosis troublesome arises from the increased number of patients with profound immunosuppression and a related increase in incidence and mortality rates of invasive fungal infections that affect these patients; a limited number of antifungal agents in clinical use; and the emergence of antifungal resistance. This phenomenon is challenging the effective management of aspergillosis, meaning that patients confronted with an invasive fungal infection from an azole-resistant isolate from *Fumigati* section will fail azole treatment and potentially die (Snelders et al., 2011). The reports on *Aspergillus* disease caused by resistance strains in individuals with no prior azole treatment suggest the emergence of azole resistance in the environment, due to the extensive use of biocides such as in agriculture practices in crop protection, or due to the use of biocides for material preservation and cleaning procedures (Verweij et al., 2016).

In fact, *Aspergillus* species can be found almost everywhere, being widespread in the environment, growing in the soil, on plants and on decomposing organic matter. Resistant *Aspergillus* strains can evolve under favorable selective environmental conditions (high moisture levels, high temperature, presence of oxygen) and azole pressure (Leendertse, 2015) (European Centre for Disease Prevention and Control, 2013). We believe that occupational settings with high levels of organic dust, such as farming, feed production, bakeries, greenhouses and waste management, can be considered hot spots for the emergence of antifungal resistance to azoles, as favorable conditions for the development of antifungal resistance are present in these settings. Thus, the characterization of antifungal resistance in these settings is critical as part of microbial surveillance strategies, as information on *Aspergillus* disease and/or risk groups should be leading for resistance management and for decisions regarding empirical antifungal therapy in specific units.

Regarding the distribution of the particles size between inside and outside of the cabins, several aspects can contribute being the fact that in confined spaces particles stick if they contact one another (agglomeration/accumulation mode which is usually the case) a possible explanation (Morawska, 2001).

Different sized particles deposit in different areas of the lung, nose and throat and may result in irritative symptoms including cough and COPD (Kirkeskov et al., 2016). Additionally, particles are also perfect carriers of other contaminants for the workers respiratory system (Viegas et al., 2014).

5 CONCLUSIONS

Workers using vehicles cabinets and the elevator as-sessed are not protected from exposure to the fungal burden and also particles. This pilot study suggests that workers carry fungi from their outdoor work in-to the vehicles cabinets and elevator. To reduce ex-posure focus should be on interventions related with good hygiene practices in vehicles cabins and eleva-tor and the improvement of the vehicles ventilation. Additionally, the equipments should be used with closed windows during circulation or when handling materials.

AKNOWLDEGEMENTS

This study would not have been possible to develop without the institutional support given by Lisbon School of Health Technology.

REFERENCES

Baatjies, R., Meijster, T., Heederik, D., Sander, I., Jeebhay MF. 2014. Effectiveness of interventions to reduce flour dust exposures in supermarket bakeries in South Africa. *Occup Environ Med.* 71(12): 811–818.

Basinas, I., Sigsgaard, T., Kromhout, H., Heederik, D., Wouters IM., Schlünssen V. 2015. A comprehensive review of levels and determinants of personal exposure to dust and endotoxin in livestock farming. *J Expo Sci Environ Epidemiol.* 25(2): 123–37.

Bünger J., Schappler-Scheele, B., Hilgers, R. Hallier E. 2007. A 5-year follow-up study on respiratory disorders and lung function in workers exposed to organic dust from composting plants. *Int. Arch. Occup. Environ. Health.* 80: 306–312.

De Hoog, G.S., Guarro, J., Gebé, J., Figueras, MJ. 2000. *Atlas of Clinical Fungi, 2nd ed.;* Centraalbureau voor Schimmelcultures: Utrecht, The Netherlands.

Douwes, J., Thorne, P., Pearce, N., et al. 2003. Bioaerosol health effects and exposure assessment: progress and prospects. *Ann Occup Hyg.* 47: 187–200.

European Centre for Disease Prevention and Control, 2013. Impact of environmental usage of triazoles on development of resistance to medical triazoles in Aspergillus spp., European Centre for Disease Prevention and Control, 2013. Stockholm.

Hawley, B., Schaeffer, J., Poole, J.A., Dooley, G.P., Reynolds, S., Volckens, J. 2015. Differential response of human nasal and bronchial epithelial cells upon exposure to size-fractionated dairy dust. *J Toxicol Environ Health A.* 78: 583–594.

Kirkeskov, L., Hanskov, D.J., Brauer, C. 2016. Total and respirable dust exposures among carpenters and demolition workers during indoor work in Denmark. *J Occup Med Toxicol.* 11: 45.

Knutsen, A.P., Bush, R.K, Demain J.G., et al. 2012 Fungi and allergic lower respiratory tract diseases. *J Allergy Clin Immunol.* 129: 280–291.

Leendertse, P. 2015. Azole fungicides, aspergillosis and preliminary "hot spots" Resistant fungi: new risk for humans.

Liebers, V., van Kampen, V., Bünger, J. 2012. Assessment of airborne exposure to endotoxin and pyrogenic active dust using electrostatic dustfall collectors (EDCs). *J Toxicol Environ Health A.* 75(8–10): 501–507.

Madsen, A.M. 2006. Exposure to airborne microbial components in autumn and spring during work at Danish biofuel plants. *Ann Occup Hyg.* 50: 821–831.

Madsen, A.M., Alwan, T., Ørberg, A., Uhrbrand, K., Jørgensen, MB. 2016. Waste Workers' Exposure to Airborne Fungal and Bacterial Species in the Truck Cab and During Waste Collection. *Ann Occup Hyg.* 60(6): 651–668.

Madsen, A.M., Zervas, A., Tendal, K. et al. 2015. Microbial diversity in bioaerosol samples causing ODTS compared to reference bioaerosol samples as measured using Illumina sequencing and MALDI-TOF. *Environ Res:* 140: 255–267.

Malta-Vacas, J., Viegas, S., Sabino, R., et al. 2012. Fungal and microbial volatile organic compounds exposure assessment in a waste sorting plant. *J Toxicol Environ Health Part A.* 75: 1410–1417.

Miaśkiewicz-Peska, E. 2011. Effect of Antimicrobial Air Filter Treatment on Bacterial Survival. *Fibres & Textiles In Eastern Europe.* 19(84): 73–77.

Morawska, L. 2001. *Environmental Aerosol Physics International Laboratory for Air Quality and Health.* Queensland University of Technology Brisbane, Australia.

Nielsen, K.F. 2003. Mycotoxin production by indoor molds* *Fungal Genetics and Biology* 39: 03–117.

Oppliger, A., Hilfiker, S., Vu Duc, T. 2005. Influence of seasons and sampling strategy on assessment of bio-aerosols in sewage treatment plants in Switzerland. *Ann Occup Hyg;* 49: 393–400.

Schenker, M. 1998. Respiratory Health Hazards in Agriculture. *American Journal of Respiratory and Critical Care Medicine.* 158(5): S1–S76.

Seedorf, J., Hartung, J., Schröder, M., et al., 1998. Concentrations and emissions of airborne endotoxins and microorganisms in livestock buildings in Northern Europe. *Journal of Agricultural Engineering Research,* 70(1): 97–109.

Sigsgaard, T., Schlunssen, V. 2004. Occupational asthma diagnosis in workers exposed to organic dust. *Ann Agric Environ Med.* 11(1): 1–7.

Snelders, E., Melchers, W.J., Verweij, P.E. 2011. Azole resistance in Aspergillus fumigatus: a new challenge in the management of invasive aspergillosis? *Science.* 80: 335–347.

Thilsing, T., Madsen, A.M., Basinas, I., Schlünssen, V, Tendal K, Bælum J. 2014. Dust, Endotoxin, Fungi, and Bacteria Exposure as Determined by Work Task, Season, and Type of Plant in a Flower Greenhouse. *Annals of Occupational Hygiene.* 59(2): 142–57.

Verweij, P.E., Chowdhary, A., Melchers, WJG, Meis, JF. 2016. Azole Resistance in Aspergillus fumigatus: Can We Retain the Clinical Use of Mold-Active Antifungal Azoles? *Clin. Infect. Dis.* 62: 362–368.

Viegas, C., Faria, T., Carolino, E., Sabino, R., Quintal Gomes, A. & Viegas, S. (2016) Occupational Exposure to Fungi and Particles in Animal Feed Industry. *Medycyna Pracy.* 67(2): 143–54.

Viegas, C., Faria, T., Gomes, A. Q., Sabino, R., Seco, A., Viegas, S. 2014. Fungal Contamination in Two Portuguese Wastewater Treatment Plants, *J Toxicol Environ Health A:* 77(1–3): 90–102.

Viegas, C., Sabino, R., Botelho, D., dos Santos, M., Quintal Gomes, A. 2015. Assessment of exposure to the Penicillium glabrum complex in cork industry using complementing methods. *Archives of Industrial Hygiene and Toxicology,* 66: 3.

Viegas, S., Almeida-Silva, M., Viegas, C. 2014. Occupational Exposure to Particulate Matter In 2 Portuguese Waste-Sorting Units. *Int J Occup Med Environ Health.* 27(5): 854–862.

World Health Organization (WHO). 1999. Hazard Prevention and Control in the Work Environment: Airborne Dust WHO/SDE/OEH/99.14

Wouters, I.M., Spaan, S., Douwes, J., et al. 2006. Overview of personal occupational exposure levels to inhalable dust, endotoxin, beta (1–>3)-glucan and fungal extracellular polysaccharides in the waste management chain. *Ann Occup Hyg.* 50: 39–53.

Occupational Safety and Hygiene V – Arezes et al. (Eds)
© *2017 Taylor & Francis Group, London, ISBN 978-1-138-05761-6*

Effect of a safety education program on risk perception of vocational students: A comparative study of different intervention methodologies

M.A. Rodrigues, C. Vales & M.V. Silva
Research Centre on Environment and Health, School of Health of Polytechnic Institute of Porto, Porto, Portugal

ABSTRACT: This study intends to analyze and compare the effect of a Safety Education Program (SEP) on vocational students' risk perception when different training methodologies are used. A SEP was designed and applied to 157 students of Vocational Education and Training (VET) programs. The sample was divided in three groups. In each group a different training methodology was applied: theory-based; demonstration-based and testimonies-based. To assess their effect on students' risk perception, the same survey questionnaire was applied two weeks before and after the SEP. Results showed a significant and important positive impact of the SEP. Comparisons among intervention methodologies showed differences in their effect on risk perception, where a greater effect was found for testimonies-based methodology. The results of this study emphasize the importance of a SEP in VET programs. However, the importance of a proper training methodology was demonstrated.

1 INTRODUCTION

Vocational Education and Training (VET) programs delivered to secondary school students are designed to prepare them to foothold in the job market, addressing the skills and employment needs. According to the Portuguese legislation (Decree-Law nº 139/2012), VET programs are essential to motivate young people to complete a program of training that can qualify them for employment. Therefore, they are seen as one of the most important ways to transit to an active life, and, simultaneously, can prepare students to continue the studies. In these programs, teaching at a vocational school alternates with practical training in a company.

VET programs are also of particular interest for occupational safety, since students can be exposed to the same risks than workers during practical training. This issue becomes even more relevant when there is evidence that young workers have an increased risk to be injured at work (see e.g. Breslin & Smith, 2005; Morassaei et al., 2013). This problem among working students was emphasized by Balanay et al. (2014). The authors found that a considerable percentage of the analyzed working students have experienced an injury at work. Also Raykov & Taylor (2013), in a study with youth apprentices, found a high incidence of injuries among youth who had participated in high school apprenticeship programs in Alberta and Ontario.

As far the concern about the increased risk of occupational accidents among young workers/students grew, explanations have emerged. Several authors as point that teens and young workers tend to have limited knowledge on occupational safety, particularly in what regards to their rights and duties, as well as about employers' responsibilities (see e.g. McClosky, 2008; Andersson et al., 2014). This was emphasized for students of VET programs by Andersson et al. (2014). Additionally, this group was also found to experience more unsafe working conditions than older workers, as well as to be willing to take risks (Breslin et al., 2007; McClosky, 2008; Lavack, 2008).

It is against this background that strategies to incorporate proper occupational safety training in the VET programs curriculum assumes particular importance. This issue is also framed in the National Strategy for Safety and Health at work 2015–2020 (Resolution of the Ministries Council nº77/2015), which emphasizes the role of education at schools in promoting the concepts of health, safety and well-being at work. However, to be effective, appropriate strategies and methodologies for training should be implemented. In view of this, the main objective of this study is to analyze and compare the effect of a Safety Education Program (SEP) on vocational students' risk perception when different training methodologies are used. These are the first results of the project "Schools Promoting of Safety", which intends to develop an intervention strategy for the schools belonging to a municipal of the North of Portugal.

2 METHODOLOGY

2.1 Sample

Data collection was carried out in two Portuguese vocational schools from the North of Portugal.

Only students from the last year of courses related to mechanics and electronics were selected. According to these criteria, a total of 157 students were involved in this study. Most of the participants were males (95.2%), and their mean age was 18.29 years old (SD = 1.47).

2.2 Study design and procedures

A SEP was designed and applied to the students by using three different types of training methodologies. In view of this, in each school, three intervention groups were defined.

To assess the effect of the SEP on students' risk perception when the different training methodologies were used, a same questionnaire (see sub-section 2.4) was administered before the SEP program and again later.

Participants were informed about the purpose of the study. It was explained that their answers would be treated confidentially and only used for the purpose of the present study.

2.3 Safety Educational Program (SEP) and training methodologies

With focus on the VET programs that the students included in this study were attending, a SEP was defined. In view of this, the designed program embraced the main risks related to handling machinery and maintenance tasks in industrial settings. The SEP was divided in four parts: (1) Introduction to industrial accidents; (2) Workers' duties and rights and employers' responsibilities; (3) Risks and control measures; (4) Workers' involvement in the improvement of occupational safety. The subjects of analysis were: mechanical risk, chemical risk and specific risks such as electrical risk and explosive atmospheres. Risks related to physical agents such as noise and vibrations were also covered, because some schools mentioned it as also important to be lectured along the sessions.

The SEP was lectured through three different training methodologies: theory-based; demonstration-based and testimonies-based. In the first methodology, theoretical contents were presented using an expositive approach supported in a power-point format presentation. On the other hand, in demonstration-based methodology, the presentation of the different risks and control measures were based on real working context images and videos. In the testimonies-based methodology testimonies from workers that have suffered an accident due to handling a machine or due to perform maintenance activities were used to explain the risks and the corresponding control measures.

The SEI was applied by the research team in 90-min lectures, with time for discussion.

2.4 Questionnaire

The questionnaire contained two main parts, which included questions covering a few important topics. In first part questions for students' characterization were presented, particularly in what regards to students' age, gender, school name, course and class name, if they knew someone who have suffered a serious work accident, as well as if they had previous training on safety and for how many hours. In the second part, questions to assess specific dimensions were included. To the aim of this study is only relevant the 7-item question to assess the students' risk perception (Table 1–items reworded as positive). The risk scenarios presented in the different items where adapted from Breslin & Smith (2005), Seo (2005) and Håvold (2010) studies. The participants were asked to assess the option that best corresponded to their level of agreement by using a five-point Likert scale (1 = Strongly disagree; 5 = Strongly agree). A pre-test of the questionnaire was performed with 10 students. They were requested to review, examine and test it.

Students were notified that the questionnaire was anonymous and that all collected data were to be used by the researchers for scientific purposes only.

2.5 Data analysis

Basic descriptive statistics were computed for all variables. Because participants could not be linked successfully due to anonymity requirements, data were treated as independent for statistical analysis. Furthermore, because variables were ordinal, non-parametric tests were applied. Therefore, Mann–Whitney U test was used to compare the rankings of each item before and after the SEP and Kruskal–Wallis test to compare differences between the three types of interventions.

The significance level was considered as $\alpha = 5\%$. Data analysis procedures were performed using the sta-

Table 1. Items included to assess students' risk perception.

---

1. *Risk Perception* (1 = Strongly disagree; 5 = Strongly agree)

1.1. All workplaces have risks.

1.2. Not comply with the safety rules is unacceptable, even when it does not bring any problem for other people.

1.3. Younger workers are more likely to suffer an occupational accident than the older ones.

1.4. Even using PPE, I can have an accident.

1.5. If dangerous parts of machines are protected, occupational accidents are more difficult to occur.

1.6. If I follow the procedures and safety rules, is less likely to suffer an occupational accident.

1.7. Smoke at the workplace carries risks.

---

tistical software package Statistical Package for Social Sciences (IBM SPSS® version 20, Inc., Chicago, Ill).

3 RESULTS AND DISCUSSION

The effect of the SEP on students' risk perception through the application of three different training methodologies was analyzed, being the results presented in Tables 2–4. This analysis was important since several authors have pointed the significance of the strategies used at the schools for safety programs (see e.g. Pisaniello et al., 2013). However, there is still limited evidence about its real impact, particularly in what regards to risk perception.

In what concerns to theory-based methodology (Table 2), statistically significant improvements in risk perception levels were observed for all the seven items. However, concordance levels remain low for the item 1.5. In general, the obtained results suggest that even through the use of traditional training methodologies, frequently pointed as less engaging (Pisaniello et al., 2013), a SEP still can have an important effect on students' risk perception.

Table 2. Risk perception level before and after the SEP for theory-based methodology.

| Item | Theory | | |
	Q1 (x±sd)	Q2 (x±sd)	P-value
1.1.	3.26 (1.16)	3.79 (0.88)	0.027
1.2.	3.58 (1.07)	4.33 (1.08)	0.000
1.3.	2.35 (1.07)	4.26 (1.07)	0.000
1.4.	3.16 (0.99)	3.65 (1.28)	0.026
1.5.	3.00 (0.84)	3.51 (138)	0.044
1.6.	3.60 (1.25)	4.30 (1.18)	0.001
1.7.	3.51 (1.01)	4.06 (1.83)	0.007

Note: Q1 = Questionnaire before SEP; Q2 = Questionnaire after SEP.

Table 3. Risk perception level before and after the SEP for demonstration-based methodology.

| Item | Demonstration | | |
	Q1 (x±sd)	Q2 (x±sd)	P-value
1.1.	3.61 (0.91)	4.75 (0.48)	0.000
1.2.	3.50 (1.15)	4.36 (0.88)	0.000
1.3.	2.54 (1.11)	4.37 (0.93)	0.000
1.4.	3.14 (1.03)	3.68 (1.04)	0.007
1.5.	2.89 (1.12)	3.09 (1.23)	0.388
1.6.	4.21 (0.89)	4.46 (0.66)	0.151
1.7.	3.48 (1.09)	4.37 (1.05)	0.000

Note: Q1 = Questionnaire before SEP; Q2 = Questionnaire after SEP.

Table 4. Risk perception level before and after the SEP for testimonies-based methodology.

| Item | Testimonies | | |
	Q1 (x±sd)	Q2 (x±sd)	P-value
1.1.	2.74 (0.89)	4.72 (0.56)	0.000
1.2.	2.91 (0.90)	4.45 (0.75)	0.000
1.3.	2.22 (0.73)	4.64 (0.48)	0.000
1.4.	2.41 (0.79)	4.28 (0.79)	0.000
1.5.	2.27 (0.70)	4.16 (0.87)	0.000
1.6.	3.97 (0.82)	4.69 (0.60)	0.000
1.7.	3.26 (0.71)	4.64 (0.61)	0.000

Note: Q1 = Questionnaire before SEP; Q2 = Questionnaire after SEP.

Table 3 presents the results of the demonstration-based methodology. Significant differences between both moments were found for five of the seven items. In fact, results showed that for the groups where this training methodology was applied, at the end of the SEP, students continued to see the importance of the use of machine protections in the same way, i.e., not as important as it was supposed. In what regards to the importance of safety procedures and rules, perceptions of students that have been subjected to this intervention were already considerable high before the SEP and remain in this way at the end of the intervention. However, it is important to note that when comparing the obtained results after the SEP with the ones found in the group where the theory-based methodology was applied, higher mean values were achieved with the demonstration, apart from the item 1.5.

Results of the effect of the SEP when the testimonies-based methodology was applied are presented in Table 4. Significant differences between both moments were found for all the items. In fact, this was the group where we could observe a greater impact of the SEP. In average, most of the students "agree" or "strongly agree" with the scenarios, even in what regards to the importance of the use machine protections.

A comparison of the results obtained after the three intervention methodologies be applied showed significant differences in their effect in four of the seven items (p < 0.05 for the items 3.1., 3.4., 3.5. and 3,7.). Higher mean rankings were achieved when the testimonies-based methodology was applied. Therefore, the presentation of workers' testimonies to explain risk scenarios and discuss control measures had a higher effect in the students' perceptions about workplaces risks, in the way as they perceived the important of the use PPE or machine protections, as well as on the risk related to some behaviors such as to smoke at the workplace.

The greatest effect of the SEP when the testimonies-based methodology was applied can be related to the students' engagement along the session (Burk et al., 2006; Burke et al., 2011; Pisaniello et al., 2013). Video materials are seen as an appropriate way for delivering safety contents at classroom settings, due to their potential to raise awareness, promote group discussion and to provide graphic illustration (Pisaniello et al., 2013). This was observed in both demonstration and testimonies-based methodologies. However, the content of the videos showed to be significant for study results. A higher engagement was possibly promoted when videos where examples of machines working with and without protections and of potential consequences (demonstration-based methodology) were replaced by videos of workers explaining the accident that they have suffered, showing the physical consequences and explain what could they have done differently (testimonies-based methodology). This last approach, may have make the students more aware about the occurrence of certain accident scenarios presented and discussed at the classrooms, making them more involved along the sessions. The importance an engaging training was emphasized by Burk et al. (2006) and later by Burke et al. (2011), mainly in what regards to safety knowledge acquisition, implications in level of safety performance and reduction of occupational accidents and injuries.

This is also a good strategy to be adopted by teachers that have little or no experience in occupational safety, in order to empower the student with the necessary knowledge and to awareness them about the occupational risks that they may face along the practical training.

4 CONCLUSIONS

This study assessed the effect a SEP using three different training methodologies. The obtained results showed, in general, a positive effect of the SEP on vocational students' risk perceptions. However, the importance of a proper training methodology was demonstrated. The use of workers' testimonies was emphasized to promote students' engagement along the sessions and consequently, to have a greater effect on their risk perceptions. Such findings highlight the importance of SEP to reduce this risk in the workplace of this group of students, not only during the practical training, but also along their job lives

The results presented in this paper are a part of a greater study; only the preliminary results are presented. In fact, and despite the importance of the results obtained and presented, there is a need for more research focused on attaining a better understanding of how occupational safety education at school settings can be made effective.

REFERENCES

Andersson, I.-M., Gunnarsson, K., Rosèn, G., Åberg, M.M., 2014. Knowledge and Experiences of Risks among Pupils in Vocational Education. Safety and Health at Work, 5, 140–146.

Balanay, J.A., Adesina, A., Kearney, G.D., Richards, S.L., 2014. Assessment of occupational health and safety hazard exposures Among working college students. American Journal of Industrial Medicine, 57, 114–124.

Breslin, F.C., Smith, P., 2005. Age-related differences in work injuries: a multivariate, population-based study. American Journal of Industrial Medicine, 8, 50–56.

Breslin, F.C., Day, D., Tompa, E., Irvin, E., Bhattacharyya, S., Clarke, J., Wang, A., 2007a. Non agricultural work injuries among youth. A systematic review. American Journal of Preventive Medicine, 21 (2), 151–162.

Burke, M.J., Sarpy, S.A., Smith-Crowe, K., Chan-Serafin, S., Salvador, R.O., Islam, G., 2006. Relative effectiveness of worker safety and health training methods. American Journal of Public Health, 96(2), 315–324.

Burke, M.J., Smith-Crowe, K., Salvador, R.O., Chan-Serafin, S., Smith, A., Sonesh, S., 2011. The dread factor: How hazards and safety training influence learning and performance. Journal of Applied Psychology, 96(1), 46–70.

Decreto Lei nº 139/2012 de 5 de julho. [Decree Law nº 139/2012 of 6 July] Diário da República, 1ª Série, nº 129. Lisboa, Portugal: Concelho de Ministros.

Håvold, J.I., 2010. Safety culture aboard fishing vessels. Safety Science 48(8), 1054–1061.

Lavack, A.M., Magnuson, S.L., Deshpande, S., Basil, D.Z., Basil, M.D., Mintz, J.H., 2008. Enhancing occupational health and safety in young workers: the role of social marketing. International Journal of Nonprofit and Voluntary Sector Marketing, 13 (3), 193–204.

McClosky, E., 2008. The Health and Safety of Young People at Work: A Canadian Perspective. International Journal of Workplace Health Management, 1(1), 41–49.

Morassaei, S., Breslin, F.C., Shen, M., Smith, P.M., 2013. Examining job tenure and lost-time claim rates in Ontario, Canada, over a 10-year period, 1999–2008. Occupational and Environmental Medicine, 70, 171–178.

Pisaniello, D.L., Stewart, S.K., Jahan, N., Pisaniello, S.L., Winefield, H., Braunack-Mayer, A., 2013. The role of high schools in introductory occupational safety education—Teacher perspectives on effectiveness. Safety Science, 55, 53–61.

Raykov, M., Taylor, A., 2013. Health and safety for Canadian youth in trades. Just Labour: A Canadian Journal of Work and Society, 20, 33–49.

Seo, D.C., 2005. An explicative model of unsafe work behavior. Safety Science, 43(3), 187–211.

Occupational Safety and Hygiene V – Arezes et al. (Eds)
© *2017 Taylor & Francis Group, London, ISBN 978-1-138-05761-6*

Ergonomic design intervention in a coating production area

M.F. Brito
DEGEIT—University of Aveiro, Aveiro, Portugal

A.L. Ramos
GOVCOPP Research Centre, DEGEIT—University of Aveiro, Aveiro, Portugal

P. Carneiro
ALGORITMI Centre, University of Minho, Portugal

M.A. Gonçalves
ISEP—Polytechnic of Porto, Portugal

ABSTRACT: The aim of this study is to redesign two workstations in a PVD coating production area, considering productivity and ergonomic aspects. Through the elimination of wastes such as unnecessary movements and transportations and by reducing the awkward postures as arm flexion larger than 45°, the productivity in the loading and unloading workstations increased 9% and 5%, respectively, and the ergonomic risk was improved from medium to acceptable. RULA was the chosen method to evaluate the ergonomic situation and anthropometric studies were performed to find the ideal ergonomic solution. This study shows the importance to consider ergonomic conditions when designing or redesigning a workstation in order to get effective productivity improvements.

1 INTRODUCTION

Due to demographic variation, fewer young workers are available and the overall number of workers will decrease. The length of absenteeism, especially due to Musculoskeletal Disorders (MSDs), increases with age (Müglich et al., 2015).

Work-Related Musculoskeletal Disorders (WMSDs) cause muscles, tendons and nerves at the joint of the neck, shoulder, elbow, wrist, finger, back, leg, etc. to be stressed and traumatised due to excessive or repetitive exertive force, awkward body posture, less resting time, cold working environment, vibration and so on (Cheol-Min et al., 2011).

With regard to Europe, the data emerging from the 5th European Survey on Working Conditions (ESWC) in 2010 (Eurofund, 2012) reports that 33% of all European workers spend at least 25% of their working time performing manual load handling. About 47% of the labour force is exposed to awkward postures during at least 25% of their working time, and over 33% of European workers perform repetitive movements of the upper limbs for almost their entire working time.

Moreover, when considering WMSDs as an occupational disease, upper limb MSDs such as hand-arm tendonitis, epicondylitis and carpal tunnel syndrome represent more than 55% of all occupational claims reported in the different insurance systems (Eurostat, 2010). It was reported by Muggleton et al. (1999) that rotator cuff tendonitis is closely associated with the upper arm abduction and forward flexion. It has been shown that, with arms raised or abducted, the blood vessels supplying the tendons of the supraspinatus muscles were compressed (Grieco et al., 1998), thus altering blood circulation. Such postures render the shoulder-arm system vulnerable to MSDs.

An ergonomic approach to the design of an industrial workstation attempts to achieve an appropriate balance between the worker capabilities and the work requirements to "optimize" both worker productivity and the total system productivity, as well as to provide worker physical and mental well-being, job satisfaction and safety. In a real world design situation, the implementation of the recommendations or guidelines needs the matching of the population anthropometry with the various components of the workstation (Das and Sengupta, 1996).

Often, in industry, the workstation is designed in an arbitrary manner, giving little consideration to the anthropometric measurements of the potential user. The situation is aggravated by the non-availability of usable design parameters or dimensions (Das and Grady, 1983a; Das, 1987). The physical dimensions in the design of an industrial workstation are of major importance from the viewpoint of production efficiency, and operator physical and mental well-being. Small changes in workstation dimensions can have a considerable impact on worker productivity, and occupational health and safety. Inadequate posture from an improperly designed workstation causes static muscle efforts, eventually resulting in acute localized muscle fatigue, and consequently in decreased performance and productivity, and enhanced possibility of operator related health hazards (Corlett et al., 1982). The aim of this work is to answer the research question: "How can be improved the workstation design of loading and unloading processes of a PVD coat production area, considering ergonomic aspects and productivity?" This case study takes place in a PVD coating production area, where workers' complaints due to shoulder pains were rising considerably. These com-plaints come mainly from the processes of loading and unloading pieces from the suspension, before and after the product entering the PVD machine, respectively. This is a repetitive job and involves several awkward postures such as: flexion of the arms above 45° (from now on "arms up"), trunk flexion, and move manually heavy suspensions. Being such a specific case study, an identical case was not found in the literature.

The paper is structured as follows: the section 2 explains the methods used to evaluate the initial situation followed by the methods used to redesign the workstation; section 3 provides a discussion of the main results and section 4 points out some conclusions and recommendations.

2 METHODS

The methodology used was the case study. According to Yin (2003), a case study should be defined "... as a research strategy, an empirical inquiry that investigates a phenomenon within its real-life context." Following this key idea, the case study, as a research methodology, helps to understand, explore or describe a given system/problem in which several factors are simultaneously involved, in a real context.

The first step was the election of a multifunctional team, including operators, to analyze the process and measure the initial situation in terms of ergonomic conditions and productivity. Then this team suggested some workstation modifications in order to improve ergonomic conditions, reduce wastes (e.g., unnecessary movements and transportations) and increase productivity. After the implementation of the suggested improvements, the team measured the productivity and the ergonomic conditions and compared them with the base scenario.

Despite the good results in the first redesign intervention, they weren't enough to achieve acceptable ergonomic risk. It was necessary to intervene again, breaking some paradigms and designing the "ideal" workstation that suited any worker with no wastes in terms of movements and transportations.

2.1 Measurement tools

RULA (Rapid Upper Limb Assessment) was the tool used to assess the postures, movements and forces exerted by the worker while performing the job, because it is especially useful for scenarios in which work-related upper limb disorders are reported.

The higher the RULA score—varies from 1 to 7, defining the action level to be taken—the higher risk associated and the greater the urgency to carry out a more detailed study and introduce modifications to the job/workstation. The scores 1 and 2 (action level 1) indicates that the posture is acceptable if it is not maintained or repeated for long periods of time. The scores 3 and 4 (action level 2) indicates that further investigation is needed. The scores 5 and 6 (action level 3) indicates that changes are required soon. The score 7 or more indicates that changes are required immediately.

The knowledge of the team in lean production was important in the achievement of the better solution in terms of productivity. The key idea of lean is "doing more with less", where less means less space, less inventory, fewer resources, among others (Womack et al. 1990).

Productivity was calculated using the number of pieces produced per hour (throughput or production rate) because it is the measure typically used in this production area, being also one of the most well-known measures of productivity in industry.

2.2 Workstation redesign

The biggest team concern was the manually suspension movement between the carpet and the table due to the effort and the awkward posture necessary to perform this task and because it involves two kind of wastes: movement and transportation. Waste means, in a lean terminology, something that doesn't add value to the product, this means something that the client doesn't pay for (Womack et al., 1990). The other concern was the elevation of the arms considering the ergonomic aspects and the tiredness accused by operators, also contributing to a loss in productivity (Figure 1).

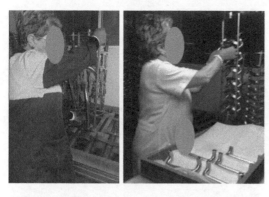

Figure 1.   Unloading workstation before improvements.

Figure 3.   Unloading workstation after improvements.

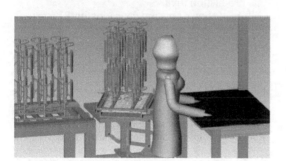

Figure 2.   Unloading workstation after improvements.

Figure 4.   Container changing.

The founded solutions for these detected problems were the following:

– Construction of a structure to place the lighter suspensions horizontally and reduce the time of arms up.
– Integration of a structure with a rotating base at the end of the machine carpet to load and unload pieces directly and eliminate the necessity of take and move manually the suspension between the carpet and the table (Figure 2).
– The new structure allows a manual adjustement of the work plan to reduce the arms flexion (Figure 2).

The Figure 3 depicts the worker in the unloading workstation after the implementation of these improvements.

The implementation of an ergonomic solution was also necessary for the container changing process.

The Figure 4 depicts the awkward posture adopted in this process. The container has an average weight of 6 kg but could rise to 9 kg maximum.

The solution was the implementation of a lift car, similar to the one in Figure 5.

Figure 5.   Lift Car.

2.2.1   *Anthropometrics studies*

Anthropometric studies were used to redesign the structure and take into account the adjustment of the workstation to the body characteristics of the operators, e.g., their stature.

In order to adjust the work plan, and eliminate the necessity of arms up above 45°, it was provided an automatism to up and down the suspension, based on the standard cycle time for producing each reference.

It was also provided an option to change from automatic to manual, when worker have difficulties to accomplish cycle time, for some reason.

The existing paradigm of the grids suspensions in rectangular shape was overcome and a round shape was elected (Figure 6). The advantage of this change is the reduction of the distance between operator, suspension and table, resulting in less movements such as trunk rotation.

The vertical amplitude of the structure was calculated based on the anthropometric database of the Portuguese population (Barroso et al., 2005): the maximum limit was calculated using the measure of floor-to-elbow of the man's 95 percentile (1159 mm) and the minimum limit was calculated by using the measure of floor-to-elbow of the woman's 5 percentile (914 mm). This structure

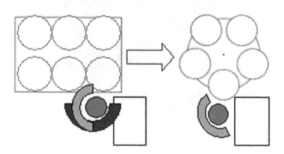

Figure 6. Suspension grid shape: rectangular vs round.

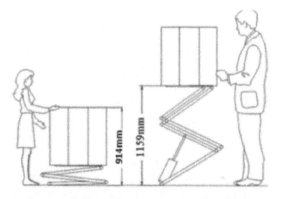

Figure 7. Structure automatized to elevating the suspensions.

also includes a rotary base to bring the suspension closer to the worker.

In Figure 7 it is possible to see that the proposed solution allows different types of workers to perform their job without elevating their arms above 20°.

Like any investment, the costs component is very important in the decision of forward or not with the project. The estimated cost to implement this solution is about 2700€.

3 RESULTS

Productivity was calculated for 23 references which represent 80% of the total quantity produced in this production area.

Table 1 shows how much productivity increased: about 9% in the load operation and 5% in the unloading operation. This difference is due to the bigger distance between the table and the carpet in the loading workstation than the same distance in the unloading workstation.

According to RULA method (McAtamney & Corlett, 1993), the most inappropriate postures before the improvements were moving the suspension and container changing (scored with 6).

The moving suspension task was eliminated and the RULA score to perform the task of changing container was reduced from 6 to 4 by the implementation of the lift car.

Table 2 and Table 3 depicts the RULA score and the percentage of time spent in each posture in

Table 1. Productivity (throughput in pieces/hour).

Workstation	Initial situation	After improvements
Loading	800	872
Unloading	900	945

Table 2. RULA score and percentage of time spent in each posture (loading workstation) before and after the ergonomic improvements.

Loading workstation	Initial situation		After improvements	
	RULA	Time	RULA	Time
Arms Flexion −20° to 20°	–	–	3	14%
Arms Flexion 20° to 45°	4	56%	4	52%
Arms Up (>45°)	5	29%	5	24%
Move Suspension	6	5%	–	–
Container Changing	6	10%	4	10%

Table 3. RULA score and percentage of time spent in each posture (unloading workstation) before and after the ergonomic improvements.

Unloading workstation	Initial situation RULA	Time	After improvements RULA	Time
Arms Extension/Flexion −20° to 20°	–	–	3	14%
Arms Flexion 20° to 45°	4	57%	4	51%
Arms Up (>45°)	5	30%	5	25%
Move Suspension	6	3%	–	–
Container Changing	6	10%	4	10%

Table 4. RULA score and percentage of time spent in each posture (loading workstation) before and after the final redesign implementation.

Loading workstation	Initial situation RULA	Time	After redesign* RULA	Time
Arms Extension/Flexion −20° to 20°	–	–	3	90%
Arms Flexion 20° to 45°	4	56%	–	–
Arms Up (>45°)	5	29%	–	–
Move Suspension	6	5%	–	–
Container Changing	6	10%	4	10%

*Estimated values.

Table 5. RULA score and percentage of time spent in each posture (loading workstation) before and after the final redesign implementation.

Unloading workstation	Initial situation RULA	Time	After redesign* RULA	Time
Arms Extension/Flexion −20° to 20°	–	–	3	90%
Arms Flexion 20° to 45°	4	57%	–	–
Arms Up (>45°)	5	30%	–	–
Move Suspension	6	3%	–	–
Container Changing	6	10%	4	10%

*Estimated values.

the initial situation and after improvements at the loading and unloading workstation, respectively. The time spent in a position implying arms up was reduced from 29% to 24% through the implementation of the horizontal structure and by lowering the work plan. The ideal posture of the arms (arms between −20° to 20°) is achieved when the worker uses the horizontal structure (14% of the time).

Table 6. RULA score summarize.

Workstation	Initial situation	After improvements	After redesign*
Loading	5	4	3
Unloading	5	4	3

*Estimated values.

In the initial situation, the weighted average was 5 for both workstations indicating that investigation and changes are required soon.

After the workstation improvements, the action level decreased from 3 to 2 means that more changes may be needed to reach the negligible level (action level 1). For this reason, another workstation redesign was performed, taking into account the anthropometric aspects and the elimination of awkward postures, i.e. trunk flexion and arms up. The team estimated that with this redesign the worker would perform 90% of their work with arms extension/flexion between −20° to 20° (Table 4 and Table 5).

After the new workstation redesign, the ergonomic risk could be reduced from the level 4 to the level 3. Although the good results, they are not enough to reach the risk level 1 – acceptable risk. The reason is the repetitiveness of the tasks. A possible solution could involve the enlargement of the job.

Table 6 summarizes the RULA score from the initial situation to the final redesign in both workstations.

Despite the demonstration made by the team of the working conditions improvements after the redesign implementation, the company decided not to proceed with the redesign due to the high investment value.

4 CONCLUSIONS

Due to the hard competition, demanding customers and competitive world that companies face, nowadays, it is very important to consider productivity measures while implementing improvements in the shop-floor. On the other hand, jobs are more repetitive leading to musculoskeletal disorders, increasing absenteeism and reducing productivity.

The conclusions of this study are limited to this case, but the authors believe that is possible to consider both aspects, ergonomic conditions and productivity, during improvements implementation.

As illustrated in the section of results, the improvements reached in the ergonomic conditions can contribute very positively for productivity increases.

The authors' opinion is that ergonomic conditions must be considered when designing/redesigning a workstation in order to get effective productivity improvements. Actually, in general, it is still difficult to implement ergonomic aspects in companies because some decision-makers do not view ergonomics as an investment, but rather as an expense.

REFERENCES

Barroso, M.P., Arezes, P.M., Costa, L.G., and Miguel, S. 2005. Anthropometric study of Portuguese workers. *International Journal of Industrial Ergonomics* 35(5): 401–410.

Cheol-Min L., Myung-Chul J. and Yong-Ku K. 2011. Evaluation of upper-limb body postures based on the effects of back and shoulder flexion angles on subjective discomfort ratings, heart rates and muscle activities. *Ergonomics* 54 (9): 849–857.

Corlett, E.N., Bowssenna, M. and Pheasant, S.T. 1982. Is dis- comfort related to the postural loading of the joints? *Ergonomics* 25:315–322.

Das, B. 1987. An ergonomics approach to the design of a manufacturing work system. *International Journal of Industrial Ergonomics* 1: 231–240.

Das, B. and Arijit K. Sengupta 1997. Industrial workstation design: A systematic ergonomics approach. *Applied Ergonomics* 27:157–163.

Das, B. and Grady, R.M. 1983a. Industrial workplace layout design: An application of engineering anthropometry. *Ergonomics* 26: 433–44.

EUROFOUND. 2012. Fifth European Working Conditions Survey. *Publications Office of the European Union, Luxembourg.*

EUROSTAT. 2010. Health and Safety at Work in Europe (1999–2007). *A Statistical Portrait. Publications Office of the European Union, Luxembourg.*

Grieco, A., Molteni, G, De Vito, G. and Sias, N. 1998. Epidemiology of musculoskeletal disorders due to biomechanical overload. *Ergonomics* 41: 1253–1260.

McAtamney and Corlett. 1993. RULA: A survey method for the investigation of work-related upper limb disorders. *Applied Ergonomics* 24: 91–99.

Muggleton, J.M., Allen, R., and Chappell, P.H. 1999. Hand and arm injuries associated with repetitive manual work in industry: a review of disorders, risk factors and preventive measures. *Ergonomics* 42: 714–739.

Müglic. 2015. Development of a database for capability-appropriate workplace design in the manufacturing industry. *Occupational Ergonomics* 24: 109–118.

Womack J.P., Jones D.T., and Ross D. 1990. The Machine That Changed the world. *Free Press, New York.*

Yin. 2003. Applications of case study research. *Sage Publications, Inc, California SA.*

Occupational Safety and Hygiene V – Arezes et al. (Eds)
© 2017 Taylor & Francis Group, London, ISBN 978-1-138-05761-6

Eye blinking as an indicator of fatigue and mental effort in presence of different climatic conditions

R.P. Martins
Research Laboratory on Prevention of Occupational and Environmental Risks (PROA/LABIOMEP), Faculty of Engineering, University of Porto, Portugal

J.M. Carvalho
DEM/CERENA, Faculty of Engineering, University of Porto, Portugal

ABSTRACT: Eye blinking pattern can be modified as a result of fatigue or mental effort or in the presence of climatic extrinsic factors such as temperature and relative humidity. The aim of this study is to compare subject's eye blinking frequency when exposed to different climatic environments and its relationship with mental effort and fatigue. A subjective experiment was conducted in a climate chamber with two different conditions of temperature and relative humidity. In this study, mental workload consisted of a cognitive task performed during 50 minutes with the simultaneous capture of eye blinking. A thermal comfort questionnaire was filled up when the subject entered the climate chamber and after the task was finished. There is indication of a slight influence of high temperature and high relative humidity on eye blinking and consequently mental effort and fatigue.

1 INTRODUCTION

Eye blinking lubricates and cleans the corneal surface and can be classified into three types: voluntary blinking that results from a conscious decision to close and open the eyelids; reflex blinking as a rapid closure movement of short duration that does not require the involvement of cortical structures; and spontaneous blinking that occurs without any external stimulus and are associated with the psycho-physiological state of the person (Stern et al. 1994, Acosta et al. 1999).

Eye blinking may also be used as a predictor of fatigue or mental effort (Stern et al. 1994, Clapp & Hively 1997, Fukuda et al. 2005) even though it is often considered a biological artifact in the Electroencephalographic (EEG) signal and in consequence is discarded in posterior EEG data analysis and interpretation. The major determinants of the Blinking Frequency (BF) are extrinsic factors, such as temperature, Relative Humidity (RH) and lighting conditions (Clapp & Hively 1997, Wolkoff & Kjærgaard 2007). However, it has to be noted, that each subject has a specific eye blinking pattern with intra- and inter-individual variability of varying degree (Stern et al. 1994, Caffier et al. 2003, Schleicher et al. 2008, Benedetto et al. 2010). The daily variation of the BF depends on a number of visual and environmental conditions (Wolkoff et al. 2005). Some

authors say that high RH reduces the BF (Tsubota et al. 1997). Other works suggests that RH below 20% may have decreasing effects on eye blinking rate (Clements-Croome 2008, Rozanova et al. 2009) and directly affect tear film stability, so a range of 40–60% RH is acceptable (Clements-Croome 2008). The increase in water evaporation may be caused by thermal factors (low RH, high room temperature), demanding tasks (concentration decreases blinking and widens the exposed ocular surface area), and individual characteristics (Wolkoff et al. 2005). An increase of 1°C in indoor air temperature above 22°C can approximately decrease the performance of office work by 1% (Clements-Croome 2008). In Tanabe et al. (2007) study, more cerebral blood flow was required to maintain the same level of task performance in the hot condition (33°C) than at a thermal neutral condition (26°C). The importance of monitoring fatigue and mental effort is enhanced by its influence in reducing the occurrence of human failures, increasing safety and productivity and to allow the evaluation of task performance (Tanabe et al. 2007, Iampetch et al. 2012, McIntire et al. 2014). There also appears to be an inverse relation between difficulty of task and BF (Caffier et al. 2003, Wolkoff et al. 2005, Wilson et al. 1994, Zheng et al. 2012, Wascher et al. 2014). The aim of this study is to compare subject's BF when exposed to different thermal environments and its relationship with mental effort and fatigue.

2 METHODOLOGY

To study the effect of moderately high temperature and high RH on task performance, eye blinking, fatigue and mental effort, a subjective experiment was conducted in a climate chamber FITOCLIMA 25000, localized at the Faculty of Engineering of the University of Porto.

10 subjects—all men; average age 26.2 ± 5.63 years; age range 18–34 years; average height 173.6 ± 8.54 cm—took part in this study. All of them were informed of the experiment's aims and conditions, and signed in an informed consent form. The climate chamber was conditioned at operative temperatures and RH, divided in two different conditions: Condition 1, 18°C and 40% RH; Condition 2, 32°C and 80% RH; both with still air. Subjects were filmed with web camera (Logitech HD Pro Webcam C920, Carl Zeiss Tessar HD 1080p). It was also used BioPlux device to monitor skin temperature, Equivital system to monitor core temperature and EEG signals were recorded to assess the similarities and differences in brain activity as a function of performance in the task—these data are not reported here.

Before entering the climate chamber subjects waited in a sedentary position for 20 minutes, until stabilizing skin temperature, and answered questions about their lifestyle. After entering the climate chamber, subjects answered a thermal comfort questionnaire designed from recommendations of ISO 10551 (1995) and reported their first thermal perception, their preference and the existence of symptoms. In this study, mental workload consisted of a cognitive task executed during 50 minutes. A Go/No-go task was used to generate low mental workload. In this task subjects had to make a motor response (go) and to withhold a response (no-go). In the end of the task, subjects needed to answer the thermal comfort questionnaire once more. In the eye blinking counting it was attempted to minimize human error by getting two investigators to manually count the blinks in the videos and any discrepancies between the two of them were solved by counting the blinks in that particular video again. In this study eye blinking is considered to be the rapid closing and reopening of the eyelid, everytime pupil is fully overlapped by the eyelid.

3 RESULTS AND DISCUSSION

3.1 Thermal comfort questionnaire

General perception of subjects about the temperature inside the Climate Chamber was indicated by a tick on a 7-point scale.

In Condition 1, after entering the climate chamber subjects reported their first thermal sensation: neutral (20%), slightly cool (50%), cool (20%), and cold (10%). Subjects preferred a thermal sensation between neutral (30%) and hot (70%). 10% of subjects presented sleepiness while 90% did not show any symptom. At the end of the cognitive task, subject's answers slightly differ from the previous answers because their perception of thermal sensation varied between slightly warm (10%), slightly cool (40%) and cold (50%). All subjects preferred a hotter thermal environment. They experienced sleepiness (40%), drowsiness (40%), chills (30%), apathy (10%), anxiety (10%) and loss of motor coordination (10%), while 20% did not show any symptom.

In Condition 2, at the entrance and at the end of the test 90% of subjects had the perception of a hot environment within the Climate Chamber, while 10% had the perception of a warm environment. Subjects preferred a cooler environment (90%), while 10% preferred a neutral environment. At the entrance, 70% subjects showed no symptom, 20% had drowsiness, 10% had shortness of breath. At the end of task subjects showed drowsiness (60%), sleepiness (40%), seizures (10%), loss of concentration (10%) and 30% of the subjects did not show any symptom.

3.2 Eye blinking

In Condition 1 (Figure 1), subject number 6 had an overall average number of blinks per minute, BF, of 23.54 ± 9.96. Over time there are oscillations with a generic monotonous increase of values between around 5 and 42 minutes then tending to stabilize. The overall standard deviation, SD, is high, reflecting the effect of the referred trend characterized by a 2nd degree polynomial regression model. The increase in the average number of eye blinks can occur, as it is the case, as a function of time on task (Stern et al. 1994, Fukuda et al. 2005).

In Condition 2 (Figure 2), the same subject had an average BF of 23.46 ± 7.47. There is a segment of higher oscillations after minute 20. At this point the subject did not experience drowsiness but at minute 40 of task he reported drowsiness, which may perhaps explain the various discontinuities observed

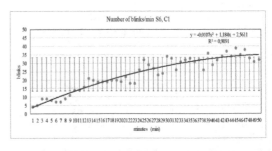

Figure 1. Condition 1: number of blinks/minute of subject 6.

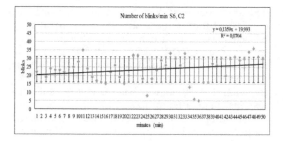

Figure 2. Condition 2: number of blinks/minute of subject 6.

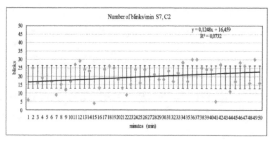

Figure 4. Condition 2: number of blinks/minute of subject 7.

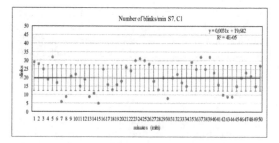

Figure 3. Condition 1: number of blinks/minute of subject 7.

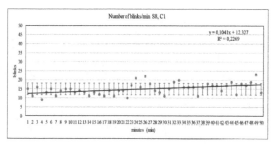

Figure 5. Condition 1: number of blinks/minute of subject 8.

after minute 20 as an attempt to counteract the state of drowsiness and maintain alertness while performing the task (Caffier et al. 2003, Wascher et al. 2014, McIntire et al. 2014). In this case a linear regression model was fitted to the data showing a much less significant trend and correlation when compared with the Condition 1 (Figure 1 graph). Although the overall SD in Condition 1 is higher than in Condition 2, there is a much higher data temporal variability, around the mean in Condition 2.

In Condition 1 (Figure 3), subject 7 had an average BF of 19.76 ± 7.52. It shows a stationary behavior, but with high temporal variability. As the subject felt sleepiness before entering inside the Climate Chamber and after finishing the task, one hypothesis for this great variability is the possibility that the occurring fluctuations might be a sign of an attempt to counteract the state of sleepiness (Caffier et al. 2003, Wascher et al. 2014).

In Condition 2 (Figure 4), subject 7 had an average BF of 19.64 ± 6.72. It is observed a not very significant overall gradual increase in eye blinking, fitted by a linear regression model. The behavior shown in both conditions is very similar, which can be a result of individual characteristics (Stern et al. 1994, Caffier et al. 2003, Schleicher et al. 2008, Benedetto et al. 2010) and the presence of sleepiness earlier in both conditions may have also concurred to the observed similarity (Caffier et al. 2003, Wascher et al. 2014).

In Condition 1 (Figure 5), subject 8 had an average BF of 14.98 ± 3.18. Data was fitted by a linear trend regression model with a small slope, showing an overall gradual linear increase in the number of eye blinks. The number of blinks obtained in the entire task was considerably lower than those obtained in previous trials. The eye blinking increase at minute 22 may be due to an attempt to counteracting the monotony of the task (McIntire et al. 2014).

In Condition 2 (Figure 6), subject 8 had an average BF of 9.64 ± 3.32, the lowest frequency obtained in all tests. There is also an overall gradual linear increase in the number of eye blinks. In both conditions it was used a linear trend regression model and it was observed a small SD. The individual variability pattern may explain the fact that eye blinking is similar in both conditions (Stern et al. 1994, Caffier et al. 2003, Schleicher et al. 2008, Benedetto et al. 2010).

The combination of the two Conditions, 1 and 2, for all subjects shows an overall similar temporal pattern, both growing from the beginning of the task until more or less the middle and stabilizing afterwards (Figure 7). Also, from the beginning of the task until more or less the middle, both Conditions have similar mean frequency of blink values, followed by a higher mean frequency of blinks in Condition 1, until the end. The influence of high temperature and high RH in the subjects in Condition 2 is apparently notorious in the early part of

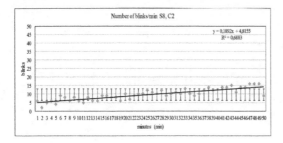

Figure 6. Condition 2: number of blinks/minute of subject 8.

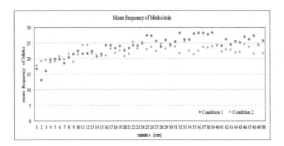

Figure 7. Conditions 1 and 2: mean frequency of blinks/minute.

the test, where the subjects probably have to adapt more significantly to changing environmental conditions and cognitive task. The adjustment period is also visible in Condition 1 in the initial part of the task, blinking continues to increase steadily over time, tending to stabilize from the middle of the task. At the final period, generally, it seems that the effect of temperature and RH of Condition 2 in mental effort lead to a decrease in eye blinking.

Bearing in mind the premise that hot environments may require an increased blood flow into the brain to maintain the same level of task performance (Tanabe et al. 2007), we can assume that it may exist a greater mental effort in Condition 2 than in Condition 1. In Condition 2 there is an increase in BF but then it stabilizes and decreases slightly at the end of the task. According to some authors (Wilson et al. 1994, Caffier et al. 2003, Zheng et al. 2012, Wascher et al. 2014), a higher mental stress or fatigue state indicates a lower blink frequency. According to the authors, it would then be expected that in Condition 1, with less mental effort, eye blinking would increase. In fact it is visible an eye blinking increase in Condition 1, which seems to occur as a function of time on task, corroborating the studies of Stern et al. (1994), Fukuda et al. (2005) and McIntire et al. (2014). As mentioned in the literature, it was confirmed that the eye blinking besides being a predictor of fatigue and mental effort, may also be influenced by several factors.

The study occurred in a laboratory environment, considered monotonous and without threatening consequences, which can allow a state of fatigue to set in quickly (Schleicher et al. 2008). Furthermore, because it is a task without different stimulus it is expected an increase in BF (Stern et al. 1994, Borghini et al. 2012), which can occur as a subject's response in an attempt to stay awake and to maintain alertness during the execution of a considered monotonous task (McIntire et al. 2014). The presence of sleepiness as one of the symptoms predominantly reported at the end of the task by subjects is, according to Caffier et al. (2003), the cause to the verified final part slight decrease in BF. It is also noted that there are considerable intra and inter-individual differences in BF as mentioned in the studies of Stern et al. (1994), Caffier et al. (2003), Schleicher et al. (2008) and Benedetto et al. (2010).

4 CONCLUSIONS

Blinking is a highly individualized activity and this was apparent in the current study namely by the fact that for some subjects and conditions there were no apparent trend and for others were observed the presence of linear or second-degree polynomial trend line regression models with distinct and varying dispersion patterns of the data along task performance.

There seems to be a slight influence of high temperature and high RH on eye blinking and consequently mental effort and fatigue. This influence may indicate a higher mental effort when the subject is in the presence of a higher temperature and RH, resulting in BF stabilization, sometimes even with a slight BF decrease. This difference is supported by thermal comfort questionnaire answers. In future studies we suggest to vary one of the climatic conditions (temperature and humidity) maintaining the other constant. In addition the use of an eye tracker instead of manually counting blinks would be helpful for increased ease and accuracy.

REFERENCES

Acosta, M.C., Gallar, J. & Belmonte, C. 1999. The influence of eye solutions on blinking and ocular comfort at rest and during work at video display terminals. *Experimental Eye Research*, 68, 663–669.

Benedetto, S., Pedrotti, M., Minin, L., Baccino, T., RE, A. & Montanari, R. 2011. Driver workload and eye blink duration. Transportation research part F: traffic psycho-logy and behaviour, 14, 199–208.

Borghini, G., Vecchiato, G., Toppi, J., Astolfi, L., Maglione, A., Isabella, R., Caltagirone, C., Kong, W., Wei, D., Zhou, Z., Polidori, L., Vitiello, S. & Babiloni, F. 2012. Assessment of mental fatigue during car driving by using high resolution EEG activity and neurophysiologic indices. *Conf Proc IEEE Eng Med Biol Soc*, 6442–6445.

Caffier, P., Erdmann, U. & Ullsperger, P. 2003. Experimental evaluation of eye-blink parameters as a drowsiness measure. *European Journal of Applied Physiology*, 89, 319–325.

Clapp, N.E. & Hively, L.M. 1997. Method and apparatus for extraction of low-frequency artifacts from brain waves for alertness detection. Google Patents.

Clements-Croome, D. 2008. Work Performance, Productivity and Indoor Air. *Scandinavian Journal of Work Environment and Health*, 34, 69–78.

Fukuda, K., Stern, J.A., Brown, T.B. & Russo, M.B. 2005. Cognition, blinks, eye-movements, and pupillary movements during performance of a running memory task. Aviat Space Environ Med, 76, C75–85.

Iampetch, S., Punsawad, Y. & Wongsawat, Y. 2012. EEG-based mental fatigue prediction for driving application. *Biomedical Engineering International Conference (BMEiCON)*, 1–5.

Mcintire, L.K., Mckinley, R.A., Goodyear, C. & Mcintire, J.P. 2014. Detection of vigilance performance using eye blinks. *Applied Ergonomics*, 45, 354–362.

Rozanova, E., Heilig, P. & Godnic-Cvar, J. 2009. The eye–a neglected organ in environmental and occupational medicine: an overview of known environmental and occupational non-traumatic effects on the eyes. Arh Hig Rada Toksikol, 60, 205–215.

Schleicher, R., Galley, N., Briest, S. & Galley, L. 2008. Blinks and saccades as indicators of fatigue in sleepiness warnings: looking tired? *Ergonomics*, 51, 982–1010.

Stern, J.A., Boyer, D. & Schroeder, D. 1994. Blink rate: a possible measure of fatigue. *Hum Factors*, 36, 285–297.

Tanabe, S.-I., Nishihara, N. & Haneda, M. 2007. Indoor Temperature, Productivity, and Fatigue in Office Tasks. *HVAC & R Research*, 13, 623–633.

Tsubota, K., Hata, S., Mori, A., Nakamori, K. & Fujishima, H. 1997. Decreased blinking in dry saunas. *Cornea*, 16, 242–244.

Wascher, E., Heppner, H. & Hoffmann, S. 2014. Towards the measurement of event-related EEG activity in real-life working environments. *International Journal of Psychophysiology*, 91, 3–9.

Wilson, G.F., Fullenkamp, P. & Davis, I. 1994. Evoked potential, cardiac, blink, and respiration measures of pilot workload in air-to-ground missions. *Aviat Space Environ Med*, 65, 100–105.

Wolkoff, P. & Kjærgaard, S.K. 2007. The dichotomy of relative humidity on indoor air quality. *Environment International*, 33, 850–857.

Wolkoff, P., Nojgaard, J.K., Troiano, P. & Piccoli, B. 2005. Eye complaints in the office environment: precorneal tear film integrity influenced by eye blinking efficiency. *Occup Environ Med*, 62, 4–12.

Zheng, B., Jiang, X., Tien, G., Meneghetti, A., Panton, O.N., & Atkins, M.S. 2012. Workload assessment of surgeons: correlation between NASA TLX and blinks. *Surgical Endoscopy*, 26, 2746–2750.

Occupational Safety and Hygiene V – Arezes et al. (Eds)
© *2017 Taylor & Francis Group, London, ISBN 978-1-138-05761-6*

Quantitative risk assessment of tower cranes based on conformity

F.O. Nunes
Instituto Superior de Engenharia de Lisboa, Portugal

P.E. Lamy
Universidad de León, León, Spain

ABSTRACT: Development of a non-statistical quantitative risk assessment methodology based on conformity assessment, i.e., compliance with legislation, applicable standards or good practice rules. The quantitative safety assessment of tower cranes is carried out based on the score of a list of items to be inspected, developed for that purpose, with specific weights in the calculation of an overall compliance level.

1 INTRODUCTION

In a typical industrial setting, workers are continually exposed to the same environment and the same hazards. A construction worker, unlike many other industrial workers, can work for several different employers and locations in the same year. The construction industry thus differs from other sectors of activity where workers are constantly confronted with new challenges and dangers as work progresses, since its operations involve complex and dynamic work environments that present new risks to their workers on a daily basis.

Because of its complex and constantly changing nature, the construction industry has high numbers of injuries and mortality rates compared to other industries (Neitzel et al., 2001).

Although there are several causal factors behind these high rates of serious and fatal accidents, many of these injuries and fatalities can be attributed to a critical equipment: the crane. However, the proportion that can be related to cranes is difficult to deduce precisely since the participation of accidents often suffers from deficiencies in classification and determination of their causes.

As a central component of many construction operations, cranes, in their various configurations, are known to be associated with a high number of fatal accidents. Numerous statistics and available studies confirm this association, and it is even suggested that cranes are involved in up to one-third of all deaths in the construction and maintenance industry (MacCollum, 1993).

It is recognized that because of their configuration and concept of operation, mobile cranes are potentially more dangerous than tower cranes. However, the use of tower cranes involves particular safety issues, as well as larger work areas that normally cover the entire yard and in some cases even crosses its borders.

A study of the causes of crane accidents (Jarasunas, 1987), concluded that, from the point of view of safety engineering, the priority is to make tools and equipment as safe as possible using the available technology. However, it is noted that hazard identification, risk control and the provision of safety information present a differentiated level of development by manufacturers and are the subject of some public policies and initiatives by safety professionals (Schulte et al., 2008; Bluff, 2014). Other studies have made specific suggestions on the training of operators (Häkkinen, 1993; Neitzel et al., 2001; Beavers et al., 2006).

In Portugal, there is no specific technical documentation of the official entities or official recognition that establishes the 'bridge' between the applicable legislation and the Harmonized Standards, for each type of work equipment, created as a national reference, integrated in the context of work equipment and having identical guidelines and assessment criteria for the various actors involved in this process.

The lack of this technical documentation means that each actor involved in this process establishes their own guidelines and evaluation criteria, which are often uncoordinated and strongly conditioned by market economic pressure, relegating to the background the proper interpretation of the legislation and the use of good practices and common sense, which should always be present.

On the other hand, there is no established system in Portugal which allows for the accreditation of entities to intervene in this area and there are no conformity assessment procedures for lifting and load handling, for tower cranes, which are harmonized with the legal requirements.

At international level, several studies exist that relate to tower cranes, but with a limited scope of action. There is, thus, still room for scientific research and development in this area to find the technical bases allowing the implementation of measures leading to conformity assessment, in an expeditious, consensual and recognized way, by the various agents involved in this process in Europe.

When they exist, statistics and accident records only provide information on the circumstances, nature, consequences, symptoms and causes nearby, and very rarely provide all the causes of the accidents investigated. Due to the incomplete and inaccurate nature of these records they do not clearly indicate the relationship between causes and effects. As such, the statistics and the investigation processes do not necessarily reflect the reality of the risks to workers' safety (Hammer, 1989; Hinze et al., 1998; Abdelhamid & Everett, 2000; Beavers et al., 2006).

The reality is that statistics suitable for a safety study involving tower cranes, are almost nonexistent. Firstly, because tower crane accidents are usually only involved in cases of serious or fatal injuries. Hence, numerous cases simply do not contribute to statistics, even if they are reported within each construction company. These cases, which may involve personal injury or property damage, constitute most crane related accidents. In addition, incidents and near misses, some of which have the potential to turn into serious accidents, are often not reported even within companies. Thus, the existing literature has concluded that the information resulting from the accidents does not provide sufficiently robust data for safety research purposes (Butler, 1978; McDonald & Hrymak, 2002).

Even when complete and accurate, information related to an accident does not always reflect the reality of effectively installed safety. One work site where a more serious accident occurred than in another does not necessarily have a worse overall level of safety. The use of accident records as a safety indicator for a single construction site is in most cases impossible because many of them do not have any reported accidents and so it is not possible to say they are safer than others with one or more accidents (Laitinen et al., 1999; McDonald & Hrymak, 2002).

2 MATERIAL AND METHODS

To evaluate safety in the use of tower cranes, the methodology to be developed should be based on the calculation of safety indices for individual sites (Beavers et al., 2006). Shapira & Simcha (2009) propose a quantitative risk assessment methodology based on a set of previously identified safety factors whose individual contribution to the assessment is weighted according to its relative importance, as assigned by a panel of experts (safety professionals and manufacturers).

An objective approach for use in each situation may also be based on direct assessments of parameters allowing the verification of the state of compliance with legislation and harmonized standards.

Normally the conformity checklists used by the inspection entities are constructed from the legislation, but merely indicating the text of the legal document regarding the applied article(s), making their application very subjective.

To avoid this situation, a weighted checklist was created, resulting from the accumulated experience of the safety inspection to tower cranes placed in service in the Lisbon area (Lamy, 2009), indicating in a systematic way, the possible items that may not be properly compliant, regarding a "reference condition".

This "reference condition" results from the interconnection of the minimum safety and health requirements applicable to this type of work equipment with the information contained in the harmonized standards (European Commission, 2016), good practices (CPA, 2008; HSE, 2008) or reference technical documentation.

The checklist presented in Table 1, where %W means the percent weight of the item in the Overall

Table 1. Tower crane conformity checklist.

ITEM$_i$		%W$_i$
1—DOCUMENTATION (9%)		
1.1	Identification plates (*)	1.80
1.2	CE marking (*)	1.08
1.3	EU Declaration of Conformity	0.90
1.4	EU Declaration of Conformity (Port.)	0.90
1.5	Instruction Manual	1.44
1.6	Instruction Manual (Portuguese)	1.44
1.7	Maintenance records	0.72
1.8	Inspection records	0.72
2—CHASSIS (8%)		
2.1	Rails Condition (*)	0.24
2.2	Rails fixings	0.40
2.3	Knockers (*)	0.40
2.4	Supports	0.40
2.5	Bearings	0.16
2.6	Support shoes	0.40
2.7	Drums (fixation)	0.32
2.8	Motor—Reducer	0.16
2.9	Brake (*)	0.48
2.10	Carcass	0.40
2.11	Screws	0.80
2.12	Ballast identification	0.20
2.13	Ballast fixation	0.20
2.14	Profiles Condition (*)	0.80
2.15	Welds (*)	0.80
2.16	Screws/connection elements (*)	0.80
2.17	Accesses	0.80
2.18	Disengagement of workspace	0.24

(*Continued*)

Table 1.   (*Continued*).

3—TOWER (8%)

3.1	Chassis (*)	0.80
3.2	Chassis connection elements (*)	0.56
3.3	Profiles condition (*)	0.40
3.4	Interconnection holes	0.24
3.5	Fire extinguisher	0.16
3.6	Welds (*)	0.48
3.7	Ladder	0.80
3.8	Body guards	0.40
3.9	Life line	0.80
3.10	Passages	0.72
3.11	Safety platforms	0.40
3.12	Body guards	0.80
3.13	Screws/connection elements of sections	0.40
3.14	Turntable	0.40
3.15	Rotation crown	0.32
3.16	Rack	0.32

4—PIVOT (3%)

4.1	Profiles condition (*)	0.45
4.2	Welds (*)	0.45
4.3	Screws/connection elements	0.30
4.4	Ladder	0.45
4.5	Body guards	0.45
4.6	Life line	0.60
4.7	Accesses	0.30

5—BOOM (8%)

5.1	Profiles condition (*)	0.80
5.2	Welds (*)	0.72
5.3	Screws/connection elements (*)	0.40
5.4	Motor—Reducer	0.16
5.5	Brake (*)	0.40
5.6	Carcass	0.16
5.7	Screws	0.16
5.8	Re-routing pulley	0.16
5.9	Pulley shaft	0.16
5.10	Car distribution profiles	0.32
5.11	Knockers	0.48
5.12	Distance/load plates	0.96
5.13	Life line	1.60
5.14	Passages	0.72
5.15	Accesses	0.80

6—COUNTER-BOOM (5%)

6.1	Profiles condition (*)	0.50
6.2	Welds (*)	0.50
6.3	Screws/connection elements (*)	0.40
6.4	Motor—Reducer	0.10
6.5	Brake (*)	0.40
6.6	Carcass	0.25
6.7	Screws	0.20
6.8	Counterweight	0.50
6.9	Re-routing pulley	0.10
6.10	Pulley shaft	0.10
6.11	Life line	1.00
6.12	Passages	0.45
6.13	Accesses	0.50

Table 1.   (*Continued*).

7—CAR (5%)

7.1	Profiles condition (*)	0.50
7.2	Welds (*)	0.50
7.3	Cylinders/rollers	0.25
7.4	Re-routing pulley	0.15
7.5	Pulley shaft	0.15
7.6	Motor—Reducer	0.10
7.7	Brake (*)	0.40
7.8	Carcass	0.25
7.9	Screws	0.20
7.10	Steel cable (*)	0.75
7.11	Cable tie in car	0.25
7.12	Fixing at the tip of the boom	0.50
7.13	Car Safety System	0.50
7.14	Maintenance basket	0.50

8—LIFTING MECHANISM (10%)

8.1	Motor—Reducer	0.50
8.2	Brake (*)	1.50
8.3	Carcass	0.50
8.4	Screws	0.50
8.5	Cable guide and pressure washer	0.60
8.6	Steel cable (*)	3.50
8.7	Cable tie	1.20
8.8	Screwdriver	0.50
8.9	Pulleys	0.80
8.10	Pulleys shafts	0.40

9—HOIST (3%)

9.1	Structure	0.30
9.2	Safety signs	0.60
9.3	Information of Safe Working Load (SWL) or Work Load Limit (WLL)	0.75
9.4	Pulleys	0.15
9.5	Pulleys shafts	0.15
9.6	Lifting hook (*)	0.30
9.7	Safety catch	0.75

10—WIRING (8%)

10.1	General power inlet	1.20
10.2	Safety device of the electric board	0.80
10.3	Contacts, fuses, transformer	0.80
10.4	Wear elements	0.40
10.5	Electrical wiring (*)	2.00
10.6	Control panel	2.00
10.7	Protection of direct and indirect contacts	0.80

11—COMMANDS AND/OR CABIN (10%)

11.1	Command identification	1.00
11.2	Condition of commands	0.80
11.3	Indication: radius & load by computer (*)	0.90
11.4	Emergency stop (*)	0.50
11.5	Indicators—lamps	0.40
11.6	Dead man system (*)	1.00
11.7	Horn	0.50
11.8	Instruction plate (Portuguese)	1.00
11.9	Load table	1.00
11.10	Glasses	0.30
11.11	Glass wipers	0.30

(Continued)

(Continued)

Table 1. (*Continued*).

11.12	Chair	0.30
11.13	Lighting	0.20
11.14	Fire extinguisher	0.40
11.15	Climatization System	0.40
11.16	Equipment Operator	1.00
12—HIDRAULIC CIRCUIT (2%)		
12.1	Sealing	0.50
12.2	Safety valves	0.30
12.3	Retention valves	0.30
12.4	Motor	0.30
12.5	Carcass	0.30
12.6	Screws	0.30
13—SAFETY SIGNALS (6%)		
13.1	Access forbidden to unauthorized persons	0.60
13.2	Identification signal near fire extinguisher	0.30
13.3	Obligatory use of the safety belt	0.90
13.4	Danger of electric shock	0.90
13.5	Several dangers	0.90
13.6	Danger of hot surfaces	0.60
13.7	Danger of suspended loads	0.90
13.8	Danger of being caught by the machine	0.30
13.9	Danger of being hit by the lifting cable	0.30
13.10	Safety signs	0.30
14—TESTS (15%)		
14.1	Lifting limit (top)	0.75
14.2	Lifting limit (bottom)	0.30
14.3	Rotational limit (*)	0.75
14.4	Limit of the car ahead	0.75
14.5	Limit of the car behind	0.75
14.6	Travel course limits (*)	0.75
14.7	Maximum load limiter (*)	1.50
14.8	Load limiter (moment) (*)	1.50
14.9	Car speed limiter (*)	0.75
14.10	Vertical speed limiter (*)	0.75
14.11	Emergency stops (*)	1.80
14.12	Weathervane system (*)	1.20
14.13	Functional tests (*)	0.75
14.14	Dynamic test (*)	0.75
14.15	Static test (*)	0.75
14.16	Aerial warning light	0.30
14.17	Sound buzzer	0.60
14.18	Anemometer	0.30
Number of items: 165		100%

Compliance Level (OCL), is exclusively for tower cranes with electric drive, horizontal boom, pivot, translation, and does not include:

o items that require measurement of parameters (noise, vibration, EMF, etc.);
o items related to the location of the cranes;
o items related to operator experience and training;
o items related to operators lift connected to the crane;
o items related to poor assembly and/or maintenance.

The compliance level of each item (Item Compliance Level—ICL) with the respective health and safety requirements is defined by the attribution of a score ranging from ICL = 0.00 (Level 1) to ICL = 1.00 (Level 5), Level 1 being for the state considered completely non-compliant and Level 5 for the full compliant state.

Each of the five compliance levels, to be interpreted by a competent person, were defined as:

- Level 5 (=1.00)—Comply satisfactorily with the inherent requirement(s) of the item. Corresponds to a conformity Condition (C);
- Level 4 (=0.75)—Slight deviation from the inherent requirement(s) of the item. Partial conformity Condition (PC);
- Level 3 (=0.50)—Important deviation from the inherent requirement(s) of the item. Partial Nonconformity Condition (PNC);
- Level 2 (=0.25)—Serious deviation from the inherent requirement(s) of the item. Nonconformity Condition (NC);
- Level 1 (=0.00)—Very serious deviation from the inherent requirement(s) of the item. Critical Nonconformity Condition (CNC).

If one of the items of Table 1 marked with asterisks is found to be a very serious condition (Level 1), the entire process is conditioned. In this situation, it is immediately classified as being on State E, as defined later.

Typically, the inspection of the tower crane with the application of the checklist takes from 90 to 120 minutes depending on its dimensions (height below 50 m, boom with car, etc.) and the number of nonconformities to be annotated.

The influence that each item i has on the Overall Compliance Level (OCL) of the tower crane, calculated between 0 and 100%, is obtained by the product of its ICL_i (0.00; 0.25; 0.50; 0.75 or 1.00) by its $\%W_i$. So, the overall compliance level is calculated by Equation (1).

$$OCL = SF\Sigma_i \left(ICL_i . \%W_i \right) \tag{1}$$

If some items cannot be applied (not applied—NA) to the crane under inspection, a Scale Factor (SF) adjusts the calculated value to the percentage scale. This factor is calculated by Equation (2), and depends of the sum of the percentage weights of all items j "Not Applied" (NA).

$$SF = \frac{100}{100 - \Sigma_j \%W_j} \tag{2}$$

Overall Compliance Level (OCL) obtained for the tower crane determines its state and the consequent actions. States are defined from A to E, according to the following criteria and consequent actions:

- State A (80 < OCL ≤ 100)—It can work. However, it is advised to resolve the reported nonconformities. 'Declaration of Adequacy to the Service' is issued with a validity period of one year;
- State B (60 < OCL ≤ 80)—Same as state A, with possible limitations, after commitment from the person in charge of the equipment of adequate resolution of the nonconformities up to 60 days;
- State C (40 < OCL ≤ 60)—Same as state B, may be necessary to recheck the equipment after resolution of the nonconformities. Operation with possible limitations and commitment of adequate resolution of the nonconformities up to 30 days;
- State D (20 < OCL ≤ 40)—It is recommended not to operate. It can be necessary to recheck the equipment after resolution of the nonconformities. 'Declaration of Adequacy to the Service' is only issued after confirmation of resolution of the nonconformities, with a validity period of six months to one year according the type of nonconformities;
- State E (0 ≤ OCL ≤ 20)—It is strongly recommended not to operate. The immediate resolution of the nonconformities is recommended. It will be necessary to recheck the equipment after resolution of nonconformities. 'Declaration of Adequacy to the Service' is issued only after confirmation of resolution of the nonconformities, with a validity period of 6 months.

3 RESULTS AND DISCUSSION

This process has a subjective component involved, but since it is detailed, that is, it involves the verification of up to 165 items in the tower crane conformity checklist, on a percentage basis, the result is consistent with the approximate safety state of the equipment.

Searching for evidence-based safety in the construction industry, Swuste et al. (2012) reports that many articles indicate that the structures and processes that are designed to ensure safety in the construction industry are poor. Authors also reported that safety management systems do not work, or are limited, business processes executed are fragmentary, it is often unclear who is responsible for safety and parties lower in the construction hierarchy tend to be saddled with the consequences.

The presented method allows for a quantitative assessment of the effective safety of each inspected tower crane. The main benefit of this methodology is to allow for an evidence-based risk assessment obtained in the mandatory inspections carried out on tower cranes. Overall, it is easy to imple-

ment with a spreadsheet and its utilization in tower crane inspections in the Lisbon area, has shown a high correlation between the *OCL* calculated from Equation (1) and the overall safety state of each inspected tower crane.

4 CONCLUSION

Mandatory inspections at times defined by legislation (initial, periodic and special inspections) must be carried out by a competent person, i.e., a person who has such practical and theoretical knowledge and experience of the lifting equipment to be thoroughly examined enabling them to detect defects or weaknesses and to assess their importance in relation to the safety and continued use of the lifting equipment (CPA, 2008). This ensures a reliable interpretation of the criteria defined for the compliance levels of the items in Table 1.

Since the model's development was tailored for quantitative compliance assessment of any individual tower crane through its Overall Compliance Level (OCL), to be carried out during the mandatory inspections, quantitative safety assessment results are in turn available.

The use of comprehensive quantitative terms also allows safety issues on construction sites to be addressed more understandably, rationally, effectively and efficiently.

REFERENCES

Abdelhamid, T.S. and Everett, J.G. (2000). Identifying root causes of construction accidents. Journal of Construction Engineering and Management. 126(1). 52–60.

Beavers, J.E., Moore, J.R., Rinehart, R. and Schriver, W.R. (2006). Crane-related fatalities in the construction industry. Journal of Construction Engineering and Management. 132(9). 901–910.

Bluff, E. (2014) Safety in machinery design and construction: Performance for substantive safety outcomes. Safety Science 66. 27–35.

Butler, A.J. (1978). An investigation into crane accidents. their causes and repair costs. Building research establishment. Building Research Station. Garston. Watford. U.K.

CPA (2008). Maintenance, Inspection and Thorough Examination of Tower Cranes: CPA Best Practice Guide. Tower Crane Interest Group (TCIG) of Construction Plant-hire Association. Reference No. TCIG 0801.

European Commission (2016). Commission communication in the framework of the implementation of the Directive 2006/42/EC of the European Parliament and of the Council on machinery. and amending Directive 95/16/EC: Publication of titles and references of harmonized standards under Union harmonization legislation. Official Journal of the European Union C 332/1.

HSE (2008). Thorough examination of lifting equipment: A simple guide for employers. HSE Books Health and Safety Executive.

Hammer, W. (1989). Occupational safety management and engineering. 4th Ed. Prentice-Hall. Englewood Cliffs. N.J.

Hinze, J., Pedersen, C. and Fredley, J. (1998). Identifying root causes of construction injuries. Journal of Construction Engineering and Management. 124(1). 67–71.

Häkkinen, K. (1993). Crane accidents and their prevention revisited. Safety Science. Volume 16. Issues 3–4. July 1993. 267–277.

Jarasunas, E.K. (1987). Crane Hazard Prevention. Hazard Prevention. March–April. 9–11.

Laitinen, H., Marjamäki, M. and Päivärinta, K. (1999). The validity of the TR safety observation method on building construction. Accident Analysis and Prevenction. 31(5). 463–472.

Lamy, P.E. (2009). Evaluación de la conformidad de grúas torre instaladas en 2009 en el distrito de Lisboa tomando como referencia la Directiva Equipos de Trabajo. Universidad de León.

MacCollum, D.V. (1993). Crane hazards and their prevention. American Society of Safety Engineers. Des Plaines. Ill.

McDonald, N. and Hrymak, V. (2002). Safety behavior in the construction sector. Research Report. Occupational Safety and Health Institute of Ireland. Dublin. Ireland.

Neitzel, R.L., Seixas, N.S. and Kyle, K.R. (2001). A Review of Crane Safety in the Construction. Industry Applied Occupational and Environmental Hygiene. 16(12). 1106–1117.

Schulte, P., Rinehart, R., Okun, A., Geraci, C. and Heidel, D. (2008). National Prevention through design (PtD) initiative. Journal of Safety Research. 39 (2). 115–121.

Shapira, A. and Simcha, M. (2009). AHP-based weighting of factors affecting safety on construction sites with tower cranes. Journal of Construction Engineering and Management. 135(4). 307–318.

Swuste, P., Frijters, A. and Guldenmund, F. (2012). Is it possible to influence safety in the building sector? A literature review extending from 1980 until the present. Safety Science. 50. 1333–1343.

Occupational Safety and Hygiene V – Arezes et al. (Eds)
© 2017 Taylor & Francis Group, London, ISBN 978-1-138-05761-6

Underground coal mine explosions: Main parameters to think about

M.K. Gökay
Department of Mine Engineering, Faculty of Engineering, Selcuk University, Konya, Turkey

M. Shahriari
Department of Industry, Faculty of Engineering, N. Erbakan University, Konya, Turkey

ABSTRACT: Safety is a serious issue in underground coal mines especially when methane is associated with the coal deposit. In deed methane and coal dust explosions are the most deadly mine disasters. More recently, many attempts have made to increase mine safety. For instance, technology has been developed to eliminate or dilute the emissim of methane both prior to and during mining. But, still a lot of work is needed to reach to an acceptable level of safety. This study is focused on parameters affecting methane explosions and to see how could it be prevented.

Keywords: Coal mines, explosions, explosion parameters

1 INTRODUCTION

Mining operations are different from other activities due to their high accident risks. Because of that mines are operated under strict workplace rules. Both surface and underground mining are at risk due to their nature of works. Due to this risky character, mine engineers have been educated and trained to understand the nature of the risk factors and being able to design safer workplaces.

Arranging mining operations and controlling the ore excavations require big engineering efforts to fulfill the tasks. As far as underground coal mines are concerned, mine engineers should control relatively big scale of work to finish their job safely. In order to prevent any mistake, engineers prefer to obey hard line rule appliance in underground coal mines. Input data uncertainties are also very common in their high risk activities. Beside, blasting operations, gas and coal dust explosion hazards make mine engineers nervous in their workplace. Because there are a lot of uncertainties associated with natural rocks and coal ore formations. Many parameters such as strength of country rocks and coal, their porosities, permeability values, gas contents, water reservoir characteristics, faults and discontinuity properties are all tested and determined in the laboratory and field to identify the design parameters and values. However, these data values might be unrealistic even the tests had been performed in international standards. In rock engineering, engineers provide excavation design or support design outputs by considering unexpected accidents due to immeasurable factors of natural rock conditions. As a matter of fact, mining workplaces have their own work conditions. In practical point of view there is no logical and scientific methods to measure all rock properties as a rock mass factors. Therefore; people in other field of works might thing that *"all the influencing parameters in mine workplaces are known and accidents are still happening. This might be occurred due to human insufficiencies"*. This type of consideration or conclusion is not true for mining workplaces. Mining workplaces are being controlled regularly by state inspectors and mine engineers are mostly very sensitive to obey workplace safety rules. It is assumed that they perceive the hazards they exposed to and if something goes wrong and cause full scale mine accident (like gas or dust explosion, collapse of tunnel, underground water burst etc.) they may lost their life as well. By knowing this fact mine engineers are forced to follow necessary precautions to prevent mine accidents. However, they also know that their efforts are not enough to prevent unexpected accidents.

2 AIM OF THE STUDY

The aim of this study is to identify the parameters affecting methane explosion in underground coal mines through literature survey. Furthermore to see how it would be possible to prevent the accidents and achieve a safer workplace.

3 COAL MINE GASES AND EXPLOSIONS

Accidents in mining sectors have been analyzed statistically by researchers and related State offices in different countries. Officially collected data provide opportunities to realize mine workplace conditions in 100 years time. Saleh and Cummings (2011) presented the deadliest coal mine disasters in USA in the 20th century based on US-National Institute of Occupational Health and Safety records. The data which were displayed in their research demonstrate how dangerous those mining workplaces were in those times. For example record for mine disaster, "Monongah No.6 and 7, WV, USA", (6th Dec. 1907) noted 362 fatalities. Countries related with coal mining have similar disaster records in their mining history. In Turkey coal mining sector had two main disasters besides six "minor" mine accidents in 2014. According to TMMOB (2014) and Spada & Burgherr (2016), 301 miners were killed in Soma (Turkey) coal mine disaster (coal mine gas accident) on 13th May 2014. After accident, total moral breakdown and shocks were recognizable at local people and in Turkey as well.

Workplace safety records in mining sector in US have been kept by US National Institute of Occupational Health and Safety. According to this institute a mine "disaster" is an event that involves five or more fatalities. Saleh & Cummings (2011) stated in their study that, the major mine disasters recorded for US in the 20th century were occurred due to methane plus coal dust explosions. Saleh & Cummings (2011) presented also the mine disaster case of "the JWR No.5 mine accident" which is located in Tuscaloosa County, Alabama (US). According to them, this mine was one of the deepest coal mine in US and due to its methane emission it was considered "very gassy underground coal mine". On September 23th of 2001, a methane explosion was occurred followed by a roof fall. Fifty five minutes later, another more powerful methane and coal dust explosion occurred which took 13 miners life (McKinney et al., 2002; Federal MSHRC, 2005; JWR, 2010). Roof rock fracturing and movements had gradually been caused roof fall in this mine (Saleh & Cummings, 2011). Collapsing roof rock formation over the coal seams damages wide range of overburden rock integrity. Methane captured in overburden rock formations then can find passageway towards the mine openings. This may happen in gradual flow rate or it may happen in sudden inrush.

Increase in methane intake to mine workplaces has sometimes been experienced in longwall coal mining also. As an excavation system, longwalls are advanced while mined out volumes are either filled or permitted to collapse. Controlled collapsing at the back side of longwalls connects the workplace areas with further deep points in overburden rock formations. Therefore mine design engineers should know methane and other gases reservoir conditions for their coal seams and country rocks. Controlled collapsing conditions and related rock fracturing patterns should be recovered to plan new longwall panels. If methane gases captured on the walls of voids and micro discontinuities of coal seams and country rocks homogeneously, these sedimentary formations should be disturbed in minimum level during coal mining activities. The risk factors for methane seepage into mine openings should be modeled by considering strength and permeability properties of coal bed formations. These engineering activities are not yet an official standard procedure in coal mine design. Without obtaining enough and meaningful data describing coal bed formation properties *(rock mechanics, chemical, physical properties)*, including discontinuity *(cleavages, fissures, joints, folds, faults etc.)* properties *(orientations, sets, spacing, openings, undulations, continuities, filling materials, weathering conditions, roughness etc.)*, underground planning of coal mines could be exposed to a high risk. This may be similar to sailing in ocean without getting weather and related boat mechanisms information in winter time. One of the main questions which should be searched for each mine gas related disasters is if responsible mine design and shift engineers had been acted according to principles of rock mechanic.

Moore (2012) reviewed related researches about coalbed originated methane gas. He mentioned the differences between gasses in rock formations (clastic gasses) and gasses originated in coalbeds (coal seams or carbon based gasses). Coal seams in general have biogenic and thermogenic methane gasses and coal ranks usually provide clue about the type of methane. Coalbeds having high rank coal (usually older coalbeds) have thermogenetic methane gasses. Moore (2012) wrote also that *"the presence of methane in coal seams has long been recognized, first as a hazard and but more lately as an economically harvestable commodity"*. Economical value of methane attracts more researchers to understand the methane origination and migration parameters. When methane origination during coalification is determined in more detail, its migration properties and reservoir characteristics are also well described. According to Moore (2012) surface of coal seam pores and joints (discontinuities) are the methane bearing locations, therefore coal seam porosity, permeability and discontinuity analyses should have been achieved to obtain coalbed methane conditions. Methane gas originated in coal beds may have emitted to atmosphere through permeable overburden rock formations. Faults, fault fillings and discontinuity parameters are all influencing factors to get permeable or impermeable rock formations. If some part of methane originated by coalification process detained and hold in coalbeds or surrounding rock formations, mine engineers have to localize their position. Firstly, it is commercial commodity for gas producing companies; secondly, it can form gassy coalbed conditions which are very dangerous for mining operations.

Su, et al. (2005) wrote about the usage of coal seam originated methane gas compounds as an energy source. They mentioned that fugitive methane, emitted from coalmines around the world, represents approximately 8% of the world's anthropogenic methane emissions that constitute a 17% contribution to total anthropogenic greenhouse gas emissions. As they presented; there are three paths where the methane has been emitted from a typical gassy mine. These paths are; "mine ventilation air (0.1–1% CH_4), gas drained from the seam before mining (60–95% CH_4), and gas drained from worked areas of the mine, e.g. goafs, (30–95% CH_4)". These researchers evaluated the conditions for usage of methane originated from coal mines; according to them this energy source can be economical after applying certain industrial gas separation technologies. In order to understand mine gas (methane) explosions and coal seam methane drainage conditions, Hemza, et al. (2009) studied on coalification and generation of methane for Upper Silesin Coal Basin, (USCB), in Check Republic. They worked on influencing parameters primarily; degree of coalification, *(rank of coal)*, basic chemical and physical properties of the coal *(volatile matter, ash content, water content)*, maceral composition of the coal and the pore structure of the coal *(permeability)*. They secondly tried to understand; processes leading to increase or decrease in the gas content of USCB coalbeds. This needs geological evolution of the coal basin after completion of the coalification process. Hemza et al. (2009) mentioned also that in the Czech part of the USCB, seams of bituminous coal were able to produce a high volume of methane during their formation considering as gas-saturated. Packham, et al. (2012) report results of field tests performed in Australia. The trial was organized to enhance gas recovery from an Australian coal mine. They wrote the objective of this field application as; "determine whether pre-drainage of methane from a coal seam can be accelerated and the residual gas content reduced to negligible levels through the use of nitrogen as an enhanced recovery agent". They concluded that nitrogen injection had accelerated methane drainage rates as they expected. They reported also some locations in underground where injected nitrogen gas could have not penetrated. When the gasses in rock masses around coalbeds are under consideration, their migration can be possible by discontinuities which have changed in geological times. That means that rock masses can be more permeable by adding new discontinuities due to tectonic forces, or less permeable by natural compaction and cementation process. Permeability is the main factor on methane content of the coalbeds and surrounding rock formations.

As mentioned earlier, some parts of methane gasses captured in coalbeds have got passageways to mine openings. Since methane is very sensitive and explosive gas, mine air should be ventilated and methane concentration must be kept under pre-described percentage. This can only be achieved through mining system planning (Gokay & Shahriari, 2016a) and effective ventilation. Wang, et al. (2015) in their study stated that malfunctioning of a ventilation system caused by a gas explosion was the primary reason for casualties in coal mines. In order to understand effects of gas explosions on mine ventilation system, they prepared "pipe model". They analyzed distribution of the methane/air mixture in pipe model of ventilation system. They compared air pressure in pipe model's branches just after the experimental methane explosions. By getting gas samples just after the explosions, they determined also their compositions.

Yin, et al. (2017) reported about human factors in underground coal mine gas explosions. They evaluated that the repeated occurrence of gas explosions, often in a similar manner and triggered by unsafe behavior, indicates inadequate changes in response in China. According to them, in an effort to further reduce these rates, the human factors associated with accident need to be addressed. Therefore they analyzed 201 significant gas explosions in China during 2000–2014. They noted that, according to *"the Chinese No.30 Order of State Administration of Work Safety (SAWS)"*, front-line mine workers are classified into 10 categories according to their jobs. According to Yin et al. (2017), eight job categories of these 10 have been related to analyzed coal mine gas explosions. They presented these categories and numbers of mine explosion cases as follows: ventilation, 127; safety inspection, 124; gas examination, 102; blasting, 76; electric, 70; tunneling, 44; mining, 19; and transportation, 4. These values demonstrate that explosive gasses emitted to the mine air should be exhausted to atmosphere immediately. In order to reach this goal; mine company should provide a good ventilation system. There should be spare machineries and side air passageways to obtain legislative ratios of air flow even in unexpected disturbances. Ventilation system contains main fans, additional underground fans and air lack doors. Automatic mine air control systems demonstrate the performance of mine ventilation system by measuring airflow rate, temperature and gases (with their concentration in the mine air). These systems can be automated together to have legislative rate of mine air in the mine. If anything goes wrong in these two systems, methane and other gasses concentration in mine air increase gradually. So mine design engineer should consider all possible circumstances in mine ventilation system to produce alternative solutions not to interrupt regulated mine air. Chinese experience show also that safe mine workplaces can be obtained through establishing deterrent mine-workplace safety inspection system. According to Gokay & Shahriari (2016b) mine inspection should be performed like shadow engineering system for each coal mine. Mine inspector should have their own input data for their inspection and they can provide their own reasoning (like the reports supplied by mine engineering consultant companies) for dangerous mine workplaces.

4 MINE GASES AND RELATED PARAMETERS

Coalbeds are sedimentary deposition of organic compounds where many chemical reactions have been continued under gradually increasing overburden pressure. Organic biogenetic gasses originated in coalbeds have "matured" during their geological evolution times. Gases originated during coalification are mostly ready to escape under pressurized conditions. If there have been no passageways for these gasses, mine openings and exploration boreholes supply artificial opportunity to run away. Mine openings and exploration boreholes are not only the way to exhaust mine gasses, they also supply atmospheric air to coalbeds. Coal itself has spontaneous combustibles in character, which would react with oxygen. Therefore mine air should be controlled for biogenetic or thermogenetic organic coalbed gasses and byproduct gasses of coalbeds spontaneous combustion. Most of these byproduct gasses are also dangerous (poisonous, explosive etc.) for mine workplaces. Atkinson & Morris (1986) defined influencing parameters of coal seam spontaneous combustion. The phenomenon of spontaneous heating in underground coal mines has mainly three factors. According to them these factors are; seam factors (such as particle size, physical properties, impurities, moisture), geological factors (such as seam thickness, faulting), and mining factors (such as mining methods, ventilation.

5 CONCLUSIONS

Coal mining has been reshaping by new opportunities aroused due to new technological devel-

opments. Mining engineers have therefore gradually redesigned their excavations according to new equipment and methods. There are also new gas policies in energy sector. Gas extracted through discontinuities and pores of rock formations could be an alternative energy source in world economy. This development has brought new technological improvements and considerations in coal mining. Coal mining companies should be more sensitive to their coalbed gas reservoirs which can be taken account saleable commodities in business. As far as the coal mining sector is concerned, methane gas in the coalbeds has been one of the main explosive sources for developing disasters. Therefore depleting coalbed gas content is very important on coal mine workplaces. Coalbeds do not cover all the gasses originated during its own coalification process. Most of the gasses might have migrated either to other positions in underground openings or exhausted to the atmosphere during geological times. Discontinuities in coalbeds and surrounding rock formations play an important role in these gas movements. Coal miners should then consider all the rock mechanics properties of coalbasin rocks together with porosity and permeability properties.

Controlling and monitoring operational parameters in underground coal mines are not enough to get safer mine workplaces. A safer coal mining starts with enhanced mine planning which covers different mine operational systems together with their demands and supplies (Gökay & Shahriari, 2016a). In order to design a safer underground coal mine, free of methane and coal dust explosion, there are many parameters which needed to be collected and tested. On the other side some parameters are unclear and there are considerable uncertainties in their test mechanism due to the nature of the environment. Due to these uncertainties not only design should be performed with extreme care, but the mine operation should be associated with a proactive and sustained mode of control to prevent the unwanted events.

REFERENCES

Atkinson, T. and Morris, R. 1986. Geological and mining factors affecting spontaneous heating of coal, *Mining Science and Technology*, 3: 217–231.
Federal MSHRC, 2005. Federal Mine Safety and Health Review Commission, 2005. Civil Penalty Proceeding, Secretary of Labor, Mine Safety and Health Administration vs. Jim Walter Resources, Inc., November 1, 2005. Docket No. SE 2003-160; AC No. 01-01322-00004. Washington, DC.
Gokay M.K. and Shahriari, M. 2016a. Design: Relation with occupational safety in mines, *8th International conference on occupational safety and health, 8–11 May 2016*, Istanbul, Halic Congress Centre, paper-0730.

Gokay M.K. and Shahriari, M. 2016b. Time to re-evaluate responsibility of mine engineers in Turkey to obtain safer workplace environment, *International Symposium on Occupational Safety and Hygiene, (ISBN 978-989-98203-6-4, Eds. Arezes, P. et al), 23–24 March 2016*, Guimaraes, Portugal, 93–95.

Hemza, P., Sivek, M., and Jirásek, J. 2009. Factors influencing the methane content of coal beds of the Czech part of the Upper Silesian Coal Basin, Czech Republic, *International Journal of Coal Geology*, 79: 29–39.

JWR, 2010. Jim Walter Resources #5 Mine Disaster, 2010. United Mine Workers of America Report. http://www.umwa.org/?q=content/jim-walters-resources-5-minedisaster, (accessed 01.03.10).

McKinney, R., Crocco, W., Stricklin, K.G., Murray, K.A., Blankenship, S.T., Davidson, R.D., et al., 2002. Report of Investigation Fatal Underground Coal Mine Explosions September 23, 2001, No. 5 Mine Jim Walter Resources, Inc. Mine Safety and Health Administration. http://www.msha.gov/fatals/2001/jwr5/ftl01c2032light.pdf (accessed 01.03.10).

Moore, T.A. 2012. Coalbed methane: A review. *International Journal of Coal Geology*, 101: 36–81.

Packham, R., Connell, L., Cinar, Y., and Moreby, R. 2012. Observations from an enhanced gas recovery field trial for coal mine gas management. *International Journal of Coal Geology*, 100: 82–92.

Saleh, J.H. and Cummings, A.M. 2011. Safety in the mining industry and the unfinished legacy of mining accidents: Safety levers and defense-in-depth for addressing mining hazards. *Safety Science*, 49: 764–777.

Spada, M. and Burgherr, P. 2016. An aftermath analysis of the 2014 coal mine accident in Soma, Turkey: Use of risk performance indicators based on historical experience. *Accident Analysis and Prevention*, 87: 134–140.

Su, S., Beath, A., Guo, H. and Mallett, C. 2005. An assessment of mine methane mitigation and utilisation Technologies, *Progress in Energy and Combustion Science*, 31: 123–170.

TMMOB, 2014. *Soma Maden Kazasi Raporu, (Report on Soma Mine Accident)*, Union of Chambers of Turkish Engineers and Architects, Ankara, Turkey, Sep. 2014, p. 64.

Wang, K., Jiang, S., Ma, X., Wu, Z., Zhang, W. and Shao, H. 2015. Study of the destruction of ventilation systems in coal mines due to gas explosions, *Powder Technology*, 286: 401–411.

Yin, W., Fu, G., Yang, C., Jiang, Z., Zhu, K., and Gao, Y. 2017. Fatal gas explosion accidents on Chinese coal mines and the characteristics of unsafe behaviors: 2000–2014, *Safety Science*, 92: 173–179.

Occupational Safety and Hygiene V – Arezes et al. (Eds)
© *2017 Taylor & Francis Group, London, ISBN 978-1-138-05761-6*

An overview of occupational diseases: Recognition and certification processes

D. Ferreira

Faculdade de Psicologia e de Ciências da Educação da Universidade do Porto, Porto, Portugal

L. Cunha

Centro de Psicologia da Universidade do Porto, Porto, Portugal

ABSTRACT: Europe has a high rate of occupational diseases and each year millions of workers are victims of work. Therefore, this study aimed to explore some cases of occupational disease in order to obtain a deeper knowledge on the topic. The results highlight the gaps of certification process: lack of clear criteria in the evaluation of disabilities; excess of time waiting for medical boards' appointments and subsequent evaluation of disease; lack of knowledge about occupational diseases and lack of monitoring by doctors and employers. Even when diseases were recognized as occupational, there was no involvement of employers in the organization of work, and difficulties in the reimbursement of costs subsisted.

1 INTRODUCTION

In the last few years greater attention has been given to working conditions and occupational risks. However, according to the V European Working Conditions Survey, "European workers remain as exposed to physical hazards as they did 20 years ago" (European Foundation for the Improvement of Living and Working Conditions, 2012). Based on the same report, 33% of workers carry heavy loads at least a quarter of their working time while 23% are exposed to vibrations. Furthermore, 46% of workers work in tiring or painful positions and repetitive hand or arm movements are a feature of work for more Europeans than 10 years ago (European Foundation for the Improvement of Living and Working Conditions, 2012). In the other hand, the number of workers who consider their health and safety threatened by work conditions decreased (European Foundation for the Improvement of Living and Working Conditions, 2012), showing that when it comes to health at work more attention is paid on the manifest, immediate and simple aspects, neglecting the importance of effects that are not obvious or immediate (Volkoff, 2011). The idea that health of most workers is satisfactory and work accidents and occupational diseases are an exception remains (Barros-Duarte & Lacomblez, 2006). This perception is justified by the fact that the concept of health continues to be defined as "absence of disease, in particular absence of recognized occupational disease" (Barros-Duarte, 2004).

In the European Union context, the term of occupational disease "covers any disease contracted as a result of an exposure to risk factors arising from work activity" (ILO, 2002) and work-related diseases are defined as having "multiple causes, where factors in the work environment may play a role, together with other risk factors, in the development of such diseases" (WHO, n.d.). In Portugal, according to national legislation, occupational diseases are those that are listed as the "lesions, functional disorders or diseases not included in the list provided since proven they are a necessary and direct consequence of the work activity carried out by workers and don't represent normal wear and tear on the body" (Decreto—Lei 98/2009 de 4 setembro, 2009).

The concept of occupational disease is a legal concept created for compensatory (not preventive) purposes, given its definition to legal and non-medical criteria (Eurogip, 2015). The compensation system for occupational diseases therefore reflects a social and political commitment (Eurogip, 2015), since the recognition of such diseases has legal and financial implications which may vary depending on the country (ILO, 2013). The systems for the recognition of occupational diseases are treated as a set of filters of different natures: conceptual—reffering to the definition of a causal relation between work and the pathology itself—intitutional, legal, social and cultural filters, and the combined effect of these filters exceeds what could be explored by a quantitative aproach (Vogel, 2014), emphasizing the need to combine statistical analysis with a qualitative approach.

The process of diagnosis and notification of an occupational disease is generally associated to physicians who are legally forced to report all cases of occupational diseases or medical conditions that could have a causal relation with work (Instituto da Segurança Social, 2015). This process is closely linked to the official lists of recognized occupational diseases of each country and its guidelines.

While the list system has the disadvantage of covering only a certain number of occupational diseases, it has the advantage of listing diseases for which there is a presumption that they are of occupational origin (ILO, 2002). Several countries of European Union—such as Portugal, Germany, France and Italy—have a mixed system for recognition of occupational diseases. This means that they have both a national list of occupational diseases and also a complementary system of recognition for the diseases not registered on the list (Eurogip, 2013). This complementary system is by nature more restrictive than the list system because the onus of proof lies with the victim and not the insurance organisation (European Commission, 2010). The approval of a notification requires a causal link to be established between the occupational activity and the disease. Therefore, the total number of recognised cases has always been considerably smaller than the number of notifications (Eurofound, 2007). In addition, many cases of occupational disease are not recognized due lack of testimony from workers, doctors and employers (Eurogip, 2015), which suggests a lack of knowledge on this matter. In fact, for the recognition of a new occupational disease, the decision-making process is often crossed by difficult negotiations between social partners (Barros & Lacomblez, 2006). The recognition of occupational diseases plays an important role in the promotion of workplace prevention (Vogel, 2014), and in the last years several countries have created initiatives to increase information regarding this subject and emphasizing the need of the report of such cases.

The insufficient orientation of society to confer visibility and vigilance to some work risk factors leads to a devaluation of such risks. Workers generally don't speak spontaneously about their work conditions and hide the effects of work on their health (Barros & Lacomblez, 2006). By having to choose between taking a job and accept certain risks and not having a job and consequently not be exposed to professional risks, job preservation and consequent exposure to risk factors tends to be more important, demonstrating the coercive effect of certain social determinants (Areosa, 2010). The underreporting phenomenon seems to be widespread, although governments, trade unions and specialists don't pronounce it equally (Eurogip, 2015). This phenomenon was mentioned in the report of the European Commission about the evaluation of the Commission Recommendation 2003/670/EC with regard of the diagnosis and recognition of occupational diseases in Europe

(Eurogip, 2015). The causes identified as responsible for this phenomenon are numerous and some are common to several countries, such as the lack of knowledge and information among doctors (especially general practitioners) regarding the concept of occupational diseases (European Commission, 2010). Doctors are sometimes criticised for taking little interest in the reporting procedure and some refrain from reporting when the chances of seeing the case recognised as an occupational disease by the insurance organisation performing recognition are small or non-existent (European Commission, 2010). Another cause of underreporting is the technical difficulty in determining the occupational origin of certain pathologies, since there is sometimes a significant latency period between the time of exposure to occupational hazards and the manifestation of the disease, or due to the interactions with extra-professional factors (Eurogip, 2016).

2 METHODOLOGY

2.1 Participants

All participants of this study were members of a non-profit association dedicated to protect the rights of victims of work accidents and occupational diseases—Associação Nacional de Deficientes e Sinistrados no Trabalho (ANDST)—aged between 16 and 80 years. This association is currently the only non-profit institution in Portugal that support people with disabilities caused by work. All participants had an occupational disease, even thought the disease was not recognized in same cases. Victims of work accidents were not considered since was not the aim of the study.

Based on the gathered information, four participants were selected by the ANDST to participate on semi-structured interviews in order to explore their professional background and their experience in occupational disease recognition processes. Of the four participants, there was a woman and three men, between 43 to 70 years old. Table 1 shows the general characterization of all the participants.

Table 1. General characterization of study participants.

Participant	Gender	Age	Occupational activity	Diagnosed disease
A	Male	43	Industrial worker	Carpal Tunnel Syndrome
B	Male	70	Metallurgical worker	Silicosis
C	Male	47	Industrial worker	Deafness
D	Female	47	Cleaner	Kienbock's disease

3 RESULTS

3.1 Actors and institutions involved in the recognition process

Factors were identified as being involved in the process of recognition of occupational disease, in addition to the protagonists themselves—occupational doctors, family physicians, private doctors or doctors from hospitals, workers' committees and employers. Occupational doctors are the most referred by participants of this study, although the verbalizations are about episodes of negligence and devaluation of participants' complaints. Family physicians and private doctors show up next, being refered as playing a role on the diagnosis of the occupational disease: *I start being treated in a private doctor in order to solve my problem faster, to keep working. Because we live in a country in which before something is done... people die first* (Participant C, industrial worker, occupational deafness). Regarding family physicians, participants emphasized the follow-up given by this doctors during the recognition process and say that in some cases it were family physicians that iniciate the recognition of occupational disease process. All participants stated that family physicians show more interest in their health problems and are more willing to assist, for example in the prescription of exams and writing medical reports. However, some participants referred that these doctors have poor knowledge about occupational diseases and that they show sometimes signs that indicates fear of retaliation in the future. These conclusions corroborate the Eurofound report (2010) which refers that occupational doctors are criticized by showing little interest on the recognition system and that family physicians are seen as having poor knowledge of the subject.

Furthermore, participants identified six institutions that are involved in this process: the ANDST, the national center of protection against professional risks (CNPRP), the national center of insurance for occupational diseases (CNSDP), a military hospital and insurance organizations.

Analyzing the content of the participants' verbalizations, we found that all participants asked ANDST for help on their recognition processes and for more information abou occupational diseases: *ANDST follow my case closely and explain things to me (...) ANDST helps me a lot, if weren't for them I would not know what to do* (Participant C, industrial worker, Occupational deafness). On the other hand, when discussing the role of CNPRP, participants evaluate negatively how is organized, existing cases of processes's documentation lost. Participants referred that the elements of this organization show little interest about workers health problems, taking only interest on medical exames and reports and not paying attention to workers complaints: *When I was called to do a surgery, I took a medical licence and then I went to the doctor to see how it was healing and I asked him about my shoulders and elbows. He answered me that the system only talked about the Carpal Tunnel Syndrome; they only care about the problems that are in the system.* (Participant A, industrial worker, Carpal Tunnel Syndrome). The CNSDP was the solution found by the Portuguese State do the problem of silicosis: *And then they create de CNSDP, and my case was solved there, not in the insurance organization* (Participant B, metallurgical worker retired, Silicosis).

Unlike insurance organizations, CNSDP insured all expenses resulting from the occupational disease and the workers were called to do annual exams to see the progression of the disease and adjust the incapacity established: *I was called to do exams every year* (Participant B, retired metallurgical worker, Silicosis). Insurance organizations are accused of neglecting workers, not giving them the necessary follow-up: *insurance companies didn't pay, or delay the payment until the person got tired. They beat workers by tiring them, they still do that nowadays* (Participant B, reformed metallurgical worker, Silicosis).

Finally, the role of trade unions in the recognition process is also refered on participants' speach. If in the past the unions didn't have enough knowledge about occupational diseases—*because in the past, unions were oblivious to everything, at that time it was impossible, we were not on 25 April. The unions were owned by the employers* (Participant B, retired metallurgic worker, Silicosis)—the scenario docs not seem to change, considering the participant's point of view: *To be honest, I am a member of the union and I didn't know many things that the ANDST explained to me* (Participant A, industrial worker, Carpal Tunnel Syndrome).

3.2 Professional risks

Physical effort and lack or inadequacy of safety equipment are the risks more often refered in the participants' speech, as well as the work intensity. The repetitiveness of work is also indicated as one of the causes of occupational disease, and there are even participants who suggest that one way of avoidying such diseases is to promote the rotation of the functions, although this does not correspond to a prevention strategy, but rather a risk management strategy. The number of verbalizations referring to the lack or inadequacy of safety equipment suggests that, although safety equipment is often available, it's not always used because difficult work: *(The safety mask) was the best at the time, but it did not solve the problem because we spent a lot of time cleaning it. And then, for example, we used to adapt it. If we unhooked it, it had a little coal in it, and we change it for cotton instead because it was a little lighter* (Participant B, retired metallurgic worker, Silicosis).

The results are consistent with recent reports referring the intensity of work and its repetitive nature as the most common occupational hazards, and which have consequences for workers' health and well-being (e.g. Eurofound, 2014). Also, the number of work hours constitutes a risk to workers' health because it involves not only physical effort but also mental effort. The excess of work hours reported by participants is commonly associated with their musculoskeletal diseases (Vogel, 2004). These diseases are also associated with other risks refered by the participants in their interviews, such as high strength required, painful posture, repetition of movements (Vogel, 2014).

3.3 *Occupational disease—Post-diagnosis*

Three participants refered that had to take medical licenses although their diseases were not always recognized as occupational: *I had been on sick leave for a "natural illness" for a couple of weeks, and it was not due to occupational disease. I had to take the antibiotic and comeback to the company* (Participant C, industrial Worker, Deafness). There is a reference to the performance medical treatments whithout medical licenses in order to guarantee a total remuneration to compensate the expenses with the treatments: *I am left with a situation that, at the end of the day, I was working and doing physical therapy at the same time (...) It turned out that I had to do physical therapy and I had to be on my work at the same time because I could not be on leave. Because I'm paid less and then I don't have the money to pay for physical therapy* (Participant A, industrial worker, Carpal Tunnel Syndrome). Two of the participants have already performed surgeries due to occupational disease, and acupuncture is referred as an alternative treatment to compensate pain resulting from the occupational disease: *And I'm trying to see if I can get the pain reduced with acupuncture because otherwise I'm going to have to fix my fist, because it's the movements that cause the pain* (Partipant D, cleaner, Kienbock Syndrome). After the recognition of occupational disease, one of the participants referred to salary consequences: *The attitude was the same, had to do the same things but I was penalized, I did not have any increase (...) If I did not work the way I worked before,, I did not have the same increase as the others* (Participant A, industrial worker, Carpal Tunnel Syndrome).

With regard to the change of functions, after the first symptoms of occupational disease, or after the beginning of the process of recognition only one participant states that he had been assigned to perform another function, although he was requested to perform the old function sometimes: *Although*

from time to time, they told me "give a hand, run from the dust as much as possible but give a hand there to the section" (Participant B, metallurgical worker retired, Silicosis). As for the other participants, they continued to perform the same function with more or less prolonged interruptions due to medical surgeries or medical casualties: *They did not change anything, quite the contrary, they even assigned me more things to do* (Participant A, industrial worker, Carpal Tunnel Syndrome).

3.4 *Recognition process*

Finally, through the analysis we were able to gather information about the process of recognition of occupational disease of each participant. It was found that all the participants dealt with the lack of interest regarding the process of recognition of occupational disease, and the waiting time from the beginning of the process until the medical committee is called, and later until the final decision is very extended: *a person waiting for a medical board to be classified, should be a faster process. Because in the end, my case started in 2008 and in 2010 I didn't even knew when I would my surgery be. Since I was diagnoted until being submitted to surgery, several years had passed. At the end of those years my health problem was agravated.* (Participant A, industrial worker, Carpal Tunnel Syndrome). In general, participants had already been involved in medical boards and could identify some flaws at this stage of the process. And they also claimed to know other cases of occupational diseases contracted in the same context. Lack of knowledge and the need for legal mesures to obtain recognition of the disease are also refered by participants, as well as reports of doctors' attempts to discourage workers of trying the recognition based on a alledged small disability: *I say this because the last time I went there, the doctor asked me what my disability was, I aswered him and he told me that was still too little. And I kept thinking, "Is it still too little?" It was still too little to change work activity, I thought, and my manager asked the same thing and told me exactly the same thing about it* (Participant A, Industrial worker, Carpal Tunnel Syndrome). The difficulty in reimbursement of expenses is also one of the problems mentioned by the participants: *It's not a fast reimbursement (...) It is a terrible mess. Summarizing and concluding, it's always to be penalized.* (Participant A, factory worker, Carpal Tunnel Syndrome).

4 CONCLUSIONS

The process of occupational disease recognition in Portugal is complex and time-consuming not always resulting in a positive decision for workers.

The time spent since the first symptoms of the disease are felt till the recognition of the disease (if recognized as occupational), can vary from months to years and in that waiting time worker's health continues to deteriorate. During this period, workers have to pay every expenses related to the disease even though they don't have an income because of their health problems, and hope that once the disease is recognized their expenses would be reimbursed. However, based on the participants' report, a disease recognized as occupational may not represent change. All participants of this study remained in the same work activity, exposed to the same occupational risks that may had been the cause of their pathology, and still end up being penalized by employers by stop receiving for example productivity prizes. As for the awaited reimbursement of the accumulated expenses, the difficulties remain and it could take months before they see any money.

It is undoubtedly urgent to give more visibility to this issue, in order to guarantee a higher level of prevention at work level, admiting that it's the right to work and the right to health at work that are in debate.

REFERENCES

Areosa, J. 2010. Riscos e Sinistralidade Laboral: um estudo de caso em contexto organizacional (Tese de Doutoramento). Instituto Universitário de Lisboa, Lisboa.

Barros Duarte, C. & Lacomblez, M. 2006. Saúde no trabalho e discrição das relações sociais. Laboreal, 2, (2), 82–92. Retirado de http://laboreal.up.pt/revista/artigo.php?id = 37t45nSU5471122785414468 81

Barros-Duarte, C. 2004. Entre o local e o global: processos de regulação para a preservação da saúde no trabalho (Tese de Doutoramento). Faculdade de Psicologia e de Ciências da Educação da Universidade do Porto, Porto.

Eurofound 2014. Psychosocial risks in Europe: Prevalence and strategies for prevention—Executive summary. Retirado de http://www.eurofound.europa.eu/publications/executive-summary/2014/eu-member-states/working-conditions/psychosocial-risks-in-europe-prevalence-and-strategies-for-prevention-executive-summary

Eurogip 2016. Première restitution 1 des points abordés lors de la conférence européenne organisée par EUROGIP le 24 mars 2016. Les débats d'Eurogip: Pathologies psychiques et travail en Europe. Paris.

Eurogip. 2015. Déclaration des maladies professionnelles: problématique et bonnes pratiques dans cinq pays européens. Retirado de http://www.eurogip.fr/fr/produits-information/publications-d-eurogip/3906-declaration-des-mp-problematique-et-bonnes-pratiques-dans-cinq-pays-europeens

European Commission 2010. Report on the current situation in relation to occupational diseases' systems in EU Member States and EFTA/EEA countries, in particular relative to Commission Recommendation 2003/670/EC concerning the European Schedule of Occupational Diseases and gathering of data on relevant related aspects. Retirado de http://ec.europa.eu/social/BlobServlet?docId=9982&langId = en

European Foundation for the Improvement of Living and Working Conditions 2012. Changes over time—First findings from the fifth European Working Conditions Survey. Retirado de http://www.eurofound.europa.eu/pubdocs/2010/74/pt/1/EF1074PT.pdf

ILO 2002. P155—Protocol of 2002 to the Occupational Safety and Health Convention, 1981. Retirado de http://www.ilo.org/dyn/normlex/en/f?p = NORMLEXPUB:12100:0::NO::P12100_INSTRUMENT_ID:312338

ILO 2013. National System for Recording and Notification of Occupational Diseases—Pratical Guide. Retirado de http://www.ilo.org/wcmsp5/groups/public/---ed_protect/---protrav/---safework/documents/publication/wcms_210950.pdf

ILO 2013. The Prevention of Occupational Diseases. Retirado de http://www.ilo.org/safework/info/publications/WCMS_208226/lang--en/index.htm

Instituto da Segurança Social 2015. Guia Pratico: Doença Profissional—Certificação. Retirado de http://www.seg-social.pt/documents/10152/24338/doenca_profissional_certificacao/3b846780–202a-4d3e-b90e-fc88a67f2cc2

Vogel, L. 2014. Un espejo deformante: apuntes históricos sobre la construcción jurídica de las enfermedades profesionales en Bélgica. Laboreal, 10 (2), 10–26. Retirado de http://dx.doi.org/10.15667/laborealx0214lv

Volkoff, S. 2011. Visibilidade. Laboreal, 7, (1), 119–121. Retirado de http://laboreal.up.pt/revista/artigo.php?id=37t45n SU5471124227839554481

Who (s/d). Occupational and work-related diseases. Retirado de http://www.who.int/occupational_health/activities/occupational_work_diseases/en/

Occupational Safety and Hygiene V – Arezes et al. (Eds)
© *2017 Taylor & Francis Group, London, ISBN 978-1-138-05761-6*

Control Banding applied to engineered nanomaterials: Short review

Andréa Pereira
PROA/LABIOMEP, Faculty of Engineering, University of Porto, Porto, Portugal

Ana C. Meira Castro
CERENA, School of Engineering, Polytechnic of Porto, Porto, Portugal

J. Santos Baptista
PROA/LABIOMEP, Faculty of Engineering, University of Porto, Porto, Portugal

ABSTRACT: Control Banding (CB) methodology is one of the approaches developed on assessment and management of the risk of hazardous substances exposure. In the last decade, several CB tools were developed regarding workers' exposure to nanomaterials. This review intends to identify CB methods already applied in occupational settings in industrial activities and laboratories with exposure to nanomaterials, aiming to help the selection of a CB tool to be applied in a specific scenario, allowing to achieve most reliable results. The results of the data base search revealed seven CB tools applied to laboratories and industrial activities. Although the similarities, these tools present differences that may affect the consistency of the resulting conclusions. CB methodologies allow to define priorities in applying adequate control measures. However, the results also indicate that workplaces' measurements and its comparison with CB results are valuable to support possible future adjustments to seek more consistent results.

1 INTRODUCTION

Nanotechnology enables many technological advances and innovations. Nanomaterials already have a major impact on a diversity of sectors, since environmental technologies, to medicine, to telecommunications and many others, being present on our day by day life (Savolainen et al., 2013). The control of potential health and safety risks related to the application of nanomaterials is a crucial to the sustainability of nanotechnology (Tc-Osh, 2013).

Engineered Nanomaterials (ENM) have specific complex properties not only in physicochemical characteristics and behavior, but also by their potential to interact with living systems (Savolainen et al., 2013).

The definition of ENM is not internationally regulated, until this date. According to the recommendation of European Commission (European Commission, 2011), a nanomaterial is a natural, incidental or manufactured material containing particles, in an unbound state or as an aggregate or as an agglomerate and where, for 50% or more of the particles in the number size distribution, one or more external dimensions is in the size range 1 nm to 100 nm. However, in specific cases and where warranted by concerns for the environment, health, safety or competitiveness, the number size distribution threshold of 50% may be replaced by a threshold between 1 and 50%. This definition has been reviewed and recently three reports intending to both clarify the definition of the term nanomaterial and to facilitate its implementation were published (Rauscher et al., 2014; Roebben, Rauscher, Sokull-klüttgen, 2014; Rauscher et al. 2015). Also the International Organization for Standardization (ISO), provided a slightly similar definition, presenting a nanomaterial as a material with any external dimension in the nanoscale (length range approximately from 1 nm to 100 nm) or having internal structure or surface structure in the nanoscale (ISO/TS 80004-1, 2015). As for Engineered Nanomaterial (ENM), the Standard ISO/TS 80004-1:2015, defines it as being a nanomaterial designed for specific purpose or function (ISO/TS 80004-1, 2015).

Control Banding (CB) is one of the approaches, applied to nanomaterials exposure risks in the last decade. (NIOSH, 2009) This strategy intends to readily get a score, allowing to stablish priorities on the implementation of measures to minimize the exposure of workers to hazardous chemicals and other risk factors in the workplace, mostly in work scenarios with limited information on hazards, exposure levels and risks (EU-OSHA, 2013).

This review intends to identify CB methods already applied in occupational settings, assessing the risk of workers' exposure to ENM. In addition, it is intended to summarize information to help the selection of a CB tool for a determined workplace scenario of application in order to achieve the most reliable results.

2 METHODS

On this review, articles selection was based on the PRISMA Statement procedure (Liberati *et al.*, 2009). The search process was performed online in 31 data bases: Academic Search Complete, Beilstein via SCIRUS (ChemWeb), Compendex, ERIC, Inspec, MEDLINE, PubMed, SCOPUS, TRIS Online, Web of Science, ACM Digital, ACS Journal, Annual Review, ASME Digital Library, BioMed Central, Cambridge Journals Online, CE Database (ASCE), Directory of Open Access Journals (DOAJ), Emerald Fulltext, Highwire Press, Informaworld (Taylor and Francis), IOP Journals, Oxford Journals, Royal Society of Chemistry, SAGE Journals Online, Science Magazine, Science direct, SIAM, Springer-Link, The Chronicle of Higher Education and Wiley Online Library. Additional sources were also used, namely: NIOSH Publications, European Agency for Safety and Health at Work, Finnish Institute of Occupational Safety and Health, Executive (HSE—United Kingdom) ISO, OECD, Official Journal of the European Union (EuroLex), Federal Office of Public Health (FOPH—Switzerland) and Google.

In order to collect all relevant published works related to the topic under review, data bases were screened with the following keywords combined: (1) "risk assessment' AND 'Control Banding' AND 'nanoparticle' AND 'Worker Exposure'; (2) 'risk assessment' AND 'Control Banding' AND 'nanoparticle' AND 'Occupational Health and Safety'; (3) 'risk assessment' AND 'Control Banding' AND 'nanoparticle' AND 'Occupational Safety and Health'. The data base search process focused on articles written in English and published between 2009 and March 2016.

3 RESULTS

The search process identified, after duplicates removed, 1309 records. After screening and selection, based on inclusion and exclusion criteria and eligibility 16 articles were elected.

This short review includes articles regarding CB methods already applied in laboratories and/industrial activities with exposure to ENP. In this review seven CB methods for assessing the risk of workers' exposure to ENP were identified. To simplify the writing and the reading process, the following acronyms and abbreviations are used, to designate the CB methods considered:

– ANSES (CB method from the French Agency for Food, Environmental and Occupational Health and Safety Method, France).
– Nanotool (CB Nanotool method, USA).

– GWSNN (Guidance Working Safely with Nanomaterials and Nanoproducts method, Holland).
– EPFL (CB method from the École Polytechnique Féderale de Lausanne, Switzerland).
– ISPESL (CB method from the Instituto Superiore per la Prevenzione e la Securezza del Lavoro, Italy).
– Nanosafer (Nanosafer method, Denmark)
– Stoffenmanager (Stoffenmanager method, Netherlands).

All CB methods identified estimate risk levels in order to determine and/or prioritize control measures. Most of these CB methods calculate risk levels based on decision matrices, considering hazard and exposure bands, with the exception of the EPFL method, whose risk level is based on a decision tree and the ISPESL method, which adds a correction factor for the equation to determine risk level.

3.1 *Control Banding methods applied in workplace environment*

Control Banding methods, applied in both laboratories and/or industrial workplaces, are present in sixteen studies. Twenty-four applications of CB methods were identified, as shown in Table 1. Six studies present the application of more than one method and the remaining ten consider only one CB method.

The case studies of CB methods application in Laboratories and in Industrial workplaces as well as the ENM involved are summarized in Table 1.

3.2 *Control Banding methods special restrictions*

Some of the CB methods possess particular restrictions that confine their applicability to laboratorial activities or to industrial workplaces with ENM manipulation. This information is summarized in Table 2.

3.3 *Advantages and limitations in the application of CB methods*

CB method application use theoretical and field information as input. This data includes characteristics of ENM such as toxicity and physical and chemical properties, workplace characteristics and workers' exposure factors, control measures involved and other factors depending on each method requirements.

3.4 *Advantages and limitations in the application of CB methods*

CB method application use theoretical and field information as input. This data includes characteristics of ENM such as toxicity and physical and chemical properties, workplace characteristics and workers' exposure factors, control measures

Table 1. Case studies focusing CB methods application in laboratories and/or industrial workplaces.

Method	Nanomaterials	Activity	References
ANSES	Metal oxides/Metals/Polymers/ Carbon nanotubes/Carbon black	Laboratory	(Groso, Meyer, 2013; Silva *et al.*, 2015; Sousa, Ribeiro, Batista, 2014)
Nanotool	Metal oxides/Metals/Nanoporous metal foams/Ceramic nanoparticles/ Carbon nanotubes/Nanocomposites/ Carbon Black/Polymers/Calcium sulphate	Laboratory	(Groso and Meyer, 2013; Melorose, Perroy, Careas, 2013; Paik, Zalk, Swuste, 2008; Silva *et al.*, 2015; Silva, Arezes, Swuste, 2015; Sousa, Ribeiro, Batista, 2014)
	Metal oxides/Nanowires/Metals/ Polymers/Quantum dots/ Nanofoams/Carbon nanotubes/ WC-coatings	Industry	(Huang, Li, Li, 2016; Liao *et al.*, 2014; Wu *et al.*, 2014; Zalk, Paik, Swuste, 2009)
EPFL	Metal oxides/Metals/Polymers/ Carbon nanotubes/Carbon black	Laboratory	(Groso, Meyer, 2013; Silva *et al.*, 2015; Sousa, Batista, Ribeiro, 2014; Sousa, Ribeiro, Batista, 2014)
GWSNN	Polymers	Laboratory	(Silva *et al.*, 2015; Sousa, Ribeiro, Batista, 2014)
ISPESL	Metal oxides/Carbon Nanotubes/ Polymers	Laboratory	(Giacobbe, Monica, Geraci, 2009; Silva *et al.*, 2015; Sousa, Batista, Ribeiro, 2014; Sousa, Ribeiro, Batista, 2014; Sousa, Ribeiro, Batista, 2013)
Nanosafer	Metal oxides	Industry	(Levin *et al.*, 2015)
Stoffenmanager	Metal oxides/Metals/Carbon nanotubes/Carbon black/Polymers	Laboratory	(Groso, Meyer, 2013; Silva *et al.*, 2015; Silva, Arezes, Swuste, 2015)
	Manufactured nano objects/Al2O3	Industry	(Barberio *et al.*, 2014; EU-OSHA, 2012)

Table 2. Particular characteristics/restrictions and references of CB methods for ENM occupational exposure assessment.

	Particular characteristics/restrictions	References
ANSES	To be used by people 'adequately qualified in chemical risk prevention'	(Groso, Meyer. 2013; Silva *et al.*, 2015; Sousa, Ribeiro, Batista, 2014)
Nanotool	Developed for preliminary qualitative risk assessment in research activities and facilities with relative small-scale use of a wide variety of MNMs[i]	(Groso, Meyer, 2013; Melorose, Perroy, Careas, 2013; Paik, Zalk, Swuste, 2008; Silva *et al.*, 2015; Silva, Arezes, Swuste, 2015; Sousa, Ribeiro, Batista, 2014; Huang, Li, Li, 2016; Liao *et al.*, 2014; Wu *et al.*, 2014; Zalk, Paik, Swuste, 2009)
EPFL	Developed for research laboratories producing and using nanomaterials	(Groso, Meyer, 2013; Silva *et al.*, 2015; Sousa, Batista, Ribeiro, 2014; Sousa, Ribeiro Batista, 2014)
GWSNN	Able to be used by non-professionals (it is intended to be a guide for employers and employees working with nanomaterials)	(Silva *et al.*, 2015; Sousa, Ribeiro, Batista, 2014)
ISPESL	–	(Giacobbe, Monica, Geraci, 2009; Silva *et al.*, 2015; Sousa, Batista, Ribeiro, 2014; Sousa, Ribeiro, Batista, 2014; Sousa, Ribeiro, Batista, 2013)
Nanosafer	Limited to down-stream use of the powder form of nanomaterials; intended to be used by EHS[ii] responsible persons in SMEs[iii].	(Levin *et al.*, 2015)
Stoffenmanager	Developed for work environments involving MNO[iv]. Applied to substances that consist of MNO with a primary size between 1 and 100 nm and/or to products with a specific surface area larger than 60 m^2/g. It is intended to be used by non-expert SMEs[v] employers and employees.	(Groso, Meyer, 2013; Silva *et al.*, 2015; Silva, Arezes, Swuste, 2015; Barberio *et al.*, 2014; EU-OSHA, 2012)

[i]MNM—Manufactured nanomaterials; [ii]EHS—Environment, Health and Safety; [iii]SMEs—Small and medium-sized enterprises; [iv]MNO—Manufactured nano-objects.

involved and other factors depending on each method requirements.

CB methods presented some advantages and difficulties while its implementation. Most authors indicate the difficulty for the user in obtaining reliable information on physical, chemical and/or toxicity of ENM (Barberio et al., 2014; Groso, Meyer, 2013; Paik, Zalk, Swuste, 2008; Silva, Arezes, Swuste, 2015; Sousa, Batista, Ribeiro, 2014; Zalk, Paik, Swuste, 2009). Some authors indicate that the shortage of reliable data and the lack of clear toxicological basis for setting nanomaterial-specific occupational exposure limits, as being a limitation for risk assessment methods application.

Also it is often referred by some authors the lack of information about the safety of nanomaterials in safety data sheets, which may include different information according to the supplier (Groso, Meyer, 2013; Silva, Arezes, Swuste, 2015).

Regarding the information to be provided by the user, it was found that each method has a specific protocol.

The method considered to require the most extensive effort in respect to information input from the user was Nanotool. However, the application of this method was considered as a simple, affordable and comprehensive way for assessing risk (Melorose, Perroy, Careas, 2013).

In contrast, EPFL was considered an easy-to-use method, requiring very little information (Groso, Meyer, 2013).

EPFL tree approach was considered (Fleury et al., 2013) as an easy tool for non-specialists' users. However, it also reported that the fusion between hazard and exposure factors caused by the tree-approach and the low level of recommendation resulting from the risk levels, may limit the scope of the method (Fleury et al., 2013).

ANSES provides by default information of safety data sheets according to the GHS system. However, this method was pointed as being used only by qualified professionals in chemical risk prevention (Groso, Meyer, 2013).

In terms of usage, Stoffenmanager was considered on a case study carried out in a small company (EU-OSHA, 2012), as being an easy to access program and easy-to-use method. It was considered one of the most complete methods to manage the nano risks when comparing to ANSES, Nanotool and EPFL (Fleury et al., 2013). However, it is referred that the increased complexity (large number of parameters and complex calculation algorithm) may limit the applicability of the method by non-specialist persons.

Regarding NM characterization, ANSES and Nanotool methods give a significant importance on its health-related properties, while EPFL, and GWSNN rather stress the analysis on their physical properties (Sousa, Ribeiro, Batista, 2014). In this context, ISEPSL is considered to be the most comprehensive risk assessment method, since it considers for analysis both health-related properties and physical properties of NM. Thus, by comparison with others, it examines more parameters (Sousa, Batista, Ribeiro, 2014).

Quantitative measurement importance (Silva et al., 2015); (Sousa, Batista, Ribeiro, 2014), concerning its contribution to determine worker's exposure was also highlighted by some authors as conducting to more reliable results.

Some case studies include quantitative measurements, using direct reading of airborne nanoparticle in the workplace environment as a complement of CB methods application (Sousa, Ribeiro, Batista, 2013; Silva, Arezes, Swuste, 2015).

However, it was found that direct measurements were not always consistent with the results obtained on CB applications (Silva, Arezes, Swuste, 2015; Sousa, Ribeiro, Batista, 2013).

4 CONCLUSIONS

According to this review, real time measurements may be very useful as input in CB methods, contributing to obtain results closer to reality. It would be important to test different CB methods integrating measurements in workplaces, in search to obtain more reliable data, and compare them in order to understand if the results are convergent or not, supporting the potential need of future adjustments.

CB methods applied to nanomaterials are well accepted and considered very useful tools for risk analysis, when facing uncertainties. However, despite the available number of CB tools for nanomaterials, yet little is known of the validity of these tools. This review leads to consider the importance of ensuring reliability of CB methods through validation and the integration of CB methods in risk management strategies.

REFERENCES

Barberio, G., Scalbi, S., Buttol, P., Masoni, P., Righi, S. 2014. "Combining Life Cycle Assessment and Qualitative Risk Assessment: The Case Study of Alumina Nanofluid Production." *Science of the Total Environment* 496: 122–31.

Belluci, S., Bergamaschi, E., Bertazzi, P., Boccuni, F., Bonifaci, G., Casciardi, S., Castellano, P. 2011. "White Book: Exposure to Engineered Nanomaterials and Occupational Health and Safety Effects". 2010th ed. Milan: INAIL.

Brouwer, D. 2012. "Control Banding Approaches for Nanomaterials." *Annals of Occupational Hygiene* 56(5): 506–14.

EU-OSHA. 2012. "Risk Assessment by a Small Company Using Stoffenmanager Nano." *Eur. Agency for Safety and Health at Work*: 1–10.

EU-OSHA. 2013. "Tools for the Management of Nanomaterials in the Workplace and Prevention Measures: Health and Safety Hazards of Nanomaterials and Exposure Routes."

European Commission. 2011. "Commission Recommendation of 18 October 2011 on the Definition of Nanomaterial." Official Journal of the European Union, Eur-Lex.

Fleury, D., Fayet, G., Vignes, A., Henry, F., Frejafon, E. 2013. "Nanomaterials Risk Assessment in the Process Industries: Evaluation and Application of Current Control Banding Methods." *Chemical Engineering Transactions* 31: 949–54.

Giacobbe, F., Monica L., Geraci D. 2009. "Nanotechnologies: Risk Assessment Model." *Journal of Physics: Conference Series* 170: 12035.

Groso, A., Meyer T. 2013. "Concerns Related to Safety Management of Engineered Nanomaterials in Research Environment." *J. of Physics: Conference Series* 429(1): 12065.

Hristozov, D., Gottardo, S., Cinelli, M., Isigonis, P., Zabeo, A., Critto, A., Van Tongeren, M., Tran, L., Marcomini, A. 2014. "Application of a Quantitative Weight of Evidence Approach for Ranking and Prioritising Occupational Exposure Scenarios for Titanium Dioxide and Carbon Nanomaterials." *Nanotoxicology* 8(2): 117–31.

Huang, H., Haijun, L., Xinyu L. 2016. "Physicochemical Characteristics of Dust Particles in HVOF Spraying and Occupational Hazards: Case Study in a Chinese Company." *J. of Thermal Spray Technology* 25(5): 971–81.

ISO/TS 80004-1. 2015. "Nanotechnologies—Vocabulary—Part1: Core Terms."

Juric, A., Meldrum R., Liberda, E. 2015. "Achieving Control of Occupational Exposures to Engineered Nanomaterials." *J. of Occ. and Environmental Hygiene* 12(8): 501–8.

Kuempel, E., Castranova V., Geraci C., Schulte P. 2012. "Development of Risk-Based Nanomaterial Groups for Occupational Exposure Control." *Journal of Nanoparticle Research* 14(9).

Levin, M. Levin, M., Rojas, E., Vanhala, E., Vippola, M., Liguori, B., Kling, K., Koponen, I., Molhave, K. 2015. "Influence of Relative Humidity and Physical Load during Storage on Dustiness of Inorganic Nanomaterials: Implications for Testing and Risk Assessment." *Journal of Nanoparticle Research* 17(8): 1–13.

Liao, H. Liao, H., Chung, Y., Lai, C., Lin, M., Liou, S. 2014. "Sneezing and Allergic Dermatitis Were Increased in Engineered Nanomaterial Handling Workers." *Industrial health* 52(3): 199–215.

Liberati, A., Altman, D., Tetzlaff, J., Mulrow, C., Ioannidis, J., Clarke, M., Devereaux, P., Kleijnen, J., Moher, D. 2009. "The PRISMA Statement for Reporting Systematic Reviews and Meta-Analyses of Studies That Evaluate Health Care Interventions: Explanation and Elaboration." *Annals of Internal Medicine* 151(4): W65–94.

Melorose, J., Perroy, R., Careas, S. 2013. "Occupational Risk Assessment of Engineered Nanomaterials by Control Banding Method in Chemistry Laboratories." *J. of Amer. Science* 1.

NIOSH. 2009. "Qualitative Risk Characterization and Management of Occupational Hazards: Control Banding (CB)." *Institute for Occupational Safety and Health* 2009–152 (Qualitative Risk Characterization and Management of Occupational Hazards: Control Banding (CB)): 118.

Paik, S., Zalk, D., Swuste, P. 2008. "Application of a Pilot Control Banding Tool for Risk Level Assessment and Control of Nanoparticle Exposures." *Annals of Occupational Hygiene* 52(6): 419–28.

Rauscher, H., Roebben, G., Amenta, V., Sanfeliu, A., Calzolai, L., Emons, H., Gaillard, C., Gibson, N., Linsinger, T., Mech, A. 2014. Luxembourg: Publications Office of the European Union. "Towards a Review of the EC Recommendation for a Definition of the Term "Nanomaterial" Part 1: Compilation of Information Concerning the Experience with the Definition". JRC.

Rauscher, H., Roebben, G., Sanfeliu, A., Emons, H., Gibson, N., Koeber, R. 2015. "Towards a Review of the EC Recommendation for a Definition of the Term "nanomaterial" Part 3: Scientific-Technical Evaluation of Options to Clarify the Definition and to Facilitate Its Implementation". JRC.

Roebben, G., Rauscher, H., Sokull-klüttgen, B. 2014. "Towards a Review of the EC Recommendation for a Definition of the Term "Nanomaterial" Part 2: Assessment of Collected Information Concerning the Experience with the Definition." JRC.

Savolainen, K., Backman, U., Brouwer, D., Fadeel, B., Fernandes, T. 2013. "Nanosafety in Europe 2015–2025: Towards Safe and Sustainable Nanomaterials and Nanotechnology Innovations Nanosafety in Europe Towards Safe and Sustainable Nanomaterials and Nanotechnology Innovations". Helsinki: Finnish Institute of Occupational Health.

Silva F., Arezes P., Swuste P. 2015. "Risk Assessment in a Research Laboratory during Sol–gel Synthesis of Nano-TiO₂." *Safety Science* 80: 201–12.

Silva, F., Sousa, S., Arezes, P., Swuste, P., Ribeiro, M., Baptista, J. 2015. "Qualitative Risk Assessment during Polymer Mortar Test Specimens Preparation—Methods Comparison." *J. Phys.: Conf. Ser* 617.

Sousa, S., Batista, J., Ribeiro M. 2014. "Polymer Nano and Submicro Composites Risk Assessment." *I. J. Working Conditions* 7: 52–67.

Sousa, S., Ribeiro M., Batista J. 2013. "Risk Assessment in Processing Polymer Nanocomposites." *Occupational Safety and Higiene*: 187–90.

Sousa, S., Ribeiro, M. 2014. "Polymeric Nanocomposites Production Risk Assessment Using Different Qualitative Analyses." *Occup. Safety and Higiene* (Table 1): 25–30.

Tc-Osh. 2013. "Priorities for Occupational Safety and Health Research in Europe: 2013–2020". European Agency for Safety and Health at Work.

UKNSPG. 2012. "Working Safely with Nanomaterials in Research & Development". The UK NanoSafety Partnership Group.

Wu, W., Liao, H., Chung, Y., Li, W., Tsou, T., Li, L., Lin, M., Ho, J., Wu, T., Liou, S. 2014. "Effect of Nanoparticles Exposure on Fractional Exhaled Nitric Oxide (FENO) in Workers Exposed to Nanomaterials." *International Journal of Molecular Sciences* 15(1): 878–94.

Zalk, D, Heussen G. 2011. "Banding the World Together ; The Global Growth of Control Banding and Qualitative Occupational Risk Management." *Safety and Health at Work* 2(4): 375–79.

Zalk, D., Kamerzell, R., Paik, S., Kapp, J., Harrington, D., Swuste, P. 2010. "Risk Level Based Management System: A Control Banding Model for Occupational Health and Safety Risk Management in a Highly Regulated Environment." *Industrial health* 48(1): 18–28.

Zalk, D., Paik S., Swuste P. 2009. "Evaluating the Control Banding Nanotool: A Qualitative Risk Assessment Method for Controlling Nanoparticle Exposures." *Journal of Nanoparticle Research* 11(7): 1685–1704.

Occupational Safety and Hygiene V – Arezes et al. (Eds)
© 2017 Taylor & Francis Group, London, ISBN 978-1-138-05761-6

Health effects on workers exposed to engineered nanomaterials: Short review

Andréa Pereira
PROA/LABIOMEP, Faculty of Engineering, University of Porto, Porto, Portugal

A.C. Meira Castro
CERENA, School of Engineering, Polytechnic of Porto, Porto, Portugal

J. Torres Costa
PROA/LABIOMEP, Faculty of Medicine, University of Porto, Porto, Portugal

J. Santos Baptista
PROA/LABIOMEP, Faculty of Engineering, University of Porto, Porto, Portugal

ABSTRACT: The huge increase in the nanotechnology industry has also resulted in an enormous growth in in the number of workers exposure to nanoparticles. Consequently, researches has been carried out regarding health and safety risks of engineered nanomaterials and the regulation of potential new hazards is a global concern. However, little is known about the impact of engineered nanomaterials in human health. This short review purpose is to identify the potential human health effects resulting from occupational exposure to engineered nanomaterials, supported on data base search. The article search revealed 5 studies on health effects performed in work-places and 14 in vitro and in vivo studies. Lung inflammatory effects, cytoxicity, oxidative stress and DNA damage were the most referred as potential health effects. However, more studies are still needed to obtain reliable data about nanomaterials effects on human health.

1 INTRODUCTION

Engineered nanomaterials are already part of an established market and are present in our daily life in many different applications. The great virtue of nanomaterials is that they possess unique chemical, physical and mechanical properties, being used in a variety of applications in different industrial branches, from the food industry to the transport services (Tc-Osh, 2013). The development of these innovative materials are very important for the competitiveness of Europe, but the enlargement of the use of nanomaterials also results in a greater number of potentially exposed workers; this exposure can be present at every stage of the nanomaterials life cycle (Tc-Osh, 2013).

Nanomaterial definition was published by a recommendation of the European Commission in 2011 indicating as a natural, incidental or manufactured material containing particles, in an unbound state or as aggregate or as agglomerate and where 50% or more of the particles in the number size distribution and one or more external dimensions is in the size range 1 nm-100 nm (European Commission, 2011). This definition has already been reviewed in three reports to clarify the definition of the term

"nanomaterial" (Rauscher et al., 2014), (Roebben, Rauscher, Sokull-klüttgen, 2014), (Rauscher et al., 2015). Also the International Organization for Standardization (ISO) published a definition for nanomaterial and manufactured nanomaterial (ISO/TS 80004-1, 2015). International Labor Office (ILO, 2015) refers that labor protection regulation of new and potential chemical hazards derived from nanotechnologies must be a concern for all countries. It also mentions that by 2020, approximately 20% of all manufactured goods around the world will be based to some extent on the use of nanotechnology and almost nothing is known about the impact of nanotechnology on health.

Therefore, different organizations, such as NIOSH (NIOSH, 2013), the European Agency for Safety and Health at Work (Tc-Osh, 2013), Finnish Institute of Occupational Health (Savolainen et al., 2013) and OECD (OECD, 2015a, b) have been developing programs concerning health and safety risks due to exposure of nanomaterials.

This review intends to identify the potential human health effects resulting from occupational exposure to Engineered Nanomaterials (ENM). With this intention, epidemiological studies concerning symptoms and/or effects to human health resulting from ENM exposure on workplaces were collected along with in vitro and in vivo studies.

2 METHOD

This short review applies the principles of PRISMA (Preferred Reporting Items for Systematic Reviews and Meta-Analysis) methodology (Liberati *et al.*, 2009), intending to summarize evidence accurately and reliably. The search process was performed online in 18 data bases: ACS Publications, BioMed Central, Elsevier, Environmental health perspectives, Informaworld (Taylor and Francis), IOP Science, MEDLINE (EBSCO), MDPI, Medscape, Nature, Oxford Journals, PubMed, Scopus, Royal Society of Chemistry, SAGE Journals Online, SpringerLink, Wiley Online Library and PLOS Medicine. Also, Google searching was included.

The data base search was conducted using a set of keywords to carry out searches for relevant studies: cohort, nanoparticle, health effect, nanomaterial, case study and epidemiological studies. These words were combined in order to form composed three-word keywords, with the words 'health effect' and 'nanomaterial' (or nanoparticle) always present. The data base search process focused on articles written in English and published between 2013 and 2016.

The data collection was performed since 2013 in order to allow more recent and updated information. The inclusion criteria defined covers cohort, cross-sectional and case studies concerning human health effects resulting from the exposure to engineered nanomaterials in work environments. Are also included toxicological studies, whose results can substantiate or confirm the results of epidemiological ones, concerning health effects or symptoms resulting from ENM exposure in occupational settings.

3 RESULTS

Data base search resulted in 484 records found and after screening and final selection was based on inclusion and exclusion criteria, resulting in 16 eligible full articles included in this review.

This review includes epidemiological studies regarding human health effects resulting from engineered nanomaterials' exposure in workplaces. To substantiate the results from the epidemiological studies found, articles with toxicological studies performed with the same ENM, were included.

The search revealed 5 epidemiological studies concerning the effects/symptoms on human health from engineered nanomaterials exposure in work environments. Also 14 toxicological studies with in vivo and in vitro tests were identified, concerning the same nanomaterials used in the epidemiological studies. Therefore, these studies were included in the systematic review in order to support the results of epidemiological studies, previously identified.

3.1 *Epidemiological studies*

Five epidemiological studies were found in the search: one cohort study, three cross-sectional studies and one longitudinal study and one case study, as presented in Table 1. These studies are related to human health symptoms/effects derived from Engineered Nanomaterials (ENM) exposure in work environments.

The cohort study (Wu *et al.*, 2014) aims to found associations between the risk level of nanoparticle exposure and Fractional Exhaled Nitric Oxide (FENO), particularly noteworthy for Nano-TiO$_2$ in the lung.

The three cross-sectional studies are related to symptoms and/or health-effects on human health of nanoparticle exposure by inhalation focused on the following: One, aims to perform non-invasively measure and evaluate the markers of oxidation of nucleic acids and proteins in the exhaled Breath Condensate (EBC) of workers and control subjects (D Pelclova *et al.*, 2015); Other (Daniela Pelclova *et al.* 2015), has the objective of detecting TiO$_2$ particles in Exhaled Breath Condensate (EBC) and urine samples to ascertain their presence and potential persistence and excretion in urine; and another (Liao, Chung, Lai, Lin, *et al.*, 2014), aspires to survey the work-relatedness of symptoms and diseases among engineered nanomaterials handling workers.

The longitudinal study (Liao, Chung, Lai, Wang, *et al.*, 2014) intends to identify sensitive and specific biomarkers related to health effects of engineered nanoparticles for potential use in nanoparticle exposure surveillance. The information of the studies, concerning the studies characteristics, are summarized in Table 1.

3.2 *Toxicological studies*

To support results on potential health effects derived from the epidemiological studies concerning engineered nanomaterials exposure in work environment, both in vitro and in vivo models were also considered. The search revealed 9 in vitro studies and 5 in vitro studies, which are included in this review.

These studies were performed with the following nanomaterials: Titanium dioxide (TiO$_2$), Silver (Ag), Silicon Dioxide (SiO$_2$) and Carbon Nanotubes (CNT), both single and multi-walled (SWCNT and MWCNT). The summary of in vivo and in vitro studies performed with the different nanomaterials is presented in Table 2.

Table 1. Characteristics of the epidemiological studies in workplaces with nanomaterials.

Author, Year	Study type	Country	NM	Size (nm)	Exposure routes	Workplaces	Sample size (number workers)	Method
(Wu et al. 2014)	Cohort study	Taiwan	Various[a]	3–200	Inhalation	NM handling factories	258 exposed 200 controls	Questionnaire; Physical examination; EBC samples; Spirometry; Control Banding nanotool
(D Pelclova et al. 2015)	Cross-sectional study	Czech Republic	TiO_2	10–100 (80%)	Inhalation	TiO_2 pigment production plant	2012: 36 exposed 2013: 32 exposed 2012/: 45 controls	Questionnaire; Physical examination; EBC sampling; Aerosol measurements.
(Daniela Pelclova et al. 2015)	Cross-sectional study	Czech Republic	TiO_2	100 (70–82%)	Inhalation	White pigment production	20 exposed 20 controls	Questionnaire; Control Banding nanotool; Aerosol measurements; Blood, urine and EBC samples; Spirometry.
(Liao, Chung, Lai, Lin, et al. 2014)	Cross-sectional study	Taiwan	Various[b]	6–200	Inhalation/ Dermal	NM handling factories	258 exposed 200 controls	Questionnaire; Control Banding nanotool
(Liao, Chung, Lai, Wang, et al., 2014)	Longitudinal study	Taiwan	Various[c]	6–200	Inhalation	NM handling factories	Baseline: 158 exposed; 6 months later: 124 exposed; Baseline: 104 controls; 6 months later: 77 controls	Questionnaire; Control Banding nanotool; Blood, urine and EBC sampling.

NM—Nanomaterial; [a] TiO_2, Ag, Fe_2O_3, SiO_2, CNT, Au; [b] TiO_2, Ag, Fe_2O_3, SiO_2, CNT.

Table 2. In vivo and in vitro studies with nanomaterials.

Ref.	NM	Size (nm)
In vivo		
(Kim et al., 2014)	TiO_2	10
(Shrivastava et al., 2014)	TiO_2	50–75
(Louro et al., 2010)	TiO_2	22
(Noël et al., 2013)	TiO_2	5–50
(Hunt et al., 2013)	Ag	10
(Kim et al., 2014)	SiO_2	10
(Nakanishi et al., 2015)	MWCNT	63
(Tian et al., 2013)	SWCNT	0,8–1,2
(Yan et al., 2015)	SWCNT	0,8–1,2
In vitro		
(Kim et al., 2014)	TiO_2	10
(Orta-Garcia et al., 2015)	Ag	<2
(Sahu et al., 2014)	Ag	20
(Kim et al., 2014)	SiO2	10
(Foldbjerg et al., 2014)	SWCNT	$31,79 \pm 8,66$

3.3 Potential health impacts of nanomaterials

The cohort study (Wu et al., 2014) concluded that FENO can be a useful indicator of broader epithelial function in addition to being an inflammatory marker for workers potentially exposed to NP. It also indicates occupational medical surveillance as a strategy for providing benefits to the individual and company for health outcomes for workers exposed to NM. This study's also found associations between the risk level of NP exposure and FENO and verified that Titanium dioxide nanoparticles (TiO_2 NP), of all NM exposed categories, was particularly noteworthy since it had a significantly increased risk on FENO.

The cross-sectional study (D Pelclova et al., 2015) focused on non-invasively measurement and evaluation of the markers of oxidation of nucleic acids and proteins in the Exhaled Breath Condensate (EBC) of workers and control subjects, concluded that the concentration of Titanium (Ti) in Exhaled Breath Condensate (EBC) of workers may serve as a direct exposure marker in workers producing TiO_2 pigment. The study also indicates that the markers of oxidative stress reflect the local biological effect of TiO_2 NP in the respiratory tract of the exposed workers.

The cross-sectional study (Daniela Pelclova et al., 2015) regarding the detection of TiO_2 particles in exhaled breath condensate (EBC) and urine samples to ascertain their presence and potential persistence and excretion in urine, concludes that the EBC collection and analysis, combined with Raman microspectroscopy, allowed to evidence the presence of TiO_2 particles/agglomerates in the respiratory tract

Table 3. Summary of results of epidemiological and toxicological studies.

Ref.	Type of study	Potential impacts on human health
Nanomaterial: Various[i]		
(Liao, Chung, Lai, Wang, *et al.*, 2014)	Longitudinal	Potential cardiopulmonary hazards
(Liao, Chung, Lai, Lin, *et al.*, 2014)	Cross-sectional	Sneezing and allergic dermatitis
Nanomaterial: Various[ii]		
(Wu *et al.*, 2014)	Cohort study	Inflammation in airways.
Nanomaterial: TiO$_2$		
(Liao, Chung, Lai, Wang, *et al.*, 2014)	Cross-sectional	Oxidative stress with possible reflections on respiratory tract
(Daniela Pelclova *et al.*, 2015)	Cross-sectional	Inflammation in airways
(Kim *et al.*, 2014)	In vitro	Cytotoxicity/Effects on Mitochondrial function and cell membrane integrity
(Kim *et al.*, 2014)	In vivo	Inflammatory effects on lungs/Lung macrophage uptake
(Shrivastava *et al.*, 2014)	In vivo	Oxidative stress on erythrocyte, liver and brain
(Louro *et al.*, 2010)	In vivo	Bio persistence and inflammatory response in liver
(Noël *et al.*, 2013)	In vivo	Lung inflammatory effects/Acute inflammatory response on BALF /Induced cytotoxic and oxidative stress effects on BALF
Nanomaterial: Ag		
(Orta-Garcia *et al.*, 2015)	In vitro	Cytotoxicity/Oxidative stress/DNA damage
(Sahu *et al.*, 2014)	In vitro	Cytotoxicity/Mitochondrial injury/DNA damage
(Hunt *et al.*, 2013)	In vivo	Induced oxidative damage on DNA
Nanomaterial: SiO$_2$		
(Kim *et al.*, 2014)	In vitro	Cytotoxicity/Effects on Mitochondrial function and cell membrane integrity
(Kim *et al.*, 2014)	In vivo	Inflammatory effects on lungs
Nanomaterial: MWCNT		
(Nakanishi *et al.*, 2015)	In vivo	Acute lung inflammation and subsequent glaucoma Histopathological changes in lungs and in lung—associated lymph nodes
Nanomaterial: SWCNT		
(Foldbjerg *et al.*, 2014)	In vitro	Induction of regulation on gene expression/Minor effects on induction of cytokine release/Reduced phagocytic ability/Effects on cell morphology/and cell cycle/Induction of necrosis/Increase of ROS levels
(Tian *et al.*, 2013)	In vivo	Inflammatory reaction in the lungs/increasing of pro-coagulant activity/Decrease of fibrinolysis capacity/ Inflammation and oxidative stress when penetrating the alveolar epithelium/Pro-thrombotic effects and promotion of development of ischemic heart disease
(Yan *et al.*, 2015)	In vivo	Inflammatory immune injury in the lung/Systemic inflammation/Vascular endothelial dysfunction/ Vascular endothelial damage/Increase of CVD risk (Thrombosis and arthrosclerosis)/Cardiovascular toxicity induced through indirect effects on vascular homeostasis.

[i] TiO$_2$, Ag, Fe$_2$O$_3$, SiO$_2$, CNT; [ii] TiO$_2$, Ag, Fe$_2$O$_3$, SiO$_2$, CNT, Au.

using non-invasive methods. This study also indicates that TiO$_2$ particles (both anatase and rutile) may persist in the lungs of exposed workers for at least several hours after the previous shift or several shifts. However, urine examination using Raman microspectroscopy produced negative results.

The cross-sectional study (Liao, Chung, Lai, Lin, *et al.*, 2014), aiming to survey the work-relatedness of symptoms and diseases among engineered nano-materials handling workers, concluded that sneezing and allergic dermatitis were significantly increased in engineered nanomaterials handling workers. It also indicates that besides allergic diseases, cardiopulmonary symptoms such as cough and angina may be used as screening tools for medical surveillance of people handling engineered nanomaterials.

The longitudinal study (Liao, Chung, Lai, Wang, *et al.*, 2014), intending to identify sensitive and specific biomarkers related to health effects of engineered nanoparticles for potential use in nanoparticle exposure surveillance, indicates that pulmonary markers, cardiovascular markers and antioxidant enzymes are possible biomarkers for medical surveillance of workers handling engineered nanomaterials. It also refers that potential cardiopulmonary hazards may occur in workers handling nanomaterials who are potentially exposed to nanoparticles; this conclusion was based at a six-month follow-up assessment and it is also mentioned that further long-term follow-up and valid design is required to confirm this possibility.

4 DISCUSSION

The impact of TiO_2 NP on human health regarding oxidative stress and inflammation in airways indicated in epidemiological studies was supported by some of the in vivo studies performed with the same nanomaterial. It was difficult to obtain a relation between the health impacts resulting from epidemiological studies with several nanomaterials and the results from in vitro and in vivo studies performed with the same nanomaterials separately. Stablishing this relation may lead to inaccurate conclusions.

5 CONCLUSIONS

More epidemiological studies in the workplace settings are needed in order to understand the symptoms and effects of nanomaterials on human health and to allow the adoption of adequate preventive measures to avoid occupational diseases that may result from engineered nanoparticles exposure.

REFERENCES

European Commission. 2011. *"Commission Recommendation of 18 October 2011 on the Definition of Nanomaterial"*. Official Journal of the European Union, Eur-Lex.

Foldbjerg, R., Irving, E., Wang, J., Thorsen, K., Sutherland, D., Autrup, H., Beer, C., 2014. "The Toxic Effects of Single-Walled Carbon Nanotubes Are Linked to the Phagocytic Ability of Cells." *Toxicology Research* 3(4): 228–41.

Hunt, P., Marquis, B., Tyner, K., Conklin, S., Olejnik, N., Nelson, B., Sprando, R. 2013. "Nanosilver Suppresses Growth and Induces Oxidative Damage to DNA in Caenorhabditis Elegans." *J. of Appl. Toxicology* 33(10):1131–42.

ILO, International Labour Conference. 2015. "Labour Protection in a Transforming World of Work—A Recurrent Discussion on the Strategic Objective of Social Protection (Labour Protection)". Geneva.

ISO/TS 80004–1. 2015. "Nanotechnologies—Vocabulary —Part1: Core Terms."

Kim, Y., Boykin, E., Stevens, T., Lavrich, K., Gilmour, M. 2014. "Comparative Lung Toxicity of Engineered Nanomaterials Utilizing in Vitro, Ex Vivo and in Vivo Approaches." *Journal of Nanobiotechnology* 12(1): 47.

Liao, H., Chung, Y., Lai, C., Lin, M., Liou, S. 2014. "Sneezing and Allergic Dermatitis Were Increased in Engineered Nanomaterial Handling Workers." *Industrial health* 52(3): 199–215.

Liao, H., Chung, Y., Lai, C., Wang, S., Chiang, H., Li, L., Tsou, T., Li, W., Lee, H., Wu, W., Lin, M., Hsu, J., Ho, J., Chen, C., Shih, T., Lin, C., Liou, S. 2014. "Six-Month Follow-up Study of Health Markers of Nanomaterials among Workers Handling Engineered Nanomaterials." *Nanotoxicology* 8(S1):100–110

Liberati, A., Altman, D., Tetzlaff, J., Mulrow, C., Ioannidis, J., Clarke, M., Devereaux, P., Kleijnen, J., Moher, D. 2009. "The PRISMA Statement for Reporting Systematic Reviews and Meta-Analyses of Studies That Evaluate Health Care Interventions: Explanation and Elaboration." *Annals of Internal Medicine* 151(4): W65–94.

Louro, H., Tavares, A., Vital, N., Costa, P., Alverca, E., Zwart, E., Jong, W., Fessard, V., Lavinha, J., Silva, M. 2010. "Integrated Approach to the In Vivo Genotoxic Effects of a Titanium Dioxide Nanomaterial Using LacZ Plasmid-Based Transgenic Mice." *Environmental and molecular mutagenesis* 51(3): 229–35.

NIOSH. 2013. *Protecting the Nanotechnology Workforce —NIOSH Nanotechnology Research and Guidance— Strategic Plan*, 2013–2016.

Nakanishi, Junko et al. 2015. "Risk Assessment of the Carbon Nanotube Group." *Risk Analysis* 35(10): 1940–56.

Noël, A., Charbonneau, M., Cloutier, Y., Tardif, R., Truchon, G. 2013. "Rat Pulmonary Responses to Inhaled Nano-TiO_2: Effect of Primary Particle Size and Agglomeration State." *Particle and fibre toxicology* 10(1): 48.

OECD. 2015. "Preliminary Guidance Notes on Nanomaterials: Interspecies Variability Factors in Human Health Risk Assessment Environment". Paris.

Orta-Garcia, S., Plascencia-Villa, G., Ochoa-Martinez, A., Ruiz-Vera, T., Perez-Vazquez, F., Velazquez-Salazar, J., Yacaman, M., Navarro-Contreras, H., Perez-Maldonado, I. 2015. "Analysis of Cytotoxic Effects of Silver Nanoclusters on Human Peripheral Blood Mononuclear Cells 'in Vitro'." *Journal of applied toxicology : JAT* 35(10): 1189–99.

Pelclova, D., Zdimal, V., Fenclova, Z., Vlckova, S., Turci, F., Corazzari, I., Kacer, P., Schwarz, J., Zikova, N., Makes, O., Syslova, K., Komarc, M., Belacek, J., Navratil, T., Machajova, M., Zakharov, S. 2015. "Markers of

Oxidative Damage of Nucleic Acids and Proteins among Workers Exposed to TiO$_2$ (Nano) Particles." *Occup. and environm. medicine* 73(2): 110–18.

Pelclova, Daniela, Zdimal, V., Fenclova, Z., Vlckova, S., Turci, F., Corazzari, I., Kacer, P., Schwarz, J., Zikova, N., Makes, O., Syslova, K., Komarc, M., Belacek, J., Navratil, T., Machajova, M., Zakharov, S.. 2015. "Raman Microspectroscopy of Exhaled Breath Condensate and Urine in Workers Exposed to Fine and Nano TiO2 Particles: A Cross-Sectional Study." *Journal of breath research* 9(3): 36008.

Rauscher, H., Roebben, G., Amenta, V., Sanfeliu, A., Calzolai, L., Emons, H., Gaillard, C., Gibson, N., Linsinger, T., Mech, A. 2014. "Towards a Review of the EC Recommendation for a Definition of the Term "Nanomaterial" Part 1 : Compilation of Information Concerning the Experience with the Definition". JRC.

Rauscher, H., Roebben, G., Sanfeliu, A., Emons, H., Gibson, N., Koeber, R. 2015. "Towards a Review of the EC Recommendation for a Definition of the Term "nanomaterial" Part 3: Scientific-Technical Evaluation of Options to Clarify the Definition and to Facilitate Its Implementation". JRC.

Roebben, G., Rauscher, H., Sokull-klüttgen, B. 2014. "Towards a Review of the EC Recommendation for a Definition of the Term "Nanomaterial" Part 2 : Assessment of Collected Information Concerning the Experience with the Definition". JRC.

Sahu, S., Zheng, J., Graham, L., Chen, L., Ihrie, J., Yourick, J., Sprando, R. 2014. "Comparative Cytotoxicity of Nanosilver in Human Liver HepG2 and Colon Caco2 Cells in Culture." *Journal of Applied Toxicology* 34(11): 1155–66.

Savolainen, K., Backman, U., Brouwer, D., Fadeel, B., Fernandes, T. 2013. "Nanosafety in Europe 2015–2025: Towards Safe and Sustainable Nanomaterials and Nanotechnology Innovations Nanosafety in Europe Towards Safe and Sustainable Nanomaterials and Nanotechnology Innovations". Helsinki: Finnish Institute of Occupational Health.

Shrivastava, R., Raza, S., Yadav, A., Flora, S.. 2014. "Effects of Sub-Acute Exposure to TiO$_2$, ZnO and Al$_2$O$_3$ Nanoparticles on Oxidative Stress and Histological Changes in Mouse Liver and Brain." *Drug and chemical toxicology* 37(3): 336–47.

Tc-Osh. 2013. "Priorities for Occupational Safety and Health Research in Europe: 2013–2020".

Tian, L., Lin, Z., Lin, B., Liu, H., Yan, J., Xi, Z. 2013. "Single Wall Carbon Nanotube Induced Inflammation in Cruor-Fibrinolysis System." *Biomed Environmental Science* 26(5): 338–45.

Wu, W., Liao, H., Chung, Y., Li, W., Tsou, T., Li, L., Lin, M., Ho, J., Wu, T., Liou, S. 2014. "Effect of Nanoparticles Exposure on Fractional Exhaled Nitric Oxide (FENO) in Workers Exposed to Nanomaterials." *International Journal of Molecular Sciences* 15(1): 878–94.

Yan, J., Lin, Z., Lin, B., Yang, H., Zhang, W., Tian, L., Liu, H., Zhang, H., Liu, X., Xi, Z.. 2015. "Respiratory Exposure to Single-Walled Carbon Nanotubes Induced Changes in Vascular Homeostasis and the Expression of Peripheral Blood Related Genes in a Rat Model." *Toxicology Research* 4(5): 1225–37.

Occupational Safety and Hygiene V – Arezes et al. (Eds)
© *2017 Taylor & Francis Group, London, ISBN 978-1-138-05761-6*

Impacts of the operation of vibratory machines on the health of workers in the rock processing industries: Systematic review

A.M. do Couto
Universidade do Porto, Porto, Portugal

B. Barkokébas Junior
Universidade de Pernambuco, Pernambuco, Brasil

ABSTRACT: Brazil is an important producer and exporter of ornamental stones, whose most significant sector is the one dedicated to the finishing activities of processing of rocks, where workers use portable motorized tools that produce high impacts on their upper limbs. Vibrations produced in the workplace show high values for both, the daily action and the daily limit levels, then that permitted by current legislation. Research indicates that the prolonged exposure to the Hand-Transmitted Vibration, by this type of tools, is associated to an increased risk of Hand-Arm Vibration Syndrome, Raynaud phenomenon, digital neuropathy and musculoskeletal disorders. This systematic review aims to study the impact of operations of vibratory machines on the health of workers in the rock processing industries and addresses the characterization of the effects and the evaluation of the risks of vibration transmitted to the hand and arm.

Keywords: risk analysis, ergonomic risks, hand-arm vibration, machine tool, marble industry, ornamental rocks

1 INTRODUCTION

Brazil is an important producer and exporter of ornamental stones such as granite, marble, limestone, quartz-based stones, slate and basalt (Estellita, et al., 2010). In this field, the sector of most significant dimension is the rock-processing one. It presents a production process that consists of operations in different sectors where there is always imminence of serious accidents, and ergonomic, physical, biological and chemical risks are observed. The sectors that stand out are the unloading and storage of plates and blocks; cutting, grinding, plate polishing and assembly of ornamental pieces (where the most dangerous conditions are present); and the loading of final products (Melo Neto, et al., 2012). In the finishing works, the workers use portable motorized tools (Gauthier, et al., 2012) such as saws, drills, grinders and polishers (Estellita, et al., 2010) that apply potential impacts on the upper limb (Palmer, et al., 2015). Research shows that the prolonged exposure to Hand-Transmitted Vibration (HTV) by this kind of tools is associated to an increased risk of Hand-Arm Vibration Syndrome (HAVS) (Xu, et al., 2016). The studies address the Raynaud phenomenon (vibration-induced white finger), digital

neuropathy and musculoskeletal disorders, such as carpal tunnel syndrome, contracture and tendinitis (Gauthier, et al., 2012; Palmer, et al., 2015). The exposure occurs in many industrial production environments, especially in manufacturing processes that involve intensive work (Bovenzi, et al., 1994; Bovenzi, 2005; Xu, et al., 2016).

The vibration exposure limits are based on specific relationships between the duration of exposure and the magnitude of vibration defined under the International Standardization Organization (ISO, 2001a; ISO, 2001b). The daily exposure to vibrations can be expressed in terms of the equivalent acceleration transmitting the same energy over a reference period of 8 h (average working day). Therefore we have A(8) (m/s^2 RMS) (Palmer, et al., 2015). For hand-transmitted vibration, action and exposure limit values are specified. The Exposure Action Value; 2,5 m/s^2 A(8), represents the daily amount of exposure above which employers must act in order to control exposure. The Exposure Limit Value—ELV; 5 m/s^2 A(8), represents the maximum amount to which a worker can be exposed during a one day period (Reynolds, 2006; Rimell, et al., 2008; Palmer, et al., 2015).

The transfer of vibrations from machine to operator can be reduced by the use of gloves or by coating the surface of the machine handle. The handle coating of hand-held motorized tools is an effective way to reduce vibrations as it is integrated into the machine and overcomes many of the problems associated with gloves (Singh, et al., 2014). Vibration Reduction (VR) gloves have also been used to reduce the risk of hand-transmitted vibration by hand-held motorized machines and tools; however, even though some of these gloves may be useful (Brown, 1990; Jetzer, et al., 2003; Mahbub, et al., 2007; Welcome, et al., 2014), there is still doubt as to its efficiency to attenuate the vibration transmitted to the fingers (Paddan, et al., 2001; Dong, et al., 2009; Welcome, et al., 2014). Some VR gloves may not be comfortable, cause hand fatigue and substantially reduce finger dexterity. They can also become a contributing factor to hand injuries due to excessive exertion (Wimer, et al., 2010; Welcome, et al., 2014). Therefore, new knowledge should be used to promote the selection of gloves suitable for operations with hand-held motorized tools, in order to assist the assessment of the risks of exposure to vibration and to support the design of better VR gloves (Welcome, et al., 2014).

This systematic review article aims to study the impacts of the operation of vibratory machines on the health of workers in the rock processing industries and addresses the characterization of the effects and risk evaluation of vibration transmitted to the hand and arm.

2 METHOD

Through the development of this review, an investigation question has been prepared to serve as a reference to the scope of the research: "what are the impacts caused by vibration on the health of machine operators in the rock processing industries." Thus, all articles presented in the integrated research of the Documentation and Information Service—SDI (Webpage of the Faculty of Engineering of the University of Porto—FEUP), between July 8th and August 29th, 2016 have been analyzed. The types of resources chosen were databases of scientific journals of all categories applied to advanced meta-research. The keywords initially chosen were: "risk analysis", "ergonomic risks", "hand-arm vibration", "machine tool", "marble industry", and "ornamental rocks" which, combined two by two, enabled the finding of 15,812. The electronic search resulted in articles from publications indexed in Annual Reviews, Nature.com, PubMed, ScienceDirect, and ScienceDirect (eJournals). With the help of the Microsoft Office® Excel program, a spreadsheet was created to list all the articles captured in the meta-search and enter their data regarding the matters of approach, structure,

content and editing details such as author, year, and country, as well as the type of study (longitudinal or transversal). The body of the spreadsheet presents, in the header, the particular topics of the study such as: Study focus, source of exposure to risk, operation/activity, type of tool used, average exposure (hours/day), vibration measuring instrument, health effects, type of cable coating of the hand-held motorized tool, use of protective equipment and standard used. Each topic presented items both related to the subject of the research and eligible to be contained in the articles considered relevant. As an item was perceived to be part of the context of an article, the information was reflected in the worksheet by marking an "X" in the square that was appropriate to the correlation of that item with the analyzed article. Filters were used to aid the classification of the articles in the face of the inclusion and exclusion criteria, thus, the articles were arranged in alphabetical order by author and submitted to evaluation. To identify them, a specific subtitle was created in order to inform their relevance in regards to the theme of the research question, where;

A: relevant article (the article is relevant to the review based on its complete text).
X: irrelevant title (the article is uninteresting to the review based on its title).
Y: irrelevant abstract (the article is uninteresting to the review based on its abstract).
Z: irrelevant text (the article is uninteresting to the review based on its complete text).
R: Repeated article.
U: Inaccessible article.

Table 1 shows the amount of articles related to the keywords and obtained in the SDI-FEUP WEB, according to the inclusion and exclusion criteria. 3900 articles which were published in years prior to 2006, and 2945 articles whose authors were not referenced were left out of the table.

Articles considered old, those published over five years ago, were discarded for potentially containing outdated data on the subject in the field of engineering. The articles were sorted in order to separate the repeated articles, of which 3916 were selected, filtered and removed from the review. The articles were then submitted to the title, abstract and full text analysis, based on four criteria: problem, intervention, results and study. The inclusion and exclusion attributes were duly justified. Analyzing the research question, the apparent problem is the vibration coming from the operation of hand-held motorized tools and the justification are the disorders caused to the health of the worker. Therefore, other risks were not included, such as noise and air pollution. In the intervention, vibration of the hand and arm were included for being

Table 1. Amount of articles by criteria obtained in the SDI-FEUP WEB resources.

Keywords	Amount of articles	A	Z	Y	X	R
Risk analysis + ergonomic risks	779	0	2	9	686	82
Risk analysis + hand-arm vibration	318	3	2	0	166	147
Risk analysis + machine tool	3097	0	1	4	1101	1991
Risk analysis + marble industry	36	0	0	0	9	27
Risk analysis + ornamental rocks	2184	0	3	12	1.314	855
Ergonomic risks + hand-arm vibration	40	0	2	1	13	24
Ergonomic risks + machine tool	1493	2	5	14	1384	88
Ergonomic risks + marble industry	47	0	1	0	7	39
Ergonomic risks + ornamental rocks	251	1	2	2	75	149
Hand-arm vibration + machine tool	180	0	1	2	37	113
Hand-arm vibration + marble industry	2	0	0	1	0	1
Hand-arm vibration + ornamental rocks	1	0	0	0	0	1
Machine tool + marble industry	117	0	0	0	100	17
Machine tool + ornamental rocks	343	0	3	0	66	274
Marble industry + ornamental rocks	126	0	0	0	18	108
Total	8967	6	22	45	4978	3916

a risk factor that is greatly present in the operation activities related to the rock processing industries; the whole body vibrations were excluded so as not to jeopardize the dimensional data effect in the research. In the results, measures to minimize the exposure to the vibrations of the population's portable motorized tools, such as policies to decrease health disorders in workers caused by equipment vibrations, were included; and the ecological and economic impacts were excluded for not being the study focus. Field and laboratory studies were included due to their scientific support, and opinion studies were considered because of their public interest. The systematic review studies were excluded for being secondary works. Table 2 presents the inclusion and exclusion criteria which were considered in the selection stage.

This way, 4978 articles were rejected by title analysis, 45 by abstract analysis and 22 by full text analysis. This systematic review gathered a total of 6 relevant articles related to the impacts on the health of operators of hand-held motorized tools. Despite this kind of tool being utilized in many industrial fields, none of these articles specifically approaches workers in the rock processing industries.

Table 2. Inclusion and exclusion criteria for articles.

Criteria	Attributes		Justifications
Problem	Inclusion	Machine vibration.	Health disorders caused by this operational attribute.
	Exclusion	Noise and air pollution.	Focus on the vibratory attribute of the equipments.
Intervention	Inclusion	Hand and arm vibration.	Present aspect in most of the activities related to the rock processing industries.
	Exclusion	Whole body vibration.	The assessment damages the dimensional effect of the research activities.
Results	Inclusion	Measures to reduce exposure to equipment vibration.	Decrease health disorders caused by equipment vibrations.
	Exclusion	Social and economic impacts.	Study focused on the health of workers.
Studies	Inclusion	Field and laboratory studies.	Scientific support for decision-making.
	Exclusion	Systematic reviews.	Features of secondary study with data collection.

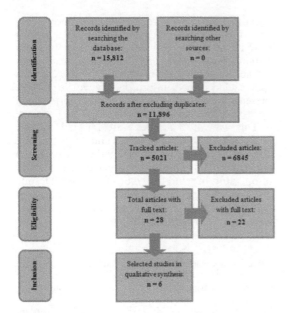

Figure 1. Flowchart of the methodological sequence of the review.

Figure 1 shows, in a synthesized way by means of a specific flowchart, the whole methodological sequence shown in this review.

3 RESULTS

This systematic review shows 6 articles related to the matter concerning the impact of vibratory machines on the health of industry workers, however, when considering the research question prepared to serve as a reference to the aim of the study, it is clear that it intends to assess the rock processing industrial sector. In the course of the research, no article was found on the disturbances caused by the vibration of portable hand-held tools in this specific factory environment, therefore the results presented in this systematic review show the impacts caused on the health of workers in industries that use hand-held vibratory machines similar to the ones used on the rock processing industries.

The relevant articles presented studies on the proposed topic in three distinct industrial areas and a laboratory test, as shown in Table 3. The relationships of these articles with the proposed theme mainly address the levels of exposure of workers to the vibrations of their instruments of work, compared to the acceptable limits proposed by the specific standards. They also address the health effects of excessive contact with hand-held motorized tools as a consequence of the vibrations transmitted to the hands and arms of these workers.

Table 3. Area of activity of the chosen articles.

Area of activity	Alonso et al., 2013	Gauthier et al., 2012	Phillips et al., 2007	Rimell et al., 2008	Singh et al., 2014	Xu et al., 2016
Mining			x			x
Construction industry	x			x		
Wood furniture manufacturing industry		x				
Laboratory tests					x	

For the analysis of the transmission of vibrations to the hands and arms performed by the tools used in the processes, all articles were based on the following standards: ISO 5349-1 (2001), Mechanic Vibration—Measurement and evaluation of human exposure to hand-transmitted vibration, Part 1: General requirements ISO and ISO 5349-2 (2001), Mechanical vibration—measurement and assessment of human exposure to hand-transmitted vibration, part 2: practical guidelines for measurement in the workplace (ISO, 2001a; ISO, 2001b). The laboratory test used a manual drilling machine to compare the different types of coatings on the handle and analyze the transmission of vibration to the wrists of the volunteers (Singh, et al., 2014).

Some studies analyzed the vibrations produced by the machines and compared with the values indicated in the current legislation. Articles that verified the exposure limits of professionals to tool vibrations in their work environments revealed high values for both the action level and the daily limit level (Phillips, et al., 2007). Table 4 shows the vibration of the tools as a function of the daily exposure during the 8 h (average working day) reference period.

Two articles dealt with harmful effects on the health of the worker due to the prolonged use of these hand-held tools. The results presented vascular, neurological and musculoskeletal symptoms (Gauthier, et al., 2012; Xu, et al., 2016). Vascular symptoms can be classified under the heading "white finger of vibration" and the prevalence of symptoms in workers using portable vibrating tools can be quite high (70% or more), depending on the type and duration of exposure (Harada, 2002; Alonso, et al., 2013). The most common neurological disorders are numbness and tingling in the fingers. Musculoskeletal damage manifests as pain in

Table 4. Values of vibrations in the analyzed tools.

Vibration A(8) (m/s²) Tool	Alonso et al., 2013	Gauthier et al., 2012	Phillips et al., 2007	Rimell et al., 2008 (máximo—mínimo)	Singh et al., 2014	Xu et al., 2016
Pneumatic Breakers			21.90	~3.7–7.1		
Hydraulic Breakers			31.00	~4,6–8.8		
Electrical Breakers	4.77		9.20	~6.7–15.0		
Diamond Core Drill				~4.4–7.0		
Plate Vibrator	4.14					
Compactor	4.14					
Power Trowel Propeller	5.90					
Demolition Hammer	5.66					
Electric Hammers	6.01			~8.0–21.0		
Hammers Drill	6.36			~5.0		6.55
Circular Saw	4.16					
Saw Stone				~1.7–19.5		
Sanders		9.80		~2.0–9.0		
Grinders Ø 115				~5.0–6.3		
Grinders Ø 125				~0.4–9.2		
Grinders Ø 150				~7.5		
Grinders Ø 230				~4.2–8.3		

Table 5. Effects of vibration in the health of workers.

Percentage of workers affected Effects on health	Alonso et al., 2013	Gauthier et al., 2012 * 8 trabalhadores	Phillips et al., 2007	Rimell et al., 2008	Singh et al., 2014	Xu et al., 2016 * 167 trabalhadores
White fingers						4,20%
Muscular fatigue						18,00%
Hand numbness						35,30%
Carpal tunnel syndrome		6,67%				14,40%
Hand pain		33,00%				7,80%
Elbow pain		33,00%				
Shoulder pain		16,70%				
Hand bulges						15,60%
Hyperidrosis Palmaris						22,20%
Wrist bulge						6,00%
Dizziness						15,00%
Headache						13,80%
Tinnitus						27,50%
Insomnia						14,40%
Memory loss						23,40%

the upper extremities (Griffin, 1998; Alonso, et al., 2013). Recent studies also point to the appearance of Carpal Tunnel Syndrome (House, et al., 2009; Alonso, et al., 2013). Table 5 shows the percentage of workers who presented symptoms of these possible diseases.

The transfer of vibrations from the machine to the operators can be reduced by the use of gloves or by coating on the surface of the machine handle. The handle coating is an effective way to diminish vibrations (Singh, et al., 2014), as shown in Table 6.

Vibration reduction gloves (VR), however, do not yet have proven efficiency against the vibration transmitted to the fingers (Paddan, et al., 2001; Dong, et al., 2009; Welcome, et al., 2014), even though some of them may be useful (Brown, 1990; Jetzer, et al., 2003; Mahbub, et al., 2007; Welcome, et al., 2014).

Table 6. Vibration x handle coating.

Vibration transmissibility (m/s²) Coating material	Uncoated	Sponge and velvet	Jute and cotton	Rubber sheets and Rexene	Cotton sandwiched between jeans cloth
Handle	1,9997	1,0635	1,0419	0,8369	0,9677
Wrist	0,7081	0,6082	0,4253	0,3621	0,3440

REFERENCES

Alonso, Mónica López, et al. 2013. Comparative analysis of exposure limit values of vibrating hand-held tools. *International Journal of Industrial Ergonomics.* 2013, Vol. 43, 3, pp. 218–224.

Bovenzi, M. 2005. Health effects of mechanical vibration. *Giornale italiano di medicina del lavoro ed ergonomia.* 2005, Vol. 27, 1, pp. 58–64.

Bovenzi, M., et al. 1994. Hand-arm vibration syndrome among travertine workers: a follow up study. *Occupational and Environmental Medicine.* 1994, Vol. 51, 6, pp. 361–365.

Brown, A.P. 1990. The effects of anti-vibration gloves on vibration induced disorders: a case study. *Journal of Hand Therapy.* 1990, 3, pp. 94–100.

Dong, R.G., et al. 2009. Analysis of anti-vibration gloves mechanism and evaluation methods Resultados da pesquisa. *Journal of Sound and Vibration.* 2009, Vol. 321, pp. 435–453.

Estellita, Letícia de Souza, et al. 2010. Analysis and risk estimates to workers of Brazilian granitic industries and sandblasters exposed to respirable crystalline silica and natural radionuclides. *Radiation Measurements.* 2, 2010, Vol. 45, pp. 196–203.

Gauthier, François, Gélinas, Dominique e Marcotte, Pierre. 2012. Vibration of portable orbital sanders and its impact on the development of work-related musculoskeletal disorders in the furniture industry. *Computers & Industrial Engineering.* 2012, Vol. 62, 3, pp. 762–769.

Griffin, M.J. 1998. Riesgos Generales, vibraciones, Enciclopedia de salud y seguridad en el trabajo. *Ministerio de Trabajo y Asuntos Sociales, Subdirección General de Publicaciones.* J.M. Stellman (Ed.), 1998, 4ª ed.

Harada, N. 2002. Cold-stress tests involving finger-skin temperature measurement for evaluation of vascular disorders in hand-arm vibration syndrome: review of the literature, (), pp. *International Archives of Occupational and Environmental Health.* 2002, Vol. 75, pp. 14–19.

House, R., et al. 2009. Current perception threshold and the HAVS Stoc-kholm sensorineural scale. *Occupational Medicin.* 2009, Vol. 58, pp. 476–482.

ISO. 2001a. ISO 5349–1. *Mechanical vibration—measurement and evaluation of human exposure to hand-transmitted vibration, part 1: general requirements ISO.* [Online] International Organization for Standardization, 2001a. [Citado em: 13 de setembro de 2016.] http://www.iso.org/iso/iso_catalogue/catalogue_tc/catalogue_detail.htm?csnumber = 32355.

—. 2001b. ISO 5349–2. *Mechanical vibration—measurement and evaluation of human exposure to hand-transmitted vibration, part 2: practical guidance for measurement at the workplace.* [Online] International Organization for Standardization, 2001b. [Citado em: 13 de setembro de 2016.] http://www.iso.org/iso/home/search.htm?qt = iso+5349–2&published = on&active_tab = standards&sort_by = rel.

Jetzer, T., Haydon, P. e D., Reynolds D. 2003. Effective intervention with ergonomics, antivibration gloves, and medical surveillance to minimize hand-arm vibration hazards in the workplace. *Journal of Occupational and Environmental Medicine.* 2003, Vol. 45, 12, pp. 1312–1317.

Mahbub, H., et al. 2007. Assessing the influence of antivibration glove on digital vascular responses to acute hand-arm vibration. *Journal of Occupational Health.* 2007, Vol. 49, 3, pp. 165–171.

Melo Neto, Rútilo P. e Rabbani, Emília R. Kohlman. 2012. Application of preliminary risk analysis at marble finishing plants in Recife's metropolitan área. *Work: IOS Press Contente Library.* 2012, Vol. 41, supplement 1, pp. 5853–5855.

Paddan, G.S. e Griffin, M.J. 2001. Measurement of glove and hand dynamics using knuckle vibration. *Proceedings of the Ninth International Conference on Hand-arm Vibration, Section, Nancy, France.* 2001, Vol. 15, 6.

Palmer, Keith T. e Bovenzi, Massimo. 2015. Rheumatic effects of vibration at work Keith T. Palmer Massimo Bovenzi. *Best Practice & Research Clinical Rheumatology.* 2015, Vol. 29, 3, pp. 424–439.

Phillips, J.I., Heyns, P.S. e Nelson, G. 2007. Rock drills used in South African mines: a comparative study of noise and vibration levels. *Oxford Journals—Medicine & Health, The Annals of Occupational Hygiene.* 2007, Vol. 51, 3, pp. 305–310.

Reynolds, D.D. 2006. New ANSI S3.34 (2.70–2006) – guide for the measurement and evaluation of human exposure to vibration transmitted to the hand. *The 41st UK Group Meeting on Human Response to Vibration.* 2006.

Rimell, Andrew N., et al. 2008. Variation between manufacturers' declared vibration emission values and those measured under simulated workplace conditions for a range of hand-held power tools typically found in the construction industry. *International Journal of Industrial Ergonomics.* 2008, Vol. 38, 9, pp. 661–675.

Singh, Jagvir e Khan, Abid Ali. 2014. Effect of coating over the handle of a drill machine on vibration transmissibility. *Applied Ergonomics.* 2014, Vol. 45, 2, Part B, pp. 239–246.

Welcome, Daniel E., et al. 2014. The effects of vibration-reducing gloves on finger vibration. *International Journal of Industrial Ergonomics.* 2014, Vol. 44, 1, pp. 45–59.

Wimer, B., et al. 2010. Effects of gloves on the total grip strength applied to cylindrical handles. *International Journal of Industrial Ergonomics.* 2010, Vol. 40, 5, pp. 574–583.

Xu, Xiangrong, et al. 2016. Occupational hazards survey among coal workers using hand-held vibrating tools in a northern China coal mine. *International Journal of Industrial Ergonomics.* 2016, Vol. 38, 9–10, pp. 661–675.

Occupational Safety and Hygiene V – Arezes et al. (Eds)
© 2017 Taylor & Francis Group, London, ISBN 978-1-138-05761-6

Environmental risks associated with falls on sidewalks: A systematic review

E.R. Araújo
University of Porto, Porto, Portugal

L.B. Martins
Federal University of Pernambuco, Recife, Brazil

ABSTRACT: Falls can compromise individuals, especially older adults, causing morbidity, mortality or reduction of activities. Their reasons vary according to each individual, and can be caused by intrinsic and extrinsic factors. This study aims to analyze extrinsic factors related to environmental causes, investigating interventions that may reduce unsafe conditions in the sidewalk environment, for the safe movement of pedestrians. Through a systematic review, following the PRISMA methodology, in data sources: ScienceDirect, IEEE, PubMed, Nature.com, e Google Scholar, 14 records were included in the study, observing that physical interventions to improve sidewalks when it comes to safety, as well as the use of digital technology, considering the characteristics of individuals, especially those with special needs, can provide a safer environment reducing pedestrian injuries.

1 INTRODUCTION

In the late 20th century, urban theorists and urban dwellers have emphasized the importance of public spaces, especially sidewalks, given their fundamental value to a democratic system and to a space of popular interaction (Ehrenfeucht & Loukaitou-Sideris, 2007). Walking and running represent the two most common forms of human walking, and eyesight provides information about the environment, which is cognitively interpreted to help shape the way pedestrians evolve in space (Miguel, 2013). Thus, the walk is considered a healthy, sustainable, accessible and economical form of transportation, being an important factor for a balanced and efficient transport system, and can be encouraged, improving the quality of the built environment. However, studies have shown that a series of infrastructure elements affects the pedestrian's ability to develop a simple walk, and they may be injured by slipping, tripping, falling or colliding with some obstacle on the sidewalk or on the shared space or if the sidewalk does not adequately accommodate people with special needs (Hunt-Sturman & Jachson, 2009), putting their safety at risk. Falls are responsible for morbidity, premature mortality and immobility, especially among the elderly (Poh-Chin et al., 2009; Rubenstein & Josephson, 2006). Elderly people are a group vulnerable to falls and landslides, due to cognitive and sensory impairments or environmental risk factors from environmental causes. The so-called "accidental falls" are triggered by environmental risks (Rubenstein & Josephson, 2006).

Studies point to the origin of falls through different intrinsic or extrinsic risk factors (King et al., 2013; Poh-Chin et al., 2009; Rubenstein & Josephson, 2006). The intrinsic causes are related to the biomechanics and psychological conditions inherent to each individual. On the other hand, the extrinsic factors are related to environmental causes and circumstantial situations. About 75% of falls occur on sidewalks (Poh-Chin et al., 2009; Lundborg & Gard, 2000), the majority caused by tripping, slipping on objects or due to irregular or slippery surfaces. This study focuses specifically on the risk of accidents and injuries to which people are exposed while walking on sidewalks, risk situations that can trigger falls in pedestrians, investigating interventions that may reduce unsafe conditions in the sidewalk environment.

Efforts to encourage an increase in the active life, with security, have been focus of studies of several areas of knowledge, mainly health professionals, engineers and architects. Effective pedestrian injury prevention strategies will reduce pedestrian injuries and increase the benefits associated with walking (Frattaroli et al., 2006). The aim is to analyze the public environment built for walking activities and investigate which variables may have a positive influence to make sidewalks safer, avoiding risks of pedestrian falls. Therefore, the intention is to synthesize the relevant knowledge, and utilized methods, providing opportunity for the development of projects and new investigation routes.

2 MATERIALS AND METHODS

Systematic searches were made, according to the PRISMA methodology requisites, using the "Database" resources" and "Scientific journals" of the SDI-FEUP (Universidade do Porto) with the keywords: "obstacles on sidewalks", "pedestrians safety", "infrastructure of sidewalks", "accident with pedestrian", "fall hazards and people with special needs", exchanged two by two. In the electronic search, carried out from June to August 2016, the articles originated from publications indexed in: SciencieDirect, IEEE, PubMed, Nature.com, and Google Scholar, those studies contained in references of articles obtained in the review.

Initially, articles dated from 2006 to 2016 were selected, since they contained data more up to date for the study; and those of previous years, those repeated, and those without access were disregarded. In this context, the following concepts were evaluated, according to Table 1.

Also, articles whose author was not informed were not considered. A first screening was performed based on the titles and abstracts, where the articles selected for this systematic review were examined in full, with the following criteria: problem, intervention, results and study. These criteria were used to identify the origin of falls occurring on sidewalks, with emphasis on the built environment, so that intervention solutions may arise to reduce these identifiable risk factors. The research elements are related to infrastructures and pedestrian perception, according to Table 2.

Table 1. Aspects approached and concepts.

Item	Concept
Human walk	Pedestrian's walking and running activity.
Urban infrastructures	Physical issues of the urban environment
Risks of accidents	Injuries and falls of pedestrians on sidewalks, for extrinsic reasons.
Risks of extrinsic falls	Environmental and circumstantial causes
Built environment	Natural environment modified by man
Environmental factors	Those related to the environment around the surface.
Surface factors	Those directly related to the surface profile.

Table 2. Inclusion and exclusion criteria for falls.

Criteria	Attributes		Justifications
Problem	Inclusion	Falls for extrinsic risks	Focused on environmental factors.
	Exclusion	Intrinsic risks	Focused on health risks.
Intervention	Inclusion	Urban sidewalks	High percentage of risks and falls.
	Exclusion	Car circulation areas	Run-over risks
Results	Inclusion	Decrease in falls	Safe transit
	Exclusion	Falls for diseases	They do not fit the context.
Studies	Inclusion	Scientific studies	Research of public interest.
	Exclusion	Systematic Reviews	Secondary study

3 RESULTS

With a total of 4926 articles selected through the keywords, 4 articles were added from references in articles collected, totaling 4930 articles. 546 duplicates were discarded, 1892 because they were less than 2007, 617 did not identify the author. Figure 1 shows, in a synthesized manner, the methodological sequence of this review.

Thus, 1875 studies were screened, excluding 1406 by title and 469 by abstract and 370 excluded by the full text of which 99 had complete texts and were qualitatively analyzed and 14 were chosen for systematic review.

In the electronic search for systematic review, 14 studies fit into the selection criteria regarding physical environmental risk factors for pedestrian falls. Articles dealing with infrastructure conditions also used the Geographical Information System (GIS), questionnaires, statistical data, on-site sample audits (terrain), as an additional tool in the scope of the investigations.

The field work, as well as the monitoring data, are used to compare the two methods, in order to verify the reliability of all, for the characterization of the built environment.

Considering this, busy streets, narrow and crowded footpaths, uneven surfaces, poor paving, commercial activities along the sidewalks, wet and slippery surfaces, highways, sidewalks, and obstacles were significant elements, resulting in unsafe pedestrian environments (Poh-Chin et al., 2009; Ping & Xiaohua, 2012).

To (Grõnqvist, 1995), the state of poor adhesion and low friction between footwear and the

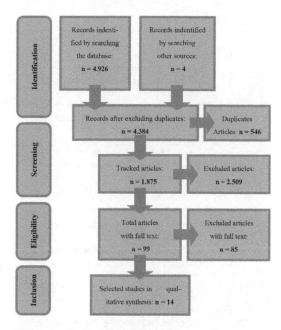

Figure 1. Methodological sequence of the review.

surface below the feet (pavement) is the main risk factor for sliding accidents. However, when stability is recovered, even if there are no falls, there may be risks of injury (Hanson et al., 1999). In order to reduce the risk of falls in wet areas during the cold, studies (King et al., 2013) have been performed replacing permeable concrete with impermeable ones, such as an anti-slip surface. It has been found that the combined permeability and frost-melting characteristics cause less frost on the surface due to the rapid absorption of the melted ice beneath the foot surface, thereby reducing the slippery effect caused by the ice. Alternatives to reduce the number of injuries due to slippery sidewalks have been used in Sweden ith 25 non-slip devices to reduce the risk of slips on ice and snow (Lundborg & Gard, 2000).

Another study (Thies et al., 2011), with a healthy, non-disabled population aged over 85 years in the United Kingdom, carried out an investigation of walk measures associated with the stability and risks of falling with surface indicators of tactile paving installed on the sidewalks. It found that despite the fact that there is no change in the walking speed of the participants, tactile paving is perceived as increasing the risk of tripping. It is also emphasized (Kobayashi et al., 2005) that these floors need a revision, considering that walking on an uneven surface affects the walking stability of people with normal eyesight, who do not need this device to travel, causing younger people and the

elderly to stumble, and also making it difficult for people with wheelchairs and pushchairs to move around. Another point emphasized by Kobayashi (2005) is that although it helps in the displacement of the visually impaired, it needs more studies because of its narrow width (300 mm) required for displacement by placing both feet when walking.

On the other hand, research has been carried out with automatic detection of tactile paving surfaces (Ghilardi et al., 2016) to support the mobility of the visually impaired. Thus, it is possible to detect images of tactile surfaces even before touching them, with a camera coupled to the user's abdomen and an audio feedback, as shown in Figure 2.

In the same way, studies with intelligent wheelchairs were carried out using images produced by a portable computer camera, allowing the user to control the chair with the movement of the face and the shape of the mouth, when perceiving an obstacle (Yeounggwang et al., 2013). Studies with digital monitoring are an important topic to help in the safe displacement of people with special needs, improving their quality of life, although the need for an improvement in the physical conditions of transit is extremely necessary to serve the different users of the transit ways.

A longitudinal study, conducted in Sidney, Australia, with 315 elderly individuals over 65 years of age, observed that people with "recurrent falls" on regular walks were 2.6 times more likely to walk, compared with Those who fell once or never had falls (Merom et al., 2015). To Poh-Chin (2009), most falls occur outdoors and on the sidewalks (73%), them being the third most common cause of deaths from unintentional injuries in

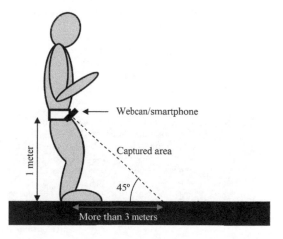

Figure 2. Capturing position (based on Ghilardi et al., 2016).

Table 3. Synthesis of the articles.

Lundborg & Gard, 2000	S	297 people over 60 years of age
	D	Evaluation of ergonomic and functional aspects in non-slip devices to reduce slips and falls on ice and snow
	V	Ice and snow
	R	Importance of making non-slip devices as safe and practical as possible
Kang, 2011	S	114 people; 15 sidewalk segments
	D	Assessment of the shared use of sidewalks by pedestrians and cyclists through video clips and interviews
	V	Independent personal, time, operational and physical variables
	R	Shows the importance of separating cyclists from pedestrians, but also provides a strategy for shared space
King, 2013	S	10 female and male young adults
	D	To verify that permeable concrete presents greater resistance to sliding than traditional concrete
	V	Traditional dry and icy concrete; permeable dry and icy concrete
	R	Use of permeable concrete to reduce risks of falling on walking surfaces
Yeounggwang, 2013	S	Experiment using 80.000 images for a one-year period
	D	Wheelchair control by the user with digital monitoring
	V	Obstacles on sidewalks
	R	Safe circulation of wheelchairs avoiding risk situations
Meron et al., 2015	S	315 inactive adults over 65 years of age
	D	Randomized clinical trials
	V	Lack of infrastructure to support walking
	R	Benefits of an adequate infrastructure on sidewalks, so as to encourage mobility
Lee, 2014	S	5 blindfold participants and 15 visually impaired
	D	One experiment using only white cane; another using only digital obstacle detector; and a third with both devices
	V	Visual impairment
	R	Increased safety for visually impaired displacement with both devices
Earl, 2016	S	2 healthy people; 3 cognitive disability; 1 moderate intellectual disability; 1 Asperger's Syndrome; 1 Hemiantopsia
	D	Eye tracking in individuals moving in a shared zone
	V	Interpretation of nonverbal signals
	R	Importance of reviewing policies for the use of shared public spaces

(Continued)

Table 3. (*Continued*)

Ghilardi, 2016	S	
	D	Tactile surface detection through digital technology and decision tree
	V	Lighting changes; occlusion, image noise and resolution
	R	The implemented approach effectively detects tactile paving with about 88.5% accuracy
Corazza, 2016	S	Physical analysis
	D	Visual inspection and registration
	V	Precarious infrastructure
	R	Sidewalk maintenance management
Hunt-Sturman, 2008	S	Pedestrians in general
	D	Review of existing hazard identification on pedestrian surfaces
	V	Inclinations and slip resistance
	R	Definition of risk score and promotion of control strategies
Thies, 2011	S	32 elderly adults
	D	Parameter analysis of gait
	V	Tactile floor roughness
	R	Verified gait imbalance
Yeounggwang, 2013	S	Use of images; laboratory
	D	Digital monitoring and wheelchair control through face movements
	V	Reflexes; image resolution and angulation
	R	Safe mobility of wheelchairs
Kobayashi, 2005	S	10 young people with normal vision
	D	Analyzes effect of tactile floors on healthy people
	V	Tactile floor roughness
	R	Shows caused harms and proposes retification
Poh-Chin, 2009	S	281 elderly
	D	Evaluates environmental factors associated with falls
	V	Infrastructure of sidewalks and obstacles
	R	Shows factors favorable to falls which need improvement studies

S = Sample; D = Design; V = Variable; R = Results.

Hong Kong. Table 3 presents the synthesis of the articles, as Table 4 summarizes the main variables cited in the articles, which make walking spaces unsafe, with risks to pedestrian injuries, and the most pointed out items were physical barriers, agglomerations and slippery floors.

Table 4. Key attributes/pedestrian safety on sidewalks.

Variables	Lundborg, 2000	Kang, 2011	Poh-Chin, 2009	Yeoungwang, 2013	Meron D. et al., 2015	Lee, 2014	King, 2013	Earl, R., 2016	Ghilardi. R., 2016	Corazza, 2016	Thies, 2011	Yeoungwang, 2013	Hunt-Sturman, 2008	Kobayashi, 2005
Non-slip devices	x											x		
Physical Barriers		x	x	x	x					x		x	x	x
Traffic volume		x		x			x			x		x		
Narrow sidewalks				x						x				
Lighting												x		
Waste on the sidewalk				x						x				
Slippery floor	x	x					x					x		
Shared use		x							x					
Tactile floor									x	x			x	
Non-slip floor								x						
Slope										x		x		
Tree roots										x				
Collision potential												x		
Slits										x				
Fall recurrence						x								
Irregular surface						x				x		x		
Permeable concrete									x					
Mobile devices						x	x				x			
Holes										x				

4 DISCUSSION

The study shows that a large proportion of falls and injuries originates from many risk factors, many of which can be modified or eliminated, and that among the most vulnerable pedestrians are the elderly, due to the decline of motor abilities, difficulty of balance linked to the changes caused by the advancement of age, thus altering their sensory, cognitive, physical and self-perception abilities. However, elements of the built environment such as obstacles, uneven surfaces, cobblestone, stones, cracks, urban furniture and slippery surfaces are environmental risk factors that require further study (Corazza et al., 2016). Some measures (Earl et al., 2016) such as shared streets, make it easier for elderly people to walk, because they do not have a curb gap, but the fact that cars, bicycles and people use the same space requires the interpretation of non-verbal communication signs, is difficult for pedestrians with cognitive problems and who do not have the ability to comprehensively process and attach importance to the visual stimulus, especially when this shared space is very full. Another aspect that conflicts with shared streets, even when tactile floors are used, is that they do not adequately meet people with impaired vision who, for the most part, have as a locomotion support some elements of the built environment, such as buildings, curb lines and the front of buildings. The lack of physical boundaries between the places for pedestrians and cars still do not work for people who need borders to identify these spaces.

In spite of some criticism of shared spaces, in China the sharing of pedestrians with cyclists is more tolerated, as research carried out by (Kang & Fricker, 2016) provided that the sidewalks are wider, and that bicycles have a controlled speed. It is believed that this tolerance is due to the customs of the place.

In this context, environmental assessment as a strategy to prevent potential risks to mobility in the built environment will greatly contribute to the public administration bodies responsible for proposing new projects or modifying existing spaces taking into account the characteristics of individuals, pedestrian perception and safety. As a consequence, we have a significant decrease in falls, and an improvement of public health.

REFERENCES

Corazza, M.V., Mascio, P.D. & Moretti, L., 2016. Managing sidewalk pavement maintenance: A case study to increase pedestrian safety. *Journal of Traffic and Transportation Engineering (English Edition)*.

Earl, R., Falkmer, T., Girdler, S. & Dahlman, J., 2016. Visual search strategies of pedestrians with and without visual and cognitive impairments in a shared zone: A proof of concept study. *Land Use Policy*, Earl, R., Falkmer, T., Girdler, S., Dahlman, J., Rehnberg, A., & Falkmer, M., pp. 327–334.

Ehrenfeucht, R. & Loukaitou-Sideris, A., 2007. Constructing the sidewalks: Municipal government and production of public space in Los Angeles, California, 1880–1920. *Geography, Journal of Historical*, 33(1), pp. 104–124.

Frattaroli, S. et al., 2006. Local stakeholders' perspectives on improving the urban environment to reduce child pedestrian injury: Implementing effective public health interventions at the local level. *Journal of public health policy*, 27(4), pp. 373–388.

Ghilardi, M., Macedo, R. & Manssour, I.H., 2016. A New Approach for Automatic Detection of Tactile Paving Surfaces in Sidewalks. *Procedia Computer Science*, Volume 80, pp. 662–672.

Grönqvist, R., 1995. A dynamic method for assessing resistance to pedestrian slip. *thesis*.

Hanson, J.P., Redfern, M.S. & Mazumdar, M., 1999. Predicting slips and falls considering required and available friction. *Ergonomics*, 42(12), pp. 1619–1633.

Hunt-Sturman, A. & Jachson, N., 2009. Development and evaluation of a risk management methodology for pedestrian surfaces. *Safety science*, 47(1), pp. 131–137.

Kang, L. & Fricker, J.D., 2016. Sharing urban sidewalks with bicyclists? An exploratory analysis of pedestrian perceptions and attitudes. *Transport Policy*, Volume 49, pp. 216–225.

King, G.W., Bruetsch, A.P. & Kevern, J.T., 2013. Slip-related characterization of gait kinetics: Investigation of pervious concrete as a slip-resistant walking surface. *Safety science*, Volume 57, pp. 52–59.

Kobayashi, Y., Takashima, T., Hayashi, M. & Fu, H., 2005. Gait Analysis of People Walking on Tactile Ground Surface Indicators. *Neural Systems and Rehabilitation Engineering. IEEE Transactions on 13*, Issue 1, pp. 53–59.

Lee, C.L., Chen, C.Y., Sun, P.C. & Lu, S.Y., 2014. Assessment of a simple obstacle detection device for the visually impaired. *Applied ergonomics*, 45(4), pp. 817–824.

Lundborg, G. & Gard, G., 2000. Pedestrians on slippery surfaces during winter—methods to describe the problems and practical tests of anti-skid devices. *Accident Analysis & Prevention*, 32(3), pp. 455–460.

Merom, D. et al., 2015. Neighborhood walkability, fear and risk of falling and response to walking promotion: The Easy Steps to Health 12-month randomized controlled trial. *Preventive medicine reports*, Volume 2, pp. 704–710.

Miguel, A.F., 2013. The emergence of design in pedestrian dynamics: Locomotion, self-organization, walking paths and constructal law. *Physics of life reviews*, Volume 10, pp. 168–190.

Ping, Y. & Xiaohua, W., 2012. Risk factors for accidental falls in the elderly and intervention strategy. *Journal of Medical Colleges of PLA*, Volume 27, pp. 299–305.

Poh-Chin, L. et al., 2009. An ecological study of physical environmental risk factors for elderly falls in an urban setting of Hong Kong. *Science of the total environment*, 407(24), pp. 6157–616.

Rubenstein, L.Z. & Josephson, K.R., 2006. Falls and their prevention in elderly people: what does the evidence show? *Medical Clinics of North America*, Volume 90, pp. 807–824.

Thies, S. et al., 2011. Biomechanics for inclusive urban design: Effects of tactile paving on older adults' gait when crossing the street. *Journal of biomechanics*, 44(8), pp. 1599–1604.

Yeounggwang, J., Hwang, J. & Eun, Y., 2013. An intelligent wheelchair using situation awareness and obstacle detection. *Procedia-Social and Behavioral Sciences*, Volume 97, pp. 629–628.

Occupational Safety and Hygiene V – Arezes et al. (Eds)
© *2017 Taylor & Francis Group, London, ISBN 978-1-138-05761-6*

The safety culture and noise level of a beverage industry

B.A.B. Nóbrega, M. Lourdes, B. Gomes & R.M. Silva
Universidade Federal da Paraíba, Paraíba, Brazil

A.M. Oliveira & D.A.M. Pereira
Universidade Federal de Campina Grande, Paraíba, Brazil

ABSTRACT: The focus of this article is to understand the company's positioning regarding safety measures and employees' point of view, analyzing the safety culture present in the beverage-bottling production line and the noise level to which workers in this line are exposed, based on the NR-15—Regulatory Standard. Semi-structured questionnaires were used to identify the safety culture. Noise measurement was performed using the Larson Dawis QUEST Dosimeter audio instrument. The results showed that the noise level in the observed sector is above the tolerance limit for 8 hours of work, according to NR 15. It can be observed that the company's safety culture for the bottling sector is inefficient. Faced with the problems identified, recommendations were made. Future work can be carried out to know the safety culture in the company, where it would explain the reason for the company's inattention to the problems.

1 INTRODUCTION

In order for organizations to survive in an increasingly competitive market, it is of great importance to create an organizational environment that offers the best conditions of installation and operation together with the professional and personal development of the worker (Neal et al., 2000; Mearns et al., 2003; Vicente, 2012). In this segment the safety culture, subcomponent of the organizational culture, reflects the characteristics of the company, the attitudes and the behavior of workers related to safety at work.

The need for Occupational Health and Safety (OHS) has always existed in all work activity, regardless of origin, culture or religion. Thus, regarding human needs, the issue of OHS maintains a prominent place, since the good performance of the activities has a direct relation with the conditions of the work environment and the safety culture, characteristic of the organizations, contributing to a good execution of the worker's activity influencing the increase of productivity. The focus of this article is to understand the company's positioning regarding safety measures and the employees point of view, analyzing the safety culture present in the beverage-bottling production line and the noise level to which workers of this line are exposed, with reference to NR-15 – Regulatory Standard.

2 THEORETICAL REFERENCE

2.1 *Noise physical agent*

According to Almeida et al. (2000); and Mello & Waismann (2004), because it is considered a frequent physical risk in almost all industrial segments, noise has received exclusivity in the approaches related to the auditory health of workers, but it has also been the subject of much research.

Noise generates innumerable implications, compromising the individual quality of life, causing hearing loss, which may be partial or total, disturbs rest, sleep and communication of human beings (Fiorini, 2004). It can cause serious problems and changes in the body such as gastrointestinal disorders, disorders related to the nervous system, such as irritability, nervousness, increased intensity of pulse acceleration, elevation of blood pressure, as well as contraction of blood vessels. (Porto et al., 2004; Rios & Silva, 2005, Moraes et al., 2014).

Occupational noise can also contribute to workplace accidents as it impairs communication, maintaining attention, concentration and memory, as well as increasing stress and excessive fatigue. (de Freitas et al., 2009). Although it has reached endemic proportions in the industrial environment, scientific studies on the natural history of workers are scarce. In Brazil it is observed that the consequences caused by noise are unknown by the majority of the population, including the most affected class by this problem, which are the workers. (Guerra et al., 2005).

In Brazil, we have Regulatory Norm NR-15, which determines unhealthy agents, limits of tolerance and the technical and legal criteria to evaluate and characterize unhealthy activities and operations. Regarding noise physical risk, the standard determines the tolerance limits for continuous or intermittent noise relating them to the respective maximum daily exposure terms that a person may be exposed during their workday, as shown in Table 1.

2.2 *Safety culture*

The concept of safety culture emerged in 1988 in the first technical report by the International Nuclear Safety Advisory Group (INSAG), which shows the results of the accident analysis of the Chernobyl nuclear power plant in Ukraine (AIEA, 1991). The errors and violations of procedures that contributed in part to this accident were interpreted as evidence of a weak safety culture in Chernobyl, in particular, and in Soviet industry in general. In this regard, the safety culture was defined as a set of basic assumptions and values, shared collectively by the members of the organi-

Table 1. Tolerance limits for continuous or intermittent noise.

Noise level dB (A)	Maximum daily permissible exposure
85	8 hours
86	7 hours
87	6 hours
88	5 hours
89	4 hours and 30 minutes
90	4 hours
91	3 hours and 30 minutes
92	3 hours
93	2 hours and 40 minutes
94	2 hours and 15 minutes
95	2 hours
96	1 hour and 45 minutes
98	1 hour and 15 minutes
100	1 hour
102	45 minutes
104	35 minutes
105	30 minutes
106	25 minutes
108	20 minutes
110	15 minutes
112	10 minutes
114	8 minutes
115	7 minutes

Source: Annex No. 1 from NR 15.

zation, that determine the structure and collective practices concerning work safety. (Hopkins, 2005; Reason, 2016).

The safety culture of an organization is the product of values, attitudes, perceptions, competencies, and behavioral patterns of individuals and groups that determine the organization's commitment, style and proficiency of managing work safety. Organizations with positive safety cultures are characterized by communication based on mutual trust, shared understanding of the importance of safety and confidence in the effectiveness of preventive measures. Thus working conditions and safety culture have intrinsic relation and it is important to analyze them together.

3 METHODS

The article is the result of a qualitative-quantitative research developed through a case study in a beverage industry, precisely in the bottling sector because it is the most affected by the incidence of noise, thus requiring an analysis of the potentialities that the level above recommended noise may cause the individual.

For the noise measurement process, the QUEST dosimeter Larson Dawis audio instrument was used, the instrument indicated by NR-15 for continuous noise measurement. From 55 employees in the beverage-bottling sector, 27 employees were evaluated, observing the amount of decibels and the exposure time of their daily activity. The readings were made in the work stations of each individual, the apparatus was placed in the waist of the worker, with the microphone affixed in the collar at the height of the right ear, during the time of measurement twenty-seven measurements of sixty seconds each were fulfilled.

In order to identify the relationship between the safety culture from both management point of view and positioning of employees with regard to working conditions and the level of knowledge about noise exposure, semi-structured questionnaires were applied for the interviews. In addition, a camera was used for the record of the analyzed site.

4 RESULTS

4.1 *Studied company*

The studied company is located in the country of Paraíba state—Brazil; it began its activities in 1973 with the production of wine. Over the years, the company has been conquering the market, updating technologically and improving the quality of productive and administrative processes. It currently has a mix of well-diversified products, which range from wines to distilled beverages. It is a prominent indus-try for the city where it is located, generating income and direct and indirect jobs, contributing to the eco-nomic development of the region. The studied sector, as already mentioned was the place of bottling.

4.2 Noise level in the bottling sector

Noise is one of the physical agents present on most factory floors, often interfering with employee pro-ductivity, compromising their physical integrity. In the studied company, noise is present in most work environments, and the bottling sector has the high-est incidence. Such a physical phenomenon comes from the filling machine and the conveyor. The conveyor, for being an old technology, presents a stronger and more intense noise, besides the noise that is caused by the bottles that hit each other as they travel through the conveyor belt along the production line.

The results obtained from the measurements made with the workers of the studied company (Table 2) showed that they are above 85 dB(A), that is, above the tolerance limit for 8 working hours, as established in NR 15 (Table 1). The amounts found were at the Visual Inspection stations 93.1 dB(A); Machine Operator 88.6 dB(A); Bottle Washing and Rinse Operator 96.1 dB(A); Bottling Machine Operator 90.4 dB(A). These data show the occur-rence of the accentuated noise level, which is directly related to the development of the task, where it has a serious and imminent risk to the worker.

Workers at this company are exposed to these noise levels during an intense workday of 8 hours. Therefore, this sector needs interventions so that future damages, such as total or partial hearing loss, among other problems are not caused.

4.3 Safety culture in the company

In view of the research carried out, it can be observed that safety culture in the company for the bottling sector is inefficient. Because the company's management is aware of the level of risk that its

Table 2. Dosimetric data of work noise.

| Activity | Noise | |
	Lavg dB (A)	Time of daily exposure
Visual Inspection	93,1	8 hours
Machine Operator	88,6	8 hours
Bottle Washing and Rinse Operator	96,1	8 hours
Bottling Machine Operator	90,4	8 hours

Source: Field research.

Figure 1. Employees working in the bottling sector. Source: Field research.

employees are exposed to, and it is known that immediate intervention is necessary, it is still pas-sive in terms of decision-making and correct prac-tices on health and safety of the company.

Employees who are exposed to high noise levels are not aware of potential problems they may be experiencing, nor are they concerned about using PPE. Most workers stated that they do not use the shell ear protector, as can be seen in Figure 1, due to the discomfort these devices cause. It was also identified that there is no effective collection by the company and the management does not clarify the danger to which they are exposed. Most workers reported feeling headaches, malaise, earaches at the end of an intense work journey; however, they did not associate with the noise level to which they are exposed. They do not know the possible dam-ages that the high level of noise can cause.

For some writers, each organization has a type of safety culture that can be classified as strong or weak (Camposa & Diasa, 2016), in this regard, the safety culture of the organization analyzed for this sector can be considered weak, since the actions, perceptions, company norms and the attitudes of all that compose it, are not directed to preserving health and physical integrity of workers, provid-ing inadequate working conditions and inducing employees to work incorrectly.

5 RECOMMENDATIONS

For protection of physical and mental health of workers, top management support to the company's department responsible for workplace safety and the involvement of all of the organization is essen-tial to the development of a healthy and proactive safety culture. Here are some suggestions.

- It is necessary that the company solve the prob-lem directly in the source, seeking to eliminate, with enclosure of the machines or even their replacement, in the case of the conveyor, because it presents an advanced life;

– Not being possible to apply the aforementioned solution, it is essential to minimize the noise in a significant way, motivating the employees to protect themselves against the agent that causes damage to the health and physical integrity, ie, providing adequate PPE (shell ear protector), in addition to training of use, custody and sanitation of them;
– Raise awareness of possible problems they may acquire if they do not use PPE correctly and safely;
– Medical control, with the accomplishment of periodic exams (biannual) and the accompaniment by the SESMT (Specialized Service in Safety Engineering and in Occupational Medicine) of the company;
– Promote events with OHS professionals for awareness through lectures, trainings, posters, allowing all those involved in the organization to obtain knowledge about the importance of developing a favorable safety culture, in which everyone gains a higher quality of life at work.

6 CONCLUSION

Considering the presented results, it was observed that both measured amounts exceed those allowed by NR-15, as the involvement of the management of the company related to the analyzed sector is passive in the face of the danger that its employees are exposed.

Such a situation requires action to be taken, first the involvement of top management by encouraging every organization to develop a culture of dynamic and efficient insurance, in which everyone has the responsibility to maintain it according to the job developed by each employee. Followed by interventions to solve the problem of noise, and the mandatory use of PPE, allying to the discipline and demand of its use, to minimize more severe damages.

In an industry that has a safety culture where everyone who compose it has an education aimed at preserving life, the benefits generated are of great proportions, since all the components of the organization always seek to develop studies and observations that indicate a state of alert against constant elimination of the causes of occupational diseases, reduction of harmful effects caused by work, prevention of worsening diseases and injuries. The consequences are to maintain workers' health and increase productivity through control of the work environment.

Future work can be carried out to know the organization's safety culture, which would explain better why the company's management inattention to the problems presented in this work. Because it is a meticulous work that would require more time, it was not possible to accomplish it at the moment. The measurement of cost benefit of replacing the machines or their enclosure are also proposed for future work.

REFERENCES

Agência Internacional de Energia Atômica - AIEA. (1991). INSAG, *Safety Culture*. 75(4). Available in: <wwwpub.iaea.org/MTCD/Publications/PDF/Pub882f_web.pdf>. Access in: 02 jan. 2017.

Camposa, D.C., & Diasa, M.C.F. (2012). A cultura de segurança no trabalho: um estudo exploratório. *Revista Eletrônica Sistemas & Gestão* [internet]. Niterói, 7(4).

de Almeida, S.I.C., *et al.* (2000). História natural da perda auditiva ocupacional provocada por ruído. Rev *Assoc Med Bras*, 46(2), 143–58.

de Freitas, A., *et al.* (2009). Alterações auditivas em trabalhadores de indústrias madeireiras do interior de Rondônia. *Rev. bras. Saúde ocup*, 34(119), 88–92.

Fiorini, A.C. (2004). Audição: impacto ambiental e ocupacional. Ferreira LP, Befi Lopes DM, Limongi SCO. *Tratado de fonoaudiologia*. São Paulo: Roca, 631–42.

Guerra, M.R. *et al.*(2005). Prevalência de perda auditiva induzida por ruído em empresa metalúrgica. *Revista de Saúde Pública*, 39(2), 238–244.

Hopkins, A. (2005). Safety, culture and risk. *CCH Australia Ltd*.

Mearns, et al. (2003). Safety climate, safety management practice and safety performance in offshore environments. *Safety science*, 41(8), 641–680.

Mello, A.P.D., & Waismann, W. (2004). Exposição ocupacional ao ruído e químicos industriais e seus efeitos no sistema auditivo: revisão da literatura. @ rq. otorrinolaringol, 8(3), 226–234.

Moraes, C., *et al.* (2014). Incidência e prevalência de perda auditiva induzida por ruído em trabalhadores de uma indústria metalúrgica, Manaus - AM, Brasil. *Revista CEFAC*, Septiembre-Octubre, 1456–1462.

Neal, et al. (2000). The impact of organizational climate on safety climate and individual behavior. *Safety science*, 34(1), 99–109.

Porto, *et al.* (2004). Avaliação da audição em frequências ultra-altas em indivíduos expostos ao ruído ocupacional. *Pró-fono*, 16(3), 237–250.

Reason, J. (2016). Managing the risks of organizational accidents. Routledge.

Regulamentadora-NR, N. (2007). NR 15–ATIVIDADES E OPERAÇÕES INSALUBRES.

Rios, A.L., & da Silva, G.A. (2005). Sleep quality in noise exposed Brazilian workers. *Noise and Health*, 7(29), 1.

Vicente, F.A.D.C.F. (2012). Gestão estratégica da segurança do trabalho na área industrial de uma usina de etanol, açúcar e energia elétrica.

Occupational Safety and Hygiene V – Arezes et al. (Eds)
© 2017 Taylor & Francis Group, London, ISBN 978-1-138-05761-6

Risk-taking behavior among drivers and its correlation with dangerousness, and sensation seeking

H. Boudrifa, M. Aissi, H. Cherifi & D. Zenad Dalila
Laboratory of Prevention and Ergonomics, University of Algiers2, Algeria

ABSTRACT: The current study aims to examine the frequency level of risk-taking behavior by drivers and how do they perceive the degree of its danger. In addition to its relation with the level of sensation seeking and so many personal characters.

A checklist about risk-taking behavior together with Zuckerman & kuhlman' test of sensation seeking and a list of personal characters were applied on a sample of 410 drivers.

Friedman ranking means test was applied to rank the items of risk taking. The results show that the items that occupy the first places according to their frequency levels reflected in general a socially adopted behavior as is shown on the first following eight items:

1. Driving fast to catch up with a delay
2. Driving after or during a dispute or hard discussion
3. Driving slowly
4. Using a mobile phone while driving
5. Being convinced that your driving is more accurate than the other drivers'
6. Feeling the ability to avoid unexpected danger
7. Not taking a rest after driving for a long time
8. Driving in a state of nervousness and high excitement

However, the items that were classified last reflect a high degree of danger, while the results of degree of danger of risk-taking behaviors show that these items occupying the first classes reflected in their behavior a set of high levels of dangerousness with high means as it is shown in the first following eight items. This implies the existence of a negative correlation between the two variables.

1. Driving under the effect of drugs
2. Driving while drunk
3. Stopping on the left of a motorway line.
4. Driving without using the car lights at night.
5. Overtaking on bends.
6. Dangerous maneuvers while driving
7. Driving carelessly near schools.
8. Getting out suddenly from a side road without taking the necessary precautions.

Obviously, the calculation of the coefficient of correlation between the frequency of risk-taking behavior and its degree of danger is estimated at (-0.359^{**}) with statistical significance at 0.01 level, while the correlation between the frequency behaviors of risk-taking and the level of sensation seeking was positive (0.307^{**}) with statistical significance at 0.01 level. Regarding the correlation between the degree of danger of risk-taking behavior and sensation seeking was also negative (-0.168^{**}) with statistical significance at 0.01 level.

The results of the study also showed that there were correlations between the frequency of risk-taking behavior, their degree of dangerousness among drivers, and sensation seeking with some individual characteristics.

Despite the fact that the means of the frequencies of risk-taking behaviors are not high, while those of the perception of the degrees of danger of the same behaviors are very high, which reflect a significant negative correlation between the two variables. Moreover, there was a significant positive correlation between risk-taking behaviors and sensation seeking while that between the latter and the perception of the dangerousness of risk-taking behaviors is negative.

Keywords: Risk-taking behavior, dangerousness, sensation seeking, personal characters, safety

1 INTRODUCTION

The vehicle has become one of the most important aspects of civilization, has provided the individual with quick and easy transportation, and contributed directly to raising the standard of living. However, it has caused disasters because of misuse, and became a problem of human suffering. Traffic accidents that are often a result of unsafe behaviors of road users, may reflect the lack of traffic education and culture. Scientific studies of behavior are willing to detect patterns and trends, which could explain the factors and motives of different patterns of human behavior (Boudrifa et al, 2010, 2012, 2014). Earlier studies have pointed out that the prevalence of risk-taking behavior among young people are higher than other age groups. This would result in a tendency of some of them to exercise some behaviors, such as driving recklessly like excessive speeding, without regard to prevailing systems and traffic rules, and acting indifferently while driving without taking into consideration signals or light marks used to keep the order of road traffic. Things that expose the driver himself and other road users to more dangers (Gonzalez & Field, 1994). Forsyth et al, (1995) indicated that risk-taking is the product of interaction of some situational characteristics and that related to the manner of the vehicle circulation, and some personality traits like risk-taking, and poor attention to the stimuli around while driving the vehicle.

Boudrifa et al, (2010), used a checklist of 150 items to measure unsafe behaviors of road users based on the third person principle. This study tool was applied on a final sample of 7058 drivers in twelve out of forty-eight districts in Algeria. They found that the frequencies of unsafe behaviors are not limited to the driver only, but expanded to cover all road users. They concluded from the high values of means that drivers realize well the danger of unsafe behaviors, despite the fact that they do not respect the traffic rules and laws in reality.

Another study found that the problem of road safety depends on changing dangerous behaviors of drivers (Elvik R; 2002), and this change depends in itself on the awareness and perception of those drivers of all dangers of road and driving. This awareness represents a principle condition for safe behavior in traffic movement in the society (Vanlaar W & Yannis G, 2006). Unsafe or dangerous behaviors are only social positions that are not a good stimulus to these behaviors as it is related to strong relation with social environment to which the individual belongs and lives (Ajzen, I., & Fishbein, 1980).

Zuckerman (1979) presented several variables on individual differences to accept risk-taking behavior, and pointed out that it may be related to the personality, sensation seeking, or new experiences, openness or excessive self-confidence, in addition to the interest rate and the desired result of risk. Furthermore, Zuckerman (2000) thinks that looking for sensual excitement is a predisposed variable with a biological basis, reflecting the differences between individuals in the maximum sensual excitement and conflicts to search for it, and the ability to withstand the positions of emotional sensation.

McClure & Turner (2003) made it clear that the risk-taking behavior could explain the high consequences of road accidents of vehicles owned by young male drivers. They tried to find out to what extent it is possible to explain the differences in risk-taking behaviors and differences in the level of accidents according to age and sex. They found that the means of males were higher than those of female in aggression, sensation seeking and accepting risk taking in general. Many studies have shown that the percentage of male taking the risk is more than the female's ratio (Kass, 1964; Lasorsa & Shoemaker, 1988; Bush & Iannoti, 1992).

It appears from the studies listed above that the issue of risk-taking behavior is very complex and interacts with several factors which have been taken from different angles in order to highlight the significant role of risk-taking in its contribution to the occurrence of traffic accidents. Hence, the question should be raised about the relation of the frequency levels of risk-taking behavior among drivers and their perception of the degree of danger with certain personal qualities or level of sensation seeking. In addition the relationship of all these variables with some individual characteristics such as age, sex, educational level, and even blood group as well as the type of profession and the daily crossed distance. That is why the aim of this study is trying to identify the frequency levels of risk-taking behavior and the perception of the degree of danger among drivers and how they interact with many of the variables mentioned above.

2 METHOD

2.1 *The study sample*

The current study sample consisted of 410 drivers, 81,2% of them are males and 18,8% are females. 42,0%, were 21 to 30 years, 29,3% of them were between 31 to 40 years, while 16,1% of them were from 41 to 50 years. In addition, 50,2% of the respondents are single and 48,0% are married. It should be pointed out that 44,9% of them have university level, while 23,9%, have an average level, and only 3,9%, have a primary level.

2.2 Study tools

Through the theoretical framework and previous studies of risk-taking among drivers, and after carrying a survey about available tools used before, it was possible to identify suitable tools to measure the variables of the present study as follows:

2.2.1 Checklist of frequency and degree of danger of risk-taking behaviors among drivers

After carrying a series of interviews and preliminary studies, a checklist of 112 items to measure risk-taking behavior of drivers was constructed. Subjects were asked to mark the frequency of each behavior (item) on a scale of five points (never, rarely, sometimes, most times, always), as well as its degree of danger on a five point scale (absolutely no danger, not dangerous, average danger, dangerous, very dangerous). The checklist was written recto-verso in Arabic and French to make it easier for drivers to answer it depending on the language they master better.

2.2.2 The sensation seeking scale

The sensation seeking scale for Zuckerman & Kuhlman (1993), was used in this study. This test consists of 19 items. Subjects need to choose the answer was yes or no. The high score indicates a high level of sensation seeking. Punctuation of the results consists of allocation of (1) one point when the answer is yes and (0) zero when the answer is no but the punctuation is reversed for the items 6 and 19. The estimation of the sensation seeking is made through five levels: 1/Very low: (00–27%); 2/Low (28–41%); 3/Average: (42–70%); 4/High: (71–84%); 5/Very high: (85–100%).

2.2.3 List of personal characters

In order to identify a set of qualities, characters, and features that may identify individual responsible for risk-taking behavior, a list of 191 personal characters was constructed. Set of features that characterizes the individual has been divided into two dimensions: the positive ones, and the negative ones. The positive dimensions are supposed not to assist the appearance of risk-taking behavior, while the negative ones refer to the negative qualities that act as promoters of risk-taking behaviors. Subjects had the following instruction: Here are some characters that describe the individual in his daily life. Please choose one of the proposals that fits you most by putting (x) in the appropriate box for each character.

2.3 Procedure

2.3.1 Distribution of the checklist

Six hundred checklists in total were distributed. Postgraduate and undergraduate students studying psychology or sociology carried out their distribution. They were paid for this task. Applicants scattered in places where drivers are expected to have free time to fill in and answer the checklists; mainly in bus and taxi drivers stations, vehicles insurances companies, vehicles' technical control stations.

2.3.2 Statistical techniques used

The Statistical techniques used in this study are:

- Means and standards deviations.
- Friedman ranking means test and Correlation.

3 RESULTS

3.1 Order of frequency of risk-taking behaviors

The data were used to obtain the means, standards deviations. Friedman ranking means test was applied to rank the items of the frequency risk taking in descending order. The results show that those occupying the first places according to their frequency levels which may reflect overall socially adopted behavior as is shown on the 20 items below, while, the items that classified last reflect a high degree of danger.

Items of risk-taking behaviors by drivers	Mean	Std.	Rank
1. Driving fast to catch up with a delay	2,76	1,169	85,90
2. Driving after or during a dispute or hard discussion	2,56	1,107	81,46
3. Driving slowly	2,39	1,020	77,06
4. Using a mobile phone while driving.	2,41	1,168	76,86
5. Being convinced that your driving is more accurate than the other drivers	2,56	1,317	76,86
6. Feeling the ability to avoid unexpected danger	2,39	1,203	74,44
7. Not taking a rest after driving for a long time	2,36	1,236	73,39
8. Driving in a state of nervousness and high excitement	2,20	,962	73,38
9. Slowing down suddenly when sensing the presence of radar	2,29	1,242	72,62
10. Slowing down suddenly	2,13	,887	72,18
11. Starting quickly when the green light comes on	2,25	1,241	71,80
12. Accelerating to try to go before the red light comes on	2,16	1,128	70,18
13. Changing direction suddenly to correct your way	2,09	,961	69,92
14. Driving on the emergency path	2,07	1,006	68,92

(Continued)

353

Items of risk-taking behaviors by drivers	Mean	Std.	Rank
15. Driving on the extreme left of the road	2,19	1,208	68,78
16. Preoccupying with what is going on edge of the road while driving.	2,07	1,008	68,59
17. Speeding in belief that it will saves you time	2,10	1,085	68,57
18. Stopping to see a road accident.	2,09	1,064	68,40
19. Not leaving enough lateral space between vehicles while driving.	2,09	1,076	68,28
20. Feeling the ability to minimize the damage of an unavoidable accident	2,11	1,129	67,30

Items of risk-taking behaviors by drivers	Mean	Std.	Rank
13. Stopping the vehicle on bends.	4,15	1,185	67,90
14. Overtaking despite the dangers of such a manoeuvre.	4,16	1,074	67,67
15. Driving carelessly in bad weather.	4,15	1,046	67,29
16. Executing dangerous manoeuvre during wedding convoy	4,12	1,123	67,25
17. Overtaking at Junctions	4,13	1,148	66,84
18. Going backward from a secondary to a main road	4,13	1,062	66,51
19. Overtaking in top of a hill	4,07	1,172	66,12
20. Excessive speed even if you are not used to	4,07	1,074	65,17

3.2 *Order of the degree of danger of risk-taking behaviors among drivers*

However, when it came to the degree of danger of the same items, the results are quite the opposite as the items occupying the first classes reflected in their behavior a set of high levels of dangerousness as it is indicated in the first following 20 items. It should be pointed out that drivers classified these items last when they were asked how frequent do they do this behaviors.

Items of risk-taking behaviors by drivers	Mean	Std.	Rank
1. Driving under the effect of drugs	4,55	1,048	81,50
2. Driving while drunk	4,53	1,065	80,29
3. Stopping on the left side of a motorway line.	4,35	1,141	74,98
4. Driving without using the car lights at night.	4,31	1,076	73,15
5. Overtaking on bends.	4,30	1,133	72,77
6. Dangerous manoeuvres while driving	4,28	1,116	72,39
7. Driving carelessly near schools.	4,30	1,013	71,97
8. Getting out suddenly from a side road without taking the necessary precautions	4,20	1,041	69,35
9. Driving in a prohibited direction.	4,18	1,136	68,94
10. Driving despite having defective brakes.	4,18	1,169	68,90
11. Getting in competition with other drivers.	4,17	1,044	68,26
12. Overtaking at the level of pedestrian crossings in residential area.	4,15	1,076	67,95

(Continued)

3.3 *Sensation seeking level among drivers*

The results of the level of sensation seeking for drivers found that subjects of the sample are distributed on the four levels of Zuckerman scale. Even though there is a considerable variation among them as fallow: high: 4.6%, medium: 36.3%, low: 15.9% and very low: 41.2%.

3.4 *Relationship between risk-taking behaviors among drivers with some variables*

A significant negative correlation between the frequency of risk-taking behavior and the degree of perception of the danger of this behavior is estimated at ($-0.359**$) with statistical significance. The results of the study also showed that there were negative correlations between the frequency of risk-taking behavior with some of the following individual characteristics: sex ($-,123*$); Age ($-,221**$); family situation ($-,154**$). However, the correlation between the frequency of risk-taking behaviors and the level of sensation seeking was positive ($0.307**$) with statistical significance. There were positive correlation between the frequencies of risk-taking behavior with some of the following individual characteristics: ownership of the vehicle ($,209**$): daily distance ($,135**$); and number of offenses ($,170**$).

However, the correlation between the degree of danger of risk-taking behavior and sensation seeking was negative ($-0.168**$). The results of the study also showed that there were positive correlations between the degrees of danger of risk-taking behavior with some of the following individual characteristics: age ($,120*$) and the number of children: ($,159**$). While there was a negative correlation with both: type of insurance ($-,125*$) and ownership of the vehicle ($-,099*$).

The results of the present study also indicate that there were negative correlations between sensation seeking and the following individual characteristics: Age (–,169**); family situation (–,114*); and the number of children (–,142**), but a positive correlation only with the year for obtaining driving licenses (,146**).

The results of the present study also indicated that there are positive general correlations between the personal characters and the frequencies of risk-taking behaviors (,322**); as well as with sensation seeking: (,270**), but negative correlation with danger of risk-taking behaviors: (–,143**). In addition, he personal characters had significant correlation with some of the following individual characteristics: Age (–,195**); family situation (–,201**); and Number of children –,113*); Type of disease (,295**); year for obtaining driving licenses (,142**); seniority in driving (–,157**); and finally type of insurance (–,105*).

3.5 The correlation of personal characters with risk-taking, degree of danger and sensation seeking

The results indicated that there are significant positive correlations of so many personal characters but mainly all the thirteen shown below with risk-taking, sensation seeking but negative correlations with the degree of danger.

Surprisingly, It can be noticed that in so many cases when there is a significant positive correlation of personal characters with risk-taking behavior the same personal character had a significant negative correlation with the degree of danger but positive with sensation seeking. Therefore, these thirteen personal characters could be used to describe the driver who has a risk-taking behavior.

Characteristics	Risk-taking	Sensation	Degree of danger
1. Impulsive	,376**	,209**	–,201**
2. Pressed	,345**	,129**	–,171**
3. Rushed	,343**	,204**	–,183**
4. Adventurer	,313**	,161**	–,177**
5. Anxious	,288**	,184**	–,102*
6. venture	,274**	,245**	–,145**
7. Indifferent	,272**	,202**	–,185**
8. Manipulator	,270**	,215**	–,148**
9. Disturber	,255**	,180**	–,142**
10. Stubborn	,253**	,205**	–,151**
11. Anarchique	,250**	,175**	–,144**
12. Rebel	,237**	,144**	–,178**
13. Aggressive	,228**	,104*	–,165**

4 DISCUSSION

The present study has shown that drivers tend to repeat risk-taking behaviors while driving, with varying degrees. The results also indicated that the most frequent risk-taking behaviors among drivers tend to express in its entirety for behaviors that are socially accepted. This is consistent with the findings of several studies (Curtis, 1960; Newcomb, 1953; Jackson, 1994). Hence, the individual has a sort of values orientation derived from the provisions issued by the individual on the human, social and physical environment. These provisions include ideas and beliefs in the behavior of the individual that is often the product of socially realise through his interaction with the community, where the individual can interpret his behavior and give it a certain justification (Abdel Fattah, 2005). Therefore, the results of the frequency of risk-taking behaviours that occupied the first ranks in the current study are acquired through the group properties, whether family or environmental surrounding. This is perhaps because doing such behaviour does not represent a risk to the individual for collective interest.

However, the results of ordering the same items of risk-taking behaviours with regard to the degree of danger came almost in reversing order to the results of the frequency. Indeed; the first item of high degrees of risk-taking behaviours (Driving under the effect of drugs) was classified last when the items were ordered in term of frequency risk-taking behaviours. Perhaps this is because the judgment on the degree of danger of a particular behaviour depends on the capability of controlling the situation, the degree of will, and the prior knowledge of the risk. Coll & Fischhoff (1981) have found that accepting the degree of risk taken voluntarily like motorcycles risks or climbing higher mountain is estimated at more than 1000 level to social risks. As for the unknown risk, it is judged by the fear that spreads and its degree of risk (Trimpop & Zimolong, 2000). Yet, the degree of awareness of the degree of risk-taking behaviour among drivers, and the frequency of these behaviours might be affected by the time factor, experience, perception quality, goals and justifications, as well as physical and mental health.

Baijonnet, Cauzard (1987) have found that the various dangers do not have the same weight (order) among respondents. The dangers of traffic system have occupied a middle rank compared to other dangers such as serious illnesses accidents, and sports competitions accidents. These results have been interpreted to the fact that the mobility is a necessity in the community, and that individuals rebuild their representations to the dangers of traffic system to make it acceptable, to reduce the mismatch of cognitive dissonance. After 10 years, Arenes & Gautier (1998) repeated the same study of representations of the dangers of traffic system at the same age groups with the same types of risks. The results showed that the dangers of traffic system ranked first among other hazards at all age groups, because traffic accidents are the risk that fears all age groups. This result was attributed to the development of preventive programs and the quality of information on the causes of deaths resulting from traffic accidents System (from: Diaz-Perez, 2003).

The results of the current study also indicated that the frequency of risk-taking behaviors among drivers has a positive correlation with the level of sensation seeking and some general personal characters of risk. This goes with the viewpoint of Zuckerman (1990) in which sensation is linked directly to risky behavior, as the individual with sensation seeking features seeks physical and social risk. Rossi & Cereatti (1992) have also indicated that the increase of the degrees obtained by individuals in the questionnaire of sensation seeking whenever the practiced activity has an objective risk. Some studies have also have shown a relationship of sensation seeking with repeat of accidents, through which it was found that this feature is not related only to the choice of activity, but it also related to adopting risky behaviours (Connolly, 1981, Rossi & Cereatti, 1992). The results of the present study also confirm the findings of Rolison, (2002), that individuals with high degree of sensation seeking are more inclined to high risk behaviour (from: Abdel Fattah Sayed, 2005).

The results of the present study has also shown that there are 13 out of 191 general personal characters, which had positive correlations over 0.20, with the frequency of risk-taking behavior. In addition to their positive correlation with sensation seeking while they all had negative correlation with the degree of danger. Therefore, these 13 general personal characters could be added to describe drivers with risk taking behaviors.

5 CONCLUSION

Perhaps it can be concluded that risk-taking behaviours among drivers is related more or less to social values in the society. This implies the need to work on improving the social values in order to minimise the different risk-taking behaviours among drivers. Hence, the society should openly refuse or reject these risk-taking behaviors just like so many behaviors in different fields that are rejected or even forbidden by the society even if they are accepted by other societies.

REFERENCES

Ajzen, I., & Fishbein, M, 1980, understanding Attitudes and Predicting Social Behavior. Prentice-Hall, Englewood Cliffs.

Ajzen, I. 1985, from intentions to actions: a theory of planned behaviour, in: Kuhl, J., Beckmann, (Eds), Action-Control: from Cognition to Behavior. Springer, Heidelberg.

Boudrifa. H, A. Bouhafs, M. Touill & F. Tabtroukia. 2010, Measurements of Unsafe Behavior of Road Users, 15th International Conference Road Safety on Four Continents, Abu Dhabi, United Arab Emirates, 28–30 March 2010.

Boudrifa. H, A. Bouhafs, M. Touill & F. Tabtroukia, 2012, factors and motives of unsafe behaviors of road users, 18th congress of The International Ergonomics Association, Recife, Brazil.

Boudrifa H. 2014, Factors and motives of unsafe behavior of pedestrians and cyclists. The proceedings Book the International symposium SHO 2014, Guimarães, from 13th to 14th February 2014. School of Engineering of the University of Minho, in Guimarães, Portugal.

Perez-Diaz C, théorie de la décision et risques routiers, Cahiers internationaux de sociologie, 2003/1, N° 114, 143–160.

Occupational Safety and Hygiene V – Arezes et al. (Eds)
© 2017 Taylor & Francis Group, London, ISBN 978-1-138-05761-6

Thermal stress and acclimation effects in military physiological performance

Joana Duarte, André Ferraz & Joana C.C. Guedes
PROA/LABIOMEP, University of Porto, Porto, Portugal

Mário Álvares
Portuguese Army, Portugal

José A.R. Santos, Mário Vaz & J. Santos Baptista
PROA/LABIOMEP, University of Porto, Porto, Portugal

ABSTRACT: This article corresponds to a case study with 6 young male military who were tested for three types of exertion: marching without load, marching with load, and running at two different environmental conditions: thermoneutral ($\pm 22°C$ and $\pm 40\%$ rh) and thermal stress ($\pm 40°C$ and $\pm 30\%$ rh), with two adaptation levels—unacclimated and acclimated to the thermal stress condition. Core temperature, heart rate, VO_2, body mass loss, and lactate were measured; the Borg Rating of Perceived Exertion (RPE) was also used. The thermal stress situation decreased the overall physiological performance and increased the rate of perceived exertion. After acclimation all participants improved their overall performance.

1 INTRODUCTION

1.1 *Scope*

The aim of human thermoregulation is to protect body processes, by establishing a relative constancy of deep body temperature, in spite of external and internal influences on it (Werner, 2010). Heat stress complications take place when heat gain is higher than the body's ability to dissipate it. This kind of matter has negative impacts on the subjects, especially in those who are directly expose to extreme environmental conditions, such as mining workers, firefighters, and military, amongst others.

Concerning temperatures, the process starts with heat stress (when temperatures rise above 37°C), then proceeds to hyperthermia (temperatures beyond 39°C) and, finally, heat stroke (higher than 40°C). Despite these tabled values, past certain temperatures (which may vary from one person to another) homeostasis cannot be sustained and death may occur (Laxminarayan, Buller, Tharion, & Reifman, 2015).

1.2 *Military's heat illness history*

Hot environments present themselves with great challenges, once heat stress may weaken human aptitude to complete complex tasks during military activities, and, in extreme cases, posing a threat to human life (Nunneley & Reardon, n.d.; Sonna, 2012).

In order to preserve homeostasis, the human body sets off physiological responses related with the rise of the core temperature (Laxminarayan et al., 2015). The main complications usually are: decrease in exercise capacity, heat injury, sunburn, *miliaria rubra* (known as "heat rash", it presents itself with a pruritic eruption), rhabdomyolysis (breakdown of skeletal muscle tissue), fluid and electrolyte loss, among others. Most of the times heat illness can be prevented with suitable training, heat acclimatization processes (one of the best strategies to prevent heat illness), and fluid availability (and adequate hydration) (Nichols, 2014).

However, these preventive strategies are not expected to eradicate heat stress completely; they ought to minimize its impact as much as they possibly can.

1.3 *Study goals*

There are records of several cases of heat stroke during military missions and even throughout the recruiting process. Heat stress greatly hinders job performance, hence it is crucial to understand the magnitude of physiological strain human being can abide, and what factors may cause tolerance so variable. The main purpose of these experiments was to evaluate the break in performance which occurs when subjects are exposed to thermal stress condition and the possible effects of an acclimation process.

2 MATERIAL AND METHODS

2.1 Sample

The participants, all Caucasian males, were randomly chosen, among militaries working for the Portuguese Army in the north region of Portugal. From an initial sample of 9 subjects, 3 were excluded: for medical, contractual and physical fitness reasons.

The participants did not present cardiac, vascular, pulmonary, or any allergic medically diagnosed disease, being considered mentally healthy, without any kind of psychological disturbances by the military center of psychology. Regular medication use was not declared.

Accordingly to the Helsinki Declaration, all participants gave written informed consent prior to their inclusion in the study.

2.2 Test protocols

The protocol was divided into three sub-protocols, occurring at two different environmental settings: the Thermoneutral Condition (NC) was set in $22.0° ± 0.5°C$ air temperature and $40 ± 2\%$ relative humidity, and the Thermal Stress Condition (SC) was set in $40.0° ± 0.5°C$ air temperature and $30 ± 2\%$ relative humidity, having the individuals two different adaptation levels: "unacclimated" and acclimated to the SC.

For each of the conditions, three tests were performed: marching without load, marching with load and running.

2.2.1 Marching Without Load (MNL)
In the marching tests the treadmill was adjusted to have 1% of inclination (in order to simulate real ground and wind friction conditions).

The test stages were: 20 minutes of adaptation (this is the only common point between the protocols), a warmup of 4 km.h^{-1} during 3 minutes and, the main experimental stage with a maximum duration of 20 minutes at 6 km.h^{-1}.

The perceived effort was estimated every 5 minutes using Borg's RPE and some physiological parameters (*2.4 Physiological monitoring*) were measured in real time (Habibi, Dehghan, Moghiseh, & Hasanzadeh, 2014).

Every subject was equipped with his uniform, consisting of a t-shirt, trousers and military boots.

2.2.2 Marching With Load (MWL)
The only difference between this protocol and the former is the load transport: each subject had to carry a backpack with the following items: tent, sleeping bag, alternative uniforms, meals, as well as belt with canteen, chargers and his weapon, creating a total load of about 29.20 kg.

2.2.3 Running
This test assessed the maximum oxygen uptake ($VO_{2\,max}$), so the treadmill was adjusted to have 0% of inclination and no maximal test duration was defined: it was expected of each subject to last, at least, 12 minutes running (Yoon, Kravitz, & Robergs, 2007; Yoshida, Chida, Ichioka, & Suda, 1986).

Previous to the test—a 5-minutes warm-up running at 8 km.h^{-1} was realized. Afterwards, the subjects had a 5-minutes recovery period, walking at 4 km.h^{-1}. The starting velocity of the test was set at 10 km.h^{-1} with increments of 1 km.h^{-1} each 2 minutes, until exhaustion.

2.3 Acclimation process

According to bibliography, the acclimation process occurred during 14 consecutive days (Nichols, 2014) where the subjects had to endure 3 hours per day inside the climatic chamber, at the defined SC, without physical effort.

2.4 Physiological monitoring

During the tests, some indicators were monitored, such as:

- Core temperature recorded, continuously, through ingestible pills, from Vital Sense;
- Heart rate through the K4-b2 equipment;
- Pulmonary gas exchange recorded breath-by-breath, also through the K4-b2 equipment;
- Lactate concentration measured by collecting a blood sample from subject's right earlobe right after exercise, at 3rd, 5th and 7th minutes after ceasing the test, through Lactate Pro;
- Dehydration through water mass loss, was calculated through the weight difference before and after each test, using a 50 g precision scale from SECA;

Some of the immediate stopping criteria defined for every protocol were:

- Core temperature above 38.5°C (Richmond, Davey, Griggs, & Havenith, 2014);
- Heart rate with a specific value for each individual, given by the formula ($HR_{max} = 220 -$ subject's age) (Bruijns, Guly, Bouamra, Lecky, & Lee, 2013);
- Respiratory coefficient higher than 1.10 (Teixeira, Grossl, De Lucas, & Guglielmo, 2014).

2.5 Climatic chamber

All of the prior mentioned tests took place inside of *Fitoclima 25000*, built accordingly to EC standards, which meets health and safety basic requirements. The subjects' performance was tested with the help of the T2100 model of General Electric treadmill, which allowed controlling velocity and exercising grade performed.

3 RESULTS

Despite the sample has 6 subjects, one of them (M06) had to be considered separately from the others because he had to take the "unacclimated" protocol twice, due to a huge mass loss that he put himself through from one year of experiment to another. However, due to calendar limitations, he first went through an acclimation process and less than one month later he performed the "unacclimated" tests.

3.1 Results for M01, M02, M05, M07 and M09

- Core temperature—increased from NC to SC for M01, M02, M07 and M09 (no sufficient data for M05). Concerning the acclimation process, it is only possible to draw conclusions for M09, in which the maximum temperature reached lower values and M07 for whom the maximum temperatures rose. For the other military this parameter did not present a linear behavior: it increased in some experiments, but decreased in others.
- Heart rate—this parameter showed significant difference between military: it increased from NC to SC for M01 and M02 and in some tests of M05 (acclimated subject protocol), however for M07 and M09 it displayed no linear behavior. The acclimation process led to lower heart rates for M01 and M09 (with the exception of running tests for both). However, for M02 and M07 heart rates rose between adaptation levels. For M05 there was not a linear behavior.
- Oxygen consumption (VO_2)—for this parameter, two calculations had to be considered: average VO_2 (marching experiments) and VO_2 max. (running experiments) (Edwards, 2013).

Concerning average oxygen consumption, the results were also variable. For M01 and M02 the body reacted differently considering adaptation level: it rose from NC to SC in the non-adapted protocol and increased in the adapted protocol. For almost every M05 tests average VO_2 lowered, for M07 it rose and M09 did not present a linear behavior. For M01, M02, M05 and M07 (for the NC tests) the acclimation process led to lower average oxygen consumption. In M09's case there was not detected a pattern.

VO_2 max displayed a linear behavior for every subject: in the "unacclimated" protocol, this parameter decreased from NC to SC but increased in the acclimated situation. With

relation to the acclimation process, M01, M02 M05, and M09 had higher VO_2 max values; however for M02 and M05 this only happened in the SC tests. Regarding M07, $VO_{2\ max}$ decreased in the SC tests despite it increased in the NC tests.
- Lactate concentration was higher from NC to SC for M01, M02, M07 and M09, with the exception of some tests. For M05 it rose in the marching tests but decreased in running experiments. The acclimation process led to lower lactate concentrations for M09, M01 in the NC tests and M02 in the SC tests. For M07 this enzyme concentration rose in almost every tests and for M05 no conclusions could be made.

However, regardless of the military, the lactate concentration was much higher for the running tests than for marching (with or without load).
- Dehydration—Body Mass Loss (BML) increase from NC to SC despite the military. This loss grew with the growth of the experiment difficulty: was higher in MWL than MNL and significantly higher in Running tests.

The acclimation process led to higher BML for M01 and most of the tests for M09. Both M02 and M07 presented lower BML and M05 results were very variable.
- Borg's Rate of Perceived Exertion this parameter did not present linear behavior for every military: in some cases the perceived exertion was higher from NC to SC, in others it led to the same values and in few, the effort perception rose from NC to SC. Despite this, concerning the acclimation process, it led to lower effort perceptions in almost every situation for every military, with the exception of M09 where in some cases there was a higher perception of exertion.

3.2 Results for M06

- Core temperature registered higher values for the SC when comparing to NC in every test and running tests had greater impact on this parameter variation. The acclimation process led lower maximum core body temperatures, except in the running tests. This occurs due the relation between useful work and temperature, caused by muscle yield.
- Heart rate tendency accompanied the core body temperature: from NC to SC heart rate increased and the acclimation process led to a better heart performance (lower heart rates).
- Oxygen consumption (VO_2) despite not varying much, average oxygen consumption was a little higher for the SC. The acclimation process led to lower values of average VO_2.

Maximal oxygen consumption change from one adaptation level to another: it increases from NC

to SC in the "unacclimated" tests but it decreases in the acclimated ones. When comparing before and after acclimation, the $VO_{2\,max}$ increased.
- Lactate concentration—the results for this parameter did not present a consistent behavior concerning adaptation level. In spite it decreased in some cases and increase in others from NC to SC. However, lactate concentrations were higher for running tests than for the marching tests.
- Dehydration—through body mass lost increased with the growing difficulty of the experiments: it was less prominent in MNL tests and hugely higher in running experiments. It is also notorious the big increase in this parameter when comparing NC with SC. When comparing acclimation levels, this parameter also led to higher values of dehydration concerning the acclimated subject protocol.
- Borg's Rate of Perceived Exertion MN—tests led to lower exertion perceptions than MWL and even Running tests. The perceived exertion rose from the NC to the SC in every experiment. The acclimation process led to lower effort perceptions.

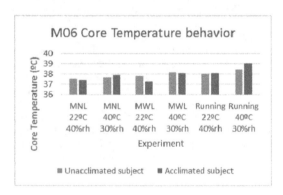

Figure 1. Core temperature variation through the different M06's tests.

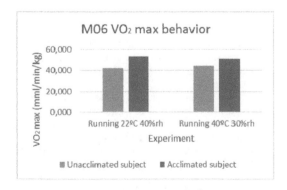

Figure 2. VO_2 variation through the different M06's tests.

Heat acclimation process also lead to the achievement of extra stages in the Running tests for almost every military, in the NC.

4 DISCUSSION

Individuals may experience reasonable dehydration during prolonged high intensity exercise, which is thought to be responsible for the increase in some physiological parameters such as: core body temperature, heart rate and pulmonary ventilation (Golbabaei, Zakerian, Dehaghi, et al., 2014; Sawka et al., 1979), coincident with the data analysis.

In hot environments there is a higher energy demand, thought to be one of the main contributes to the core body temperature increase (Dimri, Malhotra, Sen Gupta, Kumar, & Arora, 1980). There is also an intensification of the oxygen consumption, explained by the decrease in the aerobic component of the oxygen supply (and consequent increase of the anaerobic one) (Dimri et al., 1980). The maximum oxygen consumption is the physiological index that best represents the maximum aerobic power. It is clear that a good performance require a high value of $VO_{2\,max}$, but not always those with the highest values of this index, are those who have a better aerobical performance (Caputo, Fernandes, Oliveira, & Greco, 2009). This parameter was the one which lead to higher behavior variations amongst the analyzed military.

Both oxygen demand and aerobic fraction also have an impact on lactate concentration, which rises due to the increase of work severity and thermal stress on the body, observed in 5 of the 6 case-studies.

Heat acclimation produces changes in the physiological parameters which allow the body to perform better (Nielsen et al., 1993). The main differences observed usually are lower heart rate, lower core temperature, increase in tolerance time, among other factors (Gupta, Swamy, Dimri, & Pichan, 1981; Nichols, 2014; Wenger, 1997). These differences were observable in almost every case, with some mentioned exceptions. The increase in tolerance time occurred indeed in almost of the military, which led to the performance of extra stages in the Running tests.

Heat acclimation training also reduces the sweating threshold, increasing the body's ability to produce sweat. The sensitivity of the sweat glands to thermal and hormonal stimuli after acclimation may increase through an increase in the receptor density for neural and humoral stimuli or an increase in the size or number of active sweat glands (Shen & Zhu, 2015), observable in almost every performed tests.

5 CONCLUSIONS

When comparing thermoneutral condition with thermal stress condition, the body performed better while exposed to the neutral situation: core temperature was lower, heart rate showed less variation, oxygen uptake was also lower and body mass loss was not intense, despite acclimation level. Lactate parameters also met what was expected (higher values for higher physical effort), regardless of M06 it did not present a linear behavior for some of the parameters, such as lactate concentration.

It was possible to observe some improvement in the performance of the subjects due to the acclimation process: a decrease in the maximum temperatures was observed, as well as a lower heart rate variation. Oxygen uptake was also lower for most cases. One of the parameter settings discussed in which the response of the organism to acclimation was assumed as consistent with the literature was the Borg's Rate of Perceived Exertion, wherein in almost all of the performed tests the subjects perceived lower efforts.

Another conclusion is that higher exercise intensity (as caring a load or running) also have major impacts on the individual's response.

Moreover, it is necessary to deepen and validate this analysis, by increasing sample size and considering other parameters in the study.

ACKNOWLEDGMENTS

The authors would like to thank to CINAMIL, INEGI and PROA/LABIOMEP all the support in the development and international dissemination of this work.

REFERENCES

Barnett, A., & Maughan, R.J. (1993). Response of unacclimatized males to repeated weekly bouts of exercise in the heat. *British Journal of Sports Medicine, 27*(1), 39–44.

Bruijns, S.R., Guly, H.R., Bouamra, O., Lecky, F., & Lee, W. a. (2013). The value of traditional vital signs, shock index, and age-based markers in predicting trauma mortality. *The Journal of Trauma and Acute Care Surgery, 74*(6), 1432–7. doi:10.1097/TA.0b013e31829246c7

Caputo, F., Fernandes, M., Oliveira, M. De, & Greco, C.C. (2009). Exercício aeróbio : Aspectos bioenergéticos, ajustes fisiológicos, fadiga e índices de desempenho.

Dimri, G.P., Malhotra, M.S., Sen Gupta, J., Kumar, T.S., & Arora, B.S. (1980). Alterations in aerobic-anaerobic proportions of metabolism during work in heat. *European Journal of Applied Physiology and Occupational Physiology, 45*(1), 43–50.

Edwards, A.M. (2013). Respiratory muscle training extends exercise tolerance without concomitant change to peak oxygen uptake: Physiological, performance and perceptual responses derived from the same incremental exercise test. *Respirology, 18*(6), 1022–1027. doi:10.1111/resp.12100

Golbabaei, F., Zakerian, S.A., Dehaghi, B.F., Ghavamabadi, L.I., Gharagozlou, F., & Mirzaei, M. (2014). Heat Stress and Physical Capacity: A Case Study of Semi-Professional Footballers, *43*(3), 355–361.

Gupta, J.S., Swamy, Y. V, Dimri, G.P., & Pichan, G. (1981). Physiological responses during work in hot humid environments. *Indian Journal of Physiology and Pharmacology, 25*(4), 339–347.

Habibi, E., Dehghan, H., Moghiseh, M., & Hasanzadeh, A. (2014). Habibi, E., Dehghan, H., Moghiseh, M., & Hasanzadeh, A. (2014). PDF—Study of the relationship between the aerobic capacity (VO2 max) and the rating of perceived exertion based on the measurement of heart beat in the metal industries Esfahan. Journal of. *Journal of Education and Health Promotion, 3*, 55. doi:10.4103/2277-9531.134751

Laxminarayan, S., Buller, M.J., Tharion, W.J., & Reifman, J. (2015). Human core temperature prediction for heat-injury prevention. *IEEE Journal of Biomedical and Health Informatics, 19*(3), 883–891. doi:10.1109/JBHI.2014.2332294

Nichols, A.W. (2014). Heat-related illness in sports and exercise. *Current Reviews in Musculoskeletal Medicine, 7*(4), 355–365. doi:10.1007/s12178-014-9240-0

Nielsen, B., Hales, J.R., Strange, S., Christensen, N.J., Warberg, J., & Saltin, B. (1993). Human circulatory and thermoregulatory adaptations with heat acclimation and exercise in a hot, dry environment. *The Journal of Physiology, 460*, 467–485.

Nunneley, S.A., & Reardon, M.J. (n.d.). Prevention of heat illness. *Medical Aspects of Harsh Environments,* 209–230.

Richmond, V.L. b, Davey, S., Griggs, K., & Havenith, G. (2014). Prediction of Core Body Temperature from Multiple Variables. *Annals of Occupational Hygiene, 59*(9), 1168–1178. doi:10.1093/annhyg/mev054

Sawka, M.N., Knowlton, R.G., Glaser, R.M., Wilde, S.W., & Miles, D.S. (1979). Effect of prolonged running on physiological responses to subsequent exercise. *Journal of Human Ergology, 8*(2), 83–90.

Sonna. (2012). Chapter 9 Practical medical aspects of military operations in the heat heat as a threat to military operations, 293–310.

Teixeira, A.S., Grossl, T., De Lucas, R.D., & Guglielmo, L.G.A. (2014). Cardiorespiratory response and energy expenditure during exercise at maximal lactate steady state [Resposta cardiorrespiratória e gasto energético em exercício na máxima fase estável de lactato]. *Revista Brasileira de Cineantropometria E Desempenho Humano, 16*(2), 212–222. doi:10.5007/1980-0037.2014v16n2p212

Wenger, C.B. (1997). Chapter 2 Human Adaptation to Hot Environments Importance of Tissue Temperature. *Work.*

Yoon, B.K., Kravitz, L., & Robergs, R. (2007). VO2 max, protocol duration, and the VO2 plateau. *Medicine and Science in Sports and Exercise, 39*(7), 1186–1192. doi:10.1249/mss.0b13e318054e304

Yoshida, T., Chida, M., Ichioka, M., & Suda, Y. (1986). Blood lactate parameters related to aerobic capacity and endurance performance. *European Journal of Applied Physiology and Occupational Physiology, 56*(1), 7–11. doi:10.1007/BF00696368

Occupational Safety and Hygiene V – Arezes et al. (Eds)
© *2017 Taylor & Francis Group, London, ISBN 978-1-138-05761-6*

Non-ionizing radiation in vertical residences of heat islands, João Pessoa, Brazil

R.B.B. Dias, L.B. Silva, E.L. Souza, M.B.F.V. Melo & C.A. Falcão
Department of Industrial Engineering, Federal University of Paraíba, João Pessoa, Brazil

J.F. Silva
University of Brasilia—UNB, Gama, Brazil

ABSTRACT: Non-ionizing radiation of extreme low frequency is considered as a possible source causing several problems to human health, among them the infantile leukemia. Residences with exposure to magnetic fields above 0.4 µT on average can be characterized as environments exposed to non-ionizing radiation. This study presents an analysis of the residents' exposure to non-ionizing radiation in residences located in heat islands, in the city of João Pessoa. Through statistical analysis of the data, it was verified that the residences analyzed had a high index of exposure to the magnetic fields of the former low frequency (>>0.4 µT in the range of 40–60 Hz), which, from the point of view of occupational health, it could lead to serious epidemiological problems in the medium and long term.

1 INTRODUCTION

Global warming is a problem that has been intensifying since the industrial revolution (18th century) to the present day, generating a great deal of concern to society. The increased air pollution due to the excessive release of greenhouse gases into the atmosphere is another factor that confirms to the increase in global temperature. A climatic phenomenon that can occur as a result of this heating and the high degree of urbanization of the towns is the formation of heat islands. In these islands, the average temperature is usually higher than in rural areas (Ryu, 2012).

The advance technological development providing facilities and benefits to humans, adding new services and possibilities demand solutions came associated with the comfort and well-being of citizens.However, they can also produce risks to the environment and to human health, and may expose your users to electromagnetic pollution (Grelier, 2014). Non-Ionizing Radiation (NIR) does not have sufficient energy to remove electrons from atoms or molecules, but may have enough energy to break chemical bonds and molecules. And there are several types of frequencies classified as non-ionizing, namely: ultraviolet, infrared, radio frequency, laser, visible light, as well as radio waves, microwave (mobile phone, household oven, etc.) and light (Petrucci, 2011).

It is common in several vertical residences present internal transformer substations to buildings, which potentially increases exposure to Magnetic Fields (MF) in the apartments adjacent to these transformers (Kendel, 2013). Over the past two decades, residential exposure to low-frequency magnetic fields (MFBF) has been linked to childhood leukemia in consistent epidemiological studies, although causality is still under investigation (Zaryabova, 2013; Su, 2014). The International Agency for Cancer Research (IARC, 2002), ranked the Extreme Low Frequency Magnetic Fields (ELF MF) as potential carcinogens.

Studies conducted in Europe (Roosli, 2011; Harcuveny 2011) have shown that the highest values of residential exposure to MF are apartments located on the first floor of buildings associated with some kind of internal electric transformer substation.

That same focus given by the authors mentioned above, this article presents a study on vertical residences city of João Pessoa, which are associated with internal transformers and located in areas classified as urban heat islands. The purpose of the article is to analyse the exhibition of the residences in the NIR, as a result of possible existing radiation in the external environment, coupled with the radiation from household utensils found in homes, in addition to transformers associated with these houses. In the end, it discusses the potential problems arising from exposure to NIR residents of these homes.

2 MATERIALS AND METHODS

2.1 Data collection process

Data collection was carried out in two residences in the city of João Pessoa, in the Tambaú and Cabo Branco neighborhoods, in areas that are defined as urban heat islands (Santos, 2012). Both monorails are located on the first floors of their respective buildings, which had internal transformer substations.

At the residence located in the Tambaú neighborhood, measurements of magnetic fields were performed in three environments: bedroom, kitchen and living room. In the residence of Cabo Branco, magnetic field measurements occurred in two environments: bedroom and living room; Taking into account the difference of area and layout between the residences, as shown in Figures 1 and 2. In both residences the transformer chambers were located at the level of the public road of the buildings.

The values of MF were collected in the microtesla unit (µT) in the range of 1 to 100 Hz.

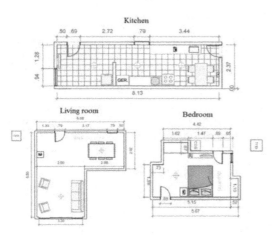

Figure 1. Layout of the Tambaú apartment.

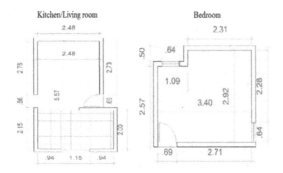

Figure 2. Dimensions of the apartment area of Cabo Branco.

Magnetic field measurements took place over a 24-hour period for each room in the residences, between October 28 and November 2, 2016, using the Spectran NF-5035 properly calibrated.

2.2 Statistical analysis

Descriptive analyzes and statistical data were performed using R Project 3.1.1 software. The Mann-Whitney and Kruskal-Wallis tests were used to compare the MF intensity values between the rooms in the Cabo Branco and Tambaú households, respectively, considering the level of significance with a value of 0.05.

The behavior of the data was evaluated for later evaluation of the accumulated probability in relation to the magnetic field intensity for the rooms in each one of the residences.

3 RESULTS AND DISCUSSIONS

The mean and standard deviation values of the magnetic field strength for each room, with the exception of the Tambaú kitchen, in the two apartments are shown in Table 1. The MF values for the kitchen of the Tambaú residence do not obey any distribution, besides Standard deviation near the mean value. The frequency of the data obtained in each room is demonstrated by the histograms of Figures 3 and 4. For the others values the behavior has lognormal distribution.

Among all the rooms observed, including the two residences, it can be seen that the highest average magnetic field intensity during the whole experiment is in the bedroom of a residence in the Tambaú neighborhood, with a value of 1.268 µT at 40–60 Hz. The highest point peak was also observed in this room, a value of 2.93 µT at 40–60 Hz. The plausible explanation lies in the fact that the location of the room is influenced by the ELF MF coming from its surroundings mainly at night, as there are other sources of heat.

Table 1. Mean and standard deviation of the magnetic field strength per room with frequency of 40–60 Hz.

Rooms	Mean (µT)	Standard deviation (µT)
Bedroom (CB*)	1,233	0,021
Living room (CB*)	1,226	0,032
Bedroom (TA**)	1,268	0,014
Living room (TA**)	1,225	0,027

*CB = Cabo Branco
**TA = Tambaú

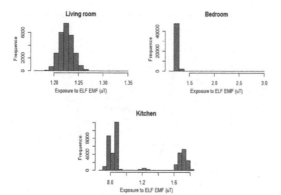

Figure 3. Frequency Histogram—Tambaú.

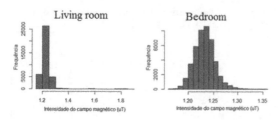

Figure 4. Frequency Histogram—Cabo Branco.

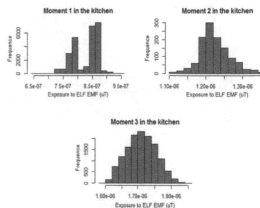

Figure 5. Frequency Histogram—Kitchen (TA).

This bedroom is exposed to ELF MF from the high voltage cables of the electric grid, as well as an external transformer, of the aerial type, that distances from the room in 10 meters. The peak value was observed at 11:58 p.m., at which time the air-conditioning tools in the room (in addition to TV and laptop) were in operation, resulting in more local MF values.

The intensity of FM in all rooms was higher than 0.4 μT in the frequencies of 40–60 Hz in the two residences. The authors (Ilonen, 2008), Hareu-veny (2011) and Grellier (2014) report that apartments located on the first floor of buildings with internal transformer substations have values higher than 0.4 μT on average in the frequencies of 50 Hz and 60 Hz, which is considered a cut bridge with respect to a residence being classified as exposed to ELF MF, according to reports and recommendations of the International Commission on Non-Ionizing Radiation Protection (ICNIRP) of the Organization (WHO) and the International Agency for Research on Cancer (IARC), as well as studies supported by the TRANSEXPO project that aims to investigate the onset of childhood leukemia related to the presence of ELF MF.

In relation to the Tambaú kitchen, the frequency distribution of the obtained data can be observed in Figure 5. The frequency histogram of the values of the field of kitchen magnetic can be divided into three moments: (1) the appliances are not in operation so it has lower values of MF (in this situation, on average, they are the lowest values when compared to the other rooms) because the environment in question is subjected only to radiation from the internal transformer to the building, causing the average value of intensity of MF in this type of situation is 0.828 μT; (2) the layout of the kitchen, as well as its location in the building, makes the external medium radiation through the low voltage conductor cables, and even external transformers have little influence in terms of MF intensity in this room. And (3) the maximum value of MF obtained in the kitchen was 1.84 μT, in the period of 11:11 hours, where practically all the appliances located in this environment were in operation. The kitchen is characterized by being the only room with values higher than 1.6 μT, and it is also the ambience where one can notice the greater presence of home appliances in operation mainly during the daytime period. Values higher than the maximum MF value of the kitchen can be observed in the Tambaú bedroom, but in a punctual way, so that they did not repeat for a significant period of time. According to Figure 5, at time 3 the average intensity of the MF in the kitchen is 1.7 μT. It can be noticed the direct increase of the intensity of the field due to the use of the household appliances present in the kitchen, which exposes the resident to periods of exposure to the NIR at a higher intensity, which can be detrimental to human health (Yithzak, 2011; Thuróczy, 2011; Zaryabova, 2013).

Figure 6 shows that the MF in the residence living room in Cabo Branco rises along the measurements up to 1.23 μT and decreases up to 1.21 μT at 40–60 Hz, whereas in the bedroom the MF starts at 1.26 μT and has been decreasing up to 1,23 μT at 40–60 Hz, remaining constant until the

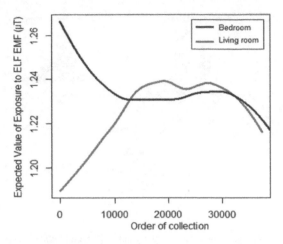

Figure 6. Exposure to the magnetic field—Cabo Branco.

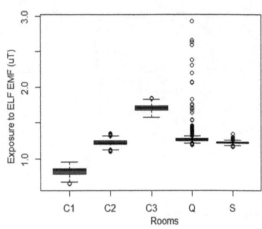

Figure 8. Comparison between rooms exposed to the magnetic field—Tambaú.

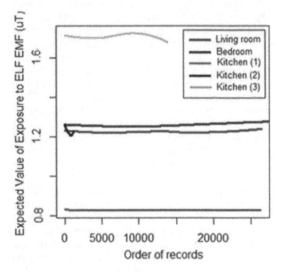

Figure 7. Exposure to the magnetic field per room—Tambaú.

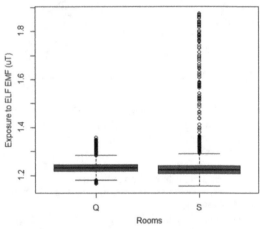

Figure 9. Comparison between rooms exposed to the magnetic field—Cabo Branco.

end of measurements. In Figure 7, the kitchen is analyzed by the three moments presented earlier. It can be noted that in the residence in Tambaú there is a significant difference in the distribution of MF data between the rooms, which was evidenced by Kruskal-Wallis test, $\chi^2 = 101769,7$; p-value $= 2,2 \times 10^{-16} < 0,05$. The Cabo Branco apartment also showed significant variation among the rooms, according to the Mann-Whitney test, with W $= 958443799$ e p-value $= 2,2 \times 10^{-16} < 0,05$.

Figures 8 and 9 exemplify this relationship between the rooms for the residence in Tambaú, where C1, C2 and C3 represent moments 1, 2 and 3 for Tambaú cuisine, respectively, mentioned previously in Figure 5. Figure 8 shows the difference of distribution of MF values, especially for the moments in which the appliances that are in the kitchen are in operation (C3). In Figure 9, the MF values of the residence in Cabo Branco, if present in a similar interval of MF intensity, however there are many spurious values that exceed this interval, mainly in the living room. One of the factors that can explain this situation is the proximity of the living room with the kitchen, because it is a residence of dimensions much smaller than the one of Tambaú, of this form the use of electrical appliances present in the kitchen can influence the measurements of MF obtained in the room for a significant period of time.

Through accumulated probability analysis it is possible to define in which intervals of magnetic field strength each room of the dwellings is understood during the time of experiment. The cumulative probability curves are shown in Figures 10 and 11, for the residences in Cabo Branco and Tambaú, respectively.

In Figure 10, the MF intensity in the residence room in Cabo Branco is representative from 1.22 µT, and becomes high at 1.23 µT, and remains constant throughout the period with probability in around 90%. The intensity of the MF in the room starts at 1.20 µT and grows over time, when at 1.24 µT it has a probability of close to 60%; At 1.26 µT exceeds 80% and, above 1.26 µT, greater than 90%.

In the Tambaú residence, Figure 11, the MF intensities in the kitchen are presented in three moments. The first rises to 0.9 µT and remains constant over time reaching a probability close to 90%. The second shows 1.2 µT with intensity up to 90%. And the third starts at 1.6 µT and

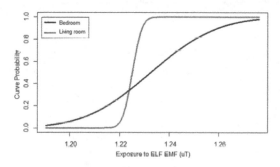

Figure 10. Cumulative probability—Cabo Branco.

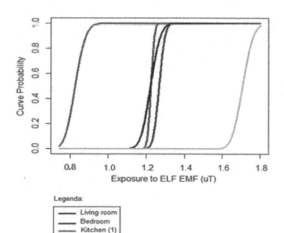

Figure 11. Cumulative probability—Tambaú.

grows up to 1.8 µT, where at 1.7 µT there is a 60% probability and at 1.8 µT, close to 90%. In the living room and in the bedroom the same intensification of the MF presented in the first moment in the kitchen.

4 CONCLUSION

A number of studies have been carried out internationally in research on the problems caused by extremely low frequency non-ionizing radiation to human health. The International Commission for the Protection of Non—Ionizing Radiation (ICNIRP) recommends that apartments submitted to extremely low frequency MF, in the range of 40–60 Hz, higher than 0.4 µT should be considered as residences highly exposed to non—ionizing radiation.

The measurements recorded throughout the experiment revealed values considered to be high and worrying regarding the occupational exposure of the human being inserted in this type of environment, since there are studies that emphasize the relationship between exposure to extremely low frequency MF and the emergence of Childhood leukemia (Ahlbom et al., 2000; Greenland et al., 2000).

REFERENCES

Ahlbom A, Day N, Feytching M, Roman E, Skinner J, Dockerty J, Linet M, McBride M, Michaelis J, Olsen JH, Tynes T, Verkasalo PK. 2000. A pooled analysis of magnetic fields and childhood leukemia. Br J Cancer 83(5):692–698.

Greenland S, Sheppard AR, Kaune WT, Poole C, Kelsh MA, 2000. A pooled analysis of magnetic fields, wire codes, and childhood leukemia. Childhood Leukemia-EMF Study Group. Epidemiology 11(6):624–634.

Grellier, J, 2014. Potencial health impacts of residential exposures to extremely low frequency magnetic fields in Europe. Environment International.

Hareuveny, R; Kandel, S, 2011. Exposure to 50 Hz magnetic fields in apartment buildings with indoor transformer stations in Israel. Journal of Exposure Science and Environmental Epidemiology.

Ilonen, K; Markkanen, A, 2008. Indoor Transformer Stations as Predictors of Residential ELF Magnetic Field Exposure. Bioelectromagnetics.

International Agency for Research on Cancer (IARC) (2002). Non-ionizing radiation, Part 1: Static and Extremely Low-Frequency (ELF) electric and magnetic fields. IARC monographs on the evaluation of carcinogenic risks to humans. Lyon: IARC Press. Vol. 80.

Kendel, S; Hareuveny, R, 2013. Magnetic field measurements near stand-alone transformer stations. Radiation Protection Dosimetry.

Petrucci, T.F, 2011. Radiações não ionizantes: causas, efeitos e prevenções.

Roosli, M; Jenni, D. 2011. Extremely low frequency magnetic field measurements in buildings with transformer stations in Switzerland. Science of the total Environment.

Ryu, Y, 2011. Quantitative Analysis of Factors Contributing to Urban Heat Island Intensity. Journal of applied metereology and climatology.

Santos, J. 2012. Campo térmico urbano e a sua relação com o uso e cobertura do solo em cidade tropical úmida. Revista Brasileira de Geografia Física.

Yitzhak, N, 2012; Time dependence of 50 Hz magnetic fields in apartment buildings with indoor transformer stations. Radiation Protection Dosimetry.

Zaryabova, V, 2013. Pilot study of extremely low frequency magnetic fields emitted by transformers in dwellings. Social aspects. Electromagnetic Biology and Medicine.

Occupational Safety and Hygiene V – Arezes et al. (Eds)
© 2017 Taylor & Francis Group, London, ISBN 978-1-138-05761-6

Experience in working shifts: The spouses/partners vision of shift workers and day workers

D. Costa, I.S. Silva & A. Veloso
School of Psychology, University of Minho, Braga, Portugal

ABSTRACT: Shift work, especially involving night work, weekends or highly valued periods of family and/or social point of view, may pose disadvantages for workers in various fields: health, organizational, family and social life. Despite the growing interest in this issue, the study of the impacts on workers' health has been emphasized in comparison to the impacts on social and/or family life. Besides, such study has favored the perspective of workers in face of third-party perspective. Bearing in mind these limitations, a research project was undertaken to evaluate the impact of shift work on family and social life alongside involving 515 participants (403 spouses/partners of shift workers and 112 spouses/partners of day workers). This paper presents and discusses the achieved results upon two open questions about the experience of working in shifts within this research project. Specifically, the answers of 178 spouses/partners of shift workers were compared with the answers of 28 spouses/partners of day workers. The results showed that according to the perspective of their spouses/partners, shift work is perceived as the most prejudicial to workers and their family and/or social life as compared to the conventional work schedule.

1 INTRODUCTION

Societies are temporally organized, largely by working hours executed by majority of citizens, since these schedules enable structuring other areas of people's life, such as family and social life (Fagan, 2001).

In the last decades, with the possibility of extending the operating time of companies up to 24 hours a day, 365 days a year, non standard work schedules have been increasing (Costa, 2003; Li *et al.*, 2014). Shift work, as a non standard work schedule, has been set as a way of work schedule of organization that allows the succession of different teams or shifts (Costa, 1997). According to the 6th European Survey of Working and Living Conditions, in 2015, 21% of the European Union workers worked in shifts and 19% performed night work (Eurofound, 2015).

This work schedule can embody advantages and disadvantages for the workers. The advantages are divided, mostly, in economic (e.g., shift allowance) and temporal reorganization (e.g., greater involvement of the father with his children when the mother works by shifts, increased free time, greater schedule flexibility) (Barnett *et al.*, 2007; Silva *et al.*, 2014). On the other hand, the disadvantages can cluster in three major areas: health, organizational and social/family life. At health level, cardiovascular, gastrointestinal and sleep problems have been found (Caruso, 2014; Costa, 2003; Prata *et al.*, 2013). In terms of organizational level, such problems have been mentioned that mainly affect the

security and the productivity in connection with the night schedule (Silva, 2012). At household and social level, the problems arise mainly due to the schedule incompatibility between the workers and their family unit along with the society, and are mainly focused on the conflicts within the marital relationships, parental and social interaction distress (Baker *et al.*, 2003; Handy, 2010). Therefore, the shift work that comprises of night work, weekend and/or highly valued periods regarding the family and/or social point of view, negatively impacts the family and social life.

Wilson *et al.* (2007) found evidences that working in a different schedule than the conventional generates greater family-work conflict. In the same direction, Simunic *et al.* (2012) concluded that conventional work schedule leads to least work-family conflict than the non-standard schedule.

Camerino *et al.* (2010) found evidences that the work schedules that involve nights and irregular schedules are more harmful for the work-family relationship than other schedules. They also found an association between the work-family conflict and the burnout, sleep and absenteeism.

Other authors (Kunst *et al.*, 2014; Mauno *et al.*, 2015) have equally found evidences that shift workers, compared with day workers, show a greater work-family conflict.

Regarding the marital relationships, Minnotte *et al.* (2015) found evidences that the work-family conflict and the family-work were negatively related to conjugal satisfaction.

On the other hand, regarding the impact of shift work on parental relationships, Rosebaum et al. (2009) found negative association between shift work and the workers' health, the conjugal relationship quality, parental interaction and family routines. The authors also found evidences that the children's behaviour was more impaired by the maternal work shift than the paternal.

Over the years, other authors, (e.g., Han et al., 2011; Gracia et al., 2016), have found negative associations between the different shift work systems and marital relationships and/or parental.

2 STUDY'S AIMS

The present study aims to help characterize the impacts that shift work has on family and social life of workers, comparing the reports of spouses/ partners of shift workers with the reports of spouses/partners of day workers. Particularly, the study presents the analysis of the participants' answers to two questions: one related to the impacts that working hours (standard vs shift work) have on parental relationships and other related with general opinions about the same schedules (the description of those questions will be resumed on the next item). The comments were made when the participants were answering a survey regarding a wider research project about the impacts of work schedules on family and social life.

Data used in this study stem from only the analysis of two open answer questions.

3 METHODOLOGY

403 spouses/partners of shift workers and 112 spouses/ partners of day workers with a nationwide geographical distribution participated in the research project. Note that all spouses/partners of the participants were Portuguese police employees. Professionals, who work in shifts, were allocated in the rotating systems that alternated between morning, evening and night shift, while police officers assigned by day work, laboured at a fixed rate and regular schedule.

Out of the 515 participants, 206 answered at least one of two open questions found in the survey, which represents an answer rate of 40%. It is worth noting that out of those 206 participants, 178 were spouses/partners of shift workers (group designated, further on, as "shift work"), whereas 28 were spouses/partners of day workers (group designated, further on, as "day work").

Data was collected through an online survey published and distributed nationally by unions of the "Public Security Police" (PSP) and the training department of the PSP National Direction. In each established contact with different organs, the

idea was reinforced several times that "the spouses/ partners of cops should be the participants". This idea was also highlighted in the surveys, as along with the assurance of confidentiality and anonymity.

The surveys were constituted by various issues of diverse nature, such as socio demographic data, professional and family situation of the participants, questions related to spouse/partner work hours and the impacts that it has on family and social life. The survey also contained two open answer questions. The first question was focused on parental relationships, and asked participants to give an example following an affirmative answer to the previous question "*If, in your view, your partner's contact with your child(ren) is low [because of your work schedule]- do you feel that this is detrimental to your child(ren) or affecting them in any way?*". On the other hand, the "non-structured" nature of the second question allows participants to make reference to every aspect that working schedules could entail for the worker and his/her family. As previous mentioned, the present study is based on the analysis of data collected in those two open answers of the questions through the content analysis (Amado, 2000).

Table 1 shows a brief characterization of those spouses/partners, bearing in mind the policeman's work schedule (i.e., spouses/partners of the participants).

Table 1. Socio demographic characterization, professional and family situation of the participants.

Variable	Shift Work Group (n = 178)		Day Work Group (n = 28)		Total (n = 206)	
	n	%	n	%	n	%
Gender						
Man	52	29.21	12	42.86	64	31.07
Woman	126	70.79	16	57.14	142	68.93
Age						
M (SD)**	40.53 (7.73)		42.21 (6.23)		40.76 (7.57)	
Status						
Single	9	5.06	0	0	9	4.37
Married	136	76.40	24	85.71	160	77.67
Union	28	15.73	3	10.71	31	15.05
Separated	5	2.81	1	3.57	6	2.91
Union Years						
M (SD)**	12.92 (7.58)		14.04 (6.57)		13.07 (7.46)	
Children's years						
Under 6 years	76	49.35	10	38.46	86	47.78
7 to 12 years	42	27.27	11	42.31	53	29.44
13 to 18 years	24	15.58	5	19.23	29	16.11
Over 18 years	12	7.79	0	0	12	6.67

*The N obtained can be different from the size of the sample (N = 206) due to the lack of values in some variables; **M (Mean), SD (Standard Deviation)

Most participants were women (68.93%), married (77.67%), with an average age of 40.76 years old (SD = 7.57). Out of the participants, who have children, over 47% have children under 6 years of age.

4 PRESENTATION OF THE RESULTS

The first question regarding the impacts that work schedule has on children obtained 137 answers,

Table 2. Frequency of obtained answers in the "negative aspects" and consequential subcategories in the open question concerning children.

Subcategories (n=151)*	Shift Work Group (n = 134)	Day Work Group (n = 17)
Coexistence time (n = 39) (e.g.: "They say, their dad has little time for them.")	35	4
Daily routines (n – 30) (e.g.: "My partner should be more present in monitoring (…) of daily routines (hygiene, etc.).")	27	3
School support/monitoring (n = 25) (e.g.: "Doesn't have time to monitor the school progress and help with homework when necessary.")	19	6
Children's development/growth (n = 18) (e.g.: "Is harmful for their growth and education.")	17	1
Leisure activities (n = 17) (e.g.: "Children want to play with their father and he has to work or rest.")	15	2
Longing the father/mother (n = 8) (e.g.: "Children miss the mother more and get sadder.")	7	1
Misunderstanding by children in relation to shift work (n = 7) (e.g.: "In a way, they always wonder "why our mother is not with us?")	7	—
Lack of patience of the police with their children (n = 4) (e.g.: "Doesn't have patience to help the son.")	4	—
Special occasions (n = 3) (e.g.: "Dad, you're never home for Christmas.")	3	—

*The n obtained in all subcategories is superior to the number of obtained answers (n = 136) due to the submission of two or more examples by some participants.

although one of them was considered invalid, as the participant didn't respond to what was asked. In turn, the second question regarding the suggestions/commentaries about the topic discussed

Table 3. Frequency responses obtained in three categories and consequential subcategories.

"Negatives Aspects" (n = 136)*

Subcategories	Shift Work Group (n = 128)	Day Work Group (n = 8)
Family life (n = 48) (e.g.: "In joke way, deep down, it is as if we were the lovers and the police his wife (…) not because they do not want to, but because they can't.")	46	2
Rigidity of hours (n = 25) (e.g.: "The government should tolerate working hours more flexible like many private companies.")	21	4
Health (n = 21) (e.g.: "It is very tiring, it takes him a lot of energy, often he can't fall asleep, against the uncontrolled alternating shifts of many years.")	21	—
Economical aspects (n = 19) (e.g.: "The shift work at least should be compensated monetarily, since it can't be extinguished because it is a necessary service 24 hours a day.")	17	2
Shift work status (n = 13) (e.g.: "Opting for greater differentiation (…) statutory (retirement age/IRS) on officers working on shifts and those working from 9 to 17h.")	13	—
Social life (n = 10) (e.g.: "In my husband's case, who works every Friday and Saturday night, except when he's on vacation, we can never organize interaction with friends and have a steady social life.")	10	—

"Positive Aspects" (n = 11)*

Subcategories	Shift Work Group (n = 0)	Day Work Group (n = 11)
Family life (n = 5) (e.g.: "We managed to combine very well working hours and family life.")	—	5
Social life (n = 4) (e.g.: "We have a much better social life.")	—	4
Security (n = 1) (e.g.: "No life danger.")	—	1
Economical aspects (n = 1) (e.g.: "Sometimes (months) makes more money.")	—	1

(Continued)

Table 3. (*Continued*).

"General Comments" (n = 21)*

Subcategories	Shift Work Group (n = 14)	Day Work Group (n = 7)
Other organizational aspects (n = 15) (e.g.: *"Distance between job and home affects family life."*)	10	5
Suggestions/opinions (n = 6) (e.g.: *"this concern is welcomed"*)	4	2

* The n obtained in all subcategories is superior to the number of obtained answers (n = 122) due to the submission of two or more examples by some participants.

obtained 122 answers. Altogether, 259 answers were analyzed.

In the first question, about the impacts of working hours in parental relationships, all answers addressed "negative aspects" (e.g., *"Children miss the mother more and get sadder."*). Therefore, a category was created with this designation. Once defined this category, subcategories were created that grouped the answers given by the participants according to different areas of impact on children (see Table 2). For each sub-category, excerpts of responses from participants are also presented that help to illustrate. It is worth noting that both participants, from the group "shift work" and the group "day work", only mentioned negative aspects, because the question was carried out accordingly.

As we can see in Table 2, the participants from the group "shift work" reported that interaction time between father/mother and children, the support and/or school monitoring, supervision on the children, daily routines, leisure activities and children development/growth were the most affected areas by shift work of their spouses/partners.

On the other hand, the participants of the group "day work" provided more examples related to interaction time between father/mother and children, as well as school support and/or monitoring.

In the analysis of the second question, regarding suggestions/comments about the working schedules, answers were found corresponding to three categories: "negative aspects", "positive aspects" and "general comments". The "negative aspects" were most reported by "shift work" group and the "positive aspects" were only reported by the "day work" group (see Table 3).

As shown in Table 3, the participants from the group "shift work" emphasized the negative aspects of the spouse/partner working schedule; the family life, health of shift workers, the schedule inflexibility and low economic compensation were mentioned as most affected areas by shift work.

In turn, the participants of the group "day work" emphasized the positive aspects of the standard working hours of the spouse/partner in family and social life.

5 DISCUSSION OF RESULTS AND CONCLUSIONS

All in all, the results are consistent with the literature. According to the perspective of the spouses/partners, the shift work entails more negative aspects for the family and social life of workers than conventional working hours. Besides family and social life, other aspects like health, economics or schedule inflexibility are cited as mark for the negative impacts of work schedules. Regarding the positive points, only spouses/partners of day workers found such outcomes.

Regarding negative impacts on workers health, only spouses/partners of shift workers found such issues, responding to previous studies (e.g., Caruso, 2014; Carneiro et al., 2015).

On the other hand, when it comes to family and social life, the results are consistent with previous studies (e.g., Camerino et al., 2010; Simunic et al., 2012) that found evidences that working on a different schedule other than the conventional can create bigger work-family conflict.

This study joins a minority of investigations that focus on the impact impacts of work hours on family and social life and highlight the necessity of carrying on the study of this issue. To conclude, the pioneer character of this study should be noted, given the shortage of investigations that cover the perspective of third party, instead of the workers themselves. In the future, for better understanding of the shift work issue, it's important to extend these opinions to other members of family unit and society.

REFERENCES

Amado, J. 2000. A técnica de análise de conteúdo. *Revista Referência* 5: 53–63.

Baker, A., Ferguson, S., & Dawson, D. 2003. The perceived value of time: Controls versus shiftworkers. *Time & Society* 12(1): 27–39.

Barnett, R.C. & Gareis, K.C. 2007. Shift work, parenting behaviors, and children's socioemotional well-being: A within-family study. *Journal of Family Issues* 28(6): 727–748.

Camerino, D., Sandri, M., Sartori, S., Conway, P.M., Campanini, P., & Costa, G. 2010. Shiftwork, work-family conflict among Italian nurses, and prevention efficacy. *Chronobiology International* 27(5): 1105–1123.

Carneiro, L. & Silva, I.S. 2015. Trabalho por turnos e suporte do contexto organizacional: Um estudo num centro hospitalar [Shift work and organizational support: A study in a hospital]. *International Journal on Working Conditions* 9: 142–160.

Caruso, C.C. 2014. Negative impacts of shitwork and long work hours. *Rehabilition Nursing* 39: 16–25.

Costa, G. 1997. The problem: Shiftwork. *Chronobiology International* 14(2): 89–98.

Costa, G. 2003. Shift work and occupational medicine: An overview. *Occupational Medicine* 53: 83–88.

European Foundation for the Improvement of Living and Working Conditions—Eurofound. 2015. *First findings: Sixth European Working Conditions Survey.* Publications Office of the European Union.

Fagan, C. 2001. Time, money and the gender order: Work orientations and working time preferences in Britain. *Gender, Work and Organization* 8(3): 239–266.

Gracia, P. & Kalmijn, M. 2016. Parents' family time and work schedules: The split-shift schedule in Spain. *Journal of Marriage and Family* 78(2): 401–415.

Han, W.J. & Fox, L.E. 2011. Parental work schedules and children's cognitive trajectories. *Journal of Marriage and Family* 73: 962–980.

Handy, J. 2010. Maintaining family life under shiftwork schedules: A case study of a New Zealand petrochemical plant. *New Zealand Journal of Psychology* 39(1): 29–37.

Kunst, J.R., Løset, G.K., Hosøy, D., Bjorvatn, B., Moen, B.E., Magerøy, N., et al. 2014. The relationship between shift work schedules and spillover in a sample of nurses. *International Journal of Occupational Safety and Ergonomics* 20(1): 139–147.

Li, J., Johnson, S.E., Han, W., Andrews, S., Kendall, G., Stradzins, L., et al. 2014. Parents' nonstandard work schedules and child well-being: A critical review of the literature. *Journal of Primary Prevention* 35: 53–73.

Mauno, S., Ruokolainen, M., & Kinnunen, U. 2015. Work–family conflict and enrichment from the perspective of psychosocial resources: Comparing Finnish healthcare workers by working schedules. *Applied Ergonomics* 48: 86–94.

Minnotte, K.L., Minnotte, M.C., & Bonstrom, J. 2015. Work– family conflicts and marital satisfaction among US workers: Does stress amplification matter? *Journal of Family and Economic Issues* 36(1): 21–33.

Prata, J. & Silva, I.S. 2013. Efeitos do trabalho em turnos na saúde e em dimensões do contexto social e organizacional: Um estudo na indústria eletrônica. *Revista Psicologia: Organizações e Trabalho* 13(2): 141–154.

Rosenbaum, E. & Morett, C.R. 2009. The effect of parents' joint work schedules on infants' behavior over the first two years of life: Evidence from the ECLSB. *Maternal and Child Health Journal* 13(6): 732–744.

Silva, I.S. 2012. *As condições de trabalho no trabalho por turnos. Conceitos, efeitos e intervenções [The working conditions in shift work. Concepts, effects and interventions].* Lisboa: Climepsi Editores.

Silva, I.S., Prata, J., & Ferreira, A.I. 2014. Horários de trabalho por turnos: Da avaliação dos efeitos às possibilidades de intervenção [Shiftwork schedules: From effect's evaluation to intervention possibilities]. *International Journal on Working Conditions* 7: 68–83.

Simunic, A. & Gregov, L. 2012. Conflict between work and family roles and satisfaction among nurses in diferent shift systems in Croatia: A questionnaire survey. *Arhiv Za Higijenu Rada i Toksikologiju* 63(2): 189–197.

Wilson, M.G., Polzer-Debruyne, A., Chen, S., & Fernandes, S. 2007. Shift work interventions for reduced work-family conflict. *Employee Relations* 29(2): 162–177.

Occupational Safety and Hygiene V – Arezes et al. (Eds)
© *2017 Taylor & Francis Group, London, ISBN 978-1-138-05761-6*

Total Productive Maintenance implementation. A way to improve working conditions

M.F. Valério
Faculdade de Ciências e Tecnologia, Universidade Nova de Lisboa, Portugal

I.L. Nunes
Faculdade de Ciências e Tecnologia, UNIDEMI, Universidade Nova de Lisboa, Lisboa, Portugal

ABSTRACT: Total Productive Maintenance (TPM) is a continuous improvement framework which includes the reduction of occupational accidents and work-related diseases as one of its goals, contributing to address companies' responsibility for the health and safety of workers. The implementation of a TPM program enables the organization to improve its performance and to manage its health and safety risks. Those benefits are achieved developing the Safety and Environment Management Pillar together with the other seven TPM pillars. The roles of Autonomous Maintenance, Focused Improvement and Education and Training pillars will be highlighted, relating them with prevention of occupational risks.

1   INTRODUCTION

Companies are embracing continuous improvement as a means to increase competitiveness while facing the increasing markets internationalization and an adverse economic conjuncture (Nunes 2015). One of the main concerns of managers is to improve productivity; however, unwary interventions to improve productivity may have a negative impact in the well-being of workers, thus resulting in hidden costs for companies.

Competition is continuously increasing and the priority given to safety and health is sometimes neglected by companies. As Goetsch notes this is the wrong option when trying to improve competitiveness and profitability and is also the wrong option from an ethical point of view (Goetsch 2011).

Ergonomics is a system-oriented approach aiming at simultaneously improve working conditions and obtain gains in productivity. Occupational safety management enables organizations to reduce the accident rate and, on the other hand, increase their competitiveness and financial performance (Fernández-Muñiz et al. 2009). Therefore, it is desirable that Ergonomics' and occupational safety principles be integrated on continuous improvement processes, such as Total Productive Maintenance (TPM) (Nunes et al. 2007) or Lean Six Sigma (Freitas et al. 2015; Nunes et al. 2012), in order to preserve (or even improve) workers' well-being while implementing measures to improve system's overall performance.

Implementing TPM is a pathway to create close relationships between good maintenance and higher productivity. Continuous improvement is a daily effort to reinforce the sense of ownership in all levels of the organization (Digalwar & Nayagam 2014).

The TPM's Safety and Environment Management pillar leads the organization to manage all the safety related issues (e.g., occupational safety and health risks, ergonomics risks, emerging risks or behavior-based safety). The present paper will address TPM as an approach to continuously improve working conditions, building safe and healthy workplaces, and, also, delivering competitive advantage between peers.

2   TOTAL PRODUCTIVE MAINTENANCE

2.1   *Origins and evolution*

TPM was born in Japan in the 1970s when manufacturers automated production lines and created the "just in time Supply Chain". This was the time when the idea of reduced or no stock was born.

However, unplanned stoppages were not compatible with this model. Therefore, unplanned stoppages turned from an unavoidable reality into an unacceptable event. This placed a big challenge which was smoothing the ups and downs of production lines' activity.

"This trend toward automation, combined with the trend toward just-in-time production, stimulated interest in improving maintenance management in the fabrication and assembly industries. This gave birth to a uniquely Japanese approach called Total Productive Maintenance, a form of productive maintenance involving all employees" (Suzuki 1994).

TPM started in the 1970's with a production-centered focus in Zero Failures, and based on five pillars. These pillars targeting the production departments are Focused Improvement, Autonomous Maintenance, Planned Maintenance, Education and Training, Early Management. In 1989, the TPM concept was extended to the entire organization involving everyone from top management to shop-floor operators and was based on eight pillars (the previous five pillars plus Quality Maintenance, Administrative and support, and Safety and Environment Management) focusing in Zero Losses. In 1997 the concepts of overall satisfaction, increased yield and costs reduction were introduced.

2.2 Key points

As proposed by Japan Institute of Plant Maintenance (JIPM) TPM should be in place in all areas and used by all people, building a culture that improves the efficiency of production systems and involves all departments in the implementation process, including development, sales and administration (Suzuki 1994). This means that all people—from top management to shop-floor workers—should be engaged.

Using a shop-floor approach, an organization that prevents every type of loss (by ensuring zero accidents, zero defects, and zero failures) was developed. Zero-loss activities are implemented based on small work teams.

As referred TPM programs consider eight pillars (or supporting activities) and has a 5S foundation, as shown on Figure 1.

5S is a structured management method whose objective is to implement specific actions to better organize and clean the workplace. It's based on teamwork and applied in sequence. The 5S refer to: Seiri (select), Seiton (sort), Seiso (shine), Seiketsu (standardize) and Shitsuke (sustain).

Figure 1.   TPM.

The teams, start the program with the first S doing the separation of what is from what isn't necessary, then applying the second S organize only the necessary items (visual inspection) in such a way that everything is easy to see and anyone can find any item and rearrange them. Cleaning the work area, it the aim of the third S, and in the fourth S the teams create the standards. Finally, in the fifth S the teams maintain the memory, carry audits and care to sustain all the achievements.

2.3 Implementation

TPM implementation follows a systematic approach in order to optimize plant and equipment performance while preserving perfect relationships between workers and equipment (Rajput & Jayaswal 2012). The implementation is performed in four phases broken down into twelve steps (Suzuki 1994) corresponding to the preparation phase (steps 1–5), introduction phase (step 6), implementation phase (steps 7–11) and consolidation phase (step 12) as shown in Table 1.

According to Campbell & Reyes-Picknell (2016) TPM is a team-based approach to organizing and working that has proven highly successful in a variety of industrial environments (e.g., automobile, light manufacturing, brewing and chemicals). The authors also note that TPM improves the way of

Table 1.   TPM implementation model.

Preparation phase	
Step 1	Formally announce the decision of introduce TPM
Step 2	Conduct introductory education and publicity campaign
Step 3	Develop a TPM promotion organization
Step 4	Establish basic TPM policy and goals
Step 5	Draft a master plan for implementing TPM
Introduction phase	
Step 6	TPM kick off
Step 7	Build a corporate constitution designed to maximize production effectiveness (develop 4 pillars: Focused Improvement, Autonomous Maintenance, Planned Maintenance, Education and Training)
Step 8	Build an early management system for new products and equipment
Step 9	Build a quality maintenance system
Step 10	Build an effective administration and support system
Step 11	Develop a system for managing health, safety and the environment
Consolidation phase	
Step 12	Sustain full TPM implementation and raise levels

working delivering good safety results, mentioning the small team as a distinctive feature of TPM and the importance of training stating that "extensive training is used to help maximize flexibility and capability while retaining safety."

3 TPM AND WORKING CONDITIONS

3.1 Safety and Environment Management pillar

Ensuring equipment reliability, preventing human error, and eliminating accidents and pollution are goals of any TPM program, through Safety and Environment Management pillar (SEM) (Setoyama 1994).

Therefore when implementing TPM, the SEM pillar leads to the creation, maintenance and improvement of safe and health workplaces (and safe and health surrounding areas) that will not be damaged by any process or procedures, achieving zero accident but also zero health damage and zero fires (Singh et al. 2013).

The SEM pillar development considers the following steps regarding Safety and Ergonomics:

Step 0 Hazard identification (considering accidents and ergonomic risk factors);
Step 1 Risk assessment;
Step 2 Defining optimal Safety and Ergonomic conditions;
Step 3 Medium level risk analysis and countermeasures implementation;
Step 4 Risk re-evaluation after countermeasures implementation;
Step 5 Establish conditions for zero accidents and zero ergonomic inadequacies;
Step 6 Control conditions for zero accidents and zero ergonomic inadequacies;
Step 7 Review and improve Safety and Ergonomic conditions.

Safety and Ergonomic risk analysis is used for finding risks by group work techniques and meetings. The groups require aware and creative people to identify the risks and suggest the best solutions. These people are normally selected among the safety committee members and maintenance engineering experts.

One of the solutions for finding safety risks is the adoption of checklists and procedures used in daily TPM visits to control any safety issues. Examples of these issues include (Jafari et al. 2014):

- Inspect wires and electrical connections;
- Sweep dust from the electrical circuits;
- Verify equipment rotating parts cleanliness and lubrication;
- Correct operation of couplings.

Highlighting the nexus between TPM and Safety, Setoyama (1994) refers some examples on how the other pillars of a full implemention TPM program also contribute to safety:

- Zero-failure and Zero-defect activities prevent faulty equipment, a common hazard source;
- Thorough application of 5S principles (as part of Autonomous Maintenance Pillar) eliminates leaks and spills and makes workplace clean, tidy and well-organized;
- Autonomous Maintenance and Focused Improvement eliminate unsafe areas;
- TPM-trained operators look after their own equipment and are better able to detect abnormalities early and deal with them promptly.

3.2 Pro-active safety in conjunction with Autonomous Maintenance

In the first step of the Autonomous Maintenance pillar, operators also have to deal with safety activities regarding zero-accident philosophy (making people understand the overriding importance of safety), training operators to anticipate risks, and avoiding having operators clean rotating parts or high places (Setoyama 1994).

Developing the Autonomous Maintenance Pillar through the seven steps requires thoroughness and continuity (Nakazato 1994):

Step 1 Initial cleaning;
Step 2 Eliminate sources of contamination and difficult to access places;
Step 3 Establish provisional cleaning and inspection standards;
Step 4 Perform general equipment inspection;
Step 5 Perform general process inspection;
Step 6 Systematize Autonomous Maintenance;
Step 7 Perform Autonomous Maintenance.

Setoyama (1994) describes the activities from the SEM Pillar that should be done during the implementation of Autonomous Maintenance Pillar to develop both pillars simultaneously:

Step 1 Detect and correct anything that might affect safety;
Step 2 Improve the cleaning and inspection activities taking in account all the safety related issues;
Step 3 Include safety procedures in provisional cleaning and inspection standards and establish individual safety routines;
Step 4 & 5 Develop equipment, safety and process competent people;
Step 6 & 7 Consolidate.

Safety features are pro-actively taken during step 2—eliminate sources of contamination prevents any future spillage or leakage occurrences contributing both to safety and health, while eliminating difficult to access places is also a big contribution to improve ergonomic conditions.

In step 3, the inclusion of safety procedures in provisional cleaning and inspection standards and the establishment of individual safety routines are a step in behavior-based safety direction.

In steps 4 & 5 all the teams are engaged in all the safety issues preventing also some of the emerging risks—the teams are co-responsible in the decisions.

After this work done in conjunction with Autonomous Maintenance, as a baseline, the SEM pillar can be fully implemented (from step 0 to step 7) as described before and benefit from the development achieved on the other pillars especially the Focused Improvement building a safety continuous improvement environment.

3.3 Zero mindset

From the formally announced decision to introduce TPM until the full TPM implementation is a journey where the zero idea is always present (refer to subsection 2.2). All the TPM process develops the psychological sense of belonging as described by Pinto (2013). A team that feels part of the organization will be proud of each activity, defending the space that is under his responsibility. Each team will be proud of his own zero activity board showing year after year the zeros achieved (different zeros in several parts of the area team) and the related work done. These zeros include "zero defects", "zero complaints", "zero accidents", "zero waste", "zero lacks in quality" (Jasiulewicz-Kaczmarek 2014).

3.4 Education and Training Pillar

As mentioned before TPM requires everyone's participation, meaning that the implementation of the Education and Training Pillar is very important. First, the identification of the competences required for each function should be done, defining the Matrix of competencies which categorizes the employees based on the following four levels:

1. Lack of theoretical and practical knowledge (need to learn);
2. Existence of theoretical knowledge and lack of practical knowledge (need to practice);
3. Existence of practical knowledge and lack of theoretical knowledge (can't teach others);
4. Existence of theoretical and practical knowledge (can teach others).

On-the-job training is a very important part of this process to achieve the expected results.

The assessment of the skill level of each team member has also to take in account Safety knowledge requirements regarding, for instance, area map, hazard map, risk assessment and safe working procedures (Borris 2006). Identically the skill level assessment should take in account the Ergonomics knowledge requirements regarding, for instance, ergonomic risk factors and ergonomic principles.

Identically every training activity, besides from the contents focused on technical aspects, must also include contents referring work safety (Jasiulewicz-Kaczmarek 2014) and workplace ergonomic setup.

3.5 Final remarks

Setoyama (1994) notes that, because of the interaction between the operator and the equipment, the operator becomes the main responsible for "his" equipment, cooperating for its correct and better use, resulting in tangible benefits—e.g., Productivity, Quality, Cost, Delivery, Safety, and Moral—and also in intangible benefits—e.g., company image, working environment, confidence, self-esteem, and bringing together maintenance and production.

The safety goals are achieved in different ways providing reduction of the global risk level, number of accidents and incidents, unsafe conditions and unsafe acts, number of injuries and ill health, and by incrementing the quantity and quality of ergonomic and healthy workplaces.

This vision is corroborated by the authors, which have developed a SEM pillar TPM implementation case study in an ice-cream production line (Nunes et al. 2007). Since this initial implementation, two more areas of zero incidents were achieved and maintained until today in the same company, increasing the morale of all the operators and teams involved.

4 CONCLUSIONS

This work discussed examples on how TPM pillars can contribute to keep the plants and equipment at the higher productive level through cooperation of all areas of organization, particularly from the Safety and Ergonomics standpoint.

Safety is "the maintenance of peace of mind" as Professor Shirose from JIPM used to say during the factory visits. Any TPM program has always to follow a strict focus on Safety to guarantee that each person is fully responsible for his equipment, safety included. In the TPM road map is mandatory that the management working together with operators in small teams brings them to be part of the solution, but the leaders should always be leaders with the example.

In the same lines, Ergonomics is key to ensure healthy and safety workplaces. The participation of management and teams in identifying the inadequacies and in proposing solutions contributes not only for the well-being but also to the overall performance of the company, namely by reducing the hidden costs resulting for instance, from absenteeism.

Finally, increasing the team's morale using the TPM program is one of the intangible benefits that become feasible because Safety and Ergonomics are present in all the implementation steps.

ACKNOWLEDGMENT

The study was funded by project Pest/EME/UI0667/2014.

REFERENCES

Borris, S., 2006. *TPM Proven strategies and techiques to keep equipment runnig at peak efficciency*, Mc-Graw-Hill Companies, Inc.

Campbell, J.D. & Reyes-Picknell, J. V., 2016. *Uptime: Strategies for Excellence in Maintenance Management* 3rd Ed., CRC Press.

Digalwar, A. & Nayagam, P., 2014. Implementation of Total Productive Maintenance in Manufacturing Industries: A Literature-Based Metadata Analysis. *IUP Journal of Operations Management*, 13(1), pp. 39–53.

Fernández-Muñiz, B., Montes-Peón, J.M. & Vázquez-Ordás, C.J., 2009. Relation between occupational safety management and firm performance. Safety Science, 47, 980–991.

Freitas, D., Nunes, V. & Nunes, I.L., 2015. Integrating Lean Six Sigma and Ergonomics—a case study P.G. Arezes P, Baptista J.S., Barroso M.P., Carneiro P., Cordeiro P., Costa N., Melo R., Miguel AS. (Ed.) *Occupational Safety and Hygiene III—Selected Extended and Revised Contributions from the International Symposium on Occupational Safety and Hygiene (SHO 2015)*, pp. 441–445.

Goetsch, D.L., 2011. Occupational Safety and Health for Technologists, Engineers and Managers, 7th Ed., Prentice Hall.

Jafari, M.M., Lotfi, R.S., Felegari H. & Ghavam A.H., 2014. The Role of Total Productive Maintenance (TPM) In Safety Improvement and Decreasing Incidents in Steel Industry. *The SIJ Transactions on Industrial, Financial & Business Management*, 2(6), pp. 278–283.

Jasiulewicz-Kaczmarek, M., 2014. Integrating Safety, Health and Environment (SHE) into the Autonomous Maintenance Activities. In *HCI International 2014 Posters*. Springer International Publishing, pp. 467–472.

Nakazato, K., 1994. Autonomous Maintenance. In *TPM in Process Industries*. CRC Press, pp. 87–143.

Nunes, I.L., 2015. Integration of Ergonomics and Lean Six Sigma. A model proposal. *Procedia Manufacturing (Proceedings of the AHFE 2015: 6th International Conference on Applied Human Factors and Ergonomics, Nevada-USA 26–30 Jul)*, 3, pp. 890–897.

Nunes, I.L., Gouveia N., Figueira S. & Cruz-Machado V., 2012. Integração da Ergonomia e da Segurança na Implementação Lean Six Sigma (Integration of Ergonomics and Safety in the implementation of Lean Six Sigma) C. Guedes Soares, A.P. Teixeira, C. Jacinto, ed. *Riscos, Segurança e Sustentabilidade (IV Encontro Nacional de Riscos, Segurança e Fiabilidade) C. Guedes Soares, A.P. Teixeira, C. Jacinto (Eds)*, 2, pp. 965–984.

Nunes, I.L., Costa, A.F., Baptista, A.F. & Valério, M.F., 2007. TPM e a Saúde e Segurança no Trabalho (TPM and the Health and Safety at Work), C. Guedes Soares, A.P. Teixeira, P. Antão (Ed.) *Riscos Públicos e Industrials*, 2, pp. 951–968.

Pinto, H.P., 2013. *TPM e desenvolvimento de pertença psicológica: análise de um caso*. Relatório de Projeto apresentado à Universidade de Aveiro para cumprimento dos requisitos necessários à obtenção do grau de Mestre em Gestão (2013). Universidade de Aveiro.

Rajput, H.S. & Jayaswal, P., 2012. A Total Productive Maintenance (TPM) Approach To Improve Overall Equipment Efficiency. *International Journal of Modern Engineering Research*, 2(6), pp. 4383–4386.

Setoyama, I., 1994. Building a Safe, Environmentally Friendly System. In *TPM in Process Industries*. CRC Press, pp. 323–349.

Singh, R., Gohil, A.M., Shah, D.B. & Desai, S., 2013. Total Productive Maintenance (TPM) implementation in a machine shop: A case study. *Procedia Engineering*, 51, pp. 592–599. http://dx.doi.org/10.1016/j.proeng.2013.01.084.

Suzuki, T., 1994. *TPM in Process Industries*, CRC Press.

Occupational Safety and Hygiene V – Arezes et al. (Eds)
© *2017 Taylor & Francis Group, London, ISBN 978-1-138-05761-6*

Emotional demands of physiotherapists activity: Influences on health

L.S. Costa
ESTESC-Coimbra Health School, Instituto Politécnico de Coimbra, Coimbra, Portugal

M. Santos
Faculdade de Psicologia e Ciências da Educação, Universidade do Porto, Porto, Portugal

ABSTRACT: The emotional demands are one of the psychosocial risk factors at work. Psychosocial risks at work are originated by working conditions, organizational and relational factors that may interact with the mental functioning and have impact on mental, physical and social health. The aim of this study is to identify the emotional demands that Portuguese physiotherapists are exposed to and relate them to their health. Results show that there are risks related to emotional demands and a correlation ($r = 0,410$, $p = 0,000$) between these risks and the health of physiotherapists. Physiotherapists are exposed to emotional demands such as, overwhelming moments, verbal aggression by the public and to have to provide an answer for the patients' suffering and difficulties. For most physiotherapists their health is moderately affected by work. A greater degree of discomfort caused by the emotional demands is associated with a health condition most affected by the work.

1 INTRODUCTION

Work-related psychosocial risks have been identified as one of the major contemporary challenges to health and safety at work (European Commission, 2010). The research followed this recognition in different ways, stating, for example, that psychosocial risks have acquired, in recent years, a greater relevance due to the evidences found in the relation between these risks and the increase of pathological processes in the workers (Gollac & Volkoff, 2000; Villalobos, 2004).

At work what makes a psychosocial health risk is not its consequences, but its cause, that is, the definition of psychosocial risk, are confined not by the costs in health, but through working conditions and organizational and relational factors. Thus, psychosocial risks are the risks that having an impact on mental, physical and social health, are originated by working conditions and organizational and relational factors and are susceptible to interact with mental functioning (Gollac & Bodier, 2011).

For the same authors, the psychosocial risk factors at work can be grouped into six dimensions: intensity and time of work; emotional demands; lack of autonomy; poor social relationships at work; ethical conflicts and job and work insecurity. Concisely, the intensity and working time factor includes subjection to rhythm constraints; unrealistic or unclear goals; polyvalence; interruption of activity, among others. Also includes, the duration and organization of working time, such as, for example, number of hours; night shift work and work-life balance. The emotional demands factor includes relationship with the public; contact with suffering; hide emotions and external violence. Autonomy comprises autonomy in the task; predictability of work and the possibility of anticipating it; monotony and boredom; use and increase of competences and work satisfaction. The social relationships factor, includes among others, recognition; social support; relationships with colleagues and with hierarchy; technical support received from superiors; appreciation of work; recognition by clients and the public, and internal violence (discrimination, harassment, etc.). Ethical conflicts comprises hindered quality and useless work. Job and work insecurity factor concerns job security, salary and career; work sustainability and change.

A European company survey of new and emerging risks (European Agency for Safety and Health at Work, 2010) shows that psychosocial risks are one of the greatest concerns to the health, social support and education sectors. Managers identify time pressure as the most important cause of psychosocial risk, followed by job insecurity and poor communication between management and workers. Among the different sectors, the biggest difference in levels of concern is dealing with difficult clients, patients, students, etc. The health sector is one of the most problematic sectors.

In fact, the provision of health care assistance is, more and more, an important pier for the societies, even due to the population aging and the existence of new pathologies and chronic conditions. However, within the health domain, the work is also almost a mission. There are many demands, particularly those which combine technical, human, ethical and even political and economic dimensions. It occurs in multiple contexts with several players (professionals from different backgrounds, patients, family members) from distinct origins and cultures (Rios, 2008). The health professionals, given the circumstances of their intervention, are exposed to a set of factors (of interaction, performance and occupation content) which are peculiar face working conditions.

Research shows the existence of psychosocial risks in these professionals, including roles overloading, many hours of work, teamwork conflicts, difficulties in work-family balance and lacking of material and human resources (Camelo, 2006; De Lange et al., 2004).

From a professional perspective health care practitioners develop a set of diversified activities with regard of the profession, the tasks, the places and type of intervention; nevertheless, for most of them, there is a set of circumstances more or less common: the pressure of having to deal, face to face, with the public, the responsibility for people, aggravated by the involvement in situations of disability, severe injuries, suffering (that they watch and inflict through the treatments, tests and procedures), physical pain and, sometimes, death. Consequently, their interactions with the patients are completely different from the interaction established by other occupational groups and they create working conditions far more demanding (Sauter & Murphy, 1995).

Since emotional demands are one of the psychosocial risk factors at work, the objectives of this study were: (1) To identify the emotional demands that Portuguese physiotherapists are subject to; (2) Relate the emotional demands to which they are exposed with the health of these workers.

2 METHODOLOGY

The INSAT (Work and Health Survey) (Barros-Duarte & Cunha, 2010; Barros-Duarte et al., 2013) was answered by 223 physiotherapists from different health care settings in Portugal. The survey is organized into different sets of questions or axis. A group of questions focuses on the work constraints and their effects on the workers: physical and environmental constraints; organizational constraints (work pace and autonomy and initiative); relational constraints (work relationships and contact with the public) and work characteristics. In these groups and for all the listed conditions, the worker is requested to indicate whether he/she is exposed

to them; for the affirmative answers the respondent shall indicate the degree of discomfort caused by such conditions. The effects of work on health may be seen through the question regarding the perception of how affected health is by the work and through a list of 24 health problems.

3 RESULTS

The sociodemographic characteristics of the 223 participants show that most physiotherapists are female (76,2%). The majority is single (54,4%) under 40 years of age (on average 33 ± 7,5) and work at the private sector (52,5%).

The conditions and characteristics of work, in the field of emotional demands, to which these physiotherapists are exposed, as well as the average of the discomfort caused by them are described in Table 1.

Physical therapists consider themselves to be more exposed to conditions and characteristics of work such as: the need to respond to the difficulties or suffering of other people (88.3%), having to bear the demands of the public (85.2%), compulsory learning (84.7%) and confrontation with tension

Table 1. Exposure to work conditions and characteristics and respective discomfort.

Work Conditions/ Characteristics	Exposure		Discomfort	
	n	%	Mean	Sd
Demands of the public	190	85,2	3,47	,943
Situations of tension in the relationships with the public	167	74,8	3,21	,930
Risk of verbal aggression of the public	143	64,1	2,91	1,076
Risk of physical aggression of the public	80	35,8	2,91	1,083
Respond to the difficulties/suffering of others	197	88,3	3,48	1,090
Always learn new things.	189	84,7	4,83	,481
With overwhelming moments	165	73,9	3,10	,973
Extremely varied	66	29,5	4,17	,985
Very complex	94	42,2	3,03	1,285
Little recognized by the colleagues	63	28,2	2,52	1,020
Little recognized by the leaders	121	54,3	2,03	,901
With which I'm not satisfied	109	48,9	1,96	1,057
In which I feel exploited	78	34,9	1,97	,946

in the relationships with the public (patients and families) (74.8%).

The perceived discomfort is, in average terms, higher for the circumstances of feeling dissatisfied with the work they perform (Mean = 1,96), in which they feel exploited (Mean = 1,97) and which is little recognized by the leaders (Mean = 2,03).

Table 2 describes the physiotherapists' perception of how much they consider their health to be affected by their work.

Most physiotherapists consider that the work they do affects their health (84.8%). On a scale of 1 to 5 the mean value of this perception is 3.29 (SD = 1,036).

The health problems most frequently mentioned by physical therapists (Table 3) are: back pain (72.2%), musculoskeletal (61.4%) and varicose veins (51.1%). The first two are considered to be caused by work by, respectively, 39% and 30.9% of the physical therapists who identify them. Varicose veins are reported to be more aggravated (33.2%) than caused by work (13.5%).

The correlation between the perception of how much health is affected by work and the discomfort with the emotional demands to which physiotherapists are exposed is positive and with statistically significant values (r = ,410, p = 0,000).

Table 2. Perception of health affected by work.

Health affected by work	n	%
Extremely	8	3,6
Very	40	17,9
Moderately	84	37,7
Little	57	25,6
Nothing	31	13,9

Table 3. Health problems identified and their relation to work.

Health problems	n	%	Caused by work	Aggravated by work
Skin	53	23,8	10,3	9,0
Respiratory	43	19,3	4,0	12,1
Musculoskeletal	137	61,4	30,9	28,3
Back pain	161	72,2	39,0	31,4
Headaches	67	30,0	8,1	14,3
Muscular Aches (Chronic)	47	21,1	12,1	7,6
Varicose veins	114	51,1	13,5	33,2
Numbness of limbs	50	22,4	7,6	13,9
Allergies	44	19,7	3,6	5,8
Nervous	18	8,1	3,6	4,0
Sleep	45	20,2	8,1	8,5

4 DISCUSSION

In this study, was questioned the perception of physiotherapists regarding the conditions and characteristics of their work related to emotional demands, in order to verify at which ones are more exposed to and which cause them more discomfort. Subsequently related them to the health problems that physiotherapists claim to have.

According to the results obtained, the emotional demands that the physiotherapists are exposed comprise working conditions such as contact with the public and some characteristics of the work. Although working in different settings, the constraints associated to the demands of the users they look to, and the tension involved, are something that physiotherapists have in common. Exposure to verbal aggression by the public, as well as the need to respond to their difficulties or suffering, also form part of their work activity.

On the other hand they have a work activity where they have to mandatorily learn new things, which is little recognized by their leaders and comprises overwhelming moments. That's a work activity where they feel exploited and with which they feel unsatisfied.

For these physiotherapists emotional demands emerge from the ongoing contact with patients, from dealing with the tensions and demands of them, from the difficulties and suffering of these patients and also from issues related to violence by them. Therefore it can be stated that emotional requirements are a psychosocial risk factor for these workers. On the other hand the presence of work characteristics such as overwhelming moments may became as an increased conditions for the emotional demands with which they have to deal. In fact, studies with Portuguese nurses reveal that the main factors of pressure at work correspond to situations of dealing with clients and overwork (Silva & Gomes, 2009; Gomes et al., 2009).

The results obtained in this study meet with a professional activity that has as one of its objectives the recovery of the physical capacity of people with orthopedic or neurological problems. Likewise, they will meet the continuous contact with people (users and families) with hard and serious clinical conditions and often with few possibilities of recovery. Therefore, also, probably with a degree of demand, and care at the relational level, which will indorse the occurrence of pressure or tensions. In clinical settings, patients have difficulty caring for themselves, are insecure and anxious. Thus, the interaction established between the professional and the user is, not infrequently, loaded with strong emotions (Dias et al., 2010).

For most physical therapists their health is affected by work, on average, moderately. Health professionals are often exposed to labor constraints that may affect their physical and mental health, and it is assumed that work characteristics lead to illness and injury (Eriksen et al., 2006; Fiabane et al., 2013). On the other hand, the demands of work seem to be the most important psychosocial factor with clear associations with subjective health complaints and need for recovery after the workday (Janssens et al., 2010). Other studies show that absenteeism due to illness is associated with tensions at work and with their psychosocial environment (Eriksen et al., 2006).

If, what has been analyzed so far, concerning emotional demands that physiotherapists are exposed to is a set of more or less predictable aspects regarding the work of these professionals, the analysis of the data concerning the discomfort that the physiotherapists perceive with it, allows to deepen its possible influence on their health and well-being.

All the constraints related to emotional demands, to which physiotherapists are exposed, are stated by them as causing discomfort. That discomfort is perceived as being greater regarding some characteristics of their work, such as: being unsatisfied with the work; a work in which they feel exploited and which is slight recognized by the leaders. The greater discomfort is, also, reported for work conditions such as, for example, being subject to verbal and physical aggression by the public. Thus, we can consider these aspects as possible risks to the health of physiotherapists.

Work in the health area exposes workers to many occupational stressors, with a particular focus on close contact with users, which can mobilize emotions and conflicts, making their workers particularly susceptible to psychic suffering and leading to work-related diseases (Rios, 2008).

In addition, the analysis carried out allows us to perceive that a greater degree of discomfort caused by the emotional demands to which the physiotherapists are subjected is associated with the perception of a state of health more affected by the work. That is, the discomfort is correlated in a positive way with the physiotherapist's perception of their health. So, not only these workers perceive that their health is affected by the work they do, as one finds an association between this perception and the discomfort that they refer by the fact they are exposed to emotional demands.

The health problems that most physiotherapists refer to are back pain, musculoskeletal and varicose veins. These problems can be associated with the physical efforts and painful postures that the physiotherapist's activity requires, but also with the strain caused by the demands they are placed upon in contact with their patients (and/or their relatives). Studies in general populations of workers (Eriksen et al., 2006) show that the combination of high physical and psychological stress is associated with a substantial increase in the risk of episodes of back pain (especially lumbar).

5 CONCLUSIONS

The emotional demands that most disturb the physiotherapists are: low recognition of their work, by the leaders; confrontation with pressure in the relationships with the patients and their families; existence of overwhelming moments at work; feel exploited at work; exposure to verbal aggression by the public and the need to respond to people's suffering or difficulties.

For most physiotherapists their health is affected by work, on average, moderately.

A greater degree of discomfort caused by the emotional demands to which they are subjected is associated with a health condition more affected by work.

REFERENCES

Barros-Duarte, C., & Cunha, L. 2010. INSAT2010—Inquérito Saúde e Trabalho: outras questões, novas relações, *Laboreal*, 6(2), 19–26, http://laboreal.up.pt/revista/artigo.php?id=48u56oTV6582234;5252:5:5292.

Barros, C., Carnide, F., Cunha, L., Santos, M., & Silva, C. 2015. Will I be able to do my work at 60? An analysis of working conditions that hinder active ageing. *Work*, 51(3), 579–90. doi: 10.3233/WOR–152011.

Camelo, S. 2006. Riscos Psicossociais Relacionados ao Estresse no Trabalho das Equipes de Saúde da Família e Estratégias de Gerenciamento (Tese de Doutoramento não Publicada). Escola de Enfermagem de Ribeirão Preto, Universidade de S. Paulo.

De Lange, A., Taris, T.W., Kompier, M.A., Houtman, I.L., & Bongers, P.M. 2004. The relationships between work characteristics and mental health: Examining normal, reversed and reciprocal relationships in a 4-wave study. *Work & Stress*, 18, 149–166.

Dias, S., Queirós, C., & Carlotto, M. 2010. Síndrome de burnout e fatores associados em profissionais da área da saúde: um estudo comparativo entre Brasil e Portugal, *Aletheia*, 32, 4–21.

Eriksen, H., Ihlebæk, C., Jansen, J., & Burdorf, A. 2006. The Relations Between Psychosocial Factors at Work and Health Status Among Workers in Home Care Organizations. *International Journal of Behavioral Medicine*, 13(3), 183–192.

European Agency for Safety and Health at Work. 2010. *European Survey of Enterprises on New and Emerging Risks: Managing safety and health at work*. Luxembourg: Publications Office of the European Union.

European Commission. 2010. *Investing in well-being at work—Addressing psychological risks in times of change*. Luxembourg: Publications Office of the European Union.

Fiabane, E., Giorgi, I., Sguazzin, C., & Argentero, P. 2013. Work engagement and occupational stress in nurses and other healthcare workers: the role of organizational and personal factors. *Journal of Clinical Nursing*, 22, 2614–2624, doi: 10.1111/jocn.12084.

Gollac, M., & Bodier, M. 2011. *Mesurer les facteurs psychosociaux de risque au travail pour les maîtriser* (Relatório do Collège d'Expertise sur le Suivi des Risques Psychosociaux au Travail), Retrived from: Collège d'Expertise sur le Suivi des Risques Psychosociaux au Travail: http://www.college-risquespsychosociaux-travail.fr/rapport-final,fr,8,59.cfm.pdf.

Gollac, M., & Volkoff, S. 2000. *Les conditions de travail*. Paris: Editions La Découverte.

Gomes, A.R., Cruz, J.F., & Cabanelas, S. 2009. Estresse ocupacional em profissionais de saúde: um estudo com enfermeiros portugueses. *Psicologia: Teoria e Pesquisa*, 25(3), 307–318.

Janssens, H., Clays, E., De Clercq, B., Casini, A. De Bacquer, D., Kittel, F., & Braeckman, L. 2014. The relation between psychosocial risk factors and cause-specific long-term sickness absence. *European Journal of Public Health*, 24(3), 428–433.

Rios, I. 2008. Humanização e Ambiente de Trabalho na Visão de Profissionais da Saúde. *Saúde e Sociedade*, 17(4), 151–160.

Sauter, S. & Murphy, L. (Eds.). 1995. *Organizational Risk Factors for Job Stress*. New York: John Wiley & Sons.

Silva, M., & Gomes, A.R. Stress ocupacional em profissionais de saúde: um estudo com médicos e enfermeiros portugueses. *Estudos de Psicologia*, 14(3), 239–248.

Villalobos, G. 2004. Vigilancia epidemiológica de los fatores psicosociales. Aproximación conceptual y valorativa. *Ciencia & Trabajo*, 6(14), 197–201.

Occupational Safety and Hygiene V – Arezes et al. (Eds)
© *2017 Taylor & Francis Group, London, ISBN 978-1-138-05761-6*

Exposure to pollutants in nightlife establishments

Vitor Custódio, João Paulo Figueiredo & Ana Ferreira
Instituto Politécnico de Coimbra, Coimbra, Portugal

ABSTRACT: Air quality and its effect on human health are one of the major problems existing in present society. The present study sought to assess people exposure to pollutants at nightclubs and find out if the pollutants in these sites have negative influences on human health due to some kind of lack with regard to the level of indoor air quality. It was collected date through measurements of concentration levels of air pollutants, in particular of Volatile Organic Compounds (VOC's), Carbon Dioxide (CO_2), Carbon Monoxide (CO), Formaldehyde (CH_2O) and particulate matter (PM_{10} and $PM_{2.5}$) of 4 nightlife establishments. It was found by the results obtained that although most of the pollutants studied were in accordance with the legislation, one of them (formaldehyde) exceeds the protection threshold with a difference statistically significant. Therefore, it is impossible to tell a good air quality inside the establishments studied.

1 INTRODUCTION

The concerns associated with the effects of air quality on public health generally take into account air pollution outside the buildings. Although in recent decades one of the great successes of the Community policy on the environment has been improving the quality of the air, showing that it is possible to decouple economic growth from environmental degradation, there are problems that persist and that must be addressed. These problems are mainly in the large population outbreaks, where is a growing development of consumption and a increased air quality degradation, thus representing a focus problematic if we consider that it is in these places that most of the population resides (APA, 2009; Ferreira, A. & Cardoso, M. 2013ab); Nakagawa, L., & Comarú, F., & Trigoso, 2010).

Taking into account that a large part of the population spends about 80–90% of their time inside buildings, these spaces become important sources of risk of air pollution. Problems related to indoor air pollution can affect different types of buildings, including homes, schools, offices, health care and other public and commercial establishments. Indoor Air Quality (IAQ) is a factor of great importance in the workplace, since it can affect the health, comfort, well-being and productivity of the occupants of a building. It is also a major factor in schools, since children are considered more susceptible to high range of polluting sources that exist in these locations. It is also noted that the occupancy rate in schools is quite higher to that which exists in service buildings Where as pollution levels in indoor spaces are much higher than in outdoor spaces and even the time usually spent by individuals inside buildings, it can be said that daily exposure to air pollutants results in large part from the inhalation of indoor air (Ferreira, A. & Cardoso, M. 2014).

At present, there is a growing demand for night entertainment establishments to fill the free time, explaining this trend by the need that people have to seek a refuge from his professional and personal life, looking for moments of relaxation, conviviality and fun. Due to this significant increase arises the concern with the quality of the spaces intended for the nightlife, being the main concern the indoor air quality. The spaces intended for the nightlife venues are more conducive to reveal lower levels of indoor air quality, which consequently present a greater risk to the health of its occupants. These types of situations are caused mainly by the huge concentration of people in a particular closed space with ventilation conditions, sometimes insufficient, and the type of materials used in the construction of buildings. According to a study conducted in the city of Braga in relation to the air pollution level caused by cigarette smoke in workplaces, it was verified that the concentrations of pollutants in nightlife venues were about 100 times higher than the concentration of pollutants observed in spaces intended for the restoration (Precious, J., Calheiros, J., López, M., & Ariza, C. 2009).

It then becomes necessary to know which air pollutants have major influence on indoor air quality.

Some air pollutants that are abroad can also be found in indoor spaces, as is the case of carbon monoxide, carbon dioxide, formaldehyde, particulates and volatile organic compounds. The particles (PM, «Particulate Matter») are identified as the most harmful air pollutant for human health in Europe (Ferreira, A. & Cardoso, M. 2013ab).

Some of the indoor air pollutants come from, in addition to the outside air, the human body, from the place over occupation, deficiencies in the ventilation system, smoke tobacco, fibre emission from building materials (asbestos, rock wool, glass wool), furniture, the use of plastics and synthetic products (paints and varnishes), the presence of carpets and curtains, gas or particles generated by the burning of fuels, chemicals and allergens (Verdelhos, M. 2011; Ferreira, A. & Cardoso, M. 2013a, b).

Various volatile organic compounds such as formaldehyde, also contribute to indoor air contamination, and may be released during use and storage of cleaning products. For an efficient control of indoor air quality must be taken into account: outside air, clean materials, mold and dampness, Heating Systems, Ventilation and Air Conditioning (HVAC) and human activities. To minimize the chemical agents present inside of buildings are necessary measures and solutions such as changes in the habits of the occupants, replacing some materials used in decoration and products used in cleaning, an adjustment of the rates of ventilation of the indoor spaces, a proper sizing of HVAC systems, use of products, materials and low-polluting equipment, proper location of vents in buildings, away from sources of pollution outside and smoking ban in indoor spaces. Exposure to Environmental Tobacco Smoke (ETS)-also called «passive smoking»—continues to be responsible for excessive morbidity and mortality in the European Union, with significant costs to society. The FTA is a complex mixture consisting of over 4000 chemicals, including air pollutants such as CO, CO_2, $PM_{2.5}$, PM_{10}, CH_2O and VOC's and more than 50 known carcinogens and many toxic agents (Verdelhos, M. 2011).

The present study had as objective to evaluate the quality of indoor air in nightlife establishments, taking into account the air pollutants that have more influence in determining the quality of the air in these places, establishing a relationship between the values obtained and the values recorded in the applicable legislation, and even compare the values obtained in the places where smoking and non-smoking are allowed. It was intended to also realize if there is any relationship between symptoms perceived by occupants of this type of premises and the effects caused by pollutants.

2 MATERIAL AND METHODS

The study was the level II (Analytical), depending on the type of observational prospective cohort study.

The study was carried out in 4 nightclubs in the municipality of Coimbra, and in 2 of them was not allowed to smoke inside, and the remaining 2 were allowed to smoke inside.

The target population were considered 25 occupants of establishments, 7 employees and 18 customers present at the time of the assessments.

For collecting date, 3 visits were carried out at different times to each facility, each of which was subjected to an analytical review, where were used 4 specific equipment of indoor air quality assessment. The environmental parameters evaluated in this study were the Temperature (T°), the relative humidity (Hr), carbon monoxide (CO), carbon dioxide (CO_2), the $PM_{2.5}$, PM_{10}, Volatile Organic Compounds (VOC's) and formaldehyde (CH_2O).

Date collection was carried out through the following facilities: Lighthouse 3016 Handheld IAQ (serial number: 110144012)—collection of quantitative date of $PM_{2.5}$ and PM_{10}; Photovac Inc. 2020 Plus, (serial number: PPXN0015)—count date collection of OVC's; PPM Formaldemeter TM (serial number: F5552)—collection of date of CH_2; Q-Trak Plus IAQ Monitor TM Model8552/8554, (serial number: 8554-01061006)—date collection CO, CO_2, T and Hr.

During the evaluations some procedures were taken into account, wherein assessments were made with a spacing of at least 2 to 3 meters from the corners of walls and windows, with a distance considered in relation to ventilation and to a height of 1.5 meters above ground level.

Also a questionnaire was carried out to workers and the few occupants of the establishments, in order to assess their perception regarding the indoor air quality in relation to the place, as well as their state of health. The survey was conducted in order to characterize the respondents with an assessment of their health condition compared to symptoms caused by pollutants.

Using the Ordinance No. 353-A/2013, of 4 December were compared the reference values set with the date collected in the process of evaluation of indoor air quality (Portaria n° 353-A/13 de 4 de dezembro; Decreto-Lei n° 118/13 de 20 de agosto).

The statistical analysis of the collected date was performed using the Statistical Software IBM SPSS Statistics 22.0 version for Windows. For evaluation, it was used a simple descriptive analysis of measures of location, dispersion and frequency, the Student's t test for independent samples and one sample and the Kruskal-Wallis test. At the level of rejection of the null hypothesis with respect to the statistical interpretation it was assumed a confidence level of 95% for a random error less than or equal to 5%. All date collected were used exclusively for statistical analysis with academic and scientific purpose, without any economic or commercial interest, being guaranteed anonymity and confidentiality with regard to information gathering.

3 RESULTS AND DISCUSSION

Was the comparison of the average levels of air pollutants for each type of establishment.

A comparison was made between the date obtained from the polluting compounds as the typology of the site allow smoking or not allow smoking. It was observed that there are no significant differences for the compounds CO, CO_2, $PM_{2.5}$ and PM_{10}, but found significant differences at the level of CH_2O and VOC's. Note that the amount of CO_2 was higher in the establishment where it was not allowed to smoke.

According to the Table 2 it can be concluded that almost all values (VOC's, $PM_{2.5}$, CO and CO_2) are significantly below the threshold ($p < 0.05$) except the value of PM_{10} that met slightly above this value in a non-significant way ($p > 0.05$), and the value of CH_2O find significantly above the reference value ($p < 0.05$). It was then compare the average levels of air pollutants depending on the time of measurement.

By analyzing the Table 3, there were no significant differences in relation to atmospheric pol-

Table 1. Comparison of the results of the average levels of air pollutants.

n = 6	Mean	SD	t	gl	p
CO Smoker	6.43	1.38	−0.23	7.79	0.82
Non-smoker	6.58	0.76			
CO_2 Smoker	418.17	84.47	−0.40	10	0.69
Non-smoker	442.33	118.22			
CH_2O Smoker	1.93	0.95	2.80	5.66	0.03
Non-smoker	0.80	0.24			
VOC's Smoker	5.25	0.40	2.61	10	0.02
Non-smoker	4.76	0.20			
PM_{10} Smoker	53.73	11.02	0.94	10	0.36
Non-smoker	47.91	10.36			
$PM_{2.5}$ Smoker	21.43	4.63	1.43	10	0.18
Non-smoker	18.21	2.94			

Student's test for independent samples.

Table 2. Comparison of the results of air pollutants with the legislated benchmark.

	Av.	SD	Ref. value	Av. Diff.	p	Parameter
CO	12	6.5	1.0	10	−3.4	0.001
CO_2	12	430.2	98.7	2250	−1819.7	0.001
CH_2O	12	1.3	0.8	0.08	1.2	0.001
VOC's	12	5.0	0.39	600	−594.9	0.001
PM_{10}	12	50.8	10.6	50	0.8	0.793
$PM_{2.5}$	12	19.8	4.0	25	−5.1	0.001

Student's test for sample 1.

Table 3. Comparison of the average levels of air pollutants depending on the time of measurement.

	Time of measurement	n	Av.	SD	p
CO	20h00 m	4	6.375	1.345	
	23h00 m	4	6.400	0.920	0.874
	1h30 m	4	6.750	1.190	
	Total	12	6.508	1.069	
CO_2	20h00 m	4	379.00	90.200	
	23h00 m	4	413.00	63.482	0.174
	1h30 m	4	498.75	115.981	
	Total	12	430.25	98.774	
CH_2O	20h00 m	4	1.108	0.814	
	23h00 m	4	1.383	0.918	0.492
	1h30 m	4	1.625	1.098	
	Total	12	1.372	0.888	
VOC's	20h00 m	4	4.800	0.163	
	23h00 m	4	5.000	0.374	0.355
	1h30 m	4	5.225	0.538	
	Total	12	5.008	0.396	
PM_{10}	20h00 m	4	47.300	10.463	
	23h00 m	4	51.225	11.076	0.491
	1h30 m	4	53.950	12.376	
	Total	12	50.825	10.640	
$PM_{2.5}$	20h00 m	4	17.958	3.242	
	23h00 m	4	19.935	4.027	0.418
	1h30 m	4	21.578	5.008	
	Total	12	19.823	4.064	

Test: Kruskal-Wallis.

lutants between different times of measurement ($p > 0.05$). Concerning the questionnaires carried out to workers and the few occupants of the establishments used the results presented were obtained. The total sample corresponds to 25 persons (18 clients and 7 employees), 18 of which are male while 7 are female. The average age of respondents is 31.56 years being the minimum value of 18 years and the maximum value of 58 years.

The following graphic represents the perception of air quality on the part of the target population in relation to places of study.

▪ Bad (4%)	
▪ Reasonable (16%)	
▪ Good (72%)	
▪ Very Good (8%)	

Graphic 1. "How would you rate the quality of the air in this place?".

Table 4. "Feel difficulty breathing in this place?".

	n	%
Yes	2	8
No	23	92
Total	25	100

Only 4% of the target population found that indoor air quality was "Bad", while 16% considered that it was "reasonable", 72% considered that the air quality was "Good".

It was possible to establish a relationship between the target population that evaluated the quality of air and the difficulty they felt breathing in those same local. Expecting that higher the air quality rating given by the respondents, smaller the breathing difficulty felt by them, it was observed that 80% of the subjects rated the air quality as "Good" and "Very Good", in the places where they were, and 92% felt that didn't have trouble breathing in these same locations. With respect to question number 9 of the questionnaire "Mark with an (X) how often you feel the following symptoms mentioned " it was found that the effect of pollutants to which the target population is exposed to is low, and that most of the symptoms never or rarely were felt by the population under study. By analysing the date obtained symptoms that showed up in greater numbers were headache, throat irritation and coughing. With regard to headache, has been mentioned as a symptom occurred occasionally for 50% of the employees, despite only 5.88% of customers mention the same symptom. The irritation in the throat is represented by occur in 50% of workers, while representing only 17.65% of the symptoms perceived by customers. Finally, the symptom described most often was the cough, and was mentioned by 75% of workers and 41.17% of customers.

With regard to Table 5, 60% of the target population showed the same symptoms on and off the premises, while 40% stated that when the senses within symptoms improve.

Table 5. "Outside of this site usually feel some of these symptoms?".

	n	%
Yes	15	60
No	10	40
Total	25	100

After analysis of the obtained results, it was found that the concentrations of pollutants were higher in establishments where smoking was allowed instead of establishments where smoking was not allowed, in particular at the level of $PM_{2.5}$, PM_{10}, CH_2O and VOC's. An explanatory hypothesis may lie in the fact that these pollutants are part of the composition of Environmental Tobacco Smoke (ETS). The VOC's and CH_2O originate essentially in constituent materials of buildings, as well as in cleaning products, paint and coatings additives such as carpets and wallpapers. As regards the resulting particles of tobacco smoke, are the $PM_{2.5}$ that most stress and producing the worst consequences to human health because they can penetrate deep into the lungs, while the PM_{10} contributing to the appearance/worsening of asthma, chronic bronchitis and respiratory tract infections. The particulate matter comes from the gas emission from automobiles and the action of the wind on the ground (Verdelhos, M. 2011; Adam, C.A., & Aciole, N.D.G. 2012; Ferreira, A. & Cardoso, M. 2013; Ferreira, A. & Cardoso, M. 2014).

On the other hand, higher values were found in establishments where smoking was not permitted with respect to the CO and CO_2 (pollutants that also make up the FTA). This may be due to the fact that the time of the assessments meet a larger number of persons within the establishments where smoking was not allowed (freeing a greater percentage of CO_2) and these same establishments are in a higher traffic flow zone (CO is a colorless and odorless, extremely toxic, resulting essentially from the greenhouse gas emissions from automobiles) (Verdelhos, M. 2011).

Regarding the different hours of measuring air pollutants, it was not recorded a statistically significant difference between them, and may be due mainly to the fact that these establishments have a regular stream of customers, and there was not a very large fluctuation in the number of people inside the properties between different times, ie, the rate of occupation and ventilation of establishments was constant.

To compare, in full, the results of the locations analyzed and the reference values, it was observed that although some pollutants were within the reference values, others were not. It should be noted that almost all values (VOC's, $PM_{2.5}$, CO and CO_2) were significantly below the threshold ($p < 0.05$) except the value of PM_{10} was slightly above that of a non-significant way ($p > 0.05$), and the value of CH_2O that was significantly above the reference value ($p < 0.05$).

Considering the results for the concentrations of air pollutants, agents could not say that the quality of the air inside the establishments studied is good, because although most air pollutants studied are in accordance with the legislation, one of them exceed the threshold of protection.

As regards the questionnaires it is important to point out that the reduced sample do not allow to obtain significant results.

The target population have the perception that the establishments studied have a good air quality.

Regarding the symptoms described by the target population and the analysis of the permanence of the same out of establishments studied, one can say that the symptoms are not related to the study establishments (most frequently described symptoms -headache, throat irritation and coughing- are common to many pathologies).

Although most air pollutants present significantly below the filing of reference ($p < 0.05$), a pollutant presents significantly above the filing of reference ($p < 0.05$) and to maintain good indoor air quality is necessary an adjustment of ventilation rates of establishments, as well as a maintenance HVAC systems regularly.

4 CONCLUSIONS

In today's society we need, increasingly, to have special regard to indoor air quality, as the places most frequented by the population are inside buildings and adjacent problems are huge risk factors for human health.

Two of the studied air pollutants, including formaldehydes and PM_{10} are not in accordance with applicable law, adversely affecting the air quality. It should be noted that other air pollutants, including VOC's, $PM_{2.5}$, CO and CO_2 are in accordance with the applicable legislation.

As it is, mainly, through the HVAC systems that you can get an improvement in indoor air quality, it is important to note that the establishment must provide an efficient HVAC system, as well as a rigorous and regular maintenance.

Considering the small number of establishments analyzed in the present study, it would be important to replicate the study in more places, in order to note this same issue. It should also replicate the number of assessments carried out for a longer period of time and at different times of the year to get a more meaningful sample. The same applies to the questionnaires.

In conclusion, the indoor air quality is an important issue for environmental health, in particular to the environmental health technician, which plays a role of the utmost importance in this field, in order to ensure good air quality, contributing to the health and well-being of the population.

REFERENCES

Adam, C.A., & Aciole, N.D.G. (2012). Formaldehyde in schools: a review. New chemistry, 35 (10), 2025–2039. http://doi.org/10.1590/S0100-40422012001000024

Agência Portuguesa do Ambiente. (2009). Qualidade do Ar em Espaços Interiores: Um Guia Técnico, 1–56.

Ferreira, A.M.C., & Cardoso, S.M. (2013ab). Estudo exploratório da qualidade do ar em escolas de educação básica, Coimbra, Portugal, 47(6), 1059–1068. http://doi.org/10.1590/S0034-8910.2013047004810

Ferreira, A.M.C., & Cardoso, M. (2013ab). Qualidade do ar interior e saúde em escolas localizadas em freguesias predominantemente urbanas, rurais e mediamente urbanas. Revista Brasileira de Geografia Médica e da Saúde—http://www.seer.ufu.br/index.php/hygeia. Hygeia 9(17): 95 115, Dez/2013, ISSN: 1980-1726.

Ferreira, A.M.C., & Cardoso, M. (2014). Indoor air quality and Health in schools. Journal Brasileiro de Pneumologia. 2014;40(3):259–268. http://doi.org/10.1590/S1806-37132014000300009.

Manuel Martins Verdelhos, V. (2011). Characterization of indoor air quality and public spaces with Permission to smoke.

Nakagawa, L., & Comarú, F., & Trigoso. (2010). Impacts on air quality and on human health of air pollution in the metropolitan region of São Paulo.

Portugal. Ministérios Do Ambiente, Ordenamento Do Território E Energia, Da Saúde E Da Solidariedade, Emprego E Segurança Social. Portaria n.º 353-A de 4 de dezembro de 2013. Diário da República. Lisboa. 3 de Dezembro, 1.ª série, n. 235, p. 6644.

Portugal. Ministério da Economia e do Emprego. Decreto-Lei nº 118, de 20 de Agosto de 2013. Diário da República. Lisboa. 26 de Julho; Série 1, n. 159, p. 4988–5005.

Precious, J., Calheiros, J., López, M., & Ariza, C. (2009). Compliance assessment of the Portuguese law of prevention of smoking in the catering sector, 22–29. Retrieved from http://repositorium.sdum.uminho.pt/handle/1822/10202.

Occupational Safety and Hygiene V – Arezes et al. (Eds)
© *2017 Taylor & Francis Group, London, ISBN 978-1-138-05761-6*

Thermal environment in the hospital setting

D. Guimarães, H. Simões, A. Ferreira & J.P. Figueiredo
ESTeSC-Coimbra Health School, Instituto Politécnico de Coimbra, Coimbra, Portugal

ABSTRACT: The purposes of this study is to evaluate the thermal conditions in a hospital setting and assess quantitatively the thermal comfort variables. Then, it will be compared the values obtained with the ones regulated by law, and it will be evaluated how health professionals—doctors and nurses. The target population is the workers of four areas of hospitalisation. Firstly, it was evaluated the quantitative data of Ta, Tr, Va and RH, as well as the metabolic rate by activity and the clothing thermal resistance. Secondly, questionnaires were made to the workers. This study showed that there were statistically significant associations when comparing health professionals wellbeing perception in that moment and the type of thermal exposure (sensation). However, when this data was associated in function of the health professionals, it was determined that their opinion varies, not always agreeing with the results from thermal exposition.

1 INTRODUCTION

Nowadays, Safety, Hygiene and Health Services at Work (SHST) conditions are a growing cause of concern since it is the main theme of some global conferences, sponsored by World Health Organization (WHO). In Portugal, in the last years, the appearance of applicable standards different fields of intervention of SHST, aiming to decrease the high levels of accidents, which still exist (Matos, 2011).

Generally, the factors of risk with an occupational origin to which workers in a hospital environment are exposed, can be classified from an etiological view in four main categories: risk factors of chemical nature, physical, biological and psychosocial.

Although, all can influence wellbeing and worker's health, this study will focus on a physical risk factor—the thermal environment (Carvalhais et al., 2011).

The theme of thermal environment and associated thermal comfort is common in the diverse legal literature produced in Portugal, acknowledging its importance to wellbeing and worker's performance (CNQ, s.d).

Studies of thermal comfort aimed to analyse and establish the necessary conditions to the evaluation and conception of a thermal environment appropriated to activities and human occupation. Also intending to establish methods and principles to an in-depth thermal analysis of a certain environment. The importance of the study of thermal comfort is based on the human satisfaction or in its wellbeing regarding the fact of feeling thermally comfortable (Moço, 2014). The sensation of thermal comfort depends on the combination and influence of many factors. These factors are divided in individual variables (type of activity and acclimatisation clothing) and environmental variable (air temperature; air relative humidity or vapour partial pressure; average radiant temperature of surrounding surfaces; and air velocity.) (Fator Segurança, s.d).

The metabolic rate, characteristic of physical activity performed, applies to the amount of heat internally produced by the body and the thermal isolation of clothing represents a barrier placed by clothing to the changes of heat with the environment. The mean radiant temperature, the relative humidity, the temperature and air velocity determine the transfer by radiation, evaporation, convection and conduction. The combined effect of all this factors establishes a comfort sensation or thermal discomfort. The clothing—responsible for thermal exchanges conditioning.

The clothing appears as a second element belonging to the component of variables of thermal comfort, representing a barrier to the heat exchanges by convection (Moço, 2014).

Thermal balance alteration results in physiological alterations, more or less uncomfortable, but bearable, considering that the warm-blood is ensured. The further the thermal environment is from the neutrality zone (warmer) or colder), the more accentuated are the physiological alterations until they reach their maximum level. Other than those limits warm-blood can't be ensured (CNQ, s.d).

Not only the bad internal environment quality, but also the activities performed outside the thermal comfort zone or the individual lack of control, in certain workplaces can intervene with work satisfaction. If the air quality is poor, alongside with uncontrolled sound, light and cooling systems in the workplace, not only fatigue but also head-aches can become a bigger inconvenience, making evident that the productivity will be affected by specific aspects of inner environmental quality (Coutinho et al., 2004).

In many cases, the thermal sensation extremes are associated to stress situations to individuals, whether by very low or very high temperatures. Displeasure sensations, physiological attrition, health disorders and heat exhaustion are some of the disorders occurring when the body tolerance margins are exceeded. This can be life-threatening if the limit of 40.6°C of the body temperature is exceeded (Moço, 2014). The commonly associated symptoms to hot thermal environment are dehydration, physical fatigue, neurological disorders and cardiovascular disorders. In the cold thermal environment are wounds, necrosis and cracked skin, exacerbation of rheumatic diseases and respiratory problems (Chagas, 2015).

Health services and, in a particular way, hospitals form quite peculiar workplaces, designed almost exclusively according to the user needs, equipped with very specific technical and organisational systems, providing its workers—whether or not health professionals—with precarious working conditions, recognized as worse than the ones verified in the majority of the remaining sectors of activity.

Paradoxically, the hospital presents risks and hazards which can represent imminent threat, causing sooner or later health problems for people, whom maintain direct daily contact and/or with this type of space. The work in a hospital environment is liable to cause health damages, which are not limited to work accidents or occupational diseases; it also adds to—often decisively—the existence of certain diseases with etiologic multifactorial origin, also known as "work-related illnesses." It also provokes the exacerbation of affections, which are etiologically independent of professional nature factors and, lastly, causes frequent situations of stress and physical and mental fatigue (Coutinho et al., 2004).

The thermal comfort evaluation combined with thermal performance simulation software is an important tool in matching climate to buildings and activities carried out on environments (Moço, 2014).

There is a progression going from state of comfort to thermal discomfort, turning into thermal stress situations, which can cause injuries or even death. When heat affects humans, the first sensation obtained is a discomfort, which increases as the heat regulating systems function to resist to the thermal aggression imposed over the body. In the meantime, it decreases working efficiency, the diseases that humans may, eventually, carry tend to rise, increases the possibility of accidents and behavior changes are evident (Leal, 2014).

This theme was chosen due to the fact that it is even more important assess issues related with Safety, Hygiene and Health Services at Work, since in performing labour activities there are numerous perils associated with thermal environment in the workplace and various variables influencing in which makes imperative the compliance with the legal and regulatory requirements to ensure the protection of safety, health and wellbeing of workers (Pinheiro, 2011).

The aim of this study is to evaluate the thermal condition in a hospital setting, taking into account all the crucial factors to this evaluation.

2 MATERIAL AND METHODS

This is an analytical, observational, cohort and cross-sectional study. The level of knowledge of this study can be classified as a level II (descriptive correlational).

This study was developed in several rooms of four medical sectors (hospitalisation) in a hospital centre, in Oporto. The target population was health professionals—doctors and nurses, with a total of 81 workers. The type of sampling wasn't probabilistic and the sampling used was convenience. This study was divided in three stages: Selecting a sample; Data collection on-site (measurement of thermal levels which workers were exposed); Conduction of a questionnaire to health professionals of the service. In this study it was evaluated individual parameters (metabolic rate per activity and clothing thermal resistance) and environmental parameters, air temperature (Ta), radiant average temperature (Tr), air velocity (Va) and relative humidity (RH), and the collection of them was conducted using the following equipment (with their respective probes): a) Delta Ohm weather station, model HD 32.1 Thermal Microclimate; b) Probe TP3275—Sensor of globe temperature; c) Probe HP32117DM—Combined sensor of wet-bulb and dry-bulb temperature; d) Probe HP3217R—Combined sensor of room temperature and relative humidity.

Using the equipment aforementioned, it was carried out seven measurements in each medical sectors, with a total of 28 measurements. All the measurements were carried out in the same time interval, between 11 o'clock and 13 o'clock. Regarding the questionnaire applied to the health professional, it was based on ISO 10551:1995—Ergonomics of the thermal environment—Assessment of the influence of the thermal environment using subjective judgement scales. The statistical processing of the data collected was carried out using the statistical software IBM SPSS Statistics, version 22 and DeltaLog10 software. Was used chi-square test. The interpretation of the statistical results was developed based on a significance level of p ≤ 0.05 with a confidence interval of 95%.

3 RESULTS AND DISCUSSION

In the following table, we tried to understand the comfort perception at the moment and its relation with the type of thermal exposure (sensation.)

Having as basis the results aforementioned, it can be determined the presence of a statistically significant association between wellbeing perception at the moment and the thermal evaluation (p < 0.05).

From the doctors who classified their wellbeing as "neither cold or hot" (n = 24), the totality of them agreed with the classification assigned analytically as "neutral." Nevertheless, of the professionals who indicated a "slightly warm" perception, the majority of the answers were consistent with the results presented analytically (slightly warm). After that, the wellbeing perception at the moment will be evaluated and associated the same condition with the analytical information (Table 2).

According to the results showed in the table above, it can be verified that there isn't a statistically significant association between the perception of wellbeing at the moment and the type of thermal exposure (sensation), according to the professional group (p > 0.05). From the doctors who classified their wellbeing as "slightly warm" (n = 18), the totality of them agreed with the classification

Table 1. Comfort perception right now and its relation with the type of thermal exposure.

Wellbeing perception at the moment		Thermal sensation			
		Neutral	Slightly warm	Slightly cool	Total
Cold	n	2	0	0	2
	%	5	0	0	2.5
Cool	n	3	0	0	3
	%	7.5	0	0	3.7
Slightly cool	n	8	0	2	10
	%	20	0	100	12.3
Neither cold nor hot	n	24	0	0	24
	%	60	0	0	29.6
Slightly warm	n	2	18	0	20
	%	5	46,2	0	24.7
Warm	n	1	15	0	16
	%	2.5	38.5	0	19.8
Hot	n	0	6	0	6
	%	0	15.4	0	7.4
Total	n	40	39	2	81
	%	100	100	100	100

Chi-Square = 82,731; df = 12; p < 0.0001.

assigned analytically as "slightly warm." However, the doctors who indicated a "hot" perception, the majority of the answers weren't consistent with the results presented analytically as "slightly warm". In the following table, it was associated wellbeing perception at the moment with the same condition with the analytical information, according to the professional group—nurse. As previously, it can be verified that there isn't a statistically significant association between the perception of wellbeing in the moment and the type of thermal exposure (sensation), according to the professional group (p > 0.05).

From the doctors who classified their wellbeing as "neither cold or hot" (n = 24), the totality of them agreed with the classification assigned analytically as "neutral" (Table 3).

Table 2. Wellbeing perception according to the professional group—doctors—and its relation with the type of thermal exposure.

Wellbeing perception at the moment: Professional group—Doctors		Thermal sensation		
		Neutral	Slightly warm	Total
Slightly warm	n	0	18	18
	%	0	46.2	45
Hot	n	1	15	16
	%	100	38.5	40
Very hot	n	0	6	6
	%	0	15.4	15
Total	n	1	39	40
	%	100	100	100

Chi-Square = 1,538; df = 2; p = 0.463.

Table 3. Wellbeing perception, according to the professional group (nurses) and their relation with the type of thermal exposure.

Wellbeing perception at the moment: Professional group—Nurses		Thermal sensation		
		Neutral	Slightly warm	Total
Cold	n	2	0	2
	%	5.1	0	4.9
Cool	n	3	0	3
	%	7.7	0	7.3
Slightly cool	n	8	2	10
	%	20.5	100	24.4
Neither cold nor hot	n	24	0	24
	%	61.5	0	58.5
Slightly warm	n	2	0	2
	%	5.1	0	4.9
Total	n	39	2	41
	%	100	100	100

Chi-Square = 6,518; df = 4; p = 0.164.

Finally, we tried to understand the thermal environment qualification and the type of thermal exposure, according to the professional group.

According to the results showed (Table 4), it can be verified that there isn't a statistically significant association between thermal environment qualification, in accordance with the health professional (doctor), in relation to the type of thermal exposure ($p > 0.05$). In the group of the doctors who qualified the environment as "a bit difficult to bear" ($n = 20$), the totality of them agreed with the classification assigned analytically as "slightly warm." Nevertheless, the doctors who qualified the environment as "quite difficult to bear" disagreed with the results presented analytically.

There isn't a statistically significant association of the thermal environment qualification, according to the health professional (nurse), in relation

Table 4. Comparison of thermal environment classification with thermal sensation in function of the professional group—doctors and nurses.

Thermal environment classification: Professional group—Doctors[a]		Thermal sensation		
		Neutral	Slightly warm	Total
Perfectly bearable	n	0	3	3
	%	0	7.7	7.5
A bit difficult to bear	n	0	20	20
	%	0	51.3	50
Quite difficult to bear	n	1	8	9
	%	100	20.5	22.5
Very difficult to bear	n	0	5	5
	%	0	12.8	12.5
Unbearable	n	0	3	3
	%	0	7.7	7.5
Total	n	1	39	40
	%	100	100	100

Thermal environment classification: Professional group—Nurses[b]		Thermal sensation		
		Neutral	Slightly warm	Total
Perfectly bearable	n	9	1	10
	%	23.1	50	24.4
A bit difficult to bear	n	17	1	18
	%	43.6	50	43.9
Quite difficult to bear	n	8	0	8
	%	20.5	0	43.9
Very difficult to bear	n	4	0	4
	%	10.3	0	9.8
Unbearable	n	1	0	1
	%	2.6	0	2.4
Total	n	39	2	41
	%	100	100	100

a) Chi-Square = 3,533; df = 4; p = 0.473; b) Chi-Square = 1,250; df = 4; p = 0.870.

to the type of thermal exposure ($p > 0.05$). From the nurses who qualified the environment as "a bit difficult to bear" ($n = 17$), the totality of them disagreed with the classification assigned analytically as "neutral." Nonetheless, the doctors who qualified the environment as "perfectly bearable" agreed with the results presented analytically.

4 CONCLUSIONS

The results of this study shows that it will be necessary to adopt some measures to improve thermal environment at the disposal of medical services.

As mentioned above, it can be concluded, as according to the equipment thermal environment at the disposal of medical services varies between "neutral" and "slightly warm". This means that the medical services offer good temperature and humidity conditions, providing wellbeing and defending worker's health by ensuring a temperature between 18°C and 22°C and a humidity between 50% and 70%. However, not always this conditions were respected, as the equipment found out that there are workplaces where the thermal environment is "slightly warm."

Although the results collected by the equipment aren't a major concern, it can't be forgotten that the data collection for this study was carried out between May and June (in the spring) and that in the different seasons of the year (specially in winter and summer) the values may change and the results can be more alarming.

In this regard, some general measures can be adopted to obtain good thermal comfort conditions for the highest possible number of workers. For instance, temperature adjustment and air renewal, which can be done in function of the workers and maintained within the convenient limits to avoid harm to human health; and structural surfaces protection as placement of suspended ceilings and thermal insulation of the walls, among others.

It would be important to reduce the duration of periods of work and/or increase the work breaks where the thermal environment is slightly warm.

It is worth pointing that, in this study, there were cases in which it was concluded that the thermal environment to which the doctors are exposed is a bit difficult to bear because it is slightly warm. Since they are in thermal discomfort, apart from decrease their productivity, since it's harmful to workers.

Hence, it is fundamental have a motorization of the factors that have influence in the thermal environment (environmental and individual) so, both workers and users and not only from the medical services, but from all the hospital centre, won't be exposed to risk situations.

REFERENCES

Carvalhais J.S, Santos J, Lourenço I, Teixeira J.P. *Conforto Térmico em Meio Hospitalar - o caso do Serviço de Esterilização*. Agência Nacional de Qualificação 2011. [Online]. Available: http://paginas.fe.up.pt/~cigar/html/documents/carvalhais.pdf. [Accessed: 20-Oct-2015]

Catálogo Nacional de Qualificações. *Ambiente, Segurança, Higiene e Saúde no Trabalho - conceitos básicos*. [Online]. Available: http://www.catalogo.anqep.gov.pt/Ufcd/Detalhe/413

Chagas D. Os riscos no ambiente laboral e os seus efeitos na saúde dos trabalhadores. 2015. Available: http://blog.safemed.pt/os-riscos-no-ambiente-laboral-e-os-seus-efeitos-na-saude-dos-trabalhadores/.

Coutinho A.S, Maria W., Santos V. Ceset - *Conforto, eficiência e segurança no trabalho*. Volume 1, Número 1, Dezembro/2004. ISSN 1806-7889

Duarte Leal A.C. *Estudo de ambientes térmicos quentes no sector da panificação: avaliação das condições de trabalho*. Tese de Dissertação de Mestrado em Gestão da Prevenção de Riscos Laborais – ISLA Leiria, 2014.

Factor Segurança. *Ambiente Térmico*. Factor Segurança. Available: http://www.factor-segur.pt/shst/docinformativos/Ambiente_termico.pdf.

Matos M.A. *Avaliação do Ambiente Térmico na Cantina do ISEC, Tese de* Mestrado em Equipamentos e Sistemas Mecânicos, Instituto Superior de Engenharia de Coimbra – 2011.

Moço S.M.O. *(Des)conforto térmico no Verão em Portugal Continental e a Perceção populacional para as alterações climáticas—comportamentos adotados aquando de vagas de calor Educação Ambiental e Sustentabilidade*. Tese de Doutoramento em Geografia, Universidade Aberta Lisboa, 2014.

Patrícia I., Pinheiro T. *Conforto Térmico e Bem-Estar numa Superfície Comercial Isolada*. Tese de Mestrado em Engenharia de Segurança e Higiene Ocupacionais – Faculdade de Engenharia da Universidade do Porto, 2011.

Occupational Safety and Hygiene V – Arezes et al. (Eds)
© 2017 Taylor & Francis Group, London, ISBN 978-1-138-05761-6

Study on injuries/diseases in workers in the municipality of Coimbra, Portugal

Marcelo Afonso, Susana Paixão, Simão Cabral, João Paulo Figueiredo & Ana Ferreira
Instituto Politécnico de Coimbra, Coimbra, Portugal

ABSTRACT: The workers of the Division of Environment, Health and Public Spaces of Coimbra Municipality are daily exposed to various triggers of disease and injury. This study aimed to analyze the prevalence and recurrence of disease and injury, as well as its surroundings in the context of work, in the workers of the Environment Division of the Coimbra Municipality. For the study of injury and disease, the period covered the biennium 2014–2015. Regarding the Results, variations were observed in terms of the age mean for the occurrence of injury and disease, as well as correlations between the hours worked in specific functions and possible disease contraction. Relapses were also observed between disease in work and injury as well as the relationships between the parts of the body that contracted injury and the season in which the injury occurred.

1 INTRODUCTION

The city of Coimbra through the Environment Division, Health and Public Spaces of the Coimbra Municipality, daily performs various actions of unsorted residues collection, cleaning and urban sanitation (Câmara Municipal de Coimbra, 2007).

Among the actions of unsorted waste collection, there is the collection of municipal solid waste, with around 10677 circuits covered and around 57,040 tons collected (on average) over the Biennium 2014–2015 (Câmara Municipal de Coimbra, 2014).

In these actions and in the actions that complement the profession, workers are subjected to situations of risk through the nature of the actions they perform. First, it is important to contextualize the nature of professions covered.

Among the various activities carried out, it is possible to differentiate which are designated for waste collection such as undifferentiated collection and selective collection of waste and the domestic objects collection, scrap and container washing. In the cleaning activities we can highlight the functions of: street cleaning, waste removal function, street sweeping, whether manual or mechanical process, elimination of herbs through the weeding or clearing processes, cleaning and dumping bins and cleaning of canine waste (Câmara Municipal de Coimbra, 2007).

The shift work and night work are also part of the working structure of the environment division workers. As for night work, this usually involves a lower level of performance and a higher frequency of accidents (Caetano & Vale, 2000). These

operational worked on average 1.411 hours (in the two years 2014 and 2015) (Câmara Municipal de Coimbra, 2014 and 2015), aware that a Portuguese worker works on average 1700.5 hours for dependent workers (OCDE, s.d.)

Thus, the performance of the employees may then be questioned depending on the stiffness of their working hours and we must notice that changes in working hours lead to changes in the performance of the employees (Caetano & Vale, 2000).

It is urgent, in a worker protection context, to minimize the exposure of the occupational risk factors, giving priority to their welfare and correct security practices, while safeguarding their health, characterizing the factors through the Professional Risk of the profession they perform. Sometimes the worker by routine work and fatigue neglects especially regarding the risks to which they are exposed in carrying out their actions (Nunes et al., 2010).

There is much speculation about how much the health and poor conditions at work can have long-term effects and be the source of occupational illnesses and health problems (CE, 2014). The World Health Organization (WHO s.d.) defines occupational diseases as any disease contracted primarily as a result of exposure to risk factors from work. These diseases are developed in the course of labor activity and have various causes, where the working environment factors can play a role, together with other risk factors in the development of certain diseases (WHO s.d.). These diseases can affect the worker, either at the time or in the course of several years, manifesting sometimes many years after the initial exposure.

The aim of this study were to analyze the prevalence and recurrence of disease and injury in the workers of the Environment Division, Health and Public Spaces of the City of Coimbra.

2 MATERIAL AND METHODS

This study refers to a statistical comparison of actual data provided by the Department of Social Development and Environment belonging to the Municipality of Coimbra. It is a descriptive and comparative study of data for disease and injury in the functions of collecting, cleaning and sanitizing. These statistics are daily observations collected by the Department of Social Development and Environment and the data related to sick leaves and disease at work, collected by the Human Resources Division of the Coimbra Municipality.

As inclusion criteria, this article takes into account the workers of the Coimbra City Council in collecting functions, cleaning and sanitizing, affected or not by injury and illness. As for exclusion criteria, this article excludes all other workers not covered by professions under study.

The data collected regarding accounting fault, personal data, professional data and the occurrence of illness and accident with injury, were provided by the Environment Division, Health and Commons under condition of anonymity.

Regarding to the data on the subject of injuries, respectively parts of the body with injury as well as the results of occupational medicine, were provided by the Human Resources of the City of Coimbra.

The sample used in this article was made up of 102 workers, in 2014, 97 male and 5 female. Within the study population, there are different categories associated with the professional function, 59 are municipal cleaners, and 31 are drivers, 1 tractor driver, 10 commissionaires and 1 mason.

Referring to 2015, there was a reduction in the quota resulting in a total of 90 workers, 85 male and 5 female. Within the professional categories are distinguished 53 municipal cleaners 27 drivers, 1 tractor driver, 8 commissioners and 1 mason.

As for age (referring to 2014), workers have an average age of 51.43 years, and the youngest employee is 33 years old and the oldest 63 years.

Data Parameters: Sex; Professional category; Age; Total Hours; Total Hours by function; Disease Occurrence; Inability determined by the Occupational Medicine; Injured Part of Body; Total of the days with Work Leave; Month of the accident with injury; Season of the accident with injury.

The statistical processing of the data collected was carried out using the statistical software IBM SPSS Statistics, version 22. The interpretation of the statistical results was developed based on a significance level of $p \leq 0.05$ with a confidence interval of 95%.

3 RESULTS AND DISCUSSION

The results were divided into two categories, the first being related to the analysis of a lesion in the second context, and with the descriptive study of disease, both in the workplace.

There were no mean differences in age between workers who suffered injury or not in either 2014 or 2015 ($p > 0.05$)

There were no mean age differences in either 2014 or 2015 ($p > 0.05$), depending on the type of disability.

Table 1. Relationship between age and injury occurrence in 2014–2015.

Injury occurrence in 2014	n	M	SD	t; df; p-value
No	63	51.97	7.496	1.055;85;0.295
Yes	24	50.00	8.501	
Injury occurrence in 2015	n	M	SD	
No	69	52.12	7.496	1.636;85;0.106
Yes	18	48.78	8.501	

Test t-Student; M: Mean; SD: Standard Deviation.

Table 2. Inability attributed by the occupational medicine in 2014 and 2015 (distribution by age).

2014	n	M	SD	Min.	Max.
No Absolute Temporary Incapacity (S/ATI)	3	54.0	6.25	47	59
Absolute Temporal Incapacity (ATI)	17	49.88	8.92	35	61
Without Incapacity	60	51.92	9.69	33	63
Total Observed Cases	80	51.56	9.37	33	63
2015	n	M	SD	Min.	Max.
No Absolute Temporary Incapacity (S/ATI)	5	49.20	10.6	35	58
Absolute Temporal Incapacity (ATI)	11	49.36	7.85	36	60
Without Incapacity	71	51.90	9.43	33	63
Total Observed Cases	87	51.43	9.24	33	63

Test Kruskal-Wallis; M: Mean; SD: Standard Deviation; Min: Minimum; Max: Maximum.

Table 3. Relationship between the distribution of body parts with injury and professional category in 2014 and 2015.

Body Parts With Injury 2014[a]		cleaner	driver	Comissioner	Tratorist	Total
Head	n	2	0	1	0	3
	%	9.5	0.0	100.0	0.0	12.5
Body	n	2	0	0	0	2
	%	9.5	0.0	0.0	0.0	8.3
Upper limbs	n	4	1	0	0	5
	%	19.0	100.0	0.0	0.0	20.8
Lower limbs	n	6	0	0	0	6
	%	28.6	0.0	0.0	0.0	25.0
Other parts	n	7	0	0	1	8
	%	33.3	0.0	0.0	100.0	33.3
Total	n	21	1	1	1	24
	%	87.5	4.2	4.2	4.2	100.0

Body Parts With Injury 2015[a]		cleaner	driver	Comissioner	Tratorist	Total
Head	n	6	0	–	–	6
	%	46.2	0.0	–	–	35.3
Body	n	2	0	–	–	2
	%	15.4	0.0	–	–	11.8
Upper limbs	n	1	0	–	–	1
	%	7.7	0.0	–	–	5.9
Lower limbs	n	2	0	–	–	2
	%	15.4	0.0	–	–	11.8
Other parts	n	2	4	–	–	6
	%	15.4	100.0	–	–	35.3
Total	n	13	4	–	–	17
	%	76.5	23.5	–	–	100.0

Chi-Square test; a) $X^2 = 13,124$; df = 12; p = 0,360; b) $X^2 = 9,590$; df = 4; p = 0,048.

We proposed to relate the presence of injury (part of the body) according to the professional category (Table 3). Both in 2014 and 2015 there was no pattern of association between the region of the affected body (injury) and the type of activity to which each worker was responsible. However, in 2014 we can verify that 25% of the lesions identified were in the lower limbs and 20.8% in the upper limbs. However, 33.3% of the injuries identified in the workers were not reported in that year. In the year 2015, only injuries were registered between two groups of workers. The lesion with the highest prevalence in this year was located in the head (35.3%) and with lower expression in the upper limbs (5.9%).

We also proposed to evaluate whether the occurrence of disease in 2014 was generally changed in 2015. See Table 4.

There was no significant change in the pattern of occurrence between the years under evaluation (p> 0.05). However, of the 73 workers who did not

Table 4. Recidivism of disease in the biennium 2014–2015.

			Disease 2015		
			Yes	No	Total
	No	n	13	60	73
		% row	17.8	82.2	100.0
Disease 2014		% column	48.1	81.1	72.3
	yes	n	14	14	28
		% row	50.0	50.0	100.0
		% column	51.9	18.9	27.7
		n	27	74	101
Total	% row		26.7	73.3	100.0
	% column		100.0	100.0	100.0

McNemar Test (p > 0.500).

Table 5. Average hours of complementary work and disease occurrence in 2014 and 2015.

Complementary works hours

Disease2014	n	M	SD	MW; p
No	71	76.03	108.64	421.000;
Yes	28	164.16	140.89	0.0001
Disease2015	N	Mean	SD	MW; p
No	13	118.46	221.76	40.500;
Yes	9	54.44	102.69	0.217

Mann-Whitney test; M: Mean; SD: Standard Deviation.

present any type of disease in 2014, 17.8% of them had disease in 2015. However, of the 28 workers with disease in 2014, the probability of having disease in 2015 was 50% and equal Probability that it does not occur.

Table 5 shows the number of average hours in complementary work and the occurrence of illness in 2014 and 2015.

As we can see in the previous table, workers who presented a disease condition in 2014 were the same ones who, on average, performed a greater number of hours of complementary work significantly compared to workers without disease (p < 0.05). In the year 2015 there was an inversion, non-significant, of the standard compared to 2014 (p > 0.05).

4 CONCLUSIONS

In conclusion, it was found in this study the influence of some factors on the occurrence of injury and Disease in the Biennium 2014–2015.

It was concluded that Age Means have little influence on the occurrence of injury, and that no parts of the body more liable to injure in different years and in similar seasons.

It was also found that there is recurrence of Disease in the two years of the study and that there was a relationship between hours worked in certain functions and their influence on the occurrence of disease.

It was also found that there was a reduction in the incidence of injuries and Disease in 2014 to 2015 as well as the level of degrees of disability assigned by the Occupational Medicine

However, understanding all factors that cause injury and other studies Diseases requires other magnitude. It is important to note the plurality of factors that can influence the objects of study.

Its redundant observation and the deeper study of these subjects are fundamental to understanding the factors that cause injury and disease, not ignoring the specific characteristics inherent to each profession.

The surrounding of the work continues to be a field to explore, the randomness of external events and their influence on the welfare of workers needs further attention.

Workers of the Environment Division, Health and Public Spaces of Coimbra City Council who have suffered injury, missed in Average 31 (Santos, 2012; Câmara Municipal de Coimbra, 2014) days together that workers who suffered Disease (not excluding that may overlap) missed in Average 32 days (Santos & Almeida, s.d.; Câmara Municipal de Coimbra, 2014).

Labor performance is inseparable from the welfare of workers. Improve working conditions reducing workers' exposure to risk factors is to improve their health and, as a result, enhance their performance at work.

REFERENCES

Brito, C. *Doenças Ocupacionais.* [Online]. Available: *https://*pt.scribd.com/doc/20704585/DO-Historia-doencas-ocupacionais. Visualizado a 6 de Agosto de 2016.

CE. Comunicação da comissão ao parlamento europeu, ao conselho, ao comité económico e social europeu e ao comité das regiões, *Melhorar a qualidade e a produtividade do trabalho: estratégia comunitária para a saúde e a segurança no trabalho 2007-2012,* Bruxelas, 21.2.2007 COM(2007)62final.[Online].Available:http://ec.europa.eu/europe2020/pdf/europe2020stocktaking_pt.pdf

Caetano A, & Vale J. *Gestão de Recursos Humanos: Contextos, Processos e Técnicas.* 2a Ed. Lisboa: Editora RH. 2000.

Câmara Municipal de Coimbra. Ambiente Saúde e Espaços Públicos [Online]. Available: http://www.cm-coimbra.pt/index.php/areas-de-intervencao/ambiente/dept-de-qualidade-de-vida/itemlist/category/108-variosCategorização das Profissões. Visualizado a 29 de Julho de 2016

Câmara Municipal de Coimbra. *Table de Horas por Função em 2014.* Divisão de Ambiente, Saúde e Espaços Públicos, Camara Municipal de Coimbra, 2014.

Câmara Municipal de Coimbra. *Table de Horas por Função em 2015.* Divisão de Ambiente, Saúde e Espaços Públicos, da Camara Municipal de Coimbra, 2015.

Câmara Municipal de Coimbra. *Table de Resíduos indiferenciados Recolhidos No Ano de 2014 em Coimbra.* Divisão de Ambiente, Saúde e Espaços Públicos, da Camara Municipal de Coimbra, 2014.

Câmara Municipal de Coimbra. *Table de Resíduos indiferenciados Recolhidos No Ano de 2015 em Coimbra.* Divisão de Ambiente, Saúde e Espaços Públicos, da Camara Municipal de Coimbra, 2014.

Câmara Municipal de Coimbra. *Table de dados Pessoais Lesão e Acidente 2014.* Departamento Recursos Humanos da Camara Municipal de Coimbra, 2014.

Câmara Municipal de Coimbra. *Table de dados Pessoais Lesão e Acidente 2015,* Departamento Recursos Humanos da Camara Municipal de Coimbra, 2015.

Lee WR. (1964) - Robert Baker: *The first doctor in the Factory Department. Part 1. 1803–1858.* Brit. J. Ind. Med, 21, pp. 85–93.

Nunes MBG. *et al. Riscos ocupacionais atuantes na atenção à Saúde da Família.* Revista Enfermagem UERJ, Rio de Janeiro, p. 204–209, abr – jun. 2010.

OCDE, *Average annual hours actually worked per Worker.* [Online]. Available: https://stats.oecd.org/Index.aspx?DataSetCode=ANHRS, Visualizado a 5 de Agosto de 2016

Santos C. *A formação em SHST versus sinistralidade laboral na limpeza urbana. Tese de Mestrado (Segurança e Higiene no Trabalho),* Escola Superior de Ciências Empresariais- Escola Superior de Tecnologia. RCAAP. 2012, 1–84.

Santos, M. & Almeida, A. *Cantoneiros: Principais riscos e fatores de riscos ocupacionais, doenças profissionais e medidas de proteção recomendadas-* Revisão Bibliográfica Integrativa. [Online]. Available: http://www.rpso.pt/cantoneiros-principais-riscos-e-fatores-de-riscos-ocupacionais-doencas-profissionais-e-medidas-de-protecao-recomendadas/

World Health Organization. *Occupational work diseases,* [Online]. Available: http://www.who.int/occupational_health/activities/occupational_work_diseases/en/, visualizado a 12 de Agosto de 2016

Occupational Safety and Hygiene V – Arezes et al. (Eds)
© 2017 Taylor & Francis Group, London, ISBN 978-1-138-05761-6

The essence of ergonomic innovation in modern manufacturing enterprises

A. Dewicka, A. Kalemba & A. Zywert

Faculty of Engineering Management, University of Technology, Poznan, Poland

ABSTRACT: The following article contains the result of copyright research being ergonomic innovation in modern manufacturing enterprises. In the first part is to see reference analysis of the essence of innovation, ergonomics, called destructive changes in terms of ergonomics and safety. In the second part of the article mentioned external and internal determinants of innovative activity of production companies. In the next part of the article has been published pilot test results qualitatively—quantitative, which was achieved in 2015 among 116 manufacturing companies, which in the years 2004 to 2014 declared conducting innovative activities in the field of ergonomics and safety. Surveys have identified the essence of innovation and the role of ergonomics in the surveyed companies. In the last part of the article is a summary of the discussed research issues.

1 INTRODUCTION

The changes in the global economy entail constant search for and implementation of innovative solutions called innovation, which in recent years have a wide range of scientific disciplines.

Innovation in enterprises are understood to be progressive in the field modification techniques and methods in the creative processes of products and services, leading to increasing the efficiency of the resources available in the organization, fulfilling the wishes and criteria for informed consumer (A. Nowak-Far, 2000).

One of the techniques aimed at the recipient-user ergonomic innovations are generated as a result of human creativity, boldness and ingenuity.

Considering the fact that the strategy of any company is based on maximizing the value and ergonomic actions may be the way to obtain greater safety, especially in the case of operator systems (Butlewski, M., and M. Slawinska, 2014; Jasiulewicz, M., and A. Saniuk, 2015), in the following paper was presented the essence of innovation, ergonomic feel that today's manufacturing companies.

2 ERGONOMIC INNOVATIONS

In the twenty-first century with the concept of innovation or innovation associated are all kinds of activities aiming at implementing the changes leading to the growth of modernity and competitiveness of companies and thus in effect to raise its value. For the modern entrepreneur innovation is the introduction of new products, implementation of new technologies and changes in production infrastructure and distribution. According to J. A. Schumpeter innovation are as creative and destructive set of operations in progress and technological development.

Innovation is considered as one of the main driving forces of business, as evidenced by one of the most popular definition of innovation, which is "in fact a deliberately designed by the human change in the product, production methods or work organization, applied for the first time in the community in order to achieve certain benefits socio-economic, meet certain technical criteria, economic and social, is the primary factor in the development of enterprises "(Brzezinski M., 2001).

As follows from the above quoted definition of targets that tend to innovate in today's enterprises are primarily economic and social, and come down to ensure conditions for the implementation of long-term development strategy, which is to lead primarily to meet the needs of demanding customers (Bogdanienko J., 2002).

One of the techniques aimed at the recipient user are innovative solutions to economic and human interaction with a particular subject of the so-called. ergonomic innovations.

Ergonomic innovation can be defined as the process of production and use "new solution" using a combination of anthropocentric, social, biotic and technical switching the parameters of objects and products in terms of size, quality, modernity and efficiency.

Keep in mind that, the advancing robotization and automation of production forces the search for solutions to replace the human decision-making process (Misztal, A., M. Butlewski, A. Jasiak, and S. Janik, 2015).

Innovation ergonomic merge achievements of many sciences humanizing working and living environment so that they are user-friendly needs psycho. Advancing robotics and automation of production and to seek solutions to replace the human decision-making processes. Measurable business benefits they bring when in a complete meet the needs of changing audience.

Innovative activity is an effective tool for preserving the competitiveness of enterprises, realized through the active implementation of the strategy management conditioned activities and relationships called determinants of innovation.

3 DETERMINANTS ERGONOMIC INNOVATION

Maturity is the appropriate level of innovative organizational culture that uses entrepreneurial sense, creativity and other abilities to create, absorb and implement innovations that are an essential element of an effective and efficient acquisition of competitive advantage (Janasz W., K. Janasz, Witness A., J Wisniewska., 2001).

Companies use a number of specific strategies, coordination and synchronization of innovative activities, among which can be distinguished: the culture of information, strategies, internal knowledge creation, strategies for local expansion of knowledge, strategies of global knowledge transfer, and organizational strategies, aimed at increasing the efficiency of methods of work organization, production, relationships and management structures by reducing administrative and transaction costs (PF Drucker, 2000).

Strategies and innovative activities of Polish enterprises determines a number of factors known as determinants of innovation that's different because of the genital-criteria can be divided into internal and external (Wziątek- A. Kubiak, Balcerowicz E., 2009).

Classification of internal and external determinants of enterprise innovation, is based on the criterion of the source of innovations, among which stands out the factors of direct and indirect impact on innovative activities. Resource direct impact on innovation include:

- human capital, their knowledge, skills and experience;
- knowledge resources, including ongoing research and scientific research;
- knowledge resources objectified (eg. Plant and equipment);
- knowledge resources (eg. Licenses, patents);
- external knowledge resources (eg. Production ties);
- commercial resources;
- and organizational resources (Janasz W., K. Janasz, Prozorowicz M., Witness A. Wisniewska J., 2002).

Among the internal factors affecting indirectly on innovation activities stands out: the financial resources, the state corporate debt, conditioning and defining a willingness to take business risks, innovative, and the size and structure of the company, affecting the possibility of funding for research. Determinants of external innovation directly and indirectly determines the environment and the environment in which the enterprise operates, including:

- consumer behavior;
- institutional conditions (eg. The state policy, legislation);
- actions and behaviors of competing operators;
- cooperation with local actors, private and central institution-I scientific research (Janasz W., K. Janasz, Prozorowicz M., Witness A. Wisniewska J., 2002).

Bearing in mind the level of materialization to these determinants of innovation include the infrastructure, materials, products, machines and devices necessary in the process of innovation, and not to these databases, patents, research, resources, business, and the overall organizational processes integrating innovative activities.

Complementarily and substitutability determinants of innovation enterprises ensures spatial and subjective level of advancement of knowledge generated and accumulated R & D, supported by capital resources and financial expenditures of enterprises.

4 MEANING ERGONOMIC INNOVATION

4.1 Research methodology

Ergonomic innovations are a multidisciplinary collection of creative activities resulting from social relationships underpinning the long-term technological development of the company.

Direct survey on the essence of innovation, ergonomics was conducted in 2015 among 116 manufacturing companies, operating in the Polish Republic, which in the years 2004 to 2014 reported to the Central Statistical Office innovative activities. Presented in the article studies are qualitative and quantitative (verified by the financial reporting of companies), and are a pilot basis to prepare for a broader study of ergonomic innovation in Polish enterprises.

The study included 116 manufacturing companies, among which 82 have declared the use of high technology production, the average of 26 and 8 medium low.

4.2 Findings

All the investigated companies declared that in the years 2004–2014 led innovative activities related to improving ergonomics and safety in plants (ergonomic innovations). Businesses were asked about the importance of (being) what was for them the implementation/running ergonomic innovation in their business entities. The questionnaire also required to answer what were the main sources of internal and external impulses to drive the development of ergonomic, and what purpose have businesses in making modernization.

Table 1 presents the essence of innovation, ergonomics, or the importance of innovation, ergonomics, called destructive changes in the area of ergonomics and safety were in enterprises. Businesses in the original questionnaire were asked, "How did you/do you assess the importance of (being) conducted ergonomic innovation in your company?".

In the opinion of entrepreneurs, among which innovation ergonomic (67.25%) had a big importance, and to 31.03% very important. Only 1.72% of the surveyed entrepreneurs declared not that conducted innovative ergonomic activity was of little importance.

The majority among the 116 surveyed companies felt the positive essence of conducting new upgrade, only two of the companies surveyed have experienced little satisfaction from the scope of conducted ergonomic innovation, which is why the next part of the questionnaire operators were asked to disclose sources of internal and external stimuli for development of systems and ergonomic measures.

The second open question entrepreneurs were invited to name sources, which initiated their businesses running ergonomic innovation. These sources are divided into internal and external, and the question in the questionnaire was "What inspired you to pursue innovation, ergonomic, please replace it."

57.75% of respondents declared that the driving force behind the development of ergonomic innovation in business entities were internal sources, among which 32.84% were top-down initiative owners, business management and human resources (50.74%), and its own R & D – 16, 42%.

To 42.25% of external sources of inspiration for innovation activities in the field of ergonomics, respondents completed the activities of competitors, suppliers, customers, suppliers, and training conferences and trade fairs run by professional consulting firms (Table 2).

In the next stage of the research entrepreneurs were asked to define the nature and purposes of deciding on innovative activities in the field of ergonomics. Table 3 were placed to answer the question of the questionnaire, which was "What were the objectives of the adopted innovative activities in your enterprise?".

According to the answers of respondents, ergonomic determinants of innovative activity has been a desire to improve the existing conditions, to meet the norms and legal standards, reduce production costs and plans for acquiring new western markets.

In the next stage of examining the importance of innovation ergonomic, entrepreneurs were asked to determine the specific location of their occurrence in economic structures. Answers to the question "Please specify the locus of innovation in Mr/Ms company." Posted in Table 4.

The next stage of research ergonomic importance of innovation in manufacturing companies

Table 1. The importance of ergonomic innovation (source: Own study based on surveys).

Response	Percentage
Very important	31,03%
Important	67,25%
It matters	1,72%
No effect	0,00%

Table 2. External sources of inspiration ergonomic innovation (source: Own study based on surveys).

Response	Percentage
Customers	22,44%
Provider	14,29%
Competition	40,82%
Collaborators	16,33%
Conferences and Fairs	6,12%

Table 3. The nature and purpose of the activities undertaken innovative (source: Own study based on surveys).

Response	Percentage
Improving working conditions	32,18%
Meeting the standards and legal standards	30,10%
Lowering the cost of production	16,26%
Winning the foreign market	21,46%

Table 4. Place of occurrence of ergonomic innovation (source: Own study based on surveys).

Response	Percentage
Organization of work	13,82%
work processes	16,85%
Working tools	19,52%
working methods	18,37%
The physical work environment	16,48%
Professional qualifications	2,84%
Operating procedures of the management	12,12%

Table 5. Interest in continuing the ergonomic innovation (source: Own study based on surveys).

Response	Percentage
Yes	93,10%
No	2,58%
No answer	4,32%

was to determine the general aspects of innovative solutions.

According to the percentage of respondents' answers to the question "How ergonomic innovations contributed to Mr./Ms company?", The innovations have contributed to:

- 27.82% modernize the machine park, which produced positive results in reducing production costs;
- 30.79% increase in efficiency and productivity;
- 32.12% reduction in the number of accidents;
- 9.27% disability compensation.

Presented in Table 4, the data reflect the practical nature and the importance of innovation ergonomic enterprises, which is why in the last stage of the research, respondents were asked to express an opinion or in the future, acting already on experience, and knowing all the pros and cons would be willing to drive further innovation development. The answers of the respondents to the latest survey question "Is the future plans Mr./Ms conducting ergonomic innovation in your company?", Presented in Table 5.

93.10% of the surveyed companies are interested in further development of innovation in the field of ergonomics, which is a prerequisite of a positive nature and importance of this interdisciplinary science in the creation of the modern market economy.

The results of the study do not exhaust the range of ergonomic analyzes the essence of innovation, which is one of the basic conditions for innovative development of the economy, lifting the rate of social life.

5 CONCLUSIONS

The condition of innovative development of the economy is appropriate policymaking creating conditions for sustainable development of the economy.

One such strategy is innovation ergonomic, found its application in the development and design of the seats, workstations, information elements and control of machines and technical equipment. Please note that the population is aging and The development of technologies useful to the elderly will therefore be a challenge that must be met, while ergonomic modeling will allow to achieve a high quality of the developed solutions (Butlewski, M., 2014).

Ergonomic innovations, as is clear from the research, helping to optimize efficiency, increase productivity, disability compensation, and prevent damage to health, which brings measurable and long-term business benefits.

Ergonomic innovation by introducing radical changes in production are one of the most important factors influencing the acquisition of qualitative and quantitative advantage in the free market economy.

REFERENCES

Bogdanienko J., W poszukiwaniu przyszłości. Zarządzanie strategiczne firmą, Wyższa Szkoła Finansów i Zarządzania w Białymstoku, Białystok, 2002.

Brzeziński M. (red.), Zarządzanie innowacjami technicznymi i organizacyjnymi, Di-fin, Warszawa 2001.

Butlewski, M., 2014, Practical Approaches in the Design of Everyday Objects for the Elderly, in L. Slatineanu, V. Merticaru, G. Nagit, M. Coteata, E. Axinte, P. Dusa, L. Ghenghea, F. Negoescu, O. Lupescu, I. Tita, O. Dodun, and G. Musca, eds., Engineering Solutions and Technologies in Manufacturing: Applied Mechanics and Materials, v. 657, p. 1061–1065

Butlewski, M., and M. Slawinska, 2014, Ergonomic method for the implementation of occupational safety systems: Occupational Safety and Hygiene Ii, 621–626 p.

Janasz W., Janasz K., Prozorowicz M., Świadek A., Wiśniewska J.: Determinanty innowacyjności przedsiębiorstw, Wydział Naukowy Uniwersytetu Szczecińskiego, Szczecin 2002.

Janasz W., Janasz K., Świadek A., Wiśniewska J., Strategie innowacyjne przedsię-biorstw, Uniwersytet Szczeciński, Szczecin, 2001.

Jasiulewicz-Kaczmarek M., Saniuk A., 2015, Human factor in Sustainable Manufacturing, Editors: M. Antona, C. Stephanidis, Universal Access in Human-Computer In-teraction. Access to the Human Environment and Culture, LNCS Vol. 9178, pp.444–455.

Misztal, A., M. Butlewski, A. Jasiak, and S. Janik, 2015, The human role in a progressive trend of foundry automation: Metalurgija, v. 54, p. 429–432

Nowak-Far A., Globalna konkurencja. Strategiczne zarządzami innowacjami w przedsiębiorstwach wielonarodowych, Wydawnictwa Naukowe PWN, Warszawa 2000.

Wziątek- Kubiak A., Balcerowicz E., Determinanty rozwoju innowacyjności firmy w kontekście poziomu wykształcenia pracowników, CASE, Warszawa 2009.

Occupational Safety and Hygiene V – Arezes et al. (Eds)
© *2017 Taylor & Francis Group, London, ISBN 978-1-138-05761-6*

Ergonomic analysis in the industry of beverages expedition sector

B.A.B. Nóbrega & L.L. Fernades
Universidade Federal da Paraíba, Paraíba, Brazil

A.M. Oliveira, D.A.M. Pereira, W.R.S. Silva & G.A.S. Biazzi
Universidade Federal de Campina Grande, Paraíba, Brazil

ABSTRACT: This study was carried out with the objective of analyzing the work station of manual dispatch of beverage boxes in the sector of disembarkation of finished products in a beverage industry. The procedure applied in this case study is based on the theoretical and methodological tool of the WEA (Work Ergonomic Analysis). In order to evaluate the level of risk that employees are subjected to when carrying out their work activities, the NIOSH (National Institute for Ocuppacional Safety and Health) lifting equation instrument was used. The results showed that the activities developed by the workers present a high probability of developing Work-Related Osteomioarticular Disorders (WROD) problems. When applying NIOSH it was noticed that the weight of the box is well above the recommended value and the values of the lifting rate present a high risk level. Recommendations have been made to improve working environment conditions.

1 INTRODUCTION

In today's scenario, an organization that seeks competitiveness needs to focus its strategies on the differentiation of its products and processes, as well as on the quality of life of employees, an essential factor for the company to achieve its objectives.

According to data available in the Statistical Yearbook of Social Security, in Brazil the food and beverages sector indicates between the years of 2012, 2013 and 2014, equivalent to 25% of the total accidents occurred in these years (Brasil, 2015a). In this way, the value positions the segment in the leading number of occurrences among the main sectors of the manufacturing industry in Brazil.

According to Fornazari et al. (2000), the imposition of intense work rhythms and prolonged journeys, coupled with incorrect postures and environments that are ergonomically inadequate to the worker, are interpreted as the cause of the commitment of his health and of his skill in the habitual tasks, predisposing him to the Development of Work-Related Osteomioarticular Disorders (WROD).

It is common to find activities that predominate the handling and manual handling of loads during movements performed in daily life, in sports and at work. The act of lifting a weight is often performed without the awareness of the mechanisms necessary for this load to be elevated or sustained (Gonçalves, 1998). The task of lifting heavy objects is a complex movement involving a large number of body structures and a decision-making system to orchestrate these structures. The relative contribution of each element of this system is of

decisive importance in the performance of the survey (Torre et al., 2005).

Many are the factors studied about the load lifting that can influence its efficiency and, thus, cause occurrence of injuries in certain parts of the body of the human being. Some of these factors are: the positioning of the joints at the beginning and during the lift, the amount of load, the speed of movement, the height at which the load is at the beginning of the lift, the presence or not of handles and their various types, the use of accessories such as the lumbar support belt, intra-abdominal pressure (Gonçalves, 1998).

The present study was developed with the objective of analyzing the work station of manual dispatch of beverage boxes in the sector of disembarkation of finished products in a distilled beverage industry located in a city of the State of Paraíba, Brazil. The procedure applied in this case study is based on the theoretical-methodological tool of the Work Ergonomic Analysis—WEA.

The WEA involves understanding, explaining and modeling real activities in order to act according to situations, it is designed to model and improve understanding of how humans act in function of work, and to highlight the active role of people in the system, Be they operators, employees, actors, managers, trainers or users. (Ribert-Van De Weerdt & Baratta, 2016)

2 METHODOLOGY

This article corresponds to a case study, descriptive and exploratory (Gil, 1991) of applied nature

(Silva & Menezes, 2001), based on the WEA method, according to Guérin et al. (2001) comes from the Franco-Belgian school of ergonomics and is based on the analysis of real work situations, which makes possible its comprehension and transformation, this method is composed of three main phases: demand analysis, task analysis and Activity analysis.

The present study was conducted in the shipping industry of a distilled beverage industry, which is composed of five employees. The monitoring of the work routine occurred between January and March of the year 2015. With the definition of the problems of the demand, the task was prescribed by the company, the activity performed by the employees, considering the working and organizational conditions in which The activity is included, making a diagnosis followed with the respective ergonomic recommendations.

The demand for this research was characterized as an induced demand, a process in which researchers intend to develop a study in order to identify problems that can be transformed into real demands from and/or authorized by the organization (Saldanha et al., 2012). In the analysis of the activity were made analyzes and interactional methods observing the real behavior of the workers in carrying out their activities, as well as the environment in which they perform it.

In order to assess the level of risk that employees are subjected to when carrying out their work activities, the National Institute for Occupational Safety and Health (NIOSH) instrument is used, proposes a Recommended Weight Limit (RWL) and the Lifting Index (LI) that Categorizes the level of injury that the operator may be exposed to. For any work situation, estimated from a set of variables that describe the conditions under which the survey is performed, in an attempt to minimize musculoskeletal overload for the worker (Oliveira, 2014). Regulatory Standard 17 (NR 17) considers the NIOSH equation adequate to indicate the weight limit (RWL) allowed for the activity (Vieira, 2012), which is calculated from the following Equation 1 and Table 1:

$$RWL = 23 \times HDF \times VLH \times LFF \times FLRT \\ \times LHQF \times VDF \qquad (1)$$

where,

3 RESULTS

3.1 Demand analysis, task analysis e activity analysis

The analysis was performed specifically for the work activity involving the loading of manual

Table 1. Components of the NIOSH equation.

Initials	Formula
HDF—Horizontal distance factor of the subject to the load	HDF 25/H H^* = Horizontal distance between feet and hands
VLH—Vertical load height factor	VLH = 1 − (0.0038*[Vc-75] Vc^* = Vertical distance between feet and hands
LFF—Lift frequency factor per minute	Look at the value in the NIOSH chart according to the frequency of lifting the load
FLRT—Factor lateral rotation of the trunk	FRLT = 1 − (0.0032*A) A^* = Torsion angle of trunk (degrees)
LHQF—Load Handle Quality Factor	Handle Vc 75 (cm) Vc > 75 (cm) Boa 1,00 1,00 Razoável 0,95 1,00 Pobre 0,90 0,90
VDF—Vertical distance factor traveled from the origin to the destination	VDF = 82 + (4.5/DC) D^* = Vertical distance traveled, corresponds to the difference in height of the load between the origin and the destination (in cm).
LI—Lifting Index	LI = Weight /LRP LI > 1.0 chance of minimal injury; 1.0 < LI > 2.0 increases the risk; LI > 2.0, will increase the risk of injury.

Source: Adapted by Brasil (2002b).

loads, in which the employees unload heavy loads (boxes loaded with beverages) by placing them on the conveyor belt, where they will proceed to the truck, manually loading them Several times a day.

Thus the demand specification is associated with the high level of physical demand, and the probable consequences that are the WRODs, which is caused by this type of activity, making development and application necessary in the most correct, safe and efficient way of work.

The task prescribed by the company for the shipping sector, which has the function of forwarding the beverage boxes to the landing truck, is described as follows: two operators climb each one on a ladder, which must be suitable for the accomplishment of the Activity in order to pick up the boxes that are at a certain height; Then deliver them to two operators, who will pass these boxes to two other employees, and these put them on the treadmill. According to the description, the task consists six employees working 8 hours. The activity have two ladders and presenting some pallets

It was observed that the activity performed is different from the task prescribed by the company, and

Figure 1. Beverages shipping industry.
Source: Research data.

Table 2. Results of application of NIOSH methods.

Get a drink carton (Origin) Results	Deliver drink box (Destination) Results
HDF = 1	HDF = 0,396825
VLH = 0,925	FAV = 0,775
LFF = 0,65 (2 per minute)	FFL = 0,65 (2 per minute)
FLRT = 0,904	FRLT = 0,776
Handle Quality: Poor	Handle Quality: Poor
VDF = 0.91	VDF = 0,895
RWL = 10, 0 kg	**RWL = 2,9kg**
LI= 2,0	**LI= 6,9**

Source: Research data.

can be seen in Figure 1, which is not made up of six employees, but consists of five employees, in which two employees climb into the boxes to pick them up and Not on their own ladders activity, after climbing into the boxes the employees pick up and throw the boxes to three employees who put them on the mat where they are following to the truck. The boxes have a weight of 20 kg and the working day is sometimes higher than the 8 hours recommended by both the company and the Brazilian labor legislation.

Employees in the analyzed sector are exposed to inadequate working conditions, which contributes to the occurrence of accidents. The work place is very small for the amount of products stored in that environment, making it difficult to perform the activity and locomotion of the employees, the treadmill has an advanced shelf life, and the way the work by these employees happens is the most worrisome factor, Because the drinks are stored in glass bottles, any wrong actions of the employees can generate breakage of the bottles and serious cuts in the workers.

3.2 Niosh

Using the NIOSH method, it was possible to determine the Recommended Weight Limit (RWL) for the activity of the two workers who pick up the boxes (origin) and deliver them (destination) to other employees to place them on the conveyor belt.

Analyzing Table 2, it can be observed that the weight of the beverage carton, which weighs 20 kg, carried by the employee in his activity is well above the RWL. In the activity of picking boxes (origin) the RWL recommends that each carton weighs a maximum of 10,0 kg, already in the activity of delivering boxes (destination) the recommended is that each carton weighs 2.9 kg. Therefore, for the activity of picking up boxes and delivering them the actual weight is at a value well above the recommended one. It was calculated the Lifting Index (LI), which indicates the level of risk for the activity, presenting such results pick up boxes (origin) LI = 2.0 and deliver boxes (destination) or LI = 6.9. The survey rates exceed the acceptable

value, this implies that currently the development of this work is unfeasible ergonomically, because according to NIOSH, LI > 2 means to say that the operator is exposed to injuries contributing to the emergence of WRODs, especially those that affect The lower back, hand, wrists, arm and forearm.

3.3 Recommendations

The excessive load of weight and bad postures adopted by the employees for the execution of the activity, the inadequate working environment, cause serious damage to the workers' physical health. Thus, some suggestions of corrective and preventive measures can be recommended, they are:

– Use of Personal Protection Equipment (EPI's) and Placing the EPCs in the places of need;
– Adoption of appropriate methods of loading to carry out the displacement and allocation of cargo at the place of destination;
– Provide training/guidance to workers with ergonomics professionals;
– Improve the prescribed task and adapt it to reality in order to preserve the physical health of employees;
– Mechanization of cargo loading activity through equipment that could replace manual unloading or adjust box weight as indicated by NIOSH; Buy a mat suitable for activity.

4 CONCLUSION

Through the Ergonomic Analysis of Work it was possible to identify a demand, diagnosing common problems in the work activity, comparing the prescribed task with the real activity analyzing the discrepancy that they present.

The results showed that the activities developed by the workers present a high probability of developing WROD problems, due to the high demands of physical effort and inadequate posture when performing the activities. Also, when applying NIOSH, it was noticed that the weight of the box is well above the recommended value and the values of the LIs present a high risk level.

Recommendations were made to improve work environment conditions, changes in posture in the act of loading the load together with the use of PPE, reorganization of the workplace. These changes would already have a significant improvement in the health and safety of the worker. However, for a new study it is necessary to use more accurate risk assessment methodologies, which allow to analyze all the employees and not only a part of them, as was done in this study, using monitoring tools.

REFERENCES

Brasil, M. T. E. (2002b). *Manual de aplicação da Norma Regulamentadora* 17.

Brasil. Ministério da Previdência Social. (2015a). *Anuário estatístico da previdência social*. Brasília. Recuperado em abril de 2016, de http://www.previdencia.gov.br/estatisticas

Fornazari, C. A. *et al.* (2000). Postura viciosa. *Revista Proteção*, v. 13 (99).

Gil, A. C. (1991). Como elaborar projetos de pesquisa. São Paulo: Atlas, 2002. *Métodos e técnicas de pesquisa social*, 5, 64–73.

Gonçalves, M. (1998). Variáveis biomecânicas analisadas durante o levantamento manual de carga. *Motriz*, 4(2), 85–90.

Guérin, F. et al. (2001). *Compreender o trabalho para transformá-lo: a prática da ergonomia*. Edgar Blucher.

Oliveira, D. R. D. (2014). Aspectos ergonômicos do levantamento manual de carga em mulheres: relação com equação de NIOSH.

Ribert-Van De Weerdt, C., & Baratta, R. (2016). Analysis of activity and emotions: a case study based investigation of an evolving method. *Le travail humain*, 79(1), 31–52.

Saldanha, M. C. W., et al. (2012). The construction of ergonomic demands: application on artisan fishing using jangada fishing rafts in the beach of Ponta Negra. *Work*, 41(Supplement 1), 628–635.

Silva E. L & Menezes E. M. (2001) Metodologia da pesquisa e elaboração de dissertação. Florianópolis, UFSC.5(6).

Torre, M. L., *et al.* (2005). Cálculo das forças internas na coluna lombar durante levantamento de carga através da dinâmica inversa. *Salão de iniciação científica* (17.: 2005: Porto Alegre, RS). Livro de resumos. Porto Alegre: UFRGS, 2005.

Vieira, J. L. (2012). Manual de Ergonomia: Manual de Aplicação da Norma Regulamentadora n 17. São Paulo: Edipro.

Occupational Safety and Hygiene V – Arezes et al. (Eds)
© *2017 Taylor & Francis Group, London, ISBN 978-1-138-05761-6*

Indoor radon in dwellings: An increment to the occupational exposure in Portuguese thermal spas

A.S. Silva & M.L. Dinis

CERENA, FEUP/UP—Centre for Natural Resources and the Environment, Faculty of Engineering, University of Porto, Porto, Portugal
PROA/LABIOMEP—Research Laboratory on Prevention of Occupational and Environmental Risk, Faculty of Engineering, University of Porto, Porto, Portugal

ABSTRACT: The main purpose of this study was to address the contribution of the exposure from non-occupational exposure to the total effective dose of workers from thermal spas. Radon indoor concentrations were measured in 14 Portuguese thermal spas and in the dwellings of 14 selected workers, one from each thermal spa. Different types of questionnaires were used to characterize the thermal spas and the dwellings. Indoor radon concentration was measured by a passive method with CR-39 detectors. The results showed that, in some cases, the non-occupational (residential) exposure to radon, and the contribution to the total effective dose, is much higher than the occupational exposure. Is addition, most of the workers have remained at the same workplace for the last 20 years, which may represent a significant risk to health due to the continuous annual doses at which they have been exposed to, over the years.

1 INTRODUCTION

The exposure to ionizing radiation from natural sources has led to a growing interest in protection the workers from others industries than just the nuclear fuel cycle. Thus, several documents have been published in the last few years addressing the protection of workers and the general public against the adverse health effects from exposure to ionizing radiation from natural sources (Mueller, 1998; Darby, 2005; Köteles, 2007; Silva et al., 2013).

The largest natural source of radiation to human exposure is radon and it is well known that radon exposure is the second-leading environmental cause of lung cancer death, after tobacco smoke (UNSCEAR, 2000; DGS, 2002; IAEA, 2003; Radolic et al., 2005; Yarar et al., 2006; Köteles, 2007; WHO, 2007; Al Zoughool & Krewski, 2009; Wiwanitkit, 2009; WHO, 2009; Nikolopoulos et al., 2010; Alberigi et al., 2011; Nikolov et al., 2012; EPA, 2013; Erdogan et al., 2013; Robertson et al., 2013; Vaupotic et al., 2013; Santos et al., 2014).

Radon is naturally present in the environment being generated in different types of rocks and soils with uranium mineralizations. Being a gas, it can migrate from places where it is produced and penetrate in confined spaces (Pereira et al., 2001). However, radon concentrations varies depending on the region, given that the distribution of uranium in rocks and soils is not uniform (Gray et al., 2009; ITN, 2010; Koray et al., 2014). Moreover, construction materials may also be a potential source of radon emanation, such as concrete, ceramic bricks, natural building materials (especially granite), gypsum and other industrial raw materials (such as gypsum), ashes, etc. (WHO, 1986).

The World Health Organization (WHO) recommended the reduction of radon in indoor air to below 100 Bq/m^3 (annual average) or, if not possible, proposed to adopt the limit of 300 Bq/m^3 and act to phase reduce this exposure (WHO, 2009), as a first step. In addition, the EU recently approved in the Directive 2013/59/Euratom, a reference value of 300 Bq/m^3 (annual average) for indoor air (Silva et al., 2014; Silva & Dinis, 2015).

The aim of this study was to evaluate the concentration of indoor radon in Portuguese thermal spas and in the dwellings of a representative worker from each thermal spa, in order to address the contribution of the exposure from non-occupational exposure to the total effective dose of the workers from these thermal spas.

2 METHODS

Fourteen thermal spas participated in the study as well as 14 selected workers, one from each thermal spa. The criteria for the workers' selection were: i) the seniority in the job; ii) the proximity of the worker's dwelling to the thermal spa; iii) the job task and the workplace within the thermal spa (preferably in balneotherapy) and iv) the number of hours spent in the dwellings.

Three questionnaires were used to characterize the thermal spas and workers' dwellings: i) the workers questionnaire; ii) the dwellings questionnaire and iii) the thermal spa's questionnaire. All questionnaires were previously validated by experts in the fields of health and geology. The aim of the questionnaires was to characterize each place in which the concentration of radon was measured (thermal spas and dwellings) and also the worker's habits that could influence the obtained results for the indoor radon concentration.

The thermal spa questionnaire addresses issues mainly related to the type of ventilation. Two different conditions were identified: i) the existence of Natural Ventilation (NV) and ii) the existence of Natural Ventilation and Mechanical Ventilation (NV/MV). The questionnaire was replied by 128 workers, which represents approximately 70% of total workers.

The indoor radon concentration was evaluated at various locations within each thermal spa: inhaler techniques room (ORL), Vapors (VP), Buvete Hall (BH), *Vichy* Shower (VS) and Thermal Pool (TP) and in the dwellings of the selected workers (living rooms) by a passive method with CR-39 detectors. The detectors were placed in each room at approximately 2 meter above the ground. After an exposure period of 45 days, the detectors were retrieved and sent to the Natural Radioactivity Laboratory, University of Coimbra, Portugal (Silva et al., 2016a).

3 RESULTS AND DISCUSSION

3.1 *Thermal spas questionnaire*

The collected data in the thermal spas' questionnaire is presented in Table 1.

Table 1. Characterization of thermal spas.

TS	N° of workers			Seniority in the workplace (years)			Task switching	TV
	M	W	Total	< 5	5–20	≥ 20		
TS1	3	8	11	3	–	8	yes	nv/mv
TS2	10	50	16	–	60	–	yes	nv/mv
TS3	nd	nd	11	11	–	–	yes	nv/mv
TS4	2	4	6	6	–	–	yes	nv
TS5	3	12	15	–	13	2	yes	nv/mv
TS6	2	13	15	–	13	2	yes	nv/mv
TS7	1	2	3	3	–	–	yes	nv
TS8	2	5	7	–	–	7	yes	nv
TS9	2	2	4	–	–	4	yes	nv/mv
TS10	6	17	23	23	–	–	no	nv/mv
TS11	nd	nd	22	22	–	–	yes	nv
TS12	12	14	16	–	16	–	yes	nv/mv
TS13	1	14	15	15	–	–	yes	nv/mv
TS14	3	15	18	–	13	5	yes	nv/mv

M—man; W—woman; nd—no data; TV—type of ventilation; nv—natural ventilation; nv/mv—natural ventilation and mechanical ventilation.

Seventy-one percent of the thermal spas has NV/MV and 29% has only NV (Table 1).

Most of the employees have been working at the same thermal spa for a period of 5–20 years and approximately 22% of the workers have been at the same workplace for the last 20 years.

All workplaces within the thermal spa have job rotation except for the TS10.

3.2 *Workers' questionnaire*

In what concerns to these questionnaires, the workers were informed about the general objectives of the study. The data provided and the results obtained are strictly confidential, becoming anonymous.

Concerning the clinical history, most of the workers do not have chronic problems, do not take medication and consider themselves healthy (Table 2).

About 21% of workers are smokers, so the risk of health effects is theoretically aggravated (Al Zoughool & Krewski, 2009).

3.3 *Workers' dwellings questionnaire*

As the presence radon will depend mostly on the rock type in the area of dwellings (geographical region) and in the area of the thermal spas, the location in the same or similar geological units may have notable effects on indoor radon levels. However, the average distance between the workers dwellings and the respective thermal spa ranges between 2 and 37 km and therefore it is not expected high variations or transitions in the geological settings at this scale (Figure 1).

Table 2. Clinical history.

TS	Chronic problems			Smoker			Family history			Healthy	
	Y	N	ND	Y	N	ND	Y	N	ND	Y	N
TS1	–	10	1	6	4	0	5	4	1	11	–
TS2	3	13	–	15	1	–	3	13	–	13	3
TS3	1	10	–	2	8	–	10	1	–	10	–
TS4	–	6	–	2	4	–	1	5	–	6	–
TS5	–	11	4	1	13	1	3	8	4	12	3
TS6	1	13	1	0	15	–	2	13	–	13	2
TS7	–	2	1	1	2	–	–	2	1	3	–
TS8	1	6	–	1	6	–	–	7	–	7	–
TS9	1	3	–	0	4	–	2	2	–	4	–
TS10	1	11	1	0	13	–	1	11	1	12	2
TS11	–	22	–	3	19	–	3	19	–	22	–
TS12	–	16	–	1	15	–	1	15	–	16	–
TS13	–	11	–	0	11	0	–	10	1	11	–
TS14	–	7	–	–	7	–	–	7	–	7	–
Total	8	141	8	32	122	1	31	117	8	147	10

Y—yes; N—no; ND—no data.

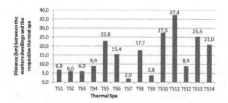

Figure 1. Distance (km) between the workers dwellings and the respective thermal spa.

Table 3. Radon concentrations in thermal (CRnTS) spa and in workers' dwellings (CRnHTS).

TS	CRnTS (Bq/m³) Location		CRnHTS (Bq/m³)	Distance (km)
TS1	ORL	3119	68	1
TS2	ORL	187	254	2
TS3	VP	465	1322	4
TS4	ORL	4335	312	11
TS5	VP	1173	1877	2
TS6	BH	1415	168	1
TS7	ORL	347	642	4
TS8	VS	360	105	3
TS9	ORL	269	714	3
TS10	AC	209	4051	11
TS11	ORL	498	257	3
TS12	VS	2873	605	5
TS13	ORL	3199	68	1
TS14	VP	398	508	9

Most of the workers live in houses, and only four have basements (workers' dwellings TS1, TS8, TS11 and TS14) that are used as garage or warehouse.

Relating to the building materials, 36% of workers' dwellings are constructed with cement and brick and 28% are constructed with granite namely at pavement level and walls of the living room and fireplace (measurement location).

The existing ventilation in dwellings is essentially natural in all areas and in all rooms, with the usual daily opening of windows (1 or 2 times a day). The dwellings' questionnaires showed that none of the selected dwellings had air conditioner. The existence of the air conditioner reduce the concentration of indoor radon.

3.4 Measurements of indoor radon concentration

Table 3 presents the indoor radon concentrations measured within both the thermal spas and workers' dwellings (living room) as well as the distance between these locations. It can be observed that, in some cases, radon concentrations in indoor air of the thermal spa are very different (see e.g., TS13) from radon concentrations in workers dwellings (Table 3).

However, indoor radon concentrations in worker's dwellings from TS3 and TS10 are higher than radon concentrations in the respective workplace and in these cases the exposure to radon will be mainly residential. The reason of these extremely high values of indoor radon concentration in workers' dwellings is due to geological settings accentuated by the fact that the house is constructed by granite materials.

The indoor radon concentrations registered in the workers' dwellings from TS1 and TS12 are lower than in the respective thermal spa. In case of worker's dwelling from TS1 the reason is mainly because the house is not constructed in granite. For worker's dwelling from TS12, although being locate in a granite zone and being constructed with granite, the ventilation appears to be sufficient to reduce the concentration of radon.

4 CONCLUSIONS

The dwellings of workers from thermal spas TS2, TS3, TS5, TS7, TS9, TS10, TS12 and TS14, are cause for concern, since they contribute significantly to radon exposure at non-occupational environment.

Regarding the questionnaires, most of the workers have remained at their workplace between 5–20 years, which may represent an additional risk to health due to the long-term annual doses at which they are exposed to, within the thermal spa. In most of the thermal spa there is job rotation. Since the estimated annual effective dose is variable within the different workplace of each thermal spa, this rotation tends to reduce the occupational exposure to radon. It was observed that for some thermal spas there is only a natural ventilation system, and it was concluded that this is not adequate to decrease the radon levels in the indoor air.

In general, a system of surveillance, monitoring and radiation protection of workers should be implemented which should include, beside other measures: the use of individual protection equipment, such as masks (when justified); promote job rotation and improve ventilation conditions within the thermal spas facilities and providing an effective ventilation system for all workplaces. In addition, inform workers how to improve their dwelling's ventilation with simple actions such as opening doors and windows. In the most serious cases, adopt architectural and constructive solutions in order to reduce the building's contact with the ground and/or implement isolation measures of radon exhalation (Silva et al., 2016b).

REFERENCES

Al Zoughool, M. & Krewski, D. (2009). Health effects of radon: a review of the literature. International Journal of Radiation Biology 85(1): 57–69. doi:10.1080/09553000802635054.

Alberigi, S., Pecequilo, B.R., Lobo, H.A., Campos, M.P. (2011). Assessment of effective doses from radon levels for tour guides at several galleries of Santana Cave, Southern Brazil, with CR-39 detectors: preliminary results. Radiation Protection Dosimetry 145(2–3): 252–255. doi:10.1093/rpd/ncr054.

DGS (2002). Vinte Anos de Diagnóstico Precoce, Edição Direcção-Geral da Saúde, Cadernos da Direcção-Geral da Saúde, N.°1. ISSN 1645-4146EPA (2013). United States Environmental Protection Agency, Consumer´s Guide to Radon Reduction, How to fix your home. URL:http://www.epa.gov/radon, EPA 402/K-10/005 | March 2013.

Erdogan, M., Ozdemir, F., Eren, N. (2013). Measurements of radon concentration levels in thermal waters in the region of Konya, Turkey. Isotopes in Environmental and Health Studies 49(4): 567–574. doi:10.1080/10256016.2013.815182.

Gray, A., Read, S., Mcgale, P. and Darby, S. (2009). Lung cancer deaths from indoor radon and the cost effectiveness and potential of policies to reduce them. BMJ 338: a3110. doi:10.1136/bmj.a3110.

IAEA (2003). Radiation Protection against Radon in Workplaces other than Mines, jointly sponsored by the International Atomic Energy Agency and the International Labour Office. Safety Reports Series No. 33, IAEA, Vienna. http://www-pub.iaea.org/MTCD/publications/PDF/Pub1168_web.pdf.

ITN (2010). Radão—um gás radioativo de origem natural. Instituto Tecnológico e Nuclear. Departamento de Protecção Radiológica. e Segurança Nuclear. URL: http://www.itn.pt/docum/relat/radao/itn_gas_radao.pps.

Koray, A., Akkaya, G., Kahraman, A., Kaynak, G. (2014). Measurements of radon concentrations in waters and soil gas of Zonguldak, Turkey. Radiation Protection Dosimetry 162(3): 375–381. doi:10.1093/rpd/nct308.

Köteles, G.J. (2007). Radon Risk in Spas? CEJOEM 13(1): 3–16. http://www.omfi.hu/cejoem/Volume13/Vol13No1/CE07_1-01.html.

Mueller, A. (1998). Handbook of Radon in Buildings: Detection, Safety, & Control. SYSCON Corporation, Brookhaven National Laboratory. ISBN-13: 000-0891168230.

Nikolopoulos, D., Vogiannis, E., Petraki, E., Zisos, A., Louizi (2010). Investigation of the exposure to radon and progeny in the thermal spas of Loutraki (Attica-Greece): Results from measurements and modelling. Science of the Total Environment 408(3): 495–504. doi:10.1016/j.scitotenv.2009.09.057.

Nikolov, J., Todorovic, N., Pantic, T.P., Forkapic, S., Mrdja, D., Bikit, I., Krmar, M., Veskovic, M. (2012). Exposure to radon in the radon spa Niška Banja, Serbia. Radiation Measurementes 47(6): 443–450. doi:10.1016/j.radmeas.2012.04.006.

Pereira, A.J.S.C., Dias, J.M.M., Neves, L.J.P.S. e Godinho, M.M. (2001). O Gás Radão em Águas Minerais Naturais: Avaliação do Risco de Radiação no Balneário das Caldas de Felgueira (Portugal Central). Memórias e Notícias, Publicações do Departamento de Ciências da Terra e do Museu Mineralógico e Geológico da Universidade de Coimbra, n.° 1, Coimbra.

Radolić, V., Vuković, B., Smit, G., Stanić, D., Planinić, J. (2005). Radon in the spas of Croatia. Journal of Environmental Radioactivity 2005, 83(2): 191–198.

Robertson, A., Allen, J., Laney, R., Curnow, A. (2013). The cellular and molecular carcinogenic effects of radon exposure: a review. International Journal of Molecular Sciences 14(7): 14024–14063. doi:10.3390/ijms140714024.

Santos, T,O., Rocha, Z., Cruz, P., Gouvea, V.A., Siqueira, J.B., Oliveira, A.H. (2014). Radon dose assessment in underground mines in Brazil. Radiation Protection Dosimetry 160(1–3): 120–123. doi:10.1093/rpd/ncu066.

Silva, A.S. & Dinis, M.L. (2015a). The presence of radon in thermal spas and their occupational implications—a review. Book chapter in: Occupational Safety and Hygiene III, Eds. P. Arezes, J.S. Baptista, M. Barroso, P. Carneiro, P. Cordeiro, N. Costa, R. Melo, A.S. Miguel, G. Perestrelo, pp. 353–355, ISBN 978-1-138-02765-7, London: Taylor & Francis.

Silva, A.S. & Dinis, M.L. (2016a). Measurements of indoor radon and total gamma dose rate in Portuguese thermal spas Book chapter in: Occupational Safety and Hygiene IV, Eds. P. Arezes, J.S. Baptista, M. Barroso, P. Carneiro, P. Cordeiro, N. Costa, R. Melo, A.S. Miguel, G. Perestrelo, pp. 485–489, ISBN 978-1-138-02942-2, London: Taylor & Francis.

Silva, A.S., Dinis, M.L., Diogo, M.T. (2013). Occupational Exposure to Radon in Thermal Spas, Book chapter in: Occupational Safety and Hygiene, Eds. P. Arezes, J. S. Baptista, M. Barroso, P. Carneiro, P. Cordeiro, N. Costa, R. Melo, A. S. Miguel, G. Perestrelo, pp.273–277. ISBN: 9781138000476, London: Taylor & Francis, 2013.

Silva, A.S., Dinis, M.L., Fiúza, A. (2014). Research on Occupational Exposure to Radon in Portuguese Thermal Spas. Book chapter in: Occupational Safety and Hygiene II, Eds. P. Arezes, J.S. Baptista, M. Barroso, P. Carneiro, P. Cordeiro, N. Costa, R. Melo, A. S. Miguel, G. Perestrelo, pp. 273–277. ISBN: 978-1-138-00144-2, London: Taylor & Francis.

Silva, A.S., Dinis, M.L., Pereira, A.J.S.C. (2016b). Assessment of indoor radon levels in Portuguese thermal spa, Radioprotection, DOI: 10.1051/radiopro/2016077 (in press).

Simões, L.M.F., Santos, J., Valente, J., Lopes, M., Borrego, C. (2007). Concentração de radão em espaços interiores da área de Viseu, 9ª Conferência Nacional do Ambiente, Aveiro, Abril, 18–20, 2007.

UNSCEAR (2000). United Nations Scientific Committee on Effects of Atomic Radiation, Sources and Effects of Ionizing Radiation. United Nations, New York, United Sales publication E.00.IX.3

Vaupotič, J., Streil, T., Tokonami, S., Žunic, Z.S. (2013). Diurnal variations of radon and thoron activity concentrations and effective doses in dwellings in Niška in Banja, Serbia. Radiation Protection Dosimetry 157(3): 375–382. doi:10.1093/rpd/nct145.

WHO (1986). Indoor air quality research. Report on a WHO meeting, Stockholm, August, 27–31, 1986.

WHO (2007). Radon and cancer. Fact Sheet Nr. 291. WHO, Geneve.

WHO (2009). WHO handbook on Indoor Radon, A Public Health Perspective, ISBN 978 92 4 154767 3. URL: http://whqlibdoc.who.int/publications/2009/9789241547673_eng.pdf.

Wiwanitkit, V. (2009). Radon in natural hot spring pools in Thailand: Overview. Toxicological & Environmental Chemistry, 91 (1): 1–4, doi:10.1080/02772240802028609.

Yarar, Y., Günaydi, T., Celebi, N. (2006). Determination of radon concentrations of the Dikili geothermal area in western Turkey. Radiation Protection Dosimetry 118(1): 78–81. doi:10.1093/rpd/nci321.

Occupational Safety and Hygiene V – Arezes et al. (Eds)
© *2017 Taylor & Francis Group, London, ISBN 978-1-138-05761-6*

Hazards identification during design phase

F. Rodrigues
RISCO, Department of Civil Engineering, University of Aveiro, Aveiro, Portugal

A. Barbosa & J. Santos Baptista
LABIOMEP, Faculty of Engineering, University of Porto, Porto, Portugal

ABSTRACT: Occupational risk prevention should be an integral part of the design process since its earliest stage to achieve a reliable level of safety. So, designers should assess the design at each stage, identifying any significant hazard, eliminating it by changing the project itself. All designs progresses from the concept to detail, but in buildings rehabilitation there is a high uncertainty of former construction processes, materials, structural stability, among others. Considering the lack of information for risk prevention in the design phase of old building rehabilitation, the characteristic of XIX Century buildings of Porto historical center were studied. Aiming to support designers in the hazards identification in an early stage of the rehabilitation design process, were developed hazard identification sheets based on different constructive processes. It can be concluded that this tools can give designers support to modify their options to eliminate/reduce the hazards during the construction and the use phase.

1 INTRODUCTION

The Construction sector has several specific characteristics such as great capital involved, intensive dynamic work, labour intensity, immobility and long life cycle of the final product, and singularity of objects that are developed in a specific institutional, economic and social context. This sector also has great fragmentation, great variety of enterprises with different dimension and specialization and lots of casual work (Vrijhoef & Koskela, 2005; Swuste et al., 2012). The sector's specificities contribute to increase the difficulty of preventive measures implementation and consequently to the increase of occupational injuries (Swuste et al., 2012). The deficient correspondence between the design and the execution phase, the lack of specificity, synergy and effectiveness between safety planning and the design and construction process are main problems, usually considered separated that effectively contribute to different levels of construction management failures along with all the construction/rehabilitation life cycle. So, design was identified by European Directive 92/57/EEC as the key role of optimising safety, which is strictly and continuously linked with constructability since the earlier design phase (Capone et al., 2014).

It is recognized that the most important contribution that a designer can make is at the concept and early design development stages of a project, when project-wide and systems for hazards and risks prevention are being considered not only for the construction phase but also for the use phase. During the longer phase of a building (use phase) several interventions of maintenance, replacement, refurbishment have to be done, and during the design phase no integrated safety measures are usually considered, neither even the reduction of exposure by designing for low frequency of maintenance actions (e.g. using low maintenance materials).

Behm & Schneller (2013), Lingard et. al (2013) and Cooke & Lingard (2011), demonstrated in their re-search the importance of looking beyond the immediate/primary circumstances of an accident to identify the failures at the upstream of construction pro-jects such as the design, client actions, education and economic environment in accident investigations (Gibb et al., 2014). Also Frijters and Swuste (2008) recognize the influence that designers, architects and structural engineers have on the health and safety of the construction sites, considering the design choices the determining factors to achieve safe construction practices and the need to integrate safety in the design objectives.

Beyond the new construction, rehabilitation of heritage buildings is a great challenge for design teams both in technical and risk prevention issues, since the early stage, during the inspections that have to be done to assess the degradation degree at the structural and non-structural level. This work aim to present occupational hazard identification sheets based on the existent anomalies in XIX century buildings of Oporto city, and on the rehabilitation solutions that have to be implemented, which can be a support to risk prevention through design.

2 DESIGN HAZARD IDENTIFICATION

2.1 *Object of study*

This study considers, as object of study, the XIX century buildings constructed in the city of Oporto, in the north of Portugal. For the development of the hazard identification sheets, research about the constructive characteristics of these buildings was made. These architectural valued buildings comprise ground floor and elevated floors, with a rectangle shape that vary from 4–7 meters of width and 15–25 meters of length. The structure is composed by large external granite stone walls (with different thicknesses). The interior floors and ceilings are in timber structure as the roof structure (Teixeira, 2004).

Heritage buildings have different structural and non-structural anomalies, that can lead to severe safety problems for all who direct or indirectly contact with the building (Holland & Montgomery, 1992). During the rehabilitation process safety is a concern since the initial phase, even during the inspection phase of the existent building. In almost the situations there is no information about the former design and of all the transformations that have been introduced along the building life cycle. Some of them imply important changes in the morphology leading to dangerous structural conditions that expose users and professionals to high risks (Cóias, 2006).

2.2 *Risk assessment during design phase*

Project management failures can occur in the aim of: project planning and subsequent construction/rehabilitation planning, inadequate project schedule, different projects incompatibility, no compliance with the risks prevention principles. These led to incorrect adoption of hazardous raw materials and dangerous constructive processes, complex architectural solutions with lower safety level, absence of execution details and lack of maintenance, and demolition procedures (Haslam et al. 2005).

The risk assessment carried out during the design phase as well as the production of documents aiming the risk prevention during the subsequent phases may constitute failures in terms of risk management, when considered as a mere paper exercise or when performed incorrectly corresponding to any specific work (Haslam et al., 2005; Fung et al., 2012). The existence of shortcomings in this area later can influence safety at workplace and the construction planning.

2.3 *Hazard identification sheets*

As in new construction, in building rehabilitation the design phase is crucial to obtain a final product with quality and safety, preserving its cultural and architectonic value and satisfying the current functional requirements. The design options (structural, architectural, materials, components and constructive systems) imply the lower or higher level of hazards exposure during the construction phase and also during the interventions during the use phase. Aiming the hazards identification during the design phase, supported on the anomalies' identification and on the technical solutions that could be adopted during this phase, hazard identification sheets have been developed for the following constructive elements: foundations, external walls, internal timber floors, roofs. These sheets point out the measures that could be implemented during the design, the construction and the maintenance phase of the building as depicted in Figure 1. In this paper the sheet that is presented is the one related with timber floors.

In this buildings, frequently the timber floor has severe structural anomalies that can be considered by two different methods, considering the floor performance: in accordance with vertical actions and the general floor, or considering the seismic actions and the whole building.

The floor reinforcement to face the vertical actions and to eliminate the existent anomalies can be obtained by three different processes (Cóias, 2006):

> Inclusion of new beams;
> Maintenance of the existent structure, with the reinforcement of the elements connections;
> Repair, partial substitution or reconstruction of some parts of the structure.

The new beams addition in the structural system is the simplest method, consisting on including new timber or metallic beams, supported on the lateral walls that have to be prepared to receive these vertical actions. The advantage of this solution is to maintain the interior free height, but the challenging of the execution of walls openings to receive the new beams is a disadvantage. When the objective is to re-store the strength of the structure it is more frequent to implement the second method: to maintain the existent structure through the junction of new pieces to fixed and reinforce the connections between the existent structure (Cóias, 2007). One of its advantages is the needless of removing any element, which is a difficult operation at the structural and occupational safety level (Roseiro, 2012). The partial repair or substitution of beams is the most arduous intervention and even the most dangerous consisting on the

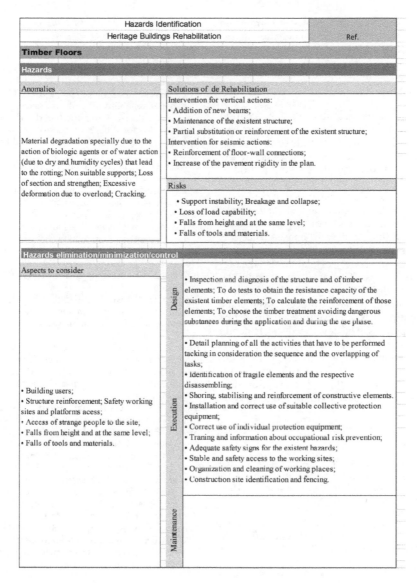

Figure 1. Hazard identification sheet for design phase—timber floors rehabilitation.

previous shoring of the structure, cutting and substitution of the damaged part by timber or steel elements.

The floor repair considering the seismic actions consists on the (Roseiro, 2012):

Reinforcement of the floor-wall connection;
Increase of the floor rigidity in the plan.

The connection between the floor structure and the walls is normally reinforced by bolts that are preached to the beams by metallic elements. When the option is to increase the rigidity of the pavement in the plan it must be avoided the significant in-crease of the structure mass that contribute to higher seismic risk for all the building. Nowadays some techniques consist on the application of pieces of wood or metal or even composite materials of fiber reinforced polymers above the existent floor (Roseiro, 2012). These hazards identification sheets sup-port the designer team, relating the anomalies identified during the inspection phase, with the different rehabilitation solutions and alert them for the occupational hazards. When applied, for each one of the constructive elements that have to be subject to maintenance actions exposing workers to hazards and risks, the sheet also include the safety measures that have to be integrated in the design, in the rehabilitation and after in the maintenance works.

3 CONCLUSIONS

Currently is strongly accepted that prevention thought design is the more reliable way to achieve high safety standards, independently of the human behavior during the execution phase. As the life cycle of a building corresponds to a large period of time in which the use phase is the longer one, it cannot be forgotten the safety measures that have to be integrated during the design phase, as for example, materials, components and constructive systems with high durability, needing in consequence less maintenance and substitution actions, and safety systems that permit safe work at roofs, facades and confined spaces. This integrated approach focused on the rehabilitation processes is intrinsically linked with the technical process given value to the occupational risk prevention in the scope of the design process. The developed sheets herein explained aims to be an easy and dynamic support for designers, giving the main highlights that will help them to think and implement the most suitable solutions envisaging the risk prevention during the next stages of the rehabilitation process.

REFERENCES

Behm, M. & Schneller, A. 2013. Application of the Loughborough Accident Causation model: a framework for organizational learning. *Construction Management and Economics* 31(6): 580–595.

Capone, P., Getuli, V. & Giusti, T. 2014. Constructability and safety performance based design: a design and assessment tool for the building process. *Automation and Robotics in Construction, Proc.* 31st *Int. Symp.* 313–320, Lithuania, 9–11 July.

Cooke, T. & Lingard, H. 2011. A retrospective analysis of work-related deaths in the Australian construction industry. *ARCOM, Proc. 27th Annual Conf.*, Reading, UK, 5–7 September.

Cóias V. 2006. Inspeçoes e Ensaios na Reabilitação de Edifício, Lisboa, IST Press.

Cóias V. 2007. Reabilitação Estrutural de Edifícios Antigos, Lisboa, Argumentum/GECoRPA.

Fung, I.W.H.; Lo, T.Y.& Tung, K.C.F. 2012. Towards a better reliability of risk assessment: Development of a qualitative & quantitative risk evaluation model (Q2REM) for different trades of construction works in Hong Kong. Accident Analysis and Prevention. 48 (0): 167–184.

Gibb, A.G.F., Lingard, H., Behm, M. & Cooke, T. 2014. Construction accident causality: learning from different countries and differing consequences". *Construction Management and Economics* 32(5): 446–459.

Haslam, R.A.; Hide, S.A.; Gibb, A.G.F.; Gyi, D. E.; Pavitt, T.; Atkinson, S. & Duff, A. R. 2005. Contributing factors in construction accidents. Applied Ergonomics, 36 (4): 401–415.

Holland, R. & Montgomery, B.E. 1992. Appraisal & Repairs of Building Structures. London: Thomas Telford

Lingard, H., Cooke, T., & Gharaie, E. 2013. A case study analysis of fatal incidents involving excavators in the Australian construction industry. *Eng. Constr. Archit. Mang.* 20(5): 488–504.

Roseiro, J.R.F. 2012. Causas, anomalias e soluções de reabilitação estrutural de edifícios antigos—Estudo de Caso. Dissertação de Mestrado, Faculdade de Ciências e Tecnologias da Universidade de Lisboa, Lisboa (in Portuguese).

Swuste, P. & Frijters, A., 2008. Safety assessment in design and preparation phase. Safety Science 46: 272–281.

Swuste, P., Frijters, A., & Guldenmund, F., 2012. Is it possible to influence safety in the building sector? A literature review extending from 1980 until the present". *Safety Science* 50(5): 1333–1343.

Teixeira, J.J.L., (2004). Descrição do sistema construtivo da casa burguesa do Porto entre os Séculos XVII e XIX. Provas de Aptidão pedagógica e capacidade científica. FAUP, Porto (in Portuguese).

Vrijhoef, R. & Koskela, L., 2005. A critical review of construction as a project-based industry: identifying paths towards a project-independent approach to construction, in CIB combining Forces.

Occupational Safety and Hygiene V – Arezes et al. (Eds)
© *2017 Taylor & Francis Group, London, ISBN 978-1-138-05761-6*

Non-ionizing radiation levels in environments with VDT of UFPB's technology center, João Pessoa, Brazil

S.L. Silva, L.B. Silva, M.B.F.V. Melo, E.L. Souza & T.R.A.L.A. Cunha
Federal University of Paraíba, João Pessoa, Brazil

ABSTRACT: Extremely Low Frequency (ELF) non-ionizing radiation has been correlated with the development of diseases such as leukemia, sleep disorders, mental disorders, and other metabolic disorders. Workers and students who make use of technology such as computers in environmental with VDT for a long period of time, may be exposed to the development of health problems, awakening to the increase to the number of studies investigating this exposure. This study presents an analysis of the magnetic field levels emitted by desktop and laptop computers in two laboratories and in a departament secretariat of the UFPB's Technology Center. It was found that one of the laboratories has high levels of non-ionizing radiation and it is because the laboratory's layout provides a concentration of computers which can increase the thermal load in the workplace.

1 INTRODUCTION

The further development of electronic technology with ample facilities and benefits made almost inevitable the use of computers, but when considering the various sources of heat in computerized environments, and some of these sources come from equipment that is being produced in smaller proportions, with configurations of their internal components closer, then it is likely that in some workplaces these environments have representative Non-Ionizing Radiation (NIR) levels. In this sense, the users of these technologies may be exposed to electric and magnetic fields, and with the more intense use of these technologies in performing various activities in various sectors, such as education, there is concern about the possible health effects on workers (Ahlborn *et al.*, 2001; WHO, 1998; Kokalari & Karaja, 2011).

Non-ionizing radiation is one that doesn't generate ionization because it lacks adequate energy for the emission of electrons, atoms or molecules. It encompasses radio waves, visible light, the infrared, the microwave, and the extremely low frequency range (Másculo, 2008). Non-ionizing radiation or electromagnetic fields around computers are Extremely Low Frequency (ELF) in the range of 3–300 Hz and Very Low Frequency (VLF), in the range of 3–30 kHz (Maisch *et al.* 1998). But the International Cancer Research Agency (IARC) in 2002 submitted a report on exposure to such fields may be related to the development of cancer. Other studies point to the onset of mental disorders (Davanipour *et al.*, 2014), sleep disturbances (Monazzam *et al.*, 2014), increased risk for leukemia (Calvante *et al.*, 2010), metabolic alterations (Lerchl *et al.*, 2015), among others.

Taking into account the published studies regarding the effects of NIR on health, and the increasing and intense use of computers in environments with new ICT, it is observed the need to investigate the magnetic fields in the workplace, as well as to analyzing factors that may interfere in the increase or decrease of non-ionizing radiation.

2 MATERIALS AND METHODS

2.1 *Magnetic field*

This study consists of measuring the magnetic field in the extremely low frequency range (3–3000 Hz) produced by the type desktop and laptop. For this frequency range, the reference levels adopted were proposed by TCO Certified Displays 7.0. (2015).

2.2 *Measuring device*

For the measurement, was used the equipment Spectran NF 5030, which allows the use of limits for the frequency range of the fields, making measurements on all coordinate axes X, Y and Z and its resulting vector.

2.3 *Characteristics of workplace*

For the measures were chosen five workplaces with VDT located at the Technology Center of the Federal University of Paraiba.

Environment 1 – Work Analysis Laboratory of the CT Technology Center (Fig. 3). It has 75 m², air-conditioned by two air-conditioning units, has a modem for wireless internet fidelity (Wi-fi), 3 printers, of which 2 have wireless system, 4 "all in one" computer and 2 desktops all of 21"; 4 stabilizers, a refrigerator, a 60 "TV and a microwave. Externally to the building about 17 m away you have a power transformer. At the same time 4 different-inch notebooks and 9 portable phones are used. It has 9 workplaces, being selected for the study only one workplace in which the technician uses a notebook of 15.6".

Environment II – Laboratory of Computational Intelligence Applied to Electrical Engineering (Fig. 4), it has 60 m², conditioned by 2 air-conditioning units, has Wi-Fi internet modem, 1 wireless printer, 8 computers 19 "desktop type and 4 14" notebooks, 1 microwave and has 14 workplaces, 2 of which were selected for the study. 12 computers of various models and 13 portable telephones are used daily. The *Workplace A* contains a 19 "computer", 1 stabilizer, and a line filter. The *Workplace B*, includes a 14 "notebook and 1 switch".

And the *Environment III* – Department of Industrial Engineering Secretary of CT (Fig. 5) has an area of approximately 30 m², has a video monitor, 2 nobreaks, 2 printers, 2 computers, 2 desktop type and 1 notebook allocated to the left side of the room, distributed by 2 jobs. In these workplace, usually 2 mobile phones are used daily. And to the right side of the room there is a microwave. In this environment, the two workplaces were analyzed. *Workplace A* has 1 21 "computer". And *Workplace B* contains a 21 "computer".

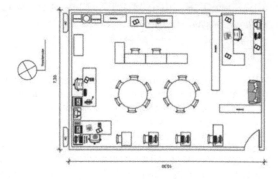

Figure 3.   Environment I Layout.

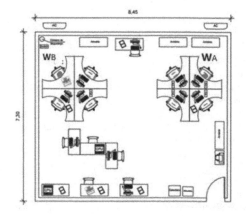

Figure 4.   Environment II Layout.

Figure 1.   Spectran NF 5030.
Font: Aaronia products.

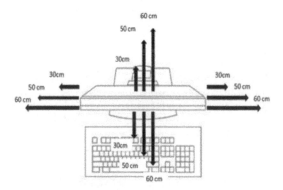

Figure 2.   Measurements of the magnetic field at distances of 30, 50 and 60 cm.

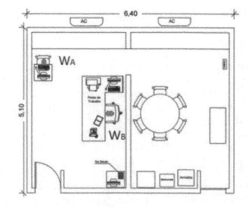

Figure 5.   Environment III Layout.

420

2.4 *Characteristics of measurements*

To perform the measurements in the three environments in the normal working period in the morning, was chosen the distances of 30, 50 and 60 cm from the computer or notebook display (Fig. 2) to the left, right, front and back, standardized according to the study of Dehagui *et al.* (2016). And was considering the usual use of the electronic devices.

3 RESULTS AND DISCUSSIONS

In the 30 cm distance measurement (Table 1), Environment I, Environment II and Environment III—Workplace B are below the limit established by TCO Certified Displays 7.0. However, Environment III—Workplace A is above the limit on the front, back and left side of the computer display, the increase of which may be due to the presence of the "nobreak" equipment.

In the measurement of 50 cm (Table 2), the Environments I and II are in compliance with the norm, as well as Workplace B of the Environment III. While Workplace A of the environment III presents with the measurements of the left and rear side of the display above the limit established by the TCO.

Table 1. Measurements under the distance of 30 cm.

30 cm	Front nT	Back nT	Left nT	Right nT
Env. I*	123.8	120.9	123.2	177.8
Env. II (A)**	175.2	190	176.3	
Env. II (B)***	174.1	170.4	171.4	
Env. III (A)	227.9	433.4	255	185.6
Env. III (B)	171.5	184.6	176.4	177.8

*Env. = Enviroment, ** (A) workplace A, ***(B) workplace B.

Table 2. Measurements under the distance of 50 cm.

50 cm	Front nT	Back nT	Left nT	Right nT
Env. I*	124.5	123.6	124.8	125.6
Env. II (A)**	175.6	183	175.5	
Env. II (B)***	174.2	171	171.2	
Env. III (A)	196.3	244.6	204.9	178
Env. III (B)	173.3	178.9	176.5	178

*Env. = Enviroment, ** (A) workplace A, ***(B) workplace B.

In the measurement of 60 cm (Table 3), we have that the Environment I, the Environment II and the Workplace B of the Environment III are below the limit of 200 nT, established by the adopted Resolution. The Workplace A of the Environment III in the back position of the display is a little above the established.

As for laterality, it is observed that in Environment I, in which the object taken for analysis was a *notebook*, a prevalence of superiority of the magnetic fields in the right side is observed, as well as in Workplace B of the Environment III. These findings are not the same as those found in the Golmohammadi (2014) study, that when measuring the magnetic field under distances of 30, 60 and 90 cm in common laptops, it found constant measurements, around 28–32 mA/m, Dehagui *et al.* (2016) performed measurements under the distances of 30, 50 and 60 cm and found for laptop measurements of the front side that are larger than the lateral and back measurements of the laptop.

As for the desktop model, most of the measurements in the back of the display are superior to the others, corroborating the findings of Dehagui et al. (2016).

The Workplace A of the environment III presented fields higher than that determined by the TCO standard adopted by Kokalaril & Karaja (2011) who observed that most of the studies of non-ionizing radiation emitted by Video Display Units (VDU) refer mainly to the requirements of TCO, which established in the low-frequency range, much stronger protection limits than the International Commission on Non-Ionizing Radiation Protection (ICNIRP).

It is also observed that in the fields measured in Environment III, its levels are higher than the other environments. It is believed that the organization of the island layout can be the determining factor of this increase, since the characteristics of equipment found in this environment in relation to the environment I are very similar. So this distribution of electronic equipment in the island layout is

Table 3. Measurements under the distance of 60 cm.

60 cm	Front nT	Back nT	Left nT	Right nT
Env. I*	124.8	124.3	123.8	124.7
Env. II (A)**	178.6	183.4	174.2	
Env. II (B)***	174.3	171.8	173.2	
Env. III (A)	185	246.3	196	177.2
Env. III (B)	174.6	180.9	176.9	178

*Env. = Enviroment, ** (A) workplace A, ***(B) workplace B.

characterized by the approximation of electronic equipment, which can concentrate the electromagnetic field and result in this increase of non-ionizing radiation. Environment II has the smallest area, and a number of electronic equipment less representative. But this area does not allow the introduction of new technologies, as it would lead to the increase of the thermal load, which could increase the level of NIR.

4 CONCLUSION

Of the analyzed workplaces, the environment that presented the smallest area concentrated a greater density of magnetic fields, although it had a smaller number of equipment than the other environments, was the enviroment III. The organization of the layout was another pre-ponderant factor in the increase of the levels of NIR, since workplace distributed in islands concentrate a greater number of equipment, which can increase the thermal load, allowing an increase of the NIR levels.

However, it was observed that at distances 30, 50 and 60 cm laptops have lower levels of NIR compared to desktops. All these equipments are close to others like printer, Wi-Fi modem, thus increasing the thermal load in the environments.

It is concluded that from the presented results, it is necessary a greater attention to the new environments equipped with new ICT, making it necessary to perform other measurements of NIR in other workplaces in different computerized environments evaluating: (1) the elements Architecture; (2) checking for the exchange of heat between man and the environment by radiation; (3) analyzing the organization of work and (4) layout from the ergonomic point of view, but taking into account the specificities of the activities performed in them.

REFERENCES

Ahlbom, A; Cardis, E; Green, A; Linet, M; Savitz, D; .. Swerdlow, A. 2001. Review of Epidemiological Literature of EMF and Health *Environmental Health Perspectives (109):* 911–933.

Calvente, I; Fernandez, MF; Villalba, J; Olea, N; Nuñez, MI.2010. Exposure to electromagnetic fields (non-ionizing radiation) and its relationship with childhood leukemia: A systematic review *Science of the Total Environment Journal.* (408): 3062–3069

Davanipour, Z; Tseng, CC; Lee, PJ; Markides, KS; Sobel, E. 2014. Severe Cog nitive Dysfunction and Occupational Extremely Low Frequency Magnetic Field Exposure Among Elderly Mexican Americans. *Br Med J Med Res* 4 *(8):* 1641–62.

Dehaghi, BF; Ghamar, A; Latifi, SM. 2016. Electromagnetic Fields and General Health: A Case of LCDs vs. Office Employees. *Jundishapur J Health Science,* 2016.

Golmohammadi, R; Ebrahimi, H; Fallahi, M; Soltanzadeh, A; Mousa Vi, SS.2014. An investigation of Extremely Low and Frequency (ELF) Electro magnetic field emitted by common Laptops *J Health Safety Work* 4 *(1):* 11–20.

IARC International Agency For Research On Cancer. 2002. *Monographs on the Evaluation of Carcinogenic Risks To Humans. Non-ionizing Radiation, Part1: Static and extr and low-frequency mely (ELF) electric and magnetic fields* Lyon France: World Health Organization. 80.

Kokalari, I; Karaja T. 2011. Evaluation of the Exposure to Electromagnetic Fields in Computer Labs of *Schools. Journal of Electromagnetic Analysis and Applications.* (3): 248–253.

Kokalaril, I; Karaja, T.2011. Evaluation of the Exposure to Electromagnetic Fields in Computer Labs of *Schools. Journal of Electromagnetic Analysis and Applications.* (3) 248–253.

Lerchl, A; Klose, M; Grote, K; Wilhelm, AFX; Spat h mann, O; Fiedler, F; Streckert, J; Hansen, V; Clemens, M. 2015. Tumor promotion by exposure to radio frequency electronic magnetic fields below the exposure limits for humans. *Bioche mical and Biophysical Research Communications.*

Maisch, D; Rapley, B; Rowland, RE; Podd, J. 1998. Chronic Fatigue Syn Drome-Is prolonged exposure to environmental level powerline frequency electromagnetic fields a co-factor to consider in treat. ment *J Austr Coll Nutr Env Med* 17*(2):* 29–35.

Master, F S.2008. Ergonomics, hygiene and safety of the three balho. In: Battle, M. (Eds.) *Introduction to production engineering.* Campus/Elsevier: Rio de Janeiro.

Monazzam, MR; Hosseini, M; Matin, LF; Aghaei, HA; Khoroabadis, H; Hesami, A. 2014. Sleep quality and general health status of employ ees exposed to extremely low fr quency and magnetic fields in the petrochemical complex *J Environ Health Sci Eng* (12): 78.

TCO DEVELOPMENT. 2015. TCO Certified Displays 7.0. 2015. Available at: < http://tcodevelopment.com/tco-certified/ > Access: <November 4, 2016>.

WHO Division of Global and Integrated Environmental Health 1998. Video Display Units (VDUs) and Human Health. FactSheet.201.

Occupational Safety and Hygiene V – Arezes et al. (Eds)
© *2017 Taylor & Francis Group, London, ISBN 978-1-138-05761-6*

Measurement of the reverberation time and sound level for improvement acoustic condition in classrooms—a case study

A. Partovi Meran & M. Shahriari
Necmettin Erbakan University, Konya, Turkey

ABSTRACT: This paper presents the results of measurements of an acoustic evaluation of lecturing rooms. The lecturing rooms located at Necmettin Erbakan University, Koyceyiz campus in Konya, Turkey. This study was focused on three classrooms representing of all lecturing rooms of the faculty of Engineering. The acoustic quality of the classrooms was analyzed according to measurements of the noise level and reverberation time. The reverberation time measured by the impulsive excitation method according to the ISO 3382-18 and ISO 3382-29. To measure internal ambient noise frequency analyzer type 2250 and frequency analysis software BZ-7223 were used. Results have been compared with reference values recommended in NASI S12.60, DIN 18041 and Turkish occupational noise control regulation. Comparison results indicate that acoustical quality of the surveyed classrooms is inadequate. It is necessary to perform an acoustical treatment on the all classrooms of campus.

1 INTRODUCTION

Education is an essential factor for developing any countries and learning quality is considered as an important indicator of a good education. On the other side, effectivity of learning process depends on quality of verbal communication between teacher and the students. In fact, excessive background noise or reverberation in classrooms interferes with speech communication and thus cause an acoustical barrier to learning. With good classroom acoustics, learning is easier, deeper, more sustained, and less fatiguing. Therefore acoustical conditions in the classroom in terms of speech transmission quality should be considered as an important task, preferably, during the design (Tang & Yeung, 2006). Building designers and planner should have in mind that environmental factors including noise conditions could affect the quality of learning and teaching process. According to Kruger and Zanin (2004):

"A poor acoustic performance of the classroom will have an effect on both the understanding by the students and the physical stress of the teacher".

In deed high levels of noise in the classroom make students tired and effect on their attention to and understanding the contents of the lectures (Zanin & Zwites, 2009). Considering what has been mentioned above, a study is needed to investigate the acoustic condition of lecturing rooms of Necmettin Erbakan University, Koyceyiz campus in Konya, Turkey. The construction of this building was finished in 2015. Therefore, a study is needed to see whether the noise level in the class rooms of the building suits the standards of acoustical performance criteria. This study is focused on the noise control of lecturing rooms of the campus. It is hoped that the results could provide useful information for improvement of acoustical quality of existing classrooms and the new classrooms during the design stage.

2 METHODOLOGY

The acoustical conditions in classrooms and effect of acoustical quality on the student and teacher health have been attracted the attention of researchers worldwide. For instance Puglisi measured the parameters affecting acoustic comfort in high-school classrooms for students and teachers ((Puglisi, 2015). Cantor Cutiva and Burdorf conducted a study to investigate effects of noise and acoustics in schools on vocal health in teachers (Cantor Cutiva & Burdorf, 2015). Shield et al. researched on the acoustic conditions and noise levels in secondary school classrooms in England (2015). To describe classroom acoustical situations, Reverberation Time (RT), Signal-to-Noise Ratio (SNR), and background noise level measurements had been used (Bradley, 1987 and Crandell et al, 1995). Also Zannin and Zwirtes measured the reverberation time, sound pressure level inside and outside the classrooms, and sound insulation parameters to evaluate the acoustic comfort of classrooms (Zannin & Zwirtes, 2009). They followed the Brazilian Standards NBR 10151 and NBR 10152 to measure ambient noise. Also to measure the reverberation time and sound insulation, the international Standards ISO 140-4, ISO 140-5, ISO 717-1, and ISO

3382 were followed. Tang and Yeung used Octave band reverberation times, background noise levels and speech transmission indices to study the acoustic condition in classrooms (Tang & Yeung, 2006).

Reverberation affects verbal-communication quality by influencing the early- and late-arriving energies, the reverberation time increases with the size of the room, and decreases with the amount of sound absorption in it (Hodgson, 2004). The RT can be measured by the interrupted noise method and by the integrated impulse response method according to the ISO 3382-18 and ISO 3382-29. Several international standards establish reference values for the RT in the classrooms. In the USA, RT is measured in the furnished and unoccupied classroom and is given as the maximum RT for mid-band frequencies of 500 Hz, 1000 Hz and 2000 Hz, and RT (ANSI S12.60, 2002). In Germany, RT is measured in the furnished and occupied classroom and that RT values represent the average in 2-octave bands including 500 Hz and 1000 Hz. This standard recommends reverberation time values for rooms with various volume as following: V = 125, 250, 500, and 750 m³ RT = 0.5, 0.6, 0.7, and 0.8 s respectively (DIN 18041, 2004). WHO (World Health Organization) recommends the value of 0.6 s for the RT in classrooms (WHO, 2001).

In this study, in order to evaluate the acoustic quality of classrooms, three different classrooms with various specifications were selected. The selected classrooms are samples of all lecturing rooms of Necmettin Erbakan University, Koyceyiz campus in Konya. The selected classrooms dimension and volumes are presented in Table 1. In the selection of rooms, parameters such as room volume, number of students, distance to outside noise sources and number of windows were considered. To verify the classrooms acoustic condition, in situ measurement of reverberation time RT and ambient noise pressure level were conducted. The Reverberation Time (RT) was measured using pistol shots method and the reverberation time module of Hand-held Analyzer type 2250 of Bruel and kjar. Hand-held analyzer type 2250 as shown in Figure 1 is a single-channel general-purpose hand-held analyzer. This analyzer with the appropriate software modules can be used as a single-channel, single-range sound level meter and frequency analyzer. Microphone type of prepolarized free field

½ inch condenser is mounted on the analyzer. For acoustic calibration, a sound level calibrator with a calibration frequency of 1 kHz and a calibration level of approximately 94 dB was used. The measurements were performed at five different points in each classroom. Sound level meter type 2250 of Bruel and kjar was employed to obtain sound pressure levels, the measurements have been done under empty and occupied conditions. In both condition, the campus was not empty, while routine activities were carried out. Figure 2 shows the height, length and width of the lecturing C3 with desk arrangement. The noise was measured at a single spot positioned at the center of the classroom with 3 min duration. The continuous equivalent sound level L_{Aeq} and its range of vibration from L_{Amin} to L_{Amax} were obtained.

Figure 1. Hand-held analyzer type 2250 during measurement.

Figure 2. Lecturing room C3 with desk arrangement.

Table 1. Dimension and volume of classrooms.

Classroom	Dimensions (m)	Volume (m³)
C1	6 × 8.2 × 4.5	221
C2	10 × 8.2 × 4.5	369
C3	13.5 × 8.2 × 4.5	498

3 RESULTS AND DISCUSSION

3.1 Internal ambient noise

To evaluate the acoustic condition of lecturing rooms, firstly, measurements were conducted inside of three empty classrooms. By frequency analyzer type 2250 and frequency analysis software BZ-7223 the following spectrum parameters were measured. L_{Aeq} (equivalent continuous sound level), L_{AFmax} (fast maximum time-weighted sound level), L_{ASmax} (slow maximum time-weighted sound level), L_{AFmin} (fast minimum time-weighted sound level), L_{ASmin} (slow minimum time-weighted sound level). During the measurements the other classrooms were engaged in normal activities and the doors and windows were kept closed.

The obtained equivalent, maximum and minimum sound levels for all classrooms in empty situation are shown in Table 2. These values compared to the comfort and acceptability parameters of ANSI S12-60 which describes the maximum comfort noise level of 35 dB (A). In Turkey according to occupational noise control regulations in school the maximum acceptable noise level is 45 dB (A). The comparison results proved that in empty classrooms the noise levels were higher than acceptable ranges. This high level of noise originates from the other rooms activities and from the ongoing construction activities around the building inside of the campus. The obtained noise values for classrooms during physical teaching activity are listed in Table 3. These sound levels related to the teacher's voice during education. The measured values indicate that noise level in classroom C3 is higher in comparison to other rooms. High noise levels affect the quality of verbal communication between teacher and students and lead to serious problems in the student's learning and intellectual development. The measurement was performed at a central spot inside the classroom during 3 minutes. The number of students during measurements was 27, 43, and 62 in classrooms C1, C2 and C3 respectively.

The spectrum noise level per frequency for class-room C3 is shown in Figure 3. The frequency analysis indicated that in the main speaking frequencies of 500–2000 Hz, the maximum noise level is 76 dB (A). When the noise levels exceed above the 60 dB (A), it disrupts the listener's concentration and attention.

Table 2. Noise levels in the empty classrooms.

Classroom	L_{Aeq} (dB)	L_{AFmax} (dB)	L_{ASmax} (dB)	L_{AFmin} (dB)	L_{ASmin} (dB)
C1	55.0	71.2	65.1	34.5	23.0
C2	58.3	75.4	70.5	36.8	29.4
C3	63.7	80.4	74.3	40.2	31.8

Table 3. Noise levels in the classrooms during lecturing.

Classroom	L_{Aeq} (dB)	L_{AFmax} (dB)	L_{ASmax} (dB)	L_{AFmin} (dB)	L_{ASmin} (dB)
C1	59.0	76.2	70.1	36.5	27.0
C2	61.3	78.8	73.1	37.9	29.2
C3	64.4	81.4	76.6	38.4	32.3

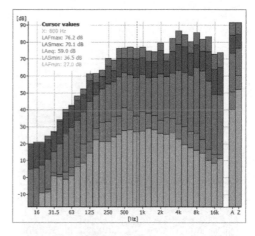

Figure 3. Noise level per frequency for classroom C3.

3.2 Reverberation time RT

Reverberation Time (RT) is the most important parameter describing the acoustic quality of a room which can be defined as the decay time for sound in a room after an interrupted noise or an impulsive excitation stops. It is the time for a 60 dB drop in the sound level. To measure the RT depending on the volume of the room it is require producing an interrupted noise, with the built-in noise generator or impulsive excitation (Schroeder Method). Impulsive excitation can be produced by a pistol or balloon. In this study the impulsive excitation method was used. Reverberation time software BZ-7227 used to measure reverberation time in 1/3- octave with furnished empty classrooms. After the analyzer started and pistol was fired, the analyzer measured and analyzed the decay and displayed the RT spectrum and decay. The RT spectrum respect to frequency shown in Figure 4.

Figure 5 shows the decay of one frequency band for the position 01 in classroom C1. RT values for all the classrooms evaluated here were higher than the 0.6 s limit recommended by the NASI S12.60 standard. The RT of the classroom C3 was the highest. The difference in the RTs of the classrooms is due to the different parallel wall surfaces that reflect the sound. Painted parallel surfaces with low sound absorption coefficients produce high RTs. The RTs measured indicated the lack of acoustic

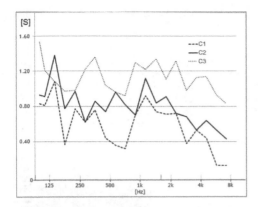

Figure 4.   RTs for classroom respect to frequencies.

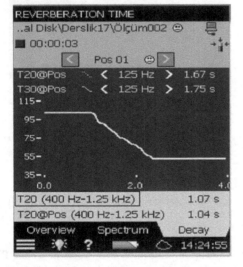

Figure 5.   The decay of one frequency band for the position 01 in classroom C1.

comfort in all three classrooms. High reverberation times lessen the intelligibility of speech. The acoustic inadequacy in the classrooms disrupts communication between students and teachers.

4   CONCLUSION

In the present study, acoustical measurements were carried out to investigate the acoustic condition of three lecturing rooms of Necmettin Erbakan University, Koyceyiz campus in Konya, Turkey. This study was focused on three classrooms representing of all lecturing rooms of the faculty of Engineering in Koyceyiz campus. The sound levels in empty situation and during teaching activities also reverberation times and frequency spectrum were measured for three classrooms. The levels of internal ambient noise measured in the classrooms were higher than the recommended values by the international standards and national regulations. This disturbing noise levels in the empty classrooms are in large part due to inadequate sound insulation in dividing walls and windows. Also in small part is due to noises originating from the construction machines working in the campus. It was informed that these activities will last for several years. The measured RT values for all three classrooms are higher than the recommended values. The main reasons of high RT values are paint material and big parallel surfaces in the classrooms.

This study results reveal the lack of acoustic comfort in the classrooms. Therefore it is necessary to perform an acoustic treatment in the classrooms. It is likely to repeat the errors and acoustic deficiencies in the classrooms under construction of the campus. Thus it is recommended to consider the acoustical condition at the architectural design stage.

ACKNOWLEDGMENTS

The authors gratefully acknowledge the Scientific Research Council (BAP) of Necmettin Erbakan University for the financial support of this research through project number 151219010.

REFERENCES

American national standard—ANSI S12.60 (2002): *Acoustical performance criteria, design requirements and guidelines for schools.* Melville.

Bradley, J. S. (1987, May). Review of methods of measuring rooms for speech. Paper presented at *the 113th meeting of the Acoustical Society of America*, Indianapolis, IN.

Cantor Cutiva, L.C., Burdorf, A. 2015. Effects of noise and acoustics in schools on vocal health in teachers. *Noise and Health* 17(74):17–22.

Crandell, C. C., Smaldino, J. J., & Flexer, C. 1995. Sound field FM amplification: *Theory and practical applications.* San Diego, CA: Singular.

DIN 18041.2004. *Acoustic quality in small to medium-sized rooms.* Germany.

Hodgson, M. 2004. Case-study evaluations of the acoustical designs of renovated university classrooms. *Applied Acoustics* 65: 69–89.

Puglisi, G.E., et al. 2015. Acoustic comfort in high-school classrooms for students and teachers. *Energy Procedia* 78: 3096–3101.

Shield, B., Conetta, R., Dockrell. J.E., Connolly, D., Cox, T. & Mydlarz, C.A. 2015. A survey of acoustic conditions and noise levels in secondary school *Noise in schools* classrooms in England. *J. Acoust. Soc. Am.* 137(1):177–188.

Tang, S.K., Yeung, M.H., 2006. Reverberation times and speech transmission indices in classrooms. *Journal of Sound and Vibration.* 294: 596–607.

World Health Organization—WHO. 2001.. Geneva.

Zannin, P.H. T. & Zwirtes, D.P.Z. 2009. Evaluation of the acoustic performance of classrooms in public schools, Applied Acoustics.70:626–635.

Occupational Safety and Hygiene V – Arezes et al. (Eds)
© 2017 Taylor & Francis Group, London, ISBN 978-1-138-05761-6

WRMSDs symptomatology in home care service workers: Risk factors associated to the work activity

D. Chagas
Higher Institute of Education and Sciences, Portugal

F. Ramalho
Higher Institute of Education and Sciences, Portugal
National Health Service, Portugal

ABSTRACT: Workers in institutional care homes for the elderly have some behaviours, due to demanding professional tasks, that contribute to the development of WRMSDs, such as moving and lifting heavy loads, incorrect postures, as well as monotonous and repetitive tasks. The objective of this study is to identify WRMSDs symptoms amidst the workers in institutional care homes for the elderly and to attempt to associate them with their daily activities. The questionnaire used was a variation of the Nordic Musculoskeletal Questionnaire (NMQ). The most common symptoms of WRMSDs are located in both the lumbar and dorsal regions (88.9%), and also in the wrists/hands (77.8%), cervical (75%) and shoulder (72.2%) regions, respectively. This study reveals that the body regions suffering from pain/discomfort can be related to the tasks that require greater physical effort.

1 INTRODUCTION

Work Related Musculoskeletal Disorders (WRMSDs) are injuries originating in professional risk factors such as overload, repetitivity and/or posture while working (DGS, 2008). Symptoms are characterized by localised pain, or pain that radiates to other body regions, discomfort, fatigue, feeling of weightness, paresthesia (numbness and/or tingling) and loss of strength (DGS, 2008).

There are several groups of factors that can contribute to the occurrence of WRMSDs: physical and biomechanical factors, organizational and psicosocial factors, individual and personal factors (EASHW, 2007).

WRMSDs are often the consequence of recurring exposure to more or less intense efforts throughout a long period of time (EASHW, 2007).

There is a high incidence of WRMSDs in institutional care home workers because they are the main support of the patients in their daily tasks. It is sometimes beyond the ability of most workers to raise and decubitus alternate the patients residing in home care institutions (Collins et al. 2006). It is always difficult to quantify the number of movements within a timeframe since it includes several factors, such as the characteristics of the patient (weight and degree of physical dependence), number of movements (positioning and raising) as well as unforeseen situations with their immediate demand and difficulty.

According to Dahlberg et al. (2004), Faucett et al. (2013), moving and lifting heavy loads, incorrect posture, as well as monotonous and repetitive tasks, are behaviours common in workers in institutional care homes for the elderly due to the demanding professional activity and they are also likely to increase the exposure to risk factors that lead to the development of such injuries. Moreover, the posture and the strength required for their labour are influenced by the conception of the work position. (Macdonald & Oakman, 2013). The physical space must be appropriate so as not to restrict movements (Alexandre & Rogante, 2000).

Several studies refer the work conditions and the risk of developing WRMSDs to which the institutional care home workers are exposed (Faucett et al. 2013; Pellissier et al. 2014; Chagas, 2016). In this perspective, it is extremely important to identify WRMSDs symptoms in institutional care home workers so as to subsequently identify and implement preventive measures, such as acquiring appropriate behaviours for moving patients and obtaining technical help (patient lifts) in order to diminish the risks associated to this activity and the occurrence of the disorder.

The objective of this study is to identify body areas more likely to develop WRMSDs amidst workers in institutional care homes for the elderly, and to attempt to associate them with their daily activities.

2 METHODOLOGY

The study took place in October 2016 and it focused on the workers of two senior care institutions (n = 37) located in the district of Leiria, Portugal. The workers were surveyed through a questionnaire. There was first a personal contact with the Technical Manager of each institution to present the objective of the study. After that, the Technical Manager received the questionnaires and passed them out to the staff. The questionnaires were answered individually and anonymously, in order to maintain the privacy of the informants. Once filled in, the questionnaires were returned to the Technical Manager. The questionnaires were collected two weeks after their delivery.

The questionnaire was adapted from the one developed by Kuorinka et al. (1987), and is divided in four sections: (i) Socio-demographic characterization (sex; age; weight; height; number the years in the current occupation; number of working hours per week; affiliation with the institution; type of schedule and number of overtime hours); (ii) characterization of the symptomatology connected to the work activity, organized by anatomical region, (reference to nine body regions: cervical; shoulders; dorsal; elbows; lumbar; wrists/hands; thighs; knees and ankles/feet) in the 12 months previous to study, in order to measure the intensity of pain/discomfort (to measure the intensity of the pain/discomfort we used an ordinal scale of 1 to 4, where 1 represents slight and 4 very intense); (iii) characterization of the work activities and their relation to the symptoms (arms above shoulders; tilt the torso; rotate the torso; arm repetitivity; hands/fingers repetitivity; exert strength with the hands/fingers; move or lift loads between 55 Kg and 70 Kg, and superior to 70 Kg); and (iv) identification of risk factors stemming from the work activity (the space where the patient is located is prepared for performing the required tasks; the height of the patient's bed is adjustable; uses auxiliary tools to move/transfer patients; number of beds made per day).

The data was collected in paper form. The IBM® SPSS® *Statistics* version 21.0 was used to analyze the collected data.

We first used a descriptive statistical analysis of the data (average, standard deviation, simple frequency, percentage) and, subsequently, we used the Chi-square test to evaluate the associations between the symptom variables and the variables of interest.

We used the Pearson coefficient correlation test to analyze the correlation. The level of significance assumed for both tests was p < 0.05.

3 RESULTS AND DISCUSSION

Of the 37 delivered questionnaires, 36 were answered, which corresponds to a response rate of 97.3%.

As for the socio-demographic characterisation, the informants are females (100%). The age range varies between 20 and 60, with an average of 32.6 years of age. The average weight of the workers is 65.6 Kg (SD = 12.1875) and the average height is 159 cm (SD = 0.0771).

Most workers have been working at the current occupation for over 10 years (33.3%) with an average weekly workload of 38.9 hours (SD = 1.4668). The most common contractual relationship is the fixed-term contact (63.9%), followed by the contract of indefinite duration (16.7%). As for scheduling practices, 86.1% of the workers work shifts while 13.9% have a regular daytime schedule. Moreover, up until 36.1% of the workers work up to 5 extra hours a week.

When characterising the symptomatology connected to the professional activity by body area in the previous 12 months, the workers in institutional care homes for the elderly have musculoskeletal complaints (Figure 1), with particular incidence in both the lumbar and dorsal areas (88.9%), as well as in the wrists/hands (77.8%), cervical (75%) and shoulder (72.2%) areas.

The causes of the prevalence of pain in the lumbar region are multifactorial, but the main causes include the workload, the handling and moving/transferring of the senior patients (Rasmussen et al. 2013; Qin et al. 2014). According to the DGS (2008), Dijken et al. (2008), the lumbar region is the most affected for workers who lift loads and perform frequent flexion movements. When applied repetitively and simultaneously, movements of compression and flexion can cause severe injuries to the lumbar region ((Rezaee & Ghasemi, 2014; Chagas, 2016). Therefore, the high number of complaints can be explained by the diverse tasks which these workers perform and

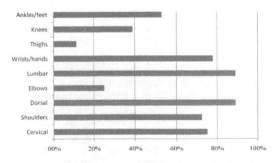

Figure 1.   Percentage of pain/discomfort by body region in the last 12 months.

which require strength, repetitive movements, both rotation and tilting of the torso, as well as a high effort demand of the upper limbs, namely while handling and moving/transferring senior patients, due to their degree of physical dependence.

As for pain/discomfort intensity, the data reveals that the dorsal and cervical region stand in the first place for the most painful anatomical region, rated between intense and very intense, while the lumbar and shoulder regions come second. The wrist/hand area stands in the first place where it comes to moderate pain (Figure 2).

Wrist tendinitis can be caused by performing repetitive movements of flexion/extension of the wrist and fingers, even when performed with light loads, or sustaining a load with an inappropriate posture (DGS, 2008).

After analyzing the professional tasks and their connection to the symptoms, it is clear that the

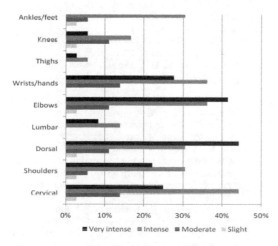

Figure 2. Intensity of pain/discomfort by body region in the last 12 months.

majority of the informants (80.6%) associate the action of lifting and moving patients weighing over 70 kg with their symptoms, as well as exerting strength with the hands or fingers (77.8%), and tilting the torso (69.4%). Raising loads has been considered as the most important factor of injury risk or illness of the vertebral column (DGS, 2008).

The literature has suggested that these workers should participate in courses about handling and moving/transferring patients as one of the strategies for diminishing the incidence of back problems (Alexandre & Rogante, 2000). Performing transfers requires specific abilities that include skill, formation and training (Collins et al. 2006).

While identifying risdk factors associated to the professional activity, 38.9%) of the informants claim that the area where the patient is located is not prepared to the execution of the movements required by their tasks. Over half the informants (61.1%) claim the bed height is nod adjustable.

The ideal situation is to have height adjustable beds, so their height can be adjusted to the task that needs to be performed (Alexandre & Rogante, 2000).

Of the sample, 58.3% of the workers claim they do not use auxiliary tools for moving/transferring patients. In other studies (Farrokhi et al. 2016), the workers do not have auxiliary tools either.

The auxiliary tools can efficiently reduce the vulnerability of a worker when dealing with heavy loads and with uncomfortable work posture which contribute to MSDs (Collins et al. 2006).

Nearly half the the workers (44.4%) claim that they make between 15 and 22 beds on a daily basis. A study by Kromark et al. (2009), in senior care institutions about 50% of the workers made over 100 beds, 30% made between 70 to 100, and 20% made less than 70.

There is a statistically significant association between the presence of symptoms in the lumbar

Table 1. The symptomatology of pain/discomfort in the different body regions and the different types of position or movements in the professional activity.

Requeriment of the tasks	Symptomatology				
	Cervical	Shoulders	Dorsal	Lumbar	Wrists/hands
Arms above shoulder height	—	$r = 0.406^*$	–	–	–
Tilting the torso	–	–	$r = 0.453^{**}$	$r = 0.453^{**}$	–
Rotating the torso	–	$r = 0.408^*$	–	$r = 0.250^*$	–
Repeatability of arms	–	–	–	–	–
Repeatability of fingers	–	–	–	–	–
Exerting strength with the hand	–	–	–	–	$r = 0.621^{**}$
Transfer of service users weighing between 55 Kg to 70 Kg	$r = 0.365^*$	–	–	–	–
Transfer of service users weight greater than 70 Kg	$r = 0.463^{**}$	–	–	–	–

Legend: r coefficient of Pearson correlation; – non significant; * statistical significance at $p < 0.05$; ** $p < 0.01$.

region and the number of years in the current occupation ($x^2 = 11.111$; $p = < 0.001$), as well as between the wrist/hand area and the number of beds made on a daily basis ($x^2 = 14.444$; $p = < 0.001$).

The presence of musculoskeletal symptomatology is related to the growing age of the workers and, therefore, with the seniority in the occupation. It is likely that there is a greater natural decline at the musculoskeletal level aggravated by the activity performed (Almeida et al. 2012).

The correlations between the symptoms related to the different positions or movements in the work activity are positive and significant for the following body areas: shoulders in relation to arms above shoulders; lumbar in relation to shifting and rotating the body; cervical in relation to moving and lifting loads between 55 Kg and 70 Kg, and superior to 70 Kg; wrists/hands in relation to exerting strength with the hands/fingers (Table 1).

4 CONCLUSION

Based on the results, it can be stated that the symptomatology of WRMSDs is a reality amidst the workers in institutional care homes for the elderly. There is a high incidence of musculoskeletal symptoms in the 12 months previous to the study, with the lumbar and dorsal regions being the most affected. This prevalence can be related with performing activities that require greater physical effort.

There is a significant relation between musculoskeletal symptoms and the different types of positions or most common movements, namely arms above shoulders, tilting and rotating the torso, lifting and moving loads between 55 Kg and 70 Kg and superior to 70 Kg, and exerting strength with the hands/fingers.

There are some methodological and procedural limitations that surfaced during this study, namely: (i) limitation of access to the informants in order to explain the objective of the study; (ii) limitations in the analysis instrument, which only measures what is in the questionnaire, leaving out relevant information.

It can be concluded that, due to the demands of the professional activity, the workers in institutional care homes for the elderly are exposed to a set of risk factors, physical, biomechanical factors and organizational. It is therefore essential to implement technical and behavioural measures in order to minimise the risks associated to this activity and the occurrence of WRMSDs.

REFERENCES

Alexandre, N.M.C. & Rogante, M.M. 2000. Movimentação e transferência de pacientes: aspectos posturais e ergonômicos. *Rev. Esc. Enf. USP* 34(2): 165–173.

Almeida, C., Galaio, L., Sacadura-Leite, E., Serranheira, F. & Sousa-Uva, A. 2012. Caraterização de LMERT em Assistentes Operacionais de um Serviço de Apoio Hospitalar. *Revista Saúde e Trabalho* 8: 131–144.

Chagas, D. 2016. Prevalence and symptomatology of musculoskeletal problems reported by home care service workers caring for the elderly. *DYNA* 83(197): 17–21.

Collins, J.; Nelson, A. & Sublet, V. 2006. Safe Lifting and Movement of Nursing Home Residents. NIOSH. Available: http://www.cdc.gov/niosh/docs/2006-117/pdfs/2006-117.pdf. [Accessed 15 November 2016].

DGS. 2008. Lesões Músculoesqueléticas Relacionadas com o Trabalho—Guia de Orientação para a Prevenção. Direcção-Geral da Saúde, Programa Nacional Contra as Doenças Reumáticas. Lisboa.

Dahlberg, R.; Karlqvist, L; Bildt, C. & Nykvist, K. 2004. Do work technique and musculoskeletal symptoms differ between men and women performing the same type of work tasks? *Applied Ergonomics* 35: 521–529.

Dijken, C., Fjellman-Wiklund, A. & Hildingsson, C. 2008. Low back pain, lifestyle Factors and physical activity: A population-based study. *J Rehabil Med* 40: 864–869.

EASHW. 2007. Introduction to work-related musculoskeletal disorders. *European Agency for Safety and Health at Work*. FACTS 71.

Farrokhi, E., Habibi, E. & Mansourian, M. 2016. Risk Assessent of Musculoskeletal Disorders Related to Patient Transfer Tasks Using the Nurse Observation Instrument Method. *Jundishapur J Health Sci* (In press):e35936.

Faucett, J.; Kang, T. & Newcomer, R. 2013. Personal Service Assistance: Musculoskeletal Disorders and Injuries in Consumer-Directed Home Care. *American Journal of Industrial Medicine* 56: 454–468.

Kromark, K., Dulon, M., Beck, BB. & Nienhaus, A. 2009. Back disorders and lumbar load in nursing staff in geriatric care: a comparison of home-based care and nursing homes. *Journal of Occupational Medicine and Toxicology* 4:33.

Kuorinka, I., Jonsson, B., Kilbom, A., Vinterberg, H., Biering-Sørensen, F., Andersson, G. & Jørgensen, K. 1987. Standardised Nordic questionnaires for the analysis of musculoskeletal symptoms. *Applied Ergonomics* 18(3): 233–237.

Macdonald, W. & Oakman J. 2013. Musculoskeletal Disorders at work: Using evidence to guide practice. *J Health & Safety Research & Practice* 5(2): 7–12.

Pelissier, C.; Fontana, L.; Fort, E.; Agard, J.; Couprie, F.; Delaygue, B.; Glerant, V.; Perrier, C.; Sellier, B.; Vohito, M. & Charbotel, B. 2014. Occupational Risk Factors for Upper-Limb and Neck Musculosqueletal Disorder among Health-care Staff in Nursing Homes for the Elderly in France. *Industrial Health* 52: 334–346.

Qin, J., Kurowski, A., Gore, R. & Punnett, L. 2014. The impact of workplace factors on filling of workers' compensation claims among nursing home workers. *BMC Musculoskeletal Disorders* 15:29.

Rasmussen, C., Holtermann, A., Mortensen, O., Søgaard, K. & Jørgensen, M. 2013. Prevention of low back pain and its consequaences among nurses' aides in elderly care: a stepped-wedge multi-faceted cluster-randomized controlled trial. *BMC Public Health* 13:1088.

Rezaee, M. & Ghasemi, M. 2014. Prevalence of Low Back Pain Among Nurses: Predisposing Factors and Role of Work Place Violence. Trauma Mon 19(4): e17926.

Occupational Safety and Hygiene V – Arezes et al. (Eds)
© *2017 Taylor & Francis Group, London, ISBN 978-1-138-05761-6*

Does workload influence the prevalence of neck pain in Portuguese physiotherapists?

A. Seixas
Universidade Fernando Pessoa, Porto, Portugal
IDMEC-FEUP, LABIOMEP, Faculty of Engineering, University of Porto, Portugal

T. Marques & S. Rodrigues
Universidade Fernando Pessoa, Porto, Portugal

ABSTRACT: Physiotherapists are professionals at risk to develop Work-Related Musculoskeletal Disorders (WRMD). Studies assessing the effect of workload in the prevalence of WRMD are lacking in the literature as little research has been done on the occupational demands of these professionals. The aim of this study was to analyze the effect of workload in the prevalence of neck pain in Portuguese physiotherapists. A cross-sectional study was conducted and, after clustering analysis, a between group comparison of prevalence in neck pain was performed. The results suggest that the prevalence of neck pain is very high in these professionals. Significant differences were found in the 7-day prevalence of neck pain and in the percentage of professionals that had to avoid normal activity in the last 12 months, with higher values in the group with higher workload, but not in the 12-month prevalence values.

1 INTRODUCTION

The term "Musculoskeletal Disorder" refers to wide number of injuries with sudden or insidious onset that might affect individuals for varying periods and constitutes a heavy burden for individuals and society (Sanders and Dillon, 2006, Sanders and Stricoff, 2006, Woolf and Pfleger, 2003). If the disorder is induced or worsened by an occupational activity or its associated circumstances, the term "work-related musculoskeletal disorder" is often used (Luttmann et al., 2003, Schneider et al., 2010). In 2014, work-related musculoskeletal disorders were responsible for 32.3% of all injuries and illnesses involving work absence, remaining the main category of injury and illness for this outcome. The median number of days away from work was 13 days (American Federation of Labor and Congress of Industrial Organizations, 2016), an increase from the previous report (American Federation of Labor and Congress of Industrial Organizations, 2015). Despite the increasing attention from researchers, the burden of work-related musculoskeletal disorders is increasing worldwide and, according to Buchbinder et al. (2015), several factors may play a role in the lack of transposition between existing guidelines and daily practice.

A physiotherapist is a professional that acts in the analysis and reeducation of movement and posture, grounding their interventions in the structure and function of the body, aiming to promote health, prevent dysfunction and helping individuals to maximize functionality and quality of life (Decreto-Lei n. 564/99, 1999). Although it would be expected, the knowledge of ergonomics, injury and rehabilitation methods that these professionals possess does not offer immunity from injury.

The reported prevalence of WRMD in physiotherapists differs within the published literature, ranging from 47.6% (Alrowayeh et al., 2010) to 91% (Cromie et al., 2000) and one of the reasons is that the analyzed time window may vary between 7 days, 12 months or even the entire working life of the worker. Several factors have been reported by these professionals as the possible cause of the WRMD, such as working in the same position for long periods, working in static postures, continuing to work while injured, performing the same task repeatedly, performing manual therapy techniques, treating an excessive number of patients in one day, lifting or transferring dependent patients, insufficient rest breaks and working in awkward positions (Cromie et al., 2000, Rozenfeld et al., 2010, Milhem et al., 2016). The neck is one of the body regions with higher prevalence of WRMD in physiotherapists with values ranging from 47.6% to 61% (Cromie et al., 2000, Vieira et al., 2016). Several factors have been reported as major contributors for to the WRMD in this body region, such as performing manual orthopaedic techniques, continuing to work when injured or hurt, performing the same task over and over, treating a large number of patients in one day and not enough rest breaks during the day (Cromie et al., 2000). Regardless of the existing research, to the best of our knowledge, the effect of workload in the prevalence of neck pain in physiotherapists has not been addressed in the literature.

The aim of the present study is to analyze the effect of workload in the prevalence of neck pain in Portuguese physiotherapists.

2 METHODOLOGY

2.1 Research design and participants

A cross-sectional method was planned for this study. Physiotherapists working in 12 private clinics in the northern region of Portugal were recruited, using a convenience sampling method after the approval of the Ethical Committee of the local University. The inclusion criteria were working as physiotherapist for more than a year.

2.2 Instruments

A two-part self-administered questionnaire was used in this research. The first part was custom made sociodemographic characterization questionnaire used to acquire information regarding gender, age, working experience, daily working hours and number of patients treated. The second part was the standardized Nordic Musculoskeletal Questionnaire used to access the prevalence of musculoskeletal complaints. The questionnaire uses a body chart that divides the human body in nine anatomical areas (neck, shoulder, elbow, hand/wrist, upper back, lower back, hip/thigh, knee and ankle/foot). For each body region 3 questions are presented: (1) if symptoms were felt within the last 12 months, (2) if, within the last 12 months, normal activity had to be avoided due to the symptoms and (3) if symptoms were felt within the last 7 days (Kuorinka et al., 1987). This questionnaire is commonly used in similar studies (e.g. Cromie et al., 2000, Vieira et al., 2016) and the Portuguese version, adapted by Mesquita et al. (2010) was used in this study.

2.3 Procedures

After institutional approval, seventy-five copies of the instruments were distributed and all participants read and signed the informed consent for this study. All the questions the individuals deemed necessary were answered and the filled questionnaires were collected by the same researcher.

2.4 Data analysis

Data was analyzed using IBM SPSS Statistics version 24. Descriptive statistics were used to characterize the study sample and interest variables. The prevalence of neck pain was calculated in two different time windows, 12 months and 7 days. A hierarchical cluster analysis was conducted to divide the study sample in groups based on the reported workload. Two variables were used in the analysis, daily working hours and daily number of patients treated. A two-cluster solution was achieved after an agglomerative approach, applying the between-groups linkage clustering method using squared Euclidean distance as measure. The chi-square test was used to compare the proportion of professionals with neck pain between the groups formed by cluster analysis. The Kolmogorov-Smirnov test was used to assess the normality of the variables distribution and, since the variables did not follow a normal distribution pattern, the independent-samples Mann-Whitney test was used to guarantee that the cluster groups were different regarding the variables used to form them. Statistical significance was set at $\alpha = 0.05$.

3 RESULTS

Out of a total of 75 questionnaires distributed, 4 (5.3%) were not returned, but the percentage of response was very high (94.7%). The 71 respondents, comprising 88.7% (n = 63) females and 11.3% (n = 8) males, had a mean age of 31.8 years (ranging between 22 and 55 years), reported an average of 9 years of practice (ranging between 1 and 30 years), an average of 7.9 daily working hours (ranging between 2 and 12 hours) and an average of 29.3 patients treated daily (ranging between 8 and 41 patients). Sociodemographic characteristics are presented in Table 1.

Although not questioning the usefulness of lifetime prevalence measures, we focused in the prevalence of neck pain within the last year and within the last 7 days as the respondents' recall of symptoms is more likely to be accurate.

Table 2 shows the results of the Mann-Whitney test regarding the comparison between the groups formed through clustering analysis. Significant differences were found in daily working hours and daily number of patients treated between the groups obtained after clustering analysis. Group 2 has a significantly higher workload than group 1. No significant differences were found for age and working experience.

The results of chi-square test for the comparison of the prevalence of neck pain and the percentage of subjects that had to avoid normal activity

Table 1. Sociodemographic characteristics of participants regarding gender, age, working experience and workload variables (daily working hours and patients treated).

Characteristics	n = 71
Gender	
male	8
female	63
Age (years)	
mean ± sd	31.8 ± 6.6
range	22–55
Experience (years)	
mean ± sd	9.0 ± 6.3
range	1–30
Daily working hours (hours)	
mean ± sd	7.9 ± 1.7
range	2–12
Daily number of patients treated (patients)	
mean ± sd	29.3 ± 7.5
range	8–41

Table 2. Between group comparison (Mann-Whitney test) for age, working experience, working hours and number of daily treated patients.

	Group 1 (n = 24)	Group 2 (n = 47)	p
Age (years)	30.0 ± 6.1	32.7 ± 6.7	0.07
Experience (years)	7.3 ± 5.7	9.9 ± 6.5	0.06
Working hours (hours)	6.6 ± 1.9	8.5 ± 1.1	< 0.01*
Treated patients (patients)	20.5 ± 5.1	33.8 ± 3.2	< 0.01*

* Significant at 95%, p < 0.05.

Table 3. Between group comparison (chi-square test) for 12-month and 7-day prevalence of neck pain.

	Group 1 (n = 24)	Group 2 (n = 47)	p
12-month prevalence (%)	83.3	83.0	0.97
Had to avoid normal activity in the last 191.312 months	4.2	38.3	< 0.01*
7-day prevalence (%)	29.2	61.7	0.01*

* Significant at 95%, p < 0.05.

because of symptoms between group 1 and group 2 are presented in Table 3.

Significant differences were found for 7-day prevalence of neck pain between groups but not for the 12-month prevalence measure.

4 DISCUSSION AND CONCLUSION

The clustering analysis approach used in this study proved adequate. Two groups were created based in two variables related to workload, the number of patients treated daily and the number of daily working hours. The groups were significantly different regarding these variables but not regarding age and working experience, reinforcing the homogeneity of both groups. To our knowledge this was the first study looking at the influence of workload in the prevalence of neck pain going beyond a descriptive analysis.

The 12-month prevalence of neck pain reported in this study was higher than in previous studies (Milhem et al., 2016). Information regarding 7-day prevalence of neck pain is lacking in the literature, however in the present study the prevalence ranged between 29.2% in the lowest workload group and 61.7% in the highest workload group.

In the present study, the sample was recruited from private clinics and according to Liao et al. (2016), the new-onset of spine related musculoskeletal disorders for physiotherapists working in clinics is 32.12%, which represents a high value. The same authors also state that physiotherapists have a 2.4-fold increased risk of developing spine related musculoskeletal disorders when compared to other health professionals.

Physiotherapists identified several risk factors associated with workload as major contributors to their WRMD in the neck in previous studies. Treating an excessive number of patients and work scheduling factors, such as overtime and length of work day, were factors rated by physiotherapists as a problem to their health in several studies (e.g. Rozenfeld et al., 2010, Shehab et al., 2003, West and Gardner, 2001, Bork et al., 1996, Cromie et al., 2000). In our study an inferential analysis was made, comparing the prevalence of neck pain between two groups differing significantly in reported workload. The 7-day prevalence of neck pain was significantly higher in the higher workload group but the 12-month prevalence of neck pain was not significantly different between groups. The information related to the number of working hours and number of patients treated daily is probably more accurate on the 7-day period because, when asked about these parameters, the respondents may have answered taking in account the present reality and not the whole year. Some workload fluctuations may have occurred during the previous 12-month period and this may have influenced the results. Significant differences were found between the proportion of subjects that had to avoid normal activity because of neck symptoms in group 1 ad group 2. Physiotherapists treat patients while they are lying in a treatment table, or seated, often while the physiotherapists' neck is in a flexion position that is sustained during long periods of time. Spending more hours and treating more patients during the day increases the time working in that position, overloading the neck. This may be related to the differences that were found between the groups in this study.

As closing remarks, the 12-month and 7-day prevalence of neck pain in the studied sample of physiotherapists is very high, highlighting that these professionals are at high risk of developing WRMD in the neck. Workload directly influenced the prevalence of neck pain in this study and attention should be paid to these factors. A high percentage of physiotherapists had to avoid normal activity because of neck symptoms within the last 12 months. Preventive and rehabilitation policies should be implemented to minimize the prevalence of WRMD in the neck among Portuguese physiotherapists.

A larger study with physiotherapists working not just in private institutions but also in rehabilitation centers and hospitals is required to strengthen our results and provide further understanding about the influence of workload in the prevalence of neck pain. Other factors related to workload should also be investigated, for instance the possible mediating effect of reported rest breaks in the reported prevalence of WRMD. Studies evaluating the effect of targeted interventions in the workplace of these professionals should also be conducted.

REFERENCES

Alrowayeh, H. N., Alshatti, T. A., Aljadi, S. H., Fares, M., Alshamire, M. M. & Alwazan, S. S. 2010. Prevalence characteristics, and impacts of work-related musculoskeletal disorders: a survey among physical therapists in the State of Kuwait. BMC Musculoskeletal Disorders, 11.

American Federation of Labor and Congress of Industrial Organizations. 2015. Report on 'Death on the Job, the Toll of Neglect: a National and State-by-state Profile of Worker Safety and Health in the United States.

American Federation of Labor and Congress of Industrial Organizations. 2016. Report on 'Death on the Job, the Toll of Neglect: a National and State-by-state Profile of Worker Safety and Health in the United States.

Bork, B., Cook, T., Rosecrance, J., Engelhardt, K., Thomason, M., Wauford, I. & Worly, R. 1996. Work-related musculoskeletal disorders among physical therapists. *Physical Therapy,* 76: 827–835.

Buchbinder, R., Maher, C. & Harris, I. A. 2015. Setting the research agenda for improving health care in musculoskeletal disorders. *Nature Reviews Rheumatology,* 11(10): 597–605.

Cromie, J., Robertson, V. & Best, M. 2000. Work—related musculoskeletal disorders in physical therapists: prevalence, severity, risks and responses. *Physical Therapy,* 80: 336–351.

Decreto-Lei N. 564/99. 1999. D.R. I Série-A 295 (21–12-1999). Diário da República. 9084.

Kuorinka, I., Jonsson, B., Kilbom, A., Vinterberg, H., Biering-Sørensen, F., Andersson, G. & Jørgensen, K. 1987. Standardised Nordic questionnaires for the analysis of musculoskeletal symptoms. *Applied Ergonomics,* 18(3): 233–237.

Liao, J.-C., Ho, C.-H., Chiu, H.-Y., Wang, Y.-L., Kuo, L.-C., Liu, C., Wang, J.-J., Lim, S.-W. & Kuo, J.-R. 2016. Physiotherapists working in clinics have increased risk for new-onset spine disorders: a 12-year population-based study. *Medicine,* 95(32): 1–6.

Luttmann, A., Jäger, M., Griefahn, B. & Caffier, G. 2003. *Preventing Musculoskeletal Disorders in the Workplace,* India, World Health Organization.

Mesquita, C., Ribeiro, J. & Moreira, P. 2010. Portuguese version of the standardized Nordic musculoskeletal questionnaire: cross cultural and reliability. *Journal of Public Health,* 18(5): 461–466.

Milhem, M., Kalichman, L., Ezra, D. & Alperovitch-Najenson, D. 2016. Work-related musculoskeletal disorders among physical therapists: A comprehensive narrative review. *International Journal of Occupational Medicine and Environmental Health,* 29(5): 735–47.

Rozenfeld, V., Ribak, J., Danziger, J., Tsamir, J. & Carmeli, E. 2010. Prevalence, risk factors and preventive strategies in work related musculoskeletal disorders among Israeli physical therapists. *Physiotherapy Research International,* 15(3): 176–184.

Sanders, M. & Dillon, C. 2006. Diagnosis of Work-Related Musculoskeletal Disorders. *International Encyclopedia of Ergonomics and Human Factors, Second Edition - 3 Volume Set.* CRC Press.

Sanders, M. & Stricoff, R. 2006. Rehabilitation of Musculoskeletal Disorders. *International Encyclopedia of Ergonomics and Human Factors, Second Edition - 3 Volume Set.* CRC Press.

Schneider, E., Irastorza, X. B. & Copsey, S. 2010. OSH in Figures: Work-related Musculoskeletal Disorders in the EU-Facts and Figures. *OSH in figures.* Luxemburg, Office for Official Publications of the European Communities.

Shehab, D., Al-Jarallah, K., Moussa, M. A. & Adham, N. 2003. Prevalence of low back pain among physical therapists in Kuwait. *Medical Principles and Practice,* 12(4): 224–230.

Vieira, E. R., Svoboda, S., Belniak, A., Brunt, D., Rose-St Prix, C., Roberts, L. & Da Costa, B. R. 2016. Work-related musculoskeletal disorders among physical therapists: an online survey. *Disability and Rehabilitation,* 38(6): 552–7.

West, D. J. & Gardner, D. 2001. Occupational injuries of physiotherapists in North and Central Queensland. *Australian Journal of Physiotherapy,* 47(3): 179–186.

Woolf, A. D. & Pfleger, B. 2003. Burden of major musculoskeletal conditions. *Bulletin of the World Health Organization,* 81(9): 646–56.

Occupational Safety and Hygiene V – Arezes et al. (Eds)
© *2017 Taylor & Francis Group, London, ISBN 978-1-138-05761-6*

Impaired sense of balance in an ergonomic evaluation of forklift operators

A. Zywert, A. Dewicka & G. Dahlke
Poznan University of Technology, Poznan, Poland

ABSTRACT: This article aims at presenting the evaluation of burden in work process of forklift operators with an important part of the study, which is pedometer testing, which aims at pointing the disturbances in the sense of balance. During their tests, the authors determined the factors that influence burden in the work process and the values of balance disturbance measured before and after work.

1 INTRODUCTION

During their work, forklift operators are exposed to numerous burdens related to the general level of vibrations, caused among others by the forklift construction and seat. Excessive burdens can be also due to working in a forced sitting position, necessity of head-turning while travelling in reverse (non-use of rearview mirrors), insufficient lighting and a large number of moving vehicles in a warehouse hall. All these factors cause that forklift operators are exposed to a variety of accidents.

2 METHODS OF DIAGNOSIS

2.1 Evaluation of physical burden

The basis for ergonomic assessment of work process is the evaluation of physical work burden. A number of different research methods can be applied to evaluate physical burden of a worker. However, the first step is to perform a chronometric study of worktime in a given job position. It is worth noting that the measurement of length of successive operations needs to be done several times for different employees. The total assessment of physical burden of a worker can be categorized into: (1) dynamic load, measured by the value of energy expenditure, (2) static load, measured by point-based estimation method, (3) repetitiveness of movements, measured by point-based estimation method, (4) hypokinesis (insufficient mobility), measured by point-based estimation method. But in many situation such a categorizing is inefficient in work description and there is a need for methodological description of operators load (Butlewski, Sławińska 2014), and method of data

aggregation (Mazur 2013). Ergonomic load model can be also insufficient in case of unknown way of action, where different factors influence operators behavior—for example age or experience, and these factors will create a need for an ethnographic approach (Butlewski 2014), which like in presented example need to be based on an experiment. Moreover companies are willing to create safe and efficient structure for developing their processes (Mrugalska, Kawecka 2014, Jasiulewicz-Kaczmarek, Saniuk, 2015). Ergonomic assessment elements should be present in addition to a safety analysis covering the environmental factors that may affect the safety of the users of technical solutions (Górny 2011, Górny, Sadłowska–Wrzesińska 2016). Maintaining logistic workers wellbeing is a key factor of system supply chain integration process (Gajšek et al. 2013).

2.2 Evaluation of mental burden

The assessment of mental burden is not simple due to individual features of every human, which directly influence his/her mental state while he/she performs operations in a given position. Mental fatigue can be measured with various indicators, among others influenced by fatigue will be (Butlewski, Dahlke, Drzewiecka 2016): (1) responsiveness to stimuli, (2) threshold tests, (3) sensorimotor activities, (4) subjective fatigue assessment, (5) the impact of fatigue on effeciency and quality of work. Additionally, mental effort can be analyzed in different aspects: physiological, mental, and the combination of both. To assess work-related mental burden in a given position, the work process must be divided into three stages, as shown below (Figure 1).

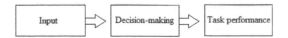

Figure 1. Stages of work process. Source: own work.

3 RESEARCH

3.1 *Subject profile*

The study was conducted on a selected group of 16 warehouse workers (1st and 2nd workhift). All are warehouse workers from both workshifts were men, 75% below 30 y.o., 12,5% between 31 and 40 y.o., 12,5%—over 41 y.o. The subjects are mostly secondary school graduates, with only 25% having vocational training. Subject workers' job consists mainly in loading and unloading of commodities with the use of forklifts (electrical and manual).

3.2 *Research methods*

The following methods have been selected from among numerous research methods used in the evaluation of burden in forklift operators: checklist, surveys, energy expenditure assessment by chronometric-tabular method, measurement of light with Sonel LXP-1 meter, measurement of microclimate with testo 415 Thermo-Anemometer, measurement of noise with the integrating sound level meter SONOPAN IM-02/m, posturography analysis with the use of pedometer. After gathering of all data the analysis was conducted, with the special insight on impaired sense of balance factors.

4 RESEARCH RESULTS

4.1 *Factors influencing operators sense of balance*

Results concluded that operators sense of balance can be impaired by:

- forklift step is located at an appropriate height,
- the majority of workers declare the performance of frequent variable movements while unloading, loading or reloading,
- due to the large size of a warehouse hall, workers drive down long travel paths (over 100 m) to pick up loads from different parts of the warehouse hall, which further results in travelling with great speed in order to fulfill the tasks,
- frequent maneuvering past other forklift operators while performing different operations (loading/unloading) requires mindfulness; 12,5% of subjects declares that they do not look back while driving,
- workers indicate the following most frequent problems related to operating forklifts: lower

back pain, neck pain, shoulder pain, hand muscle pain, leg muscle pain, wrist pain, deterioration of sight, swollen legs, mental fatigue.

For energy expenditure assessment using chronometric-tabular method, a senior warehouse employee was selected, with adequate qualifications, internal authorisations for forklift driving and valid medical examination. The work of a forklift truck operator was rated as fairly heavy in the company where reasearch was conducted. Based on chronometric study, energy expenditure of forklift operator was calculated, amounting to 2977 to 4243 [kJ] (715–994 kcal). The amount of energy expenditure is assessed by comparing the calculated values with the data in work burden classification table. Based on the chronometric job position study of forklift truck operator, the energy expenditure was calculated, which was 715 to 994 kcal per workshift. This result allows to assess the performed work as light to fairly light in the work burden classification.

The warehouse hall where light measurements were performed is 12,2 m high and is lit by 826 fluorescent lamps. To perform light measurements, measuring points were designated along travel paths in the 168 m × 180 m warehouse hall. The number of measuring points was calculated with the following formula:

$$p = 0,2 \cdot 5^{\log d} [m] \tag{1}$$

where p = maximum distance between measuring points; d = shorter dimension.

After designating the number of measuring points and presenting the result to the warehouse manager, the distance between the measuring points had to be shortened to 2 meters. The measurements were conducted in the middle of travel path of the vehicles, which could have endangered the safety of both workers and subject. The light measurement results with the use of Sonel LXP-1 meter are shown in Table 1.

The light intensity measurement shows that lighting in workplace is correct, amounting to minimum 150 1x (according to lighting requirements for interiors contained in PN-EN 12464-1:2012 standards).

The measurement of microclimate was conducted in spring between 1 pm and 1.30 pm, with the temperature outside of circa 15°C. The humidity in warehouse hall was the same as outside because there are no sources of humidity in spring in that hall. In summer the hall is cooled by mechanical extract ventilation, while in winter it is heated with the use of heaters. The results of air flow velocity and temperature measurements at forklift operator work area are shown below (Table 2).

Table 1. Light intensity measurement results.

No.	Light intensity min [lx]	max [lx]
1	302	574
2	390	526
3	498	620
4	449	524
5	360	443
6	389	523
7	403	640
8	444	637
9	401	768
10	251	472
11	158	401

Source: Own work.

Table 2. The results of Thermo-Anemometer measurements.

No.	Air flow velocity [m/s]	Temperature [°C]
1	0,07	23,0
2	0,13	21,0
3	0,12	20,7
4	0,1	20,0

Source: Own work.

The analysis of the conducted tests shows that the microclimate in the tested area is moderate, and the temperature value complies with the Ministry of Labour and Social Policy Regulation of September 26, 1997 and amounts to no less than 14°C. The noise measurement with the integrating sound level meter was taken on May 19, 2016 at 12:30 at the forklift operator work area at the subject's head level. The sound level falls within the range of 68 dB to 74 dB, consequently no actions need to be undertaken in order to lower the sound level in the work environment for the noise exposure with the reference to 8-hour daily working time amounts to 80 dB (according to the Decree of Ministry of Economic Affairs and Labour from August 5, 2005).

4.2 Posturography analysis with the use of pedometer

The measurements were conducted with a pedometer equipped with 2304 tension sensors that enable readings at the frequency of 200 Hz on shifts in foot pressure. The device also allows for the determination of the centers of gravity of the feet and the entire body, as shown in Figure 2 (Dahlke et al. 2016).

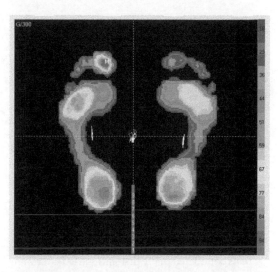

Figure 2. Screenshot from the WIN-POD program to support pedometer. The color codes reflect levels of foot pressure against the base of support. The white curves in the center of the chart and the two curves near the middle of the feet depict sways in the center of gravity. Source: (Dahlke et al. 2016).

Table 3. Description of measurements. Source: own work.

Assessment number	Description
Measurement 1 and 1A	A static assessment; the subject's eyes were open; their gaze directed at a plain white wall ca. 1.5 m away from subject;
Measurement 2 and 2A	A static assessment; the subject's eyes were closed;
Measurement 3 and 3A	Posturography assessment (measured over a 30 s interval at the frequency of 200 Hz); the subject's eyes were open; their gaze directed at a plain white wall ca. 1.5 m away from subject;
Measurement 4 and 4A	Posturography assessment (measured over a 30 s interval at the frequency of 200 Hz); the subject's eyes were closed.

Four measurements per subject were taken before and after work. The individual measurements are described in Table 3.

Measurements 1, 2, 3 and 4 were performed before work. Measurements 1A, 2A, 3A and 4A were made in the same conditions as measurements 1, 2, 3 and 4 but were conducted after work. The test results were analyzed statistically and displayed in tables and graphs. The pedometer and the Win-pod computer software can be used to collect data on, inter alia:

- Area of left and right foot [cm²],
- Pressure on left and right foot [%],
- Length of center of gravity trace [mm],
- Area of center of gravity trace [mm²],
- Average speed of center of gravity trace Q [mm/s],
- Deflection speed along X and Y axis of center of gravity [mm/s],
- Average deflection shifts of center of gravity along X and Y axis (X_{av}; Y_{av}) [mm].

The tests were conducted during the first and second workshift on November 5, 2015. The number of tested subjects was respectively:

- 1st shift: 6 people;
- 2nd shift: 10 people.

Tests conducted in the first shift started after 9 am, so the workers were exposed to work-related factors. The test results for the second shift, which began before work, have also been shown for comparison.

The analysis of the values of foot trace areas in workers, has shown a drop in the mean values after work (for both 1st and 2nd shift workers) (Figures 3 and 4).

An interesting trend of body weight distribution can be observed in Figures 5 and 7. Left-sided assymmetry of body weight distribution is shown.

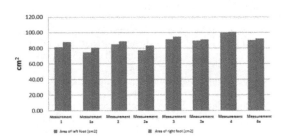

Figure 3.   Airthmentic mean values of the surface areas of left and right foot in 1st shift workers. Source: own work.

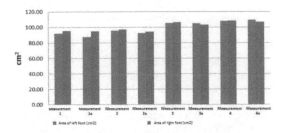

Figure 4.   Arithmetic mean values of the surface areas of left and right foot in 2nd shift workers. Source: own work.

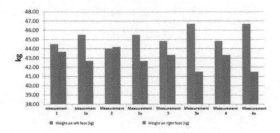

Figure 5.   Arithmetic mean values of body weight distribution on foot in 1st shift workers. Source: own work.

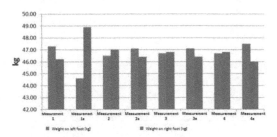

Figure 6.   Arithmetic mean values of body weight distribution on foot in 2nd shift workers. Source: own work.

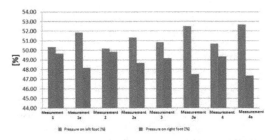

Figure 7.   Arithmetic mean values of body weight percentage distribution on foot in 1st shift workers. Source: own work.

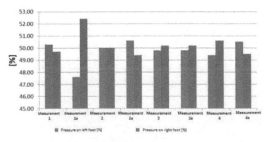

Figure 8.   Arithmetic mean values of body weight percentage distribution on foot in 2nd shift workers. Source: own work.

This trend increases towards the end of shift. The same regularity was observed in second shift workers, although it is less visible. (Figure 6 and 8).

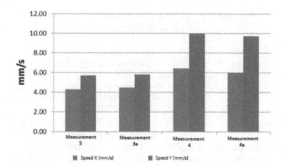

Figure 9. Artithmetic mean values of the average speed of trace in axes X (transverse) and Y (longitudinal) through the centers of gravity of 1st shift workers. Source: own work.

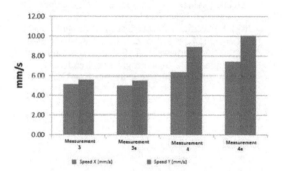

Figure 10. Artithmetic mean values of the average speed of trace in axes X (transverse) and Y (longitudinal) through the centers of gravity of 2nd shift workers. Source: own work.

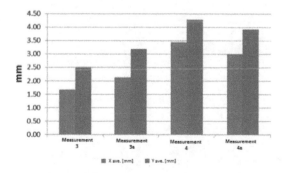

Figure 11. Artithmetic mean values of the average deflection of trace in axes X (transverse) and Y (longitudinal) through the centers of gravity of 1st shift workers. Source: own work.

The analysis of results in Figures 9 to 12 shows that the value of subjects' body deflection in the anterior-posterior direction is greater than in the medial-lateral direction.

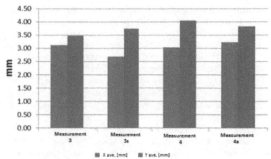

Figure 12. Artithmetic mean values of the average deflection of trace in axes X (transverse) and Y (longitudinal) through the centers of gravity of 2nd shift workers. Source: own work.

The analysis of test results confirmed a considerable drop in the balance system performance when subjects' eyes were closed. It has been observed in numerous studies. (Dahlke et al. 2015). It can be detrimental to the stability and reliability of workers in case of insufficient lighting in both the task area and immediate surroundings.

5 CONCLUSIONS

This paper has addressed the problem of the evaluation of burden in work process of forklift operators, with particular focus on the sense of balance. Analysis of the conducted research indicates that the burdens of forklift operators are mainly related to head-turning while driving in reverse −44% of the working time (due to load on the fork operator was driving back much longer then without a load driving forword). Another problem for forklift drivers were: changing microclimate (thermal discomfort especially in winter) and the incompatibility of the work tool with the unique anthropometric features of a worker. During posturography analysis no considerable influence of turning positions was observed; however, regarding the special character of the research, it should be repeated in various seasons of the year, with different transport load and work tasks.

REFERENCES

Butlewski, M., and M. Slawinska, 2014, Ergonomic method for the implementation of occupational safety systems: Occupational Safety and Hygiene Ii, 621–626 p.
Butlewski, M., 2014, Practical Approaches in the Design of Everyday Objects for the Elderly, in L. Slatineanu, V. Merticaru, G. Nagit, M. Coteata, E. Axinte, P. Dusa, L. Ghenghea, F. Negoescu, O. Lupescu, I. Tita, O. Dodun, and G. Musca, eds., Engineering Solutions

and Technologies in Manufacturing: Applied Mechanics and Materials, v. 657, pp. 1061–1065.

Dahlke, G. Exposure to noise of communities residing near slow-traffic public roads. A practical example. Sho2015: International Symposium on Occupational Safety and Hygiene, p.p 91–93, 2015.

Dahlke, G., Butlewski, M., & Drzewiecka, M. (2016). Impact of exposures to environmental factors on sense of balance stimulation. Occupational Safety and Hygiene IV, pp. 243–248.

Gajšek B., Grzybowska K., Rosi B., Semoli B., The Understanding of the 'Logistics Platform' Concept in Poland and Slovenia, w: The International Conference on Logistics & Sustainable Transport, 2013.

Górny A., 2011, The Elements of Work Environment in the Improvement Process of Quality Management System Structure, [in:] W. Karwowski, G. Salvendy (eds.), Advances in Human Factors, Ergonomics, and Safety in Manufacturing and Service Industries, pp. 599–606, CRC Press, Taylor & Francis Group, Boca Raton (ISBN: 978-1-4398-3499-2).

Górny A., Sadłowska–Wrzesińska J., 2016, Ergonomics aspects in occupational risk management, [in:] P. Arezes, J.S. Baptista, M.P. Barroso, P. Carneiro, P. Cordeiro, N. Costa, R. Melo, A.S. Miguel, G. Perestrelo (eds.), Occupational Safety and Hygiene, SHO 2016, p.p 102–104, Portuguese Society of Occupational Safety and Hygiene (SPOSHO), Guimares (ISBN: 978-989-98203-6-4).

Jasiulewicz-Kaczmarek M., Saniuk A., 2015, Human factor in Sustainable Manufacturing, Editors: M. Antona, C. Stephanidis, Universal Access in Human-Computer Interaction. Access to the Human Environment and Culture, LNCS Vol. 9178, pp. 444–455, DOI 10.1007/978-3-319-20687-5_43.

Kawecka-Endler, A., Mrugalska, B., 2014. Humanization of work and environmental protection in activity of enterprise. In: Masaaki Kurosu (ed.), Human-Computer Interaction. Applications and Services.16th International Conference, HCI International 2014, Heraklion, Crete, Greece, 22–27 June 2014, Proceedings, Part III, LNCS 8512: 700–709.

Mazur A., Application of fuzzy index to qualitative and quantitative evaluation of the quality level of working conditions, in: HCI International 2013—Posters' Extended Abstracts, International Conference, HCI International 2013, Las Vegas, NV, USA, July 21–26, 2013, Proceedings, Part II, Springer-Verlag Berlin Heidelberg, 2013, pp 514–518, ISBN: 978-3-642-39475-1.

Misztal, A., M. Butlewski, A. Jasiak, and S. Janik, 2015, The human role in a progressive trend of foundry automation: Metalurgija, v. 54, p.p 429–432.

Occupational Safety and Hygiene V – Arezes et al. (Eds)
© *2017 Taylor & Francis Group, London, ISBN 978-1-138-05761-6*

"HugMe"—Validation of a prototype of an inclusive toy for children

Rosária Ferreira & Demétrio Matos
School of Design, IPCA, Barcelos, Portugal

Filomena Soares
Algoritimi Research Centre, University of Minho, Guimarães, Portugal

Vítor Carvalho
School of Technology, IPCA, Barcelos, Portugal and Algoritimi Research Centre, University of Minho, Guimarães, Portugal

Joaquim Gonçalves
School of Technology, IPCA, Barcelos, Portugal

ABSTRACT: Toys should be firstly approached according to each child's interest. Regarding children with special needs, it is indispensable to analyse the way they physically reach the toy, their interaction and difficulties/limitations. So, a toy that is specifically designed for children with special needs is a fundamental piece in their development and quality of life, also being an auxiliary tool for therapists to use in therapy sessions. Based on that, this article presents a proposal of an inclusive toy, named "HugMe", which was especially designed for children with special needs. The results from the first tests are presented which allow to infer its application in game activities with children with special needs and to identify possible improvements of the toy.

1 INTRODUCTION

Playing is important, in order for the children to develop their social, emotional and physical cognitive abilities (Isenberg & Quisenberry, 2002).

The toy is directly associated to the action of playing, having a high experimental value, which allows the children to acquire multiple abilities and skills (Pereira, 2009). This becomes a reflex of how child is part of the society, showing his/her interaction with the toy (Brougère, 2004). So, an inclusive toy for children with special needs, which could also be used as a more technical auxiliary technique, will have great value. Playing is used by therapists and technicians as a way to stimulate the child in reaching an objective, regarding his/her rehabilitation, or abilities.

The development of the "HugMe" toy project has already had an initial phase, which had, as its main goal, establishing a contact with professionals who deal with children with special needs on a daily basis. (Ferreira, Matos, Carvalho, & Soares, 2016).

Throughout the project, different aspects, regarding the child, were taken into account, in order to ensure a high level of protection and safety (Oficial et al., 2009).

1.1 Theoretical assumptions

Having conception ergonomic as an objective, instead of correction ergonomic, all the actions that have been followed from the beginning of the project were taken according to the necessities of what is intended to design. The work was done regarding the interactions of use, taking user-product safety into account, usage comfort, and its efficiency, in a way to try and solve the problems and focus on the optimization of the interaction (Fig. 1).

When designing a product, specifically a toy for children with special needs, it is necessary to assure

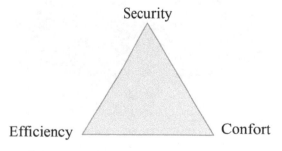

Figure 1. Optimization of the interaction.

its safety, and reduce the risks for any potential damage. So, the implementation of good designing practices and the consideration of anthropometric measurements for the design of the toy becomes a core issue (Saptari & Poh Kiat, 2013).

In this process, it was necessary to obtain certain references about children's anthropometric measurements, and also about any auxiliary devices they can use (ex: wheelchair), in order to obtain the ideal toy, in accordance to the user's anatomy. Measurements regarding children aged 3 to 6 years old were taken into account, and stipulated for the toy (Tilley & Wilcox, Stephen B. Associates, 2002). The measurements that were taken into account were the width and perimeter of the waist, the distance from the waist to the knee and the total length of the arm.

The distance between the toy and the child is one of the main focus of the project, because, in most adapted toys, the available buttons which promote the interaction with the toy are far from the child. This distance is a key factor for children with posture control problems and for those who have slow, uncoordinated movement patterns (Abreu et. al., 2015), (Heide et. al., 2005).

2 TOY PROTOTYPE –"HUGME"

The toy in Figure 2 is based on the principles that the toy should be attractive, with different colours and textures; intuitive; able to be used in different positions; able to be adjusted, both in volume and difficulty level of the activities; able to be used by children of different ages and/or levels of development, and always having in mind that it should have a cheap price in comparison to other toys for children with special needs (Directiva 2009/48/CE).

The toy has a 3-part structure, in which each component has a specific function.

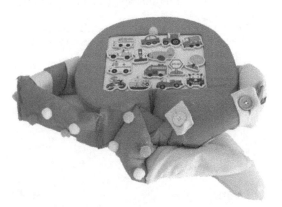

Figure 2. Prototype "HugMe".

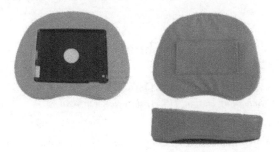

Figure 3. Toy base.

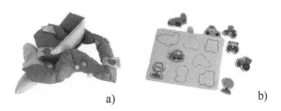

a)   b)

Figure 4. a) "Arms" b) Activity.

The "base" (Fig. 3) is the main structure, and the piece which supports the other two. Its components are a pillowed body with an inclination, necessary for the child to remain in the most correct posture. It takes into account the possible motor difficulties of the child, being possible to place the toy in different configurations: on a chair, on a table or on the floor. It is in this "base" that the activities and/or the tablet will be fitted.

The arm (Fig. 4a) is attached in the rear part of the "base", and has different textures and fine movement activities along its body. It is an extra element for stimuli. Consequently, the toy is always closer to the child, which is not the case on most adapted toys. The "arms" are fitted in a simple way, their extremities are made of Velcro and their length is malleable, making them easy to fit.

The activities piece (Fig. 4b) is a key piece which has the possibility of fitting different modules with diverse activities, designed for the different needs of the children. Also, it allows the simple exchange between activities. It also has a design which enables the fitting of a tablet, which increases the number of activities and interaction potential the child may have with the toy.

The toy can be used in different positions, being adapted to the child's daily posture, both in his/her most comfortable or in any posture correction context in a therapy session. Thus, it can be used on the floor, on a table, or on the wheelchair, due to the malleable "arm" which is adaptable to multiple scenarios, being also able to be easily removed, if necessary.

The toy hygiene requirements were also considered, which is why all the toy parts that can be touched by the child can be washed and disinfected. This allows for the use of the toy by different children, in an institution, for example.

3 PROTOTYPE VALIDATION TESTS

This section presents the description of the prototype validation tests, including the characterization of the participants, the tools and equipment, design procedures, and the results and discussion.

3.1 *Participants*

The information was collected using a 21-element sample, with children ranging between 3 and 6 years of age, also adding 4 more elements which do not fit the stipulated age group, with the intent of studying the possible need of widen the age group for the toy.

There was not any distinction, neither on gender nor in the child's pathologies, such as Autism; Cerebral Paralysis (affected motor part); DiGeorge Syndrome; left hemiparesis; global development delay, Sensorial Integration Dysfunction. The usability tests have been carried out in three different associations: APAC (in Portuguese "Associação de Pais e Amigos de Crianças"), APPC (in Portuguese "Associação do Porto de Paralisia Cerebral") and 7SENSES. The tests took place in therapy sessions.

3.2 *Equipment and tools*

The equipment that was used in the tests was the prototype, considering its parts, and the pieces that are used in the activity (Fig. 5).

Figure 5. Prototype usability tests.

3.3 *Design and procedures*

The objective of the tests was to understand if the use of the toy can have a positive effect in children with special needs.

The main focus of the tests was the base, the arms, and the way they interacted with the activity. The activity was designed specifically for these tests, in order for the children to have a game associated with it, making it possible to analyse the situation in a real scenario.

A questionnaire was designed in order to infer the performance of the toy. The questionnaire was fulfilled by the parents and the therapists who accompanied the children, as the children could not express themselves. In case they were able to give feedback, their answer was considered.

The questionnaire consisted of 4 specific parts:

- child's characterization;
- toy level of effectiveness;
- efficiency;
- satisfaction level.

3.4 *Results and discussion*

The questionnaire considered a Likert scale: Full agree-5; Agree-4; Neutral-3; Disagree-2; Full disagree-1.

3.4.1 *Effectiveness level*
The effectiveness level was used to try to understand how the child reacts to the toy, to analyse its shape and dimensions.

Table 1 presents the answers obtained accordingly to the dimensions of the different parts of the toy.

Regarding the toy structure, taking the 3 parts into account, it was necessary to understand if the dimensions were adequate. By analysing Table 1, it can be concluded that there is a consent on the dimensions, however, regarding the "base", the results are almost evenly distributed throughout the scale, which leads to the conclusion that this factor should be considered in future prototype studies.

Next, the arm fixation issue was approached, namely, if it was simple to apply, which had a 52% agreement and 28% full agreement. Posteriorly,

Table 1. Percentage of answers according to the dimensions of the different parts of the toy.

Adequate dimension of the constituent parts:

	5	4	3	2	1
Base	28%	24%	28%	20%	0%
Arms	4%	44%	28%	16%	8%
Activity	56%	44%	0%	0%	0%

the question whether or not the "arms" textures were stimulating for the child, received less positive results: 48% neutral and 28% disagree that the child felt stimulated by the textures that were presented.

It was also verified if the material that was used in the prototype caused irritation, in which 68% of the inquired fully disagreed that it had happened.

The possibility of having the option of fixing the tablet, as a way to have an additional activity on the toy, was also surveyed, and 64% of the respondents fully agreed, which shows the importance of the table for children (Table 2). Nevertheless, 44% of the inquired agreed, and 20% fully agreed (64% total) that the children showed more interest in the activity than with the tablet. The positive answers for "the children presented more interest in the tablet" totaled 52% (level 4 and 5).

Regarding the toy ability to adapt to different positions, 72% of the inquired fully agreed on this matter.

3.4.2 *Efficiency level*

The aim of the efficiency level is to know the level of interaction between the child and the toy, which allows for the identification of the possible need for external help, as well as eventual difficulties on using the product.

Depending on the child, it was either necessary, or not, to use external help in order to fit the toy (fitting the "arms" around the child) and/or an explanation of how the toy worked (Table 3). However, 60% of the respondents were able to use the toy, indicating full agreement.

When they were questioned about the ability of the toy to allow for the development of new ways of playing, 32% of the inquired fully agreed, while 28% were doubtful about the matter, having a neutral answer.

Table 2. Percentage of answers about the possibility of being able to fix the tablet and comparison of the interest between the tablet and the activity.

	5	4	3	2	1
Possibility of fixing the tablet	64%	16%	12%	4%	4%
Presented more interest in the tablet	28%	24%	36%	4%	8%
Presented more interest in the activity	20%	44%	32%	0%	0%

Table 3. Percentage of answers regarding the toy's efficiency level.

	5	4	3	2	1
Was able to use the toy	60%	32%	4%	0%	4%
Need help or explanation	8%	28%	24%	12%	28%

3.4.3 *Satisfaction level*

Regarding the satisfaction level, this parameter allows for the verification of the level at which the toy could reach its goals, achieving or going beyond the initial expectations.

Initially, the adequacy of the toy to the age group of the child was approached, and 68% of the inquired agreed that the toy is useful to the children of different ages.

Additionally, in order to understand if the toy could be used by the child for long periods of time, using different activities and having positive effects on him/her, 44% of the inquired agreed, and 20% fully agreed.

In the last part of the questionnaire, the inquired had the chance to openly state their opinions and suggestions about the toy. There was a good amount of comments on the fact that the toy could also be used for adults with reduced mobility, mainly on the upper limbs, and use the tablet to communicate or just for leisure. This solution was considered to be of added value to their lives.

4 CONCLUSION AND FUTURE WORK

The "HugMe" prototype is an inclusive toy, a solution for the problems that children with special needs face. Thus, it is possible for children to be able to play without feeling the awkwardness and problems that come with using toys that were designed for typically developing children.

After analysing the results of the usability tests that were carried out, a higher understanding on how the designed toy can adapt to children is achieved. However, it will be subjected to improvements and future studies. As possible improvements, parts as the "arm" can be evaluated, regarding the textures along its length, which did not spark the desired interest on the children. The size of the base is also an issue that must be improved, in order to be better adaptable to the anthropometric measurements of the children of the intended age group. Additionally, the possibility of the toy being used by adults must also be considered, especially on the use of the tablet.

The "HugMe" toy is still a developing project, with a high potential. In the future, the development of a pack of activities that can be applied to the toy is also part of the plan.

ACKNOWLEDGEMENTS

We would like to thank the children, the therapists and assistants of children with special needs for their collaboration with the research team.

This work has been supported by COMPETE: POCI-01–0145-FEDER-007043 and FCT—

"Fundação para a Ciência e Tecnologia" within the Project Scope: UID/CEC/00319/2013.

REFERENCES

Abreu, A.R., Arezes, P.M., Silva, C., & Santos, R. (2015). A case study of product usability of a pelvic device used by children with neuromotor impairments. Procedia Manufacturing, 3(Ahfe), 5451–5458. http://doi.org/10.1016/j.promfg.2015.07.677

Brougère, G. (2004). Brinquedos e companhia. (Cortez, Ed.). São Paulo.

Ferreira, R., Matos, D., Carvalho, V., & Soares, F. (2016). "HugMe" Development of an inclusive toy: first insights. In 2º International conference of the portuguese society for engineering education. Vila Real.

Heide, J.C.V.A.N.D.E.R., Fock, J.M., Otten, B., & Stremmelaar, E. (2005). Kinematic Characteristics of Postural Control during Reaching in Preterm Children with Cerebral Palsy, 58(3), 586–593. http://doi.org/10.1203/01.pdr.0000176834.47305.26

Isenberg, J.P., & Quisenberry, N. (2002). A Position Paper of the Association for Childhood Education International PLAY: Essential for all Children. Childhood Education, 79(1), 33–39. http://doi.org/10.1080/00094 056.2002.10522763

Oficial, J., Do, C.E., Europeu, P., Conselho, D.O., Europeia, C., Europeu, S., Euro, P. (2009). Directiva 2009/48/ce do parlamento europeu e do conselho de 18 de Junho de 2009 relativa à segurança dos brinquedos. Jornal Oficial Da União Europeia, 1–37.

Pereira, M.L.D. (2009). Design Inclusivo—Um Estudo de Caso: Tocar para Ver—Brinquedos para Crianças Cega e de Baixa Visão. Universidade do Minho Escola de Engenharia.

Saptari, A. & Poh Kiat Ng. (2013). The Importance of Child Anthropometry in Child Product Designs The Importance of Child Anthropometry in Child Product Designs Adi Saptari, Poh Kiat Ng and Mohd Musa Mukyi. ResearchGate, (January). http://doi.org/10.13140/2.1.4006.2726

Tilley R., A., & Wilcox, Stephen B. Associates, H.D. (2002). The Measure of Man and Woman: Human Factors in Design, Revised Edition. (J.W. & Sons, Ed.) (revised ed). New York.

Occupational Safety and Hygiene V – Arezes et al. (Eds)
© 2017 Taylor & Francis Group, London, ISBN 978-1-138-05761-6

Review of the state of knowledge of the BIM methodology applied to health and safety in construction

A.J. Aguilar Aguilera
Master in Construction Management and Security, University of Granada, Granada, Spain

M. López-Alonso
Department of Construction and Engineering Projects, University of Granada, Granada, Spain

M. Martínez-Rojas
University of Granada, Granada, Spain

M.D. Martínez-Aires
Department of Building Construction, University of Granada, Granada, Spain

ABSTRACT: The construction sector is one of the most dangerous sectors in the world. Notwithstanding the crisis that has recently affected construction, accident rates in this sector are among the highest in all the industrial sectors. However, traditional methods for security management, based on documents and 2D plans, continue to be used. Building Information Modelling (BIM) has emerged as a fundamental tool in the architecture, engineering and construction industry. Application of BIM in construction projects is advantageous in that it allows the project to be controlled from the outset, with prevention measures applied right from the design stage. The aim of the present study is to carry out a review of the state of knowledge of the BIM methodology and safety in construction works. To this end, publications have been analysed based on variables such as year of publication, country and applied technology.

1 INTRODUCTION

The construction sector is one of the most dangerous in the world, largely due to its special characteristics. One of every five fatal accidents (21.4%) in the EU in 2013 happened in the construction sector (EUROSTAT, 2013). Given the high number of fatalities in the sector, it is evident that safety management throughout the whole construction process must be improved. One of the main problems affecting the industry is the use of traditional methods both in the design of the projects and the execution of works. The methods used are based on 2D plans or drawings employed to identify potential risks and establish preventative measures (Hadikusumo & Rowlinson, 2002).

This trend has begun to change over recent years, especially in Anglo-Saxon countries, with the implementation of Building Information Modelling (BIM) methodologies as a fundamental tool in the Architecture, Engineering and Construction (AEC) industry. (Bryde et al., 2013). BIM, as a methodology, involves the creation, management and storage of information about the properties and characteristics of the different parts of a construction. This does not only refer to geometric or visual properties, but also includes other aspects. Furthermore, it requires participation and collaboration between the different agents involved in the project (Eastman et al., 2008).

The BIM model is a digital representation of all the physical and functional characteristics of a building, a database of reliable information which can be consulted throughout the service life of a construction, from design to demolition. The concentration of all information and data on a project in one single model offers a global vision of a project and leads to better coordination between all involved. Problems which may appear later during construction work can be anticipated and studied with time, eliminating possible interactions and risks which could lead to accidents or injuries. (Rodrigues & Alves, 2015).

The growing application of BIM in the AEC industry is changing approaches to safety in construction (Zhang et al., 2013). BIM can potentially improve prevention from the design stage whilst offering support to architects, engineers and builders to improve worker safety during the construction process (Ku & Mills, 2010).

This study offers a review of the state of knowledge of research into the application of BIM methodology in improving health and safety in construction.

2 METHODOLOGY

The methodology used to carry out this study is based on a systematic literature review on the BIM methodology applied to safety and health. The review process is highlighted in Figure 1.

2.1 Selection of journals

The first step was to identify the sources of information to be consulted. For the review, only journal articles with the Journal Citation Report (JCR) were considered. No other academic publications or conference publications were taken into account.

2.2 Literature search

For the literature review three search criteria were employed. The articles were identified through token (key words);

1. "BIM" OR "Building Information Modelling"
2. "Safe" OR "Risk" OR "Hazard"

Criteria 1.- The articles have been identified using the token (1) together with (2) in the title and/or abstract and/or key words.

Criteria 2.- The articles have been identified using the token (1) in the main body of the article and the token (2) in the title and/or abstract and/or key words.

Table 1. Search results.

Search criteria	N° articles
Token (1) and (2) in Title and/or Abstract and/or Keywords	74
Token (1) in full text; Token (2) in Title and/or Abstract and/or Keywords	91
Out search criteria	8
Elimination of duplications	18
Total	155

Criteria 3.- At this stage certain articles which did not meet the previous search criteria were included. These were identified using references from the articles themselves or through a search for the most prominent authors in this field. Table 1 shows a summary of the results obtained from the search.

2.3 Selected articles

In total 173 articles were obtained. After eliminating the duplications, 18 in total, 155 articles were identified.

Subsequently, following a detailed analysis, 113 articles were excluded as, although they met the search criteria, these articles did not deal with safety and BIM. In total 42 articles were included for analysis.

2.4 Article coding

Article coding was carried out depending on (1) title, (2) year of publication, (3) journal title, (4) country, (5) research topic and (6) applied technology.

3 RESULTS OF LITERATURE REVIEW

The results were analysed based on the date of publication, country and technology applied in BIM. The analysis is discussed in the following sections.

3.1 Year profile of publications

Figure 2 shows the historical evolution of the BIM methodology applied to safety in construction.

The first reference dates back to 2009. The number of publications is especially low up until 2013 when the number of publications begins to increase gradually.

For this reason, two distinct periods have been chosen. The first, between 2009–2012, represents 11% of the total publications, whilst the second period, from 2012–2016* (until July 2016), accounts for 89%.

From 2013 there is a marked increase in the amount of research into safety and BIM.

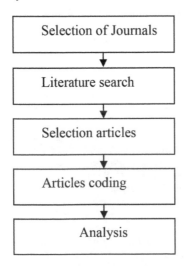

Figure 1. Flow/process review.

3.2 Publications distributed by country

Distribution was analysed according to country of reference of the author and the main institution where research was carried out. Following the analysis, research articles were collected from 11 different countries (Figure 3).

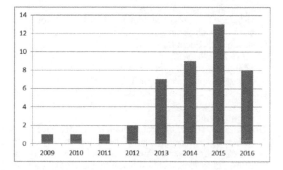

Figure 2. Distribution of publications by year (until July 2016).

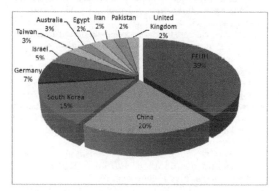

Figure 3. % Publications distributed by country.

The country with the highest number (38%) of the articles is the United States, followed by China with 19% and South Korea with 14%. The rest of the countries offer between 2 and 7% of publications.

From a global point of view, although most of the articles come from the United States, Asia is the continent with the highest number of research projects, making up 50% of the total.

3.3 Publications distributed by technology

Graph 9 shows the distribution of technology applied across the publications. As for the most commonly used technologies, it is worth noting that almost 50% of the publications apply either algorithms (24%) or 4D simulation (22%).

These two technologies are followed by Real Time Location Systems (RTLS), rule checking, game engine, Virtual Reality (VR) and ontology, each accounting for between 7 and 14% of the total publications.

The last group denotes the use of information retrieval and 3D object design, used in 2% of publications.

Algorithms are used in risk identification (Sacks et al., 2009; Kim et al., 2016b) and evacuations in case of an emergency (Abolghasemzadeh, 2013; Li et al., 2014; Li et al., 2015a).

4D simulation is used in risk identification (Guo et al., 2013; Zhou et al., 2013; Kang et al., 2013) and workspace congestion (Moon et al, 2014a; Kassem et al., 2015; Moon et al., 2014b; Choi et al., 2014).

Real Time Location Systems are used to gain knowledge in real time of the workers on a building site, generally to identify danger areas or machinery collision risks. For example, RFID employed to

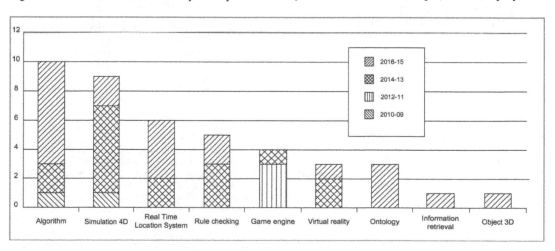

Figure 4. Publications distributed by technology.

449

identify danger areas (Kim, et al., 2016a), and GPS to avoid collisions between workers and machinery (Cheng & Teizer, 2013; Golovina, et al., 2016).

Rule checking is used in risk identification, specifically, the risk of falls from height (Zhang, et al., 2013; Zhang, et al., 2015a; Qi, et al., 2014) and collapse (Wang, et al., 2015).

The game engine and VR publications are geared towards the immersion of professionals and workers in virtual worlds in order to improve knowledge on safety (Albert, et al., 2014; Li, et al., 2012; Li, et al., 2015b; Perlman, et al., 2014) and simulations evoking emergency evacuations in case of fire (Rüppel & Schatz, 2011).

Ontology is used to organize and manage information related to safety (Wetzel & Thabet, 2015; Zhang, et al., 2015b; Ding, et al., 2016).

Kim et al. (2015) apply an information retrieval system to revisit accidents in the hope of avoiding potential future similar accidents.

4 CONCLUSION

Since 2013 the research which introduces BIM methodology in the management of safety on construction works has increased significantly.

The country with the highest number of publications is the USA, one of the leading countries in the implementation of BIM in the construction sector. It is worth noting that China has the second highest number of publications.

Among the main advances regarding the use of BIM applied to safety are automation in the identification of risks and application of preventative measures, identification of workspace congestion, real time information on safety and creation of virtual spaces to improve both learning about and management of safety.

Among the research analysed the main trend is towards the use of algorithm and (4D) simulation.

REFERENCES

Abolghasemzadeh, P., 2013. A comprehensive method for environmentally sensitive and behavioral microscopic egress analysis in case of fire in building. *Safety Science* 59:1–9.

Albert, A., Hallowell, M.R., Kleiner, B., Chen, A. & Golparvar-Fard, M. 2014. Enhancing Construction Hazard Recognition with High-Fidelity Augmented Virtuality. *Journal of Construction Engineering Management* 140(7).

Bryde, D., Broquetas, M., & Volm, J. M. 2013. The project benefits of building information modelling (BIM). *International Journal of Project Management* 31(7): 971–980.

Cheng, T. & Teizer, J., 2013. Real-time resource location data collection and visualization technology for construction safety and activity monitoring applications. *Automation in Construction* 34: 3–15.

Choi, B., Lee, H.S., Park, M., Cho, Y.K. & Kim, H. 2014. Framework for Work-Space Planning Using Four-Dimensional BIM in Construction Projects. *Journal of Construction Engineering and Management* 140(9).

Ding, L., Zhong, B., Wu, S. & Luo, H., 2016. Construction risk knowledge management in BIM using ontology and semantic web technology. *Safety Science* 87: 202–213.

Eastman, C., Teicholz, P., Sacks, R. & Liston, K. 2008. *BIM Handbook: a guide to building information modeling for owners, managers, designers, engineers, and contractors*. New Jersey: Jhon Wiley & Sons, Inc.

EUROSTAT, 2013. Accidents at work statistics, Luxembourg: Office for Official Publication of the European Communities: Eurostat.

Golovina, O., Teizer, J. & Pradhananga, N., 2016. Heat map generation for predictive safety planning: Preventing struck-by and near miss interactions between workers-onfoot and construction equipment. *Automation in Construction*.

Guo, H., Li, H. & Li, V., 2013. VP-based safety management in large-scale construction projects: A conceptual framework. *Automation in Construction* 34: 16–24.

Hadikusumo, B. & Rowlinson, S. 2002. Integration of virtually real construction model and design-for-safety-process database. *Automation in Construction* 11: 501–509.

Kang, L. S., Kim, S.-K., Moon, H. S. & Kim, H. S., 2013. Development of a 4D objectbased system for visualizing the risk. *Automation in Construction* 31: 186–203.

Kassem, M., Dawood, N. & Chavada, R., 2015. Construction workspace management within an Industry Foundation Class-Compliant 4D tool. *Automation in Construction* 52: 42–58.

Kim, H. et al., 2015. Information Retrieval Framework for Hazard Identification in Construction. *Journal of Computing in Civil Engineering*, 29(3).

Kim, H., Lee, H.S., Park, M., Chung, B., & Hwang, S. 2016a. Automated hazardous area identification using laborers' actual and optimal routes. *Automation in Construction*. 65: 21–32.

Kim, K., Cho, Y. & Zhang, S. 2016b. Integrating work sequences and temporary structures into safety planning: Automated scaffolding-related safety hazard identification and prevention in BIM. *Automation in Construction*. 70: 128–142.

Ku, K., & Mills, T. (2010). Research needs for building information modeling for construction safety. In *International Proceedings of Associated Schools of Construction 45nd Annual Conference*, Boston, MA.

Li, H., Chan, G. & Skitmore, M., 2012. Visualizing safety assessment by integrating the use of game technology. *Automation in Construction* 22: 498–505.

Li, N., Becerik-Gerber, B., Krishnamachari, B. & Soibelman, L., 2014. A BIM centered indoor localization algorithm to support building fire emergency response operations. *Automation in Construction* 42: 78–89.

Li, N., Becerik-Gerber, B., Soibelman, L. & Krishnamachari, B., 2015a. Comparative assessment of an indoor localization framework for building emergency response. *Automation in Construction* 57: 42–54.

Li, H., Lu, M., Chan, G. & Skitmore, M., 2015b. Proactive training system for safe and efficient precast installation. *Automation in Construction* 49: 163–174.

Moon, H., Dawood, N. & Kang, L., 2014a. Development of workspace conflict visualization system using 4D object of work schedule. *Advanced Engineering Informatics* 28(1): 50–65.

Moon, H., Kim, H., Kim, C. & Kang, L., 2014b. Development of a schedule-workspace interference management system simultaneously considering the overlap level of parallel schedules and workspaces. *Automation in Construction* 39: 93–105.

Perlman, A., Sacks, R. & Barak, R., 2014. Hazard recognition and risk perception in construction. *Safety Science* 64: 22–31.

Qi, J., Raja, R. A., Olbina, S. & Hinze, J., 2014. Use of Building Information Modeling in Design to Prevent Construction Worker Falls. *Journal Computing Civil Engineering* 28(5).

Rodrigues, F. & Alves, A. 2015. Contribution of BIM for hazards' prevention through design. *Occupational Safety and Hygiene III*. London:Balkema, pp. 67–69.

Rüppel, U. & Schatz, K., 2011. Designing a BIM-based serious game for fire safety evacuation simulations. *Advanced Engineering Informatics* 25(4): 600–611.

Sacks, R., Rozenfeld, O. & Rozenfeld, Y. 2009. Spatial and Temporal Exposure to Safety Hazards in Construction. *Journal of Construction Engineering Management*. 135(8): 726–736.

Wang, J., Zhang, S. & Teizer, J., 2015b. Geotechnical and safety protective equipment planning using range point cloud data and rule checking in building information modeling. *Automation in Construction* 49: 250–261.

Wetzel, E. M. & Thabet, W. Y., 2015. The use of a BIM-based framework to support safe facility management processes. *Automation in Construction* 60: 12–24.

Zhang, S. et al., 2013. Building Information Modeling (BIM) and safety: Automatic safety checking of construction models and schedules. *Automation in Construction*, 29: 183–195.

Zhang, S., Sulankivi, K., Kiviniemi, M., Romo, I., Eastman, C.M. & Teizer, J. 2015a. BIM-based fall hazard idetification and prevention in construction safety planning. *Safety Science* 72: 31–45.

Zhang, S., Boukamp, F. & Teizer, J., 2015b. Ontology-based semantic modeling of construction safety knowledge: Towards automated safety planning for Job Hazard Analysis (JHA). *Automation in Construction* 52: 29–41.

Zhou, Y., Ding, L. & Chen, L., 2013. Application of 4D visualization technology for safety management in metro construction. *Automation in Construction* 34: 25–36.

Occupational Safety and Hygiene V – Arezes et al. (Eds)
© 2017 Taylor & Francis Group, London, ISBN 978-1-138-05761-6

Analysis of whole-body vibrations transmitted by earth moving machinery

M.L. de la Hoz-Torres
Master in Construction Management and Security, University of Granada, Granada, Spain

M. López-Alonso
Department of Construction and Engineering Projects, University of Granada, Granada, Spain

D.P. Ruiz Padillo
Department of Applied Physics, University of Granada, Granada, Spain

M.D. Martínez-Aires
Department of Building Construction, University of Granada, Granada, Spain

ABSTRACT: Whole-body vibration presents a physical risk which affects the health of workers in the construction sector. Heavy equipment vehicles, such as earth moving machines, are widely used in this sector, and machine operators are regularly exposed to Whole-Body Vibration (WBV). The study presents five case studies. The aim of this study is to identify and measure WBV emissions and daily exposure levels. Three earth moving machines were analysed whilst performing different tasks. The results were obtained by measuring actual operations.

Vibrations were measured in accordance with the EU Directive on physical agents (vibration) 2002/44/EC and ISO 2631-1: 1997 Mechanical vibration and shock, evaluation of human exposure to whole-body vibration.

1 INTRODUCTION

Exposure to mechanical vibration is a problem which affects thousands of workers. The industrialisation of production processes and the mechanisation of the workplace have eliminated many of the risks which previously existed, however these phenomena have also led to a higher exposure to risks caused by physical agents, including vibrations (INSHT, 2014).

Exposure to vibrations is one of the ergonomic features of the workplace which poses a risk factor for the apparition of work-related Musculoskeletal Disorders (MSDs), both in the neck and upper extremities (Punnett & Wegman, 2004).

Millions of workers in different sectors all over Europe suffer work-related MSDs, making this the most common complaint amongst workers in Europe (45% of the total job-related illnesses (EUROSTAT, 2004). It affects workers' quality of life and costs the companies dearly (EU-OSHA, 2016).

There is epidemiologic evidence for the link between MSDs and Whole-Body Vibration (WBV). If exposure values exceed the limits established in legislation, the risks to which the workers, particularly machine operators, find themselves exposed are high. (Wilder & Pope, 1996).

MSDs caused by vibrations can develop over time and cause job-related osteoarticular or angioneurotic illnesses. They can also cause serious pain which forces the worker to take sick leave and even to seek medical help. (EU-OSHA, 2016).

Aware that workers' health is a fundamental requisite for productivity and economic development (WHO, 2007), strategies and regulations have been developed to control and prevent damage caused by vibrations (INSHT, 2014). Several European countries have tools put in place to evaluate exposure to vibrations using information collated in a database.

The present study has come to light due to the importance of determining the risks that construction workers are exposed to. The analysis makes it possible to determine levels of exposure to vibrations of the earth moving machine operators in the construction sector, a situation which effects companies, manufacturers, as well as the actual workers.

The aim of this project is to determine whether earth moving machine operators, during a normal working day, are exposed to levels of WBV which exceed those set out in the legislation. This paper present five different cases studies. To this end in-situ measurements were taken from two mixed backhoe loaders on wheels and a midi excavator with a pivoting cab on tracks. The data on WBV levels provided by the manufacturers as well as that found on databases were also analysed.

2 MATERIAL AND METHODS

2.1 Legislation

In European legislation Directive 2002/44/EC, of the European parliament and of the Council of 25 June 2002, has been drawn up, on the minimum health and safety requirements regarding the exposure of workers to the risks arising from physical agents (vibration) (sixteenth individual Directive within the meaning of Article 16 of Directive 89/391/EEC).

All Member States of the EU have transposed this directive to their national legislation. In Spain this was done through the Royal Decree 1311/2005, of 4th November, concerning the protection of the health and safety of workers against risks related to or possibly related to exposure to mechanical vibrations.

In this legislation, the minimum provisions for the protection of workers against risks to their safety and health related to, or possibly related to, exposure to mechanical vibrations are set out (European Commission, 2007).

2.2 Equipment used

The equipment chosen to measure vibrations is SVAN 106, manufactured by SVANTEK. The measurements made by this machine comply with ISO 2631-1, 2 & 5 and ISO 5349. It allows simultaneous measurements to be taken using two triaxial accelerometers.

The machines chosen for the project are two mixed backhoe loaders on wheels and a midi excavator with a pivoting cab on tracks (see Table 1).

2.3 Sample collection

Once the earth moving machines were chosen for the study, the next step was to analyse the manufacturers' instruction manuals to obtain information relating to WBV transmission. This data was compared to information available on national and international databases namely: "Banca Dati Vibrazioni Corpo intero" (Laboratorio di Sanità Pubblica dell'Azienda Sanitaria USL Toscana Sud Est), "Databases for Vibration Machines" (Umeå University) and "Base Vibra" (INSHT, 2014).

Based on this data the typical activities that workers performed during their working day were determined.

Table 1. Machines selected for measurement.

Type	Manufacturer	Model	Year	Wheels or tracks
Backhoe loader	JCB	3CX4C	2003	Wheels
Backhoe loader	JCB	3CX4C	2007	Wheels
Excavator	Takeuchi	TB175	2006	Tracks

The work cycle includes "the operation or series of operations repeated", with operation being "an identifiable activity for which a representative vibration magnitude measurement is made, being a combination of a type of work and a working condition" (AENOR, 2004). The strategy for measuring the continuous operations throughout the working day was established.

2.4 Calculation of vibration exposure

Following European Directive 2002/44/EC (2002), the parameter used to evaluate WBV is the effective energy level or equivalent continuous acceleration ($A_{eq,t}$) in acceleration units m/s^2.

The European Directive allows exposure to vibrations to be measured using two different methods, both defined in the ISO 2631-1:1997 standard. The equivalent daily exposure A(8) is expressed as equivalent continuous acceleration over an eight-hour period, calculated as the highest (rms) value (see equation 1).

$$A_i(8) = k_i \cdot a_{wi} \sqrt{\frac{T_{exp}}{T_0}} \qquad (1)$$

where a_{wi} is the effective value of acceleration weighted in frequency according to the orthogonal axes x, y, z.; T_{exp} is the exposure time; T_0 is the reference time of 8 hours; k_i is the factor of multiplication (UNE-ISO 2631-1:2008).

The value A(8) is calculated as the maximum of the value in the three directions (see Equation 2).

$$A(8) = \max\left[Ax(8), Ay(8), Az(8)\right] \qquad (2)$$

The other method is the Vibration Dose Value (VDV), defined as a cumulative dose based on the fourth root of the acceleration to the fourth power and its unit is m/s$^{1.75}$ (European Commission, 2007) (see equation 3).

$$VDV = \left\{ \int_0^T (a_w(t))^4 \, dt \right\}^{\frac{1}{4}} \qquad (3)$$

where $aw(t)$ is the frequency-weighted instantaneous acceleration; T is the total measurement period in seconds. Daily VDV is the highest value among the calculated values in the three directions (see Equation 4).

$$VDV = \max\left[VDV_{exp,x,i}, VDV_{exp,y,i}, VDV_{exp,z,i}\right] \qquad (4)$$

This method is defined in the UNE-ISO 2361-1standard as being more sensitive to vibration peaks by using the fourth power instead of the second power of the acceleration time history as the basis for average (AENOR, 2008).

The established limit values are the daily Exposure Limit Values (ELV), which, when standardised to an eight-hour reference period, shall be 1.15 m/s², or a vibration dose value of 21 m/s$^{1.75}$;

Finally, the daily Exposure Action Value (EAV), standardised to an eight-hour reference period, shall be 0.5 m/s² or, at the discretion of the Member State concerned, a vibration dose value of 9,1 m/s$^{1.75}$.

3 RESULTS

3.1 Analysis of the data

Once measurement data has been obtained it is possible to calculate the workers' exposure levels.

Using this information, it is then possible to determine whether the levels exceed those indicated by the manufacturer in their instruction manual. Subsequently this will be compared to exposure limits laid out in the legislation (EAV; ELV).

3.1.1 Mixed backhoe loader 3CX4C JCB
With the 3CX4C machine, manufactured in 2007, two measurements were taken (see Table 2): movement on tarmac (Case 1); earth moving and displacement (Case 2).

Two measurements were also taken with the 3CX4C machine, manufactured in 2003 (see Table 3): movement; cutting and pruning on agricultural land (Case 3, Case 4).

3.1.2 Takeuchi TB175 excavator
With the Takeuchi TB175 manufactured in 2006 one measurement was taken (see Table 4): demolition of a dwelling in an urban area (Case 5).

Table 2. Results obtained from the 3CX4C (2007).

Sample	A(8) (m/s²)	VDV (m/s$^{1.75}$)
Case 1	0.163	6.012
Case 2	0.541	11.817

Table 3. Results obtained from the 3CX4C (2003).

Sample	A(8) (m/s²)	VDV (m/s$^{1.75}$)
Case 3	0.304	6.974
Case 4	0.319	11.548

Table 4. Results from the TB175 (2006).

Sample	A(8) (m/s²)	VDV (m/s$^{1.75}$)
Case 5	0.239	9.506

4 DISCUSSION

4.1 Information provided by manufacturers

The manufacturers of the JCB 3CX4C machines stipulate in their instruction manual that, for a cycle based on tests including excavation and work with the loader (earth), the normalised value for a reference period of 8 hours A(8) is 0.28 m/s². The emission value in VDV is not indicated in the instruction manual.

The results obtained from the JCB 3CX4C exceeded in the samples Case 2, Case 3 and Case 4 the vibration emission values published by the manufacturer.

Bearing in mind that the daily value A(8) provided by the manufacturer refers to a work cycle of earthwork, only sample Case 2 is comparable. In this case, the sample more than doubles the figure provided by the manufacturer.

The manufacturers of the Takeuchi TB175 indicate in their manual that vibration emissions should not exceed the value of acceleration 0.5 m/s². As no type of work cycle is indicated, it can be assumed that vibration emissions should never exceed this level. The manufacturer does not indicate the VDV.

The results obtained from the Takeuchi TB175 (Case 5) do not exceed the limits published by the manufacturer.

It is not always possible to use the vibration emission values indicated by the manufacturer as the conditions in which these were measured is not always specified. Emission values, when associated with different operations, are not comparable, therefore cannot be used to carry out an evaluation.

4.2 WBV data bases

A search was conducted on the following international WBV databases: BaseVibra, INSHT España (INSHT, 2016), Department of Public Health and Clinical Medicine, Occupational and Enviromental Medicine. Universidad de Umea (Umea, 2016) and Laboratorio di Sanità Pubblica dell'Azienda Sanitaria USL Toscana Sud Est. Italia (Azienda, 2016).

The search of 3CX JCB produced 4 results in INSHT database and 2 results in Azienda database (see Figure 1). There wasn't any result in Umea database.

The search of Takeuchi wasn't produced any result.

4.3 Exposure limit values

Measurements between EAV and ELV limits set out in the Physical Agents Directive (Directive 2002/44/EC, 2002), are samples Case 2, Case 4 and Case 5.

The results of A(8) and VDV for sample Case 2 exceed the EAV (see equation 5).

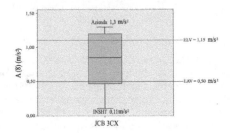

Figure 1. 3CX JCB results on WBV database.

$$\text{Case 2} \quad \begin{aligned} A(8) &= 0.541\, m/s^2 > 0.5\, m/s^2 \\ VDV &= 11.817 m/s^{1.75} > 9.1 m/s^{1.75} \end{aligned} \quad (5)$$

However, in samples M4 and M5 only the VDV exceeds the EAV (see equations 6 and 7).

$$\text{Case 4} \quad VDV = 11.548\, m/s^{1.75} > 9.1\, m/s^{1.75} \quad (6)$$

$$\text{Case 5} \quad VDV = 9.506\, m/s^{1.75} > 9.1\, m/s^{1.75} \quad (7)$$

This is because the VDV is more sensitive to shocks over short time periods than the mean square root of the A(8) calculation.

The levels are not above the ELV in any of the operations.

4.4 Preventative measures

As established in Directive 2002/44/EC, in any event, workers shall not be exposed above the exposure limit value.

If the exposure limit value is exceeded, the employer shall take immediate action to reduce exposure below the exposure limit value (Directive 2002/44/EC, 2002).

The employer is under obligation to carry out health surveillance on those workers who are, or might be, exposed to values exceeding the EAV. Furthermore, the employer must comply with what is established in terms of training, information, participation and consultation of the workers (Directive 2002/44/EC, 2002).

5 CONCLUSIONS

Not all manufacturers indicate the vibration emission levels of their earth moving machines. When the value is indicated in the instruction manual, in no cases can this exceed the ELV or EAV stipulated in Directive 2002/44/CE, of the European parliament and of the Council of 25 June 2002, on the minimum health and safety requirements regarding the exposure of workers to the risks arising from physical agents.

The manufacturers are obliged by the Machinery Directive to provide vibration emission values only in terms of the RMS value of acceleration, making it difficult to perform evaluations employ-

ing the VDV method from the information provided by the manufacturer.

Of the three cases where the VDV exceeded the EAV, only one exceeded the value A(8). This is due to the fact that the VDV calculation is more sensitive to shocks over short time periods than over long.

If the evaluation of the earth moving machine operator's workplace was limited to the calculation of A(8) during different operations, without taking into consideration the VDV calculation, operations which exceed EAV could possibly be considered safe.

REFERENCES

AENOR, 2004. UNE-EN 14253:2004+A1 Vibraciones mecánicas. Medidas y cálculos de la exposición laboral a las vibraciones de cuerpo completo con referencia a la salud. Guía práctica. AENOR.

AENOR, 2008. UNE-ISO 2631-1:2008 Mechanical vibration and shock. Evaluation of human exposure to whole-body vibration. AENOR.

Azienda, 2016. Base de datos de vibraciones del Laboratorio di Sanità Pubblica dell'Azienda Sanitaria USL Toscana Sud Est. Italia. Disponible en: http://www.portaleagentifisici.it. Acceso: Junio de 2016.

Comisión Europea, 2007. Guía de buenas prácticas no vinculante para la aplicación de la Directiva 2002/44/CE, sobre las disposiciones mínimas de seguridad y de salud relativas a la exposición de los trabajadores a los riesgos derivados de los agentes físicos (vibraciones). Luxemburgo: Oficina de publicaciones UE.

EU-OSHA, 2016. Trastornos musculoesqueléticos. [En línea] Available at: https://osha.europa.eu/es/themes/musculoskeletal-disorders [Último acceso: 31 Mayo 2016].

EUROSTAT, 2004. Work and health in the EU. A statistical portrait, Luxembourg: Office for Official Publication of the European Communities: Eurostat.

INSHT, 2014. Aspectos ergonómicos de las vibraciones, Madrid: Instituto Nacional de Seguridad e Higiene en el Trabajo (INSHT).

INSHT 2016. Instituto nacional de Seguridad e Higiene en el trabajo de España. (BaseVibra—Base de datos "Vibraciones mecánicas". Disponible en: http://vibraciones.insht.es:86/findcuerpo.aspx. Acceso: Junio de 2016.

Laboratorio di Sanità Pubblica dell'Azienda Sanitaria USL Toscana Sud Est, s.f. Guida uso Banca Dati Corpo Intero. [En línea] Available at: http://www.portaleagentifisici.it/fo_wbv_guida_uso_bancadati.php?lg=IT [Último acceso: 5 Junio 2016].

Organización Mundial de la Salud, 2007. Salud de los trabajadores: plan de acción mundial – 60.ª ASAMBLEA MUNDIAL DE LA SALUD.

Punnett, L. & Wegman, D.H., 2004. Work-related musculoskeletal disorders: the epidemiologic evidence and the debate. Journal of electromyography and kinesiology, Issue 14, pp. 13–23.

Umeå University, 2016. Databases for Vibration Machines. Recuperado el 06 de 11 de 2016, de http://www.vibration.db.umu.se/Default.aspx?lang=en

Wilder, D. & Pope, M., 1996. Epidemiological and aetiological aspects of low back pain in vibration environments. An update. Clinical Biomechanical., 11(2), pp. 61–73.

Occupational Safety and Hygiene V – Arezes et al. (Eds)
© 2017 Taylor & Francis Group, London, ISBN 978-1-138-05761-6

Mental workload of nurses at an intensive care unit, João Pessoa, Brazil

R.L. Silva & L.B. Silva
Federal University of Paraíba, João Pessoa, Brazil

ABSTRACT: The level of mental workload was evaluated in nurses working in an Intensive Care Unit in the city of João Pessoa/Brazil, using NASA-TLX and a complementary questionnaire to trace the socioeconomic profile of the participants, containing important issues for greater Knowledge of the sample. After the data were collected, these were tabulated and, later, the NASA-TLX score was calculated. Descriptive analyzes of the data were performed. It was found that the mental load of the workers was low in view of the low Level of Frustration, since the motivating feelings of the professional choice reaffirm the humanitarian character of love, donation and satisfaction that are perpetuated during professional practice. It is important to conduct research with greater depth, capable of investigating and analyzing the physiological responses of the human body to the demands of mental load, from the neuroergonomic perspective.

1 INTRODUCTION

For Motter et al. (2015) the concept of workload is derived from studies of work psychology and activity ergonomics. The same has been important to clarify questions related to the physical and mental health of the worker. Antonelli (2011) complements this concept by stating that workload is a quantitative and/or qualitative measure of the level of motor, physiological and mental activity required to perform a job.

Mental load can be considered as a brain/mindset that modulates the human performance of perceptual, cognitive and motor skills, and its evaluation is an important theme in research and practice in ergonomics (Kramer & Parasuraman, 2007).

Amin et al. (2014) states that in a hospital environment, mental effort can be generated by the need to meet the patient's needs of nursing services or to interact with patients or relatives related to the most intense emotional aspects of life. Therefore, it is assumed that this environment can have an effect on the workload. The workload of nurses in the Intensive Care Units (ICU) is worrisome, mainly due to the nature of the work and the criticality of the patients' results, which strengthens the accomplishment of this article.

2 THEORETICAL REFERENCE

2.1 *Mental load of work and ergonomics*

Mental workload is generally understood as a field of interaction between the demands of the task and the capacity for human achievement. This is a consequence of the worker performing the task itself, taking into account all the complexity when performing an activity. Thus, the mental load is not only derived from work, but also from other factors extrinsic to the task, such as: individual, sociocultural, intellectual capacity, age, level of education, vocational training, learning, experience (noise, heat and toxic) (Cardoso & Gontijo, 2012).

For DiDomenico & Nussbaum (2011) the evaluation of the mental workload is an important aspect in the conception and evaluation of an occupational task. Efficient performance requires concentration on task-related information and suppression of extraneous stimuli to avoid overloading of information.

Excessive levels of mental load can cause errors or delay information processing, especially if the amount of information presented exceeds processing capacity (DiDomenico & Nussbaum, 2011).

For a better understanding of the mental load, a brief description is necessary about some concepts and definitions of terms commonly used and associated with mental load, namely: psychic load, mental load and cognitive load (Cardoso, 2010).

Cardoso & Gontijo (2012) define psychic load as affective aspects present in the work, relating to how the worker can be affected with the work that he/she performs. Cognitive load refers to the loads arising from the cognitive demands of tasks. The use of memory, perception, attention, concentration, reasoning, and decision making are task related. The mental load contemplates psychic and cognitive aspects covering the concepts of the psychic and cognitive load.

Due to the nature of the activities that involve the nursing professional, it provides assistance to the sick or in situations of illness, frequently encountering situations of restlessness, suffering, degradation and death (Ferreira & Ferreira, 2014).

2.2 *Mental workload assessment*

For Cain (2007), the main reason for measuring workload is to quantify the mental cost of task execution to predict operator and system performance. This measure provides insight into how far increased job demands can lead to unacceptable performance, with the ultimate goal of better working conditions, intuitive design of the workstation for more effective procedures.

Cardoso (2010) & Cardoso & Gontijo (2012) establish the categories and definitions of the methods used for this purpose. Among them we have: Primary Task Measures that relate directly to the performance of the task itself; The Multiple Tasks Measures that measures the level of the load through the use of two tasks, a primary, more sophisticated task, and a second task less sophisticated and with the level of load already known; Physiological Measures that measure physiological responses related to responses to changes in levels of mental loads; And finally, Subjective Measures, these seek the subjective responses to the experiences related to the mental work load, often administered through questionnaires applied at the end of the task accomplishment.

The subjective measures are the most used to measure the mental work load, because the level of wear or mental load is associated with the capacity of the worker to perform his or her job. Some of these techniques attempt to address individual differences by developing weights for the proportion of scales for a global workload calculation that are operator specific, such as NASA TLX (Cardoso, 2010; Cardoso & Gontijo, 2012; Cain, 2007).

NASA TLX is a method that has proven to be effective in measuring the level of mental load in workers. This tool has been developed by Hart & Staveland (1988) as a multidimensional rate process that provides an overall workload score based on a weighted average of evaluations in six subscales (Hart & Staveland, 1988):

- Mental Requirement: amount of mental and perceptual activity that the task requires (thinking, deciding, calculating, remembering, looking, searching, etc.);
- Physical Demand: Quantity of physical activity that the task requires (pulling, pushing, turning, sliding, etc.)
- Temporal Demand: Time pressure level felt. Ratio between the time needed and the available time.

- Performance: To what extent does the individual feel satisfied with the level of performance and performance at work;
- Effort: Degree of mental and physical effort that the subject has to perform to obtain his level of income;
- Level of Frustration: To what extent does the subject feel insecure, stressed, angry, disgruntled, while performing the activity.

In the work mental load score, this questionnaire is divided into two sections: sources of loads (or weights) and the magnitude of loads (or assessments).

- Sources of Charges (Weights): The weights refer to diagnostic information about the nature of the workload imposed on the task. There are 15 possible pairs of comparisons within the 6 scales or dimensions, each pair is represented on a card. The subject circulates the pair member on the card that most contributes to their workload on the task. The number of times each factor is chosen is marked. They can be selected 0 times (no relevance) or 5 (more important than some other factor). A combination of different weights is obtained for each task. The same combination of weights can be used in different versions of the same task if the contributions or if the 6 factors for the workload are reasonably similar.
- Load Magnitude (Ratings – Fees): The second requirement is to obtain a numerical rate for each scale that reflects the magnitude of that factor in a given task. Subjects respond by scoring each scale in the desired position. Each scale has a 12 cm line divided into 20 equal parts anchored in bipolar descriptions. The twenty-first part marks each division of the scale from 0 to 100, of 5 by 5. If the subject marks between two marks, the right value will be used.
- Weighted average weighting and calculation: The overall workload score for each subject will be computed by multiplying each assessment by the weight given by the subject for each factor. The sum of the weights evaluated by each task is divided by 15 (the sum of the weights).

Daily nursing professionals are exposed to a large number of factors that contribute to the mental and psychic load, some of which are inherent in the work itself, or to the work organization itself. Mental load and psychic load are associated and may result in occupational stress situations (Ferreira & Ferreira, 2014).

Considering that the quality of health care is directly related to patient safety, it can be considered that not only the quality of the professionals involved, but also the quantity and workload are relevant for the proper development of nursing actions. Studies have shown that a 0.1% increase in the patient/nurse ratio leads to a 28% increase in the rate of adverse events (Serafim, 2015).

Kang et al. (2016) state that the incidence of adverse effects to patients may increase when the number of these, allocated to each nurse, is high and the nursing workload increases. These events occur within a hospital, which compromises the patients' health, in addition to generating significant financial burden for the health system.

Most of the studies aimed at assessing the nursing workload are performed with several scales directed to the evaluation of the physical load as Therapeutic Intervention Scoring System (TISS), PRN-80, Nursing Activities Score (NAS), among others (Serafim, 2015). These instruments evaluate several aspects—patient severity, physical workload of nurses, hours spent in nursing care for each patient—therefore, it is necessary to investigate the mental workload of these professionals and their effects on nursing practice and patient care.

3 MATERIAL AND METHODS

3.1 Characterization of the study

This is a cross-sectional study with a quantitative approach, aiming to analyze the mental workload in nursing professionals. The study population was composed by the nursing team that perform their work activities in the Adult Intensive Care Unit of an emergency hospital in the city of João Pessoa/Brazil. A questionnaire was initially applied to acquire the socioeconomic profile of the study participants. Then, to analyze the mental load of these individuals, the NASA TLX questionnaire was used.

3.2 Statistical analysis of data

After collecting the data regarding the socioeconomic questionnaire, these were tabulated in Excel spreadsheet.

In the calculation of the weighted average for NASA-TLX, certain scores were considered according to Chile (2016):

- Low Mental Load Level: 0 to 500;
- Medium Mental Load Level: 501 to 1000;
- High level of mental load: >1000.

After this calculation, descriptive analyzes were performed.

4 RESULTS

A total of seven nursing professionals participated in the study, of which 71.4% (n = 5) of the total female sample and only 28.6% (n = 2) were males aged 29 to 52 years. The majority have 5–10 years of work in the area, 71.4% (n = 5); 14.3% (n = 1) over 10 years and only 14.3% (n = 1) less than 5 years of performance.

Regarding the weekly workload, the majority reported working between 30 and 40 weekly hours (n = 4, 57.2%); Up to 30 hours per week (28.5%, n = 2) and only 14.3% (n = 1) worked more than 40 hours per week. An important fact is that 71% of the sample has more than one labor bond, and this same percentage of participants had a mixed work scale, that is, it works at day and night schedules. The majority (n = 4, 57.2%) work between 5 and 10 years in the hospital. The results of the NASA-TLX analysis are presented in Table 1.

It can be observed that all had a low score of mental work load, being able to be influenced by the low Level of Frustration. This is explained by Fidlarczyk & Silva (2009) in affirming that the motivating feelings of professional choice reaffirm the humanitarian character of love, bestowal, and satisfaction that have been perpetuated during professional practice. Only one of the respondents had a high level of frustration.

Another factor that may explain this low mental demand in the study sample is the time (years) of service, where 57% work in the profession between 5 and 10 years. This finding is endorsed by Sapata (2012) because in his study of coping strategies, ie, cognitive and behavioral efforts to deal with situations of harm, threat or challenge when a routine or automatic response is not available, Which in nurses with less than 1 year of age experience more stress generated by the psychological environment, that is, by situations of death or suffering of patients, by inadequate preparation to deal with the emotional needs of the patient and the family, due to lack of Peer support, and uncertainty in treatments; As well as by the hospital social

Table 1. Results of the mental load of nurses working at the Intensive Care Unit, obtained by NASA–TLX.

Questionnaire	1	2	3	4	5	6	7
MD	25	100	55	25	100	100	100
PD	50	70	55	25	100	60	100
TD	80	55	60	25	55	100	100
LP	40	5	45	5	5	40	5
EL	50	100	50	100	100	100	50
LF	35	25	80	5	5	100	15
Weighted score	233	365	303	211	358	440	421

Legend: MD: Mental Demand; PD: Physical Demand; TD: Temporary Demand; LP: Level of Performance or Performance; EL: Effort Level; LF: Level of Frustration.
Source: Research Data (2016).

environment, that is, by conflicts with doctors or conflicts with other nurses and with bosses. Within the same perspective, it is emphasized that the less experienced stress is associated with professional instability and earned salary.

There is yet another point that may have influenced a low mental demand in these nurses: the weekly workload. Most of the interviewees (57%) work between 30 and 40 hours a week. Dalri et al. (2014), when analyzing the relationship between nurses' hours and physiological stress reactions, concluded that although the majority of nurses performed their duties for more than 36 hours a week, they did not physiologically present high stress responses. Such workers dealt with conflicts in vertical and horizontal relationships between professionals, family members and patients.

The fact that 57% of the sample work in more than one job does not influence the increase in mental workload. This corroborates with the studies of Sapata (2012) that found that it is not significant that stress-generating situations have more than one employment relationship.

5 CONCLUSION

The participants of this research presented a low level of mental workload, explained by several factors, the one with the greatest impact being the motivational feelings of professional choice that reaffirm the humanitarian character of love, donation and satisfaction that were perpetuated during the practice professional level, leading to a low level of frustration.

It is important to carry out research with greater depth, capable of investigating and analyzing the physiological responses of the human body to the demands of mental workload, but from the perspective of neuroergonomics, understanding better the relation between the human brain and the brain—performance of the man during the performance of his activities.

REFERENCES

Amin et al. 2014. Measuring Mental Workload in a Hospital Unit Using EEG—A Pilot Study. *Proceedings of the 2014 Industrial and Systems Engineering Research Conference Y. Guan and H. Liao, eds.*

Antonelli et al. 2011. Avaliação da carga de trabalho físico em trabalhadores de uma fundição através da variação da frequência cardíaca e análise ergonômica do trabalho. *Ação Ergonômica.* V. 6. N. 2, 18–23.

Cain, B. 2007. A review of the mental workload literature. *Defense Research and Development Canada Toronto. Human System Integration Section.* Recovered from: www. handle.dtic.mil/100.2/ADA474193.

Cardoso, M.S. & Gontijo, L.A. 2012. Avaliação da carga mental de trabalho e do desempenho de medidas de mensuração: NASA TLX e SWAT. *Gestão da Produção,* São Carlos, v. 19, n. 4, p. 873–884.

Cardoso, M.S. 2010. *Avaliação da carga mental de trabalho e do desempenho de medidas de mensuração: NASA TLX e SWAT.* Masters dissertation. Federal University of Santa Catarina - Graduate Program in Production Engineering - Master's Degree in Ergonomics. Florianópolis, Santa Caratina, Brazil.

Chile. Ministério del Trabajo e Previsión Social (2016). Método NASA – TLX. Recovered from: http://www. campusprevencionisl.cl/appergo/nasatlx/nasa-tlx.pdf.

Dalri et al. 2014. Carga horária de trabalho dos enfermeiros e sua relação com as reações fisiológicas do estresse. *Revista Latino-Americana. Enfermagem.* V 22. N 6. Pág. 959–65.

DiDomenico, A. & Nussbaum, M.A. 2011. Effects of different physical workload parameters on mental workload and performance. *International Journal of Industrial Ergonomics.* V. 41. pág. 255–260.

Ferreira, M. & Ferreira, C. 2014. Carga mental e carga psíquica em profissionais de enfermagem. *Revista Portuguesa de Enfermagem de Saúde Mental,* ESPECIAL 1.

Fidlarczyk, D. & Silva, L.R. 2009. Os sentimentos dos enfermeiros sobre ser enfermeiro. *Journal of Nursing UFPE on line,* Volume 3, Número 3.

Hart, S.G. & staveland, L.E. 1988. Development of NASA-TLX (Task Load Index): Results of empirical and theoretical research. In: Hancock, P.A. & Meshkati, N. (Eds.). *Human mental workload.* Amsterdam: North Holland, p. 139–183.

Kang et al. 2016. Nurse-Perceived Patient Adverse Events depend on Nursing Workload. *Osong Public Health Res Perspect.* V. 7. N. 1. Pág 56–62.

Kramer, A.F. & Parasuraman, R.. 2007. Neuroergonomics: Applications of Neuroscience to Human Factors. In: Cacciopo, J.T. & Tassinary, L.G.; & Berntson, G.G. *Handbook of Psychophysiology.* Reino Unido: Cambridge University Press.

Motter et al. 2015. O que está à sombra na carga de trabalho de estivadores?. *Revista Produção Online,* vo.15, n. 1, p. 321–344.

Sapata, A.F.R. 2012. Stress e estratégias de *coping* em enfermeiros: estudo comparativo entre Portugal e Espanha. Masters dissertation. Lusófona University of Humanities and Technologies. Faculty of Psychology. Lisbon, Portugal.

Serafim, C.T.R. 2015. *Eventos Adversos relacionados à Gravidade e Carga de Trabalho de Enfermagem em Unidade de Terapia Intensiva.* Masters dissertation. Paulista State University Júlio de Mesquita Filho. Medical School. São Paulo, São Paulo, Brazil.

Occupational Safety and Hygiene V – Arezes et al. (Eds)
© *2017 Taylor & Francis Group, London, ISBN 978-1-138-05761-6*

Application of statistical tools to the characterization of noise exposure of urban bus drivers

L. Pedrosa
PROA/LABIOMEP, Faculty of Engineering, University of Porto (FEUP), Porto, Portugal

M. Luisa Matos
LNEG, National Laboratory of Energy and Geology, S. Mamede de Infesta, and FEUP, Porto, Portugal

J. Santos Baptista
PROA/LABIOMEP, Faculty of Engineering, University of Porto, Porto, Portugal

ABSTRACT: Workplace noise is a global scale problem affecting workers of almost all activities. Driving urban buses is one of those activities. In order to investigate the relationship between noise and city driving variables, in this work were analysed twelve drivers, four models of buses, four routes with different characteristics and executed four trips in each route. The measurements were done in continuously along all driving time. The equivalent continuous sound level (L_{Aeq}) was obtained from observation periods with a mean time around 150 min. The data were submitted to a statistical treatment with principal components analysis, Kolmogorov-Smirnov and variance analysis. Based on the records resulting from monitoring performed, it was concluded that the L_{Aeq} values are dependent on the type of bus and on the route, with no clear evidence of the influence of the human factor (driver) on the observed noise levels.

1 INTRODUCTION

The noise in the workplace is a global problem affecting workers worldwide. Noise is not only a source of discomfort for workers, but also the main cause of work-related hearing loss (Freitas, et al., 2013). Noise is recognized as a relevant health problem. Its effects are, increasingly, considered a major public health problem (World Health Organization., 2015). In an occupational point of view, exposure to noise has been also associated with numerous detrimental health effects (Berglund, Lindvall, & Schwela, 1999).

In the EU-27, one third of the workforce, around 60 million workers, are exposed to excessive noise (OSHA, 2009). In the United States, it is estimated that 22 million workers (17%) are exposed to noise levels considered to be hazardous (Tak, Davis, & Calvert, 2009), (Deshaies, et al., 2015).

Legally, is considered that hearing loss is minimal to exposures bellow 80 dB (A) and increases with the growth of noise level (Lutman & al, 2008). However, the effects of noise are not limited to hearing damage. This kind of exposure can also lead to increased fatigue, stress, sleep disturbances, cardiovascular problems and even tinnitus.

In the workplace, one of the more negative potential effects of the noise level is that it overlaps to audible warnings, impairing communication and also increasing the risk of accidents (OSHA, 2009). The noise induced hearing loss was recorded in 2001 as the fourth most common occupational disease in the EU-12. Fourteen million workers in EU-27, or 7% of total, believe that their work affects their health causing hearing disorders. Actually, the incidence rate of hearing disorders is of 11.5 cases per 100 000 workers (OSHA, 2009). A recent research (EU-OSHA, 2011) developed through a questionnaire in order to evaluate the perception of drivers regarding the safety hazards and health (OSH), identified noise as one of the factors that contributes to accidents. Bus drivers are responsible in a daily basis by transport million people between their homes and work and get them back safely.

In studies carried out with the aim of evaluating the level of exposure to noise in buses (L_{Aeq}), it is unanimous that the L_{Aeq} depends on many variables, in which it is possible to highlight factors such as: the type and dimensions of the bus, bus route, type of driving, number of passengers and, bus age.

The most of the times, the noise source is the engine influenced by the changes in speed, the existence of automatic speeds, the use of the accelerator and the number of pauses (Nassiri, Ebrahimi, Monazzam, Rahimi, & Shalkouhi, 2014; Nadri, et al., 2012; Universitas, 2010). In these studies, the location of the engine is also considered important since drivers are exposed to a higher noise level in

Table 1. Characteristics of the buses involved in the study.

Bus type	Brand	Engine	Door	Year	Gearbox	Bus capacity		
						Seated	Standing	Wheelchair
Mini Bus	Volkswagen	Front	Front and rear	2012	Manual	15; 15	5; 2	0; 1
Double-decker	MAN	Rear	Front, middle and rear	2011	Automatic	91; 91	36; 33	0; 1
Articulated	VOLVO	Rear	Front, middle and rear	2010	Automatic	48	99	1
Standard	Mercedes	Rear	Front and rear	2000	Automatic	34	61	1

Table 2. Summary of noise monitoring.

Types of bus	Minibus	Double-decker	Articulated	Standard
N° of buses monitored	3	9	10	10
N° of drives evaluated	12	12	12	12
Routes evaluated	2	2	2	2
N° of travels by route	4	4	4	4
N° of travels	4*12	4*12	3 + 4*11	4*12

Table 3. Code assigned to the original data.

Character	Identified variable
1st	Driver
2nd	Route
3rd	Type of bus
4th	Travel direction

buses where the engine is located at the front of the vehicle (Zannin, Diniz, Giovanini, & Ferreira, 2003; Bruno & Zannin, 2010; Nassiri, Ebrahimi, Monazzam, Rahimi, & Shalkouhi, 2014; Silva & Correia, 2010). In addition to all those factors, Mohammadi (2015) refers, as an important aspect, the lack of adequate vehicle maintenance. In order to analyse the real results of the implemented protective measures are presented in this article the preliminary results about the relationships between noise and city driving variables in a big company of the north of Portugal.

2 MATERIAL AND METHODS

To reach the objectives, four different type of buses have been analysed (Table 1), traveling in real conditions by different routes. A total of 115 hours of measurements, corresponding to an average of 2:30 hours per route were monitored. The summary of the sampling performed can be observed in Table 2.

All studied busses were equipped with an air conditioning system. Each one of the four different

routes was tested four times by the same bus (two trips in one direction and two trips in the other).

For this article were processed data obtained during the first 15 minutes of each monitored trip, using all the combinations between the considered variables. The next step was the encoding of the variables with a four digits code built as is shown in Table 3.

Coding examples:
– ac21 - Driver a, in route c, in the Double-decker bus, the trip being done on going direction;
– bb13 - Driver b, in route b, in the Minibus at the return trip.

After variables codification, statistical data treatment were done using the following steps:

1. Determination of the basic statistics of each variable (variable is understood as the code applied to each tetranomial: driver/route/bus/trip). Were tried to have a one-dimensional characterization of the data analysis;
2. Application of a Principal Components Analysis (PCA) with the objective of identify the relational structure between the variables;
3. Verification of the theoretical probabilistic model (normal distribution) adjustable to the data, using a Kolmogorov-Smirnov adjustment test;
4. After analysing the required assumptions for its application, a Variance Analysis (ANOVA) was carried out to investigate the influence of the different factors on the homogeneity of the data, that is, which factor influences the monitored noise level.

3 RESULTS AND DISCUSSION

3.1 Results obtained by performing histograms and basic statistical analysis

The histograms were done directly with the data in dBTrait 5.1 Software. Analysing the set of monitored data, these were very effective in terms of visual analysis, since similarities were easily identified between the histograms form and the Gaussian distribution (Figure 1).

However, in some situations, it is not so clear to assume the normality of the records (Figure 2).

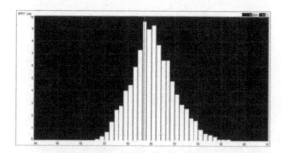

Figure 1. Distribution of L_{Aeq}, monitoring of *kc3*.

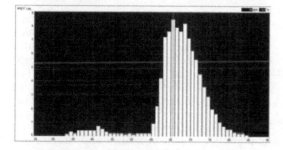

Figure 2. Distribution of L_{Aeq}, monitoring of *dd4*.

So, subsequent analysis were necessary by applying Kolmogorov-Smirnov test in order to clarify, for each situation, the effective adjustment of the data to the distribution settings.

Basic statistics analysis allows for a better and more sustained interpretation of the variability of the registers, contributing to the characterization of the behaviour of equivalent continuous sound level for each of the different combinations of "driver/route/bus/trip". This study presents, however, some limitations since it only allows a one-dimensional observation of the phenomenon under study, saying nothing about the influence of the different factors on the L_{Aeq} values.

3.2 Principal Component Analysis (PCA)

For the application of PCA, a Matrix of original data with 191 columns and 300 lines was made. The criterion for retention of the factorial axes to be interpreted obeyed, on the one hand, the known Kaiser criterion (eigenvalue associated with each factorial axis superior to the unit) and on the other hand to the rate of inertia transported (variability explained) in each axis (Bandalos & Boehm-Kaufman, 2008). There is awareness that the percentage of explanation for each axis is low, but this fact is justified by the profusion of variables that are simultaneously under analysis. It was considered that a variable would be correlated with the factorial axis and in this way susceptible of being interpreted in that axis, whenever its coordinates had a value equal or superior to 0.5. Variables projection in the first factorial plane (F1, F2) are presented in Table 4 and Table 5.

There is a tendency for the aggregation of variables to be established as a function of route *b*, carried out by Bus 1 (Mini Bus), regardless of the driver.

It can be seen that in the group of variables projected on axis 1, the negative half-axis, the factors of association are the route *d*, the type of bus 3 (articulated) and the driver *d*. In the study carried out by Portela, (2010), there was an association with strong correlation (Pearson) among L_{Aeq} produced by articulated buses.

3.3 Kolmogorov-Smirnov (KS) adjustment test

Following the univariate statistical analysis, KS adjustment tests were performed, considering two levels of significance (alpha values $\alpha = 0.05$ and $\alpha = 0.01$) that gave rise to two values of critical deviation (critKS = 0.079 and critKS = 0.094). The use of two levels of significance results from the observed proximity, in some of the tests, between the observed maximum deviation (maxKS) and the maximum allowable deviation (critKS).

The KS test made possible to identify, for some events, that they did not reveal the origin of a population with normal distribution (H_0 rejection), when tested with an alpha value of 0.05, but that,

Table 4. Projection of the variables in the first factorial plane (F2).

Projection in plane	Variable	description
Positive half-axis	*kc23*	Driver k on route *c*, driving the Double-decker bus, in second one way
	fd21	Driver f on route *d*, driving the bus two floors in the first one way
	cd44	Driver c, route *d*, standard bus, on the second return way
Negative half-axis	*gb14*	Driver g on route *b*, driving the Mini Bus on the second return trip
	hb12	Driver h on route *b*, driving the Mini Bus, on the first return trip
	jb12	Driver j on route *b*, driving the Mini Bus on the first return trip
	ed21	Driver e on route *d*, driving the bus Double-decker in the first one way

Table 5. Projection of the variables in the first factorial plane (F1).

Projection in plane	Variable	description
Positive half-axis	*lb12*	Driver l on route *b*, driving the Mini bus, on the first return trip
	ed24	Driver e on route *d*, driving bus Double-decker on the second one way
	cd43	Driver c on route *d*, driving the standard bus, on the second one way
Negative half-axis	*da13*	Driver d on route *a*, driving the Mini Bus on the second one way
	lb13	Driver l on route *b*, driving the Mini Bus on the second one way
	lc34	Driver l on route *c*, driving the articulated bus, on the second return trip
	ad34	Driver a on route *d*, driving the articulated bus on the second return trip
	dd31	Driver d on route *d*, driving the articulated bus, on the first one way
	dd33	Driver d on route *d*, driving the articulated bus, on the second lone way
	ed31	Driver and on route *d*, driving the articulated bus on the first one way
	dd44	Driver d on route *d*, driving the standard bus, on the second return trip

when the level of significance of 0.01 was used, they were already identified with normal distribution (non-rejection of H_0).

3.4 *ANOVA tests*

As a result of previous statistical treatments, a reasonable understanding of the variables under study behaviour was achieved at this research stage. As conclusion of the previous statistical analyses and once the assumptions for the application of the ANOVA are guaranteed, on different matrices of input, homogeneity tests of the samples were done. The initial application of those tests had the objective of verifying the influence of the different buses on the generated sound levels (H_0—The continuous sound levels produced by the different buses are homogeneous). So, the data were grouped by bus type and by driver, to which a mean value of monitored sound level is associated, taking into account the first 15 minutes of each trip.

Table 6. Results of Fisher-Snedecor test to ANOVA.

Variation source	SQ	gl	MQ	F	P value	F critical
Between groups	52.69	3	17.56	5.540	0.0026	2.816
Within groups	139.50	44	3.17			
Total	192.20	47				

Significance level = 0.05; *SQ* – Sum of squares; *gl* – degrees of freedom; *MQ* – mean square; *F* – Observed value of the quotient between the sources of variation.

Considering the results obtained, and comparing the value of *F* (5.540) with the *critical F* value (2,816), it is verified that *F* is higher, so the hypothesis (H_0) is rejected, that is, buses influence the generated sound level. In the study by Nassiri (2014), using the ANOVA test, was found that the bus type had a significant effect on the noise level inside the bus. Another study, carried out by Portela (Portela & Zannin, 2010), also shows that the location of the motor influences the equivalent noise level. However in other study Nadri (2012) found no significant differences in the variables that influence noise levels monitored in four types of buses.

Another application of ANOVA was to verify the influence of drivers on sound levels generated when driving different buses (H_0—The continuous sound levels produced by different drivers are homogeneous). Were obtained values of F (1,998), lower than the critical value F (2.060), so were not rejected the hypothesis (H_0), that is, drivers do not influence the generated sound level.

The application of the ANOVA Test was possible because it was verified that the data follow a normal distribution. In the studies identified in which the authors use the ANOVA Test (Lopes, et al., 2012), (Nassiri et al., 2014), (Bruno and Zannin 2010) and (Nadri, et al., 2012), nothing is mentioned in relation to the distribution of the data.

4 CONCLUSIONS

With the application of the KS test it was found that, for a significance level of 0.05, in a total of 191 events, 88% (167) did not reject the null hypothesis, that is, data can be considered as coming from a population with a normal distribution. In doing the same test now for a significance level of 0.01, the near unanimity of non-rejections of the null hypothesis (96%). This finding opens the way to future studies in which the probability path, through the probability density function and/or the distribution function, can be used to predict the behaviour of the L_{Aeq} variable.

From the application of PCA were found a strong positive association between the variables of 13, lb13, lc34, ad34, dd31, dd33, ed31 and dd44, to which values are associated lower levels of noise. It was verified that in this group of variables the route *d*, the type of bus *3* (articulated), also associated to the driver *d*. In this group of variables, some of them (lb12, ed24 and cd43), projected in Axis 1, positive half-axis, are negatively associated. The analysis of the group of variables gb14, hb12, jb12 and ed21, projected in the Axis 2, negative semi-axis, allows to infer a tendency for the variables corresponding to the route *b*, carried out in the bus *1* (Mini Bus) present a behavioural similarity that associates, all this regardless of the driver.

Were concluded that the monitored sound levels have some dependence on the type of bus and route, rather than the human factor.

With the application of the ANOVA Test to the obtained results, it can be concluded that the Driver does not influence noise levels, but the level of noise depends on the type of bus.

ACKNOWLEDGMENTS

The authors thank to Masters in Occupational Safety and Hygiene Engineering (MESHO), Faculty of Engineering, University of Porto (FEUP), their full support and international dissemination of this work.

REFERENCES

Bandalos, D., & Boehm-Kaufman, M. (2008). Four common misconceptions in exploratory factor analysis. (T. &. Francis, Ed.) *Statistical and Methodological Myths and Urban Legends: Doctrine, Verity and Fable in the Organizational and Social Sciences.*, pp. 61–87. Obtido de ISBN 978-0-8058-6237-9.

Benzécri, J. P., & et al. (1973). *L'Analyse des Données.* Paris: Dunod.

Berglund, B., Lindvall, T., & Schwela, D. H. (1999). Guidelines for Community Noise. *World Health Organization (WHO)*, pp. 159.

Bruno, P., & Zannin, P. H. (2010). Analysis of factors that influence noise levels inside urban buses. *Journal of Scientific & Industrial Reserch, 66*(9), 684–687.

Deshaies, P., Martin, R., Belzile, D., Fortier, P., Laroche, C., Leroux, T., ... Picard, M. (2015). Noise as an explanatory factor in work-related fatality reports. *Noise Health*, pp. 294–299.

EU-OSHA. (2011). *https://osha.europa.eu/.* doi:10.2802/5632

Fiúza, A. (2003). *Aquisição e Análise de Dados.* Faculdade de Engenharia da Univesidade do Porto.

Freitas, L. C., Parreira, A., Baptista, C., Frade, F., Marçal, J. E., Ferreira, P., ... Marques, V. (2013). *Manual de Segurança e Saúde do Trabalho.* Lisboa: Edições Lusófanas.

Gelman, A. (2005). Analysis of Variance—Why it is More Important Than Ever. (I. o. Statistics, Ed.) *The Annals of Statistics, Vol. 33*, pp. 1–53. doi:DOI 10.1214/009053604000001048

ISO. (1990). ISO 1999. *Acoustic—Determination of occupational noise exposure and noise induced hearing impairment.* Geneve: International Organization for Standardization.

ISO. (2000). ISO 7029. *Acoutics. Statistical distribution of hearing thresholds as a function of age.* Geneve: International Organization for Standardization.

Lopes, A. C., Otowinz, V. G., Lopes, P. M., Lauris, J. R., & Santos, C. C. (2012). Prevalence of noise-induced loss in drivers. *International Archives of Otorhinolaryngology, 16*(4), pp. 509–514. doi:10.7162/S1809-97772012000400013

Lutman, M. E., & al, e. (2008). *Epidemiological evidence for the effectiveness of the noise at work regulations.* Health and Safety Executive..

Mohammadi, G. (2015). Noise exposure inside of the Kerman urbam buse: measurements, drivers and passengers attitudes. *Iranian Journal of Health, Safety & Environment, 2*(1), 224–228.

Nadri, F., Monazzam, M. R., Khanjani, N., Ghotbi, M. R., Rajabizade, A., & Nadri, H. (2012). An Investigation on Occupational Noise Exposure in Kerman Metropolitan Bus Brivers. *International Journal of Occupational Hygiene, 4*(1), 1–5.

Nassiri, P., Ebrahimi, H., Monazzam, M. R., Rahimi, A., & Shalkouhi, P. J. (2014). Passenger Noise and Whole-Body Vibration Exposure—A Comparative Field Study of Comercial Buses. *Journal of Low Frequency Noise, Vibration and Active Control, 33*(2), 207–220.

OSHA. (2009). *Novos Riscos Emergentes para a Segurança e Saúde no trabalho.* Luxemburgo: Serviço das Publicações Oficiais das Comunidades Europeias.

Portela, B. S., Queiroga, M. R., Constantini, A., & Zannin, P. H. (2013). Annoyance evaluation and the effect of noise on the health of bus drivers. *Noise & Health, 15*(66), pp. 301–306.

Silva, F., & Correia, F. (2010). Bus passager's Noise Exposure Assessment in Itajuba, Brazil. Em S. I.-I. SPA (Ed.), *39 th International Congress on Noise Control Engineering*, (p. 10). Lisboa.

Stephens, M. A. (Sep. de 1974). EDF Statistics for Goodness of Fit and Some Comparisons. (L. Taylor & Francis, Ed.) *Journal of the American Statistical Association, Vol. 69*, pp. 730–737. doi:DOI: 10.2307/2286009

Tak, S., Davis, R. R., & Calvert, G. M. (2009). Exposure to hazardous workplace noise and use of hearing protection devices among US workers-NHANES, 1999–2004. *Am J Ind Med*, pp. 358–371.

Trombetta, Z. P. (2006). Occupationsl noise in urban buses. *International Journal of Industrial Ergonomics*, 901–905. doi:10.1016/J.ergon.2006.06.014

Universitas, C. (2010). Saúde, Segurança e Higiene no Trabalho. *Instituto da Mobilidade e dos Transportes Terrestres, IP.*

World Health Organization. (2015). *Occupational and community noise. Fact sheet No. 258. Geneva.* Obtido em 9 de 8 de 2016, de http://www.who.int.

Zannin, P., Diniz, F., Giovanini, C., & Ferreira, J. (2003). Interior noise profiles of bus in Curitiba. *Transportation Research Part D*, 243–247.

Occupational Safety and Hygiene V – Arezes et al. (Eds)
© 2017 Taylor & Francis Group, London, ISBN 978-1-138-05761-6

Mental health and well-being among psychologists: Protective role of social support at work

C. Barros, C. Fonte, S. Alves & I. Gomes
University Fernando Pessoa, Porto, Portugal

ABSTRACT: Work is a significant part of an individual's life that affects mental health and well-being. The practice of psychology can be demanding, challenging and emotionally exhausting which makes it susceptible to occupational health risks. The purpose of the present study was to investigate workplace social support predictors of well-being among psychologists, using the Mental Health Continuum—Short Form (MHC-SF) and the Health and Work Survey (INSAT). Consistent with the literature, the present study showed that social support in a work environment can have a protective role in psychologists' mental health and well-being. These results highlight the importance of carrying out management practices that promote supportive relationships in the workplace.

1 INTRODUCTION

1.1 Mental health, well-being and work activity

The World Health Organization (WHO) defines mental health as a positive state that is defined as *a state of well-being in which the individual realizes his or her own abilities, can cope with the normal stresses of life, can work productively and fruitfully, and is able to make a contribution to his or her community* (WHO, 2004, p 12). This definition builds on two longstanding traditions in studies on a life well lived (Deci & Ryan, 2008; Ryff, 1989): the hedonic tradition and the eudaimonic (Keyes, 1998). The hedonic tradition focuses on subjective well-being – measurement of satisfaction with life and positive emotions (Diener et al., 1999). The tradition of eudaimonia focuses on psychological well-being (Ryff, 1989), and social well-being (Keyes, 1998) that considers optimal psychological functioning in life and reflects the extent to which individuals view themselves as functioning well in life (Keyes, 2002). Ryff (1989) developed a model of psychological well-being that comprises six dimensions (self-acceptance, personal growth, purpose in life, positive relations with others, autonomy, and environmental mastery) regarding the challenges that individuals encounter as they strive to realize their potential. Keyes (1998, 2002) proposed a model of social well-being with five dimensions (social integration, social contribution, social coherence, social actualization, and social acceptance) and focuses on the individuals' evaluations of their public and social lives. Taking both the hedonic and the eudaimonic approaches into account, mental health can be defined as the presence of subjective, psychological, and social well-being (Keyes, 2002), in accordance with the definition of the WHO (2004).

Work is a significant part of an individual's life that affects mental health and well-being. The achievement of well-being is one key factor of a positive work environment. Although psychologists' well-being has often been conceptualized in ways that highlight each individual's responsibility to maintain their well-being, the work context may well affect the experience of compassion, fatigue, and burnout (Thompson et al. 2014), because the practice of psychology can be challenging and emotionally exhausting (Barnett et al. 2007, Lee et al. 2011).

1.2 Occupational health among psychologists

Psychologists face a number of challenges and stressors that place them in risk. Work activity is characterized by intense emotional demands that require a close relationship with people, which makes it susceptible to occupational stressors (Moreno-Jiménez et al. 2006, Spiendler et al. 2015).

Despite individual self-care, certain behaviors can reduce the risk of occupational health problems (Benedetto & Michael 2014) making understanding how working conditions may contribute to health and well-being. Greater amount of time spent with clients, greater length of time spent working (long work hours), lack of workplace support, less control over work activities, and over-involvement with clients can increase the risk of health problems (Mckim & Smith-Adcock 2014).

However, work atmosphere (Thompson et al. 2014), coworker support (Ducharme et al. 2008), and social support from supervisors and colleagues might interfere in the protection of their own well-being (Thompson et al. 2014). In fact, workplace relational hierarchy plays an important role in mental health: quality relationships can help manage stress and negative emotions (Saint-Hilaire 2008).

It is accepted that work is healthy for individuals and contributes to overall well-being if the working conditions are ensured (Arandjelovic 2011, European Agency for Safety and Health at Work 2013, Ilies et al. 2015) in the sense that all work should provide workers the possibility of having an active role in their conduction, reflecting the ability of acting by themselves and on their work (Clot, 2008).

This approach takes a comprehensive, holistic and integrative perspective on occupational health focusing on well-being at work (Barros et al. 2015, Barros-Duarte & Lacomblez 2006, Silva et al. 2016). A number of studies suggest that social support at work is mainly beneficial for people under stress (Stansfeld & Candy 2006, Visweswaran et al. 1999). This protective effect of social support must receive more attention in studies on occupational health as a well-being predicting factor. Occupational health is a multidimensional phenomenon comprising the interaction of physical, psychological, and social dimensions. It can affect workers at cognitive and behavior levels as well as in their physical and psychological health (Ilies et al. 2015, Ilmarien 2009, Molinié & Leroyer 2011).

According to this perspective, this study aims to explore the mental health and well-being among psychologists, and to analyze the working conditions—work relational constraints – which seem to have an important role in well-being.

2 METHOD

2.1 Participants

One hundred ninety-seven psychologists ranging in age from 22 to 58 ($M = 36.41$; $SD = 7.39$) participated in this study. Seventy-two percent of the participants were female ($n = 141$), and 28% were male ($n = 56$). The overwhelming majority of respondents came from the Northern region of Portugal (96%) and represented various areas of specialization that included school psychology, clinical psychology, forensic psychology, work psychology, and counseling psychology. Participants also represented a considerable range in their years of experience, from those who only have one year of practice to others who had been working as a psychologist for 32 years ($M = 7.36$; $SD = 6.53$). Seventy-seven percent of the participants were employed either under permanent or fixed-term contracts, and 20% were working as independent contractors (3% failed to indicate their employment status). Over half of the practitioners were working on a full-time basis (72%), in jobs without schedule flexibility (70%).

2.2 Measures

Mental Health Continuum—Short Form (MHC-SF). Developed by Keyes (2005), this self-reported scale consists of 14 items rated on a 6-point Likert scale, ranging from 1 (never) to 6 (every day). This instrument includes three sub-scales that measure emotional, psychological and social well-being (3, 6 and 5 items, respectively). The total score (ranging from 14 to 84) has been used as a measure of positive mental health, where higher scores indicate *flourishing* mental health. The MHC-SF has been translated into Portuguese, and it has been shown to have good internal reliability (Cronbach's α coefficients for the total scale as well as sub-scales were all above 0.80; Fonte et al. in press, Monte et al. 2015).

Health and Work Survey (INSAT—Inquérito Saúde e Trabalho). INSAT is a self-reported questionnaire comprising 154 items organized in seven sections, that measure working conditions, health and well-being, and the relationship between them (Barros-Duarte & Cunha 2010). Regarding the purpose of the present study, only a subset of the second section (related to work characteristics and conditions) was used. This subset measures work relational constraints from 23 items, organized in two domains: work relationships (17 items) and dealing with the public (6 items). For each item, participants are asked to identify if a specific situation is present or absent (using a dichotomous scale 'yes' or 'no') and to indicate the degree of discomfort if present (rating on a 5-point Likert scale, ranging from 1 'absence of discomfort' to 6 'maximum discomfort'). In terms of psychometric properties, INSAT has been found to have good internal consistency, in a Rasch PCM analysis, with a reliability coefficient > 0.8 (Barros et al. in press).

2.3 Procedure

The study protocol was approved by the Ethics Committee of Fernando Pessoa University prior to the conduct of the investigation. Psychologists were recruited through the snowball method and data collection was conducted for those who gave informed consent to participate in this research, according to their availability. Participation was voluntary, and confidentiality and anonymity was guaranteed. The instruments were handed out together with a response envelope in which to return the questionnaire to the researchers.

2.4 Statistical analysis

Descriptive statistics were used to determine the relative frequency of the nominal variables and the central tendency parameters for scale variables (mean, standard deviation and median). Point-Biserial correlations were calculated to determine the association between well-being measurements (scale variables) and work relational constraints (nominals variables that are binary variables with only two possible values: 0-no; 1-yes). For those variables that reached significant associations, a simple linear regression to explain the relationship between the dependent variable (well-being) and the independent variable (work relational constraints) was conducted.

The data were analyzed by means of the Statistical Program SPSS (Statistical Package for the Social Sciences) for Windows, version 22.0. The significance level adopted was $p \leq 0.05$.

3 RESULTS AND DISCUSSION

3.1 Descriptive statistics

In relation to well-being, we found higher well-being scores in terms of total score of well-being (MHC-SF), emotional well-being, psychological well-being, and social well-being (cf. Table 1), which suggest that these psychologists have good mental health according to World Health Organization definition (WHO, 2004).

These results shows that psychologists can positively cope with the normal stresses of professional life, work productively and fruitfully, and are able to make a contribution to society. It can also be highlighted that psychological well-being has high scores in the sample which means that psychologists view themselves as functioning well in life (Deci & Ryan, 2008).

Regarding the work relational constraints, Table 2 shows the frequency distribution of the answers to the specific items that include the alternatives "yes" and "no". We can mainly identify that the majority of the participants reported having contact with the public in their professional activity, in agreement with the characterization of the work in psychologists that require a close relationship with people, which makes it susceptible

Table 1. Characterization of well-being in the sample.

MHC-SF	n	M	SD	Range
Total well-being score	196	64.15	10.24	14–84
Emotional well-being	197	14.24	2.67	3–18
Psychological well-being	197	29.20	4.84	6–36
Social well-being	196	20.74	4.47	5–30

Table 2. Characterization of work relational constraints in the sample.

	Yes		No	
	n	%	n	%
Work relationships				
Frequently needing help from colleagues but not always getting it	94	47.7	103	52.3
Frequently not having help from colleagues when I need it	30	15.2	167	84.2
It is rare to exchange experiences with other colleagues to better performed the work	17	8.6	180	91.4
Not having my opinion taken into consideration to the running of the department	8	4.1	189	95.9
Not being recognized by the managers	31	15.7	166	84.3
Not having recognition by colleagues	7	3.6	190	96.4
Impossible to express myself	7	3.6	190	96.4
Verbal aggression	37	18.8	160	81.2
Physical aggression	24	12.2	173	87.8
Sexual harassment	19	9.6	178	90.4
Moral harassment	21	10.7	176	89.3
Threat of job loss	18	9.1	179	90.9
Sexual discrimination	14	7.1	183	92.9
Age discrimination	14	7.1	183	92.9
Race or nationality discrimination	13	6.6	184	93.4
Physical and mental disabilities discriminations	10	5.1	187	94.9
Sexual orientation discrimination	9	4.6	188	95.4
Dealign with the public				
Direct contact with the public	180	91.4	17	8.6
Endure the demands of the public	157	79.7	40	20.3
Deal with situations of tension in the relationship with the public	150	76.1	47	23.9
Be exposed to verbal aggression from the public	105	53.3	92	46.7
Be exposed to physical aggression from the public	73	37.1	124	62.9
Being exposed to the suffering of the others	155	78.7	42	22.1

to occupational stressors (Moreno-Jiménez et al. 2006, Spiendler et al. 2015).

3.2 Relationships between work relational constraints and well-being

The correlation analysis revealed a significant inverse relationship between psychologist perception of "not being recognized by the management" and the total well-being score ($r_{pb} = -.197; p \leq .001$), emotional well-being ($r_{pb} = -.197; p \leq .001$), and psychological well-being ($r_{pb} = -.188$; $p \leq .001$). That is, psychologists who perceived a lack of recognition by their managers reported less total well-being, less emotional well-being, and less psychological well-being. Another inverse relationship was demonstrated between the psychologist's perception of "frequently needing help from colleagues but not always getting it" and emotional well-being ($r_{pb} = -.153$; $p \leq .05$). The psychologists who report "frequently needing help from colleagues but not always getting it" have less emotional well-being. No other significant correlation was observed.

These results suggest that work can be healthy for individuals and contributes to overall well-being if the social support by colleagues and hierarchy is ensure (Arandjelovic 2011, European Agency for Safety and Health at Work 2013, Ilies et al. 2015).

3.3 Predicting well-being

Research questions about the influence on well-being of these work relational constraints were estimated using linear regression analyses. The results revealed that "not being recognized by the management" and "frequently needing help from colleagues but not always getting it" were significant predictors of the total well-being score [$F(2, 195) = 6.659$; $p \leq .05$], and of social well-being [$F(2, 195) = 5.224$; $p \leq .05$]; this model explains 6.5% and 5.1% of the variance, respectively. When emotional and psychological well-being was regressed based on the variables "not being recognized by the management" and "frequently needing help from colleagues but not always getting it", the latter was not found to be a significant predictor of either emotional or psychological well-being. The exposure to "not being recognized by the management" explains 4.7% of the variance in emotional well-being [$F(2, 196) = 4.831$; $p \leq .05$], and 5.3% of the variance in psychological well-being [$F(2, 196) = 5.396$; $p \leq .05$].

These results are in accordance with other studies of the role of social support in the process of work impact in individual's mental health and well-being (Viswesvaran et al. 1999). In fact, social support, social relations (Gollac & Bodier 2011), coworker support (Ducharme et al. 2008), and social support from supervisors and colleagues can be pointed as protective factors on mental health and well-being (Thompson et al. 2014), and can help manage stress and negative emotions (Nielsen et al 2010, Saint-Hilaire 2008).

4 CONCLUSIONS

In conclusion, there is convincing evidence of mental health benefits of social support in general.

Specifically, the results found in this study show that work relational conditions may operate as protective factors in mental health and well-being of psychologists. This indicates the importance to establish a professional environment of sharing, peer support and supervisors recognition.

Creating a professional climate supportive of cooperative relationships can improve the work activity of psychologists even when the practice is demanding, challenging, and emotionally exhausting.

Although the body of evidence on social support analyzed in this article is fairly compelling, a study with a larger sample of psychologists can provide more evidence about its impact on mental health and well-being, and thus contribute to better practices for promotion and protecting good mental health among these professionals.

New research in this area will have important implications for the understanding of mental health and well-being determining factors. Such knowledge will serve to strengthen the supportive aspects of psychological practice, which will contribute to the individual, families, and larger society.

REFERENCES

Arandjelovic M. 2011. A need of holistic approach to the occupational health developing (in Serbia). *International Journal of Occupational Medicine and Environmental Health* 24(3): 229–40.

Barnett, J.E., Baker, E.K., Elman, N.S. & Schoener, G. R. 2007. In pursuit of wellness: The self-care imperative. *Professional Psychology: Research and Practice* 38(6): 603–612.

Barros, C., Carnide, F., Cunha, L., Santos, M. & Silva, C. 2015. Will I be able to do my work at 60? An analysis of working conditions that hinder active ageing. *WORK: A Journal of Prevention, Assessment & Rehabilitation* 51(3): 579–590.

Barros, C., Cunha, L., Baylina, P. & Oliveira, A. in press. Development and validation of health and work survey based on Rasch model among Portuguese workers. *International Journal of Occupational Medicine and Environmental Health.*

Barros-Duarte, C. & Cunha, L. 2010. INSAT2010 - Inquérito Saúde e Trabalho: Outras questões, novas relações. *Laboreal* 6(2): 19–26.

Barros-Duarte, C. & Lacomblez, M. 2006. Santé au travail et discrétion des rapports sociaux. *Perspectives interdisciplinaires sur le travail et la santé* 8(2): 1–17.

Benedetto, M. & Michael, S. 2014. Burnout in Australian psychologists: Correlations with work-setting, mindfulness and self care behaviours. *Psychology, Health & Medicine* 19(6): 705–715.

Clot, Y. 2008. *Travail et pouvoir d'agir*. Paris: PUF.

Deci, E.L. & Ryan, R.M. 2008. Hedonia, eudaimonia, and well-being: An introduction. *Journal of Happiness Studies* 9(1): 1–11.

Diener, E., Suh, E.M., Lucas, R. & Smith, H.L. (1999). Subjective well-being: Three decades of progress. *Psychological Bulletin* 125(2): 276–302.

Ducharme, L. J., Knudsen, H. K. & Roman, P. M. 2008. Emotional exhaustion and turnover intention in human service occupations: The protective role of coworker support. *Sociological Spectrum* 28: 81–104.

European Agency for Safety and Health at Work 2013. *Well-being at work: Creating a positive work environment*. Luxembourg: Publications Office of the European Union.

Fonte, C., Ferreira, C. & Alves, S. in press. Estudo da saúde mental em jovens adultos. Relações entre psicopatologia e bem-estar. *Psique*.

Gollac, M. & Bodier M. 2011. *Mesurer les facteurs psychosociaux de risque au travail pour les maîtriser*. Collège d'Expertise sur le Suivi des Risques Psychosociaux au Travail, http://www.college-risquespsychosociaux-travail.fr/rapport-final, fr, 8,59.cfm.pdf.

Ilies R., Aw, S. & Pluut, H. 2015. Intraindividual models of employee well-being: What have we learned and where do we go from here? *European Journal of Work and Organizational Psychology* 24(6): 827–38.

Ilmarien J. 2009. Work ability—A comprehensive concept for occupational health research and prevention. *Scand J Work Environ Health* 35(1):1–5. 10.5271/sjweh.1304.

Keyes, C. L. M. 2005. Mental illness and/or mental health? Investigating axioms of the complete state model of health. *Journal of Consulting and Clinical Psychology* 73(3): 539–548.

Keyes, C.L.M. 1998. Social well-being. *Social Psychology Quarterly* 61(2): 121–140.

Keyes, C.L.M. 2002. The mental health continuum: From languishing to flourishing in life. *Journal of Health and Social Behavior* 43(2): 207–222.

Lee, J., Lim, N., Yang, E. & Lee, S. M. 2011. Antecedents and consequences of three dimensions of burnout in psychotherapists: A meta-analysis. *Professional Psychology: Research and Practice* 42(3): 252–258.

Mckim, L. & Smith-Adcock, S. 2014. Trauma counsellors' quality of life. *Int J Adv Counselling* 36: 58–69.

Molinié, A-F. & Leroyer, A. 2011. Suivre les évolutions du travail et de la santé: EVREST, un dispositif commun pour des usages diversifiés. *PISTES* 13(2).

Monte, K., Fonte, C. & Alves, S. 2015. Saúde mental numa população não clínica de jovens adultos: Da psicopatologia ao bem-estar. *Revista Portuguesa de Enfermagem de Saúde Mental* 2: 83–87.

Moreno-Jiménez, B., Meda-Lara, R.M., Morante-Benadero, M.E., Rodríguez-Munõz, A. & Palomera-Chávez, A. 2006. Validez factorial del inventario de burnout en una muestra de psicólogos Mexicanos. *Revista Latinoamericana de Psicologia* 38(3): 445–456.

Nielsen, K., Randall, R., Holten, A-L. & Rial-González, E. 2010. Conducting organizational-level occupational health interventions: What works? *Work & Stress* 24(3): 234–59.

Ryff, C.D. 1989. Happiness is everything, or is it? Explorations on the meaning of psychological well-being. *Journal of Personality and Social Psychology* 57(6): 1069–1081.

Saint-Hilaire F. 2008. *Mental health at work: The relational side*. Québec: Université Laval.

Silva, C., Barros, C., Cunha, L., Carnide, F. & Santos, M. 2016. Prevalence of back pain problems in relation to occupational group. *International Journal of Industrial Ergonomics* 52: 52–58.

Spiendler, S., Carlotto, M., Ogliari, D. & Giordani, K. 2015. Estressores ocupacionais em psicólogos clínicos brasileiros. *Psicogente* 18(33): 104–116.

Stansfeld, S.A. & Candy, B. 2006. Psychosocial work environment and mental health—A meta-analytic review. *Scand J Work Environ Health* 32: 443–62.

Thompson, I., Amatea, E. & Thompson, E. 2014. Personal and Contextual predictors of mental health counselors' compassion fatigue and burnout. *Journal of Mental Health Counseling* 36 (1): 58–77.

Viswesvaran, C., Juan, I., Sanchez, J. & Fisher, J. 1999. The role of social support in the process of work stress: A meta-analysis. *Journal of Vocational Behavior* 54: 314–334.

World Health Organization. 2004. *Promoting mental health: Concepts, emerging evidence, practice* (Summary Report). Geneva: WHO.

Occupational Safety and Hygiene V – Arezes et al. (Eds)
© *2017 Taylor & Francis Group, London, ISBN 978-1-138-05761-6*

The application of TRIZ and SCAMPER as innovative solutions methods to ergonomic problem solving

A. Kalemba, A. Dewicka & A. Zywert
Poznan University of Technology, Poznan, Poland

ABSTRACT: The increasing complexity of design problems causes the demand for complex tools to solve them. However, in many cases the evolutionary methods of solving problems have ceased to have an evolutionary character, and thus are insufficient. Hence the interest in heuristic methods for solving problems that are related to ergonomic design. Two tools, TRIZ and SCAMPER, used for achieving innovative solutions to solve problems of an ergonomic nature were analyzed. The results of the analysis were compared with the results of the group that did not use any tools supporting innovative design. Both of the used methods, TRIZ and SCAMPER, are applicable to ergonomic problem solving. It was found that even partially carried out methods aiding innovative design made it possible to obtain solutions with a much greater potential to achieve a high ergonomic quality compared with the traditional approach to ergonomic design.

1 INTRODUCTION

Design is an ergonomic problem and regardless of the undertaken context of the design work, one will certainly need to consider the future needs of the user, as well as those of the non-default user, such as a serviceman. Let proof of this be the fact that ergonomic design includes any of the areas of human functioning, from complicated mega-systems, consisting of a network of interconnected human-technical systems (Butlewski & Tytyk 2012), to basic tools. Thus it can be said that if we are discussing the design process, it cannot be unergonomic (Butlewski 2013, Butlewski 2014). The dehumanized approach, which assumes that technology is the most important part, and other aspects, including those concerning human functioning with the proposed technical product, will be resolved "later," does not work. Moreover, taking into account the needs of the user, and therefore ergonomic design, requires a number of organizational changes within the company performing them. One of the most important and at the same time hardest, changes is an alteration of organizational culture and acceptance of the participation of employees in designing and implementing new solutions (Mrugalska & Kawecka-Endler 2012, Gabryelewicz et al. 2015). It should be noted that ergonomic design results from pragmatic reasons; it allows for attaining better outcomes in terms of quality at the predicted cost of production (Drożyner et al. 2011). Studies show that the human factor criteria are effectively incorporated into the design process in the early stages. This means that the most efficient ergonomic design exists when heuristic methods have the most application. However, very often thoughtless replication of solutions occurs with just a few changes of a corrective character in terms of workstations, including those for people with disabilities, making it impossible to achieve a higher level of ergonomic quality (Butlewski et al. 2014). There is ample evidence that the focus on human needs not only brings an advantage of better health conditions for the user, but also a greater effectiveness of interaction of ergonomic solutions with the human that uses them. Ergonomic design goes hand in hand with obtaining a better quality of the working environment and safety at work, and these are in general dependent on the quality management system, for example, through the use of TQM (Górny 2011). We are dealing with a system of connected components where the ergonomic quality of the working conditions or used products can affect the overall quality of both the manufacturing and service systems.

The use of ergonomic solutions and those complying with safety rules is crucial to success of a commercial entity (Górny 2012). However, attaining a greater ergonomic quality as compared to existing solutions requires increasing effort and the application of new methods. Hence the interest in heuristic methods, which have the potential to overcome many of the problems faced by experts in ergonomics. Many of these problems (lack of ergonomic quality) still exist, despite the existence of modern design programs, which are very helpful in system modeling but still do not solve all design aspects (Zabłocki et al. 2006).

Once the problem is not "whether to use" ergonomic design—the question is now how to do it and how to do it effectively. Most methods used require a change of an evolutionary nature, which does not provide an adequate "profit," while at the same time exposing one to having to achieve solutions by trial and error. The design structure can be described by the following steps (De Medeiros & Batiz, 2012):

- Initial planning: analyzing the nature of the project, formulation of questions for the product design planning;
- Data capture: gathering necessary information and checking for possible alternatives;
- Selection of alternatives: stage of selecting one of the developed alternatives;
- Implementation: developing the selected alternative;
- Evaluation: evaluating the impact of the project and setting the necessary corrections.

However, within the Conceptual Phase, which is before Selection of alternatives, measures are foreseen which employ tools such as (De Medeiros & Batiz, 2012, p. 1022): anthropometric tables; anthropometric dummies, 2D/3D software. This means that the solutions achieved will be based on previously devised solutions, which will be adapted only ergonomically. The difference here is considerable because ergonomic design done sufficiently early will allow for achieving solutions that are not available in the later decision-making stages. Hence the interest in heuristic methods, which allow for achieving nonstandard solutions, significantly increasing the level of the ergonomic quality of solutions, and are an important source of innovation in the early stages of design (Yilmaz & Seifert 2011).

2 METHOD

TRIZ and SCAMPER are examples of heuristic methods, which are used to overcome the limitations of thought and to open the designer to new areas that were previously overlooked (Butlewski 2012, Królak & Butlewski 2016). A feature of heuristic methods and techniques is introducing conditions that positively stimulate our inventiveness. These two heuristic tools have been analyzed during the solving of problems of an ergonomic character.

2.1 TRIZ

TRIZ, the algorithm of inventive problem solving, is a technique based on years of research by Henry Altszuller in the area of solving technical problems. This method is designed to direct the search for solutions, in order to minimize unnecessary effort during the search by setting the vector for the search. The TRIZ method consists of three main steps. Each of the stages is assigned sub-stages, which guide the user to the appropriate thinking track.

A detailed scheme of the proceedings in the method is as follows (Altszuller 1972, p. 101): 1. Analytical stage; 1.1. Task presentation; 1.2. Imagining the ideal solution; 1.3. Determining what is the obstacle to achieving the ideal solution (i.e., identifying the contradiction); 1.4. Determination of the cause of contradiction; 1.5. Determination of the terms of removal of the obstacles (determination of terms of contradiction elimination); 2. Operational stage; 2.1. Explore the possibility of any changes in the object itself (or machine or technological process—for example changing the dimensions, shape, material); 2.2. Explore the possibility of division of the object into independent groups; Establishment of a "weak" group; Establishment of a "necessary and sufficient" group; Division of the object into equivalent groups; Division of the object into different functional groups; 2.3. Explore the possibility of any changes in the external environment of the object; for exchanging the parameters of the environment; 2.4. Explore the possibility of any changes to the neighboring cooperating objects; Determine the relationship between the previously independent objects involved in the performance of the same work; The elimination of one object at the expense of the transfer of its functions to another; Increase the number of objects, simultaneously operating in a limited space, at the expense of the use of the free reverse side of the surface; 2.5. Search for patterns in other fields of technology (once the given contradiction has been eliminated there); 2.6. Return to the starting point (in case of uselessness of aforementioned activities) and widen assumptions, that is, transition to others, which are more general and will allow for the consideration of additional aspects; 3. Synthetic stage; 3.1. Make changes to the shape of the object (the new nature of the machine should be accompanied by a new shape); 3.2. Make changes to other objects associated with the given; 3.3. Make changes in the methodology of application of the object; 3.4. Examine the degree of applicability of the invented principle to resolve other technical issues; The presented method has an iterative character and often in order to achieve the desired effect it is necessary to analyze its stages repeatedly at different levels of generality. It should be noted that the same algorithm TRIZ is only part of a broader methodology known as TRIZ, and within it there are rules and methods which also assist in innovation in technical design—including so-called "inventive tricks".

2.2 SCAMPER

SCAMPER is an acronym derived from the first letters of the names of the steps one can take towards creating a concept of a new solution in terms of a product (substitute, combine, adopt, modify/distort, put to other purpose, eliminate, rearrange/

reverse) (Butlewski 2012, p. 97). Examples of some questions within SCAMPER in brackets example questions (Michalko 2006):

1. Substitute (What materials or resources can you substitute or swap to improve the product?; What other product or process could you use?; What rules could you substitute?).
2. Combine (What would happen if you combined this product with another, to create something new?; What if you combined purposes or objectives?).
3. Adapt (How could you adapt or readjust this product to serve another purpose or use?; What else is the product like?; Who or what could you emulate to adapt this product?; What else is like your product?).
4. Modify (How could you change the shape, look, or feel of your product?; What could you add to modify this product?).
5. Put to Another Use (Can you use this product somewhere else, perhaps in another industry?; Who else could use this product?;).
6. Eliminate (How could you streamline or simplify this product?; What features, parts, or rules could you eliminate?; How could you make it smaller, faster, lighter, or more fun?).
7. Reverse (What would happen if you reversed this process or sequenced things differently?; What if you try to do the exact opposite of what you're trying to do now?).

Questions in an extended version (above are merely examples) were asked to the study group.

2.3 Participants

The participants of the experiment were a group of 53 engineers (safety engineering) during their last semester of post-graduate studies. The research group previously passed courses of ergonomics and ergonomic design, and during undergraduate studies has gained the right to work in the service of OSH. The control group consisted of 29 engineers (safety engineering) with the same rights as the above group, but immediately after their presentation of their thesis work—a group during the first semester of their post-graduate studies. In the first group the amount of women and men was similar with 25 women out of 53 participants in the research group, whereas in the control group, 18 women out of 29 participants. In connection with the occurrence of mixed groups it was decided not to identify the differences resulting from the gender of the participants of the experiment.

2.4 Procedure

The study on the research group was preceded by a training in the operation of the two methods

(SCAMPER and TRIZ). The research group was divided into two groups and an exercise was given to make improvements in terms of ergonomics to: (could choose) computer workstation, apparatus for locomotion by people with disabilities (this was intended to be a wheelchair, but everyone could choose any transport device for people with disabilities). The persons participating in the experiment due to their education had the appropriate qualifications for the designing of workstations and equipment for people with disabilities (objects with such issues were part of the curriculum). The second attempt was based on the fact that the group with the task of implementing the SCAMPER method in the first exercise was asked to use TRIZ method and vice versa. This time, however, participants were asked to solve a problem of a general technical nature—under consideration was the improvement of a city bike's structure (in any direction desired by the user—to improve the safety, convenience, or comfort).

The author and animator of the experiment clearly indicated the present technical contradictions that had to be considered during the analysis. Participants in all groups carried out the task in pairs. Each couple received a record of the given method and had been asked to examine the different stages to reach a solution that meets the principles of ergonomics (the first attempt), or the technically optimal solution (the second attempt) through the use of innovative solutions.

The control group was informed that their task is to design an ergonomic vehicle for people with disabilities on the basis of their knowledge and experience, but also using their imagination and creativity (this remark had been repeatedly articulated by the instructor). Students in the control group had anthropometric atlases available for use (sections regarding the sitting and standing positions).

After conducting the research an analysis was made of the innovativeness of the proposed solutions. For this purpose the experience of the author was applied (based on the current knowledge of available design solutions). The proposed solutions were classified as: found in already commonly used solutions, rarely seen solutions and prototypes, and innovative custom solutions (the last was evaluated for technical feasibility of implementation and rationality). The applied classification is admittedly subjective, but according to the author was the most effective way to evaluate the proposed solutions. It is not possible to categorize the solutions because using heuristic methods yields a very broad, diverse, and variable range of solutions which makes them further unclassifiable. The classification was justified by the author's knowledge in the field of applied solutions for people with disabilities, as well as ergonomic workstations.

3 RESULTS

When solving the problem of an ergonomic character both research groups achieved better results in terms of solutions than the control group. However, there were significant differences between the groups. The group applying the SCAMPER method received dozens of custom solutions, and three that could be described as innovative, one of which met the criteria of rationality. The solution in question was a walker—walking stick, for which the load due to the pressure of a leaning person will bring about a block on the moving elements, but after the relief of the load will easily move—which is a typical technical contradiction and ergonomic problem at the same time. The TRIZ method, against the author's predictions, brought significantly worse results—found only 9 custom solutions (most of them of an irrational character). It should be noted that a repeated solution was treated as one, and the standard solutions were not counted, nor were they evaluated for ergonomic correctness. The widely interpreted questions from SCAMPER method thus yielded more and subjectively better solutions. It was also noted that during the experiment using the SCAMPER method, even misunderstood by the users questions allowed the experiment to obtain positive results and did not change the answers to the following questions, while the more formalized TRIZ method, during a misunderstanding of the questions or commands (especially in the early stages of the method), made it impossible to achieve good solutions.

On the other hand, not surprising were the results obtained by the control group, which demonstrated that in case of no application of the method to support innovative solutions, any proposed ergonomic solutions will be purely structural and based on the design of commonly known and previously used solutions.

Despite the previously mentioned command for the application of innovative solutions, none of the participants in the control group suggested deviating from standard solutions. The reason seems to be in this case the transfer of data from the anthropometric atlas, which encouraged the experiment participants to reproduce the anthropometric characteristics in the technical object, while not allowing for an analysis of more general issues of design decision-making.

Quite interesting seem to be the general conclusions on the use of SCAMPER and TRIZ methods during the solving of ergonomic problems. Targeting of ergonomic solutions itself has led to a significant focus on the "adaptation" of a technical object to man, but in many cases it does not specify how this adaptation is supposed to manifest. This is in the author's opinion a typical dysfunction of educational activities in the field of ergonomics and workstations in Polish educational institutions.

Table 1. The criteria for the application of SCAMPER and TRIZ methods for ergonomic problems.

Critrion/Method	SCAMPER	TRIZ
type of ergonomic problem	general or specific	clearly defined
number of generated solutions	high	small
required knowledge of methods	small	high
effort	small	high
consistency	not required	required

The carried out and described above experiment allows for the identification of the main differences in terms of the ergonomic application of both of these methods, which are presented in Table 1.

Table 1 contains the differences between the two methods. No attempt was made to assess the ergonomic quality of solutions obtained through both methods due to the large thematic variety of solutions obtained (participants had to some degree flexibility in the formulation of the problem).

The second attempt, during which the study participants were supposed to use the TRIZ and SCAMPER methods in a simple technical problem led to the conclusion that this time the first and the second groups obtained similar results. An average of about 8 ideas for solutions to the technical problem chosen for the group employing TRIZ, while about 12 for SCAMPER. In this case it was also decided not to compare the quality of solutions obtained, however, based on a subjective assessment of their technical state it can be said that some of them were innovative and have a chance to be applied in practical solutions. Quite surprising was the poor application of the TRIZ method to a problem of an ergonomic nature, while good results were obtained with this method in the development of a solution to a simple technical problem. While it is difficult to measure the ergonomic effects of the application of both of these methods, it is due to the much greater emphasis in the TRIZ method on the analysis of the problem that gives a better understanding of the complex relationship between the human—technical object (Chulvi et al. 2012).

4 CONCLUSION

Heuristic methods have the potential to overcome many of the problems faced by experts in ergonomics. Many of these problems (lack of ergonomic quality) still exist, despite the existence of modern design programs, which are very helpful in system modeling but still do not solve all design aspects. Human factor criteria are effectively incorporated into the design process in the early stages, i.e. until the end of the preliminary design. This means that the most efficient ergonomic design exists during when heuristic methods have the most application.

5 DISCUSSION

The carried out analysis did not allow to unambiguously determine the effectiveness of a particular method for ergonomic design (although quantitative analysis indicates the superiority of the SCAMPER method). It's true that it is known that the SCAMPER method can be used for solving ergonomic problems, where the methodological approach to the design problem does not give the expected results, and in order to find the solution one needs to mentally "test" the range of solutions (Butlewski 2013). However the carried out attempt made it possible to demonstrate the need for the application of methods to support innovation in ergonomic design, and the inadequacy of the TRIZ method in its original form to the general discussion of ergonomic problems. It should be noted that the method itself, and even the ability for creative thinking requires more motivation (Amabile 1998), hence the results obtained depend on the determination and motivation of the persons implementing them.

It seems reasonable to make an attempt to construct a dedicated method for ergonomic activities, which would overcome the technical-human problems in the same way like TRIZ and SCAMPER allow for overcoming technical problems.

REFERENCES

Altszuller H., Algorytm wynalazku, Wiedza Powszechna, Warszawa; 1972.

Amabile, T.M. How to kill creativity. Harvard business review; 1998, pp. 76–87.

Butlewski M., (2013) Heuristic Methods Aiding Ergonomic Design, Universal Access in Human-Computer Interaction. Design Methods, Tools, and Interaction Techniques for eInclusion, Lecture Notes in Computer Science Volume 8009, pp 13–20.

Butlewski M., Tytyk E., The assessment criteria of the ergonomic quality of anthropotechnical mega-systems; p. 298–306, [in]: Advances in Social and Organizational Factors, Edited by Peter Vink, CRC Press, Taylor and Francis Group, Boca Raton, London, New York; 2012, ISBN 978-1-4398-8.

Butlewski, M. (2014) Practical Approaches in the Design of Everyday Objects for the Elderly, In Engineering Solutions and Technologies in Manufacturing, pp 1061–1065. DOI: 10.4028/www.scientific.net/AMM.657.1061.

Butlewski, M., A. Misztal, E. Tytyk, and D. Walkowiak, 2014, Ergonomic service quality of the elderly on the example of the financial market: Occupational Safety and Hygiene Ii, 579–583 p.

Chulvi, V., Sonseca, Á., Mulet, E., Chakrabarti, A. Assessment of the relationships among design methods, design activities, and creativity, Journal of Mechanical Design, Transactions of the ASME, 134 (11); 2012.

Dahlke G., Horst W. M., Preventing musculoskeletal disorders by creating a database. Maximum time of power grip of the hand, in: Horst W., Dahlke G. [eds.], Musculoskeletal disorders. Causes, consequences, prevention, Poznan University of Technology, Monograph, Poznań; 2009 pp. 135–164, ISBN: 978-83-7143-859-2.

De Medeiros, I.L., Batiz, E.C. The inclusion of ergonomic tools in the informational, conceptual and preliminary phases of the product design methodology; 2012, Work, 41 (SUPPL.1), pp. 1016–1023.

Drożyner P., Mikołajczak P., Szuszkiewicz J., Jasiulewicz-Kaczmarek M., Management standardization versus quality of working life, M.M. Robertson (Ed.): Ergonomics and Health Aspects, HCII 2011, LNCS 6779, pp. 30–39 © Springer-Verlag Berlin Heidelberg; 2011.

Gabryelewicz I., Sadłowska-Wrzesińska J., Kowal E. 2015. Evaluation of safety climate level in a production facility. Procedia Manufacturing: 6th International Conference on Applied Human Factors and Ergonomics (AHFE) and the Affiliated Conferences: 6211–6218, Elsevier. B.V.

Górny, A., Ergonomics and Occupational Safety in TQM Strategy, In: Š. Hittmár (ed.). Theory of Management 4: The Selected Problems for the Development Support of Management Knowledge Base. Žilina: University Publishing House; 2011.

Górny, A., Ergonomics in the formation of work condition quality. Work: A Journal of Prevention, Assessment and Rehabilitation; 2012, 41, 1708–1711.

Królak, P., & Butlewski, M. (2016). Application of the TRIZ method in design oriented to the various needs of people with disabilities. Occupational Safety and Hygiene IV, 275.

Michalko, M. Thinker Toys. A handbook of creative? thinking techniques. Berkeley, CA: Ten Speed Press; 2006.

Mrugalska Beata, Kawecka-Endler Aleksandra, Practical application of product design method robust to disturbances, Human Factors and Ergonomics in Manufacturing and Service Industries; 2012, Vol. 22, Iss. 2, s. 121–129.

Yilmaz, S., Seifert, C.M. Creativity through design heuristics: A case study of expert product design, Design Studies, 32 (4); 2011, pp. 384–415.

Occupational Safety and Hygiene V – Arezes et al. (Eds)
© *2017 Taylor & Francis Group, London, ISBN 978-1-138-05761-6*

Diagnosis of occupational hygiene and safety of work in an industry of the footwear sector

J.C.M. Cunha, M.B.G. Santos, R.S. Carvalho, P.B. Noy & M.B. Sousa
Federal University of Campina Grande, Campina Grande, Paraíba, Brazil

ABSTRACT: The economic importance of footwear companies in Brazil and the concern about risks to the health and safety of workers in the sector motivated this study that aimed to analyze the occupational hazards to which the workers of a footwear industry in the city of Campina Grande-PB/Brazil are exposed. The methodology used included bibliographical research, application *in locu* and *in situ* verification sheets, and measurements using an hygro decibelimeter luximeter type instrument in order to quantify the thermal, acoustic and luminous aspects of the environment. After analyzing the quantitative data collected, it was possible to verify that the variables noise, ambient temperature and illuminance were above the values prescribed by the specific legislation. Therefore, aiming at a work environment conducive to the development of production activities without compromising workers' health and well-being, it is suggested that greater monitoring and control of the mentioned environmental variables be undertaken.

1   INTRODUCTION

The economic importance of footwear industries in Brazil is due to the diversity of manufacturing processes, the multiplicity of labor involved and the different sizes and locations of the companies.

Brazil is the third largest producer of footwear in the world and ranks fourth in the global consumer category (IEMI, 2015).

According to the Brazilian Service of Support to Micro and Small Companies of Paraiba (2016), the footwear industry in the Northeast Region is the fastest growing in the country in recent years and ends up being responsible for a significant part of the production national.

In Brazil, according to data from the Ministry of Social Security, according to the Statistical Yearbook of Labor Accidents (2014), the number of occupational accidents, according to the status of registration and reason, according to the National Classification of Economic Activities, considering the CNAE code 15.3 for the manufacture of trousers, there are a total of 4,496 accidents at work, being: 1,955 typical accidents; 934 traffic accidents; 245 diseases of work; and 1,362 without work accident communication recorded.

Considering the relevance of the footwear sector, especially in the Northeastern Brazilian region, and the concern with risks to the health and safety of workers, which is a hallmark of entrepreneurs and trade unions in the sector, this research aimed to analyze the occupational risks to which are exposed workers of a footwear industry of the city of Campina Grande-PB/Brazil are exposed in their jobs.

2   THEORETICAL FOUNDATION

The existence of risk of accident is intimately and strongly linked with routine and life in companies and organizations, persisting in all the processes involved. That is, from the idea of launching a new product or service to the final result of the same (WELLER, 2008).

Although the footwear sector is not the record holder of accidents in Brazil, it occupies the 32nd position in the ranking of accidents by sector of activity, according to the Statistical Yearbook of Social Security (2014) and deserves some attention.

In a study to identify, quantify and evaluate the occurrence of work accidents in small and medium-sized companies in the footwear industry, Salim (2007) pointed out that in typical accidents, the most frequently affected part of the body corresponds to the "upper limb", with 85.5% of the records, with emphasis on the "finger" and "hand" subdivisions. As for the causative agents, the most important are "equipment, accessories or tools", followed by "imbalance, effort, twisting or badly" and "chemical", which account for 79.2%, 8.0% and 3.1% of occurrences. And the most frequent functional categories in the accidents were: "multipurpose worker in shoemaking", "machine shoe seamstress".

Regarding the ergonomic risks and the emergence of occupational diseases in the footwear sector, Lourinho *et al* (2011) point out the presence of risk factors such as poor posture, performing excessive or repetitive force, static overload presence or dynamic, with inadequate as determinants of high ergonomic risk and that may contribute to the onset of work-related musculoskeletal disorders.

The Ergonomics Booklet in the Footwear Industry (2011), elaborated by the Joint Tripartite Commission of Ergonomics, composed of representatives of the employers, through the Brazilian Association of Footwear Industries; of the Federal Government, through the Ministry of Labor and Employment; and workers, believes that disturbances in thermal comfort can cause functional changes and that excessive heat leads to fatigue, drowsiness, reduces the readiness of responses and increases the probability of failure, directly affecting yield, production capacity and increase of workers' accident and illness rates.

According to the Occupational Health and Safety Manual for the Footwear Industry (2002), local climatic conditions, together with the characteristics and the manufacturing process, are responsible for impairing thermal comfort. And in order to achieve comfortable conditions in the productive environment it is necessary to consider factors such as ambient temperature, ventilation, air circulation, lighting and noise levels, as well as the construction system.

3 CHARACTERIZATION OF THE COMPANY

This study was carried out in a limited liability company of the footwear sector, located in the city of Campina Grande—PB/Brazil, founded in 1988. The industry is responsible for the shoe production line, having as activities by CNAE 15.33-5 referring to the manufacture of footwear made of synthetic material, and CNAE 15.31-9 referring to the manufacture of leather shoes, both with risk degree 3.

The industry operates with a total of 176 workers working throughout the factory, without turning at the workstations, with production activities performed in the morning and afternoon shifts.

4 MATERIALS AND METHODS

This study was classified as exploratory and descriptive by means of bibliographic research in scientific articles, books, websites, legislation, etc. Analyzes related to the qualitative and quantitative aspects of the environment were also carried out.

The qualitative data were collected through on-site and on-site application of check sheets used to verify disagreements regarding regulatory standards, in particular NR 05—Internal Commission for the Prevention of Accidents—ICPA; NR 06—

Personal Protective Equipment—PPE; NR 08—Buildings; NR 10—Security in Facilities and Services in Electricity; NR 12—Machinery and Equipment; NR 15—Unhealthy Operations and Activities; NR 17—Ergonomics; NR 23—Fire Protection; and NR 26—Safety Signs; as well as the technical norms of the Paraiba Fire Department and the technical standard ABNT NBR ISO/ICE 8995-1: Lighting of work environments—Part 1: Interior.

The quantitative analysis was performed by monitoring data related to the production environment, whether related to the acoustic, light and thermal aspects, by collecting noise data (dB), luminance (Lux), relative humidity of the air (%), ambient temperature (°C), in the workstations of the sector of cut and in the sector of homogenization of the rubber because these sectors are judged as deficient as the mentioned environmental variables.

Therefore, the quantitative data were collected in the morning from 07:30 h to 10:30 h (considering the official Brasilia time) on October 07, 2016, using a type instrument hygro decibelimeter luximeter of the brand Instrutherm model TDHL-400. Although it is a short time interval, we had to rely on the results obtained in it, because this was the schedule that the company offered us to carry out the measurements. At this hour, the noise measurements were performed, the equipment being positioned as close as possible to the worker's ear canal, performing three consecutive measurements in the fifteen minute interval, completing a cycle of four measurements per hour. In relation to the light variable, two point measurements were performed at four points uniformly distributed in the work area of the operator in each work station analyzed and the mean of these were used for analysis of the results, in which the illuminance measurement was positioned the equipment 0.70 cm above the floor early in the morning at 07:35 h, and at the end of the morning at 10:20 h. The measurements of ambient temperature and relative humidity were carried out at 07:30 h and another at 10:30 h in the workplace, in the measurement of the ambient temperature was placed the measuring device near the center of mass of the operator of the workstation of the rubber homogenization sector.

5 RESULTS AND DISCUSSIONS

5.1 Qualitative evaluation

The company has one work safety technician since it has a degree of risk 3, and this technician works in conjunction with ICPA, as required in NR 5, both regularly constituted and operating.

It was observed that in all sectors of the industry there is the use of ear protectors, in the sector of homogenization it becomes mandatory the use of gloves, and in the sector of sewing the use of masks.

The company has complete material for First Aid and offers its workers all the training required by the technical standards of the Fire Department. There is also a sound alarm system and fire brigade for personnel protection, although there is still no automatic combat installed and fire hydrants. Color signaling is used to alert areas requiring more attention by workers, as well as the protection of moving parts of machines handled during the production process.

As for the buildings, a straight foot above three meters, as specified by NR 8, was observed, thus ensuring good air circulation in the work environment and providing thermal comfort to employees.

The electrical installations of the company meet the recommendations of the NR 10, designed and constructed in order to prevent electric shocks or any similar nature of accident, since the electrical installations machines and equipments were positioned so that they do not have contact with water, shielding existing electrical installations in areas where there is a risk of explosion.

On machines and equipment it has been observed that the minimum spacing of 0.60 m to 0.80 m as specified in NR 12 is being respected. All areas are properly signaled and the main circulation routes have a width of 1.20 m and they find clear. All machines have more than one immediate stop device located in regions that the operator can operate in his work place and that can be triggered by another person in an emergency, none of which is located in a hazardous area of the machine.

In addition, no disagreements were observed with the regulatory norms cited, nor were alarming differences observed in the items observed.

5.2 *Analysis of the identification and recognition of risks*

Two workstations, cutting sector and rubber homogenization sector, were analyzed in this study because it was believed that these would be the company's critical points regarding noise, ergonomic risks and accidents, as highlighted in the methodology.

The first analysis was performed in the cutting sector, where the employee performs the cutting of the foam from the side of the sneaker.

The noise of this work station is caused by the actual operation of the machine, and will depend on the rigidity of the material and its thickness when it is cut.

The sector has a satisfactory ambient temperature, according to the perception of the workers, due to the structure of its building composed by a high ceiling and *cobogos* on the walls, thus allowing a continuous circulation and renewal of natural air.

And in the sector there is also complementation of ventilation through the existence of artificial ventilation, through the use of ventilators, arranged in pairs in each production sector, and the artificial renovation of the air through the exhaust fan.

The second work station studied belongs to the rubber homogenization sector, in which this process results in a reduction of the processing time and vulcanization with better properties through a mixture of resin of various types to rubber.

To work in this workstation, it is essential that the worker has at least five years of experience and undergoes training because of the risk of an accident. In addition to exposure to agents of ergonomic risk, noise and heat.

In this work sector, according to workers' opinion, there is considerable thermal discomfort, since the machine responsible for the homogenization of rubber is a source of heat. However, there is a natural ventilation system due to the structure of the high-rise building and *cobogo*, and it also has artificial ventilation, through the use of fans in order to reduce the temperature of the environment relative to this work station.

Based on the NR 17, it can be verified that the company under study respects the criteria to preserve the physical health of the employee, since an ergonomic study had already been carried out in the workstations, so that, in addition to maintaining physical health, employees feel even more motivated, given the concern of managers and managers with their well-being. It can be observed, for example, that the workers responsible for the development of activities in the homogenization sector that perform manual transports of non-light loads, have training for the handling of these in order to safeguard health and prevent accidents.

The panels and tables in the work area of the studied sectors have height that allow a proper posture for the development of the activity without compromising the health of the operator, as well as easy reach and visualization. However, an aggravating factor for the health of the worker in the ergonomic aspect, it is the execution of repetitive activities during the operation of the activities in the sector of cut, in which it is proposed as a mitigating measure the application of a system of rotation between the workers.

It can be verified that even in the event of an accident, in no case is there any need for intervention, since the machinery and equipment already have the safety devices, fixed parts and moving parts required by NR 12, which follow international standards. In addition, the rubber preparation and homogenization cylinder, for example, has several emergency stop devices, both those required by the standard and the additional ones proposed by the manufacturer.

5.3 *Quantitative evaluation of environmental variables*

5.3.1 *Ambient temperature*

In Graph 1, can observe the values of ambient temperature measured in two different schedules. There is a considerable increase in the temperature between the two schedules, mainly due to the heat generated by the operation of the machine that is located in the workstation under study.

According to observations made at the study site, it can be verified that this variable does not present thermal discomfort, according to the evaluation of the workers, since the right foot of the factory meets the conditions of comfort, safety and health, guaranteeing adequate ventilation, air renewal and avoiding excessive sunshine, as specified in NR 08.

As only the ambient temperature and the relative humidity of the air were measured in this study, it is only possible to make an approximation to what is disposed in NR 17 on effective temperature (which involves the interaction of the ambient temperature with the relative humidity and the velocity of the air) and the measured ambient temperature. Therefore, it can be verified that although there is no discomfort on the part of the worker, they should be monitored in order to investigate if the elevated temperature to which they are exposed during their work period does not affect their health and welfare.

Graph 2 shows the percentage of relative air humidity measured at two different times and shows that in both cases the minimum value of NR 17, of 40%, was obeyed.

5.3.2 *Noise*

In Graphs 3 and 4 we can observe the mean levels of noise measured at these times, taking into account the tolerance limit of 85 dB established by NR 15 for 8 hours of work.

Considering these noise level values, it is possible to calculate the noise dose (D), the Equivalent Level of Noise (Neq) and the standardized exposure level (NEN):

As can be seen, with respect to the Normalized Exposure Level (NEN) of noise at the press work-

station the value found was 101.33 dB (A). This level has a maximum recommended time of exposure to the worker for up to 45 minutes in every 1 hour of work, according to the annex I of NR 15. The working day of the worker of this work station lasts 8 hours, the use of attenuated ear protectors of at least 16 dB is used in order to reduce the effects caused by noise.

According to NR 15, in its Annex II, the tolerance limit for impact noise will be 130 dB (linear). In the intervals between the peaks, the existing noise should be evaluated as continuous noise. As all values of noise levels obtained at the cylinder workstation were lower than 130 dB, they were considered and evaluated as continuous noise.

The Normalized Exposure Level (NEN) was found to be 100.36 dB (A), as set out in Table 1,

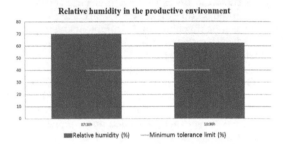

Graph 2.   Relative air humidity.

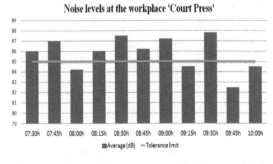

Graph 3.   Noise levels at the workplace 'Court press'.

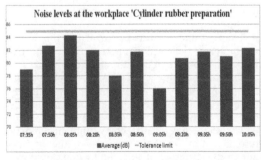

Graph 4.   Noise levels at the workplace 'Cylinder rubber preparation'.

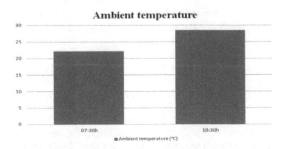

Graph 1.   Ambient temperature at the 'Rubber preparation cylinder'.

Table 1. Results of the calculation of Doses, Neq and NEN.

Workplace	Dose (D)	Equivalent Noise Level (Neq) in dB (A)	Normalized Exposure Level (NEN) in dB (A)
Court press	0.43	101.33	101.33
Cylinder	0.34	100.36	100.36

where the recommended time for this value is 45 minutes according to NR 15. As the worker is exposed to this noise level during the working day, which lasts 8 hours, then at least 15 dB attenuating ear protectors must be used during the working day in an attempt to Noise and protect the worker from the effects caused by it.

5.3.3 Lighting

In the environment of the homogenization sector, an average of 1,073.75 can be observed, obtained through the arithmetic average of the four points measured in the two schedules, in accordance with ISO/IEC 8995-1, for the environments for cutting, finishing and inspection in the sludge company must have 750 Lux (value corresponding to illuminance maintained which is the value below which the average illuminance of the specified surface should not be reduced).

When performing the measurements of luminance in the cutting sector, an average of 1.073.6 was obtained through the four points measured in the two schedules, according to ABNT NBR ISO/CIE 8995-1, for shoe manufacturing sector in the leather company, must have 500 Lux (value corresponding to the illuminance maintained which is the value below which the average illuminance of the specified surface should not be reduced).

Although there is an oversized level of illumination in the studied environments, the lighting is adequate for the execution of the activities assigned to each production sector, with the purpose of better illuminating the production system, since it is a detailed process, thus providing visual safety at work and meeting the needs for the performance of the activity in an efficient manner. However, attention should be paid to the fact that this super-measurement can negatively interfere with the health of the worker.

6 CONCLUSIONS

Among the types of environmental risks (risk of accident and ergonomic risk) and physical risks (ambient temperature, noise and lighting), what was most critical was noise, since it presented high values that exceeded the limits of tolerance established by NR 15. To this end, it is recommended that the company always seek to purchase hearing protectors that can ensure the optimum degree of attenuation of the noise in order to adjust it to the value within the tolerance limits established by NR 15, and always advise their workers to make the correct use of PPE in order to avoid exposure to doses of noise considered unhealthy according to the norm.

Therefore, it is expected that studies, analyzes and improvements will continue to be made in the company in order to always keep it a working environment that considers the safety, health and well-being of its workers.

REFERENCES

Brazilian Association of Technical Standards—ABNT. *Technical Standard ABNT NBR ISO/IEC 8995-1*: Workplace Safety—Part 1: Interior. 2012. Available in: <http://pt.slideshare.net/micheleferracioli/nbr-8995-1-iluminacao>.

Brazilian Service of Support to Micro and Small Companies—SEBRAE. *The footwear industry in the Northeast Region is the fastest growing in the country*. 2016. Available at: <http://www.sebrae.com.br/sites/PortalSebrae/ufs/pb/cursos_eventos/setor-calcadista-pb,a5a2ba18deda8410VgnVCM2000003c74010aRCRD>

Lourinho, Mayra Guasti, et al. Risks of musculoskeletal injury in different sectors of a footwear company. *Physiotherapy and Research*, São Paulo, v. 3, n. 18, p.252–257, July 2011.

Market intelligence. *Indicators of the footwear sector*. 2016. Available at: <http://www.iemi.com.br/indicadores-do-setor-calcadista/>.

Ministry of Labor and Employment. *Norma Regulamentadora (NR) 17—Ergonomics*. Atlas Legislation Manual. 59th. Edition, 2006b.

Ministry of Labor and Employment. *Regulatory Standard (NR) 15—Unhealthy Activities and Operations*. Atlas Legislation Manual. 59ª. Edition, 2006b.

Ministry of Labor and Social Security. Company of Technology and Information of the Social Security. *Statistical Yearbook of Labor Accidents 2014*. Brasília: MPS/DATAPREV, 2014. Available at: <ftp://ftp.mtps.gov.br/portal/acesso-a-informacao/AEAT201418.05.pdf>.

Ministry of Labor and Social Welfare. Company of Technology and Information of the Social Security. *Statistical Yearbook of Social Security 2014*. Brasília: MPS/DATAPREV, 2014. Available at: <http://www.previdencia.gov.br/wp-content/uploads/2016/07/AEPS-2014.pdf>.

SALIM, Celso Amorim (Coord.). *Technical Report 1—Shoe Industry*: Research on work accidents in micro and small industrial companies in the footwear, furniture and clothing sectors. Brazil: Fundacentro, 2007.

Social Service of Industry Regional Department of São Paulo—SESI/SP. *Manual of safety and health at work for footwear industry*. São Paulo: 2002. 297p.

___. *Ergonomics booklet in the footwear industry*: guidelines for occupational safety and health/ABICALÇADOS; FETICVERGS; Ministry of Labor and Employment. New Hamburg: Feevale, 2011. 96p.

Occupational Safety and Hygiene V – Arezes et al. (Eds)
© *2017 Taylor & Francis Group, London, ISBN 978-1-138-05761-6*

Influence of different backpack loading conditions on neck and lumbar muscles activity of elementary school children

S. Rodrigues, G. Domingues, I. Ferreira, L. Faria & A. Seixas
Universidade Fernando Pessoa, Porto, Portugal

ABSTRACT: The purpose of the present study is to analyze the influence of four backpack loading conditions (0, 10, 15 and 20% of body weight) on the electromyographic activity of neck and lumbar muscles of elementary school children. 14 Elementary school children participated in the present study, between the ages of 6 and 10. Bilateral sternocleidomastoid, rectus abdominis and erector spinae muscles activity were recorded using Surface Electromyography (sEMG), while tested for the effect of four randomized backpack loading conditions. The results showed an increase in the activity of both right cervical erector spinae muscle and right rectus abdominis, while right erector lumbar spinae decreased in the presence of heavy loading. 20% body weight backpack causes the most significant muscular changes, however statistically significant differences were also found at 15% of body weight conditions. Analysis of the present findings suggest the choice of 10% body weight as the upper limit recommended backpack load for schoolchildren.

ClinicalTrials.gov Identifier: NCT02725645.

Keywords: Surface electromyography, children, backpack, backpack safety

1 INTRODUCTION

1.1 *Backpack safety*

Backpacks are a popular and practical way for children and teenagers to carry schoolbooks. When used correctly, they are able to distribute the weight of the load among some of the body's strongest muscles.

However, backpacks that are too heavy or are worn incorrectly can cause problems for children and teenagers. Improperly used backpacks may injure muscles and joints. This can lead to severe back, neck and shoulder pain, as well as posture problems (Schor and Pediatrics, 1995).

There is an extensive debate on the literature on the choice of 10%, 15% or 20% of body weight cutoff point for safe weight of school backpacks (Brackley and Stevenson, 2004, Hong and Brueggemann, 2000, Mackenzie et al., 2003, Schor and Pediatrics, 1995). It is also suggested that even loads corresponding to 10% of body weight appear to be sufficient to cause changes in posture (Ramprasad et al., 2010).

1.2 *Safety and prevention*

When choosing for the Right Backpack, parents should be looking for wide, two padded shoulder straps, padded back and with waist straps. Even the backpack itself should be light on load. The rolling backpack could be an option if students need to carry heavy loads. In order to prevent injury when using a backpack, students should use both shoulder straps and tighten the straps so that the pack is close to the body and pack light (Schor and Pediatrics, 1995).

1.3 *Aims*

The purpose of this study is to determine whether different backpack loading conditions influences neck and lumbar muscle activity of elementary school children.

2 METHODS

2.1 *Subjects*

14 elementary school children participated in the study, 9 female and 5 male (mean age 8.93 ± 0.73), with backpack mean weight 3.92 ± 0.93 Kg (result of one week evaluation), which represents 11,58% of their average body mass. However analysis of the maximum backpack weight would represent 18,73% of the child body mass.

In terms of athropometric characteristics, the abdominal skinfold mean were 13.79 ± 7.15 mm and the body mass 35.00 ± 7.92 Kg. Obese subjects were withdrawn from the study.

All students whose parents/legal guardians have given permission for their participation were included. On the other hand, all those whose parents/guardians were not authorized to participate or who were not at school on the day of the evaluation were excluded.

2.2 Instruments

The myoelectric activity was recorded using bio-PLUXresearch, a device that collects signal from sensors located on the skin. The channels were of 12 bit, with a sampling frequency of 1000 Hz. The processing was done offline using MATLAB software.

A Tanita brand scale and a SECA stadiometer were used to measure and weigh the children and weigh their backpacks.

2.3 Procedure

After contacting the school, the parents of the children under study were invited to sign the informed consent, authorizing the participation of the children in the study.

Data were collected from the participants, where the anthropometric characteristics of each individual were recorded. Body height and abdominal skin fold were assessed according to the protocol recommended by the International Society for the Advancement of Kinantropometry (ISAK) (Norton et al., 1996).

The evaluation of the different myoelectric activation pattern (according to the loading characteristics, 0, 10, 15 and 20% of the body weight) occurred on the same day. The order of application of the loads was randomized so that bias of the succession of loads was avoided. The loaded conditions were achieved using usual school objects (books, notebooks, cases, etc.). Between each presentation of loads a rest period of 5 minutes was given.

The weight to be transported ranged from 0 to 20% of body weight, mostly because several studies have shown the relative weight of backpacks of school-age children varies mainly among these values (Forjuoh et al., 2003, Pascoe et al., 1997), 15% of body weight was refered as the weight normally carried (Motmans et al., 2006).

Each participant was studied individually for a period of 30 minutes. Initially the skin was prepared and standard Ag/Cl electrodes were placed on the muscle belly of the sternocleidomastoid, erector of the cervical spine, rectus abdominis and erector of the lumbar spine according to Cram and Criswell (2011). An electrode (the reference) was positioned over a bone prominence.

The resting myoelectric activity was then recorded during the application of the 4 loads under study. The loads to be transported were prepared and evenly distributed in the backpack, as well as the straps adjusted for each student.

For the processing of electromyographic data, after applying a Butterworth filter and transformation in millivolts, a root mean square analysis was performed on the data.

The backpacks of the children under study were weighed daily for 5 consecutive days to determine a weekly mean of the backpack weight.

2.4 Statistical procedures

The Statistical Program for Social Science (SPSS) software for Windows, version 22, was used.

The mean and standard deviation of the myoelectric activity was considered for statistical purposes. Since the data was not normally distributed, non-parametric statistics were used to compare the means of repeated measurements using the Friedman's Two-Way Analysis of Variance by Ranks.

3 RESULTS

The next graph shows the electromyographic activity of the lumbar erector spinae muscles, during different backpack loading conditions (Figure 1).

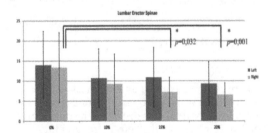

Figure 1. Route mean square mean and standard deviation of the electromyographic activity of lumbar erector spinae muscles, during different backpack loading conditions, with reference to statistical significance of the 15% and 20% of the body weight, when compared to no load condition.

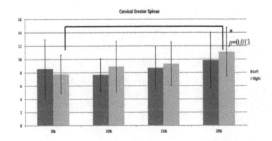

Figure 2. Route mean square mean and standard deviation of the electromyographic activity of cervical erector spinae muscles, during different backpack loading conditions, with reference to statistical significance of the 20% of the body weight condition, when compared to no load.

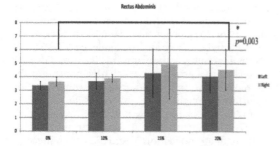

Figure 3. Route mean square mean and standard deviation of the electromyographic activity of rectus abdominis muscles, during different backpack loading conditions, with reference to statistical significance of the 20% of the body weight, when compared to no load condition.

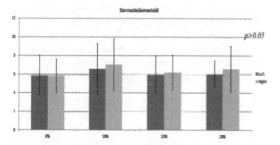

Figure 4. Route mean square mean and standard deviation of the electromyographic activity of sternocleidomastoid muscles, during different backpack loading conditions, with reference to absence of statistical significance between test conditions.

The next graph shows the electromyographic activity of the cervical erector spinae muscles, during different backpack loading conditions (Figure 2).

The next graph shows the electromyographic activity of the rectus abdominis muscles, during different backpack loading conditions (Figure 3).

The next graph shows the electromyographic activity of the sternocleidomastoid muscles, during different backpack loading conditions (Figure 4).

4   DISCUSSION

Backpacks have been referred as an adequate and safe way to carry loads, while maintaining stability (Ramprasad et al., 2010), however, the recommended limit for weight carried in the backpack remains a matter of considerable discussion in the literature. Some authors, such as Brackley and Stevenson (2004) and Hong et al. (2008) recommend that the maximum weight to be carried by children in backpacks should range between 10% and 15% of body weight. However, Mackenzie et al. (2003), demon-strated that transporting a backpack with more than 15 to 20% of body weight induces the appearance of musculoskeletal symptoms. Nevertheless, according to Ramprasad et al. (2010), even 10% of body weight may predispose to postural changes, namely increased tension in the lumbar muscles.

In the present study, the mean backpack recorded weight, during one week, corresponded to 11.58% of the children's body weight. Using 10 to 15% of body weight as a reference (Brackley and Stevenson, 2004, Hong et al., 2008, Mackenzie et al., 2003), the average load transported by the participants might be considered as adequate, although it should be noted that when the average of each individual child is analyzed, there is at least one participant who carries 18.73% of his body weight during the week, which alerts us to the need for health education in schools.

The results of the study suggest that there are significant differences in the electromyographic activity of the erector of the cervical spine, erector of the lumbar spine and rectus abdominis of the right side. The activity of the right erector of the cervical spine increases with increasing load and a statistical significant difference was found between 0 and 20% of body weight. Kim et al. (2008), in their study also observed that when carrying a backpack the activity of the erector of the cervical spine also tends to increase. The literature also suggests that as a consequence, a significant increase in the anterior projection of the head occurs as far as an increased craniovertebral angle (Ramprasad et al., 2010). The craniovertebral angle is the angle formed by a horizontal line drawn through the spinous process of the seventh cervical (C7) vertebra and a line joining the spinous process of C7 vertebra with the tragus of the ear (Lau et al., 2009).

With the increase of load in the backpacks a subsequent displacement of the center of gravity in relation to the base of support takes place. In order to counterbalance the created extension moment, anterior projection of the head occurs (Goh et al., 1998). In this sense, the increased activation of the erector of the cervical spine found in the present study seems to be related to the attempt to respond to the anterior projection of the head, created to cope with the increased load in the backpack, especially when transporting 20% of the body weight. In the present study, there are only significant changes in the activity of the right erector of the cervical spine, possibly due to the fact that the excess of load induces a rotational component of the head, caused by the transfer of load to the right side.

Regarding the activity of the right erector of the lumbar spine, it is verified that it reduces with the increase of load, with a significant variation between 0 and 20%, and 0 and 15% of the body weight. The results observed in the erector of the lumbar spine show the same tendency of the results presented by Habibi (2009) and Motmans et al. (2006). While in the study of Al-Khabbaz et al. (2008) no significant differences were found between different loading conditions. In biomechanical terms, it can only be suggested that the children from the present study tended to project their center of gravity backwards, as the load increased, with an increase in the response of the rectus abdominis muscle to control the positioning of the spine. Similarly, in Al-Khabbaz et al. (2008) study, even though there were no significant differences, the posture pattern adopted by the sample seems to corroborate the data of the present study, with an increase in trunk extension associated with increased rectal abdominal activity. In order to counterbalance the moment of extension of the trunk created by the increase of the load. Habibi (2009), in order to explain the alterations in the electromyographic activity of the erector of the lumbar spine and rectus abdominis muscle, suggested that these muscles play an important role in trunk extension and resistance against extension due to increased load. The electromyographic activity of the erector of the lumbar spine and rectus abdominis respond to the application of different weights, thus a maximum activation level is found in the abdominal rectus, when a maximum load (20%) is applied (Habibi, 2009). In summary, the extensor muscles have to withstand a bending movement of the trunk, without load, because the center of gravity of the upper body is located at the front of the lumbosacral joint. In a back loaded situation, the center of gravity of the torso shifts backward creating an extension moment. In order to counterbalance the weight on the back occurs a slope forward of the trunk, together with an increased trunk flexor activity as a compensation mechanism, which reduces the activity of the erector of the lumbar spine (Pascoe et al., 1997).

In accordance, in the right rectus abdominis a significant increase of its activity was observed between 0 and 20% of body weight, in order to resist the posterior projection of the center of gravity associated with the increase of the posterior load. In fact the literature suggests that the abdominal rectus activity also increases significantly with increasing backpack weight (Goh et al., 1998, Al-Khabbaz et al., 2008). Motmans et al. (2006), also verified that there is an asymmetry between the activity of the right and left rectus abdominis.

In fact, an increase in the activity of the abdominal rectus, bilaterally, would be expected to cope with the posterior displacement of the center of gravity under loaded conditions and as a result of the extra extension of the trunk (Goh et al., 1998). This would be the expected way of compensating for the effect of backpack weight, through trunk flexion (Pascoe et al., 1997), combined with the simultaneous co-activation of the erector of the lumbar spine, in order to stabilize and compensate for trunk displacement (O'Sullivan et al., 2002). However, in the present study, an asymmetric tendency of increased rectus abdominis activity towards the right side may be related to a greater tendency of support over the dominant limb when under loaded conditions, by transferring weight to the preferred side.

Finally, as in the present study, Devroey et al. (2007), demonstrated that the electromyographic activity of sternocleidomastoid did not change significantly with increasing backpack load, possibly due to the anterior projection of the head.

5 CONCLUSIONS

There seems to be an asymmetrical activation of the axial muscles, with preference towards the right side of the body, in a loading condition.

No alteration was recorded in the sternocleidomastoid muscles (right and left) or in left spine muscles, with increase in carried load.

The results showed an increase in the activity of both right cervical erector spinae muscle and right rectus abdominis, while right erector lumbar spinae decreased in the presence of heavy loading.

20% body weight backpack causes the most significant muscular changes, however statistically significant differences were also found at 15% body weight conditions. In this sense, the data from the present study seem to corroborate the choice of 10% of the body weight as the recommended maximum limit for the weight of the school children backpacks. It is important though, to emphasize the need for more studies to observe if these patterns of compensation are consistent in different age groups.

REFERENCES

Al-Khabbaz, Y. S., Shimada, T. & Hasegawa, M. 2008. The effect of backpack heaviness on trunk-lower extremity muscle activities and trunk posture. *Gait & posture,* 28, 297–302.

Brackley, H. M. & Stevenson, J. M. 2004. Are children's backpack weight limits enough?: A critical review of the relevant literature. *Spine,* 29, 2184–2190.

Cram, J. R. & Criswell, E. 2011. *Cram's Introduction to Surface Electromyography,* Jones & Bartlett Learning.

Devroey, C., Jonkers, I., De Becker, A., Lenaerts, G. & Spaepen, A. 2007. Evaluation of the effect of backpack load and position during standing and walking using biomechanical, physiological and subjective measures. *Ergonomics,* 50, 728–742.

Forjuoh, S., Little, D., Schuchmann, J. & Lane, B. 2003. Parental knowledge of school backpack weight and contents. *Archives of disease in childhood,* 88, 18–19.

Goh, J., Thambyah, A. & Bose, K. 1998. Effects of varying backpack loads on peak forces in the lumbosacral spine during walking. *Clinical Biomechanics,* 13, S26-S31.

Habibi, A. 2009. Weight varying effects of carrying schoolbags on electromyographic changes of trunk muscles in twelve-year old male students. *Archives: The International Journal of Medicine,* 2, 314–319.

Hong, Y. & Brueggemann, G.-P. 2000. Changes in gait patterns in 10-year-old boys with increasing loads when walking on a treadmill. *Gait & posture,* 11, 254–259.

Hong, Y., LI, J.X. & Fong, D. T.-P. 2008. Effect of prolonged walking with backpack loads on trunk muscle activity and fatigue in children. *Journal of Electromyography and Kinesiology,* 18, 990–996.

Kim, M., Yi, C., Kwon, O., Cho, S. & Yoo, W. 2008. Changes in neck muscle electromyography and forward head posture of children when carrying schoolbags. *Ergonomics,* 51, 890–901.

Lau, H. M. C., Chiu, T. T. W. & Lam, T.-H. 2009. Clinical measurement of craniovertebral angle by electronic head posture instrument: a test of reliability and validity. *Manual Therapy,* 14, 363–368.

Mackenzie, W. G., Sampath, J. S., Kruse, R. W. & Sheir-Neiss, G. J. 2003. Backpacks in children. *Clinical orthopaedics and related research,* 409, 78–84.

Motmans, R., Tomlow, S. & Vissers, D. 2006. Trunk muscle activity in different modes of carrying schoolbags. *Ergonomics,* 49, 127–138.

Norton, K., Olds, T. & Commission, A. S. 1996. *Anthropometrica: A Textbook of Body Measurement for Sports and Health Courses,* UNSW Press.

O'Sullivan, P. B., Grahamslaw, K. M., Kendell, M., Lapenskie, S. C., Möller, N. E. & Richards, K. V. 2002. The effect of different standing and sitting postures on trunk muscle activity in a pain-free population. *Spine,* 27, 1238–1244.

Pascoe, D. D., Pascoe, D. E., Wang, Y. T., Shim, D.M. & KIM, C. K. 1997. Influence of carrying book bags on gait cycle and posture of youths. *Ergonomics,* 40, 631–640.

Ramprasad, M., Alias, J. & Raghuveer, A. 2010. Effect of backpack weight on postural angles in preadolescent children. *Indian pediatrics,* 47, 575–580.

Schor, E. L. & Pediatrics, A. A. O. 1995. *Caring for Your School-age Child: Ages 5 to 12,* Bantam Books.

Occupational Safety and Hygiene V – Arezes et al. (Eds)
© *2017 Taylor & Francis Group, London, ISBN 978-1-138-05761-6*

Risk perception associated with noise-induced hearing loss. State of the art and design of interactive tools for workers training and awareness

L. Sigcha, I. Pavón & L. Gascó
Instrumentation and Applied Acoustics Research Group (I2A2), Universidad Politécnica de Madrid, Spain
Campus Sur UPM, Madrid, Spain

J.E. González & J. Sánchez-Guerrero
Master's Degree in Acoustic Engineering in Industry and Transport of the Universidad Politécnica de Madrid
Máster Universitario en Ingeniería Acústica en la Industria y el Transporte, Alumno, ETSI Industriales,
Universidad Politécnica de Madrid, Madrid, Spain

D.M. Buitrago
Facultad de Ingenierías, University San Buenaventura, Sede Medellín, San Benito, Colombia

ABSTRACT: Workplace noise exposure and noise-induced hearing loss are closely related to worker's attitudes about the use of hearing protection devices. Previous studies have shown that workers are unwilling to use them for different reasons: The lack of knowledge of the noise effects in hearing health, difficulties to use them correctly, discomfort, etc. With the aim of improving the workers' attitudes towards the use of hearing protection devices, a bibliographic study was made about the factors that influence the worker's behavior about the use of hearing protectors, as well as their knowledge of the associated risks. After analyzing the background and current needs, a user-friendly interactive tools were developed to raise workers' awareness of this issue. It is presented a review of the literature on the use of hearing protectors, an analysis of the different approaches commonly used for training and awareness of workers on the risks of noise-induced hearing loss, and finally the tools developed as a proposal to improve the hearing loss risk perception among workers.

1 INTRODUCTION

Noise-induced hearing loss is an important public health problem. Exposure to noise in the workplace causes direct and well-known effects such as hearing loss, but also indirect effects that are difficult to establish with the current assessment systems, such as loss of concentration and productivity or the increase of work accidents due to the absence of alarms and warning signs (Pavon, 2015).

People who work in noisy environments, and therefore suffer from exposure to high noise levels tend to underestimate the risk of noise to their hearing. Although in many situations the use of hearing protection is mandatory, many workers are reluctant to use it, arguing for discomfort or other causes (Buitrago, 2015).

One of the most historically effective security measures has been the implementation of hearing conservation programs. Such programs are known with this terminology in the United States. The minimum requirements for a hearing conservation program are (McReynolds, 2005):

– Monitoring and recording the exposure to noise of workers.

– Provide notification to workers about the noise exposure.
– Provide adequate hearing protection devices.
– Provide instructions in the fit and using hearing protection devices.
– Provide training regarding the effects of noise through employee training programs, including access to training materials and information.
– Provide employees with the results of periodic audiometric tests and information on the meaning of such tests.

The European Union addresses each of the previous headings, through the text of Directive 2003/10/ CE, although it does from a somewhat different perspective. The evaluation of noise exposure in the workplace is collected in the Article 4.1 (Determination and assessment of risks), which refers to the measurement of noise levels workers are exposed. The Article 4.4 provides for the preservation of records. The notification of exposure to employee noise is provided in the Article 8.

There are some differences between the two approaches, mainly about the hearing protection devices. In USA the hearing protection devices are a fundamental part of the hearing conservation programs, since 2 of its 6 points are related to hearing protection. In contrast, in the EU, the collective protection measures take precedence over individual protection, with the use of hearing protectors being considered as the last option to be used in the absence of other ways to prevent the risks of noise exposure.

Directive 2003/10/CE briefs in the Article 6 about the use of hearing protection devices, and the Article 8 about the training and information to provide to workers about the use of hearing protectors, the nature of the risks of noise exposure, the measures taken in the workplace to reduce risks, results of noise assessments and measurements, correct use of hearing protectors, how to detect and report symptoms of hearing loss, etc. Finally, the article 10 reports how to perform the preventive audiometric control.

With the information about the hearing conservation programs and the actions to be taken according to Directive 2003/10/EC, the need for training and information for employees is highlighted, in topics like: the risks, the noise control measures, individual health monitoring, correct use of hearing protectors and safe working practices.

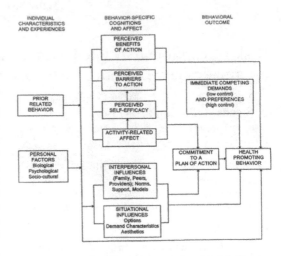

Figure 1. Pender Health Promotion Model (Pender, 1996).

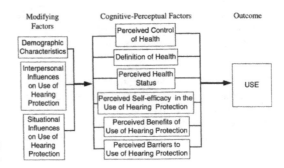

Figure 2. Health Promotion Model adapted to the use of hearing protection (Lusk et al., 1997).

2 RISK PERCEPTION OF HEARING LOSS DUE TO NOISE EXPOSURE AND WORKERS ATTITUDES

Occupational noise exposure is a common risk present in the workplace, so there are many attitudes and habits that are deeply rooted among exposed workers. In order to minimize the risk of hearing loss among workers, it is very important to know their attitude to hearing protection, risk perception and behavior.

In 1975 Nola Pender published "A Conceptual Model for Preventive Health Behavior", laying the foundations for the study of how individuals make decisions about taking care of their own health, identifying factors that influence the decision making and individual's actions to prevent diseases (Pender, 1975).

The updates of the Health Promotion Model propose that by modifying sociological, personal or social factors, it is possible to influence cognitive factors such as perceived self-efficacy. The Pender Health Promotion Model (Fig. 1) has been adapted to different areas, such as the use of hearing protection (Fig. 2).

Several studies have used the Health Promotion Model to predict behaviors that promote the individual health, and explain the modification of behaviors specifically related to the use of hearing protection devices. In (Lusk et al., 1995) a study was made to identify which factors influence in the rate of use of the hearing protection devices.

Based on a self-assessment questionnaire for workers exposed to noise, the most influential factors related to the use of hearing protection devices are:

– The personal perception of the health control.
– The perceived self-efficacy of hearing protection use.
– The perceived benefits of using hearing protection devices.
– The perceived barriers and difficulties found in the use of hearing protection devices.

Similar results were found in (Arezes & Miguel, 2005), where the self-efficacy was identified as the main predictor of hearing protection use. The increase in the use of hearing protection devices in industrial environments depends of the improvements in the perception of risk by workers and the removal of barriers and obstacles.

The perception of individual risk has been identified as a fundamental indicator of risk behavior (Diaz et al., 2000, Gendon & McKenna, 1995), so the perceived risk of noise-exposed workers is an important topic in which it can be developed strategies for improvement.

In 2001 a study was made in the United Kingdom (Hughson et al., 2002), involving 280 workers from 18 companies in different sectors. Initially, a survey was made to the workers to know their risk perception, their attitude and behavior related to hearing protection. Subsequently, different actions were taken in four selected companies. The actions were designed specifically for each company, including all the typical aspects of a hearing conservation program (McReynolds, 2005). Next, another survey was made and there was an improvement in the knowledge and risk perception; and an increase in the rate of hearing protection use. One of the main conclusions is that the traditional methodology used in companies to train and report on noise at work was not very efficient:

– Only 27% of newly recruited employees were trained in hearing conservation.
– When information posters were used, only 46% of the employees remembered it.
– In the cases where information leaflets were used, only 39% acknowledged having been informed.

Although 100% of the workers had been trained in the use of hearing protection, two-thirds of the workers did not remember being trained and only a fifth remembered being informed about the limitations of the effectiveness of the protectors.

In the case of audiometries, 83% acknowledged having been subject to audiological tests and having received the results. Employers identify that the most appropriate time to give training is to coincide with the periodic audiometric test, because the worker has a greater facility to relate the risk and its effect on hearing health from an individual and personal point of view.

Regarding the use of hearing protectors, it was found that there are a higher rate of use in situations where:

– Workers understand the physiological effects of exposure to noise.
– Noise levels are higher.
– The noise levels are constant (there is no fluctuation of levels).
– When work or task is routine.
– Where the company demonstrates a commitment to hearing conservation.
– Where there is support from coworkers.

The main conclusion of this study is that workers with more training and awareness are able to identify the risks and take actions to minimize them.

The use of hearing protectors is a factor that depends on the perception of the risk by the worker or individual, or in other words the ability of the worker to identify if a particular situation

or exposure constitutes a risk to his health. The identification of these situations and exposure to noise in these environments depends on different factors such as educational level, age, work experience, among others (Arezes & Miguel, 2006).

The most representative predictors of the use of HPDs were shown in a review made by Costa & Arezes in 2013, finding many factors of interest to predict the use of HPDs (Costa & Arezes, 2013).

The study made by (Gorski, 2014) shows the usefulness of an interactive tool in hearing conservation programs. This work proposes the use of a tool for training in the use of hearing protectors in order to improve the effectiveness of their use, as well as to serve as a monitoring tool to evaluate the change in the risk perception and their attitude.

Other recent works show different initiatives based on applications for mobile devices related to the diagnosis of hearing loss. An example of this is the development of an interactive game application for performing audiometries for children (Yeung et al., 2015), a perfectly transferable approach to training and informing workers about occupational noise.

3 DESIGN OF INTERACTIVE TOOLS FOR TRAINING AND AWARENESS ABOUT THE RISK OF HEARING LOSS

Based on the previous evidence, it has been identified the need to develop interactive training tools for workers and professionals in occupational hazards, providing the user with a realistic idea of the estimation of hearing loss induced by the noise in the work environment, using the variables of workers hearing loss and loss due to the insertion of hearing protection devices.

The most prolific initiatives that have been made until now for training and informing are the next:

– Training sessions.
– Informative sessions, mainly for workers.
– Training videos (3M Soluciones Auditivas).
– Posters (OMS, 2015)
– Training guides (CCOO, 2008).
– Hearing loss simulation software (NIOSH,, Liedtke, Starkey, 3M E-A-RFIT)

There are an increasing to use methodologies based on participatory tools that allow to experience and feel the implications of a given exposure.

With the aim of equipping workers with tools that allow them to train and report risks, but at the same time with an entertaining and participative way, some applications have been designed.

In 2014 a tool was designed to determine the audibility of alarm signals (Gasco, 2014). This application has the objective of training technicians in occupational risk prevention to choose a

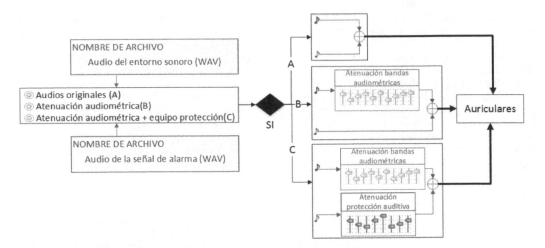

Figure 3. Simplified block diagram of the software Audible emergency signals.

Figure 4. Simplified block diagram of the software Audible emergency signals.

suitable hearing protection device for each situation (Fig. 3). The developed software allows to evaluate the need for hearing protection based on the existing noise level, the noise level of the alarm signals (which must be perceived), the hearing protector used and the current hearing loss of the worker.

In 2015 CEP software was developed for the training of workers in the use of hearing protection. The tool estimates hearing loss induced by: noise exposure, age (presbycusis) and takes into account hearing protection. The application allows loading audio files and listen, how a particular sound would be perceived by modifying noise, age-induced hearing loss and adding the attenuation provided by hearing protection devices (Buitrago, 2015), (Fig. 4).

4 CONCLUSIONS

Based on the review of the literature published, it has been identified and verified that some of the training and information strategies that are performed for: Occupational noise, hearing protection and the risk of hearing loss are not being effective. Although training and information are key aspects of the hearing conservation programs and they are covered by Directive 2003/10/EC, the objectives are generally not accomplished and there are possibilities for improvement.

Specifically:

– It has been found through literature review, workers and technicians in occupational risk prevention are not really aware of the consequences of the exposure to occupational noise for the health and safety.
– The methodologies for training, informing and raising awareness about exposure to occupational noise and associated risks can be improved, since too much information is transmitted to workers, but the approaches and methodology are not being effective.

It is proposed to train workers, employers and technicians in occupational risk prevention using interactive tools, where they can simulate (and perceive) how they would listen in the future according to certain conditions of exposure and hearing protection use.

Two software applications have been developed, one of them is aimed to the training of technicians in occupational risk prevention to assess the need for hearing protection based on: the existing noise level, the noise level of alarm signals (that must be perceived), the hearing protector used and the worker's hearing loss.

The second application was designed for training workers in the use of hearing protection devices, which allows to hear how a certain sound would be perceived by modifying: The noise-induced hearing loss, the age and the attenuation provided by hearing protectors.

Future research topics will be based on using this kind of tools to assess if through their use, there is an improvement in the knowledge of the risks and if this causes positive effects on the behavior and attitude towards the work noise, increasing the use of hearing protection devices.

ACKNOWLEDGEMENTS

This work has been developed in the course Industrial and occupational noise of the Master's Degree in Acoustic Engineering in Industry and Transport of the Universidad Politécnica de Madrid.

http://www.i2a2.upm.es/masteracustica/

The tools shown here are the result of work done by the students of the subject-course and two master's degree projects.

The funding required to carry out to this paper has been provided by the Research Group on Instrumentation and Applied Acoustics (i2a2) of the Universidad Politécnica de Madrid.

http://www.i2a2.upm.es/

REFERENCES

3M, *E-A-Rfit™ Dual-Ear Validation System*. [Online]. Available: http://www.3m.com/3M/en_US/worker-health-safety-us/safety-equipment/hearing-conservation/hearing-protection-fit-testing/?WT.mc_id=www.3m.com/earfit [Accessed 11/18 2016].

3M™SOLUCIONES AUDITIVAS, *¿Cuál es tu sonido favorito? Formación en protección auditiva*. [Online]. Available: https://youtu.be/ErHGeBKpy_Y.

Arezes, P.M. & Miguel, A.S. 2005. Hearing protection use in industry: The role of risk perception, *Safety Science*, vol. 43, no. 4, pp. 253–267.

Arezes, P.M. & Miguel, A.S. 2006. Does risk recognition affect workers' hearing protection utilization rate? *International Journal of Industrial Ergonomics*, vol. 36, no. 12, pp. 1037–1043.

Buitrago, D. 2015. Predicción de pérdida auditiva, como herramienta para la formación de trabajadores y profesionales en riesgos laborales, basado en las normas ISO 1999, ISO 4869 e ISO 9612, MSc Thesis. Universidad Politécnica de Madrid.

CCOO 2008. *El ruido, un daño silencioso Guía práctica*. [Online]. Available: http://www.fsc.ccoo.es/comunes/recursos/17629/pub12008_guia_sobre_el_ruido_en_el_entorno_laboral.pdf.

Costa, S.; Arezes, P. M. 2013. On the nature of hearing protection devices usage prediction; *Occupational Safety and Hygiene*. Arezes et. al. (eds). Taylor and Francis.

Diaz, Y.F. 2000. *Predicting employee compliance with safety regulations, factoring risk perception*, ProQuest Dissertations Publishing.

Directive 2003/10/EC, of European Parliament and Council of 6 February 2003 on the minimum health and safety requirements regarding the exposure of workers to the risks arising from physical agents (noise) (Seventeenth individual Directive within the meaning of Article 16(1) of Directive 89/391/EEC).

Gasco, L. 2014. *Audibilidad de Señales de Emergencia*, Grupo de Investigación en Instrumentación y Acústica Aplicada.

Glendon, A.I., Clarke, S. & McKenna, E. 2016. *Human safety and risk management*, Crc Press.

Górski, P. 2014. Model of Interactive System for Training in the Proper Use of Hearing Protection Devices, *Archives of Acoustics*, vol. 39, no. 1, pp. 11–15.

Hughson, G.W., Mulholland, R.E. & Cowie, H.A. 2002. *Behavioural studies of people's attitudes to wearing hearing protection and how these might be changed*, Health and Safety Executive, United Kingdom.

Liedtke, M., *Hearing impairment calculator*. [Online]. Available: http://www.dguv.de/ifa/praxishilfen/noise/gefaehrdungsbeurteilung-laermmessung-unterweisung/software-berechnung-von-hoerschwellenverschiebungen/index.jsp [Accessed 11/18 2016].

Lusk, S., Ronis, D. & Hogan, M. 1997. Test of the health promotion model as a causal model of construction workers' use of hearing protection, *Research in nursing & health*, vol. 20, no. 3, pp. 183–194.

Lusk, S.L., Ronis, D.L. & Kerr, M.J. 1995. Predictors of Hearing Protection Use among Workers: Implications for Training Programs, *Human Factors: The Journal of the Human Factors and Ergonomics Society*, vol. 37, no. 3, pp. 635–640.

McReynolds, M.C. 2005. Noise-induced hearing loss, *Air Medical Journal*, vol. 24, no. 2, pp. 73–78.

NIOSH 2002. Mining Product: HLSim—NIOSH Hearing Loss Simulator. [Online]. Available: http://www.cdc.gov/niosh/mining/works/coversheet1820.html [Accessed 11/18 2016].

OMS 2015. *"Escuchar sin riesgos!"*. [Online]. Available: http://www.who.int/topics/deafness/safe-listening/es/ [Accessed 11/18 2016].

Pavón, I. 2015. *Webinar: Nuevas tecnologías para la valoración de la exposición al ruido ocupacional*. [Online]. Available:http://prl.ceoe.es/es/contenido/webinar-nuevas-tecnologias-para-la-valoracion-de-la-exposicion-al-ruido-ocupacional [Accessed 11/18 2016].

Pender, N. 1996. *Health Promotion Model-Diagram*. [Online]. Available [Accessed 11/18 2016]: https://deepblue.lib.umich.edu/handle/2027.42/85351.

Pender, N.J. 1976. A conceptual model for preventive health behavior, *[Kango kyōiku] Japanese journal of nurses' education*, vol. 17, no. 11, pp. 710.

Starkey Hearing Technologies, *What hearing loss sounds like* [Online]. Available: http://www.starkey.com/hearing-loss-simulator [Accessed 11/18 2016].

Yeung, J.C., Heley, S., Beauregard, Y., Champagne, S. & Bromwich, M.A. 2015. Self-administered hearing loss screening using an interactive, tablet play audiometer with ear bud headphones, *International journal of pediatric otorhinolaryngology*, vol. 79, no. 8, pp. 1248–1252.

Occupational Safety and Hygiene V – Arezes et al. (Eds)
© 2017 Taylor & Francis Group, London, ISBN 978-1-138-05761-6

Psychosocial risks in radiographers work

A. Neves, D. Durães, A. Saraiva, H. Simões & J.P. Figueiredo
ESTeSC-Coimbra Health School, Instituto Poltécnico de Coimbra, Coimbra, Portugal

ABSTRACT: The current tendencies in promoting Health, Safety and Hygiene in the Work Place include not only Physical Risks, Chemicals and Biologicals in the working environment, but also Psychosocial Risks which influence the physical and mental well-being, of a worker. Professional health workers, including Radiographers, are highly exposed to this risk. Our propose is to Evaluate Psychosocial Risks factors among Radiographers and seeks the relationship between social-demographics, social-professional and physiological with those same risks, using the COPSOQ to collect the data, and also a Check list for self-evaluation of the Hygiene and Safety Conditions in the work place. Dimensions groups in the COPSOQ, reveal a statistic differences between the dimension "social relationship and leadership" and the variable gender. Since the multiple transformations that occurred in the laboring world and, by different constraint still taking place, Psychosocial Risks in the work-place is still a concern subject of professional risks.

1 INTRODUCTION

The emergence of new Occupational Health risks, fundamentally the Psychosocial Risks, led to a growing awareness of the importance and necessity of its prevention (1).

Thus, the current trends in promoting Hygiene and Safety at Work (HSW) not only include the physical, chemical, ergonomic and biological risks of work environments, but also the multiple psychosocial factors and the way these factors influence the Physical and mental well-being of the professional (2).

Gollac & Bodier define Psychosocial Risks as being the health risks created by work through social and psychic mechanisms (3). These are not only individual in character, but are also associated with economic expectations, human relationships and their emotional aspects. Currently, Psychosocial Risks such as Stress, Burnout and Mobbing are the challenge of Occupational Safety and Health (4). It is then considered as an interaction between the individual and the environment in which it operates and, consequently, attempts are made to deal with the problem (5).

The first author (Bernard) to approach this concept, considering that the physical threats to the integrity of the organism evoke contradictory responses to the threat (6). Other author (Seyle) explains Stress as the non-specific result of a bodily requirement from a physical, mental or emotional nature (7). Stress-inducing causes are varied. People react differently to stressors, which may be psychosocial, physical, chemical and biological (8).

The most frequent symptoms are low satisfaction and involvement with work, tension, anxiety, depression, psychological fatigue, frustration, irritability and Burnout (9).

Radiographers are largely exposed to Stress and its consequences due to the closed work environment, the frequency of unforeseen and urgent situations that require quick and effective action, lack of time and resources, interpersonal conflicts, contact with patient suffering and work shift (10).

Burnout, "a state of physical, emotional and mental exhaustion caused by long-term involvement in situations of high emotional demands in the workplace, conceptualized as a syndrome of emotional exhaustion, depersonalization and lack of personal fulfillment at work" usually caused by a combination of very high expectations and chronic Occupational Stress, especially in the professions that focus on service delivery (11; 12).

2 MATERIAL AND METHODS

The convenience sample of this study is non-probabilistic of 40 Radiographers to performing functions in the health units A, B and C of public units (two) and private (one) from the central region of the country. Data collection was carried out through a socio-demographic and socio-professional questionnaire, the "Copenhagen Psychosocial Questionnaire—COPSOQ", a medium version, adapted and validated for the Portuguese population, and also using a Self-Check Check List of the HSW Conditions.

In the first phase, the questionnaire under study incorporates the characterization of a personal nature includes information about the sociodemographic variables (gender, age, marital status, academic qualifications) and socio-professional characteristics (type of labor bond, managerial functions, years of service, functions). The second part concerns the characterization of Radiographers in the perception of social support and self-concept.

COPSOQ emerged in a recent format, reformulated in 2007, consisting of 8 dimensions and 25 sub-dimensions. The items of these sub-dimensions are answered on a Likert-type scale, with five response options.

Finally, in a second phase, we used a Check-List of self-assessment of the conditions of HSW.

There was a concern to carry out the data collection in the study places during working hours and according to the availability of the professional, between March 1 and May 17, 2016. The statistical treatment of the data referring to this study was done through the Statistical Package for the Social Sciences (SPSS) program in 21.0 version.

3 RESULTS

Data showed that the large percentage of elements, 60%, works in the health unit A. 15.5% in health unit B and 27.5% in health unit C.

Regarding gender, the sample is composed mostly of women, 77.5% (n = 31) and 22.5% (n = 9) men.

The mean age for both males and females is 42 years, with a standard deviation of 9.58 and 8.02 respectively, being the minimum 28 years and the maximum of 65 years old.

Regarding the marital status and with regard to the totality of the sample, it is verified that the marital status/union is predominant (65%), followed by the group of singles (20%). In the female gender we can see that the married/de facto civil status prevails with 71% and then the divorced/separated state with 16.1%. In the masculine gender it is verified equality in the married state/domestic partnership and single, with 44.4%.

Regarding academic qualifications, most of them have a superior level degree (65%), with a significant number of Radiographers with a Master's degree (22.5%), and 12.5% with a Bachelor degree.

Analyzing statistics on service duration, the average does not vary according to gender, being 19 years of professional practice, with standard deviations of 9.09 in the masculine gender and 9.68 in the feminine gender.

Regarding the type of employment relationship, a large number of Radiographers have a definitive link (57.4%), followed by a permanent contract (31.5%) and a fixed term contract (11.1%). As for the managerial functions, only a small part of performs in these areas (12.5%). This situation occurs in both genders, 11.1% in men and 12.9% in women.

Subsequently, an analysis was made for the correlations between the variables gender, academic qualifications and professional linkage and the COPSOQ size sets.

Regarding the sociodemographic variables, some characteristics were grouped. In the academic qualifications, Radiographers were grouped with Bachelors to the Graduates, designated by 1st Cycle and the Masters designated by 2nd cycle. On the other hand, the type of definitive link to work was added to the contract without term.

With regard to the 75 percentile, as recommended by the authors of the used instrument, after the stratification in sub dimensions, the percentage frequency table was elaborated relative to the number of elements belonging to each sub-dimension. This tripartite division assumed a "traffic light" interpretation by means of the impact on health, where green corresponds to a favorable situation for health (< 2.33), yellow to an intermediate situation (2.33 to 3.66) and red to risk for health (> 3.66).

By analyzing the results of the Chi-Square tests, it was found that the items that stand out with a high risk of being harmful to health are the "Cognitive Requirement" (74.4%), "Transparency of Work Paper Role" (89, 7%), "Rewards" (92.3%), "Labor Conflicts" (82.1%), "Social Support of Colleagues" (60.0%), "Social Work Community", "Quality of Leadership" (47.5%), "Self-efficacy" (62.5%), "Meaning of Work" (95.0%) and "Job Satisfaction" (56.4%).

With a moderate risk of being harmful to health, we highlight the "Quantitative Requirement" (55.0%), "Sleeping Problems" (47.5%), "Predictability" (62.5%). "Social Support Superior" (47.4%), "Horizontal Trust" (90.0%) and "Vertical" (53.8%), "Justice and Respect" (52.5%), "Commitment to the Workplace" (0%), "Occupational Insecurity" (52.5%), "Conflicts of Work/Family" (47.5%), "Burnout" (55.0%) and "Stress" (67.5%).

With a low risk of being harmful to health, we stand the "Influence at Work" (45.0%), "Depressive Symptoms" (45.0%) and "Offensive Behaviors" (100.0%).

Regarding gender, the results of the Chi-Square tests show that there is a statistically significant difference between the male gender and the female gender in the "Social Relations and Leadership" dimension, namely in the sub-dimensions "Transparency of the Worked Paper Role" (p = 0.004) and "Social Support of Superiors" (p = 0.005). Thus, we can affirm that feminine Radiographer perceive, in average, more Psychosocial Risks related to the set "Social Relationships and Leadership". For the remaining sets of dimensions, the gender difference is not statistically significant.

Regarding academic qualifications, it was found that there were no statistically significant differences between the 1st Cycle and the 2nd Cycle, in any group of dimensions of Psychosocial Risks.

For the Check-List regarding HSW Conditions applied in the health units under study, a total of 19 different diagnostic rooms were evaluated: Conventional Radiology (RC); Computed Tomography (CT); Mammography; Angiography/Hemodynamics and Magnetic Resonance (MRI). Most of the rooms correspond to RC rooms (37%), followed by TC rooms (21%), Angiography/Hemodynamics (16%), Mammography (11%) and, at last, the MRI room (5%). Its completion was based essentially on the observation and analysis of each room by the investigators (when diagnostic tests were not in progress) as well as on the collection of information from the Radiographers.

In relation to the General Conditions of HSW, it was verified that all health units had a policy statement that reflected the organization's commitment to Occupational Health and Safety, with regular audits and/or safety inspections. However, the analysis of occupational accidents is not always done. All Radiographers assume the knowledge of each type of risk (chemical, physical, biological, ergonomic and psychosocial), as well as their preventive measures.

Workplace conditions are favorable in all health facilities, since they have structural stability, stable plain floors and continuous ceilings. The sanitary facilities and locker rooms are easily accessible, however they are not separated by gender. The ventilation is sufficient, continuous and well distributed, with regular maintenance. With respect to noise and vibrations it has been found that there are logically only noisy environments in the MRI rooms. The thermal environment is adequate, and the thermal stress evaluation is normally performed. Also the maintenance of the lighting system is carried out, being this uniform and localized. However, only in the Mammography room of health unit B (5%) if was verified the presence of emergency lighting.

The electrical installations of all the health units are in good general condition, without overloading the outlets and avoiding humidity or flammable material. One note: the electrical boards do not have safety rules attached.

With regard to the chemicals present in the TC rooms, Angiography/Hemodynamics they are stored in their own place, properly labeled and with the safety data sheets available. Although the professionals received information on the use of PPE (lead apron, gloves, lead glasses, among others), only in 58% was their use verified. Regarding fire prevention and protection, only the RC rooms of health unit B (11%) have fire extinguishers available.

In none of the rooms, the existence of a plan and/or emergency plan was confirmed, as well as any type of information related to this item. It was also verified that only one of the RC rooms of the health unit B (5%) presents official pathways routes and exits of emergencies. Moreover, in any room there is no sign of any type of emergency, obligation, prohibition or warning.

Finally, in relation to ergonomics, it was found that the work space was adequate in 95% of the rooms. However, only one of the TC rooms of health unit A (5%) presents a reduced working space compared to the number of health professionals.

4 DISCUSSION

As a pioneer study, there are no studies that allows comparing the results obtained with regard to the perception of the Psychosocial Risks of Radiographers, thus limiting the discussion to the other health professionals. According to the European Agency for Safety and Health at Work, Psychosocial Risks, especially Occupational Stress, is one the most commonly reported causes of illness by workers, affecting more than 40 million people across the European Union. A European survey on new and emerging risks (ESENER) has shown that occupational accidents and musculoskeletal injuries and stress are the main concerns (14) (15).

Occupational Stress is often reported as one of the main concerns of health managers (14). Figures from different Member States of the European Union show that Occupational Stress in the health sector is more common in Slovenia (60%), followed by Greece and Latvia, where Stress was reported by 54% of professionals. (15) (16) (17). However, according to the European Labor Research and scientific literature, violence and harassment also predominate in the health sector. In 2005, the threats of physical violence were reported by 14.6% of workers, while actual physical violence was felt by 8.4% of workers. Harassment was reported in 7.8% of workers, with sexual harassment reported by 2.7% (15) (16) (17).

COPSOQ results show that the sub-dimensions with a value greater than 3.66 correspond to health risk, those with a value lower than 2.33 correspond to a favorable situation for health and the sub-dimensions in which the value is between 2.33 and 3.66 correspond to an intermediate health situation.

In the present study, it was found that the subdimensions that stand out with a high risk of being harmful to health are the "Cognitive Requirement", "Transparency of Work Paper Carried Out", "Rewards", "Labor Conflicts", "Social Support of Colleagues", "Social Work Community", "Quality of Leadership", "Self-Efficacy", "Meaning of Work" and "Job Satisfaction". Compared with national averages for health professionals, it was found that Radiographers are exposed to other risks that are not included in national averages ("Rewards", "Labor Conflicts", "Social Support of Colleagues", "Quality of Leadership" and "Job Satisfaction"). However, the national averages also indicate worse result the sub-dimension "Vertical Trust" which is not verified in the present study (18).

With a moderate risk of being harmful to health, we highlight the "Quantitative Requirement", "Sleeping Problems", "Predictability", "Social Support of Superiors", "Horizontal and Vertical Trust", "Justice and Respect", "Commitment to Workplace Instability", "Work/Family Conflicts", "Burnout" and "Stress". These results aren't against the national averages in the subdimension "Vertical Trust". On the other hand, in the national averages it was verified that the subdimensions "Influence in the Work", "Rewards", "Labor Conflicts", "Social Support of Colleagues", "Quality of Leadership", "Labor Satisfaction" and "Depressive symptoms present an intermediate risk to health" (18).

With low risk of being harmful to health stands out the "Influence in the Work", and "Offensive Behaviors". These results only agree with the national averages in the sub-dimension "Offensive Behaviors", and it is verified that Radiographers are in a favorable situation in more sub-dimensions compared to national averages for health professionals (18).

Concerning the perception of Psychosocial Risks, the difference between the Radiographers masculine and feminine gender is very small, with average values very closed. However, Radiographers female have a greater perception of these in the dimension "Social Relations and Leadership" compared to the male gender, constituting the only statistically significant difference. These results confirm the data obtained by Silva and Gomes in studies conducted in Portugal and other countries report that women tend to experience higher levels of Occupational Stress related to relationships at work, professional careers, overwork, salary and family problems (14) (19).

Regarding the academic qualifications, it was verified that there are no statistically significant differences between the 1st and 2nd Cycle in any group of dimensions of Psychosocial Risks. Our results are in agreement with other studies, where

it is verified that health professionals with Post-Graduation or Masters report higher levels of exhaustion, depersonalization, total Burnout value and work-family conflict (19).

Regarding the type of professional relationship, it was verified that there are no statistically significant differences between the definitive labor bond and the fixed term contract in any group of dimensions of Psychosocial Risks. Our results are in agreement with other consulted studies where they reveal Stress factors in the different types of contractual relationship, being the professionals with more unstable work contracts the ones that reveal more problems related to overwork, professional involvement, professional instability, Salary and socio-professional status. Effective professionals show fewer problems in terms of depersonalization compared to temporary professionals (19).

So, the data now obtained suggest the need for further studies that can test the relationship between Psychosocial Risks and health professionals, especially Radiographers.

Finally, it is essential not to consider the relevance of the development of intervention strategies, which seek to better promote HSW conditions in each context of work

5   CONCLUSIONS

Psychosocial Risks constitute a theme that, in view of the multiple transformations that have occurred in the world of work, due to different constraints, continue to be a concern in terms of occupational risks.

This study allowed us to verify that Radiographers are in a favorable situation for health in more sub dimensions, compared to the national averages for health professionals.

As a future implication, it is important to study the factors that are at the origin of Psychosocial Risks and whose presence can be detected in a suitable manner, as well as the creation of tools for the intervention and transformation of the HSW Conditions.

REFERENCES

[1] Coelho, Aguiar, et al. Riscos Psicossociais no Trabalho. 2009.
[2] Duarte, Vitor. Riscos Psicossociais no Trabalho dos Enfermeiros.
[3] Costa, Lúcia e Santos, Marta. Fatores Psicossociais de Risco no Trabalho: Lições Aprendidas e Novos Caminhos. Porto: RICOT, 2013. ISSN 2182-9535.
[4] Coelho, João. Gestor Hospitalar e Prevenção de Riscos Psicossociais no Trabalho. Lisboa: LDA, 2011. ISBN: 978-989-8463-16-6.

[5] Sacadura-Leite, E. e Uva, A. Stress Relacionado com o Trabalho. Lisboa: s.n., 2007.

[6] Melo, Sandra. Stress relacionado com o trabalho e burnout em técnicos de radiologia. Lisboa: s.n., 2012.

[7] Camelo, Silvia. Riscos Psicossociais relacionados ao eStresse no trabalho das equipes de saúde da família e estratégias de gerenciamento. Ribeirão Preto: s.n., 2006.

[8] Ribeiro, Miguel. Riscos Psicossociais no Trabalho dos Enfermeiros. Viseu: s.n., 2012.

[9] Montanholi, Liciane, et al. EStresse: fatores de risco no trabalho do enfermeiro hospitalar. Uberaba: s.n., 2006.

[10] Lima, J.A. utilização de equipamentos de protecção individual pelos profissionais de enfermagem— práticas relacionadas com o uso de luvas. Minho: s.n., 2008.

[11] Coelho, João. Gestão Preventiva de Riscos Psicossociais no Trabalho em Hospitais no Quadro da União Europeia. Porto: s.n., 2009.

[12] Fonte, Cesaltino. Adaptação e validação para português do questionário de Copenhagen Burnout Inventory (CBI). Coimbra: s.n., 2011.

[13] Costa, Ricardo. Níveis de Burnout nos estudantes da licenciatura em Optometria e Ciências da Visão da Universidade da Beira Interior. Covilhã: s.n., 2013.

[14] Agência Europeia para a Segurança e Saúde no Trabalho. Relatório do observatório de risco europeu. Previsão dos peritos para riscos psicossociais emergentes relacionados com a segurança e saúde ocupacional. Bélgica: s.n., 2007.

[15] Agência Europeia para a Segurança e Saúde no Trabalho. Inquérito Europeu das Empresas de riscos novos e emergentes (ESENER). Gestão da segurança e saúde no trabalho. 2010.

[16] Agência Europeia para a Segurança e a Saúde no Trabalho. Relatório do observatório de risco europeu. OSH em números: Stress no trabalho— factos e números. Luxemburgo: s.n., 2009.

[17] Organização Mundial de Saúde. Protecção dos serviços de saúde dos trabalhadores No.9. PRIMA-EF Orientação sobre o Quadro Europeu para a Gestão de Risco Psicossocial. Um Recurso para o empregador e os representantes dos trabalhadores. Reino Unido: s.n., 2008.

[18] Silva, Carlos. Copenhagen Psychosocial Questionnaire: Portugal e países africanos de língua oficial portuguesa. Portugal: Análise Exacta, 2000.

[19] Silva, M., Gomes, A. Stress Ocupacional em profissionais de saúde: um estudo com médicos e enfermeiros portugueses. Estudos de Psicologia. Minho: s.n., 2009.

Occupational Safety and Hygiene V – Arezes et al. (Eds)
© 2017 Taylor & Francis Group, London, ISBN 978-1-138-05761-6

Management of the activity of the safety coordinator

Cristina M. Reis
School of Sciences and Technology, UTAD, Vila Real, Portugal

C. Oliveira
Instituto Politécnico de Viana do Castelo, Escola Superior de Tecnologia e Gestão, Portugal

A.A.M. Márcio
PERITS Engenharia, Braga, Portugal

J. Ferreira
Instituto Politécnico de Viana do Castelo, Escola Superior de Tecnologia e Gestão, Portugal

ABSTRACT: Following the entry of Decree-Law n. 273/2003, 29th October, began to emerge a concern in the safety coordinator activity, that is managing is activity in safety work. Therefore, it is intended to establish one methodology that supports safety coordinators, on how they should perform their duties and at the same time ensuring all legal obligations.

The choice of this theme relates to the wide range of literature available on the Department of Health and Safety at Work and ways of implementation, this thesis is an attempt to be c a document that can help the Safety Coordinator in carrying out their duties.

1 INTRODUCTION

1.1 *Construction safety and health management*

Increasingly, today the management of works is one of the most important aspects in the construction sector. And in this matter, the most varied decisions are implied for the project to be carried out.

Construction is an economic activity with its own specificities, characterized by a great diversity that follows:

- From customers with a demand that goes from the State/Autarquias, to the particular that intends to self-build, from the large multinational companies to the small traditional promoters;
- Of projects, where each work presents different characteristics and specificities;
- products that cover both traditional housing and more complex works (roads, dams, etc.);
- productive operations, where the final product results from the interaction between various specialties with differentiated degrees of demand and technology;
- Of technologies, as a result of intervention in an undertaking of various specialties and of the

coexistence of new production technologies with the old ones.

It is a very demanding and varied sector, in which there are diversified works. Often, it is difficult to coordinate all of them, and as such there are laws, rules and obligations that are intended to regulate the sector and that are thoroughly pursued so that the whole process of building develops in a more organized way. Within the sector, one of the facades that is often neglected or placed in the background, is safety and health at work.

1.2 *Legal framework*

Decree-Law no. 273/2003, of 29th October, establishes general rules for planning, organization and coordination to promote safety, hygiene and health at construction sites and transposes Directive n. Council Directive 92/57/EEC of 24th June concerning the minimum safety and health at work prospects for construction sites.

The activity of the safety coordinator in terms of obligations is outlined in Decree-Law no. 273/2003, of 29 October, however given the importance of this activity it will be pertinent to analyze the way in which it manages its activity.

2 DEVELOPMENT

2.1 Conditioning safety coordinator activity

Within the universe of occupational health and safety, the coordinator has the moral and legal obligation to prevent the occupational risks of his workers. It must therefore ensure that its workers have conditions of safety, hygiene and health, irrespective of their employment relationship.

However, this obligation to prevent is always conditioned by an obligation on the part of the employer. The motivation to fulfill an objective can be seen as a constraint for the planning and prevention of work safety. More than a product, the construction is defined as a process developed in three phases:

- Phase of the conception: that implies the technical definition regarding the construction and its implantation;
- Phase of the organization: which involves the drawing up of specifications and the negotiation of proposals for the implementation of the project;
- Execution phase: where the site preparation, installation of the site and the construction work are carried out.

Safety coordination is the set of actions carried out with the aim of reducing damages and losses caused in companies, so it is essential that the safety elements are integrated into the preparation of business plans.

2.2 Health and safety plan

The need arises to draw up a Health and Safety Plan (HSP). The HSP is a dynamic document that will be updated if necessary, starting its preparation during the project phase and will be concluded with the final reception of the project.

The HSP describes all aspects relevant to safety and health, to be taken into account in the execution phase of the work. This is the main instrument for the prevention of inherent risks, with the aim of minimizing the risks of accidents and incidents, contributing to an increase in the safety of workers during the execution phase of the work and also to the subsequent users, In the exploration phase.

Thus, the HSP, as well as being fundamental, is one of the requirements within the framework of the regime resulting from the transposition of Community Directive no. 92/57/EEC into Portuguese domestic law.

This legislation, in addition to making prevention at the project level compulsory, imposes the existence of coordination and planning of safety and health at the design and construction stages and defines the line of responsibilities of each

actor, with the first responsibility of the construction owner.

The construction owner is a person, singular or collective, on whose behalf the work is performed. Its responsible for developing the safety coordination system, ensuring the preparation of the HSP, technical compilation and prior communication; To appoint the Safety Coordinators in the Project Phase (CPP) and Construction Phase (CCP) phase.

Within the scope of its obligations, the contractor is responsible for the communication of the HSP to the contractor and the other actors in the execution of the work and in the Safety Coordination system, as well as the referral to the Labor Inspection of the Previous Communication. Designers must ensure project safety and contractors and subcontractors must ensure the safety and health of their workers.

2.3 Prior communication (in accordance with article 15, Decree-Law no. 273/2003 29th October

The prior communication is an administrative act, which consists of the communication by the executing entity to the regulatory body for the Conditions of Work, ACT, informing it of the opening and implantation of its building site and consequently the beginning of work in the work to which it refers.

1. The owner of the work must first communicate the opening of the building site to the Working Conditions Authority (ACT) when it is foreseeable that the execution of the work involves one of the following situations:
   a. A total period of more than 30 days and, at any time, the simultaneous use of more than 20 workers.
   b. A total of more than 500 working days, corresponding to the sum of the working days provided by each worker.
2. The prior communication referred to in the previous number must be dated, signed and indicate:
   a. The complete address of the construction;
   b. The intended nature and use of the work;
   c. The developer, the author or authors of the project and the executing entity, as well as their respective domiciles or headquarters;
   d. The supervisor or supervisor of the work, the project safety coordinator and the safety coordinator on the construction site, as well as their respective homes;
   e. The technical director of the contract and the representative of the executing entity, if he is appointed to remain in the building site during the execution of the work, as well as

the respective domiciles, in the case of public works contract;

f. The person in charge of the technical direction of the work and its domicile, in the case of private work;

g. The dates scheduled for the start and end of the works at the building site;

h. An estimate of the maximum number of self-employed and self-employed workers who will be present at the same time at the building site or of the days of work of each worker, depending on whether the prior communication is based on a) or b) of paragraph 1;

i. Estimation of the number of companies and self-employed workers operating at the building site;

j. Identification of subcontractors already selected.

3. Prior notification shall be accompanied by:

a. Declaration of the author or authors of the project and the project safety coordinator, identifying the work;

b. Declarations of the executing entity, the on-site safety coordinator, the supervisor or supervisors, the technical director of the contract, the representative of the executing entity and the person in charge of the technical management of the work, identifying the construction site and the dates foreseen for the beginning and end of works.

4. The contractor shall notify ACT of any alteration of the communication elements referred to in points a) to i) within forty-eight hours and shall at the same time give notice thereof to the on-site safety coordinator and Entity.

5. The contractor shall communicate the updating of the elements referred to in paragraph 2 j) to the ACT on a monthly basis.

6. The executing entity shall affix copies of the prior communications and their updates to the site in a conspicuous place.

3 ACTIONS FOR RISK PREVENTION

3.1 *Action plan existing constraints*

The project must be adjusted to the conditions that will be found in the space reserved for its execution. This is an essential measure so that there are no surprises to the "posteriori" and to prevent accidents that may occur throughout the execution phase of the work.

In-depth knowledge of the deployment site helps identify and confirm constraints that interfere with the execution of the work. These conditions can create risk conditions that, detected in advance, can be prevented or even eliminated in some cases.

The survey of these constraints, allows to write down all the elements that will interfere with the enterprise. Situations can be analyzed on a case-by-case basis so that the best solution for each one can be adopted.

Particular attention should be paid to work which has direct interference with the public road, in particular with possible affected services. All risks involved with the existence of these infrastructures will oblige the contractor to carry out a cadastral survey and to record all the elements that may interfere with the execution of the work and the site itself.

3.2 *Action plan for building site constraints*

The organization of traffic routes and the deployment of safety signs on a site must be defined in the light of a number of factors not only linked to production, but also to the commercial sector, human resources, maintenance of equipment and, of course, safety and rescue in the event of a major accident.

There is a need to create a Signaling and Circulation Plan for the building site. This plan must be drawn up on the site of the building site.

Some situations are highlighted:

– Obligation to use personal protective equipment;
– Prohibition of entry of unauthorized persons;
– Location of construction site facilities (office, cafeteria, medical post);
– Prohibition of approach to dangerous areas;
– Warning of danger of falling objects (construction entrances must be protected by suitable cover);
– Indication of the places where fire fighting devices (extinguishers and hydrants) are located.

At the construction site, safety, hygiene and health at work measures should also be adopted, and the responsible person of the building site must take the necessary initiatives to:

– Keep the construction site in good order and in an adequate state of health;
– Guarantee the correct movement of the materials, in particular through equipment adapted to each case;
– Maintain and control facilities and equipment prior to commissioning and at regular intervals during operation;
– Define and organize storage areas for materials, in particular hazardous substances;
– Safely collect hazardous materials used;
– Store, dispose of or dispose of waste and debris;
– Determine and adapt, depending on the evolution of the building site, the actual time to be devoted to the different types of work or phases of work;

3.3 Signaling and circulation plano on the building site

The plan of signaling and circulation in the building site must have two types of signaling:

– Safety and health signage;
– Traffic signs.

The safety signage directly targets the individual (worker or visitor) and comprises a vast group of signs.

– Metallic plates that combine different symbols and colors with specific meaning (rescue and emergency signals, signs of firefighting equipment and signs of general information, warning signs, prohibition and obligation);
– Acoustic signals;
– Light signals;
– Signals.

In the case of metallic plates, the colors to be used shall be those listed in the following table in accordance with Ordinance 1456-A of 1995.

3.4 Collective protection plan

The collective protection plan has as its first objective to guarantee and ensure the implementation of collective protection equipment on site. This equipment must be adequate and depending on the risks that workers may be exposed to. Collective protection equipment is intended to protect all workers at the construction site and should always be superimposed on individual protection in accordance with the application of the general principles of prevention. The contractor or the executing entity must submit to the developer the collective protection plan, which must then be approved by the developer or if there is a safety coordinator for the construction phase.

For its approval, the contractor must conscientiously analyze the construction site's design, the design of the contract and the constructive methods and processes to be used in order to identify the foreseeable risks with the aim of preventing them. The collective protection plan will be implemented with priority over individual protections.

3.5 Individual protection plan

They should be used where existing risks can not be limited or avoided by technical means of collective protection, or by processes of work organization.

The conditions for their use (as well as their duration) are determined according to the severity of the risk, the frequency of exposure of the worker to the risk and the characteristics of his/her position.

There are several Individual Protection Equipment (IPE). However, we must distinguish IPE from mandatory use of IPE from temporary use. The former must be used by all workers at the building site (eg protective helmets and boots with a steel toe and insoles) and IPE for temporary use, will be used by workers according to the type of task to be performed. Workers have to be informed of the risks against which the Personal Protective Equipment seeks to protect them.

Training on their use should also be ensured and if there are any doubts on the part of the workers, organize safety exercises to elucidate them.

4 SAFETY COORDINATOR

Considering the obligations set forth in Decree-Law no. 273/2003, it can be said that the Safety Coordinator on Project phase (SCP) has the following functions:

– Ensure cooperation between designers;
– Ensure that designers fulfill their obligations;
– Require the preparation of a safety and health plan in the design phase;
– Advise the owner whenever necessary;
– Ensure the preparation of the Technical Compilation.

4.1 SCP mission

Its main task, which is defined in the Construction site Directive No 92/57/EEC, is the follow-up of the implementation of the General Principles of Prevention. Also defined in the Directive, there is a need to strengthen coordination between the different actors, both in the preparation phase of the project and in the construction phase.

This directive also provides for the obligation to apply the general principles of prevention during the drawing up of projects to the owner or his representative, in particular as regards architectural options and technical options. However, most countries of the European Union assign this obligation to project authors through transpositions into the national law of each country.

The safety coordinator in the design phase has the obligation to coordinate or ensure that project developers comply with and implement the general principles of prevention with regard to accidents and occupational safety.

4.2 General Prevention Principles (GPP)

The coordinator should stipulate a hierarchy of general principles of prevention. The following is a series of measures that show this hierarchy in a simple way. However, it should be noted that none of these measures will be more or less important than the others. They serve only as a line of measures to follow, starting and beginning by what seems most obvious.

The general principles of prevention (PGI) are laid down in the Framework Directive 89/391/EEC of 12th June, which are:

1. Avoid risks;
2. Assess risks that can not be avoided;
3. Combat risk at source;
4. Adapt work to man, in particular as regards the design of workplaces, as well as the choice of work equipment and working and production methods, with a view, in particular, to alleviating monotonous work and Work and reduce their effects on health;
5. Take into account the stage of development of the technique;
6. Replace what is dangerous so that it is free from danger or less dangerous;
7. Plan prevention with a coherent system that integrates technology, work organization, working conditions, social relationships and the influence of environmental factors on the job;
8. Give priority to collective prevention measures in relation to individual protection measures;
9. Give appropriate instructions to workers.

5 GOOD COORDINATION AND MANAGEMENT PRACTICES

Throughout these years of research into hygiene and safety legislation, there are several examples of a good safety coordination management applied on construction. The most recent example was the construction of the Marão tunnel, where despite the work dimension, the high risk and the number of workers involved, it ended without any kind of fatal accident. Another complex construction that was followed, was the new bridge over the Lima river, in the Jolda zone, including the respective accesses. It was applied several safety measures through adequate safety coordination management. As a result of this performance, the productivity resulting levels were excellent, minimizing the labor loss ratio and the economic costs resulting from accidents.

It is possible to mention several examples of good coordination practices such as the Baixo Sabor dams, Foz Tua, Caniçada, all of them with high and specific risks, where, through a good implementation of the safety plan, the number of accidents was minimized.

6 CONCLUSIONS

At the end of this research, some relevant conclusions are presented. The first conclusion is related to the competencies necessary to carry out the role of the Safety Coordination with special emphasis on SCP. In this sense, it is important to mention the lack of regulation on the profile of this function.

Another of the issues to mention are the factors that may influence the implementation of the Safety and Health Coordination in the design phase. Thus, although direct interveners at the project stage interact with each other, it is possible to distinguish some issues that help to understand the difficulties in implementing this prevention function.

First, highlight the poor participation of the Project Safety Coordination during the design phase. This may occur for a variety of reasons, namely it may be a consequence of your non-appointment by the Construction Owner.

For reasons of lack of information, or even disregard of his obligations, sometimes does not make such appointment. It can also be associated with the lack of knowledge of the role and importance of the development of this function during that phase. It is thus important that the developer is informed of his/her competencies.

Another of the reasons that seems to go hand in hand with the lack of prevention is that the designers do not correctly consider issues related to the constructability and maintenance of the building and refer them to the executing entity.

In the same sense, the risk assessment process is rarely applied. However this process is an added value for the project and, when applied, seems to result in benefits for all parties involved in the design and execution of works.

As such, it is of special interest and relevance that the safety coordinator be a member involved from the beginning of the entire design process. It can thus avoid risks or control them, which can also prevent monetary losses on the part of those involved in the execution. As can be seen in the course of this study, fewer risks mean fewer accidents, and as such fewer stops and delays in execution times, which in turn mean less spending in monetary terms.

REFERENCES

- Decree-Law no. 273/2003, 29/10
- Directive no. 89/391/CEE de 12/6
- Directive no. 89/656/CEE de 30/11
- Decree-Law no. 348/93 de 1/10
- Directive no. 89/391/CEE de 30/11
- Decree-Law no. 331/93 de 25/9

- Directive no. 92/58/CEE de 24/6/92
- Decree-Law no. 141/95 de 14/6
- Ordinance no. 988/93 de 6/10
- Reis, Cristina M.; Oliveira, Carlos; Marcio, Mieiro; Santos, Cristina. 2014. Different transposed legislation analysis in safety matter for construction sites: Three study cases. In Occupational Safety and Hygiene II, ed. Arezes, P., Baptista, J.S., Barroso, M.P., Carneiro, P., Cordeiro, P., Costa, N., Melo, R., Miguel, A.S., Perestrelo, G.P., 529–532. ISBN: 978-1-138-00144-2. London: CRC Press Taylor & Francis Group.
- Reis, C.M.; Oliveira, C.; Pinto, D.; Ferreira, J.; Mieiro, M.; Silva, P.. 2015. "Health and safety plans analysis", Occupational Safety and Hygiene III—Selected Extended and Revised Contributions from the International Symposium on Safety and Hygiene, III: 447–452.
- Reis, Cristina M.; Oliveira, Carlos; Marcio, Mieiro; Machado, Teresa. 2016. Management of the activity of the Safety Coordinator. In Occupational Safety and Hygiene IV, ed. Arezes, P., Baptista, J.S., Barroso, M.P., Carneiro, P., Cordeiro, P., Costa, N., Melo, R., Miguel, A.S., Perestrelo, G.P., 519–522. ISBN: 978-1-138-02942-2. London: CRC Press Taylor & Francis Group.

Occupational Safety and Hygiene V – Arezes et al. (Eds)
© 2017 Taylor & Francis Group, London, ISBN 978-1-138-05761-6

Control Banding—Qualitative risk assessment system for chemical handling tasks: A review

P.E. Laranjeira & M.A. Rebelo
CIICESI, ESTG, Politécnico do Porto, Felgueiras, Portugal

ABSTRACT: Several of chemical substances in commerce have no established occupational exposure limits. In the absence of established limits, employers and workers often lack the necessary guidance on the extent to which occupational exposures should be controlled. Control Banding (CB) is a qualitative strategy for assessing and managing hazards associated with chemical exposures in the workplace. The conceptual basis for CB is the grouping of chemical exposures according to similar physical and chemical properties, processes/handling, and exposure scenarios. The utility of qualitative risk management strategies such as CB has been recognised by a number of international organizations. The different CB models published reveals varying levels of complexity and applicability. This review attempts to capture the state-of-the-science of CB as reflected in research and practice.

1 INTRODUCTION

The majority of chemical substances in commerce have no established Occupational Exposure Limits (OELs). In the absence of established OELs, employers and workers often lack the necessary guidance on the extent to which occupational exposures should be controlled.

A strategy to control occupational exposures that may have value when there are no relevant OELs is known as Control Banding (CB). CB is a qualitative strategy for assessing and managing hazards associated with chemical exposures in the workplace. The original CB model was developed within the pharmaceutical industry; however, the modern movement involves models developed for non-experts to input hazard and exposure potential information for bulk chemical processes, receiving control advice as a result.

2 MODELS OF CONTROL BANDING

2.1 COSHH Essentials (United Kingdom)

The pioneering control banding tool was COSHH Essentials which was developed in the United Kingdom by the HSE to guide compliance with the COSHH regulation. At the time of its release, COSHH Essentials was the most developed strategy for chemical risk assessment. It selected the appropriate control band based on four elements:

- Type of task;
- Chemical hazard band;
- Volatility or potential of the chemical to become airborne;
- Quantity of chemical used.

The resulting risk management level (control band) had one of the following control approaches:

- Control Approach 1: General ventilation. Good standard of general ventilation and good working practices;
- Control Approach 2: Engineering control. Ranging from local exhaust ventilation to ventilated partial enclosure;
- Control Approach 3: Containment. Containment or enclosure, allowing for limited, small scale breaches of containments;
- Control Approach 4: Special. Seek expert advice.

Internet link:
http://www.coshh-essentials.org.uk/

2.2 Risk potential hierarchy (France)

The French-derived strategy used information from Material Safety Data Sheets (MSDSs) and chemical labels to assign a hazard class to the chemical based on three criteria:

- Hazard classification;
- Frequency of use;
- Quantity used.

The frequency and quantity of use are combined to give an exposure potential, which is then combined with the hazard classification, which results in a numerical score for the process. That score corresponded to one of three priority bands: elevated, middle or weak.

Although this strategy is a simple and clear triage method, internal validation studies indicated it overestimated in 19% of cases and underestimated in 1%. It was determined this strategy should be used in conjunction with another method.

Internet link:
http://www.inrs.fr

2.3 Chemical Management Guide (Germany)

The Chemical Management Guide tool was developed by Germany with the intent of assisting third world countries in risk assessment.

The early version of this control banding tool was called the GTZ Chemical Management Guide which had three steps:

- The first step identified portions of the process where inefficient storage, handling, use, and disposal were observed. These portions were "hot spots";
- The next step was a detailed chemical inventor;
- The final step used one of the following control strategies: basic risk assessment, description of control strategies, MSDSs, safety phrases for hazardous substance, or symbols for labeling hazardous materials.

This tool seemed to be too sophisticated for many small companies, but could be beneficial for larger enterprises with more MSDSs on site. This strategy expanded to an easy-to-use workplace control scheme for hazardous substances called EMKG as a means to provide practical guidance for workplace risk assessment in small and medium companies. After the information from the MSDSs was applied to the workplace, EMKG could derive control strategies to minimize inhalation and dermal exposures.

EMKG was similar to COSHH Essentials, but EMKG assessed dermal exposure. Like COSHH Essentials, EMKG provided non-regulatory guidance for controlling exposure, but was well supported by regulations.

An enhanced version of EMKG (EMKG 2.2), connects substances with legal OELs. The user began the risk assessment with the OEL which aligned to a hazard band and then two possible control schemes. For hazard groups above the OEL, the employer must perform workplace measurements, but for hazard groups below the OEL, the employer could waive workplace measurements.

Internet link:
http://www.baua.de/de/Themen-von-A-Z/Gefahrstoffe/EMKG/Software.html

2.4 Stoffenmanager (Netherlands)

The Stoffenmanager tool developed in The Netherlands was a compilation of the preceding control banding tools.

The tool calculated a risk score using a similar chemical classification as COSHH. Stoffenmanager "supports the requirements for maintaining inventory of hazardous substances, assessing and controlling risk in a risk inventory, obtaining a plan for control measures, making instruction sheets for the workplace, and helping to store chemicals according to guidelines".

This tool was generic, but the Dutch plan to modify the tool to be industry specific.

Internet link:
https://www.stoffenmanager.nl/

2.5 KjemiRisk (Norway)

Norwegian oil industry was integral in the development of KjemiRisk which used chemical and process information such as:

- Physical properties;
- Handling methods;
- Personal barriers;
- Duration of exposure;
- Frequency of task.

Chemicals were grouped into one of five health hazard categories, and processes fall into one of fifteen predefined common tasks. All the information listed above was used in the risk assessment which was divided into two phases: potential risk and final risk.

The risk was influenced by the reliability and effectiveness of the control measures in place, and the task's potential for illness to the lungs, internal organs, and skin. When used by production managers or safety and health generalists, KjemiRisk was a rough risk assessment tool, but when used by an industrial hygienist it becomes an expert tool.

Internet link:
https://chemirisk.ohs.no

2.6 Regetox and SOBANE (Belgium)

The Belgium tool Regetox was a two-stage assessment strategy which was developed in response to the European Chemical Agents Directive and combines COSHH Essentials and the Estimation and Assessment of Substances Exposure (EASE) model. This directive required companies to assess and manage chemical risks in the workplace. To reduce the number of chemicals requiring assessments, and thereby reduce cost, the first phase was used to rate chemicals based on the risk and the quantity used. Only those products with a rating of medium or high move forward to the second phase which used COSHH Essentials or EASE.

Regetox attempts to address the challenge of evaluating mixtures, and exposure where the process 11 generates the contaminants. A feasibility study of the tool which was done at two facilities revealed lacking or inadequate MSDSs. In one case, the model failed to identify a need for improvement. Overall, Regetox was believed to be helpful for companies not prepared to comply with the European Chemical Agents Directive. However, it did require training, planning, and involvement from employers.

A second tool developed out of Belgium is the Screening, Observation, Analysis, Expertise (SOBANE) method. This strategy consists of four phases, the first of which was the screening phase. The purpose of the screening stage was to identify the major problems on the worksite, and solve the simple problems immediately. The more complex problems identified in the screening phase are examined in greater detail, with the help of a nine page tool, during the observation phase. Those problems remaining after the previous two stages are analysed by a safety and health practitioner. Finally, in the Expertise phase, an expert designs more complex solutions, as needed.

Internet link:
http://www.regetox.be/

2.7 Semi-Quantitative Risk Assessment (Singapore)

The foundation of the Semi-Quantitative Risk Assessment (SQRA) tool is a three stage method for exposure evaluation: monitoring personal exposure, selecting exposure factors, and estimating exposure. SQRA was developed in Singapore; however, it has been compared to COSHH Essentials.

SQRA consisted of five control levels that paralleled COSHH Essentials control strategies:

– The first control level in SQRA was classified as negligible risk and level two was low risk with suggested periodic reassessment and personal air monitoring. These first two levels in SQRA were comparable to the general ventilation control strategy in COSHH Essentials;
– Level 3 (medium risk) parallels the engineering control strategy of COSSH Essentials. This third level suggested controls and reevaluation of the process every three years. Training and personal air monitoring may be necessary;
– The fourth level of SQRA was high risk and indicates a need engineering controls, personal air monitoring, training, personal protective equipment and reassessment after all controls

are in place. The guidance of this fourth level aligns with the containment control strategy of COSHH Essentials;
– The final level of SQRA was very high risk and directs the user to seek guidance from a specialist, which paralleled the "special" control strategy of COSHH Essentials.

2.8 Korean control toolkit (Korea)

Korea Occupational Safety and Health (KOSH) Agency developed a web-based tool for small and medium companies called the Korean Control Toolkit (KCT).

The tool was a semiquantitative assessment strategy which produced guidance on controlling hazards associated with 12 specific chemicals. Like SQRA, KCT was a modification of the COSHH Essential. Users selected one of the twelve chemicals from a menu online and then entered information related to the chemical use and workplace conditions (quantity used, duration and frequency of use, and physiochemical properties).

The KCT then used the COSHH Essential algorithm to give a risk grade, band class, and process-specific control guidance. The user then selected the specific control tool and the tool provided process-specific control guidance.

KOSH has plans to expand coverage to include a total of thirty chemicals in KCT. An additional element of the KCT project is modifying MSDSs to better communicate health hazards using less technical terminology.

3 GENERAL LIMITATIONS OF CONTROL BANDING TOOLS

The most significant challenge in the development and implementation of control banding tools is the accuracy of the decision log. Under prescription of controls can lead to hazards resulting in serious injuries, illnesses, or deaths. In contrast, over prescription of controls can result in unnecessary expense. The variables that must be considered in validation studies of control banding strategies are:

– Prediction of Exposure: The exposure data used to validate the decision log methodology needs to characterize the entire range of exposure for the work process under consideration. The measurements must account for interworker, intraworker, and interworkplace variation. Repeat measurements on each worker are required. For processes with engineering controls, the use of those controls needs to be discontinued, assuming acceptable risk, for testing purposes.

– Hazard Prediction: The toxicological endpoint of a chemical is typically used to define the specific risk level of that chemical. However, risk levels do not capture the relative severity of different chemicals in the same risk level. Where toxicological data, such as with OELs and TLVs, exist, they can be used to further assess the hazard.
– Control Recommendations: The recommendations from control banding strategies can be compared to those of experts, preferably for scenarios where the exposure and hazard predictions of experts have already been tested and determined to be accurate.
– Training: Control banding tools are intended for non-experts in the field of industrial hygiene. Therefore, an understanding of the tool and the methodology of the strategy is vital to proper implementation. The training on how to use the tool should be evaluated to ensure that it is appropriate for the target audience, offered frequently by a credible source, and effective. Because most control banding tools are accessible online, delivering and measuring effectiveness of training remains a challenge.
– Control Implementations: Follow up is needed to validate proper implementation of controls, and periodic evaluation of controls is required to maintain effectiveness.

4 CONCLUSION

Control Banding (CB) strategies offer simplified solutions for controlling worker exposures to constituents often encountered in the workplace.

CB models are, essentially, a simplification of scientific information into a format that is accessible to the multitudes, tending to provide safe-sided judgment. A possible interpretation of this is that control banding is inherently designed to secure workplace safety by compensating for its insufficient exposure information with safe-sided judgment criteria and by requiring experts' intervention in high-risk cases.

Control banding could be widely and effectively utilised, especially by employers in small enterprises, provided that the above models are pre-acknowledged. However, to this aim, it is essential to establish institutional mechanisms for facilitating employers' access to expert advice on this matter.

Furthermore, the need for a more complete analysis of CB model components and, most importantly, a more comprehensive prospective research process is still necessary to better understand its strengths and weaknesses.

REFERENCES

Balsat, A., J. de Graeve, and P. Mairiaux: A structured strategy for assessing chemical risks, suitable for small and medium-sized enterprises. Ann. Occup. Hyg. 47(7):549–56 (2003).

Farris, J.P., A.W. Ader, and R.H. Ku: History, implementation, and evolution of the pharmaceutical hazard categorization and control system. Chemistry Today 24:5–10 (2006).

Garrod, A., and R. Rajan-Sithamparananadarajah: Developing COSHH essentials; dermal exposure, personal protective equipment and first aid. Ann. Occup. Hyg. 47(7):577–588 (2003).

Jackson, H.: Control banding–Practical tools for controlling exposure to chemicals. Asian-Pacific Newsletter 9:62–63 (2002).

Jones, R.M., and Nicas M.: Margins of safety provided by COSHH Essentials and the ILO Chemical Control Toolkit. Ann. Occup. Hyg. 50(2):149–156 (2006).

Kromhout, H.: Design of measurement strategies for workplace exposures. Occup. Environ. Med. 59: 349–354 (2002).

Maidment, S.C.: Occupational hygiene considerations in the development of a structured approach to select chemical control strategies. Ann. Occup. Hyg. 42(6):391–400 (1998).

Russell, R.M., S.C. Maidment, I. Brooke, and M.D. Topping: An introduction to a UK scheme to help small firms control health risks from chemicals. Ann. Occup. Hyg. 42(6):367–376 (1998).

Swuste, P., and A. Hale: Databases on measures to prevent occupational exposure to toxic substances. Appl. Occup. Environ. Hyg. 9:57–61 (1994).

Occupational Safety and Hygiene V – Arezes et al. (Eds)
© *2017 Taylor & Francis Group, London, ISBN 978-1-138-05761-6*

Prevalence and incidence of upper-limb work-related musculoskeletal disorders at repetitive task workstations in a dairy factory

A. Raposo, J. Torres da Costa & R. Pinho
PROA/LABIOMEP, Faculty of Medicine, University of Porto, Porto, Portugal

J. Santos Baptista
PROA/LABIOMEP, Faculty of Engineering, University of Porto, Porto, Portugal

ABSTRACT: work-related musculoskeletal disorders (WMSDs) are a leading work-related health issue. Work-related disability and illness are common in a wide range of activities. The association of the musculoskeletal disorders and work possibly have a temporal cause-effect relation. This study investigated the prevalence and incidence of upper-limb work-related musculoskeletal disorders in a factory dairy products. The study was carried out between 2010 and 2014 among the 166 workers belonging to the cheese sector. Two evaluations were done, one in 2011 and another in 2014 and the study is limited to a total of 134 workers who were evaluated in both. It was found a total symptoms prevalence of upper limb musculoskeletal disorders of 38.1% and 31,1% in the first and second evaluation, respectively. Respect to the incidence of musculoskeletal disorders from the first to the second evaluation, the value found was 8.2% These values justifies the need to undertake preventive measures, reducing the injuries and consequently costs associated with these disorders.

1 INTRODUCTION

Work-related musculoskeletal disorders (WMSDs) are common among workers from different areas of activity. In addition, WMSDs are the leading cause of injury, illness work-related disability, leading to absenteeism and loss of productivity in developed countries (Baldwin 2004).

The association of the musculoskeletal disorders and work possibly has a temporal cause-effect relation, but prevention and the early diagnosis are not valorized. This is due to the absence of information able to identify economic sectors and working conditions that increase the risk factors, but also because there is no complete knowledge of the true "size" of the problem (Costa 2015). The risk factors are intensified due to some organizational characteristics such as long working hours and repetitive tasks (Nikpey et al. 2013).

The prevalence of WMSDs is reported to be usually high (Alrowayeh et al. 2010, Akodu et al. 2010). In fact, existing data provided by international organizations like European Agency for Safety and Health at Work (EASHW) point to a growing trend in the prevalence of Musculoskeletal Disorders in Europe, making them one of the most important causes of long-term absenteeism.

Respect to the incidence of WMSDs, many studies and systematic reviews have shown that in activities with high physical demand as abnormal force, sustained abnormal posture, vibration and also repetitive movements the total incidence is higher than the general working population (Dale et al. 2015).

Considering all this problems, the aim of this study is investigate the prevalence and incidence of upper-limb work related musculoskeletal disorders in a factory dairy products, LACTOGAL, *Produtos Alimentares* S.A, and contribute to the development of prevention and intervention strategies to reduce these disorders.

2 MATERIAL AND METHODS

This study was developed in a food factory specialized in dairy products and its derivatives, LACTOGAL, in Oliveira de Azeméis, Portugal, in a specific production sector: cheese sector, which start activity in 2008. The company has a total of 620 employees and the study was carried out between 2010 and 2014 among all the 166 cheese sector workers. Two evaluations were done, one in 2011/2012 and another in 2013/2014 and the study is limited to a total of 134 workers who were evaluated in both.

The clinical history, as well as the occupations, whether professional, or not, outside the context of the factory, were collected for all participants in the study. The data were collected, in both evaluations, using several methods:

a. Survey to assess symptoms of upper limb musculoskeletal disorders: Nordic Questionnaire validated for the Portuguese population (Serranheira et al. 2003), with assisted response;
b. Physical examination with observation of the upper limb by an orthopedist; the medical examination took place at the company's facilities during normal working time, and was performed within the validity period of the Nordic Questionnaire. This analysis determined the existence of disease, whenever presented positive semiological data for one or more of the following diagnoses: Tendinitis, Tenosynovitis, Compressive Syndrome, Algodistrophy, Neuropraxia, Stiffness, Contracture, Arthritis/arthrosis, Joint instability and synovial cyst.
c. Risk assessment in the workplace through the application of the RULA—Rapid Upper Limb Assessment method (McAtamney & Corlett 1993);
d. Risk assessment in the workplace through the application of OCRA—Occupational Repetitive Actions method (Occhipinti 1998);
e. Imaging of the upper limb by radiology and ultrasonography, to complement the information collected in the survey and the examination performed by the orthopedist;
f. Biomechanical analysis based on three-dimensional images of the production process.

For the purpose of the present study, data collected from application of the Nordic Questionnaire, and from the clinical examination (orthopedic examination), in both evaluations, were used.

2.1 Data analysis

Descriptive statistics were performed using SPSS version 22 to estimate the prevalence and incidence of WMSDs. Frequencies and cross-tabulations were used to compare musculoskeletal disorders prevalence between demographics and work history. Chi-square tests were also used to assess these relationships with statistical significance evaluated at $\alpha = 0.05$, so $p < 0.05$ was considered as statistically significant associated.

3 RESULTS

3.1 Demographic characteristics and working profile of participants

From the 134 participants of the study, 41 (30.6%) were male while 93 (69.4%) were female, aging from 20 to 61 years old. The mean value of age was 32.42±7.6 years old, and the mean values of height, weight and body mass index (BMI) were, 1.65±0.09 m, 67.63±13.5 kg and 25.99±7.63 kg/m² respectively (Table 1).

Respect to working profile, 80% of the participants worked less than 8 years in the factory, and 94% of workers are right-handed.

3.2 Prevalence and Incidence of musculoskeletal disorders

In the first participants evaluation the total symptoms prevalence of upper limb musculoskeletal disorders was 38.1%. By segment, the upper limb have a prevalence of 18.7% in shoulder, 11.2% in elbow and 23.9% in wrist/hand (Table 2).

The point prevalence in the second evaluation was 31.3% and respect to shoulder, elbow and wrist/hand was 17.1%, 8.2% and 18.7% respectively (Table 3).

It was found that of the workers who presented complaints of WMSDs in the first evaluation period, 11.5% presented the same complaints in the second evaluation period and 88.5% changed the complaints (either no longer had complaints, or had new complaints, or accumulated with other complaints). In particular, 19.4% have no longer complaints of WMSDs, and regarding segments

Table 1. Demographic characteristics of participants.

Variable	Mean ± standard deviation
Age	32.42 ± 7.6 years old
Height	1.65 ± 0.09 m
Weight	67.63 ± 13.5 kg
BMI*	25.99 ± 7.63 kg/m²

* Body Mass Index.

Table 2. Prevalence of upper limb musculoskeletal disorders, in first evaluation.

Total prevalence	38.1%
Shoulder prevalence	18.7%
Elbow prevalence	11.2%
Wrist/Hand prevalence	23.9%

Table 3. Prevalence of upper limb musculoskeletal disorders, in second evaluation.

Total prevalence	31.3%
Shoulder prevalence	17.1%
Elbow prevalence	8.2%
Wrist/Hand prevalence	18.7%

the complaints disappear in 9.7% respect to shoulder, 8.2% respect to elbow and 12.7% respect to wrist/hand.

Comparing the values of the WMSD perceived symptoms prevalence based on Nordic Questionnaire with the prevalence of WMSD clinical findings it was verified that the latter was inferior, with 37.5% and 26.1% of workers having a positive diagnosis, for the first and second evaluation, respectively (Table 4).

The relationship between the perceived symptoms prevalence of musculoskeletal disorders and the variables participant's gender, age group, Body Mass Index (BMI), years of service in the company years of service in the cheese sector was assessed by Chi-square test (Table 5). Prevalence of musculoskeletal disorders was not significantly associated with participant's gender, Body Mass Index (BMI) and age but of those with WMSDs complaints, 74.5% were female, 54.9% were in normal weight (considering BMI normal values between 18.5 and 24.9) and 41.2% belong to the age group of 19–30 years old. The only significant associations verified was with the years of service in the company (p = 0,007) and years of service in the cheese sector (p = 0.01). Of those with WMSDs complaints, 54.9% were workers of cheese sector since its opening.

Respect to the incidence, in this study was used the incidence proportion that is a measure of the risk of disease, and it is defined as a ratio between the number of new cases of upper limb WMSDs during a specified period and the size of population at the start of period. So, the incidence value achieved from the first to the second evaluation was 8.2%. Regarding the segments, the incidence was 8.2% in shoulder, 5.2% in elbow and 8.2% in wrist/hand (Table 6).

Table 4. Prevalence of clinical findings and perceived symptoms.

	Prevalence	
	Perceived symptoms	Clinical findings
First evaluation	38.1%	37.5%
Second evaluation	31.3%	26.1%

Table 5. Relationship between the perceived symptoms prevalence of WMSDs and the accessed variables, in first evaluation.

Variable	p value
Gender	0.315
Age group	0.196
Body Mass Index	0.672
Years of service in company	0.007*
Years of service in cheese sector	0.010*

* Significant value (p < 0.05).

Table 6. Incidence of upper limb musculoskeletal disorders, from first to the second evaluation.

Total incidence	8.2%
Shoulder incidence	8.2%
Elbow incidence	5.2%
Wrist/Hand incidence	8.2%

1 DISCUSSION

The purpose of this study was to determine the prevalence and incidence of upper limb work related musculoskeletal disorders among cheese sector factory workers. The prevalence of WMSD was 38.1% and 31.3% in the first and second evaluation, respectively. This results agree with other studies of the upper limb WMSDs in manufacture companies with repetitive task workstations like the study of Punnet et al. (2004) that reported a prevalence of 31%, and the study of Roquelaure et al. (2002) who reported prevalence also about 30%.

Regarding a systematic review of the upper-limb work related musculoskeletal disorders, the value found in this study is superior of annual prevalence ranges (from 0.14 to 14.9) (Costa et al. 2015).

However, comparing the prevalence values of this study with studies of the musculoskeletal disorders in another types of working population including also non-repetitive tasks the prevalence finding values are below of the values reported on those studies all above 50% (Parot-Schinkel et al. 2012, Rasotto et al. 2015, Mesquita et al. 2010).

Findings from this study reveal that complaints upper limb WMSDs were major female with 74.5%, and belonging to the age group of 19–30 years old (41.2%) which is a reflection of the workers in the study population. It should be noted that a high prevalence of musculoskeletal disorders in women is in accordance with other studies carried out in the general population (Hagberg et al. 1995, Silverstein et al. 2009).

Respect to the most frequent area of upper limb WMSDs complaints were observed that is in wrist/hand, probably as a result of being the most requested zone in the lifting movements of all workstations.

The small decrease in the prevalence rates from the first to the second evaluation, might be a result of new working strategies adopted by the workers like diminished muscular strain applied in tasks or a result of some ergonomic improvements implemented in the workstations between evaluations. One of the improvement implemented was the rotation of tasks in the workstations, so that each worker could spend less time in the same task. This can explain the disappearance of 19.4% complaints of upper limb WMSDs (by segments the complaints disappear in 9.7% respect to shoulder, 8.2% respect to elbow and 12.7% respect to wrist/hand)

Comparing the values of the WMSDs perceived symptoms prevalence based on Nordic Questionnaire with the prevalence of WMSDs clinical findings it was verified that the latter was slightly lower, with 37.5% and 26.1% of workers having a positive diagnosis (first and second evaluation respectively). These results also support other studies (Mehlum et al. 2009, Paarup et al. 2012, Roquelaure et al. 2006) in which perceived symptoms and clinical findings were in disagreement. In the present study there is only a slightly disagreement, probably due to the cases considered as non-work-related by the clinical experts or maybe the definition and criteria for WMSDs used by both participants and experts. Thus, this may indicate that neither perceived symptoms nor clinical examination can be used as a stand-alone diagnostic test.

Respect to the incidence of musculoskeletal disorders from the first to the second evaluation, the value of 8.2% was in accordance with other studies who reported 7% (Roquelaure et al. 2002) and ranges of annual incidence from 0.08 to 6.3% (Costa et al. 2015). However if compared with some studies related with other activities, the value found is lower than value achieved for example in health-care staff of Nursing Homes, which was about 15% (Pelissier et al. 2014).

Although high, perhaps it was expected a higher incidence value due to the very repetitive tasks. The improvements implemented by the company in the workstations after the first evaluation, may contribute to the not so high incidence value.

Despite the large turnover of workers in the company, it was observed that the only significant associations between upper limb WMDSs and the variables gender, age group, body mass index (BMI), years of service in the company and years of service in the cheese sector was with the years of service in the company (p = 0,007) and with the years of service in the cheese sector (p = 0.01, which may confirm the relation of WMDSs with work tasks.

5 CONCLUSIONS

This study shows that there is a significant prevalence and incidence of WMSD symptoms presented by workers in workstations with repetitive tasks.

Although the results obtained in this study are consistent with some similar studies, there are discrepancies with other studies. For these differences, are most likely to contribute the different work activities, the conditions under which they occur, the different methodologies, the subjectivity inherent at answers given in the surveys relative to the perception of the worker and associated to his psychological characteristics.

This results underlined, the importance of having WMSDs education and prevention programs, including the need of effective and routine analysis of health for the prevention of these disorders and for prevent a rapid evolution of the disease.

The conclusions of this study reaffirm the importance of establishing appropriate and consensual diagnosis criteria that allow a better comparison of the results.

REFERENCES

Alrowayeh H.N., Alshatti T.A., Aljadi S.H., Fares M., Alshamire M.M., Alwazan S.S. 2010. Prevalence, characteristics, and impacts of work-related musculoskeletal disorders: a survey among physical therapists in the State of Kuwait. BMC *Musculoskeletical Disorders.* Jun 11:116.

Akodu, A. et al. 2015. Work-related musculoskeletal disorders of the upper extremity reference to working posture of secretaries. *South African Journal of Occupational Therapy*, D 45(3): 16–22.

Baldwin M.L. 2004. Reducing the costs of work-related musculoskeletal disorders: targeting strategies to chronic disability cases. *J Electromyogr Kinesiol* 14: 33–41.

Costa J.T., Baptista J.S., Vaz M. 2015. Incidence and prevalence of upper-limb work related musculoskeletal disorders: A systematic review. *Work* 51(4): 635–44.

Dale, A. et al. 2015. Comparison of musculoskeletal disorder health claims between construction floor layers and a general working population. *Occupational And Environmental Medicine*, 72(1): 15–20.

Hagberg, M., Silverstein, B., Wells, R., Smith, M, Hendrick, H., Cararyon, P., & Perusse, M. 1995. Identification, measurement and evaluation of risk. In: Kuorinka, I., Forcier, L. (Eds.), *Work Related Musculoskeletal Disorders (WMSDs): A manual for prevention.* Taylor&Francis. London.

McAtamney, Lynn, & Corlett, E Nigel. (1993). RULA: a survey method for the investigation of work-related upper limb disorders. *Applied Ergonomics*, 24 (2): 91–99.

Mehlum, I. S., Veiersted, K. B., Waersted, M., Wergeland, E., & Kjuus, H. 2009. Self-reported versus expert-assessed work-relatedness of pain in the neck, shoulder, and arm. *Scand J Work Environ Health*, 35(3): 222–232.

Mesquita, Cristina Carvalho, Ribeiro, José Carlos, & Moreira, Pedro. 2010. Portuguese version of the standardized nordic musculoskeletal questionnaire: cross cultural and reliability. *J Public Health*, 18: 461–466.

Nikpey, A., Ghalenoei, M., Safary Variani, A., Gholi, Z., & Mosavi, M. 2013. Musculoskeletal disorders and posture analysis at workstations using evaluation techniques." *Journal Of Jahrom University Of Medical Sciences* 11, n° 3: 16–23.

Occhipinti, E. 1998. OCRA—a concise index for the assessment of exposure to repetitive movements of the upper limbs. *Ergonomics*, 41:9: 1290–1311.

Paarup, HM, Baelum, J., Manniche, C., Holm, JW., & Wedderkopp, N. 2012. Occurrence and co-existence of localized musculoskeletal symptoms and findings in work-attending orchestre musicians-an exploratory cross-sectional study. *BioMed Central*, 5.

Parot-Schinkel, E., Descatha, A., Ha, C., Petit, A., Leclerc, A., & Roquelaure, Y. 2012. Prevalence of multisite musculoskeletal symptoms: a French cross-sectional working population-based study. *BMC MSD*, 13(1): 122.

Pelissier C, Fontana L, Fort E, Agard JP, Couprie F, Delaygue B, Glerant V, Perrier C, Sellier B, Vohito M, Charbotel B. 2014. Occupational risk factors for upper-limb and neck musculoskeletal disorder among health-care staff in nursing homes for the elderly in France. *Ind Health*. 52(4): 334–46.

Punnet, L., Gold, J, Katz, JN, Gore, R, & Wegman, DH. 2004. Ergonomic stressors and upper extremity musculoskeletal disorders in automobile manufacturing: a one year follow up study. *Occup Env Med*, 61: 668–674.

Rasotto C, Bergamin M, Simonetti A, Maso S, Bartolucci GB, Ermolao A, Zaccaria M. 2015. Tailored exercise program reduces symptoms of upper limb work-related musculoskeletal disorders in a group of metalworkers: A randomized controlled trial. *Manual Therapy* 20(1): 56–62.

Roquelaure, Yves, Mariel, J, Fanello, S, Boissière, J-C, Chiron, H, Dano, C., Penneau-Fontbonne, D. 2002. Active epidemiological surveillance of musculoskeletal disorders in a shoe factory. *Occup Environ Med*, 59: 452–458.

Roquelaure, Y, Ha, C, Leclerc, A, Touranchet, A, Sauteron, M, Melchior, M,. Goldberg, M. 2006. Epidemiologic surveillance of upper-extremity musculoskeletal disorders in the working population. *Arthr Rh*, 55(5): 765–778.

Serranheira, Florentino, Pereira, Mário, Santos, Carlos Silva, & Cabrita, Manuela 2003. Auto-referência de sintomas de lesões músculo-esqueléticas ligadas ao trabalho (LMELT) numa grande empresa em Portugal. *Rev. Saúde Ocupacional*: 37–47.

Silverstein, B., Fan, J., Smith, C., Bao, S., Howard, N., Spielholz, P., Viikari-Juntura, Eira. 2009. Gender adjustment or stratification in discerning upper extremity MSD risk? *Sc J Work Env Health*, 35(2): 113–126.

Occupational Safety and Hygiene V – Arezes et al. (Eds)
© 2017 Taylor & Francis Group, London, ISBN 978-1-138-05761-6

Particle exposure levels in cement industries

D. Murta, H. Simões, J.P. Figueiredo & A. Ferreira
ESTeSC-Coimbra Health School, Instituto Politécnico de Coimbra, Coimbra, Portugal

ABSTRACT: The cement industry is based on long cement manufacturing processes in which are inevitably released particles and compounds such as free crystalline silica that is present in the raw materials of cement. Therefore, the objective set for this study was to analyse the exposure levels to inhalable particles and free crystalline silica in respirable fraction and the impact that this exposure can have on operators of these plants, evaluating the evolution in the 3 factories where samples were taken and evaluating the similarity and differences between them.

Developing this case study, it was collected the concentration data from inhalable particles and crystalline silica in respirable fraction at 31 workstations in 3 cement plants in order to obtain a representative sample of plants to be analyzed. The data correspond to the years 2008, 2009, 2011, 2013 and 2015.

According to the literature, repeated exposure to particles and exposures above the exposure limit values specified in the national standard NP 1796:2014 may lead to the development of occupational diseases such as pneumoconiosis, chronic obstructive pulmonary disease or silicosis (due to the presence of silica in raw materials used in these industries).

Analysing the concentrations of the agents tested, it can be concluded that the plants studied are very similar, only showing significant differences only relative to the crystalline free silica concentrations in respirable fraction at 2015.

1 INTRODUCTION

The statistics of recent years indicate that the number of workers with occupational diseases have been increasing, and this has attracted the attention of concerned researchers because of the related health issues the impact on the life quality of the workers.[1] This is a cross-reality to all areas, each with different occupational risks involving the tasks of each profession.

According to the European Agency for Safety and Health at Work (EU-OSHA), safety at work consists in the prevention of all that can compromise the health of any worker in the course of their profession either due to work accidents or occupational diseases.[2]

Several previous studies show that the cement industry is an industry with a high risk of occupational diseases, more specifically, respiratory tract disorders due to high exposure to particles suspended in air with high penetration capacity of the human respiratory system.[3–7]

The cement production is a long process that begins with the extraction of raw materials (various types of rock such as: limestone) in large dimensions that are sent to the crusher for an initial size reduction. The extracted raw materials are transported to the factory where begins the crude preparation process (material created by mixing and doing the homogenization of raw materials originally extracted). The crude is then routed to the ball mill where it is again homogenized and subjected to some chemical corrections and a drying operation until it becomes a flour following to storage where is stored up to the production of clinker. In the production of clinker, the same is pre-heated and sent to a furnace where they various chemical reactions happen to 1450°C. After this the clínquer is rapidly cooled in a grate cooler. Later the clinker is stored and transported to cement hoppers where it is mixed gradually with additions for cement production (plaster, fly ash and steel slag). Finally, the cement is created, bagged and palletized or loaded directly in bulk in different industrial and dispatched transport.[8–10]

During this entire cement production process, dust particles are released in large quantities, constituting the main danger for all workers in these factories anywhere in the production process.[3]

When the air is inspired also small particles in there are inspired. These due to the small dimension and mass, they are suspended in the air and can be dragged in the natural air currents.[11] The respirable particles, depending on their dimension, can deposit throughout the respiratory system and only the smallest ones can penetrate until the end of the respiratory system.[12]

In the industrial hygiene, is used the concept of fraction to divide particles according to size and ability to penetrate the human body. The national standard NP EN 481:2004[16] has established standardized values for these same fractions.[12]

From a risk prevention point of view for exposed workers, these fractions are relevant in respiratory system in which the particles can cause damage, setting the limit values specifically for each fraction. In the case of crystalline silica, damage to health is taken at the alveoli level so only the respirable fraction is seen to exposure assessment.[12] According to the literature on the subject, workers in cement industries are exposed to amounts of dust particles that can lead to disorders such as nasal discharge and congestion, allergic rhinitis, or even more serious disorders such as pneumoconiosis, chronic obstructive pulmonary disease, increase the probability of contracting tuberculosis or lung cancer. As a result of the presence of silica in the raw material constituting the cement is still likely that the inhalation of this compound can lead to diseases such as silicosis, which has shown a common occupational disease among workers in the construction and cement industries[6,12–14].

The dustiness associated with the cement manufacturing are still dermatological disorders such as prurience, allergies, burns or blisters and gastrointestinal disorders such as diarrhea or hemorrhoids.[6]

After a review of literature on the effects that operators in cement plants are subject because of the exposure to particles and free crystalline silica, it was found that although the rules that determine the exposure limits (VLE) for these contaminants were force since 1983, through the NP 1796 standard (currently in its 5th edition (NP 1796: 2014),[15] there are few studies on exposure to dust and free crystalline silica in this type of industries in Portugal This standard establishes the exposure limit for respirable particles at 10 mg/m^3 and for free crystalline silica at 0,025 mg/m^3. Thus, defined the objective of this study was to evaluate the levels of exposure to inhalable particles and free crystalline silica in respirable fraction and still establish a relationship between the exposure levels in the factories and the possibility of developing occupational diseases.

2 MATERIAL AND METHODS

The conducted study began in the academic year 2015/2016. The data colleting ended up in April.

In this study was determined the concentrations of particles of inhalable and respirable fractions, in particular, of free crystalline silica in the latter fraction and analysed using the Portuguese standard NP 1796:2014 to check if the values exceeded the exposure limit values.

The universe of study were the employees of three cement plants in the central region of Portugal, having a sample represented by 190 workers of 31 work posts. The people to whom the study is addressed are the responsible for the safety of the workers.

The different work posts were sampled for determination of total or inhalable particles (PT) and crystalline free silica in the respirable fraction (SC) requiring the analysis of respirable particles (PR). Sampling points represented in this study were designed to provide a significant representation of existing jobs in the factories since many of these jobs are occupied by mobile workers circling the respective factory and often changing activities performed during the day. The data collected from different samples correspond to different years (2008, 2009, 2011, 2013 and 2015). This data was collected by the same company over time.

The study carried out was a level II study (descriptive-correlative) with retrospective cohort nature since it was intended to relate the PT and SC exposure values between plants and compare the evolution of plants from 2008 to 2015.

To collect the data, it was asked to the responsible entities from the 3 factories to provide the reports of the determination of PT and SC tests with the values collected in the samples of all the years that had made such an assessment. All determinations were made by a company certified for the purpose by the national accreditation authority (IPAC- Portuguese Accreditation Institute). The collection of the data from 2015 was conducted in collaboration with the company that provides this service in 2 of the 3 factories.

According to the reports of all the previous sampling and of the sampling made in collaboration, these were made in accordance with the methods presented by the National Institute for Occupational Safety and Health (NIOSH), being performed as follows:

Sampling of total particles (inhalable fraction)— Sampling constant flow (1 L/min and 2 L/min) through the PVC membrane filter (Ø = 37 mm), pre-weighed, made in proximity of the airways according to the NIOSH method 0500: 1994.[17]

Sampling of free crystalline silica in the respirable fraction—Sampling constant flow (2.2 L/min) through silver membrane filter (Ø = 25 mm), pre-weighed, made in proximity of the airways, by equivalent procedure of NIOSH method 7500: 2003.[18]

Laboratory analysis of filters collected in the samples was performed by the laboratory of the same company that made the sampling, which is also certified for this process. The analysis was performed by a X-ray diffractometer method.

For data analysis, we created a database on statistical analysis program SPSS Statistical 22 (IBM-SPSS), in which was determined average exposure to PT, PR and SC and was determined the exposure values of the different plants and then compared for each year. It was also created comparative graphs with the evolution of each plant in the years 2013 and 2015. The statistical tests used for data analysis were the variance analysis nonparametric tests of Kruskal-Wallis and multiple nonparametric comparison test of Dunn.

Analysing the comparative data from an evolutionary perspective, it was excluded the values of the years 2008, 2009 and 2011since the data from these years were only related to plant 1 and there was no means of comparison against the other plants analysed. Thus they used the values of the years 2013 and 2015 to compare the situations of the plant in recent years, and to compare the evolution of measurements.

3 RESULTS

The sample used in the study consisted of 31 sampling points in 3 different cement plants in central Portugal. In these places were taken 626 samples of SC and 197 of PT. The results of the 5 years of data are as follows:

An initial analysis of 1 table values shows that there are significant differences in variation of SC values depending on the different cement plants in the study (p = 0.001).

Applying the multiple comparisons test of Dunn, the factories 1 and 3 showed no significant differences in variation of SC values in the study

(p = 0.280). However, the factory 2 shows significant differences in their values from the remaining factories since according to the same test has lower levels of significance (p. (Factory 1–2) = 0.002 and p. (factory 2–3) = 0.004).

Applying the Kruskal-Wallis test to the values of Table 2 to compare the mean values of each factory and determine whether they are similar, it was detected that the values are similar since the test of significance level was set at p. = 0.474.

Since the calculations of asymmetry and kurtosis with the respective standard errors (skewness/standard error; Kurtosis/standard deviation) are always higher than 1.96, it is shown that the data distribution was negative and leptokurtic asymmetric.

In order to make a more detailed comparative analysis, it was used the Kruskal-Wallis test to analyse the average concentrations of contaminants per year at each plant to realize on each year, only using the years 2013 and 2015 (because the years 2008, 2009 and 2011 had no data from factories 2 and 3). There were no significant differences between plants.

The following tables present the results of contaminant concentrations per year and per factory:

According to the results of Table 3 gave a significance value that allows to realize that in the year 2013 there were no significant differences between the SC concentration of 3 plants.

However, it can be observed that in 2015 exists a significant difference between SC concentrations of factories (p. = 0,001), using the multiple comparison test of Dunn it can be seen that the factory 3 differs in relation to the concentrations of the factories 1 and 2 having a value of p. = 0.000.

Table 1. Analysis of SC values by factory.

	Factory 1		Factory 2		Factory 2	
No of Samples	345		163		118	
Average Concentration	0,013		0,017		0,014	
Confidence range of 95%	Lower Limit	Upper Limit	Lower Limit	Upper Limit	Lower Limit	Upper Limit
	0,12	0,015	0,014	0,019	0,011	0,016

Table 2. Analysis of PT values by factory.

	Factory 1		Factory 2		Factory 2	
No of Samples	104		53		40	
Average Concentration	2,742		3,411		3,575	
Confidence range of 95%	Lower Limit	Upper Limit	Lower Limit	Upper Limit	Lower Limit	Upper Limit
	2,136	3,348	2,456	4,368	2,107	5,044

Table 3. Analysis of SC concentrations per year at each factory.

Year	Factory	Average	Standard deviation	N
2013	1	0,013	0,011	62
	2	0,020	0,018	81
	3	0,016	0,016	60
	Total	0,017	0,016	203
2015	1	0,020	0,021	79
	2	0,015	0,013	82
	3	0,012	0,013	58
	Total	0,016	0,017	219

Table 4. Analysis of PR concentrations per year at each factory.

Year	Factory	Average	Standard deviation	N
2013	1	1,029	0,760	62
	2	0,929	0,985	81
	3	0,877	0,862	60
	Total	0,944	0,883	203
2015	1	1,729	1,589	79
	2	1,278	1,021	82
	3	1,052	0,891	58
	Total	1,381	1,255	219

Table 5. Analysis of PT concentrations per year at each factory.

Year	Factory	Average	Standard deviation	N
2013	1	2,256	2,289	24
	2	3,006	2,653	28
	3	4,135	5,762	20
	Total	3,070	3,715	72
2015	1	3,452	2,665	23
	2	3,865	4,213	25
	3	3,017	3,071	20
	Total	3,476	3,389	68

By applying the Kruskal-Wallis test to the values of Table 5 it was found that in the year 2013 there were no significant differences between the PR concentrations of three plants being the most significant value (p. = 0.093) this year. However, in 2015 there is similarity of concentrations between the factory and the remaining 2 plants with higher significance 0.05 (p. factories 1–2 = 0.276 and p. factories 1–3 = 0.508) but presenting a significant difference between the concentrations of factory 3 relative concentration will plant 1 having a significance value less than 0.05 (p. = 0.011).

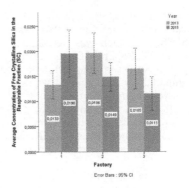

Graphic 1. Comparison of average values of concentration of SC in the years 2013 and 2015.

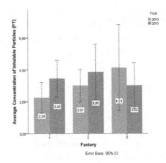

Graphic 2. Comparison of average values of concentration of PT in the years 2013 and 2015.

To represent the values of total particles, the Table 5 shows the average concentrations of each factory in the years of 2013 and 2015.

Applying the Kruskal-Wallis test to the values of the Table 5, it was obtained a significance value that demonstrate that in the years of 2013 and 2015 there were no significant differences between the concentrations of the PR values in the 3 factories, since the values of p. value were superior than 0.05 (p. (2013) = 0.493 and p. (2015) = 0.674) which demonstrate that the 3 factories are very similar.

In order to demonstrate the evolutionary perspective of the two main topics (SC and PT), it was created 2 graphs that demonstrate de values of the average concentrations of the years 2013 and 2015 in each factory.

4 DISCUSSION

This type of industry were often similar in regard to PT, PR and SC exposure since the cement manufacturing process is a process of inevitable release of dust, this being the main occupational hazard in this industry.[3]

When analysing with more detail it was observed that the factories, even obtaining very similar values, demonstrated different developments between the years of 2013 and 2015, having the factory 1 and 2 presenting a negative development as the average value obtained was higher in 2015 than the value of the year 2013 and the factory 3 presenting a positive development as the average value decreases from year 2013 to 2015.

By analysing the past few years developments graph, it can be observed the positive development of the factories 2 and 3, decreasing the SC concentration average while the factory 1 increased the exposure value (average concentration at which workers were exposed) and increased the number of sampling points that went beyond the VLE for exposure to SC, with 2 sampling points to exceed the VLE in 2013 and 5 sampling points exceeding in 2015 (keeping the 2 points in the initial risk and worsening exposure in 3 more).

The negative developments in the exposures to SC or PT can be explained by several hypotheses. One hypothesis is risk behaviour of some workers, such as the behaviour of "blow" clothing before each work break or the use of dry cleaning methods in their workplaces, keeping windows and doors open on loaders machines among other behaviours that can increase the presence of dust in the air. Another hypothesis to justify is the change of activities during the days of the measurements between the years (2013 and 2015) because as mentioned earlier, the workers are in constant motion, having to change the activities during the same working day.

A limitation of the study conducted, similar to the limitations of other published studies, is the inability to draw definitive conclusions about the maximum concentration in the air particles that are associated with diseases because of lack of necessary data.[19] The existence of a VLE gives us an indication to the acceptability of the risk. Comparing the value of daily exposure worker with the VLE it can assessed to which the worker is exposed.[12]

If an exposure higher than the VLE is considered a high level of exposure, each sampling point which exceeded the VLE of SC or of PT represent a risk of developing occupational diseases, existing several studies that prove the relationship of exposure to dust and the occurrence of such diseases.[3,7,12,20,21] According to the literature, repeated exposure to dust may lead to rhinitis and chronic bronchitis[19] and even with short periods of high exposure to silica can lead to the development of silicosis.[22]

It should be developed one or more control measures to reduce them to levels considered safe. Taking controlling as the elimination or reduction of dust in factories, to concentrations that cannot result in damage to the health of the workers, or the elimination or limitation of exposure to such agents.[14]

The exposure to dust control measures can be divided into two parts: measures concerning work environment and measures concerning individual workers,[14] prioritizing them in this respetive order according to the general principles of prevention. Some measures examples for the work environment are having the work areas with high levels of ventilation, installation of dust collection points or having regular cleaning of the factory using a vacuum system. As individual measures it can be used, for example, the reduction of the exposure, lowering the time of work to most exposed workers, use of personal protective equipment (protective respiratory mask, glasses and gloves) as well as training and information of the workers on the effects of silica and dust on their health, formation on which health behaviours to adopt and the importance of these behaviours to prevent diseases.[12,20]

5 CONCLUSIONS

This study allowed to relate the exposures of dust and free crystalline silica in the cement industry in the central region of Portugal, and concluded that the factories presented here had many similarities at the level of PT and PR concentrations, differing slightly on the SC concentration in the last year of assessment made. It was also observed that there were different developments of the concentrations of each agent in these factories over the last two years of measurements, concluding that the factory 3 was the only one that shown a positive development in relation to inhalable particles and free crystalline silica in the respirable fraction.

Noticing that there were many jobs that exceed the VLE of the PT, PR and SC in the last two years was also taking in consideration the fact that this over-exposure is related to various diseases in the lung and possibly dermatologic, and so it's necessary to intervene at the level of dust exposure prevention both a collective level and individual level.

In conclusion, it would be valuable in a national level if it were possible to establish the exact positions of work or manufacturing areas in this type of industry suffering from more exposure to dust, since it was not possible such findings in this study due to a data failure. However, the existence of value in the national standards[15] that limits the legal exposure to which a worker is exposed is already a measure that helps to protect these workers and perhaps encourage employers to be more careful with the situation and to increase reduction measures of these exposures.

REFERENCES

[1] Santos P, Martins P. Qualidade De Vida No Trabalho: Contribuições Dos Programas De Saúde E Segurança [Internet]. 2016;3(1):35–44. Available from: file:///C:/Users/Utilizador/Downloads/2989-8693-1-SM.pdf

[2] EU-OSHA. Safety [Internet]. 2013 [cited 2016 Jun 2]. Available from: https://oshwiki.eu/wiki/Safety#Defining_safety

[3] Neghab M, Choobineh A. Work-related Respiratory Symptoms and Ventilatory Disorders among Employees of a Cement Industry in Shiraz, Iran. J Occup Health. 2007;49:273–8.

[4] Nordby KC, Fell AKM, Not H, Eduard W, Skogstad M, Thomassen Y, et al. Exposure to thoracic dust, airway symptoms and lung function in cement production workers. Eur Respir J. 2011;38(6):1278–86.

[5] Ribeiro FSN, Oliveira S, Reis MM dos, Silva CRS da, Menezes MAC, Dias AEX de O, et al. Processo de trabalho e riscos para a saúde em uma indústria de cimento. Cad Saúde Pública. 2002;18(5):1243–50.

[6] Manjula R, Praveena R, Clevin R, Ghattargi C, Dorle A, Lalitha D. Effects of occupational dust exposure on the health status of portland cement factory workers. Int J Med Public Heal [Internet]. 2013

[7] Maury MB, Blumenschein RN. Produção de cimento : Impactos à saúde e ao meio ambiente. Sustentabilidade em Debate. 2012;75–96.

[8] CIMPOR. Processo de fabrico de cimento [Internet]. [cited 2016 Jun 1].

[9] Secil. Processo de fabrico de cimento. 2006 [cited 2016 May 20].

[10] Secil. Processo Produtivo [Internet]. 2003 [cited 2016 May 20]. p. 1–28. Available from: http://www.secil.pt/default.asp?pag = proc_fabrico.

[11] Ventura DD, Francisco E, Ctcv S. Ventilação Industrial no Sector da Cerâmica [Internet]. UC; 2011. Available from: https://estudogeral.sib.uc.pt/bitstream/10316/20115/1/Daniel_Ventura_2006116085_2011_rf.pdf.

[12] CTCV. Guia de Boas Práticas para a Redução da Exposição à Sílica Cristalina Respirável na Indústria Cerâmica [Internet]. apicer. Apicer; 2012 [cited 2015 Nov 18].

[13] NEPSI. Acordo relativo à protecção da saúde dos trabalhadores através da utilização e manuseamento de sílica cristalina e produtos contendo sílica cristalina [Internet]. 2006 [cited 2016 Nov 18]. Available from: http://www.nepsi.eu/agreement-good-practice-guide/agreement.aspx.

[14] Ferreira V, Lu O. Avaliação e controle da exposição ocupacional à poeira na indústria da construção. Ciências & Saúde Coletiva. 2003;8:801–7.

[15] Instituto Português da Qualidade. NP 1796:2014—Valores Limite de Exposição a agentes químicos. Instituto Português da Qualidade; 1983.

[16] Instituto Português da Qualidade. NP EN 481 (2004) -Atmosferas dos locais de trabalho: definição do tamanho das fracções para medição das partículas em suspensão no ar. Instituto Português da Qualidade; 2004.

[17] National Institute for Occupational Safety and Health. NIOSH 0500:1994. Centers of Disease Control and Prevention; 1994.

[18] The National Institute for Occupational Safety and Health. NIOSH 7500:2003. Centers of Disease Control and Prevention; 2003.

[19] Uk Health and Safety Executive. EH75/7—Portland Cement Dust: Hazard assessment document. 2005;

[20] NEPSI. Guia de Melhores Práticas para a protecção da saúde dos trabalhadores através do correcto manuseamento e utilização da sílica cristalina e produtos relacionados [Internet]. 2006 [cited 2015 Nov 18]. Available from: www.nepsi.eu.

[21] Rafeemanesh E, Alizadeh A, Afshari Saleh L, Zakeri H. A study on respiratory problems and pulmonary function indexes among cement industry workers in Mashhad, Iran. Med Pr. 2015;66(4):471–7.

[22] Nogueira DP, Certain D, Brólio R, Garrafa NM, Shibata H. Ocorrência de silicose entre trabalhadores da indústria cerâmica da cidade de Jundiai, SP. Rev saúde Pública SPaulo. 1981;15:263–71.

Occupational Safety and Hygiene V – Arezes et al. (Eds)
© 2017 Taylor & Francis Group, London, ISBN 978-1-138-05761-6

Safety control evaluation of food storage in mass caterer

A.L. Baltazar
ESTeSC-Coimbra Health School, Dietetics and Nutrition, Instituto Politécnico de Coimbra, Coimbra, Portugal

J.P. Figueiredo
ESTeSC-Coimbra Health School, Exact Sciences, Instituto Politécnico de Coimbra, Coimbra, Portugal

A. Ferreira
ESTeSC-Coimbra Health School, Environmental Health, Instituto Politécnico de Coimbra, Coimbra, Portugal

ABSTRACT: The catering business provides food to people and covers all sectors of society. Proper food storage will help maintain the quality and safety of products and is crucial for the safety of the meals produced in these establishments. This study was conducted in mass caterer companies, covering a period of four years. The objective of this work is to evaluate the evolution of food safety conditions in storage in small mass caterer companies. Storage conditions were evaluated according to hygiene, food handling practices and HACCP documentation, using an assessment tool created for this work. After the evaluations, an action plan was put into work to improve the food safety conditions of the establishments. Following these results, we can conclude that monitoring was crucial to improve food storage, not only in terms of increasing the catering team's knowledge but also in terms of cementing good practices in food handling in this phase.

1 INTRODUCTION

Mass caterer establishments are restaurants, canteens, schools, hospitals and catering enterprises in which, in the course of a business, food is prepared to be ready for consumption by the final consumer (Regulation (EU) n°1169/2011)

The mass caterer sector has grown in recent decades and several factors have been identified such as increased number of individuals living in urban areas, distance from home/work, increased percentage of women in the workplace, increased financial power and dietary concerns (Medeiros *et al.*, 2012) (Baptista & Linhares, 2005).

The economic activity of catering in Portugal embraces (I) the preparation and sale activities of food products for consumption, usually on site or in other establishments that do not produce those products; (II) the activities of preparation of meals or dishes delivered and/or served at the place determined by the customer for a specific event; (III) the supply activities and, where appropriate, the preparation of meals and drinks to well-defined groups of people, like public collectivities (hospitals, schools, elderly places, etc.). It includes canteens and military spaces; it also includes provision of meals based on a contract for a given time period; (IV) the sales activities of drinks and small meals for consumption on the premises with or without

spectacle. The catering sector is mostly located in the capital, Lisbon, roughly 36% (Bank of Portugal, 2011).

The European food laws introduced a new concept in the food market, "from farm to fork", by designing a cross accountability to all stakeholders in the food chain (Veiros *et al.*, 2009). The catering sector assigns a very important role to entrepreneurs, considering them primarily responsible for food safety (APHORT, 2008).

It is estimated that millions of people have had a foodborne disease at least once. Health agencies associate these numbers with the consumption of meals in restaurants or mass caterer (Medeiros *et al.*, 2012).

The European Union (EU) has created legal tools to ensure food hygiene in the sector, as well as official entities in charge of controlling and inspecting establishments to ensure public health (Veiros *et al.*, 2007).

The need and obligation to produce safer food goes to the inevitable implementation of effective food safety systems along the entire chain of production, shipping and distribution, namely a system based on the principles of HACCP methodology (Hazard Analysis and Critical Control Points) (Sun & Ockerman, 2005).

This preventive system (Regulation (EC) n.°852/2004) requires a strategic approach of the stages of production/distribution, based on the identification of inherent hazards such as biological, chemical and physical hazards.

The HACCP system is a preventive system resulting from the application of scientific and technical principles. It is an essential tool for identification and analysis of Critical Points (CP) at different stages of the process, while allowing the establishment of the necessary means to control these points and apply preventive monitoring. The HACCP system stands for proactivity instead of reactivity (corrective approach) (Forsythe & Hayes, 1998).

Food safety control can be applied to any stage or type of food systems for corrective actions in case problems arise. Food storage requires optimal conditions to achieve the desired material characteristics and extended shelf life. Understanding food properties and their phase and state transitions and impact on food processing and storage is fundamental (Roos & Drusch, 2016).

It is very rare for food to improve in quality with storage (maybe with the exception of wine and cheeses).

Food texture, flavor, color, and nutritional composition can all change, and their rates of change are often regulated by the conditions they are stored in. If products are removed from cool or cold storage, they often deteriorate more rapidly and so their best-before dates, or use-by dates, are shorter and designed to allow consumption prior to spoilage (Tanner, 2015).

This study aims to evaluate the importance of monitoring in mass caterer establishments, in this case restaurants, in terms of food storage practices (cold and room temperature).

2 MATERIAL AND METHODS

The aim of this study was to evaluate food storage in restaurants according to food safety procedures. The method applied to this assessment is demonstrated in Table 1.

The time schedule of the stages was: a month between the first and the second stage and between second stage and the remaining three months. After the fourth stage, it would start again in the third stage, for four years.

The companies selected for this study were small and medium enterprises, according to their economic volume invoicing (up to 10 million euros), because they are the ones with lack of economic resources to invest in food safety and represent the largest portion of mass caterers in Portugal.

The public catering companies were selected according to the following criteria:

- Portuguese economic activity code in Portugal for mass caterer (financial Portuguese code related with company activities)

Table 1. Study action plan.

Stage	Action plan
1	Diagnosis inspection (checklist) and HACCP data collection
2	Diagnosis report
	Training Action "Hygiene & Food Safety – Public Catering"
3	Inspection/Audit (using checklist)
4	Inspection report
	Training Action "Treatment of non-compliance—inspection report"
	Improvement Plan Application

- Turnover (up to 10 million euros)
- Geographic area (Leiria district)
- Consulting companies in food safety interested in participating in the study

Out of 40 mass caterer companies were eligible, and 22 remained in the study for the four-year period.

The data collection instrument for the inspection was a checklist, created by food safety technicians after a pre-test, organized in three modules divided into specific topics (items).

In this article, storage stage will be analyzed from the data collected. Storage was evaluated separately, according to temperature specifications. The at room temperature inspection focused on: separation of food and non-food; stock type (First in, first out (FIFO)/First to expire, first out (FEFO)); labeling/products identification; products without minimal conditions of consumption; materials and facilities and equipment hygiene. In cold storage, inspection covered: defrosting conditions; stock type (FIFO/FEFO); freezing procedure; labeling/products identification; packaging; products and materials without minimal conditions; facilities and equipment hygiene; temperatures and refreezing conditions.

After every evaluation, the catering team was trained by the experts in food safety that participated in the study. The training plan's aim was to make the team aware of the non-conformities detected and the corrective measures.

The results of the study were subject to statistical analysis, namely, Fisher's exact test and Wilcoxon's T test. For the statistical Inference, we took into account a 95% confidence level for a random error of up to 5%. The specialized software for treatment of analytical data was the IBM SPSS Statistics version 24.

The results were analyzed by score percentage (0–100) represented by the storage items in the checklist towards all inspection criteria used in all the meals production stages (reception, storage,

Table 2. Time evaluation on storage improvement measures.

Food Storage n = 22		M	SD	Dif. M ± DP	Z; p-value
All Food Storage	Diagnostic	78.79	19.37	−13.64 ± 19.68	−2.714; 0.004
	Last evaluation	92.42	17.61		
Storage Room Temperature	Diagnostic	63.64	16.78	−18.94 ± 25.35	−2.817; 0.005
	Last evaluation	82.58	21.50		
Cold Storage	Diagnostic	62.63	9.42	−14.14 ± 24.77	−2.428; 0.015
	Last evaluation	76.77	24.22		

Wilcoxon Test; M: Mean; SD: Standard Deviation; Dif. M ± DP: Difference Mean ± Standard Deviation.

preparation and distribution). Since it was a continuous study, the data analyzed was based only in comparing the average scores provided between the diagnostic evaluation (first) and the last evaluation (fourth), to have a summary knowledge of the change and improvement of the food safety practices during the study.

3 RESULTS AND DISCUSSION

The results (Table 2) between the diagnosis and the last evaluation (fourth) in food storage showed a significant mean improvement in scores of adequate condition from diagnosis to the final evaluation (z = −2.714, p = 0.004). That means that there was an improvement at all levels in food safety practices.

With regard to room temperature storage, the scores obtained between diagnosis and the fourth evaluation showed a significant increase in compliance (z = −2.817, p – 0.005) for the 22 companies monitored in the 4 year period, 72.7% of the companies improved and 3 regressed in their classification.

In cold storage, the scores obtained showed a significant increase in compliance (z = −2.428; p = 0.015), as 50% of the companies improved and 4 regressed in their classification (18.2%).

The worsening of some catering companies in storage practices was due to lack of professionals, inadequate premises conditions, high turnover of catering teams, and insufficient equipment or in poor condition.

However, the overall enhancement of mass caterer companies was very satisfactory.

4 CONCLUSIONS

The primary causes of food poisoning in catering are cross contamination between raw and cooked foods; insufficient heating; keeping food at room temperature for extended periods of time; contamination by infected food handlers and contamination by inadequately cleaned equipment (Ko, 2015).

The proper storage of food minimizes potential contamination and growth of microorganisms.

The methods used in this study allowed us the possibility of assessing food handlers in food storage practices and of investigating how the action plan was useful in changing behaviors towards food safety.

The results of the diagnostic evaluation/inspection highlighted significant gaps in knowledge, attitudes and practices of safe food handling in storage. The areas of high concern were the lack of adequate storage (e.g.: food products were in the danger zones), multiple freeze thaw cycles, and thawing of frozen food at room temperature. These gaps were resolved with risk-based training of food handlers in the facilities, using appropriate training aids to encourage understanding and assurance in the application of food safety principles in their day-to-day operations. The training interventions covered appropriate storage temperatures, food identification and thawing of frozen foods, because these were the main nonconformities found.

In this work, the action plan was linked to a change of habits to obtain a sustained improvement in food safety. It is important to refer that, after continuous and systematic training, with frequent inspection processes, food storage in mass caterers has improved.

REFERENCES

APHORT (2008). Code of Good Practices in Food Safety. Portuguese Association of Hotels, Restaurants and Tourism, Portugal (published in Portuguese only).

Bank of Portugal (2011). Sector Analysis of accommodation, catering and alike. Studies of central balances. Lisbon, Portugal (published in Portuguese only).

Baptista, P. & Linhares, M. (2005). Hygiene and Food Safety in Catering: Volume I—Initiation. Forvisão. Guimarães, Portugal (published in Portuguese only)

Boundless, 2015. Food Spoilage by microbes. Boundless Microbiology. 21 July 2015.

Forsythe SJ & Hayes PR (1998) Food Poisoning and other foodborne hazards. In: Forsythe, SJ & Hayes, PR (eds). *Food hygiene, microbiology and HACCP*. Third edition. Gaithersburg, MD: Aspen Publication pp. 21–85.

Ko, W. (2015). Food suppliers' perceptions and practical implementation of food safety regulations in Taiwan. Journal of Food and Drug Analysis, Volume 23, Issue 4, Pages 778–787.

Medeiros, CO.; Cavalli, SB.; Proença, RPC. (2012). Human resources administration processes in commercial restaurants and Food Safety: the action of administrators. International Journal of Hospitality Management 31, 661–674.

Regulation (EC) No. 852/2004. The European Parliament and the Council of 29th April 2004.

Regulation (EC) No. 1169/2011. The European Parliament and the Council of 25th October 2011.

Roos H & Drusch S (2016) Chapter 9—Food processing and storage. In: Roos H & Drusch S (eds). *Phase Transitions in Foods*. Second Edition. Academic Press pp 315–355.

Tanner, J. (2015) Food Quality, Storage, and Transport. Reference module in food science. Elsevier.

Veiros, MB.; Proença, RPC.; Santos, MCT.; Rocha, A.; Kent-Smith, L. (2007). Proposta de check-list hígio-sanitária para unidades de restauração. Alimentação Humana, 13(3), 51–61. (published in Portuguese only).

Veiros, MB.; Proença, RPC.; Santos, MCT.; Kent-Smith, L.; Rocha, A. (2009). Food safety practices in a Portuguese canteen. Food Control 20, 936–941.

Sun, Y. & Ockerman, HW. (2005). A review of the needs and current applications of hazards analysis and critical control point (HACCP) system. Food Control 16, issue 4, 325–332.

Occupational Safety and Hygiene V – Arezes et al. (Eds)
© 2017 Taylor & Francis Group, London, ISBN 978-1-138-05761-6

Risk perception and hearing protector use in metallurgical industries

I.C. Wictor, A.A. de P. Xavier & A.O. Michaloski
Federal Technological University of Paraná, Ponta Grossa, Brazil

ABSTRACT: Although hearing protectors are defined as a temporary solution, they are often widely employed as the only measure against noise exposure. However, it is also known that unless workers wear the hearing protector continuously, their effectiveness will be very low. In this regard, there are some surveys that show that workers do not always wear their protectors properly and consistently while exposed to noise. The purpose of this article is to present the results of an investigation about noise perception, relating with noise exposure in working environment and the hearing protectors' use. Was consider that noise perceptions effects can support minimizing risk and improve industries safety politics. The study sample was carry out in 5 metallurgical industries and 243 workers from Parana—Brazil. The survey data was collected and analyzed by Statistical Package for the Social Sciences (SPSS). The questionnaire results showed that workers are exposed at high noise levels, in increasing risk of developing Noise-Induced Hearing loss (NIHL). About the hearing protection use, it can be evaluated that the use is more effective in companies with a higher level of noise exposure. The perception of risk plays a fundamental role, which predicts the use of hearing protectors; therefore, the perception of the work environment, regarding the lower or higher risk, can be directly linked to the use of hearing protectors. The companies with more rigorous safety procedures also indicate a greater report of effective use of hearing protectors.

1 INTRODUCTION

Workers exposed to high noise levels are at risk of noise-induced hearing loss—NIHL (Ahmed et al., 2001). The occupational noise level is a permanent concern in all regions, being the major cause for the incapacitating deafness in the world (Reddy et al., 2012), is a public health problem with many social and economic consequences (Lie et al. 2015), and despite the imposed regulations and standards, occupational hearing loss persists.

For many industries, the actions of hearing conservation are summarized in the application of hearing protectors. According Stephenson et al. (2011), the implementation of an effective hearing conservation program should be established after determining the factors that substantially influence the real use of hearing protection by workers. The program and the measures that will be taken should be planned together, with all the company's staff, so that the actions are punctual, and really are effective in protecting workers.

Sensorineural hearing loss is a result of exposure to high noise levels that is linked not only to exposure time, but also to noise characteristic (frequency, intensity), nature (continuous or floating noise) characteristics that may affect the degree of hearing deficiency. In addition, the level of hearing loss tends to increase with age, however, the trend line is higher in workers exposed to high occupational noise, and becomes even more significant if there is no continuous use of hearing protectors (Araujo, 2002; Hunashal & Patil 2012; Whittaker et al. 2014).

In this way, companies must invest in efficient hearing conservation programs in order to promote a safer and more comfortable working environment. Educational training programs and training are also important (Stephenson & Stephenson 2011, Bockstael, et al. 2013) for the awareness of workers. The organizational climate is essential to promote the effective use of auditory protectors (Lusk, et al. 1998, Arezes & Miguel 2005), as well as rigid and well-structured policies and practices on safety and auditory conservation among all employees.

2 HEARING PROTECTORS USE

A personal hearing protection device (or hearing protector) is an acoustic barrier to protect the ear and reduce the level of airborne sound that reaches the eardrum (Miranda 2003). The main purpose of hearing protectors is to reduce, to an acceptable level, excessive levels of noise. These devices are easily implemented, as they are low cost methods that minimize hearing loss by continuous exposure to high intensity noises.

When workers are exposed to excessive noise levels, administrative or engineering control are recommended to reduce its. Therefore, when these techniques are not available immediately, the equipment can be used, but this type of solution should not be considered definitive, due to the intrinsic characteristics of the protectors, such as poor comfort, difficulty in verbal communication (Gerges, 1992).

Some authors point out barriers about comfort to the use of the protectors (Melamed et al., 1996; Davis et al., 2009; Byrne, 2011). Some authors indicate the relationship between hearing protection use and risk perception in the work environment (Rabinowitz et al., 2007) and indicate that the most effective use of hearing protection in the workplace is more related in places where noise exposure is higher. According them, the perception of risk is higher in these environments.

Arezes and Miguel (2006) suggests that supervision helps to improve the use of hearing protection, but does not lead to increased perception of risk. According to the authors, perception of risk is also quite high in companies with more rigid safety policies, although somewhat lower than in industries with higher levels of exposure.

3 MATERIALS AND METHODS

3.1 Data collection procedures

To develop an analysis about the risk perception and the hearing protection use, 243 workers from four medium-sized and one large metallurgical enterprises in Parana State—Brazil were interviewed. Company size was measured by the number of employees, one of the most common measures. The European Commission defines micro enterprises as those with 0–9 employees, small businesses with 0–49 employees, medium-sized companies with up 250 employees and large enterprises with 250 or more persons employed (Laforet 2013).

One of the main points for choosing the sample of employees to participate was to determine the areas with a higher incidence of noise. The selection of the sample considered the noise levels reported by the company, since it was intended that the sample was exclusively composed of workers exposed to noise above 85 dB (A), (above the levels allowed for daily exposure without protection established by Brazilian Legislation).

The data collection was performed in four steps: (1) selection of the workers (which were working in environment with level up than 85 dB (A). (2) Interview (first, general data were collected: age, sector of working, service time in the same sector and educational level. (3) Collection of

noise levels (informed by company) (4) Interview (main questionnaire/employees).

To select the sample study, some inclusion and exclusion criteria were applied: Inclusion—agree to participate voluntary in the survey; be working in the same sector at least 1 year. Exclusion—working in environment with levels below 85 dB (A), according Brazilian regulatory limits, which limit value is 85 dB (A) for eight hours, and the maximum dose of 100% were used. In sequence, data collected were analyzed.

The data were analyzed through descriptive statistics. The Levene Test was used as it is recommended to evaluate if the variances of a single metric variable are equal between groups. Variance Analysis—ANOVA is used to verify if there is a systematic difference between the means of results (Vieira & Ribas, 2011, p. 69)

To test the significance of the ANOVA results, the Tukey test was opted, since it is considered "one of the most robust deviations from normality and homogeneity of variances for large samples". (Maroco, 2014, p. 133).

This study was approved by the National Commission of Ethics in Research (CONEP) under the number 53661315.4.0000.5547. All selected workers in this research signed a voluntary commitment agreement.

3.2 Interviewees and questionnaire

At first, for identification of the companies and future comparisons, other general questions were asked, such as activity sector, number of employees and noise levels.

One of the researchers conducted the interviews personally in the months from March to June 2016. As agreed, the enterprises names will not be revealed, so letters here identify them.

In the employees' questionnaire asked about individual perception, the individual perception of risk, the individual perception of aspects related to hearing protection, safety culture and organizational factors and risk behavior. The questions about risk perception were made using a Likert scale questionnaire, this method was developed by Arezes (2002), and the ranking perception answer used ranges from 1 = no risk to 5 = too much risk, and, 1 = totally disagree to 5 = totally agree.

Individual risk perception is an important antecedent for risk behavior (Diaz & Resnick, 2000; Glendon & Stanton, 1995) the way that workers perceive the risks they are exposed can be an important factor for a better understanding of risk management (Arezes & Miguel, 2005).

The Risk Behavior Assessment verifies workers' risk behaviors, such as actions that violate safety rules and procedures, as well as items related to non-use of hearing protectors. The risk behaviors were analyzed between companies, in order to verify differences between them and the level of noise exposure.

The questionnaire applied to workers provides four questions that assess the individual perception of risk treated as follows: sources of risk; knowledge about noise; perception of self-efficacy; and perception of protection.

4 RESULTS

4.1 Interview questions and answers

The general data shows the population profile that was found in the sample study.

Table 1 shows the distribution of the 5 companies surveyed that were identified through letters (A) through (E), followed by the total number of employees, the number of the sample that was surveyed, followed by information that refers to the average age of employees of each company, and the working time in the same sector (in years) in each company.

It is important to note that the data related to the working time are related to the time the employee works in the same sector - and not the total working time in the company. This criterion was adopted to better evaluate the time (years) of exposure to noise that, the workers are exposed to.

It is possible to observe that the companies participating in the study have different sizes. Concerning the age of the workers and working time in the respective companies, it possible to verified that, companies (A) and (B) have a longer period of employment, between 8–9 years of work and the average age is also higher.

In the same table are reported the types of hearing protectors used by workers in each company. The use of the ear protector was reported by 100% of the workers surveyed in the companies.

Note that, plug type is the most used, and in company D there is only one option of hearing protection equipment.

However, it is analyzed that a significant number of workers are exposed to occupational noise over 85 dB (A) for more than 6 years.

In a survey carried out in metallurgical companies in Brazil, Guerra (2005) found that the prevalence of cases suggestive of Noise Induced Hearing Loss—NIHL rises from six years of activity in the company, compared to workers with shorter working hours.

4.2 Occupational noise exposure

According NR-9 (MTE, 2011) establishes the obligation to elaborate and implement the Environmental Risk Prevention Program—which according to the Standard considers the physical agents (noise, vibration, pressure, temperature, and radiation), chemical and biological, depending on their nature, concentration or intensity, can cause harm to the worker's health.

Due to the obligatoriness, annually the companies make the measurements in the levels of noise in the workplace. Therefore, the noise level was obtained through information from the Program for the Prevention of Risks and Accidents—PPRA of the company surveyed.

The results of company noise levels are presented on Table 2.

We can observe that, in company (A) and (D), noise levels are very high and there are many workers exposed in this environment. In the other hand, company (E) has lower noise levels comparing with the other.

Referring to the use of hearing protectors, a brief analysis is important. The company (E) has the lowest noise level among those surveyed; however, this company offers options for protectors, 44% of employees of this company use HPD earmuffs. Already in company (D), where a considerable number of workers are exposed to levels above 90 dB (A) there is no option of hearing protector provided to workers.

4.3 Risk perception

The risk behavior was evaluated, so it was possible to verify differences between companies and the level of noise in which the worker is exposed.

Through the ANOVA and Levene results it can be verified that there is a systematic difference ($p < 0.001$) between the means of risk behavior

Table 1. General data of the companies interviewed.

Company	Employees number	Sample number	Age average	Sd	Service time	Sd	HPD ear plug	HPD earmuffs
A	385	58	40.8	11.1	8.9	6.0	79%	21%
B	80	43	37.6	12.5	7.9	6.8	93%	7%
C	720	89	38.9	11.5	7.1	7.3	87%	13%
D	165	19	31.5	8.4	4.3	4.5	100%	0%
E	220	34	34.5	9.7	4.6	5.4	56%	44%

Table 2. Percentage of workers exposed to noise levels exceeding 85 dB (A).

Noise levels dB (A)	A	B	C	D	E
85–86	10%	0%	24%	0%	68%
87–89	28%	40%	48%	16%	26%
90–92	26%	58%	25%	84%	6%
93–98	36%	2%	3%	0%	0%
Sample	58	43	89	19	34

Table 3. Results for "my colleagues do not usually wear protectors" compared to noise levels.

Noise level dB (A)	N	alpha = 0.05	
		1	2
Above to 90	62	**1.76**	
87 to 88	47	1.87	
89 to 90	84	2.17	2.17
85 to 86	50		**2.68**
Sig.		0.344	0.159

Table 4. Results for "my colleagues do not usually wear protectors" compared to different companies.

Company	N	alpha = 0.05	
		1	2
C	89	**1,8**	
A	58	1,97	
D	19	2	
B	43	2,4	2,4
E	34		**2,88**
Sig.		0,282	0,494

among companies. The lowest risk behavior was identified in company (A) and company (D), and the highest risk behavior index in company (E) and (C).

It was also found that there is a significant difference ($p < 0.001$) between the means of results between the noise exposure levels of each worker and their responses on the risk behavior. In conclusion, workers exposed to higher noise have lower risk behavior, while workers exposed to lower noise tend to be at higher risk behavior.

The results showed that there is a higher risk behavior in workers who are exposed to noise levels between 85 dB (A) and 86 dB (A), and there is a lower risk behavior in workers exposed to noise above 90 dB (A).

4.4 Hearing protection use

Some research has shown that few workers use hearing protection devices throughout their working time (Williams, et al. 2004; Ahmed 2012). Therefore, in this research the verification between the self-report of the use and the effective use was performed, for this, two questions were analyzed: "I do not always use the protectors as it should" and "My colleagues do not usually use protectors".

It was found that the variation ($p < 0.001$) is significant for the question "my colleagues do not usually wear protectors". The differences were analyzed through the Tukey Test, in order to identify in which groups the differences are located.

Table 3 shown the results, it is possible to notice that the report of hearing protector use of colleagues is greater in the environment with noise above 90 dB (A) and being smaller in the environment between 85 and 86 dB (A).

The same way, was analyzed the differences between companies. There were differences between the report of hearing protectors use and the report of use by colleagues among the different companies. The results are presented on Table 4.

The company (C) presented smaller difference between the two answers. It is understood that the responses of self-report of use and the use by colleagues are practically equivalent. A close result also recorded by company A. It is important to mention that both companies (A) and company (C) recorded high noise level.

The company (E) registered a greater discrepancy between the conflicts of the two responses, thus, the hearing protector use is not similarly related to the reports that the colleagues answered. At this point, it is important to note that the company (E) has the lowest levels of noise among the companies surveyed, and it can be evaluated that the use is more effective in companies with a higher level of exposure to noise. According Arezes & Miguel (2005) the perception of risk plays a fundamental role, which predicts the use of a hearing protector, therefore, the perception of the work environment, with regard to lesser or greater risk, can be directly connected with the hearing protector use.

5 CONCLUSIONS

Understanding workers' perceptions, the safety culture of a workplace, and attitudes are important factors in assessing safety needs. The perception about the risk addressed through the questions, seek mainly to understand the dimension of the perception of risk and to relate among the companies surveyed.

There were differences in risk behavior among companies surveyed. The company (E) registered the highest risk behavior index, in this company the lowest noise was found (between 85 and 86 dB (A)) when compared to the other companies surveyed.

This factor can strongly influence the perception of "non—risk". Another important consideration is that, the company (E) 41% of employees makes use of the hearing protector earmuffs, which can improve the perception of hearing protection.

Already in company (A) was identified lower risk behavior among the companies surveyed.

In order to confirm this evaluation, it is possible to relate the behavior of risk with the level of exposure to noise of the workers.

In this perspective there was a significant difference between the means, and it was observed that the lowest risk behavior index is located in workers who are exposed to noise levels higher than 90 dB (A), while the highest risk behavior index was found in the group of workers exposed to noise levels between 85 and 86 dB (A).

We conclude in this research that, in an occupational environment where high levels of noise are recorded, workers tend to comply more strictly with the use of hearing protectors, and the annoying level of noise can motivate workers to use the equipment if compared with workers in areas with less exposure to noise.

REFERENCES

Ahmed, H.O., et al. 2001. *Occupational noise exposure and hearing loss of workers in two plants in eastern Saudi Arabia.* Annals of Occupational Hygiene 45 (5): 371–380.

Ahmed, H.O. 2012. *Noise exposure, awareness, practice and noise annoyance among steel workers in United Arab Emirates.* Open Public Health Journal 5: 28–35.

Araújo, S.A. 2002. *Perda Auditiva Induzida Pelo Ruído em Trabalhadores de Metalúrgica.* Rev. Bras. Otorrinolaringol, v.68, n. 1, pp. 47–52.

Arezes P.M.F.M. 2002. *Percepção do Risco de Exposição Ocupacional ao Ruído.* 240 f. Tese (doutoramento em Engenharia de Produção) Escola de Engenharia da Universidade do Minho.

Arezes, P.M.; Miguel A.S. 2005. *Hearing protection use in industry: The role of risk perception.* Safety Science 43(4): 253–267.

Arezes, P.M.; Miguel., A.S. 2006. *Does risk recognition affect workers' hearing protection utilisation rate?* International Journal of Industrial Ergonomics 36(12): 1037–1043.

Bockstael, A., et al. 2013. Hearing *protection in industry: Companies' policy and workers' perception.* International Journal of Industrial Ergonomics 43(6): 512–517.

Byrne, D.C., et al. 2011. *Relationship between comfort and attenuation measurements for two types of earplugs.* Noise and Health 13(51): 86–92.

Davies, H.W., et al. 2009. *Occupational noise exposure and hearing protector use in canadian lumber mills.* Journal of Occupational and Environmental Hygiene 6(1): 32–41.

Diaz, Y., Resnick, M. *A model to predict employee compliance with employee corporate's safety regulations factoring risk perception.* In: Proceedings of the IEA2000/HFES2000 Congress, vol. 4, pp. 323–326.

Gerges, S.N.Y. 1992. *Ruído. Fundamento e Controle.* 2ª Edição. Florianópolis: Editora Imprensa Universitária UFSC.

Glendon, A.I., Stanton, N.A. e Harrison, D 1995 *Factor Analyzing a performance shaping concept questionnaire.* In Contemporary Ergonomics: ergonomics for all, ed. S.A. Robertson, Taylor and Francis, London, pp. 340–345.

Guerra, M.R. 2005. *Prevalência de perda auditiva induzida por ruído em empresa metalúrgica.* Revista de Saúde Pública—USP. 39 (2): 238–44.

Hunashal, R.B.; Patil, Y.B. 2012. *Assessment of Noise Pollution Indices in the City of Kolhapur, India.* Procedia—Social and Behavioral Sciences, v. 37, n. 0, pp. 448–457.

Laforet, S. 2013. Organizational innovation outcomes in SMEs: Effects of age, size and sectors. *Journal of World Business,* 48(4), pp. 490–502.

Lie, A., et al. 2015. *"Occupational noise exposure and hearing: a systematic review."* International Archives of Occupational and Environmental Health 89(3): 351–372.

Lusk, S.L., et al. 1998. *Use of hearing protection and perceptions of noise exposure and hearing loss among construction workers."* American Industrial Hygiene Association Journal 59(7): 466–470, 1998.

Maroco, J. 2014. *Análise estatística—com utilização do SPSS.* 6 ed. Lisboa: Edições Sílabo, 990 p.

Melamed, S., et al. 1996. *"Usefulness of the protection motivation theory in explaining hearing protection device use among male industrial workers."* Health Psychology 15(3): 209–215.

Ministério do Trabalho e Emprego. 2011. *Norma Regulamentadora nº 9.* Brasília Ministério do Trabalho e Emprego.

Miranda, E.F.V. 2003. *Avaliação experimental e numérica da atenuação sonora de protetores auditivos para ruído impulsivo.* Tese. Universidade Federal de Santa Catarina—UFSC.

Rabinowitz, P.M., et al. 2007. *Do ambient noise exposure levels predict hearing loss in a modern industrial cohort?* Occupational Environment Medicine 64:53–59.

Reddy, R.K., et al. 2012. *Hearing protection use in manufacturing workers: A qualitative study.* Noise & Health 14(59): 202–209.

Stephenson, M.R., et al. 2011. *"Hearing loss prevention for carpenters: Part 2—Demonstration projects using individualized and group training."* Noise and Health **13**(51): 122–131.

Stephenson, C.M.; Stephenson, M.R 2011. *Hearing loss prevention for carpenters: Part 1-Using health communication and health promotion models to develop training that works.* Noise & Health 13(51): 113–121.

Vieira, P.R. da C. Ribas, J.R. 2011. *Análise Multivariada com o uso do SPSS.* Rio de Janeiro: Editora Ciência Moderna Ltda.

Whittaker, J.D. et al. 2014. *Noise-induced hearing loss in small-scale metal industry in Nepal.* The Journal of Laryngology & Otology, v. 128, n. 10, pp. 871–880.

Williams, W., et al. 2004. *Hearing loss and perceptions of noise in the workplace among rural Australians.* Australian Journal of Rural Health 12(3): 115–119.

Occupational Safety and Hygiene V – Arezes et al. (Eds)
© 2017 Taylor & Francis Group, London, ISBN 978-1-138-05761-6

Standards of heat stress—a short review

T.F.O. Galvan, A.O. Michaloski & A.A. de P. Xavier
Federal Technological University of Paraná, Ponta Grossa, Brazil

ABSTRACT: Guidelines and standards are set to limit heat stress in work environments, all over the world. This paper summarizes the revision of published articles that used Wet Bulb Globe Temperature—WBGT index, Required Sweat Rate – SWreq, and Predicted Heat Strain—PHS, and were able to verify their application. To accomplish this task, a systematic literature search was performed on databases. The articles selected were tabulated and statistically analyzed. Results showed that WBGT was the most used index for the assessment of thermal conditions. The PHS model was normalized only in 2004, which may justify the fact that there are not yet many studies using this index. The SWreq is an index that has been replaced by PHS, so it was expected that this index would no longer be used after of 2004.

1 INTRODUCTION

Ergonomics, in the industrial context, can be evaluated based on several criteria; one of them is the thermal conditions at workplace. Heat stress can affect the productivity and health of workers, athletes, and other people (Epstein & Moran 2006).

The existence of extreme hot conditions in many work environments may cause a serious negative effect on the health and safety of employees (Fogleman et al. 2005). Guidelines and standards are set to limit heat stress in work environments, all over the world.

Freitas & Griegorieva (2015) searched on the literature to identify and classify all developed indices about human thermal climate conditions, and they identified 162 of them. Because there were so many indices involved in that process, the choice of a single index for the assessment of heat stress is very difficult.

However, the usage of standard indices became a facilitating option. The International Organization for Standardization—ISO has many standards for ergonomics of the thermal environment, including the specific standards of heat stress such as: 1) ISO 7243:1989—Hot environments - Estimation of the heat stress on working man, based on the WBGT-index (wet bulb globe temperature); 2) ISO 7933:1989—Hot environments - Analytical determination and interpretation of thermal stress using calculation of required sweat rate; and, 3) ISO 7933:2004—Ergonomics of the thermal environment—Analytical determination and interpretation of heat stress using calculation of the predicted heat strain.

In the past, ISO 7933:1989 was used for the calculation of the required sweat rate—SWreq, although, in 2004, the standard was revised, and from then on, the predicted heat strain—PHS is used for the assessment of heat stress instead.

The research reported on the article aims to verify the application of these three indices in published articles and analyze them statistically.

2 RESEARCH METHODOLOGY

2.1 Search strategy

A systematic literature search was performed on the Science Direct and Web of Science databases. No temporal restriction was imposed.

Three searches were made in both databases. The following combinations of terms were used in each search: WBGT AND heat stress; Required Sweat Rate AND heat stress; Predicted Heat Strain AND heat stress.

2.2 Criteria for inclusion and exclusion

The criteria for inclusion consisted in having the full text available online, and observing that the content had to be related to the studied topic and suitable to the intended objectives.

Duplicated papers, papers with unidentified author, and papers not published in English felt into the exclusion criteria.

2.3 Analysis of results

The results were analyzed through descriptive statistics.

3 RESULTS AND DISCUSSION

The first step was to search the databases to obtain the available articles related to the studied topic. After that the exclusion criteria was applied to select relevant articles to work with. Then, the relevant, available articles had their title, abstract and keywords analyzed. Finally, and, if necessary, relevant articles had their full text analyzed.

The results of this search are presented on Table 1.

The selected articles were tabulated according to their characteristics such as year of publication, journal where published and type of article. The articles were classified according to their type: If they were a review, a case study, an experiment/laboratory/chamber, an analysis of data, a development/proposal of model, and others.

3.1 WBGT results

One hundred and twenty five articles considering WBGT index were selected. They were tabulated according to the chosen methodology. Figure 1 shows that the majority of articles found were case studies, followed by experiment/laboratory/chamber articles, and, then, review articles.

The majority of the case studies occurred in surroundings such as construction, steel/iron, glass, and agricultural industries. There are, also, many case studies about urban climate and sports.

Despite the fact that the origin of the WBGT index is usually traced to 1957 (Alfano et al. 2014), the greatest amount of publications using WBGT index is recent, as observed in Table 2. The years with the most publications are 2016, 2015 and 2008. Most part of these articles use WBGT as an index for the assessment of heat stress in workplaces.

The oldest publication found dated from 1967. Beginning in 1987, several articles were selected from mostly every year until 2016.

Another interesting result from this search is the fact that some of these articles were published in sports journals. This occurs because WBGT is the common heat limit reference recommendation for sports guidelines (Larsen et al. 2007).

3.2 Required Sweat Rate Results

Nine articles considering Required Sweat Rate Results were analyzed and tabulated according to the same methodology. As a result, Figure 2 shows

Table 1. Search and analysis results.

Index Articles/base	WBGT		SWreq*		PHS**	
	B1	B2	B1	B2	B1	B2
Search Results	347	230	12	65	10	42
Avaliable	138	94	12	27	10	21
Selected	56	69	7	2	2	7
Total	WBGT		SWreq*		PHS**	
Selected Articles	125		9		9	

*Abbreviation of the term Required Sweat Rate.
**Abbreviation of the term Predicted Heat Strain.
***B1 corresponds to Science Direct and B2 corresponds to Web of Science.

Table 2. Years with greater number of publication on WBGT articles.

Year	Articles
2016	14
2015	14
2008	11
2007	7
2012	6
2006	6
2001	5

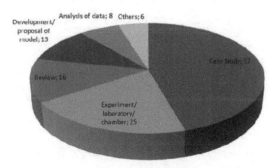

Figure 1. Amount of articles on WBGT according to their types.

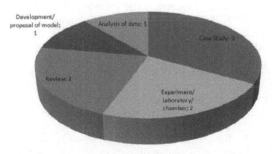

Figure 2. Amount of articles on SWreq according to their types.

Table 3. Years with greater number of publication on SWreq articles.

Year	Articles
2000	3
2001	2
1999	2
1991	1
1982	1

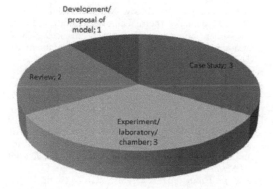

Figure 3. Amount of articles on PHS according to their types.

Table 4. Years of publication on PHS model articles.

Year	Articles
2002	2
2016	1
2012	1
2010	1
2007	1
2006	1
2001	1
2000	1

that the majority of articles analyzed were case studies, followed by the experiment/laboratory/ chamber, and, then, the review category.

Table 3 demonstrates that the largest number of publications occurred before 2004, the same year that happens the revision of ISO 7933, when the SWreq was replaced by the PHS model.

3.3 *PHS results*

Nine articles were, also, selected concerning PHS results. They were tabulated according to the same methodology. Figure 3 demonstrates the amount of published articles according to their type.

Table 4 shows that there were articles from 2000 until 2002, just before the revision of ISO 7933. In addition, it shows that there are newer publications (2016, 2012 and 2010).

4 CONCLUSIONS

The results obviously demonstrate that WBGT is the most used ISO index for the assessment of thermal conditions both for the monitoring of thermal conditions in the laboratory or chamber, and for the evaluation of indoor and outdoor environments. Although, several case studies using the WBGT are about athletes and sports, most of the studies are about the assessment of workplaces.

Despite the fact that WBGT index is still being widely used, there are some studies after 2004 revision with the application of the PHS, including study with the comparison between index WBGT and PHS (Holmér, I. 2010). The PHS model was normalized only in 2004, what may justify the fact that not many studies have yet been published using this index.

The SWreq is an index that has been replaced by PHS, so it was expected that this index would no longer be used after the year of 2004.

After presenting the results of this short review, it is clear that further research is needed. The amount of articles about SWreq as well as PHS researches demonstrates that the terms of search can be expanded in order to obtain more results. It is, also, important to perform this type of research in different bases to achieve more journals. These points should be considered in future researches.

REFERENCES

Akatsuka, S., Uno, T. & Horiuchi, M. 2016. The Relationship between the Heat Disorder Incidence Rate and Heat Stress Indices at Yamanashi Prefecture in Japan. *Advances in Meteorology* 2016: 1–11.

Al-Ashaik, R.A., Ramadan, M.Z., Al-Saleh, K.S. & Khalaf, T.M. 2015. Effect of safety shoes type, lifting frequency, and ambient temperature on subject's MAWL and physiological responses. *International Journal of Industrial Ergonomics* 50: 43–51.

Alfano, F.R.A., Malchaire, J., Palella, B.I. & Riccio, G. 2014. WBGT Index Revisited After 60 Years of Use. *The Annals of Occupational Hygiene* 58: 955–970.

Assunta, C., Ilaria, S., Simone, D.S., Gianfranco, T., Teodorico, C., Carmina, S., Anastasia, S., Roberto, G., Francesco, T. & Valeria, R.M. 2015. Noise and cardiovascular effects in workers of the sanitary fixtures industry. *International Journal of Hygiene and Environmental Health* 218: 163–168.

Borg, D.N., Stewart, I,B. & Costello, J.T. 2015. Can perceptual indices estimate physiological strain across a range of environments and metabolic workloads when wearing explosive ordnance disposal and chemical. *Physiology and Behavior* 147: 71–77.

Budd, G.M. 2001. Assessment of thermal stress—the essentials. *Journal of Thermal Biology* 26: 371–374.

Budd, G.M. 2008. Wet bulb globe temperature WBGT its history and its limitations. *Journal of Science and Medicine of Sport* 11: 20–32.

Buonanno, G., Frattolillo, A. & Vanoli, L. 2001. Direct and indirect measurement of WBGT index in transversal flow. *Measurement* 29: 127–135.

Chan, A.P.C. & Yang, Y. 2016. Practical on-site measurement of heat strain with the use of a perceptual strain index. *International Archives of Occupational and Environmental Health* 89: 299–306.

Dianat, I., Vahedi, A. & Dehnavi, S. 2016. Association between objective and subjective assessments of environmental ergonomic factors in manufacturing plants. *International Journal of Industrial Ergonomics*. 54: 26–31.

Ducharme, M.B. 2006. Heat stress of helicopter aircrew wearing immersion suit. *Industrial Health* 44: 433–440.

El Hachem, W., Khoury, J. & Harik, R. 2015. Combining Several Thermal Indices to Generate a Unique Heat Comfort Assessment Methodology. *Journal of Industrial Engineering and Management* 8(5): 1491–1511.

Epstein, Y. & Moran, D.S. 2006. Thermal Comfort and the Heat Stress Indices. *Industrial Health* 44: 388–398.

Fogleman, M., Fakhrzadeh, L. & Bernard, T.E. 2005. The relationship between outdoor thermal conditions and acute injury in an aluminum smelter. *International Journal of Industrial Ergonomics* 35(1): 47–55.

Freitas, C.R. & Grigorieva, E.A. 2015. A comprehensive catalogue and classification of human thermal climate indices. *International Journal of Biometereology* 59: 109–120.

Fujii, R.K., Horie, S., Tsusui, T. & Nagano, C. 2008. Effectiveness of a head wash cooling protocol using non-refrigerated water in reducing heat stress. *Journal of Occupational Health* 50: 251–261.

Gosling, C.McR., Gabbe, B.J., McGivern, J. & Forbes, A.B. 2008. The incidence of heat casualties in sprint triathlon: The tale of two Melbourne race events. *Journal of Science and Medicine of Sport* 11: 52–57.

Grundstein, A., Williams C., Phan, M. & Cooper, E. 2015. Regional heat safety thresholds for athletics in the contiguous United States. *Applied Geography* 56: 55–60.

Hólmer, I. 2010. Climate change and occupational heat stress: methods for assessment. *Global Health Action* 3 (0).

Inaba, R. & Mirbod, S.M. 2007. Comparison of subjective symptoms and hotprevention measures in summer between traffic control workers and construction workers in Japan. *Industrial Health* 45: 91–99.

International Organization of Standardization (ISO). 1989a. ISO 7243:1989—Hot environments—Estimation of the heat stress on working man, based on the WBGT-index (wet bulb globe temperature).

International Organization of Standardization (ISO). 1989b. ISO 7933:1989—Hot environments—Analytical determination and interpretation of thermal stress using calculation of required sweat rate.

International Organization of Standardization (ISO). 2004. ISO 7933:2004—Ergonomics of the thermal environment—Analytical determination and interpretation of heat stress using calculation of the predicted heat strain.

Kampmann, B., Bröde, P., Schütte, M. & Griefhan, B. 2008. Lowering of resting core temperature during acclimation is influenced by exercise stimulus. *European Journal of Applied Physiology* 104: 321–327.

Kikumoto, H., Ooka, R. & Arima, Y. 2016. A study of urban thermal in Tokyo in summer of the 2030s under influence of global warming. *Energy and Buildings* 114: 54–61.

Konstantinov, P.I., Varentsov, M.I. & Malinina, E.P. 2014. Modeling of thermal comfort conditions inside the urban boundary layer during Moscow's 2010 summer heat wave (case study). *Urban Climate* 10: 563–572.

Kralikova, R., Sokolova, H. & Wessely, E. 2014. Thermal Environment Evaluation According to Indices in Industrial Workplaces. *Procedia Engineering* 69: 158–167.

Larsen,T., Kumar, S., Grimmer, K., Potter, A., Farquharson, T. & Sharpe, P. 2007. A systematic review of guidelines for the prevention of heat illness in community based sports participants and officials. *Journal of Science and Medicine of Sport* 10: 11–26.

Lee, K.L., Chan, Y.H., Goggins, W.B. & Chan, E.Y.Y. 2016. The development of the Hong Kong Heat Index for enhancing the heat stress information service of the HongKong Observatory. *International Journal of Biometereology* 60: 1029–1039.

Leithead, C.S. 1967. Prevention of the Disorders Due to Heat. *Transactions of the Royal Society of Tropical Medicine and Hygiene* 61(5): 739–745.

Liang, C., Zheng, G., Zhu, N., Tian, Z., Lu, S. & Chen, Y. 2011. A new environmental heat stress index for indoor hot and humid environments based on Cox regression. *Building and Environment* 46: 2472–2479.

Lin, Y., Chang, C., Li, M., Wu, Y. & Wang, Y. 2012. High temperature indices associated with mortality and outpatient visits: Characterizing the association with elevated temperature. *Science of the Total Environment* 427–428: 41–49.

Maiti, R. 2008. Workload assessment in building construction related activities in India. *Applied Ergonomics* 39: 754–765.

Matsuzuki, H., Ayabe, M., Haruyama, Y., Seo, A., Katamoto, S., Ito, A. & Muto, T. 2008. Effects of heating appliances with different energy efficiencies on associations among work environments, physiological responses, and subjective evaluation of workload. *Industrial Health* 46: 360–368.

Maurya, T., Karena, K., Vardhan, H., Aruna, M. & Raj, M.G. 2015. Effect of heat on undergraound mine workers. *Procedia Earth and Planetary Science* 11: 491–498.

Miller, V.S. & Bates, G.P. 2007. The Thermal Work Limit Is a Simple Reliable Heat Index for the Protection of Workers in Thermally Stressful Environments. *The Annals of Occupational Hygiene* 51(6): 552–561.

Moran, D.S., Epstein, Y. 2006. Thermal comfort and the heat stress indices. *Industrial Health* 44: 399–403.

Moran, D.S., Pandolf, K.B., Shapiro, Y., Heled, Y., Shani, Y., Mathew, W.T. & Gonzalez, R.R. 2001. An Environmental Stress Index (ESI) as a substitute for the Wet Bulb Globe Temperature (WBGT). *Journal of Thermal Biology* 26: 427–431.

Nag, P.K., Nag, A. & Ashtekar, S.P. 2007. Thermal limits of men in moderate to heavy work in tropical farming. *Industrial Health* 45: 107–117.

Ohashi, Y., Ihara, T., Kikegawa, Y. & Sugiyama, N. 2016. Numerical simulations of influence of heat island countermeasures on outdoor human heat stress in the 23 wards of Tokyo, Japan. *Energy and Building* 114: 104–111.

Palella, B.I., Quaranta, F. & Riccio, G. 2016. On the management and prevention of heat stress for crews onboard ships. *Ocean Engineering* 112: 277–286.

Parsons, K. 2006. Heat stress standard ISO 7243 and its global application. *Industrial Health* 44: 368–379.

Pérez-Alonso, J., Callejón-Ferre, A.J., Carreño-Ortega, A. & Sánchez-Hermossila, J. 2011. Approach to the evaluation of the thermal work environment in the greenhouse construction industry of SE Spain. *Building and Environment* 46: 1725–1734.

Pryor, R.R., Bennet, B.L., O'Connor, F.G., Young, J.M.J. & Asplund, C.A. 2015. Medical Evaluation for Exposure Extremes Heat. *Wilderness & Environmental Medicine* 26: S69-S75.

Pyke, A.J., Costello, J.T. & Stewart, I.B. 2015. Heat strain evaluation of overt and covert body armour in a hot and humid environment. *Applied Ergonomics* 47: 11–15.

Rowlinson, S., YunyanJia, A., Li, B. & ChuanjingJu, C. 2014. Management of climatic heat stress risk in construction: A review of practices methodologies and future research. *Accident Analysis and Prevention* 66: 187–198.

Saha, R., Dey, N.C., Samanta, A. & Biswas, R. 2008. A comparison of cardiac strain among drillers of two different age groups in underground manual coal mines in India. *Journal of Occupational Health* 50: 512–520.

Santamouris, M., Mihalakakou, G., Patargias, P., Gaitani, N., Sfakianaki, K., Papaglastra, M., Pavlou, C., Doukas, P., Primikiri, E., Geros, V., Assimakopoulos, M.N., Mitoula, R. & Zeferos, S. 2007. Using intelligent clustering techniques to classify the energy performance of school buildings. *Energy and Buildings* 39: 45–51.

Sinclair, W.H., Crowe, M.J., Spinks, W.L & Leicht, A.S. 2008. Thermoregulatory responses of junior lifesavers wearing protective clothing. *Journal of Science and Medicine of Sport* 11: 542–548.

Suzuki-Parker, A., Kusaka, H. & Yamagata, Y. 2015. Assessment of the Impact of Metropolitan-Scale Urban Planning Scenarios on the Moist Thermal Environment under Global Warming: A Study of the Tokyo Metropolitan Area. *Advances in Meteorology* 2015: 1–11.

Tanaka, M. 2007. Heat stress standard for hot work environments in Japan. *Industrial Health* 45: 85–90.

Taylor, N.A.S. 2006. Challenges to temperature regulation when working in hot environments. *Industrial Health* 44: 331–344.

Wang, Y., Gao, J., Xing, X., Liu, Y. & Meng, X. 2016. Measurement and evaluation of indoor thermal environment in a naturally ventilated industrial building with high temperature heat sources. *Building and Environment* 96: 35–45.

Yamamoto, S., Iwamoto, M., Inoue, M. & Harada, N. 2007. Evaluation of the effect of heat exposure on the autonomic nervous system by heart rate variability and urinary cathecolamines. *Journal of Occupational Health* 49: 199–204.

Occupational Safety and Hygiene V – Arezes et al. (Eds)
© 2017 Taylor & Francis Group, London, ISBN 978-1-138-05761-6

LEED certification in a commercial enterprise in Brazil

G.M. Zellin, A.S.C. Fernandes & L.A. Alves
Department of Civil Engineering—CEFET, RJ, Rio de Janeiro, Brazil

J.L. Fernandes
Department of Production Engineering—CEFET, RJ, Rio de Janeiro, Brazil

ABSTRACT: The advance of construction on a global scale contributes to the increase of the impacts caused to the environment. Thus environmental certifications in buildings aim to work, developing goals to be followed in favor of the environment, minimizing environmental impacts and making the enterprise in the process of certification, sustainable and cleaner from its construction to operation. The focus of the work is the LEED (Leadership in Energy and Environmental Design) certification, presenting and analyzing the implementation of level 1 certification of a commercial enterprise located in Barra da Tijuca, Rio de Janeiro—Brazil. At this level, fewer sustainable actions are involved to implement the certification, and new credits can be added to reach a higher level. However, even adopting LEED level 1 certification, the company has implemented appropriate certification actions, thus contributing to a more sustainable construction industry.

Keywords: LEED Certification, Construction, Management Systems

1   INTRODUCTION

The environmental impacts generated by construction are a global problem. The United States Green Building Council—USGBC (2009) states that in the United States buildings alone account for 65% of electricity consumption, 30% of raw material use, 30% of waste generated annually, 30% Of the greenhouse effect and 12% of drinking water consumption.

Brazil faces the same reality, according to data from the Brazilian Council for Sustainable Construction—CBCS (2015) 75% of the natural resources extracted in Brazil are destined for the civil construction sector and the main environmental impacts of Brazilian buildings are: 80 million tons of waste generated per year in its construction and demolition; Consumption of 21% of the country's drinking water; Consumption of wood often illegally extracted; Noise, dust and erosion generated by construction sites; Waterproofing of soil; Consumption of 50% of the electric energy in the country. Based on the sustainability tripod, which covers social, environmental and economic issues, a sustainable building, called a green building, is designed to increase the efficiency of the use of natural resources (water, energy and materials) through constructive measures and procedures and to reduce the impacts of construction and

minimization of impacts on people's health and the environment (Krygiel and Nies, 2008).

With the aim of minimizing as much of these impacts as possible. Therefore, the object of study of this work is to analyze the certifications, mainly the LEED Certification.

For the Brazilian Council for Sustainable Construction (2015), within the global sustainability theme in civil construction, there are essential subthemes to be addressed that will be expressed in items 1.1 to 1.12.

1.1   *Ethics and formality of companies and service providers*

Business ethics is related to an ideal and fundamental posture for sustainability. The formality deals with the obligatory fulfillment of the Laws, Regulations and Norms with the purpose of guaranteeing the solidity, usefulness and quality of the works.

1.2   *The importance of the stages of conceptualization, planning and design of constructed spaces*

In the phases of conceptualization, planning and design of constructed spaces is where the best opportunity to implement actions that lead to the quality, rationality and efficiency of the building during its period of use is found. The concept that promotes the synergistic integration of concepts, projects, technologies and design allows to create spaces built with a high degree of performance and that will attend to their form, function, flexibility, simplicity of use and ease of maintenance to the interest of users and investors.

1.3   Integration

Integrate areas of knowledge, professionals, construction processes and technologies from the conceptualization of the project and include data on available infrastructures and social, cultural, environmental and economic characteristics of the region. This is certainly the most relevant theme of the whole process in view of the complexity and opportunities not yet explored in the construction sector.

1.4   Life cycle analysis (systemic)

For a better understanding of the rational use of our natural resources, we must understand the life cycle of the products and verify the best use of inputs considering the relation of efficiency or scarcity, the primary and potential uses of each resource and the alternatives.

1.5   Urban planning and infrastructure

Urban planning and adequate infrastructure are synonymous with efficiency, competitiveness and wealth for cities, states and countries. Due to the lack of planning and effective public policies, the issue of infrastructure is becoming a problem that afflicts the entire planet.

1.6   Energy efficiency, alternative and renewable sources

Humanity is extremely dependent on energy. In this context, there are three opportunities: to work on demand with rational and energy conscious use, avoiding waste and providing mobility and comfort for people; Massively adopt economizing technologies; And, thirdly, in the medium term, non-renewable energy should make room for renewable energy.

1.7   Water and sanitation

Fundamental to life, water must be used rationally. After its consumption, the effluent must pass through sanitation actions, returning to the environment with quality characteristics similar to or higher than those withdrawn from nature. The great challenge lies in equating the rational consumption with the health, the comfort and the pleasure that

the contact with the water offers us. For this, it is necessary to make use of the technologies and systems already created and always innovate.

1.8   Materials

The Brazilian Council for Sustainable Construction argues that there is no good material or bad material, but material suitable for the purpose. For this analysis, one should consider formality, quality, durability in use, acoustic performance, thermal performance, reliability, ease of use and maintenance, among other requirements specific to the application. And always meet the safety, needs and wishes of those who will use, operate and maintain.

1.9   Construction place

Safety, health and training of workers, maintenance of equipment and machinery, mitigation of pollution and noise, and the organization of the work and the production system are important factors in the implementation of the construction. The safety and health of workers appear first, they must take precedence over factors such as economic, terms and customer service, since the integrity of human life must always occupy a prominent place.

1.10   Housing of social interest

Aspects such as quality of life, able to provide greater productivity of families in both work and learning, should be considered. Quality is an important aspect, which must be considered, because in cases of poor construction quality, durability is no longer met and public spending, which subsidizes families, becomes inefficient. In addition, the delivery of housing must be provided by water treatment and sanitation systems, essential for the health of the residents.

1.11   Public policies

In many cases, government intervention is important in order to establish public policies that consider:

Minimum parameters and indexes of quality and efficiency of the construction systems with regard to the quality, reliability, durability and comfort of the built environment;

- Policies for the rational use of natural resources;
- Economic policies to stimulate more efficient building systems throughout life, making technology viable in cases with a higher cost of implementation;
- Policies to encourage innovation and technological development.

1.12 Integration of agents for sustainability

Throughout the life cycle of the enterprises, the various agents involved in the construction chain for their work and knowledge have great potential for integration to add sustainability to the enterprises, neighborhoods and cities.

2 LEED CERTIFICATION

The LEED System (Leadership in Energy and Environmental Design), the central theme of this study, was created by the USGBC (United States Green Building Council), one of the most relevant bodies that develops the theme of sustainable construction in the world. It was designed to classify commercial, institutional and residential buildings being they new or existing based on accepted USGBC (2009) environmental and energy principles.

The insertion of LEED worldwide can be verified through Table 1 with the ranking of countries with the highest number of registered projects.

2.1 LEED in Brazil

The first application for registration of an enterprise for LEED in Brazil occurred in 2004, and since then the growth rate in the registration number is extremely high.

The first building to obtain LEED certification in Brazil was an agency of Real Banc (now Santander Banc) located in the city of Cotia—São Paulo, in August 2007. Since then the country has 313 certified enterprises (GBC BRAZIL, 2015).

The high growth in the number of registrations in Brazil is not distributed evenly in all regions of the country. According to data from the GBC

Table 1. Ranking of countries with the highest number of projects registered for LEED certification.

Country	Number of registered projects
USA	53908
Canada	4814
China	2022
India	1883
Brazil	1058
United Arab Emirates	910
Turkey	477
Germany	431
Korea	279
Taiwan	149
Sweden	197

Source: Adapted from USGBC (2016)

Brazil (2015), the state of São Paulo concentrates 53.8% of the records, being the Southeast the region with the highest number of records as presented in Table 2 with the distribution of records and certifications by state of Brazil.

Brazil is the 4th country in the world in terms of LEED certifications.

2.2 LEED system types

According to Hernandes (2006) with the release of version 2.1 of the system in November 2002, LEED underwent minor changes in its bureaucratic process and also became LEED-NC (LEED—New Construction). As of 2004, the program began to categorize and differentiate its system according to the projects, types and purposes of the buildings. The following categories have been created:

- LEED NC—New Constructions and Major Renovations;
- LEED EB—Existing Buildings: Operation and Maintenance;
- LEED CS—Core & Shell;
- LEED CI—Commercial Interior;
- LEED ND—Neighborhood Devolpment;
- LEED Healthcare;
- LEED Schools;
- LEED Home;
- LEED Retail.

2.3 Distribution of LEED in Brazil

Even LEED having nine different system types, according to the GBC Brazil (2015), the LEED NC and LEED CS systems account for 38% and 43.4%, respectively, for certification registrations in Brazil. For this reason, this project will be deepened in these two categories of the system. The distribution can be verified in Table 3.

The weighting of points in the LEED version 3.0 system was based on the environmental impact assessment system of The American Environmental Agency (EPA) and the NIST (National Institute of Standards and Technology). The two systems together give a solid approach to

Table 2. Registration by state.

Certifications in Brazil	Values (%)
PI [2]	53.8
RJ [192]	18.3
PR [73]	6.9
Other	21.0

Source: GBC Brazil (2015)

Table 3. Registering the LEED system by category.

LEED System	Values (%)
LEED CI [71]	6.8
LEED CS [456]	43.4
LEED EB_OM [77]	7.3
LEED NC [399]	38.0
Other	4.5

Source: GBC Brazil (2015)

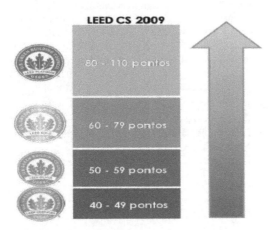

Figure 1. The four certification levels. Source: USGBC (2009).

the determination of the values of each credit (USGBC, 2015).

According to USGBC (2009) the credit weighting follows the following parameters to maintain consistency and ease of use, being able to receive four different levels of certification (Figure 1).

3 CASE STUDY

Insertion of LEED Core and Shell certification into a commercial venture under construction. In this modality it certifies all the envoltória of the enterprise, its common areas and internally the air-conditioning system and elevators. It is used when you will market the built areas. As previously described, it must meet all the certification prerequisites and obtain the minimum credit necessary to be certified.

The company chose to target the level of certification—called Certificate, which is the first level, needing to get between 40 and 49 points. During this first stage of the project, in the case under construction, the company chose to meet 6 credits and 1 prerequisites. Table 4 presents the goals panel of

Table 4. LEED certification goal for the enterprise.

Prerequisites and Credits	
Sustainable Terrain	Possible points: 21
Rational Water Use	Possible points: 2
Energy and Atmosphere	Possible points: 10
Materials and Resources	Possible points: 6
Indoor Air Quality	Possible points: 4
Project Innovation	Possible points: 3
Regional priorities	Possible points: 3
Goal:	49

Source: Internal company file

the project, showing all the prerequisites and credits that will be met.

The current phase of the project comprises the construction of five commercial buildings, located in Barra da Tijuca, in the city of Rio de Janeiro—Brazil. Of these 05 buildings, one will be built following the concepts of sustainable construction of the Leadership in Energy and Envrionmental Design (LEED) certification. Block 01 consists of a pavement with two rooms of 693.42 and 753.03 m², eight type pavements with four rooms each of 507.68 m² on average in addition to the ground floor and two basements intended for parking.

3.1 Analysis of the implementation of the LEED certification in the enterprise THE CITY by the company PDG

The consulting company in the implementation of the LEED certification in the venture stipulated with the company the goal of punctuation to be followed by the same that we can verify in Table 4 of this work, giving to the enterprise the level of certification, CERTIFICATE. However, by analyzing some of the credits that were left out of the goal and its potential to enter the goal in order to aggregate points to obtain a higher level of certification, one can analyze that the credits below could enter without major complications to the company. Thus, collaborating even more with regard to the sustainability and marketing of the enterprise. Here are the potential credits to exemplify the idea:

- SSc 4.4—Parking area
Preferential vacancies for drivers who offer a ride to people with the same destination, the enterprise certified. Score: 2 points
- WEc2—Innovative Wastewater Technologies
The goal is to reduce the use of drinking water by 50% through water conservation devices such as sink and flush sensors or use of non-potable water

in some devices such as sanitary basins. In addition to locally treating 50% of the waste water and reusing them. Score: 2 points
• WEc3—Reduction of water consumption
Reducing water consumption by 40% can achieve the highest credit score, for this, it is necessary to employ strategies to reduce consumption. Considering that mandatory prerequisite 1 is similar to credit, but decreasing 20% of consumption, we can understand that they will already have a way to be followed. It would be less complicated to implement this credit. Score: 4 points.

With the implementation of these 3 more credits, we can verify that we have already acquired 8 points beyond the target, which would give us the sum of 57 points and the SILVER level, which would already be a great advance for the company.

4 CONCLUSION

Through the case study of the implementation of LEED certification in the commercial enterprise it was possible to verify in practice the sustainable techniques required in certification. According to the consulting company the venture is in compliance with the requirements of LEED, on the certification label. It was also possible to observe that the company was not very bold about the certification goal, remaining in the first level instead of implanting more credits to reach higher levels. Even so, the venture, succeeding in everything until the end of the project, could be one of the projects certified in Brazil, gaining credibility and helping the environment.

REFERENCES

CBCS. 2015. Action Guidelines. Available in: <http://www.cbcs.org.br/_5dotSystem/userFiles/Sobre%20CBCS/CBCS_Diretrizes%20de%20Acao_rev1.pdf> Access in: 10/10/2015.

GBC BRAZIL. 2015. Available in: <http://www.gbc-brasil.org.br>. Access in: 12/18/2015.

GSA. General Service Administration. Assessing Green Building Performance. 2008. A Post Occupancy Evaluation of 12 GSA Buildings. Washington DC. 20f. Available in: <http://www.gsa.gov/portal/category/21083> Access in: 12/05/2015.

HERNANDES, T.Z. 2006. *LEED-NC as a system of evaluation of sustainability: a national perspective?* 2006. Dissertation (Master in Architecture)—Faculty of Architecture and Urbanism of the University of São Paulo, São Paulo, 2006. Available in: http://www.teses.usp.br/teses/disponiveis/16/16132/tde-28032009–111851/pt-r.php. Access in: 11/26/2015.

KRYGIEL, E.; NIES B. Green BIM. 2008. Successful Sustainable Design with Building Information Modeling. Indianapólis. Wiley Publishing Inc.

USGBC. 2009. LEED Reference Guide for Green Building Design and Construction: For the Design, Construction and Major Renovations of Commercial and Institutional Buildings Including Core & Shell and K-12 School Projects. Washington DC: USGBC. Available in: <http://www.usgbc.org/sites/default/files/LEED%202009%20RG%20BD+C-Supplement_GLOBAL_10_2014_Update.pdf.> Access in: 10/03/2016.

USGBC. 2015. Available in: <http://www.usgbc.org>. Access in: 12/05/2015.

Zellin, G.M. 2016. Environmental Certifications in Buildings in Brazil—Study on Certification LEED in a Commercial Enterprise. 2016. Final Project of Civil Engineering Course—Celso suckow Federal Center of Technological Education of Fonseca, Rio de Janeiro, 2016.

Occupational Safety and Hygiene V – Arezes et al. (Eds)
© 2017 Taylor & Francis Group, London, ISBN 978-1-138-05761-6

Chemical risks of plant protection products—preventive measures

A.L. Baltazar, A. Ferreira, A. Lança, D. Barreira, J. Almeida & T. Neves
ESTESC, Coimbra Health School, Saúde Ambiental, Instituto Politécnico de Coimbra, Coimbra, Portugal
ESTESC, Coimbra Health School, Dietética e Nutrição, Instituto Politécnico de Coimbra, Coimbra, Portugal

ABSTRACT: Plant protection products, called pesticides, promote the protection of plants and all agricultural production. There are several constituents of plant protection products that can present chemical risk to health and the environment. Therefore, the need to regulate the way these products are handled and marketed came to be, resulting in the Law 26/2013 of April 11, allowing the regulation of distribution activities, sales and application of these products for professional use, defining the procedures for monitoring and use of plant protection products. It is pretended, through a bibliographical research, to evaluate the impact that the plant protection products have on the environment and the health of those who handle them, verifying if they can lead to the development of occupational diseases or accidents at work. It is concluded that the application of these products should be viewed responsibly, since phytopharmaceuticals have contributed to the increase of occupational accidents and diseases due to the continuous exposure to pesticide derivatives.

1 INTRODUCTION

Applying a plant protection product aims to solve a particular plant health problem. The realization of this goal depends on several factors, which cannot be forgotten in any way. Incorrect application, in addition to wasting product, may cause additional problems in the crop, contaminate the applicator and the environment. (3)

Plant protection products are therefore an indisputable instrument in agricultural production, but their application involves exposure to serious risk factors for human health, both at the application stage and for public health and the environment. It is in this context that studies have been developed to evaluate the effects of the use of these products on agricultural production, both regarding the target population/workers or the final recipient/consumer of the product. These studies consider the elimination of their use as much as possible, with the risks associated being more or less severe depending on the way they are used. (13)

The preparation of the syrup requires specific care by the operator. At this stage, in addition to handling undiluted concentrated products, there are a variety of operations, such as mixing and filling of the spray tanks, which require extra attention. Preparing the syrup is a fairly responsible operation that should only be performed by qualified persons. It is necessary to ensure that there are no people or animals near the place where the syrup is prepared and take all precautions so that no errors or accidents might occur, with negative consequences for the quality of the treatment, the operator and the environment. (3)

Thus, for a good use of a pesticide, the farmer must know what pest or disease to combat, so that the technicians can inform him about the product to be used, its form of its application, the dose, the frequency, methods and equipment to be used. (4)

Immediately after the application of plant protection products, it is important to pay attention to three distinct aspects: to know and to put into practice the conditions expressed on the label regarding periods of re-entry in the treated crop, safety intervals, maintenance, cleaning of the application material, the Personal Protection Equipment (PPE), and the operator hygiene. (3)

Situations in which the victim is exposed accidentally to a plant protection product may present an increased risk to the rescuer. The victim should only be approached if there are safety conditions to do so. It is essential to avoid contact with these products, in particular by using gloves and other forms of body protection and not inhaling the vapors released by it. The goal of emergency treatment is to reduce the effects of intoxication, as soon as possible. Do not lose time looking for an antidote, unless it can be obtained immediately. Provide emergency transport to the hospital, accompanying the victim, whenever possible, with the packaging or the label. (8)

In fact, these products, being a production factor, clearly have the great benefit of contributing to an increase of harvests by reducing yields, improving agricultural product quality and efficiency in various tasks, but the great majority of chemical products, they have inherent in themselves a certain negative charge, consequence of the greater or less toxicity and the characteristics of each one. It is therefore confirmed that, in addition to the benefits associated with its use, there might also be dangers to human and animal health and unacceptable impacts on the environment, facts which are needed to know and minimized. (14)

2 PHYTOPHARMACEUTICAL PRODUCTS

With the growing demand and use of plant protection products in the agricultural sector, and given the dangers inherent in their use, as well as the producers handling them, there was a need to implement Law No 26/2013 of April 11. The purpose of this document is to regulate the distribution, sale and application of plant protection products for professional use and adjuvants of plant protection products, defining procedures for monitoring the use of plant protection products. (2) On the other hand, Decree-Law No. 101/2009 of May 11 aims to regulate the non-professional use of plant protection products in the domestic environment, establishing conditions for their authorization, sale and application. (1)

Law No. 26/2013 of April 11 covers agricultural/forestry, urban/leisure, communication and confined areas. These products must state on their label phrases such as "this product is intended for use by farmers and other plant protection product applicators", "to avoid risks to humans and the environment to comply with the instructions" and "to keep out of the reach of children". All companies under this law must have specific training required, exclusive storage facility, and all products must be sold in exclusive establishments and licensed for this purpose. (7)

Law No. 101/2009 of 11 May has therefore dominion of the domestic environment, such as indoor plants, gardens or family gardens. This law applies to vegetable gardens with areas less than 500 m², productions intended for self-consumption not subject to residue control. This law covers ready-to-use products, packaging with capacity or weight of 1 L or 1 Kg, packaging with the words "unprofessional use" and "bottom plant line", as well as child-resistant fastening. This Law does not require any kind of mandatory training nor installation of warehouse and the places destined to the sale of these products not require licensing. (7)

2.1 Phytopharmaceuticals

Plant protection products are therefore all products intended to control organic organisms which are considered to be harmful in the agricultural sector. They are subdivided into (12):

- Fungicides (reach fungi)
- Herbicides (reach the plants);
- Acaricides (reach the mites);
- Rodenticides (reaching rodents);
- Molluscicides (reaching slugs and snails);

- Nematodecides (reach nematodes);
- Growth regulators;
- Biostimulants (potentiates endogenous factors);
- Adjuvants (used in mixtures with insecticides to attract flies);
- Attractive (used in mixtures with insecticides, facilitating the contact of the pests with the insecticide syrup);
- Repellents (used alone or as adjuvants of these products, in order to repel certain vertebrate organisms);
- Insecticides (reach insects)

The latter are further subdivided into synthetic organic insecticides, synthetic inorganic insecticides, botanical insecticides and biological agents. Since this article aims to assess the chemical risks associated with the application of plant protection products in public health and the environment, we will address only the most harmful ones, namely organic synthetic insecticides. These are organochlorines, organophosphates, carbamates and pyrethroids, all of which have harmful effects on the environment and public health. (11)

Organochlorines have the disadvantage of degrading very slowly in the environment, thus leading to their gradual accumulation in the ecosystem. Therefore these pesticides are persistent in the environment with tendency of accumulation in the tissues of the alive beings. (11)

Concerning the aquatic environment, insecticides such as organophosphates can contaminate the environment by their discharging or discharging effluents into the water, by infiltrating the soil (contaminating groundwater) or by pulverization. Carbamates in this medium can contaminate water in the same way as organophosphates, which have the particularity of rapidly decomposing into an aquatic environment. (11)

In addition to the harmful effects of organophosphates in the aquatic environment, it may be hazardous to the health of the workers applying and to others who may come in contact with them, and intoxication by the substance may lead to direct health problems. The dispersion of the organophosphates through their spraying can be caused by winds spreading over areas of 1 km to 2 km, causing a serious risk of contamination. (11)

2.2 Phytopharmaceuticals classified as possibly carcinogenic to humans

With all these public health problems arising from the application of pesticides, World Health Organization (WHO) found five plant protection products classified as possible carcinogens for humans (10):

1. Glyphosate
It is the most widely used herbicide worldwide and has recently been classified as possibly carcinogenic to man.

2. Malathion

It is an insecticide, used in agriculture and also in domestic environments for pest disinfestation. Using it professionally, it can increase the likelihood of prostate cancer.

3. Diazinon

It is also another insecticide that is used both in agriculture and domestically, for this pesticide, as for the other substances indicated, the evidence is still limited, however this is related to the increased risk of lung cancer.

4. Parathion

It is an insecticide, considered as a possible carcinogen. Restrictions were imposed in some countries on their use, being used in agriculture to eliminate pest infestation.

5. Tetrachlorovins

It is an insecticide, similar to Parathion, in that it is considered a carcinogen and is also subject to restrictions in certain countries.

It is important for countries and their governments to take into account WHO indications in order to restrict or prohibit the use of certain pesticides that pose a potential hazard to human health and also to the environment. These substances are not only harmful to those who handle them or to those who consume foods treated with these products, since the substances end up spreading in the environment, by the air and earth, contaminating the water and the soil.

The toxicity of pesticides depends fundamentally on their chemical composition and the concentration in which they occur. The risk of poisoning to humans depends on the toxicity of the active substance, the time they are exposed to the pesticide, the conditions under which they are handled and applied, atmospheric/environmental conditions, and in particular how they come in contact with the organism, either by inhalation, ingestion or by direct skin contact. (4)

In order to avoid intoxication of the worker it is important that a correct evaluation of the risks is made, so that an appropriate selection of preventive measures to adopt in that situation is performed.

It is recommended that the application of plant protection products is carried out, preferably away from habitations and animal facilities, always keeping in mind the direction of the wind and the height of the application, always using the specific materials for handling the substances. Packaging of products must be handled with care, ensuring their integrity. (4)

3 WORKER PROTECTION

To ensure worker protection, appropriate Personal Protective Equipment (PPE) and work clothing must be worn.

In the selection of this type of clothing and equipment we must take into account the information that must be indicated on the pesticide label, written information on the characteristics and conditions of use, conservation of personal protective equipment (provided by the respective vendor) and quality assurance equipment, as well as the CE certification. In order to select PPE and clothing there should be a special attention to the protection of the eye, nose, mouth, hands, body, feet and legs (Figure 1). (4)

Eye protection should be worn when wearing protective goggles, and since these are toxic and irritating products, it is advisable to have a more careful and frequent hygiene.

For protection of the nose and mouth, a mask should be used when the pesticide label indicates this and the filters should be replaced according to the manufacturer's recommendations. This care is essential to avoid inhaling gases or dust.

For hand protection, use suitable gloves and comply with EN 374—protective gloves against chemicals and micro-organisms. After handling pesticides, always wash hands. At the end of work, gloves should be washed inside and out when handling or applying toxic pesticides, and especially when in high concentrations.

For body protection, a protective suit suitable for pesticide application should be used, it can be cotton with sleeves and wide legs (but adjusted at the wrists and ankles), easy to carry and dry.

For protection of the feet and legs (either during preparation of the product or in its application, or in subsequent contact with treated plants) rubber boots should be used, avoiding an unprotected space between the boots and pants. (4)

Figure 1. Individual Protection Equipment depending on the toxicity of the product. (Source: "Autoridade para as Condições de Trabalho. 2014. Utilização de Pesticidas Agrícolas. pag 10").

4 COLLECTIVE PROTECTION

However, we must not neglect collective protection, where simple measures such as the use of exhaust systems for fumes and gases, can be crucial to guarantee the workers' safety when preparing the products. (5)

Although it is important to protect farmers from exposure to these products, we can't ignore the environment. And consequently emerged an automated system developed by a company that offers solutions for the protection of crops, seeds and biotechnology, and environmental sciences. (6)

On a farm there is a significant number of operations which can result in a potentially contaminated effluent, even if carried out in accordance with Good Agricultural Practices, representing a risk of pollution to the environment. The operations with the highest risk associated are the handling and preparation of the products (concentrated effluent) and the washing of the application equipment (high volume effluent). This automated system aims to reduce the risk of contamination and pollution of water by phytosanitary effluents, as well as contribute to the reduction of the volumes of water spent in operations by the dissemination of Good Agricultural Practices. (9)

This is a practical system that recreates what happens naturally in nature, (6) but also added the possibility of controlling and optimizing factors that influence microbiological activity (humidity, air, light), and also safety without risk of contamination, because it's a watertight system isolated from the environment. (9)

Maintenance is performed by the farmer through well-known operations (example: it may be necessary to ventilate the substrate, add moisture to the substrate, or add organic matter). It works all year, for any type of farm or culture. Scientific studies demonstrate that the system is efficient in the degradation of all types of plant protection products (the degradation time obviously depends on the active substance under consideration). (6)

The automated system responds to everyone's concerns about water quality preservation, as well as the needs of an increasing number of farmers and farm entrepreneurs who certify their holdings, either by individual entities, or in the environmental certification of their farms. The implementation of this system also has the benefit of being able to contribute to the disclosure and application of Good Agricultural Practices in our country. (9)

5 CONCLUSION

Plant protection products are a key tool in the agricultural sector today, enabling the sustainable development of production. However, we know that there are several risks to public health and the environment, as they should be sold, distributed and applied according to Law nº 26/2013 of April 11.

It lays down all the principles and measures necessary to reduce the risk in the application of plant protection products.

It is essential that the applicators of these products have training to enable them to handle plant protection products safely and responsibly. To do this they must be aware of the danger of the products, that is, how to interpret the label of the packaging in its entirety, in order to prepare the syrup in a safe and correct manner. In addition they should know all the personal protective equipment to be used in relation to the product to be applied, as well as the correct way to handle it. (4)

REFERENCES

[1] Assembleia da República. Lei n.º 101/2009 de 11 de Maio. Diário da República. I série. Páginas 2806–2809. Portugal. 2013.

[2] Assembleia da República. Lei n.º 26/2013 de 11 de Abril. Diário da República. I série. Páginas 2100–2125. Portugal. 2013.

[3] Associação Nacional e Europeia de Proteção das Culturas. 2007. Manual Técnico de Produtos Fitofarmacêuticos.

[4] Autoridade para as Condições de Trabalho. 2014. Utilização de Pesticidas Agrícolas.

[5] Autoridade para as Condições de Trabalho. Abril de 2015. Segurança e Saúde no Trabalho no Setor Agro-Florestal.

[6] Bayer Phytobac. Bayer Phytobac—sistema de tratamento de efluentes para a exploração agrícola desenvolvido em Portugal em parceria com a Tomix. Portugal: 2015 [updated 2015; cited 2016 20 december]; Available from: http://www.bayercropscience.pt/internet/empresa/artigo.asp?menu = &id_artigo = 756&seccao = 70

[7] DGAV. Produtos Fitofarmacêuticos de Uso Profissional e Não Profissional. 2015.

[8] Direção-Geral de Agricultura e Desenvolvimento Rural. 2010. Código de conduta nos circuitos de distribuição e venda de produtos fitofarmacêuticos.

[9] Engª Maria do Carmo Romeiras. PHYTOBAC—Um sistema simples e prático para prevenir a contaminação da água. Portugal: 2010 [updated 2010; cited 2016 20 december]; Available from: http://www.bayercropscience.pt/internet/empresa/artigo.asp?id_artigo = 562&seccao = 70

[10] IARC Monographs Volume 112: evaluation of five organophosphate insecticides and herbicides. Worth Health Organization and International Agency for Research on Cancer. 2016. Volume 112. Available from: https://pedlowski.files.wordpress.com/2015/03/monographvolume112.pdf

[11] M. Anastasila, N. Luciano. A química dos pesticidas no meio ambiente e na saúde. Revista Magaio Académico. Jan/Jun 2016; V. 1, 55–58.

[12] Ministério da Agricultura, do Mar, do Ambiente e do Ordenamento do Território. DGAV. Guia dos Produtos Fitofarmacêuticos—Lista de Produtos com venda autorizada. 2012.

[13] Paixão, S. Ferreira, A. Lança, A. & Leitão, A. (2016a). Gestão de Riscos em Aplicadores de Produtos Fitofarmacêuticos, 1–7.

[14] Simões, J. S. 2005. Utilização de Produtos Fitofarmacêuticos na Agricultura.

Occupational Safety and Hygiene V – Arezes et al. (Eds)
© 2017 Taylor & Francis Group, London, ISBN 978-1-138-05761-6

Effect of occupational activity on ambulatory blood pressure behavior in university teachers

J. Pereira, A. Teixeira & H. Simões
ESTeSC-Coimbra Health School, Instituto Politécnico de Coimbra, Coimbra, Portugal

ABSTRACT: Hypertension (HBP) is a key risk factor for cardiovascular diseases, being strongly associated with behavioral and environmental aspects of living. Professional activities, amongst others that take place throughout the day, are responsible for important Blood Pressure (BP) variations and may increase it. This study aims at ascertaining the blood pressure profile and variation in teachers, during a typical teaching session.

Ambulatory Blood Pressure Monitoring (ABPM) was performed in a cohort of 21 university teachers during a typical professional day, comprising the following periods: 24-hour period, day period, night period, morning period, 2 hours before class, during class, 2 hours after class, aerobic exercise period and 1 hour after exercise period.

Teachers demonstrated higher BP during the occupational activities (137.71/88.57 mmHg) compared to the period before (128.81/82.43 mmHg) and after the class (132,38/85, 19 mmHg) ($p < 0.05$). It was found that systolic BP has the greatest variability across the considered activities and time periods. In a gender analysis, men had higher systolic BP compared to women (141.55 mmHg/133.50 mmHg, respectively), and demonstrated greater variability across activities.

The results clearly demonstrated the existence of important variations in BP due to different daily activities. The occupational period produced a significant increase in the different components of BP and heart rate. The long-term effects of repeated exposure to this increase in BP related with the occupational contexts remains to be demonstrated.

1 INTRODUCTION

Blood pressure varies from moment to moment as a response to the different activities and emotions experienced by individuals. Hypertension (HBP) is considered an exaggerated rise in Blood Pressure (BP), above 140/90 mmHg for systolic blood pressure and diastolic blood pressure, respectively. (1) HBP can progress without symptoms for years and, if left untreated, be a cause of major complications and death. (2)

In most cases, HBP has no apparent cause, although several aspects contribute to its manifestation, such as age, weight, smoking, sedentary lifestyle and ethnicity. Hence, HBP is associated with other conventional cardiovascular risk factors. A minority of cases of HBP have a secondary and identifiable cause. (3) The diagnosis of HBP is a key aspect in the management of this major cardiovascular risk factor, and should be made precociously. Ambulatory Blood Pressure Monitoring (ABPM) is a method through which multiple indirect BP measurements are obtained over a period of 24 consecutive hours and during the daily activities. (4) This method is essential for the diagnosis and treatment monitoring of HBP patients, and is also a fundamental tool for

the understanding of circadian and seasonal variations of BP, and to explore how occupational contexts affect the cardiovascular physiology.

Several studies have shown a strong incidence of HBP and coronary heart disease associated with occupational stress, strengthening the idea that occupational activity affects the health of individuals and could be, in some instances, seen as a cardiovascular risk factor. (5, 6). For this reason, ABPM has been shown to be a valuable method in the study of the relationship between occupational activity and the behavior of BP during work. (7).

The main objective of this study was to investigate the BP variation in university teachers, during a daily routine involving periods of teaching in classroom.

2 MATERIAL AND METHODS

A cross-sectional study was conducted in a Superior Education Institution, including teachers of both genders, with a full-time professional link with the institution. A 24 hour ABPM was performed to all teachers that voluntarily agreed to participate in the study. Demographic data was also collected to all participants.

The ABPM was performed non-invasively, with a validated device and according to a specific protocol: the ABPM monitor was programmed to measure BP every twenty minutes from 6 am to 10 pm, and every thirty minutes from 10 pm to 6 am, during a 24 hour period. The monitoring was made during a regular working day, without any compromise of routine daily activities. The participants were asked to do a 30 minute walk during the day period so that BP under aerobic exercise could be assessed.

All information was digitally collected and statistically analyzed with the SPSS software (Statistical Package for the Social Sciences, version 16.0; IBM, USA).

A descriptive statistical analysis was performed. The statistical tests used were: for continuous variables with a normal distribution, the repeated measures ANOVA, and the t Student test for independent or pairwise samples; for non-parametric continuous variables, the Friedman test, the Mann-Whitney test and the Wilcoxon test; for categorical variables, the Qui-square test, with Fisher's correction whenever appropriated. The level of significance used for the interpretation of the tests was $p < 0.05$.

The study complied with all the ethical dispositions for human research. Confidentiality of the data was warranted. All participants gave their informed and written consent. This study has no commercial or financial interest.

3 RESULTS

The cohort consisted of 21 teachers, 10 were female and 11 were males, aged between 26 and 52 years. More than half presented overweight (66.6%) and family history of hypertension (52.4%). Almost half of the cohort indicated regular physical activity habits (47.6%). Alcohol consumption during meals (19%) and smoking (9.5%) were observed in a small proportion of the participants. Systolic Blood Pressure (SBP), Diastolic Blood Pressure (DBP) and mean Pulse Pressure (PP) were situated within normal values in the majority of the cohort during the 24-hour period [SBP (90.7%), DBP (81%) and PP (85.7%)], as well as Heart Rate (HR).

Considering the BP profile during the occupational period, as compared with the before and after periods, an increase in BP was depicted. In fact, there was an increase in SBP during classes compared to the previous and subsequent periods ($p < 0.01$), and the same was true for DBP ($p = 0.037$). This behaviour was independent of gender for SBP ($p = 0.139$), although the mean increase in DBP was greater in males as compared with females ($p = 0.034$). Similar pattern was depicted for PP, HR and mean BP.

To assess the effect of mild exercise on BP of the participants, a 30-minute walk period was recommended to the participants. Curiously, SBP was significantly reduced after this period ($p = 0.023$) in all the participants, indicating that exercise could be an effective tool to promote well-being and cardiovascular health, as expected from the available evidences.

Considering other aspects of BP variation furing the 24 hour period in the cohort, it was found that participants with HBP and normotension had a similar dipping profile ($p = 0.586$) and similar mean 24h PP ($p = 0.08$).

4 DISCUSSION

Hypertension is considered an important risk factor for cardiovascular diseases, and constitutes a major public health problem in Portugal. Professional activities, as well as other activities that take place throughout the day, actually BP. Therefore it seemed pertinent to study how the occupational context in university teachers affects their BP during a typical working day.

The results showed a significant increase in BP during the occupational period, so that the higher SBP and DBP identified during the 24 h monitoring period were in fact observed during this period. These results are in line with previous reports which showed that professions that are characterized with high mental workload or intellectual engagement, or that are related to feelings of anxiety and mental exhaustion, provide greater increases in BP so that BP tends to be higher during the exposure to the occupational ecology as compared with the BP at home or during other daily routines. (7) Other studies found that activities that require mental effort (such as teaching) deeply influence BP variability. (8, 9) Other researcher stated that the sense of responsibility towards the people with whom one works may also be directly related to the increase in BP. (10).

Our results identified an increase in all components of BP, such as PP and MAP, as well as HR, during the occupational context, thus indicating a clear impact of the professional activity over the cardiovascular system. The increase in PP during the teaching session could be somehow related with an increased risk for cardiovascular events, as PP has been strongly associated with cardiovascular events. In fact, Nobre & Coelho (2003) stated that hypertension and increased PP are associated with higher cardiovascular risk. (11) Also, it was previously observed a five-fold greater risk of cardiovascular events in individuals with increased ambulatory PP (>53 mmHg), and Muxfeldt & Salles (2008) concluded that individuals with hypertension and an increase in PP values are more predisposed to cardiovascular risk.

Considering the effect of gender in the variation of BP during the occupational period, a trend to greater increases of BP was observed for males as compared with females, particularly considering DBP. Others have found no difference in the BP variation during professional activities in regards to gender (14), although others also identified greater BP reactivity in men during occupational contexts (e.g. 15), which might reflect a greater sympathetic drive in men workers.

Additionally, our data identified benefits of aerobic exercise in this population, producing a significant decrease in SBP. This finding is in line with other study that reported that BP reduces with physical activity in both hypertensive and normotensive individuals. (16). Furthermore, we found no significant differences between hypertensive individuals with abnormal BP drop and hypertensive individuals with a normal dipping profile. Notwithstanding the cohort was mostly normotensive (81%), a proportion of 58,8% presented an abnormal dipping in nocturnal BP. Several studies found a significant and independent association of an abnormal dipping pattern with major cardiovascular events and overall target organ damage in hypertensive populations. (11, 12, 13, 17, 18, 19, 20).

5 CONCLUSIONS

The study provided further evidence for the characterization of BP behaviour during the occupational moments in university teachers. An increase in all components of BP and also HR was depicted during the teaching tasks, clearly highlighting the influence that the occupational context has over the hemodynamic physiology of the teachers. Whether a prolonged exposure to this effect has long-term effects over the cardiovascular system, and whether in could contribute to the appearance of HBP in normotensive persons remains to be demonstrated in longitudinal prospective studies.

Limitations: Small cohort, with predominantly low cardiovascular risk participants; objective measures of cholesterol levels, glycaemia and other relevant parameters were not available; the information compiled through the questionnaire relied exclusively on the participant's sincerity and recollection.

Future investigations: A study with a larger cohort, followed in a longitudinal and prospective design should provide further evidence, particularly regarding the eventual association of the exposure-dependent increase in BP in the long-term. The inclusion of further physiologic, psychological and social parameters should provide a more effective evaluation of the effects of this profession on the variation of BP according to the physiological response of each individual to a given task.

REFERENCES

[1] Corrêa, A. Confiabilidade Metrológica no Sector da Saúde no Brasil, Estudo de um Caso: Qualidade Laboratorial na Saúde Pública e Controle Metrológico de Equipamentos Médico-Hospitalares. [Dissertação]. Rio de Janeiro. Pontifícia Universidade Católica do Rio de Janeiro; 2001.

[2] Costa, J. N. A. Hipertensão Arterial e o Local de Trabalho. In: Falcão L M, editors. Clínica e Terapêutica da Hipertensão Arterial. Lisboa: Lidel; 1997: p 15–19.

[3] Trigo, M.; Coelho, R.; Rocha, E. Factores de Risco Clássicos e Sócio-Demográficos da Doença das Artérias Coronárias: Revisão da Literatura. Sociedade Portuguesa de Psicossomática 2001 Jul-Dez; 2 (3): p. 239–262.

[4] Nobre, F. Monitorização Ambulatorial da Pressão Arterial (MAPA). Medicina 1996; 29: 250–257.

[5] Mendes, A. A Hipertensão Arterial e o Local de Trabalho. In: Falcão L M, editores. Clínica e Terapêutica da Hipertensão Arterial. Lisboa: Lidel, 1997: p. 21–33.

[6] Robazzi M. L., Veiga E. V., Nogueira M. S., Hayashida M. & Ruffino M. C. Valores depressão arterial em trabalhadores de uma instituição universitária. Ciência Y Enfermeria 2002; (8): 57–65.

[7] Couto H. A., Vieira F. L., & Lima E. G. Estresse ocupacional e hipertensão arterial sistémica. Revista Brasileira de Hipertensao 2007; 2(14): 112–115. 10

[8] Mattos C. E., Mattos M. A., Toledo D. G. & Filho A. G. Avaliação da Pressão Arterial em Bombeiros Militares, Filhos de Hipertensos—Através da Monitorização Ambulatorial da Pressão Arterial. Universidade Federal do Rio de Janeiro e Corpo de Bombeiros Militar do Estado do Rio de Janeiro 2006 Fevereiro; 6 (87): 741–746.

[9] Schnall P. L., Landsbergis P. A., Warren K., & Pickering T. G. Relation between job strain, alcohol and ambulatory blood pressure. Hypertension 1992; 19: 488–494.

[10] Serra, A. V. O stress no trabalho. In: A. V. Serra, editores. O Stress na vida de todos os dias. Coimbra: 2002.

[11] Nobre, F.; Coelho, E. B. Três Décadas de MAPA—Monitorização Ambulatorial da Pressão Arterial de 24 horas: Mudanças de Paradigmas no Diagnóstico e Tratamento da Hipertensão Arterial. Arquivo Brasileiro Cardiológico 2003 Março; 4 (81): 428–34.

[12] Noll C. A., Lee E. N., Schmidt A., Coelho E. B. & Nobre F. Ausência de queda da pressão arterial entre os períodos de vigília e sono. Revista Brasileira de Hipertensão2001; 8: 468–471.

[13] Muxfeldt, E. S. & Salles, G. F. Pulse pressure or dipping pattern: which one is a better cardiovascular risk marker in resistant hypertension? 5 de Abril de 2008. [0 ecrans].

[14] Egeren, L. F. The relationship between job strain and blood pressure at work, at home and during sleep. *Psychosomatic Medicine* 1992; 54 (3): 337–343.

[15] Reckelhoff, J. F. Gender differences in the regulation of blood pressure. Hypertension 2001; 37: 1199–1208.

[16] Arroll B. & Beaglehole R. Does physical activity lower blood pressure: a critical review of the clinical trials. Journal of clinical epidemiology 1992; 45: 439–447.

[17] Silva A. M., Monteiro A., Nogueira B., Cardoso F., Maldonado J., Morais J., et al. Hipertensão arterial na prática clínica. Lisboa: 2006.

[18] Ohkubo T., Hozawa A., Yamaguchi J., Kikuya M., Ohmori K., Michimata M., et al. Prognostic significance of the nocturnal decline in blood pressure in individuals with and without high 24-h blood pressure: the Ohasama study. *Journal of Hipertension* 2002; 20: 2183–2189.

[19] Brandão A. A., Pierin A., Amodeo C., Giorgi D. M., Mion D., Nobre F., et al. III Diretrizes para uso da Monitorização Ambulatorial da Pressão Arterial, I Diretrizes para uso da Monitorização Residencial da Pressão Arterial. Revista da Sociedade Brasileira de Hipertensão 2001; 4 (1): 6–18.

[20] Gomes, M. Normatização dos Equipamentos e Técnicas para Realização de Exames de Mapeamento Ambulatorial de Pressão Arterial (MAPA) e de Monitorização Residencial da Pressão Arterial (MRPA). Arquivo Brasileiro Cardiológico 2003; 80: 225–34.

Occupational Safety and Hygiene V – Arezes et al. (Eds)
© 2017 Taylor & Francis Group, London, ISBN 978-1-138-05761-6

A proposal for a project plan on quality, human resources and stakeholders for events in the tourism area

G.R. Freitas
DCC/NPPG, Politecnic School—Federal University of Rio de Janeiro, Rio de Janeiro, Brazil

J.L. Fernandes
Department of Production Engineering—CEFET, RJ, Rio de Janeiro, Brazil

A.S.C. Fernandes & L.A. Alves
Department of Civil Engineering—CEFET, RJ, Rio de Janeiro, Brazil

ABSTRACT: The present work presents a strategic planning using the tools SWOT, Porter and Balanced Scorecard as subsidies for the conduction of the National Seminar of Tourism and Culture. This analysis will bring qualitative and quantitative benefits to the organization of an event, such as cost reduction, greater control under your project, timelines and less wear and tear on the organizers of this event.

Keywords: project plan on quality, stakeholders, SWOT, Tourism

1 INTRODUCTION

The event can be a simple business meeting or even a large technical and scientific congress. According to Britto et al. (2011) much more than a successful event, party, communication language, public relations activity or even marketing strategy, the event is the sum of efforts and actions planned with the goal of achieving defined results with your audience—target.

Arthur Bormann, one of the authors who emerged in the 1940s, from the Berlin school, defines tourism as a set of journeys aimed at pleasure or commercial, professional or similar reasons, during which it is absence of habitual residence. (Barretto, 2000, pp. 23–24). Although it does not have a single definition for tourism, it is relevant to note that, in accordance with the World Tourism Organization/United Nations Recommendations on Tourism Statistics, they define it as: "The activities that people perform during their travels and stay in places other than those living for a period of time less than a consecutive year for leisure, business and other purposes" (Conceito Turismo, 2015).

1.1 Description of the event to be proposed in the project

This paper will present an analysis about the First National Seminar on Tourism and Culture. The 1st National Tourism Seminar will be held at Casa Rui Barbosa, located in Botafogo/Rio de Janeiro—Brazil, where around 300 people will participate in this event.

The event will be attended by the academic community and tourism professionals, who will discuss the relations and intersections between tourism and culture. In this first edition, which took place on May 10 and 11, 2016, due to the Olympic year, the theme will especially address the effects of mega-venues on tourist flows, highlighting opportunities and challenges for Rio de Janeiro. In the seminar, the following thematic axes of the four working groups will be described as follows:

a. GT 1—Heritage, Hospitality and Tourism: WG 1 (Working Groups) aims to explore the issue of cultural heritage, considering the material and immaterial dimensions, as well as its relevance for the construction of cultural identities. It is still necessary to establish a bridge with the studies of hospitality, reflecting on its manifestations in the field of tourism.
b. GT 2—Mobility and Tourism: In addition to encompassing studies related to the transport universe, WG 2 aims to stimulate the debate about the Paradigm of New Mobility. Extrapolating the bodily mobility, essential for tourism, will be considered other categories, as for example: 1) the physical movement of objects;

2) imaginative mobility; 3) virtual mobility; and, 4) communicative mobility. The interface of this paradigm with tourism studies will be especially valued.

c. GT 3—Major Events, Tourism Management and Planning: In the tourism planning and management bias, WG 3 proposes an analysis of the Great and Mega-events, their positive and negative impacts, as well as the opportunities and challenges they represent for Tourism. It intends to develop a re-flexion about the big events that the city of Rio de Janeiro hosted in the last decade, considering also the preparations for the Olympics in 2016. Finally, the WG intends to temper the dynamics of the Actors involved, such as public authorities, companies and the communities involved.

d. GT 4—Faces of the modern phenomenon of leisure: GT 4 aims to scrutinize the modern phenomenon of Leisure, and Tourism is only one of its various manifestations. It is intended to explore, conceptually and empirically, the various tensions inherent in the field, as well as its relevance to contemporary Western society.

Tables 1 and 2 show the schedule of the event I National Seminar on Tourism and Culture:

The Casa de Rui Barbosa Foundation, located in Rio de Janeiro—Brazil, has its origins in the museum-library instituted in 1928 by President Washington Luis, the House of Rui Barbosa. In 1966, the institution had its legal personality changed by Law No. 4,943, in order to better fulfill its purposes of development of culture, research and teaching, as well as the dissemination and worship of the work and life of Rui Barbosa. (Casa Rui Barbosa, 2016).

Table 1. Schedule of Day 1—Event I National Seminar on Tourism and Culture.

DIA 1	Location: Auditorium Casa de Rui Barbosa Foundation
	09 h–09 h30: Accreditation
	09 h30–10 h: Welcome Coffee
	10 h–10 h30: Opening of the Pales-tra Event (Lia Calabre)
	10.30–12.30: "Reflections on the intersection between culture and tourism"
	12 h30–14 h: Lunch
	14 h–15 h30: First session of presentation of scientific papers
	3:30 p.m. to 4 p.m.: Coffee Break
	16 h–17 h30: Second session of presentation of scientific papers
	17 h30–18 h: Closing of the First Day of Seminar

Table 2. Schedule of Day 2—Event I National Seminar on Tourism and Culture.

DIA 2	Location: Course room Casa de Rui Barbosa Foundation
	09 h–09 h30: Accreditation
	09 h30–10 h: Welcome Coffee
	10 h–10 h30: Opening of the Pales-tra Event (Lia Calabre)
	10.30–12.30: "Reflections on the intersection between culture and tourism"
	12 h30–14 h: Lunch
	14 h–15 h30: First session of presentation of scientific papers
	3:30 p.m. to 4 p.m.: Coffee Break
	16 h–17 h30: Second session of presentation of scientific papers
	17 h30–18 h: Closing of the First Day of Seminar

Source: Author.

2 ASPECTS OF THE STRATEGIC ANALYSIS OF THE EVENT I NATIONAL SEMINAR ON TOURISM AND CULTURE (RESULTS)

2.1 SWOT analysis

The SWOT analysis will be demonstrated through Table 3, which will address the points of Strength, Weakness, Opportunities and Threats, identified from the analyzed event.

Through the SWOT Analysis, there are factors that can be considered as relevant in the strategic planning of the 1st National Seminar on Tourism and Culture. In the analysis of threats there was a need to attract sponsors and supporters to provide resources and materials for the Events, or implement some system that provides actions for certificates, make a creation of books to attract sponsorship and support. Keeping the theme of the event interesting is not only sharpen the interest of the participants and implement more marketing campaigns aimed at reaching a target audience, offering small compensations such as: differentiated service with innovating solutions for events and implement an online confirmation system along with the intrascription system to give participants feedback via email.

2.2 Studying the competitiveness of the market from the five PORTER forces

For the event I National Seminar on Tourism and Culture, another strategic tool will also be analysed, which is the analysis of the five forces of PORTER. This analytical tool is used to understand competition between companies as well as in industry and can also guide event managers and their entire marketing decision-making. (Allen, 2008) Figure 1. will

Table 3. SWOT analysis.

	FACTORS INTERNAL	FACTORS EXTERNAL
FAVORABLE FACTORS	FORCES 1. Teachers of high credibility in federal colleges; 2. Interesting theme for the 2016 Olympic year; 3. Free event to the public; 4. Event with full inscriptions.	OPPORTUNITIES a. Growth in the event market; b. To arouse the interest of the participants; c. Only seminar in the year for tuition; d. Strengthening the image of the institution.
UNFAVORABLE FACTORS	WEAKNESSES 1. Event does not provide certification to the participants; 2. Few means of dissemination; 3. Number of limited participants; 4. There was no consignature of the participants at the time of registration.	THREATS a. Lack of interest of participants; b. No one's attending speaks to you; c. Organizations with new events in the market; d. New potential competitors.

Source: Author.

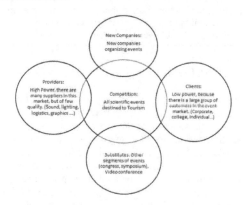

Figure 1. Analysis of the Five Forces of PORTER. Source: Author.

analyze PORTER's Five Forces at the First National Seminar on Tourism and Culture.

Through the Protests Analysis, it is important to establish a relational strategy of establishing suppliers to ensure continuity of supply at reasonable prices. Although some event suppliers can have a lot of power, relationship marketing strategies help you deal with this imbalance of forces.

2.3 Balanced Scorecard (BSC)

The strategic elements of the event can not be dissociated from the institution that promotes it and sought to reflect the entity's mission, vision and values. The entity (Casa Rui Barbosa Foundation) has as its mission the development of culture, research and teaching, especially the dissemination and worship of the work and life of Rui Barbosa. In this way, the institution can contribute to the knowledge of cultural diversity and to the strengthening of citizenship, ensuring the implementation of the other policies of the Ministry of Culture. (Site 28).

Strategy Definition—In general, institutions have difficulties in defining a clear and well-defined strategy (Niven, 2007). This thesis is also defended by Kaplan and Norton (2000), who point out that difficulties in the definition of strategies are very common in organizations, since it is rare to find entities that align the strategy with the Leadership of your products and services and with the customer.

Based on the above elements, they have been defined as strategies that will contribute to the accomplishment of the objective, mission, vision and values of the event, the following initiatives are aligned with Casa Rui Barbosa's objectives:

Encourage the participation of students, teachers, researchers, tourism professionals and others interested in broad dissemination in different media;

Signing partnerships with public agencies and other private institutions to publicize and sponsor the event;

Foster debate and scientific production through sections of presentation of national and international academic productions;

Facilitate the process of enrolling researches and participants through the creation of an eletronic site or through the creation of registration systems;

Ensure the correct flow of information through the organizing committee of the event to organize the communication between all the coordinators of the event, aiming to homogenize the procedures of relationship with the interested parties (Controlling Governors, Sponsors, Participants, etc.) and Improve the image of the event and institutional;

Measure satisfaction and obtain feedback through the elaboration and application of opinion and satisfaction research with supporters/sponsors, speakers and participants through electronic means.

After presenting the strategies of the event, the strategic objectives and their interactions are presented.

Table 4. Integration between strategies.

	INTEGRATION
SWOT	Improve marketing, encourage marketing campaigns by promoting differentiated services; Make more campaigns of events; Form or have a marketing management team; Capture more sponsors and supporters; Adopt good practices of event management in the organization's culture; Partner with public and private competing companies; Have more flexibility with the number of participants; Implement a system to confirm participants' registrations, giving feedback via e-mail to subscribers; Have a database of all the speakers; Train the organizing team with training (when necessary); Further enhancing your team's know-how in event management;
PORTER	Through the competitive market analysis carried out through the PORTER tool, it is understood that the event market is very competitive. And for this competitive strategy to go live, the other aggregate tools were SWOT and BSC analysis.
BSC	The strategic map of the event was based on the results of the SWOT and PORTER analyzes, in order to create value for the event and the institution. Manage competent, committed, qualified people; Gar-antir internal and effective communication; Have an adequate information system for decision making; Have database system; Have online registration system giving feedback; Strengthen communication with stakeholders and teams; Bus-car excellence in management; Guarantee the necessary resources; Manage sustainable aspects correctly; Fulfill the mission of the event; Promote the event in various media; Signing partnerships with public and private companies; Strengthen the image of the institution and the credibility of the event; Ensuring the satisfaction of participants, supporters and sponsors; Increase the number of event participants and sponsors.

According to Kaplan and Norton (2004), the strategic map represents the missing link between the formulation and execution of the strategy. Its construction is guided by the following premises:

a. Financial Perspective—to be financially successful, as we should be seen by our shareholders; Promote the event in various media and establish strategic partnerships with public and private companies.
b. Customer Perspective—to reach our vision, how we should be seen by our customers; Strengthen the image of the institution and the credibility of the event; Increase the number of participants and sponsors of the event.
c. Perspective of Internal Processes—to deepen our vision, how we will maintain our ability to change and improve; Strengthen communication with stakeholders and teams; Run the production of the event with quality and efficiency. D. Learning and Growth Perspective—to satisfy our shareholders and customers, how will we sustain the ability to improve and change; Manage and ensure effective internal communication of the people involved in the Event.

2.4 *Integrations between the SWOT, PORTER and Balanced Scorecard*

The integrations between the SWOT, Porter and BSC analyzes will be possible to visualize the integrations between the strategies approached in this work, which will be demonstrated in Table 4.

3 CONCLUSIONS

The integration of the strategy with the company's business strategy involves aspects of the company's culture and values and refers to the importance of the date to the issues in the company's strategic planning.

In order for all the strategic actions raised here to be implemented, they must be inserted as culture within the company, as well as good practices in event or project management, must be part of the organization's day-to-day values. This is a task that, although it should be led by the project manager, also depends on the support of the entire team participating in the project.

The integration between the strategies of the organization aims to maintain the communication and synchronization of everything that is happening in a way that everything fits perfectly.

REFERENCES

Allen, J. et al. 2008. Organization and Management of Events. 3rd edition. Elsevier Publishing Company.

Barretto, M. 2000. Introduction to the Study of Tourism. 8th edition. Papirus Publishing Company.

Britto, J. et al. 2011. Strategies for Events 2nd edition expanded and updated. Publishing Company Aleph.

House Rui Barbosa. 2016. Available at: http://www.casaruibarbosa.gov.br/. Access in: February/2016.

Kaplan, R. and Norton, D. P. 2000. Organization Oriented to the Strategy. Campus Publishing Company.

Niven, P. R. 2007. Balanced Scorecard step-by-step: Raising performance and maintaining results. Rio de Janeiro: Qualitymark7.

Tourism Concept. 2015. Available at: http://en.slideshare.net/cursotiat/01-conceitos-turismo. Access in: February/2016.

Occupational Safety and Hygiene V – Arezes et al. (Eds)
© 2017 Taylor & Francis Group, London, ISBN 978-1-138-05761-6

Occupational exposure to chemical agents released on cooking processes at professional kitchens

A. Ferreira, A. Lança, D. Barreira, F. Moreira, J. Almeida & T. Neves
ESTESC, Coimbra Health School, Saúde Ambiental, Instituto Politécnico de Coimbra, Coimbra, Portugal

ABSTRACT: In the catering sector, there are several risks related to the tasks performed in the kitchen. The smokes produced in the confection of food, where fats and oils are burned an it consists on the most common risk, also known as Cooking Oil Fumes. The cooking processes releases harmful components to the Work environment and then to the workers lungs. The thermal environment is also a factor that influences the performance of the worker's activities, and represents an extremely important role in improving working conditions. This study aims to evaluate how the cooking smokes influence the quality of working environment and the thermal environment, and consequently the health of the workers. It was conducted an in-depth research on the components released and their consequences. As conclusion the inhalation and the contact with released pollutants like these are related to the appearance of several pathologies, such as cancer.

1 INTRODUCTION

The improvement of air quality over the last decades has been one of the great achievements of Community policy on the environment, showing that it is possible to dissociate economic growth from the degradation of the environment. However, there are problems that persist and need to be solved. Concerns related to the effects of air quality on public health, generally consider the air pollution only outside the buildings. The Restoration sector is constantly expanding and represents an important source of employment in the services sector, employing around 7.8 million people in the European Union. The working conditions of this sector presents a number of risks associated to physically demanding work, exposure to high noise levels, work performed in hot or cold environments, falls, cuts, burns, exposure to hazardous substances and psychosocial risks related to the ergonomics conditions, job requirements, working hours and autonomy, among others.(7)

Tasks performed in a kitchen abound in risk factors/occupational hazards, however, some of these are not obvious to the worker, as is the case of exposure to cooking fumes/vapors (COFs— cooking oil fumes).(8) Cooking at temperatures of 240°C to 280°C, oils and fat acids break down into volatile products that may differ in quality and quantity, depending on the source and the purity of the oil. High temperatures lead to the emission of large amounts of fumes from the oil used and they can cause eye irritation; and the emission of

a wide variety of toxic agents, some of which are potentially carcinogenicand mutagenic. The particles emitted during the cooking activity contain a variety of toxic agents, such as polycyclic aromatic hydrocarbons, which, depending on temperature and humidity, may exist in the gases and particles generated at high temperature during food confection mode.(6) Taking into account the toxicity of agents released in the confection of food, this article has as its main objective to characterize the conditions and substances to which the workers are exposed in the performance of their daily functions, also demonstrating that when exposed to these same agentes, they can trigger the development of considerably serious diseases. To develop this article was conducted an extensive literature search of papers and magazines which were crucial to its progress.

2 AIR QUALITY

Kitchens are places where food is prepared, there is daily cleaning of equipment and facilities. The quality of food in restaurants is closely related to it's structural and environmental conditions, and there are factors such as indoor air quality that can compromize food quality.(7) During some of these processes, heat generation, water vapor, chemical substances, smoke, and odors affecting indoor air quality and the working conditions of workers are observed.(6) Because of that, is necessary to guarantee the extraction of the chemical substances and agents, as well as the high thermal loads generated during all the activities involved. it is possible to fight this by venting the indoor air, and renewing it with fresh and clean air. Ventilation may be mechanical, natural or combined, but the first option will be the right one because it makes control easier and more reliable, guaranteeing constant inflation and extraction.(5)

2.1 Indoor air quality

Currently, people spend most of their time indoors, in these indoor spaces both the number of pollutants and their concentration are generally much higher than in the outside air.(2)

We know that pollutants like—carbon monoxide, carbon dioxide, ammonia, sulfur oxide and nitrogen—are produced inside the building by building materials based on organic solvents, cleaning materials, mold, human metabolism, and man's activities, such as cooking, or washing and drying clothes. Such pollutants compromise workers' health and income from work.(4)

A combination of factors can influence indoor air such as the climate, building ventilation system, sources of pollutants in the interior, the exchange of pollutants with the exterior, the number of occupants of the building and the activities of these on-site. Also, thermal parameters such as temperature and humidity are also factors that influence the emission of pollutants such as Volatile Organic Compounds (VOCs) and Polycyclic Aromatic Hydrocarbons (PAHs), and the development of microorganisms.(6)

In the preparation of meals, workers are exposed to several dangerous substances contained in the smoke or vapors released, resulting either directly from confection processes or from cleaning and dis-infection products used or even from industrial washing machines. These substances can be VOC's, HAP's, fats and water vapor.(5)

2.1.1 Fats

The fat may be in the liquid or gaseous state. In the extraction process, filtration and separation must be used, since their accumulation in the ventilation networks may increase the risk of fire, as well as more frequent maintenance for cleaning.(5) A very clear example is the animal fat that in contact with high temperatures, when grilled, releases about 200 VOCs, some of them of toxicological nature with carcinogenic and mutagenic properties.(5)

2.1.2 Volatile Organic Compounds (VOCs)

Volatile Organic Compounds are a group of chemicals, with or without odor, which contain organic compounds of carbon and at a pressure of 101,325 kPa and evaporate at a temperature of 250°C or less. (3) These substances may originate from varnishes,

paints, fossil fuels that have not been or were only partially burned, or released directly by food confection. Exposure to these agents may have adverse effects on human health and may therefore be considered toxic or even carcinogenic and exposure to them should be avoided.(5) VOCs include a number of chemical compounds such as formaldehyde and benzene, which may have short or long term effects. The concentration of VOCs found in indoor air is about ten times greater than that of out-door air. (6) In the manufacturing processes volatile organic compounds such as acetic aldehyde, pentene, pentanol, cyclopentane, alkylbenzene, benzene, butadiene, isomers of octane, octene, n-heptanal, diene, unsaturated hydrocarbons and particularly serious, we emphasize the aldehydes of short chain. (3) Exposure to VOCs can be by ingestion, breathing or by dermatological absorption. The effects vary de-pending on their concentration, which increases greatly indoors and without air circulation.(5)

2.1.3 Polycyclic Aromatic Hydrocarbons (PAHs)

PAHs are chemical substances composed of two or more benzene rings that have their main origin in the incomplete combustion of organic matter. These substances may originate from the incomplete burning of fossil fuels, waste, among other organic substances such as tobacco or confectionery.(5) PAHs are usually colorless, white, or light greenish-yellow. (3) PAHs were one of the first carcinogens to be identified and are formed be-cause carbon or fuel is not converted to CO and CO_2 in the combustion process in professional kitchens.(3) Concern over the existence of these substances in professional kitchens has been increasing, and several studies over the last 30 years have reported an increased risk of cancer among cooks and other catering workers.(5)

2.1.4 Effect of exposure to fats, VOC's, and PAH's on the health of the workers

There are many factors that determine whether an individual exposed to PAH contracts any disease, what type and severity. The health effects of exposure to a hazardous substance depend on the dose, duration, as discussed, whether other chemicals are present, and varies with age, sex, nutritional status, genetic tendency, lifestyle, and health status. In humans, studies have shown that long-term breathing or skin exposed to mixtures containing PAHs and other compounds may also develop cancer.(3) Prolonged exposures to VOC's can cause kidney and liver effects, resulting in increased levels of serum enzymes, slight cellular changes and metabolism, in higher concentrations this may result in irritation of the eyes and respiratory tract, such as nausea and loss of balance.(5)

Many mutagenic and carcinogenic compounds have been identified in cooking oil fumes and previous investigations have also shown that cooks may have an increased risk of developing long cancer.(10)

Several Epidemiological studies were made in many kitchens who show an elevated incidence of cancers among non-smoking women with long-term exposure to cooking oil-fume and an excessive bladder cancer rate among cooks exposed to kitchen air.(9)

3 THERMAL ENVIRONMENT

The thermal environment is one of the most important parameters to be characterized in order to ensure comfort and improve working conditions.(5) Thermal comfort is not exact and does not involve unique thermal conditions. The temperature of the work place affects the productivity of the workers, besides, that the conditions of thermal comfort also decrease the number of accidents. When the interior temperature in professional kitchens is too high (above 28°C), productivity and overall comfort rap-idly decrease. (3) According to EASHW (European Agency for Safety and Health at Work) the ideal temperature for the kitchens is between 20°C and 22°C, so that above 24°C, a productivity loss of around 4% for each degree above this value begins to occur. The risks increase significantly from 26°C, with a decrease in concentration, power, and work capacity decreases.(5)

3.1 Thermal confort

Thermal comfort can be evaluated from the combination of different factors, both physical and individual.(3) Thermal comfort is a non-linear concept, dependent on various personal factors or the sur-rounding environment. At a personal level, one must consider the clothing used, the metabolism, the energy expended to perform a certain task and the physical constitution of the worker. The environment around the person will have as main influences, the air temperature, average radiant temperature, humidity, and the speed of air circulation.(5)

3.2 Effect of high temperatures on the human body

A hot thermal environment can be defined in which the body must resort to alternatives in order to keep the thermal balance null.(5) High temperatures have a very significant impact on humans. When passing the comfort zone there is a lack of satisfaction and bad feeling on the part of the individual in the space in which it is.(5) With the rise in temperature, a series of events occur at the level of the organism, a cardiac and circulatory system overload be-gins to exist. Long periods of exposure to high temperatures may involve disorders such as thermal fatigue.(5) The temperature of the human body is usually between 36 and 37°C. When body temperature exceeds these values, the body initiates certain physiological mechanisms to counteract this increase, it reacts through the circulation of blood to the skin, causing the temperature of the skin to increase, and release to the outside the excess heat. With muscles producing work, less blood is available to circulate to the skin and it is not possible to release the heat produced. The body temperature, when increasing and not being diminished by sweating, causes the body temperature to increase more and more and the worker goes into thermal stress. (3, 5) A detailed analysis of the influence of the environment on thermal stress always requires knowledge of four basic parameters already mentioned: air temperature, average radiant temperature, air velocity and humidity.(3) Thermal discomfort can also be caused by heating or cooling located on a particular part of the body. The most common factors of discomfort are the asymmetry of average radiant temperature, air currents, vertical air temperature differences and cold or hot pavements.(3)

3.3 Thermal environment in professional kitchens

Until very recently, it was not considered that such a prolonged and intense exposure, such as kitchens workers, can have a serious impact on the quality of life of these people, jeopardizing their safety and health. This may be due to emissions of malignant particles as well as the thermal conditions to which these workers are subjected. In this way, governmental norms and impositions have been developed, so that employers can offer good working conditions to their professionals.(5) In the specific case of kitchens, thermal comfort is something that is impossible to keep constant, being considered as areas more precarious the areas next to appliances with strong heat emissions.(5)

Figure 1. Source: Carneiro PMCMF. Ambiente Térmico e Qualidade do Ar em Cozinhas Profissionais. Dissertação. Coimbra. Fac-uldade de Ciência e Tecnologia da Uni-versidade de Coimbra; 2012.

There are many different measures of protection against heat, however to carry out their application a detailed characterization of the thermal environment of the place in question must be carried out. These measures may affect the sources of heat, the working environment or individual protection measures. (3)

3.4 Prevention measures to avoid discomfort and thermal stress in kitchens

A system of rotating tasks and rest periods should be adopted, and the same worker should not be exposed to high temperatures for long periods of time, alternating with cooler places. This will prevent and relieve thermal stress. Hydration is also something that should be a constant concern on the part of the workers. Drinks containing caffeine or alcohol should be avoided. The physical work must be controlled and the transport of high loads should be done by means of manual trolleys. Thus, heavier physical work should also be scheduled for cooler times. The clothing used should be lightweight, comfortable and made of cotton, to ensure air circulation and sweat evaporation. (5) There must be provision of personal protective equipment, such as gloves, aprons, among others, whenever necessary. Exhaust systems should be installed above high heat production sites, such as stoves, fryers, grills, among others. The most effective way to reduce excessive heat and vapours is through a ventilation system with displacement ventilation. With this form of ventilation the air introduced at ground level, at a reduced speed and at a slightly lower temperature than the local environment, causes the warm air and contaminants produced to rise to the ceiling where they are extracted. The displacement ventilation also has the advantage of energy efficiency and a quiet operation.(3) The training and awareness of workers will undoubtedly be the main measure to be taken. A worker who is subject to these thermal environments should be able to recognize the signs and symptoms of problems related to exposure to high temperatures. They should also know the factors that influence the thermal sensitivity at the individual level and should be aware of the plan adopted to avoid problems and health risks.(5)

4 CONCLUSION

In kitchens, workers are always exposed to risks, some bigger than others but can potentially lead to future injuries and potentially very serious diseases. The factors of the indoor air quality of the kitchens and the thermal environment are undoubtedly the risks present in this area. It is important that soon in the construction of the establishments the concern is to ensure that there is adequate ventilation to the type of establishment that is being created, following the legislation, unfortunately there is often such a concern to save, that they prefer to spend in other places, somewhat devaluing the ventilation and coditions for the workers. We can see that workers can be exposed to numerous substances that become extremely harmful, especially if in contact with them systematically over the years. To reduce contact with harmful substances, ventilation and artificial exhaust systems should be available in such a way that workers do not inhale them or at least ensure a reduction in exposure. The Ventilation System will also help keep the room at a stable ambient temperature, ensuring thermal comfort for the worker, guaranteeing greater safety and efficiency in the performance of their tasks. Above all, it is extremely important to ensure good conditions for workers, making them healthier and more satisfied. It is necessary that the workers are properly educated and informed about the risks and substances to which they are exposed, for this should be scheduled recurrent training over time.

REFERENCES

[1] Agência Portuguesa do Ambiente. Ar. Portugal: 2016; [updated 2016; cited 2016 1 November]; Available from: https://www.apambiente.pt/index.php?ref=16&subref=82.
[2] Agência Portuguesa do Ambiente. Qualidade do Ar Interior. Portugal: 2016; [updated 2016; cited 2016 1 November]; Available from: https://www.apambiente.pt/index.php?ref=16&subref=82&sub2ref=319.
[3] Baptista FM. Ventilação de Cozinhas Profissionais (Ambiente Térmico e Qualidade do Ar). Dissertação. Coimbra. Faculdade de Ciência e Tecnologia da universidade de Coimbra; 2011.
[4] Carmo, A. T., & Prado, R. T. A. (1999). Qualidade do Ar Interno, 35.
[5] Carneiro PMCMF. Ambiente Térmico e Qualidade do Ar em Cozinhas Profissionais. Dissertation. Coimbra. Faculdade de Ciência e Tecnologia da universidade de Coimbra; 2012.
[6] Coelho PIC. Exposição Aos Compostos Orgânicos Voláteis—Trabalhadores Em Cozinhas Escolares. Lisboa. Instituto Politécnico de Lisboa Escola Superior de Tecnologia da Saúde de Lisboa. 2014
[7] Ferreira D, Rebelo A, Santos J, Sousa V, Silva MV. Estabelecimentos de Restauração e Bebidas: Estudo sobre a Qualidade do Ar Interior em Cozinhas. International Symposium on Occupational Safety and Hygiene 2012: 189–191.
[8] Santos M, Almeida A. COFS (Cooking Oil Fumes). Revista Portuguesa de Saúde Ocupacional online 12 de Outubro de 2016; volume 2, 1–2. Available from: http://www.rpso.pt/cofs-cooking-oil-fumes/
[9] Singh A, Kamal R, Mudiam MKR, Gupta MK, Satyanarayana GNV, Bihari V, Shukla N, Khan AH, Kesavachandran CN. B-ON. 12 February 2016.
[10] Svendsen K, Jensen HN, Sivertsen I, Sjaastad AK. Exposure to Cooking Fumes in Restaurant Kitchens in Norway. Oxford Journals. 17 December 2001. Available from: http://annhyg.oxfordjournals.org/content/46/4/395.short.

Occupational Safety and Hygiene V – Arezes et al. (Eds)
© 2017 Taylor & Francis Group, London, ISBN 978-1-138-05761-6

Bacteria load a surrogate to assess occupational exposure to bioaerossols?

A. Monteiro, M. Santos & T. Faria
Environment and Health Research Group (GIAS), Escola Superior de Tecnologia da Saúde de Lisboa, ESTeSL, Instituto Politécnico de Lisboa, Lisboa, Portugal

C. Viegas
Environment and Health Research Group (GIAS), Escola Superior de Tecnologia da Saúde de Lisboa, ESTeSL, Instituto Politécnico de Lisboa, Lisboa, Portugal
Centro de Investigação e Estudos em Saúde Pública, Escola Nacional de Saúde Pública, ENSP, Universidade Nova de Lisboa, Lisbon, Portugal

S. Cabo Verde
Centro de Ciências e Tecnologias Nucleares, Instituto Superior Técnico, Universidade de Lisboa, Lisbon, Portugal

ABSTRACT: In feed production, pig farms and swine slaughterhouses, the workers are exposure to bio-aerosols. The aim of this study was to assess occupational exposure to bacteria load (air and surfaces) in one feed production industry, in two pig farms and in one swine slaughterhouse. In addition, the authors intended also to analyze the results together with the air fungal load already reported. Air samples were collected through an impaction method, while surface samples were collected by the swabbing method. The pig farms demonstrated to be the setting with the highest bacterial load. Bacteria load should not be a surrogate to assess occupational exposure to bioaerosols, since bacterial and fungal load didn't present the same load trend, mainly in the feed industry and in clean and dirty slaughterhouse.

1 INTRODUCTION

In recent decades, the importance of bioaerosols in occupational and housing contexts has been emphasized, due to the adverse effects they cause on human health and consequent impact on public health (Douwes et al. 2003; Badri 2014). Bioaerosols are a heterogeneous mixture of biological agents, airborne particles including bacteria, fungi, viruses and their by-products, endotoxins and mycotoxins which in addition to infectious diseases can cause allergic reactions, irritant or toxic (Oppliger, 2014).

Bacterial cells or their fragments present in the inhaled air cause health problems, mainly pulmonary or cutaneous diseases (Miaśkiewicz-Peska & Łebkowska, 2011). In addition, the presence of gram-negative bacteria is an increased risk of endotoxin production, which can lead to acute and toxic inflammation of the lungs (Miaśkiewicz-Peska & Łebkowska, 2011), which has been recognized as an important factor in the etiology of occupational lung diseases, including asthma (Douwes et al. 2003).

Fungal spores are complex agents that may contain multiple hazardous components. Health hazards may differ across species and strains because fungi may produce different allergens and also mycotoxins, and some species can infect humans (Eduard 2009).

The animal feed industry has been reported to be a setting with significant bio-contamination and some of the microorganisms involved in respiratory problems have been detected in this environment, namely several gram-negative bacteria whose toxic effects are derived from the inhalation of endotoxins (Awad 2003; Kim et al. 2009). Also the presence of fungal load in feed production (Halstensen et al. 2013; Viegas et al., 2016a), has been considered a factor contributing significantly to these symptoms (Swan & Crook 1998). Moreover, some studies in last years suggested pig farm activities associated with a variety of respiratory symptoms among farmers (Heederik et al. 1991).

In slaughterhouses, the biological risk is present not only from the direct or indirect contact with animal matter (feces, innards, feathers), but also from the exposure to bioaerosols (Viegas et al. 2016b).

In addition, according to several authors (Stetzenbach, Buttner & Cruz, 2004; Viegas et al. 2016c) surfaces analysis complements microbiological characterization of the air and is used in order to identify contamination sources.

The aim of this study was to assess occupational exposures to bacteria air and surfaces load in one feed production industry, in two pig farms and in one swine slaughterhouse. In addition, the authors intended also to analyse the results together with the air fungal load already reported (Microbiota air load analyses—3.2 section).

2 MATERIALS AND METHODS

2.1 Assessed settings

All places were assessed in July 2015 during a normal working day. One feed production, two pig farms and one swine slaughterhouse were selected, all located in Setubal district, Portugal.

The sampling sites selected were chosen considering the local where the workers spent the majority of time by during their occupational activity. In the slaughterhouse, the selected points for the air sampling focus on two distinct zones, the clean and the dirty zone (Table 1). In all the analyzed workplaces none of the workers used respiratory protection devices.

2.2 Sample collection, preparation and analyses

Air samples were collected based on conventional methods. The amount of collected air ranged from

Table 1. Sampling sites selected from each setting.

Setting	Sampling sites
Feed Production (FP)	Bagging
	Control
	Warehouse
	Pharmacy
Pig farms (PF)	Gestation (in parks, in saddles)
Gestation	Batteries
	Maternities
	Fatten
	Quarantine
Swine Slaughterhouse (SW)	
(Clean zone)	Packing
	Chorizos packing
	Labeling
	Production greenhouse
	Production room
	Cut
(Dirty zone)	Gut room
	Bleeding swine
	Cutting swine

25 L (from feed production), 100 L (from dirty zone in slaughterhouse and in the pig farms) to 250 L (from silage in the feed production, in the clean zone in slaughterhouse and outdoor). Air samples were collected through the impaction method with a flow rate of 140 L/min (Millipore air Tester, Millipore—Billerica, Massachusetts, USA). Tree different culture media were used in order to enhance the selectivity for bacterial and fungal growth: Tryptic Soy Agar (TSA) supplemented with nystatin (0.2%) used for mesophilic bacterial population and Violet Red Bile Agar (VRBA) for bacteria belonging to the Enterobacteriaceae family (e.g. coliforms—Gram-negative bacteria).

Samplers were placed at a height of 0.6–1.5 m above the floor, approximately at the breathing zone level, and as close as possible to the worker. Outdoor samples were also performed to be used as reference. Surface samples were collected by swabbing the same indoor sites with a 10 by 10 cm square stencil, disinfected with 70% alcohol solution between samplings, in line with the International Standard ISO 18593 (2004) requirements.

All the collected samples were incubated at 30°C and 35°C for 7 days (bacteria) (mesophilic bacteria and coliforms, respectively). After laboratory processing and incubation of the samples, quantitative colony-forming were obtained.

Fungal burden assessment procedures are well explained at Viegas et al. (Viegas et al., 2016a, b; Viegas et al., submitted).

2.3 Data analysis

Statistical software SPSS V22 was applied for statistical analysis. The data analysis was performed using univariate descriptive statistics with frequency (n; %), median and graphical representations appropriate for the nature of the data.

3 RESULTS AND DISCUSSION

3.1 Bacteria load

Total bacterial load ranged from <4 CFU/m³ (pharmacy to 150 CFU/m³ (warehouse) in the feed production; >10000 CFU/m³ in all sampling places in the two pig farms; and from 400 CFU/m³ (production room) to 9920 CFU/m³ (bleeding swine) in swine slaughterhouse. Airborne coliform load ranged from <4 CFU/m³ (pharmacy) to 20 CFU/m³ (bagging) in the feed production; <4 CFU/m³ (in all most places in pig farm 1 and quarantine in pig farm 2) to 5880 CFU/m³ (batteries in pig farm 2) in pig farms; and <4 CFU/m³ (gut room, bleeding swine and production room) to 352 CFU/m³ (labeling) in swine slaughterhouse.

The presence of bacteria in the air was verified in most of the sampled places, which could be explained by the fact that bacteria are microorganisms found in nature, and in the skin, in particular by human desquamation and in activities such as talking, coughing and sneezing, which lead to the projection of bacteria from the upper respiratory tract into the air (Terkonda 1987).

There are any national guidelines to impose limit values for bioaerossols exposure neither for bacteria load. However, some studies (Goyer et al. 2001) have been carried out in order to propose guidelines for eight hours of work indicating 10000 CFU/m³ for total bacteria and 1000 CFU/m³ for Gram-negative bacteria for agricultural and industrial settings. Pig farms were the setting that presented values exceeding these limits for both total and Gram-negative bacteria.

Regarding the surfaces sampling, the total bacterial load ranged from 2×10^4 CFU/m² (pharmacy) to 3×10^7 CFU/m² (warehouse) in the feed production; 5×10^5 CFU/m² (quarantine in pig farm 1) to 3×10^9 CFU/m² (gestation in saddles in pig farm 2) in the pig farms; and 1×10^5 CFU/m² (packing) to 2×10^4 CFU/m² countless (gut room) in swine slaughterhouse.

In a total of 23 samples, Gram-negative bacterial isolates were detected on 14 surfaces. The load of Gram-negative bacteria on surfaces ranged from $<1 \times 10^4$ CFU/m² (control) to 6×10^5 CFU/m² (warehouse) in the feed production; $<1 \times 10^4$ (gestation in parks and in saddles, and quarantine in pig farm 2) to 6×10^5 CFU/m² (maternity in pig farm 1) in pig farms; and $<1 \times 10^4$ CFU/m² (gut room, bleeding and cutting swine, packing, labeling) to 1×10^7 CFU/m² (production room) in swine slaughterhouse. The presence of Gram-negative bacteria in indoor air may be due to human activity (Zhu et al. 2003).

Almost all samples collected (22 total bacteria samples and 17 gram-negative bacteria samples out of the 23 samples for each) collected presented higher values of airborne bacterial concentration were detected in indoor than the outdoor, which could indicate that the contamination is originated in the indoor environment (Kuo et al. 2008).

3.2 *Microbiota air load analyses*

Regarding feed industry in pharmacy no bacteria counts were obtained, probably due to antibiotics aerosols presence. However, fungal contamination was observed and surpassing the guideline proposed by World Health Organization (WHO) (maximum value of 150 CFU/m³) not only in pharmacy, but in all sampling sites (Figure 1).

Concerning both pig farms assessed, was observed that in the pig farm 1 (Figure 2), where bacteria load was countless in 4 out of 5 samples collected, fungi presented the same trend with higher load than the other pig farm where bacteria load was also with lower counts. Interestingly, also Gram-negative bacteria presented the same trend that fungi load in the same pig farm with higher counts in the pig batteries (Figure 2). All sampling sites from both pig farms presented values surpassing the guideline proposed by World Health Organization (WHO) (maximum value of 150 CFU/m³) (Figures 2 and 3).

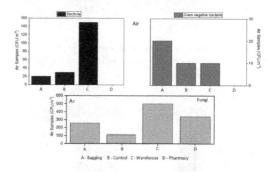

Figure 1. Microbiota (bacteria and fungi) air load from feed industry. Fungal load data adopted from Viegas et al. submitted.

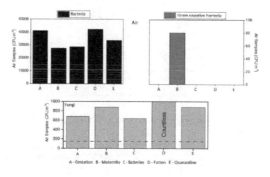

Figure 2. Microbiota (bacteria and fungi) air load from pig farm1. Fungal load data adopted from Viegas et al. submitted.

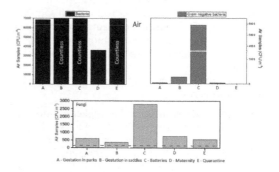

Figure 3. Microbiota (bacteria and fungi) air load from pig farm 2. Fungal load data adopted from Viegas et al. submitted.

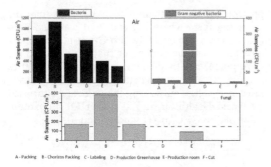

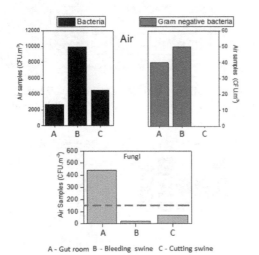

Figure 4. Microbiota (bacteria and fungi) air load from clean slaughterhouse. Fungal load data adopted from Viegas et al. 2016b.

Figure 5. Microbiota (bacteria and fungi) air load from dirty slaughterhouse. Fungal load data adopted from Viegas et al. 2016b.

Regarding clean slaughterhouse (Figure 4) was possible to observe that although bacteria load was lower than in dirty slaughterhouse in all sampling sites (Figure 5), regarding the fungal load in clean slaughterhouse presented one more sampling site than dirty slaughterhouse surpassing the WHO guideline.

From microbiota graphical analyses several observations can be done:

– Fungal load was observed in higher counts than bacteria in the assessed feed industry. This can be due to raw material that can be very prone to fungal infection (Straumfors et al. 2015).
– Pig farms were the settings with the highest fungal and bacteria load and this situation is already reported in several studies (Attwood et al. 1987; Portengen et al. 2005; Radon et al. 2002).

– Fungal contamination didn't decrease from the dirty to the clean slaughterhouse as was observed regarding bacteria counts. Several different environmental variables can be influencing microbiota and we should consider the possibility for the applied biocides in clean slaughterhouse being more efficient for bacterial than for fungal load. In addition, fungi are more resistant to lower-refrigerated temperatures and this can be an advantage in the clean slaughterhouse (Tournas 1994; Viegas, Meneses & Viegas 2016d).

The bacterial community in the agricultural and animal production bioaerosol exposures is not characterized well enough in order to understand the etiology of occupational diseases. Hence the study of compositions and dynamics of airborne microbial communities in the agricultural environment is fundamental for the exposure assessment and the implementation of preventive measures (Martin et al., 2013). Therefore, further studies should include de identification of bacterial isolates to assess the prevalence and persistence of bacterial species in these settings. Previous works have reported the prevalence of Gram-positive bacteria, such as *Bacillus* spp. and *Staphylococcus* spp., in the airborne bacterial community of agricultural industries (Martin el al. 2013).

4 CONCLUSIONS

All the studied settings indicated presence to airborne microbial load. However, pig farms were the workplaces with the highest bacterial and fungal load.

Bacteria load is not a surrogate to assess occupational exposure to bioaerosols, since bacterial and fungal load didn't present the same load trend, mainly in the feed industry and in clean and dirty slaughterhouse.

This highlights the potential occupational exposure of this animal-related industrial settings and the need of preventive and control measures to mitigate the microbiological risk. Workers should use personal protective equipment for the respiratory tract and work uniform, and increase the frequency of hygiene of the installations.

ACKNOWLEDGEMENTS

This study would not have been possible to develop without the institutional support given by Lisbon School of Health Technology and Associação Para o Desenvolvimento de Conhecimento e Inovação (POLITEC & ID).

REFERENCES

Attwood, P. et al. 1987. A study of the relationship between airborne contaminants and environment factors in Dutch swine confinement buildings. *Am. Ind. Hyg. Assoc. J.* 48: 745–751.

Awad, A.H.A., 2003. Evaluation of bio-aerosols at an animal feed manufacturing industry: A case study., (November 2016).

Badri, R.M. 2014. Identification and Characterization of Bacteria Air Pathogens from Homes in Some Areas of the Baghdad City Material and Method, 2(6): 384–388.

Douwes, J. et al. 2003. Bioaerosol health effects and exposure assessment: Progress and prospects. *Annals of Occupational Hygiene*, 47(3): 187–200.

Eduard, W. 2009. Fungal spores: A critical review of the toxicological and epidemiological evidence as a basis for occupational exposure limit setting. *Crit. Rev. Toxicol.*, 39: 799–864.

Goyer, N. et al. 2001. Bioaerosols in the workplace: Evaluations, Control and Prevention Guide. p. 83.

Halstensen, A.S. et al. 2013. Exposure to Grain Dust and Microbial Components in the Norwegian Grain and Compound Feed Industry., 57(9): 1105–1114.

Heederik, D. et al. 1991. Relationship of airborne endotoxin and bacteria levels in pig farms with the lung function and respiratory symptoms of farmers. *Int Arch Occup Environ Health*, 62(8): 595–601.

Kim K-Y, Kim H-T, Kim D, Nakajima J, H.T. 2009. Distribution characteristics of airborne bacteria and fungi in the feedstuff-manufacturing factories. *Journal of Hazardous Materials*, 169(1–3): 1054–1060.

Kuo N-W, Chiang H-C, Chiang C-M 2008. Development and application of an integrated indoor air quality audit to an international hotel building in Taiwan. *Environ Monit Assess*, 47:139–147.

Martin, E. et al. 2013. Microbial exposure and respiratory dysfunction in poultry hatchery workers. *Environ. Sci. Processes Impacts*, 15: 478–484.

Miaśkiewicz-Peska, E. & Łebkowska, M. 2011. Effect of antimicrobial air filter treatment on bacterial survival. *Fibres and Textiles in Eastern Europe*, 84(1): 73–77.

Oppliger, A. 2014. Advancing the Science of Bioaerosol Exposure Assessment. *Ann. Occup. Hyg*, 58(N): 661–663.

Swan, J.R.M. & Crook, B. 1998. ORIGINAL ARTICLES. pp. 7–15.

Terkonda, P.K. 1987. Sources and effects of microbiological indoor pollutants. In *Indoor Air '87, Proceedings of the 4" International Conference on Indoor Air Quality and Climate*. Berlin: Institute for Water, Soil and Air Hygiene, p. 713–717.

Tsapko, VG, et al. 2011. Exposure to bioaerosols in the selected agricultural facilities of the Ukraine and Poland—a review. *Ann Agric Environ Med*, 18 (1): 19–27.

Viegas, C. et al. 2016. Slaughterhouses Fungal Burden Assessment : A Contribution for the Pursuit of a Better Assessment Strategy. *Int. J. Environ. Res. Public Health*, 13(297): 1–11.

Zhu, H., Phelan, P. E., Duan, T., Raupp, G. B., Fernando, H.J.S., Che, F. 2003. Experimental study of indoor and outdoor airborne bacterial concentrations in Tempe, Arizona, USA. *Aerobiologia*, 19: 201–211.

Occupational Safety and Hygiene V – Arezes et al. (Eds)
© 2017 Taylor & Francis Group, London, ISBN 978-1-138-05761-6

Health risks for the public and professionals exposed to sewage wastewaters: A review on legislation and regulatory norms in Brazil

Scarllet Dos Santos & Tomi Zlatar

Research Laboratory on Prevention of Occupational and Environmental Risks (PROA/LABIOMEP), University of Porto, Porto, Portugal

ABSTRACT: Sewage wastewaters represent a high health concern for developing and underdeveloped countries. In Brazil, illegal sewage disposal is represents serious direct and indirect consequences for public and wastewater professional's health. The aim of this study is to contribute with a review on Brazilian legislation, norms and health risks for the public and professionals exposed to sewage wastewaters. The web-site of the World Health Organization, Google Scholar and PubMed were screened using keywords "sewage" and "wastewater", while for screening Brazilian legislation and norms, the Ministry of Labour and Employment of the Federative Republic of Brazil and Presidential Palace were screened using keywords "sewage system", "environmental health" and "sewage/wastewater operators". The identification process resulted with different microbial and chemical risks, 5 legislation and 2 norms. In order to resolve the health challenges posed by sewage there is a need for more organized structural wastewater solutions, treatment, rising public concern, and water sampling on and inspections on regular basis.

Keywords: Illegal sewage disposal; Environmental health; Occupational safety and health; WHO; Recife

1 INTRODUCTION

Although most ancient systems of wastewater management date to 1500 BC, and its constant development through the centuries, it continued to be a challenge as the number of inhabitants in the cities were increasing considerably (Wiesmann, Choi, & Dombrowski, 2006). In the middle of the 19th century, wastewater produced in the fast-growing industrial regions and cities was discharged directly into rivers and canals, as well as the soil below the toilets at the courtyards of tenement blocks. Frequently, the drinking water pump was located directly next to toilets. Therefore, it was no wonder that cholera illnesses often occurred, especially in large cities. In London, a large canal was constructed in 1865–1868 along the river Thames (based on a method used in Rome 2000 years earlier), which received most of the countless wastewater streams which were previously disposed directly into the river (Wiesmann et al., 2006). The extremely polluted river Thames (Reich, 1871) resulted with a need for investigating and developing wastewater treatment processes, taking samples and measuring pollutants (Wiesmann et al., 2006).

Today in developing and underdeveloped countries, sanitation services are still extremely deficient and in many cases non-existent, leading to the spread of diseases and occurrences of deaths, especially among children (Bos, Carr, & Keraita, 2009). In Latin America 91% of population has access to water in the household, while for sanitation (excreta and wastewater management) coverage is 77%, where, moreover, even greater deficits persist in rural areas. At the same time, countries face the need to control risks associated with industrialization and unplanned development in large cities. In Brazil, the coverage of water supply and sanitation in 2013 was 82.40 and 37.50% respectively (Americas, 2007).

One study on the impact of sanitation deficiencies on public health conducted in Brazil from 2001 to 2009showed that due to inadequate sanitation there were on average 13,449 deaths per year (1.31% of total deaths) (Teixeira, 2014). The annual average of compulsory notification on cases to diseases was 466,351 cases, with an expense of R$ 30,428,324.92 with medical appointments in this period (Teixeira, 2014). The origins could be found in poor socio-economic conditions, lack of water resources availability and low awareness on water use (Costa Dos Santos & Benetti, 2014). In addition to the public health risks, there are occupational risks to consider, especially for sewage wastewater workers. During regular work they are exposed to toxic chemical and microbial risks, which lead to health problems and sometimes even deaths (Brown, 1997).

An example of urban wastewater management complexity Recife, according to the City Hall of Recife, only 30% of a total of 220 km² of the city's territory has a public sewage collection network (Recife, n.d.), with represents 88% of households from which 9% have wells or springs without pipeline.

Only 42.9% of households are connected to the general sewage network or rainwater network (characterizing the absolute separator system), 46.6% use septic and rudimentary septic tanks and about 8% discharge their untreated effluent in the drainage networks (storm water tunnels), compromising drainage structures. In addition to the unavailability of sewage collection services, there is a lack of supply of garbage collection services, aggravated by the population lack of organizing garbage, causing serious problems in the incorrect disposal of household waste (Recife, n.d.).

In areas without sanitary sewage systems the connections are made illegally to drainage networks (Mattiello & Farias Cunha, 2009), which due to the impermeability of the soil often results with flooding and therefore disseminates contaminated water throughout the city and spread waterborne diseases (Tucci E.M., 2008). According to one study on waterborne diseases, in 2012, Recife had on each 10.000 inhabitants, a number of 161 persons infected with waterborne diseases (Programa cidades sustentáveis, n.d.).

The highest rates of illegal disposal are in peripheral areas, created by illegal housing and "favelas," where the garbage collection service either does not exist or is not used properly by residents. Often the garbage is irregularly disposed in rivers and streams, on empty lands and streets. The garbage further on result with undesirable effects, silting rivers and streams, obstructing drainage networks and contaminating green areas. Further on, accompanied with bad smell, proliferation of flies, cockroaches and rats, it is representing serious direct and indirect consequences for public health. Even household waste that is traditionally laid out on the sidewalks, waiting for collection, is often scattered by rains, and ends up in drainage networks, drastically reducing its drainage capacity (Ábalos et al., 2012).

Investments in basic sanitation in underdeveloped countries did not accompany the population growth of large cities, which explains the unequal access to basic sanitation infrastructure and services. Thus, like the great cities of the country, Recife expresses low rates of basic sanitation (Starling et al., 2005).

The aim of this study is to contribute with a review on health risks for the public and professionals exposed to sewage wastewaters. Further on, to review the Brazilian legislation on the sewage wastewater issue, and its directives toward protection of human health and improvement of the system. Finally, to give suggestions to minimize health risks for public and professionals exposed to wastewaters.

2 METHODOLOGY

The process of creating the database was divided into searching methods. The first searching method was used in order to find articles on health risks of humans exposed to sewage wastewaters, for which the web-site of the World Health Organization (WHO), Google Scholar and PubMed were screened using keywords "sewage" and "wastewater" regarding publications, fact sheets, guidelines and news releases in all WHO regional sites, with special attention on the "Americas WHO regional site" (WHO, 2006, 2016).

The second searching method was used in order to find Brazilian legislations and norms regarding wastewaters, for which the web-sites of the Ministry of Labour and Employment of the Federative Republic of Brazil and Presidential Palace were screened using keywords "sewage system", "environmental health" and "sewage/wastewater operators".

References were managed using the Mendeley 1.15.3.

3 RESULTS

The identification process in the screening of the WHO website resulted with microbial and chemical risks illustrated in Tables 1 and 2.

Table 1.   Microbial risks for health of humans exposed to sewage wastewaters by the World Health Organization.

	Pathogens	Type of waterborne pathogenic bacteria	Transmission
Microbial risks (waterborne infectious disease)	bacterias	typhoid bacterium; vibrio cholera bacillus; shigella sp.; pathogenic and enterohaemorrhagic E. coli; yersinia anterocolitica	Ingestion (Drinking contaminated water or eating uncooked shellfish from the contaminated water)
	viruses	hepatitis A; hepatitis E vírus; norovirus; sapovirus; rotavirus	Inhalation of aerosol droplets
	protozoans	giardia intestinalis; entamoeba histolytica; cryptosporidium parvum	Transmitted on dirty hands
	helmithes	dracunculus medinensis; schistosoma spp.	Through a carrier (insect)

Table 2. Chemical risks for health of humans exposed to sewage wastewaters by the World Health Organization.

	Types	Source	Health effects	Transmission
Chemical risks	Arsenic	natural	cancer	Ingestion (Drinking contaminated water or eating uncooked shellfish from the contaminated water)
	Fluoride	natural	dental and crippling skeletal fluorosis	
	Lead	corrosion of lead pipe	neurological	
	Pesticides	agricultural use and spills	variable effects	
	Nitrate and nitrite	agricultural and sewage	infant deaths	
	Radon	natural geology to indoor air and some groundwaters	cancers	
	Sulfates	natural	causing temporary diarrhea to non-residents	

Table 3. Brazilian legislations and norms regarding wastewaters and its protection of health.

		Topic	Considerations
Legislations	Presidency of the Federative Republic of Brazil, year 1943, Decree-Law No. 5,452/43 – Consolidation of Labour Laws (Presidência da República do Brasil, 1943)	Regulates individual and collective relations to work of the unhealthy or dangerous activities	No risk related to sewage systems mentioned
	City Hall of Recife, year 1995, Law No. 16.004/95 – Code of Municipal health (Prefeitura da Cidade do Recife, 1995)	Regulates standards of access to the collection and treatment system of sanitary sewers, addressing the promotion, protection and recovery of environmental health	Risks related to exposure of the population to untreated sewage: waterborne diseases; Dissemination of animal vectors of diseases
	Presidency of the Federative Republic of Brazil, year 2007, Law No. 11.445/07 – National Guidelines for Basic Sanitation (Governo do Estado de Pernambuco, 1998)	Establish standards for the provision of basic public sanitation services	
	Government of the State of Pernambuco, year 1998, Decree No. 20.786/98 – Health Code of the State of Pernambuco (Presidência da República do Brasil, 2007)	Regulates the inspection and sanitary control of the collection and destination of sanitary sewage	Risks related to exposure to physical, chemical, biological and ergonomic agents of sanitation professionals
	Presidency of the Federative Republic of Brazil, year 2016, Law No. 11.445/16 (Presidência da República do Brasil, 2016)	Complements guidelines for public basic sanitation services by determining the maintenance of drainage networks	Risks related to exposure of the population to untreated sewage dumped irregularly into drainage channels
Norms	Ministry of Labor and Employment, year 1978, NR 13 – Boilers, pressure vessels and pipelines (Ministério do Trabalho e Emprego, 2014)	Directs minimum inspection requirements for public sewage treatment and sewerage pipelines for the safety and health of workers	Risks related to exposure of workers to biological agents present in raw sewage
	Ministry of Labor and Employment, year 1978, NR 15 – Unhealthy Activities and Operations (Ministério do Trabalho e Emprego, 1978)	Defines unhealthy operations and establishes specific remunerations according to the degree of unhealthiness	Risks related to biological agents present in sewage (galleries and tanks)

The identification process in the screening of the Brazilian legislation and norms regarding wastewaters and its protection of health resulted with 5 legislation and 2 norms, which are illustrated in the Table 3.

4 DISCUSSION

The problem of misconnections to separate sewage systems (accidental or deliberate) remained a challenge even in developed and well organized urban areas in the UK (Ellis & Butler, 2015), therefore there is a need for a multi-aspect approach in order assure environmental, health and water quality standards.

In the Table 3 the Brazilian legislations refer to the standards and guidelines that should be followed in order to avoid or mitigate the health risks posed in contact with sewage. However, there is still a growing effort to universalize basic sanitation. From a historical perspective, public policies were not able to reach the universalization of public services in quality basic sanitation, which would contribute in improvement of the living conditions, reduce social inequalities and increase environmental quality of the country (Ministério das Cidades, 2013). This problem results in population exposed to sewage drained throughout the streets and with illegal connection of sewages to drainage networks.

Another concern is the frequent flooding caused by excessive waterproofing of public areas due to a growing urban sprawl, together with occupation of riverside areas and heavy rains. The situation is aggravated by large number of waste disposed on streets in an inadequate way. And in cases of intense rainfall, overflows can be caused, affecting the health of the residents in the city.

Microorganisms may seep through some soils for long distances until they reach a body of surface water or groundwater. Leaking septic tanks, inadequate latrines and contaminated drainage networks may contaminate nearby drinking water. Some soils, such as sand-stone, are effective at filtering microorganisms, but coarser and fractured soils may allow transport of pathogenic organisms for long distance and depth (WHO, 2006) which might be later on inhalated by passengers and pose a serious risks.

As it is illustrated in Tables 1 and 2, there are many health risks both for public and professionals exposed to sewage wastewaters. Microbial and chemical risks can be both transmitted through ingestion by drinking contaminated water or eating uncooked shellfish from the contaminated water. Further on, microbial risks represent an even higher risk for contamination, as it can be inhalated through aerosol droplets.

Sewage wastewater workers work directly exposed to raw (untreated) sewage, either in maintenance in the collection networks or in the treatment plants. As a result, they are exposed to complex mixtures of toxicants including pathogens, heavy metals, chlorinated organic solvents like chloroform, dichloroethane, perchloroethanol, other solvents (benzene, toluene), aldehydes, nitrosamines, pesticides, dyes, polychlorobiphenyls, and polycyclic aromatic hydrocarbons(Al Zabadi et al., 2008). Studies have shown that wastewater treatment may generate aerosols, so sewage workers may be at increased risk of cancer or adverse birth outcomes (Brown, 1997).

The Brazilian Ministry of Labor and Employment (Ministério do Trabalho e Emprego, 2002) specify preventive measures for sewage wastewater workers routinely involved in processes at treatment plants or maintenance of the catchment network in every urban city network:

– companies should provide adequate conditions for personal hygiene, including bathing at the end of the working day
– provision of uniforms for daily exchange, with hygiene at the company's expense (personal protective equipment), in addition to the provision of locker rooms with individual lockers
– measures taken to avoid exposure to contaminants using personal protective equipment, training and instructions on the severity of health risks are also required.

In comparison to the Brazilian legislation, the Directive from the European Parliament and of the Council (Directive 2000/54/EC of the European Parliament and of the Council of 18 September 2000 on the protection of workers from risks related to exposure to biological agents at work, 2000) gives regulations that have been implemented into national legislation as minimum requirements. Some Member States have introduced Codes of Practice and guidelines for safe handling of biological agents. By the directive, it is set that employers have to keep records including information about exposure and health surveillance of workers likely to be exposed (European Agency for Safety and Health at work, 2003).

Further on, the Directive requires the employer to:

– assess the risks posed by biological agents
– reduce the risk to the workers (eliminate or substitute; prevent and control exposure; inform and train workers
– provide health surveillance as appropriate.

The USA National Institute for Occupational Safety and Health (NIOSH) give a more detailed guideline for controlling potential risks for workers at wastewater treatment plants (National Institute for Occupational Safety and Health (NIOSH), 2002). The guideline gives detailed instructions on how to prevent work-related illnesses, specifying steps toward basic hygiene, appropriate protective equipment, hygiene stations, training and toward minimizing occupational exposures.

For now, it is well documented that sewage wastewaters pose high healthy risks for the public and professionals, and that there is a need for urgent, well organized approach to this challenge. There is a need for assessing different sources on public and occupational preventive guidelines and develop a more structured and easy applicable guideline for Brazil. Further studies should be conducted on drainage networks in Recife and other cities in Brazil. More studies should be conducted on recorded diseases through the past years in order to reach more consistent conclusions and increase public awareness on this problem.

5 CONCLUSIONS

Sewage wastewaters represent a high health risk for developing and underdeveloped countries. In order to resolve the health challenges posed by sewage there is a need for more organized structural wastewater solutions, treatment, rising public concern, and water sampling on and inspections on regular basis. In developing and underdeveloped countries sewage microbial and chemical risks represent a serious concern as they are present in most of the city parts in drainage networks, which is affecting all the city population by aerosol droplets and with every coming of intense rainfall. Finally, sewage wastewater workers, due to their job description, are constantly highly exposed to microbial and chemical risks, therefore need to use adequate protective equipment and following safety procedures.

REFERENCES

Ábalos, F., Sulimam, F., Mosseri, I., Ota, N., Farina, R., Filho, K.Z., ... Porto, M.F.A. (2012). Gestão de Resíduos Sólidos e Impactos sobre a Drenagem Urbana Sumário.

Al Zabadi, H., Ferrari, L., Laurent, A.-M., Tiberguent, A., Paris, C., & Zmirou-Navier, D. (2008). Biomonitoring of complex occupational exposures to carcinogens: the case of sewage workers in Paris. BMC Cancer, 8(2007), 67. http://doi.org/10.1186/1471-2407-8-67

Americas, M. of H. of the. (2007). Health Agenda for the Americas 2008–2017. Security, (June 2007). Retrieved from http://www.paho.org/English/DD/PIN/Health_Agenda.pdf

Bos, R., Carr, R., & Keraita, B. (2009). Wastewater Irrigation and Health. Wastewater irrigation and health.

Assessing and mitigating risk in low-income countries. Routledge. http://doi.org/10.4324/9781849774666

Brown, N. (1997). Health hazard manual: Wastewater treatment plant and sewer workers. Manuals and User Guides, 2, 1–54.

Costa Dos Santos, D., & Benetti, A. (2014). Application of the urban water use model for urban water use management purposes. Water Science and Technology, 70(3), 407–413. http://doi.org/10.2166/wst.2014.229

Directive 2000/54/EC of the European Parliament and of the Council of 18 September 2000 on the protection of workers from risks related to exposure to biological agents at work (2000).

Ellis, J.B., & Butler, D. (2015). Surface water sewer misconnections in England and Wales: Pollution sources and impacts. Science of the Total Environment, 526, 98–109. http://doi.org/10.1016/j.scitotenv.2015.04.042

European Agency for Safety and Health at work. (2003). Biological agents. Safety And Health, (September). Retrieved from https://osha.europa.eu/en/tools-and-publications/publications/factsheets/41

Governo do Estado de Pernambuco. (1998). Decreto no 20.786/98 – Código Sanitário do Estado de Pernambuco.

Mattiello, R., & Farias Cunha, G. (2009). Análise Da Problemática Da Contaminação Por Efluentes Domésticos Dos Cursos De Água Que Afluem Ao Campus Da Ufsc, Florianópolis, SC. Universidade Federal de Santa Catarina.

Ministério das Cidades. (2013). Plano Nacional de Saneamento Básico. Ministério Das Cidades, 172. Retrieved from http://www.mma.gov.br/port/conama/processos/AECBF8E2/Plansab_Versao_Conselhos_Nacionais_020520131.pdf

Ministério do Trabalho e Emprego, B. (1978). Nr 15 – Atividades E Operações Insalubres. Portaria MT 3124, de 08/06/1978, 4(15). http://doi.org/10.1007/s13398-014-0173-7.2

Ministério do Trabalho e Emprego, B. (2002). Manual de Procedimentos para Auditoria no Setor Saneamento Básico.

Ministério do Trabalho e Emprego, B. (2014). NR 13 - Caldeiras, vasos de pressão e tubulações. Dou, (13), 22. Retrieved from http://portal.mte.gov.br/data/files/FF80808145B26962014600A0AF41169F/NR-13 (Atualizada 2014).pdf

National Institute for Occupational Safety and Health (NIOSH). (2002). Guidance for Controlling Potential Risks to Workers Exposed to Class B Biosolids, (July), 1–7. Retrieved from httpwww.cdc.gov/niosh/docs/2002-149/pdfs/2002-149.pdf

Prefeitura da Cidade do Recife, B. (1995). Lei no 16.004/95 – Código Municipal de Saúde, 1–44.

Presidência da República do Brasil. (1943). Decreto-lei no 5.452/43 – Consolidação das Leis do Trabalho.

Presidência da República do Brasil. (2007). Lei no 11.445/07 – Diretrizes Nacionais para o Saneamento Básico.

Presidência da República do Brasil. (2016). Lei no 11.445/16.

Programa cidades sustentáveis. (n.d.). Doenças de veiculação hídrica—Recife, PE. Retrieved November 20, 2016, from http://indicadores.cidadessustentaveis.org.br/br/PE/recife/doencas-de-veiculacao-hidrica

Recife, P. do. (n.d.). Doencas deveiculação. Retrieved September 26, 2014, from http://indicadores.cidadessustentaveis.org.br/br/PE/recife/doencas-deveiculacao-hidrica

Reich, O. (1871). Der erste und zweite Bericht der im Jahre 1968 in England eingesetzten. Dtsch Vierteljahresschr. Offentl. Gesundheitspfl, 3, 278–309.

Starling, F.A., Kutianski Romero, G.F., Sousa, G.M., Machado, G.M., Tavares Nascimento, W., & Carreira, W. (2005). Influência do Saneamento Básico na Saúde Pública de Grandes Cidades. ESCOLA POLITÉCNICA DA USP.

Teixeira, J.C.E. Al. (2014). Estudo do impacto das deficiências de saneamento básico sobre a saúde pública no Brasil no período de 2001 a 2009. Eng. Sanit. Ambient., 19(1), 87–96. http://doi.org/10.1590/S1413–41522014000100010.

Tucci E.M., C. (2008). Textos para discussão cep al • ipea. TEXTOS PARA DISCUSSÃO CEPAL • IPEA.

WHO. (2006). Health As Pects of Plumbing. Health Aspects of Plumbing.

WHO. (2016). Water and health. Retrieved November 15, 2016, from http://www.who.int/en/

Wiesmann, U., Choi, S., & Dombrowski, E.-M. (2006). Historical Development of Wastewater Collection and Treatment. In Fundamentals of Biological Wastewater Treatment (pp. 1–23). Weinheim, Germany: Wiley-VCH Verlag GmbH & Co. KGaA. http://doi.org/10.1002/9783527609604.ch1

Occupational Safety and Hygiene V – Arezes et al. (Eds)
© 2017 Taylor & Francis Group, London, ISBN 978-1-138-05761-6

Noise propagation emitted by the pile driver in building sites inside the urban zone

E.M.G. Lago & P. Arezes
University of Minho, Guimarães, Portugal

B. Barkokébas Junior
University of Pernambuco, Recife, Pernambuco, Brazil

ABSTRACT: The cities evolve and that leads to consequences to the environment. Progress entails prosperity and well-being; however, it can also reduce the quality of life for the population and cause environmental degradation. This paper will showcase the surveying of noise from the pile driver in a building site and the study of its propagation, checking its suitability with Brazilian legislation. The pile driver's levels of sound pressure were measured; a chart for cross-reference was created from the equipment; the noise levels surrounding the site were analysed and compared to the legislation. By analysing the site and one of the equipments it is already possible to state that the measured values are already very elevated and the value that comes to the environment added to the one which already exists is above the legislation threshold for a predominantly residential, mixed zone.

1 INTRODUCTION

The cities evolve and that leads to consequences to the environment. Progress entails prosperity and well-being; however, it can also reduce the quality of life for the population and cause environmental degradation.

This evolution is based on technology, being noticeable by the presence of machines and equipments that generate noise, which is present in urban environments causing noise pollution, which, in turn, can be generated by traffic, nocturnal environments, industries, car horns, whistles, loudspeakers, animals, public demonstrations, sports, schools, neighbourhood sounds, building sites both on normal and on street levels, and so forth.

That noise generated by all those sources affects the environment and causes problems for the neighbouring population. Exposed citizens can suffer the consequences, like being disturbed, which affects reasoning, oral communication, politeness and wellbeing, thus limiting the human potential. Socioeconomic growth itself is affected by the lack of capacity to comprehend and react to noisy urban, industrial and leisure environments, made worse by the high population density. Following the 2010 demographic census (IBGE, 2013), around 84% of the population lives inside urban environments in Brazil; 73,1% in the northeast, and in Pernambuco, 76,5%, with 1.537.704 million of inhabitants living in the urban area of Recife.

Prolonged exposure to noise affects the individual on many levels (Gerges, 2000), being hearing loss the most meaningful, with disturbances like accelerated heart rates and narrowing of the blood vessels, the increase of blood pressure, heart overload, causing changes in hormonal secretion, muscular tensions, among others, arising from such exposure.

The effects surface through behaviours, changing the performance of individuals in work, possibly causing absenteeism. Among those effects are nervousness, mental fatigue, stress, irritability caused by mental and emotional hardships, and social conflicts.

According to WHO (2007), the non auditory effects originated by exposure to excessive noise are many, being known that they depend on the intensity, frequency, duration, time of exposure, etc. Even if away from the source, noises are disturbing and cause impact on human beings such as fatigue, lack of concentration, sleep disturbance, cardiovascular problems, among others (Dani & Garavelli, 2001).

This paper will showcase the surveying of noise from the pile driver in a building site and the study of its propagation, checking its suitability with Brazilian legislation.

2 METHOD

The neighborhood of Boa Viagem, in the city of Recife, was chosen because nowadays it is the most populous (>122.000 hab.), and has a high density of population (±163 hab/ha).

The main factor responsible to such a high population density is the neighborhood's intense verticalization both for commercial and residential ends. The neighbourhood of Boa Viagem leads the preferences for habitation, concentrating 22,64% of its total area in construction.

In 1996, Boa Viagem had 43% of its habitation units in properties with more than 10 floors, in 2003 it was possible to ascertain that 4,2% of the total of the 57% of habitation units had more than 20 floors, a number that grew, in 2014, to 7,8% (SEPLAN/PCR, 2016).

Steps were taken to measure the levels of sound pressure of pile drivers used in the foundation phase; they were done approximately 30 centimetres away from the equipment, in order to obtain the level of noise from the functioning equipment. For that phase, the time took measuring each one of the equipments was of approximately 15 minutes.

For the second phase, a chart for cross-reference was created from the equipment, which aimed to measure various dots close to the machine, with the goal of tracing acoustic curves. The first dot was in the equipment, the subsequent were situated apart from each other through the mapped chart, which varied from 5 to 10 metres (inside the limit of the site's terrain), creating dots in a blueprint, feeding the SURF software, to verify the propagation of the emitted noise, and its reach. A survey\questionnaire was also developed and applied in order to classify the site being researched.

The third phase of the collection of data was the measurement of noise, this time in the terrain surrounding the building site. The measurement points relevant to the data gathering were defined; the front side of the site, the opposing side, and at the street corners surrounding it. Measurements were done with the noise-generating pile drivers in function, for the intention was to verify the noise effectively reaching the neighbourhood and already added to the natural noise. After measurement, the values were compared to the levels allowed by the Brazilian regulation (Norma Brasileira—NBR 10151—"Evaluating noise in Inhabitated Areas, Aiming for the Comfort of the Community.".

With that, it was possible to develop a questionnaire based on the criteria of scales of altitude, aimed at the inhabitants/workers of the surrounding properties, addressing the goal of verifying the interface between the noise generated by the sites and the nuisance felt by those who live and work around the place, however this paper will not show this questionnaire, which is an integral part of the author's doctorate thesis and is nearing completion.

The pile driver, equipment used to lay deep foundations in big constructions (depending on the method with which it is possible to lay a stake on the soil), may be of many types, like precast in concrete, wood, metal and other materials. These equipments are basically made of a tower which raises a hammer (or other kinds of weight that falls on the stake through gravitational energy, or through a hydraulic hammer) aiming to embed the stake inside the soil. In the first case, there is a cable suspending the hammer, activated by a crane, and in the second there is a hydraulic pump which injects oil in the circuit, forcing the hammer up and down. The sampled pile driver was of the hydraulic type (as per Picture 1), and the embedded stakes were of the metallic kind.

The measurements of the equivalent level of noise were made according to the recommendations of NBR 10151, which establishes that the equipment should have resources to measure the Equivalent Sound Pressure Level, in decibels in the "A" (dBA) scale, in "fast" mode.

The two pile drivers of the site were measured. For each point, measurements in a standard height which varied from 1,20 m to 1,50 of the horizontal scope, and at least 2,00 m from any other reflective surfaces like walls, sidings, etc were made.

The measurements followed the mapped chart in the interior and exterior of the site and in determined spots, such as in the equipment, in front of the site and in front of the buildings on both sides.

In order to make the measurements, a QUEST audiometer, model 2900, serial number CDB 040027, was used alongside a sound level calibrator type QC-10/QC-20 from the same brand. The microphone was protected with a windscreen, as per the NBR 10511 guidelines.

Picture 1.　Site with pile driver.

Each reading took an average 10 minutes, preceded by the calibration of the sound pressure level device before and after and measurements in each spot.

With the measured data of each equipment, the chart of noise propagation was mapped with the help of the graphic program SURF® version 8.0 (2002), then tracing the level curves regarding this propagation, being thus possible to map the reach of the noises.

For a better understanding, the propagation area was highlighted in the map, delineating and pointing where the noise levels obey the maximum limits established by the environment office in the city of Recife. The yellow lines indicate noise equal or above 85dB(A), White lines indicate noise between 65dB(A) and 85dB(A) and the Green lines show noises below 65dB(A).

3 RESULTS

To characterize the company being researched, a survey\questionnaire was applied and answered by the site's manager.

The questionnaire was elaborated in two parts, the first aimed to showcase the company, its context in the sector and in the sampled region; the firm has been active for more than 30 years in the real estate market and, in building sites, has safety programs with the OHSAS certification already implemented. For the second part, the focus was on the building itself, which is of the commercial type, and has 39 floors.

Concerning the existence of noise emission controls for the machines and equipment used in the building sites during the execution of the work, the company affirms that there is none, be it at source level or during propagation.

It is possible to verify that 60% of the buildings of the company have 30 or more floors and are located at roads deemed as important, or main roads.

The sampled site was found during its foundation phase, and all of the work related to quantification had begun with the measurement of the

Sound Pressure Level (NPS) of the pile driver equipment, previously determined and classified.

Table 1 presents the measured NPS (Leq) of each pile driver. The methodology of measurement for all points followed the directives of NBR 10151.

Picture 2 shows the location of the site, the limit of the construction areas (this particular site covered the entire block) having its main front side directed at the Ministro Nelson Hungria street with Maria Carolina street. Picture 2 also indicates the position in which the equipment was found during data gathering.

Table 2 informs the phases of building, the equipment and the NPS values measured, also showing the values of environmental noise around the site. Environmental noise contemplates the working equipment, movement and activities in the vicinity.

Measurements took place in the same spots A, B and C indicated in Picture 2 during the morning after work had started (between 7:00 hs and

Table 1. NPS of the pile drivers.

Pile Drivers	NPS dB(A)
1	99,6
2	102,6

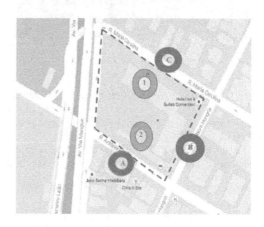

Picture 2. Site location.

Table 2. Measured Values/Site (Source and environmental).

Building Site (Commercial Type)			Environmental Noise—NPS dB(A)			
Work phase	Equipment	NPS dB(A)	Local	Morning	Midday	Afternoon's End
Foundation	1-Metallic Stake	99,6	A—Av. Antônio Falcão Frente da obra	73,6	79,6	82,4
	2-Metallic Stake	102,6	B—Street Min. Nelson Hungria (Left lateral)	70,9	72,9	82,7
			C—Street Maria Carolina (Site backgrounds)	67,9	71,7	81,5

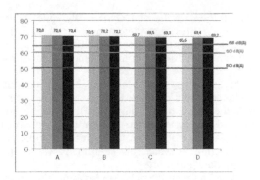

Chart 1. Values measured compared to the legislation.

Figure 1. Map of the isolines of propagation.

9:00 hs), during midday (between 11:30 hs and 12:30 hs), that is, by the end of the morning during traffic, and at the end of the afternoon (between 16:30 hs and 18:00 hs).

The intention was to evaluate the worst moments of the day, checking values that reach the environment, the site's vicinities, in order to verify the perception of those who answered the questionnaire applied at this point.

It is possible to verify through Chart 1, originated from Table 2, that the levels of noise measured during any time of the day are above the values allowed by legislation.

And then, the isolines showing the propagation of the noise generated by the equipment.

4 CONCLUSIONS

Conclusions are still partial, considering the author is still in the completion phase of the thesis. The research for the thesis involves 32 equipments of 7 types in 10 distinct building sites, and the added application of 500 structured surveys\ questionnaires.

However, by analysing just one site and one of the equipments it is already possible to state that the values measured from the equipment are very elevated and the value that reaches the environment, when added to an already existing noise, is above the levels determined by legislation for a mixed, predominantly residential zone.

Noise is not the only cause for complaints coming from the neighbourhood when it comes to civil construction worksites. Other elements, like the increased traffic of cargo trucks, the constant production of suspended particles, among others, were detected. However, noise is the most affecting, because it is present during the whole construction process. By this study, it was possible to verify the high level of noise produced by the workplace and how much it affects its workers and the nearby community.

According to the analyses carried in this paper, it was sought out to make recommendations which may contribute to the reduction of sound pressure levels produced by workplaces. In order to attain better results of noise control through every step of the construction, three steps must be taken while planning for the construction: mitigation in the source, layout changes and the building of barriers to minimize propagation.

REFERENCES

Dani, A.; Garavelli, S.L. (2001). *Principais Efeitos da Poluição Sonora em Seres Humanos*. Revista Universa, 9(14), 659–678.

Gerges, Samir Nagi Yousri (2000). *Ruído: fundamentos e controle*. 2ª Ed. Florianópolis, SC: NR Editora.

IBGE—Instituto Brasileiro de Geografia e Estatística. *Características da população brasileira*. Consultado em 14.05.2013, disponível em http://www.ibge.gov.br/home.

IBGE—Instituto Brasileiro de Geografia e Estatística. *Geociência/geografia/cartogramas/mesoregião*. Consultado em 14.09.2016, disponível em http://www.ibge.gov.br/home.

PCR—Prefeitura da Cidade do Recife. *Perfil dos bairros do Recife*. Secretaria de Planejamento. Consultado em 03/05/2015, disponível em http://www.recife.pe.gov.br/pr/secplanejamento/inforec/bairros/php.

WHO World Health Organization (WHO: 1995) *Guidelines for Community Noise*. Consultado em 17/11/2013, disponível em http://www.euro.who.int/eu/health

WHO World Health Organization (WHO: 2007). *Night Noise Guidelines for Europe*. Consultado em 17/11/2013, disponível em: http:/www.ec.europa.eu/health/ph_projects/2003/acti-ion3/docs/2003_08_frep_en.pdf.

Occupational Safety and Hygiene V – Arezes et al. (Eds)
© *2017 Taylor & Francis Group, London, ISBN 978-1-138-05761-6*

Fragmented occupational identities. A study on Portuguese and British contact centre workers

Isabel Maria Bonito Roque
Centre for Social Studies, Faculty of Economics, University of Coimbra, Coimbra, Portugal

ABSTRACT: According to Marxian theory, there is a general trend to reduce workers to an undifferentiated mass, who can be easily replaced—cybertariat, precariat or proletariat. Work is deeply connected with the subjective dimension of the human being, a person's occupation is one of the most important delineators of social identity. In that sense it becomes relevant to analyze if contact centres are offering new opportunities for career development or if they are enacting more routinized, deskilled and devalued forms of work, preventing workers from building an occupational identity. Between 2015 and 2016, a longitudinal analysis was conducted through biographical interviews undertaken with former and present Portuguese and British contact centre workers. It was concluded that the majority of workers, especially the Portuguese ones, engage on acceptance and resignation behaviors, whilst the British are more involved in trade unionism and resistance.

Keywords: Contact Centres, Identity, Psychodynamics of Work, Psychopathology of Work

1 INTRODUCTION

1.1 *Research aims*

In the academic literature there is a lack of analysis concerning psychosocial factors between Portuguese and British contact centre workers. In that sense, there is the urge for studying the way how routinized, flexible and precarious work affects these workers. The study was conducted in Portugal and Great Britain with the support of a Short-Term Mission of the Project "Dynamics of Virtual Work", supervised by Professor Ursula Huws. In that sense, between 2015 and 2016, twenty semi-structured and biographical interviews were conducted with present and former Portuguese and British contact centre workers. The profile of the interviewees consisted mostly of highly educated individuals aged between 21 and 60 years old, mostly female. The contact was established through the help of the supervisor, but also by email and through friendships that the author established whilst working for several years in several Portuguese contact centres. The main concepts used in the interviews sought to obtain information about the workers' household, parents' social class, their marital status, material possessions, educational level, employment situation, number of years worked in contact centres, their unionization situation, level of precariousness, alienation and happiness/fulfillment with their work. They were also questioned about their career experiences, their aims and future expectations.

1.2 *British and Portuguese societies*

After the 2008 crisis, contact centres became one of the easiest and fastest ways for insertion in the labor market, especially for the highly skilled, unemployed and feminine workforce (Roque, 2010). In Great Britain the failure of Fordism gave place to a neoliberal orthodoxy resulting in programmes of deregulation, privatization and withdrawal of the state from many areas of social provision leading to the flourishing of contact centres (Harvey, 2007; Fisher, 2009). Most British contact centres deal with charity and sales services, employing predominantly women, older people who lost their longtime jobs, students and seasonal workers, like musicians and actors, who seek a "guaranteed" wage to pay for their bills. On the 6th April 2011, Portugal signed the Troika's Memorandum which mandated various austerity measures (i.e., fiscal consolidation through spending cuts and revenue increases) and implemented various structural reforms (i.e., longer-term changes targeting aspects of the economy's operations, such as state-owned enterprises, financial sector regulation and labour market rules and regulations) (Gurnani 2016). Austerity changed peoples' lives in such a way that the rate of immigration and unemployment increased tremendously. These significant changes in work organization led to a profound disruption of social relations at work, contributing to mental and physical vulnerability of the workers, as well as their dignity and identity (Dejours, 2013).

2 CONTACT CENTRES

2.1 *The digital revolution of the 21st century*

The digital revolution of the 21st century allowed work to be performed in network and in constant connection. In this sense, new technologies of information and communication were disseminated, bringing new challenges for the world of labor, mainly social, economic and technological, involving many advanced economies (OECD, 1998). It encouraged qualification, mobility, flexibility and the creation of work, leading to a learning society focused on the production and exchange of knowledge in a continuous learning process. New Toyotist and post-Taylorist production models were implemented, based on flexible organizational structures and networks, appealing to a sense of initiative, autonomy and versatility, as well as a new labor relationship of self-employment and the replacement of the employee by after-hours timetable service providers (Kovács, 2002). Tertiarization, feminization and dissemination of flexible forms of employment are significant trends in the recent evolution of employment in Europe. The development of the tertiary sector allowed professions associated with the branches of telecommunications, environmental engineering and informatics to be announced as professions of the future (Kovács & Casaca, 2004).

2.2 *The relevance of contact centres*

According to Bergevin et al. (2010) contact centers started when the telephone became a common household device. Established organizational arrangements were challenged by the growing power of the integration of communications and computer technologies, encouraging the creation of contact centres (Boddy 2000). Contact centres perfectly illustrate the relationship between the 21st century technologies and the 19th century labor conditions, enacting Toyotist flexibility and managerial Taylorism control techniques over the worker (Antunes & Braga, 2011). Contact centres are corporations where the lean production model is employed with flexible and specialized work and lowest investment in workers, subcontracting through temporary work agencies with a high level of job rotation (Kovács, 2002). Though labor flexibility comprehends any area, it is in companies like contact centres that workers are more exploited. Contact centres are a symbol of the modern service economy where services are available all around the clock, being deliverable from almost any spot (Paul & Huws, 2002). According to the neoliberal system, these companies dislocate to foreign countries where costs are lower and labour is cheaper.

The location of contact centres varies widely in different parts of the world, reflecting diversities in technology and culture (Bonnet, 2002).

3 PSYCHOPATHOLOGY OF WORK

3.1 *Customer Service Representative (CSR)*

Contact centre workers are knowledge workers (Drucker, 1959) who perform abstract/immaterial work (Marx, 1990). They organize and redirect the information, performing virtual delivery of products, keeping and managing the relationship between capital and clients of the service sector. CSR's are reduced to mere gestures and tasks that do not allow them to develop their mental capabilities, being considered mere extensions of the computer, leading to a total absence of passion and autonomy at work. The worker is restrained to a cubicle in a room, having access to a shared keyboard, chair, headset and mouse, where there is the absence of proper cleaning and regulated room temperature. The worker has also to be fast, attentive, friendly, emotionally balanced, flexible and needs to be able to deal with unexpected situations; the training is very occasional, mostly virtual and in the attendance seat; the teams are formed by 20 to 30 workers, managed by one supervisor. There are also occasions of moral, gender and racial harassment, where female workers are screamed at. Career progression is very scarce or even null, being the role of supervisor the highest one that a CSR can achieve. In most cases, this leads to the lack of a sense of belonging to the company. According to Brophy (2009), the neotaylorist mode of production provides the workers with low wages, high stress, precarious employment, rigid management, draining emotional labor and a pervasive electronic surveillance. Quoting a female British CSR:

> "It's almost like you were at the sweatshops, sometimes they expected you to do so many calls per day, they expect you to do the donations and the rate that you are supposed to get on them was quite very intense. If you didn't reach the target they would be on your back every hour until you get it." (Female Charity CSR, 55 years of age, November 2015)

Taylor and Bain (1999) have identified the contradiction between the quantitative and qualitative dimensions of the labor process in contact centres, which cannot be resolved, creating an assembly-line in the worker's head. So, we can see that in contact centres there is a high level of stress, emotional exhaustion and job insecurity which leads to the vulnerabilization of the worker. The worker's identity is deconstructed through the non-identification with the role of Customer Service Representative, the impossibility of constructing a professional career leading to the absence of an occupational identity and a feeling of belonging to the company (Huws, 2003). In this sense, a process of subjectification takes place, creating a "cybernetic personality" where the non-equivalence of professional and educational skills, adapted to the tasks and to the wage, result in a process of "status discord", i.e., disqualification (Kosugi, 2008). In this sense, and being considered as the fastest developing form of e-work, contact centres are conceived as "information processing factories" or "modern-day sweatshops", providing images of call handlers chained to cage-like workstations by their headphones (Paul & Huws, 2002:71).

3.2 Cybertized biographical pathways

Twenty biographical interviews were conducted with Portuguese and British men and women, with five female workers and five male workers from both countries. In the Portuguese case there were two trade union representatives, a man and a woman, and in the British we had two men and two women. In this sense, it becomes relevant to interview them not only due to the fact that they are contact centre workers but because they engage in resistance practices to the company's power, searching for the defense of their dignity and personal identity. In the Portuguese case, 45% of the interviewees were between 31 and 39 years old and 50% were between 41 and 56 years old. The majority of workers, about 45%, lived with their parents, 35% had a marital status and 20%, lived alone. So we can see that their degree of social emancipation was very low, mainly as a consequence of the low wages (in the majority of cases, Portuguese contact centre workers receive the minimum wage). About 40% of the workers had a degree, while 35% had a masters' degree. About 80% had a job but 15% were unemployed, after being fired from a contact centre. The majority of them, 70%, worked from two to four years and 25% of the cases worked in a contact centre for more than fifteen years. Around 75% never achieved more than the Customer Service Representative role/status, mainly in the inbound customer service. In trade unionism, we can see that 65% of the interviewees were unionized, revealing a sign of hope for the improvement of their working conditions. Around 90% of the workers felt precarious and 85% alienated, revealing a 70% level of unhappiness concerning the work executed on a daily basis. Nevertheless, 30% mentioned having a certain degree of happiness by answering calls, only in the sense that they were helping other people.

In the British case, five men and women were interviewed and the majority, about 60%, was between 40 and 61 years old. Only three of the ten interviewees lived with their parents, revealing higher levels of social emancipation, engaging more often in several working experiences throughout their professional careers when compared to Portuguese workers. Only 40% of the workers were in a contact centre for more than ten years, whilst 60% had several seasonal jobs during their professional careers. Only 20% of the interviewees were not unionized, while 80% belonged to trade unions, mostly as representative officers. Concerning the workers' subjectivity, around 80% felt precarious and alienated being a CSR and 40% felt happy working there and helping others with information.

In order to proceed to a more profound analysis, two biographies, one from each country, will be presented on the next lines. The first one is of a 42 years old Portuguese woman who was single and lived with her mother in Coimbra. She was born in Angola but went to Coimbra at a very young age. Besides having an academic degree in Sociocultural Animation, she had several precarious jobs such as working at a factory, being a military, a cook, a supermarket worker and a sales promoter. In 1995 she joined a Portuguese Inbound Contact Centre in Coimbra and fifteen years later she became a supervisor. She has also been a union header for the past ten years. Besides being a supervisor she feels no satisfaction in the work she does nor in the wage she receives, being subjected to extra hours and having her contract (besides from being uncertain) always established with the temporary agency, quoting:

> "Before I became a supervisor I was happy because I took it as a professional evolution that doesn't exist. It's just a status. I consider myself a precarious person. Nowadays, I think that the majority of people are precarious but they don't realize that. Being precarious is being disposable. Basically we are machines."

The second biography is represented by a 32 year old Indian male worker who was single and lived alone in a rented flat in London. He was born in India but came to the UK to study at the age of 20. He had no access to a British Passport because, according to the British labor law, he was not eligible for it, nor for the dole. He finished his degree in music and worked in several precarious jobs like a plastics and a dairy factory, worked two years in care house with the elderly, at the National Cross also for two years, was a cab driver, also worked in an outbound charity fund contact centre with a zero hours contract and nowadays works at an events' company. Quoting:

> "I think that working in a contact centre can definitely create a precarious mind, especially if you work fulltime or in sociable hours. You just want to shut down and to be as mindless as possible

because the work isn't sociable at all. You are being treated like a machine and you are just literally acting as a command voice. You are always speaking to someone, always calling. It is always just work, work, work and your breaks are very, very short."

Besides the physical degradation that contact centre workers experience, like deterioration of vision, musculoskeletal disorders connected to work, hearing loss, skin and respiratory diseases, and also back and kidneys problems, psychic suffering is the most recurring situation amongst customer service representatives (Barreto, 2001). Psychosomatic illnesses, like stress and burnout, the permanent irritability and anxiety, chronic fatigue, body ache, insomnia, nausea, anxiety, restlessness, irritability and depression, due to odd working hours and stress, can lead to social phobia, social isolation, fear, apathy, culminating in depression, drug dependency and absenteeism (Roque, 2016), as reported by the Portuguese Call Centre Workers' Trade Union leader. He also mentioned that medical leaves are extremely frequent and when people return to the company they are dismissed or given an inferior working position.

4 PSYCHODYNAMICS OF WORK

4.1 *Emotional labor*

The CSR is forced to do emotional work to attract and retain customers making use of their emotions (Hochschild, 1983). There is the commodification of emotions in the form of customer service, i.e., the "fusion" of their commercialised self with their real, "private sphere" self (Brook, 2009). Women are best suited for contact centre work because of their presumed communication and social abilities, in particular their capacity to smile down the phone (Taylor & Bain, 1999). In this sense, service workers have to sell their feelings in the service industry on a daily basis, rearranging their emotions and their entire personality becomes a commodity of emotions. There is the subordination of the Self to the needs of the company, the ability of disguising what one feels, pretending to feel what one does not feel (surface acting) or the manipulation of internal thoughts and feelings, involving deceiving oneself as much as deceiving others (deep acting) (Hochschild, 1983).

4.2 *Resistance or resignation*

The previous situations can be considered as resistance or resignation strategies. There is the creation of a cyber and mechanized self (identity) by deleting the emotions, so that any aggression becomes

harmless to their ears and stress management is carried out in the shortest time possible to engage and keep customers (Taylor and Bain, 1999; Roque, 2016). Even though the majority of Portuguese workers feel unhappy with the work they do, they prefer to engage in resignation and conformity logics, postponing their dismissal in search for a better job and/or a healthier work environment (Roque, 2010). There is the sense of fear in losing their jobs. In England, people are more prone to have several jobs throughout their working lives. In this sense, some of these workers instead of joining trade unions or engaging in resistance logics prefer to suffer in silence, some of them seeking psychological help or just taking drugs:

> "But there were a number of people in the call centre who took some things like Aderall, you know like drugs to try and focus while they were there and make sales and so that they could make their money and not work there anymore and I think that in a long time it would have damaging side effects you know, taking stimulants in a contact centre." (British Sales Contact Centre Worker, 30 years of age, November 2015)

The lack of recognition is directly connected to "motivation-satisfaction (or desire-pleasure) and works like one of the elements of workers' psychic discharge" in the sense of avoiding suffering and illness, transforming suffering into pleasure (Dejours, 1994). Quoting a female inbound CSR:

> "I used to say that the cleaning lady has a profession, she is an employee of a company and dedicates herself to the cleaning business. A contact centre operator does not have a profession or a career, only works at a contact centre. I think that no one there feels like having a profession but just works at a contact centre. I am nothing, there's no sense of identity, there' isn't a feeling of career." (Portuguese Inbound Contact Centre Worker, 36 years of age, January 2015)

In contact centres, especially in the Portuguese case, the role of the Customer Service Representative is not recognized as a profession. It is a high stress occupation.

5 CONCLUSION

The analysis of the present study allows us to conclude that these individuals, mostly on the Portuguese case, prefer to hold on to something they initially plan as temporary but eventually becomes permanent (Roque, 2010). In this sense, they stop looking for another job because they are convinced that this situation is good, given the grim job market, and it makes no sense for them to look outside their comfort zones. According to Standing (2011), they form a growing army connected to the precariat, a symbol of globalization, electronic life and alienated work. Huws (2009) conceives contact centre workers as the cybertariat who deal with cyber work, embracing insecure jobs which do not allow them to build an occupational identity or desired career. Drawing upon the interviews it can be concluded that the number of qualified women is higher compared to men in both countries. In the British case men and women who lost jobs after a working life, mainly miners, fishing workers and secretaries, found a new occupation. According to the Communication Workers' Union representative, there is a higher rate of women unionized and involved in social movements, since they tend to spend more years working in contact centres. The banking/financial sector displays a higher male rate, while the telecommunications' sector is predominantly female in both countries. Nevertheless, in Britain workers experience zero hours contracts and a higher rate in charity contact centres is registered. In Portugal there are more workers in the inbound customer support service with short-term contracts and seasonal mass dismissals. So, analyzing these biographies it can be seen that contact centre services are very important in the Portuguese and British labour market. Nevertheless, even though it represents an easy entry into the labor market it offers no stability, career opportunities, nor the construction of an occupational identity for the majority of workers.

REFERENCES

Antunes, R.; Braga, R. (orgs) (2009) Infoproletários: degradação real do trabalho virtual. São Paulo: Boitempo.

Barreto, F. (2001) O sofrimento psíquico e o processo de produção no setor de telefonia: tentativa de compreensão de uma atividade com caráter patogênico. (Dissertação de Mestrado). Faculdade de Engenharia de Produção, Universidade Federal de Minas Gerais, Belo Horizonte, MG.

Bergevin, R, Kinder, A., Siegel W. and Simpson, B. (2010) Call Centres for Dummies. Mississauga, Ontario: John Wiley and Sons Canada.

Boddy, D. (2000). "Implementing interorganizational IT systems: lessons from a call centre project". Journal of Information Technology, 15, p. 29–37.

Bonnet, N. (2002). "The establishment of call centres in France: between globalism and local geographical disparateness". NETCOM 16, p. 75–78.

Brook, P. (2009) "The Alienated Heart: Hochschild's "emotional labour" thesis and the anti-capitalist politics of alienation", Capital & Class 98, pp. 7–31.

Brophy, E. (2009) "Resisting Call Centre Work: The Aliant Strike and Convergent Unionism in Canada." Work Organisation, Labour and Globalisation, 3(1): 80–99.

Dejours, C. (1994) Psicodinâmica do trabalho. São Paulo: Atlas.

Dejours, C. (2013) "A sublimação, entre o sofrimento e prazer no trabalho". Revista Portuguesa de Psicanálise, 33(2), 9–28.

Drucker, P. (1959), Landmarks of Tomorrow, Harper, New York, NY.

Fisher, M. 2009. Capitalist Realism: Is There No Alternative? Winchester: Zero Books.

Gurnani, S. (2016) The Financial Crisis in Portugal: Austerity in Perspective, The Libraries Student Research Prize. Paper 9.

Harvey, David (2007) A Brief History of Neoliberalism. Oxford: Oxford University Press.

Hochschild, Arlie R. (1983) The Managed Heart: Commercialization of Human Feeling. Berkeley: University of California Press.

Huws, U. (2003) The Making of a Cybertariat. Virtual Work in a Real World. Monthly Review Press, New York; The Merlin Press, London.

Huws, Ursula (2009) Working at the Interface: Call Centre Labour in a Global Economy. 31 Jul 2009 In: Work Organisation, Labour and Globalisation. 3, 1, p. 1–8.

Kovács, I. (2002) As Metamorfoses do Emprego: Ilusões e Problemas da Sociedade de Informação. Oeiras: Celta Editora.

Kovács, I., Casaca, S. (2004) "Formas flexíveis de trabalho e emprego no sector das tecnologias de informação e comunicação" Atas das comunicações apresentadas ao V Congresso Português de Sociologia, Sociedades Contemporâneas, Reflexividade e Ação, Universidade do Minho, Braga, Portugal.

Kosugi, R. (2008) "Escape from work: freelancing youth and the challenge to corporate Japan". Melbourne: Trans Pacific Press.

Marx, K. (1990) Capital. A Critique of Political Economy. Volume One (trans. Ben Fowkes). Penguin Books, London.

OECD (1998) 21st Century Technologies, Promisses and Puerils of a Dynamic Future. Paris: OECD Publishing.

Paul, J.; Huws, U. (2002) 2nd draft report for the Tosca Project, Analytica Social and Economic Research Ltd, August.

Roque, I. (2010) "As linhas de montagem teleoperacionais no mundo dos call centres". Coimbra: Faculdade de Economia a Universidade de Coimbra, 2010. Dissertação de Mestrado.

Roque, I. (2016), "Riscos Psicossociais em Contact Centres Portugueses", Revista Segurança Comportamental, X, 14–16.

Standing, G. (2011) The Precariat: The new dangerous class". London: Bloomsbury Academic.

Taylor, P. & Bain, P. (1999) "An assembly line in the head: Work and employee relations in the call centre", Industrial Relations Journal, Blackwell Publishers, Oxford, UK.

Turner, B.S. (2006) Vulnerability and Human Rights. University Park, PA: The Pennsylvania State University Press.

Occupational Safety and Hygiene V – Arezes et al. (Eds)
© 2017 Taylor & Francis Group, London, ISBN 978-1-138-05761-6

An integrated risk assessment and management for the rehabilitation of a university department

L.A. Alves, E.G. Gravatá & A.S.C. Fernandes
Department of Civil Engineering—CEFET, RJ, Rio de Janeiro, Brazil

J.S. Nobrega
Department of Industrial Engineering—Veiga de Almeida University, Brazil
DVSST, Federal University of Rio de Janeiro, Brazil

E.G. Vazquez
Department of Construction Engineering—Federal University of Rio de Janeiro, Brazil

ABSTRACT: This research deals with the design of an analysis and risk management. It aims to present the case study of the rehabilitation of the department of construction, from a university. At a first stage the project and the events are described, raising problems and risks, followed by the analysis of major risks, causes and consequences. The second step relates to the construction of a flowchart for the risks and the possible interventions. This work is intended to serve as a model to resolve doubts and prevent future errors found in similar works.

1 INTRODUCTION

1.1 Risk management and analysis

According to OHSAS 18001 (BSI, 2007), risk is a combination of the probability of the occurrence of a dangerous event, explosion of a lesion or disease that can be caused or exposed. Galante (2015) defines risk as an event or condition that, if occurred, may have a positive or negative effect on some of the project objectives. Risk management encompasses the process of identifying, advising, troubleshooting, monitoring and reporting risks (Zeleňáková & Zvijáková 2016). Vose (2008) states that risk analysis aims to help managers better understand the risks and opportunities that can emerge during the project life cycle and to evaluate the options available for their control.

The concept can also be defined as the combination of probability or frequency of the event characterized as a risk, analyzing its impact and severity (Rausand & Utne, 2009 and Rausand and Hoyland, 2011). The process consists on the identification of risks given that the objectives are well defined, involving the project team and stakeholders, and an assessment of environmental, organizational culture and project management plan (Forteza et. al. 2016).

A qualitative risk analysis consists on fitting for high, medium or low probability and high, medium or low impact, as show in Table 1. A qualitative assessment of the probability of a risk event and the impact it would produce can be made by

assigning descriptions to the magnitudes for each event (Vose, 2008). For this type of scoring system, the higher the score, the greater the risk. An event in areas 1 and 2 should have an action plan containing mitigating and preventive measures.

A more detailed analysis can be done by the Preliminary Risk Analysis (PRA). For this analysis, the frequency and consequences of each risk should be studied, to calculate a risk degree that allows the team to work on the highest degrees first. The frequency categories are A for extremely remote, B

Table 1. Qualitative risk analysis.

Impact				
High	1	2		2
Medium	0	1		2
Low	0	0		1
	Low	Medium		High
	Probability			

Table 2. Risk degree classification.

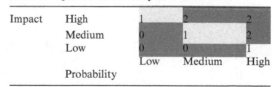

		Frequency				
		A	B	C	D	E
Severity	IV	2	3	4	5	5
	III	1	2	3	4	5
	II	1	1	2	3	4
	I	1	1	1	2	3

for remote, C for improbable, D for probable and E for frequent. The severity of the consequences is classified as I for negligible, II for acceptable, III for critical and IV for catastrophic. After assessing the classifications, the risk degree can be calculated as show in Table 2, where risk degree 1 is classified as negligible, risk degree 2 as minor, risk degree 3 as moderate, risk degree 4 as serious and risk degree 5 as critical (Technical Standard NR-18, 2011).

1.2 Case study

Aiming for readjusted spaces, necessary to better serve the expectations of teachers, students and staff of a department in a University, a rehabilitation was proposed. The scope of the project seeks to meet the demand of new needs for the current view of cooperation between faculty and students and to create a living space for the practice of research and extension, the construction of a multifunctional space following the requirements of a meeting room, monitoring attendance and a work space for scientific research. This aims to improve the facilities of the department and to increase the level of productivity and user satisfaction. Consists on improving facilities with modernization, the use of water saving devices and the provision of a water tank to attend possible disruptions on water supply.

The implementation of the project began in architecture plan definition, use and occupancy of the new space and layout study. Figures 1, 2, 3 and 4 show,

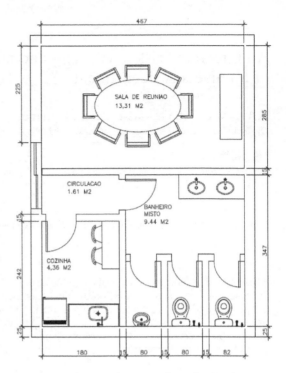

Figure 2. First round review (schematics drawings).

respectively, the original use of the department area, first and second round review, in chronological order, and the as-built result, result of the need to layout change due to lack of an as-built plant for the latter design. Revisions suffered its modifications after brainstorming meetings between faculty, staff and trainees of the department, except for the final revision. The project had as basic premises to transform the space where the bathroom was located into a multifunctional space and the other bathroom in a double use, however individualized for each gender. Moreover, in the kitchen, an improvement of facilities was expected, transferring the deposit a top floor, also designed, and allocating the electric board to its internal part, creating a safe route escape in case of fire by electrical failure, because the original place was blocking the only entrance/exit of the department.

As the most important restrictions for the rehabilitation was the 25 thousand BRL budget limit, it is important to emphasized the need to keep the doors and the space assigned to the kitchen and the available space previously fixed without possibility of expanding external measures.

The next step consists on the selection of contractors. A descriptive spreadsheet of services and specification of materials and equipment to be used was prepared and delivered to three companies. The choice of the firm was due to lowest cost and deadline, as shown in Table 3. In the services

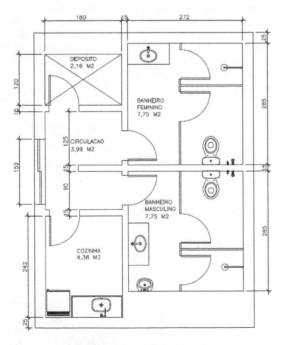

Figure 1. Original use (schematics drawings).

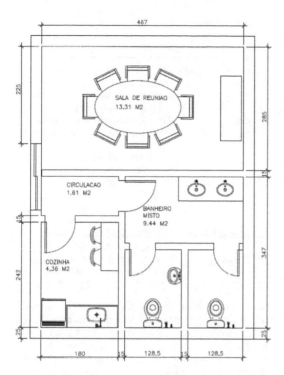

Figure 3. Second round review (schematics drawings).

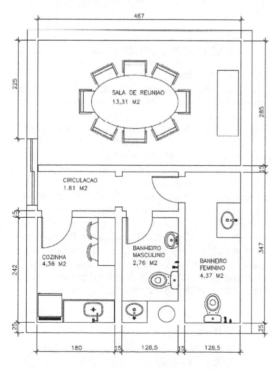

Figure 4. As-built results (schematics drawings).

Table 3. Conditions proposed by the companies.

Company	Budget	Deadline	Payment conditions
Co. 1	BRL 40.560,00	35 dd	50% on signing 25% 15 days 25% final
Co. 2	BRL 19.700,00	15 dd	30% on signing 30% thought 40% final
Co. 3	BRL 17.820,00	15 dd	To be agreed
Co. 4	BRL 20.900,00	30 dd	To be agreed

performed by the company, it is extremely important to point out that the purchase of ceramics and sanitary equipment would be charged for the client, to cheapen the cost of the work.

2 RISK ANALYSIS AND IDENTIFICATION

2.1 Identification of probable risks

The first imminent risk observed on the case study was the choice of layout. The cause is attributed to the missing plants of the original use of the department. As for consequences, project reviews can be cited, as well as delay on deadline and readjustment of decisions taken on the planning phase.

Another risk would be the wrong ceramic material delivered. As this is a task done by the client, the cause could be attributed to a mistake on the order taking from the company responsible for the delivery. This resulted on delay on the deadline and legal issues for the university since the company charged the wrong material, and the university refused to pay.

Executions errors are one of the most recurrent risk on this type of project. The causes vary from the incorrect installation of the toilet, wrong fixation of the mirror on the bathroom, clogged toilet due to sanitary installation errors, wrong connection of the reactor wires. All this causes have as consequence the need to redo the work, extra expenses, and materials waste.

Work incident also occurs in a large scale. For this study, the cause is pointed out as a broken equipment within the field team and dropping of tools from the counter.

Finally, bad quality materials caused by buying from a lower quality supplier as a need to lower the budget has severe consequences, such as leaking, i.e. for hydraulic equipment, and the need to exchange the material, raising the final cost of the project.

2.2 Analysis and intervention proposal

For a complete assessment of the problem, a Preliminary Risk Analysis (PRA) was carried out for the rehabilitation of this case study. This analysis seeks to highlight the consequence event from the probable risks, its causes, future consequences, severity and risk degree. Table 4 describes the respective cause and their Frequencies (F) and Table 5 shows each consequence associated with the cause and their Severity (S). As observed all events are connected by cause and consequence. The risk degree for the PRA is classified as show in Table 2, on the introduction, and presented on Table 6.

After completing the PRA, it can be concluded that the greater risks, classified as serious and critical, could be avoided. It is important to note that

Table 4. Description of causes and frequencies.

Events	Causes	F
Excessive expenses	Execution errors	E
	Incorrect analysis	C
Need for replacement	Wrong handling of material	D
	Work incident	C
	Mistaken deliverables	A
Wrong ceramic tile	Wrong code by supplier	A
	Communication error	B
Changes in layout	Missing original as-built	D
Legal issues	Wrong ceramic tile	A

Table 5. Description of consequences and severity.

Causes	Consequences	S
Execution errors	Over budget	III
Incorrect analysis		
Wrong handling of material	Extended deadline	III
Work incident		
Mistaken deliverables		
Wrong code by supplier	Legal issues for the university	IV
Communication error		
Missing original as-built	Extended deadline	III
Wrong ceramic tile	Inconvenience	II
	Legal budget	I

Table 6. Preliminary risk analysis.

Events	Frequency	Severity	Risk
Excessive expenses	E	III	5
	C		3
Need for replacement	D	III	4
	C		3
	A		1
Wrong ceramic tile	A	IV	2
	B		3
Changes in layout	D	III	4
Legal issues	A	II	1
	I		1

Table 7. Preventive or mitigating measures.

Events	Causes
Excessive expenses	Daily monitoring and training of the team
	Review and analysis of a check-list by more than one person
Need for replacement	Team training, business valuation
	Observe internal standards, regular trainings
	Conference of material or invoice upon receipt of material
Wrong ceramic tile	Pre-analysis of the supplier, require a copy of the order
	Bureaucratize the system by logging all iterations
Changes in layout	Analysis of all systems before designing the project
Legal issues	Conference of material or invoice upon receipt of material

the risks classified as negligible cannot be tolerated (i.e. wrong tile) because their consequences had a catastrophic effect. Thus, Table 7 presents the proposals for preventive or mitigating measures for each event that occurred, as a simplified version of the fault tree build for this analysis. A fault tree starts with the outcome and looks at the ways it could have been arisen. The results presented on Table 7 can be used to avoid these types of risks in future works.

3 FINAL CONSIDERATIONS

Based on the data obtained, it is possible to conclude the importance of an adapted and reliable methodology that allows the professionals of the security area to ensure that the work environment has the risks identified and properly controlled.

The application of the PRA in the evaluation of the development of the plumbing works of the Department of Civil Construction is simple in the identification and prevention of accident risks by determining the categories of risk and the preventive measures before the operational phase. The wrong tile was considered the main risk, since this fact caused several damages for the execution of the project. By adopting the preventive measures proposed in the work, it is expected that the occurrence of the risks highlighted in the work will decrease considerably on further applications.

REFERENCES

Forteza, F.J., Sesé, A., Carretero-Gómez, J.M. 2016. CONSRAT: Construction sites risk assessment tool. *Safety Science* 89: 338–354.

Galante, E.B.F. (1st ed.) 2015. *Princípios de gestão de riscos*. Brazil: Ed. Appris.

Rausand, M. & Hoyland, A. (2nd ed.) 2011. *HAZOP—Hazard and Operability study: System reliability theory; Models, statistical methods and applications*. New York: John Wiley & Sons.

Rausand, M. & Utne, I.B. 2009. Product Safety—Principles and practices in a life cycle perspective. *Safety Science* 47(7): 939–947.

The British Standard Institution, 2007. OHSAS 18001: *Occupational Health and Safety Management*.

Vose, D. (3rd ed.) 2008. *Risk analysis: a quantitative guide*. New York: John Wiley & Sons.

Work Inspection Secretary, 2011. Brazilian Technical Standard NR-18: *Condições em meio ambiente de trabalho na indústria da construção*.

Zeleňáková, M. & Zvijáková, L. 2016. Risk analysis within environmental impact assessment of proposed construction activity. *Environmental Impact Assessment Review* 62: 76–89.

Occupational Safety and Hygiene V – Arezes et al. (Eds)
© 2017 Taylor & Francis Group, London, ISBN 978-1-138-05761-6

Development of innovative clean air suits to increase comfort and simultaneously decrease operating room infections

P. Ribeiro, C. Fernandes, C. Pereira & M.J. Abreu
2C2T-Centre for Textile Science and Technology, Department of Textile Engineering, University of Minho, Azurém, Guimarães, Portugal

ABSTRACT: Comfort is a growing concern in medical textiles and specifically in medical clothing industry and must be considered when it's necessary to develop new products.

Specifically, in the surgical garment, gowns and clean air suits, is important to consider comfort as part of the requirements for any personal protective equipment or medical device, since the products are formulated to meet the primary requirement: the protection, according with the European standard EN 13795:2011+A1:2013: Surgical drapes, gowns and clean air suits, used as medical devices for patients, clinical staff and equipment. General requirements for manufacturers, processors and products, teste methods, performance requirements and performance levels.

Measuring comfort is difficult and this paper shows how to quantify comfort parameters using a thermal manikin for this specific PPE. For this study was made a quantitative analysis on two new developed clean air suits.

The results show that textile materials can provide a good comfort experience for the user and not compromise their function as a barrier material against operating room infections. With this study was possible to ensure a clean air suit with good protection, comfort and ergonomics features with the right balance between protection and comfort.

1 INTRODUCTION

In the present day, society expect much more from clothing than just to satisfy their basic needs (Das and Alagirusamy 2010). Textile Clothes are used as protective equipments for the body and they can protect the user in many different working areas. The healthcare sector uses several protective equipments and most of them are from textiles. Over the last years, a new range of materials and clothes to protect medical professionals and patients has emerging, improving the performance of medical clothing. The use of nonwoven based single-use materials like SMS (spundbond/meltblown/spundbond) ou spunlace that present a good resistance to liquid penetration, bursting and tensile strength or the common use of reusable laminated textile structures that offer good resistance to microorganisms and bacteria with bioactive or antimicrobianic properties, making the clothing system a complete barrier to both professional and patient (Anand *et al.* 2006).

Clean Air Suits (CAS) are considered Class I medical devices according to the definition and classification rules of the Medical Devices Directive 93/42/EEC, amended by 2007/47/EC (Council Directive 93/42/EEC). Clean air suit is defined as a

"suit intended and shown to minimize contamination of the operating wound by the wearer's skin scales carrying infective agents via the operating room air thereby reducing the risk of wound infection" by EN 13795:2011+A1:2013 (CEN 2011). The standard EN 13795 presents general performance requirements concerning properties which require assessment in CAS like resistance to microbial penetration, microbial and particle matter cleanliness, linting, bursting strength and tensile strength (CEN, 2011). As further characteristic of medical clothing, EN 13795 takes in consideration the comfort of the users but only as an informative note, in annex D. According the annex D thermophysiological comfort of a garment will depend on thermal resistance, air permeability, water-vapour resistance and drapeability (CEN 2011).

Inside of the Operating Room (OR), thermal comfort of medical clothing apparel is a very important parameter, since the lack of comfort can lead to thermal stress that influence the physic and psychological conditions of the surgeon, as the ability to maintain constant vigilance and concentration, which the correct surgical procedure is dependent (Hohenstein 2011, Abreu *et al.* 2015). Thermal comfort of the user depends on thermal properties and its adjustment to the environmental conditions in the OR during the surgery, among many other factors like design, size and fabric characteristics (Cho *et al.* 1997). Extremely insulating and low absorbent medical apparel will result in an increase of skin temperature, leading to a greater moisture accumulation between the professional skin and clothing. To overcome this situation, surgical clothing needs to satisfy some requirements; they should be comfortable, breathable, loose fitting, keep the user in cool conditions and allow heat exchange changes between the body and environment (Fanger 1973, Song 2004).

Comfort can be understood as the set of sensations, body responses and reactions to several interactions of clothing. Comfort it's a pleasant state of physiological, psychological and physical harmony between the human and the environment. For health professional can achieve thermal comfort, the mean skin temperature should be between 33°C and 34°C and sweating or chills should not occur.

In this way, the aim of the presented study was to compare thermal properties between two types of clean air suits, reusable and single-use, using a thermal manikin in a controlled environment.

The two clothing systems, reusable or single-use, must answer to primal function: protection. The ideal clothing system must ally protection and comfort all in one to allow the user to perform his job without any risks of thermal stress. To do so, comfort must be extended to all range of properties and characteristics such us sizing and design, textile materials choice and be focus on the user needs (Ribeiro *et al.* 2014).

The thermal manikin is an instrument for measuring the thermal insulation (I_T) of garments and clothing ensemble. It is considered to be one of the most useful tools for evaluating thermal comfort of overall clothing systems, and the most studies about thermal clothing insulation occur with the use of thermal manikins. In comparison to other methods for measuring thermal properties of clothing, thermal manikin studies allow to investigate fully assembled clothing in the way these garments are supposed to be used (multi-dimensional), however without any influence of subjective interpretation of human testing or simply physical testing of the materials (bi-dimensional) (Vidrago & Abreu 2012).

To calculate I_T there are three calculation methods – *serial, parallel and global*. As referred by Cho *et al.* (1997) and Oliveira (2008) the *global* method is the only one that is valid, despite of the common use of the *serial* and *parallel* calculation methods because it is the one less susceptible to thermal significant variations. In this study we use the three methods to compare two different clean air suits and see the significant differences and variations between calculation methods (Huang 2012).

2 MATERIALS AND EXPERIMENTAL METHODS

2.1 Clothing system

Disposable and reusable clean air suits were used in this study, Table 1. The reusable clean air suit (CAS-1) is made of a laminated structure with polyester (to the outside) and cotton (to the inside) with cotton cuffs. As for the single-use (CAS-2) we used the common SMS material, once it is the main material used for single-use clothing system. Both clothing systems were one piece design.

2.2 Thermal manikin and adiabatic chamber

Thermal manikins provide a good estimate of the total dry heat loss from the body and the distribution of heat flow over the body surface. In a standard environment, these measures can be used to describe the thermal characteristics of clothing (Oliveira 2008).

The thermal insulation of clothing ensemble was measured using a thermal manikin, Figure 1, with controlled skin surface temperature that simulates the wear. This thermal manikin, called "Maria", has a woman's body; its size and configurations are similar to an adult woman. It is divided in 20 thermally independent sections and only sense dry heat transfer. Thermal manikin, positioned 0.1 m from

Table 1. Tested clothing systems.

Ref/Type CAS	Design	Composition (%)	Weight (g/m²)	Thickness (mm)
CAS-1- Reusable	Jumpsuit	50%PES/ 50%CO	139	0.28
CAS-2- Single-use	Jumpsuit	100% SMS	35	0.23

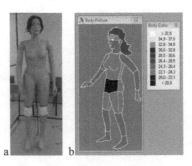

Figure 1. Thermal manikin: photo of nude thermal manikin used on tests (a) and software body picture of thermal manikin and color legend of body segments temperature (b).

Table 2. Operating Theatre conditions tested in the adiabatic chamber.

	Conditions in the adiabatic chamber			
Temperature (°C)	20°C	20°C	24°C	24°C
Relative Humidity (%)	40%Rh	60%Rh	40%Rh	60%Rh

the floor, was kept standing with their legs and arms held in vertical position without any motion. The skin temperature was set, and during the test period maintained at $33 \pm 0.1°C$. Thermal manikin operates with the constant skin temperature regulation mode. The constant skin temperature regulation mode was used (Cho *et al.* 1997).

The tests were conducted in an adiabatic chamber where ambient conditions characteristic of an operating theatre were simulated. The climatic chamber can achieve temperatures around 15°C to 35°C and relative humidity around 35% to 85%. The test conditions performed simulate the temperature and relative humidity inside an Operating Room, according to Table 2:

Data acquisition started after achieving stationary conditions and did not take more than 60 min. During the acquisition period, the heat flux and the skin temperature of each body part record every minute.

The three thermal insulation calculation methods—the serial, the global and the parallel—were considered to calculate thermal insulation (I_T) in order to compare the different test methods and the different clothing systems behaviour. Calculation of the Thermal insulation (I_T) according to global, parallel and serial method.

3 RESULTS

The results presented were related to two clothing systems: a reusable clean air suit, made of 50%PES/50%CO and a single-use clean air suit, made of SMS material. Table 3, 4, 5 and 6 shows the values of thermal insulation of the thermal manikin nude (reference) and dressed with each suit, for the four OR simulations of temperature (°C) and relative humidity (%).

According with Nilsson (1997) referred by Oliveira (2008) *serial* method present the highest values of thermal insulation once the differences between *serial and parallel method* increase with non-uniform insulation. So for this we can expect higher values for I_T when used *serial* method. The difference between these two methods can amount to 30% or more. Holmér (2001) also referred by Oliveira (2008) said that *serial* method overestimates the insulation values.

As we can see, for the temperature of 20°C and relative humidity between 40% and 60% the clothing system with the best thermal insulation values is CAS-1, that present the lowest values of thermal insulation in all three calculationg methods. As expected the values of thermal insulation for reusable CAS-1 are lower. This clothing system is made of reusable textile materials and the heat flux is well conducted through this materials that is translated to the sensation of freshness/coolness by the user. Higher the value of thermal insulation, more the user will heat and sweat, so lower values of thermal insulation are always best. Although CAS-1 presented the best results and the lowest

Table 3. Results of thermal insulation for all the conditions T = 20°C and 60% Relative Humidity, and the three calculation methods.

	T 20°C Hr 60%		
	Global method	Serial method	Parallel method
Nude			
Ia [m2.°C/W]	0,092	0,106	0,092
Ia [m2.K/W]	0,096	0,106	0,092
Clo	0,620	0,680	0,592
CAS 1			
It [m2.°C/W]	0,130	0,153	0,129
It [m2.K/W]	0,136	0,153	0,129
Clo	0,878	0,985	0,830
CAS 2			
It [m2.°C/W]	0,153	0,189	0,152
It [m2.K/W]	0,160	0,189	0,152
Clo	1,031	1,221	0,983

Table 4. Results of thermal insulation for all the conditions T = 20°C and 40% Relative Humidity, and the three calculation methods.

	T 20°C Hr 40%		
	Global method	Serial method	Parallel method
Nude			
Ia [m2.°C/W]	0,111	0,124	0,111
Ia [m2.K/W]	0,116	0,124	0,111
Clo	0,749	0,797	0,715
CAS 1			
It [m2.°C/W]	0,170	0,199	0,169
It [m2.K/W]	0,177	0,199	0,169
Clo	1,140	1,286	1,088
CAS 2			
It [m2.°C/W]	0,183	0,220	0,182
It [m2.K/W]	0,191	0,220	0,182
Clo	1,233	1,422	1,174

Table 5. Results of thermal insulation for all the conditions T = 24°C and 60% Relative Humidity, and the three calculation methods.

	T 24°C Hr 60%		
	Global method	Serial method	Parallel method
Nude			
Ia [m2.°C/W]	0,110	0,126	0,109
Ia [m2.K/W]	0,117	0,126	0,109
Clo	0,753	0,811	0,705
CAS 1			
It [m2.°C/W]	0,158	0,190	0,156
It [m2.K/W]	0,167	0,190	0,156
Clo	1,078	1,226	1,009
CAS 2			
It [m2.°C/W]	0,174	0,218	0,173
It [m2.K/W]	0,185	0,218	0,173
Clo	1,192	1,409	1,117

Table 6. Results of thermal insulation for all the conditions T = 24°C and 40% Relative Humidity, and the three calculation methods.

	T 24°C Hr 40%		
	Global method	Serial method	Parallel method
Nude			
Ia [m2.°C/W]	0,089	0,097	0,089
Ia [m2.K/W]	0,095	0,968	0,089
Clo	0,612	0,625	0,573
CAS 1			
It [m2.°C/W]	0,161	0,182	0,160
It [m2.K/W]	0,172	0,182	0,160
Clo	1,110	1,171	1,034
CAS 2			
It [m2.°C/W]	0,178	0,202	0,177
It [m2.K/W]	0,188	0,202	0,177
Clo	1,213	1,305	1,141

I_T values, both clothing systems present very close values which means that in an OR situation they would serve to the user as well.

4 CONCLUSIONS

The aim of this study was to compare thermal insulation of two clean air suits with the same design made of different materials: reusable (woven fabric) and single-use (nonwoven fabric) using a thermal manikin using different calculation methods. Thermal manikins provide a good estimate of the total dry heat loss from the body and the distribution of heat flow over the body surface. Both measures

are used to describe the thermal characteristics of clothing. The results obtained with the three calculation methods of the thermal insulation showed that values of the *serial* method were always the highest, as expected and the *parallel* method always present the lowest values. *Global* method is less susceptible to significant variations and it should be used for calculating thermal insulation of clothing for all manikin control modes, especially for thermal comfort regulation mode (Cho *et al.* 1997).

The difference between wearing CAS-1 or CAS-2 are not very significant, however CAS-1 presented better results of I_T and best performance due to heat loss, meaning that CAS-1 materials have a good influence to the thermal behaviour of the clothing system.

We can conclude that several factors can influence thermal insulation of clothing such as the fabric characteristics, drapeability, sensitive touch, fit, sizing and others. So, comfort is a requirement that have to be consider when talking about clothing systems to operate as medical devices (non-active) or personal protective equipment in order to get to the user a good comfort experience.

ACKNOWLEDGMENTS

This work is supported by Portuguese National Funding, through FCT—*Fundação para a Ciência e a Tecnologia*, on the framework of project UID/CTM/00264/2013.

REFERENCES

Abreu, I.M., Ribeiro P., and Abreu, M.J. 2015. "Comparison of different medical clothing used in operating rooms (OR's)—the importance of thermal comfort at work". In Arezes et al. (Ed.) Occupational Safety and Hygiene III, London, Taylor & Francis Group, 2015, pp.47. ISBN 978-1-138-02765-7.

Abreu, M.J., Abreu, I., Ribeiro, P. 2014. Thermo-physiological behavior of Single Use Scrub Suits using a Thermal Manikin. EGEMEDITEX—2nd International Congress on Healthcare and Medical Textiles. Izmir, Turkey.

Anand et al. 2006. Medical textiles and biomaterials for healthcare. Woodhead Publishing. ISBN: 978-1-85573-683-2.

CEN 2011 "EN 13795:2011+A1:2013, Surgical drapes, gowns and clean air suits, used as medical devices for patients, clinical staff and equipment. General requirements for manufacturers, processors and products, test methods, performance requirements and performance levels".

Cho, J.-S., Tanabe, S.-I. & Cho, G. 1997. Thermal Comfort Properties of Cotton and Nonwovens Surgical Gowns with Dual Functional Finish. Applied Human Science, 16, 87–95.

Council Directive 93/42/EEC of the European Parliament Council of 14 June 1993 concerning medical devices. Official Journal L 169, 12.7.1993, pp. 1.

Das, A; Alagirusamy, R. 2010. Science in Clothing Comfort. Woodhead Publishing. ISBN: 978-18-456-9789-1

Fanger, P. O. 1973. Assessment of man's thermal comfort in practice. British Journal of Industrial Medicine, 30, 313–324.

Hohenstein Institute, Disposable or reusable clothing in the operating theatre?—The brain prefers reusable. Operating Theatre Journal, September 2011, Issue 252, p.4.

Huang, J. 2012. Theoretical analysis of three methods for calculating thermal insulation of clothing from thermal manikin. The Annals of Occupational Hygiene. 56 (6): 728–35. DOI: 10.1093/annhyg/mer118

ISO 9920:2007, Ergonomics of the thermal environment—Estimation of thermal insulation and water vapour resistance of a clothing ensemble.

Oliveira, V., Gaspar, A.R., Quintela, D.A. 2008. Measurements of Clothing Insulation with a Thermal Manikin Operating Under the Thermal Comfort Regulation Mode—Comparative Analysis of the Calculation Methods. European Journal of Applied Physiology, Vol. 104 (4), pp. 679–688. DOI: 10.1007/s00421-008-0824-5

Ribeiro, P, Abreu, I. Abreu, M.J. 2014. Design and Development of Innovative Non Active Medical Devices: Clean Air Suits. Conference 7th International Textiles, Clothing & Design Conference—Magic World of Textiles. Croatia. ISSN: 1847-7275.

Song, G. 2011. Improving comfort in clothing. Woodhead Publishing. ISBN: 978-1-84569-539-2.

Vidrago, C. Abreu, M.J. Optimization of the thermal comfort behaviour of bed linen using different softening formulation in Proceedings of the TRS2012—The 41st Textile Research Symposium, 2012, University of Minho. ISBN: 978-972-8063-67-2.

Occupational Safety and Hygiene V – Arezes et al. (Eds)
© *2017 Taylor & Francis Group, London, ISBN 978-1-138-05761-6*

Identify minimum training requirements for the chainsaw operators

M.C. Pardo Ferreira, J.C. Rubio Romero, A. Lopez-Arquillos & F.C. Galindo Reyes
University of Málaga, Málaga, Spain

ABSTRACT: The chainsaw can be considered one of the most dangerous work equipment in the field of occupational health and safety. In Spain, currently, there is no minimum specific and compulsory training for the chainsaw operators. In order to detect the current deficiencies in the spanish educational materials aimed or likely to be used as support in the training of this workers, the contents of the main Spanish education and training materials available from Andalusia were analyzed. The only contents that appear in all analyzes performed with a frequency higher than 50% are Preventive Measures, PPE and Chainsaw Risks. On the other hand, some of the less mentioned were contents focus on work in height, stacking, pruning and Environmental Safety. In conclusion, it is necessary to continue completing the contents of the documents available for the training and information of workers unifying them in a single manual. Emphasizing that it is important to establish minimum mandatory training for chainsaw operators, filling the legal gap that currently exists in Spain.

1 INTRODUCTION

The chainsaw can be considered one of the most dangerous work equipment in the field of occupational health and safety (Robb et al, 2014; Enez et al, 2014). The chainsaw is used in different sectors of activity such as agricultural, forestry, construction or gardening. Although it is a working equipment especially used in the forestry sector, so most of the studies about accidents with chainsaws focused on this sector. In Spain, the accident rate in the forestry is significantly higher than in other sectors considered to be of high risk, revealing that forest work is the most dangerous, with loggers being the most exposed (Robb et al, 2014; Albizu et al., 2013; Lefort et al., 2003; Bell, 2002; Peters, 1991). In semi-mechanized logging operations, a great majority of accidents are usually caused by chainsaws, above all in the logging phase (Nieuwenhuis and Lyons, 2002; Neely and Wilhelmson, 2006; Shaffer and Milburn, 1999; Axelsson, 1998; Peters, 1991; Albizu et al., 2010).

Risk perception is of vital importance for workers, as it causes them to expose themselves unduly to unacceptable risks, and so is an important element in training and preventive measures. (Albizu et al., 2013). Thus, it is essential that workers receive good training that includes as much the necessary theoretical knowledge as practical exercises on work techniques.

In Spain, currently, there is no minimum specific and compulsory training for the chainsaw workers. Only the basic training in safety and health defined by Law 31/1995 on the prevention of occupational hazards is required. It is true that there is voluntary training, but because of the magnitude of the risk to which workers are exposed, it is necessary to regulate the minimum training accurately. For a significant reduction of serious and fatal accidents, a mandatory training should be extended to all chainsaws users (Cividino et al., 2015).

The first step is identify minimum training requirements for the chainsaw operators To continue determining the state of the materials and documents available for the training of these workers by analyzing what the contents are currently included in the training materials of the chainsaw workers and which should be included but do not appear.

Focusing on the theoretical knowledge that the chainsaw workers must receive, as main sources of materials and resources available to implement a preventive programme, a great quantity of resources and materials are available for workers and companies in Internet. (Albizu et al., 2013). Therefore, in this study have been studied all the materials available in Andalusia through the internet for the training of chainsaw workers, both in paper and in digital format.

The aim of this paper is to determinate what contents are currently included in the spanish educational materials such as guides, guidelines, manuals and documents aimed or likely to be used as support in the training of workers using the chainsaw as work equipment in any sector, in order to detect the current deficiencies. The results obtained will help the development of new common standards both on a European and international level.

This study is developed by the University of Malaga as partner, along with other entities from various regions throughout Europe, of the project VET-SAFETY proyect 2014–2017, "Vocational education & training standards in agriculture, forestry & environmental safety at heights". The project is supported by the EU Erasmus + programme & ABA International.

2 METHODOLOGY

Initially, a search via internet was performed in Spain in order to find the documents available in Spanish and focused on the use of chainsaws in the workplace. Subsequently, the documents found were classified according to their main characteristics. Then the content of each was analyzed according to the items in a checklist, previously designed. Finally, the results were analyzed.

The search was conducted using Google and Google Books as search engine. The main manufacturing chainsaw site (Husqvarna, Stihl, ECHO...) and bookshop sites were also consulted. The keywords used were: chainsaw, felling, forestry work, working at heights, winches, pruning, chainsaw safety and health, occupational risk prevention chainsaw, chainsaw PRL, work with chainsaw and chainsaw workers. Combinations of these keywords were also used. The search was conducted from Malaga in Andalusia (Spain) during the month of June 2016. Some documents found were not available online and were purchased as textbooks.

Thus, 61 documents written in spanish language were found. Four groups or types of documents were established according to the characteristics of the documents:

– Textbooks (T) on paper, large size, between 92 and 268 pages, and oriented training specialties or professional qualification linked to the chainsaw.
– Guides and Manuals (GM) focused on training and information for workers and produced by government agencies, manufacturers, organizations or associations. Many of these documents serve to support teaching.
– Fact Sheets (FS) characterized by its short length, between 2 and 15 pages, and focused on developing specific aspects.
– Technical Instructions (TI) developed within the field of systems management risk prevention in enterprises.

The information sources used to design the checklist, with the main contents to be analyzed in the documents, were three: Chainsaw qualifications levels, promoted by ABA for harmonisation of skills qualifications worldwide and with the support of the European Union; Professional qualification in forestry activities, promoted by Ministry of Education, Culture and Sport of the Government of Spain (R.D. 108/2008) and PhD Thesis of Albizu (2012) on the diagnostic safety the forest harvesting operations in Castilla y León (Spain). The checklist developed included the main contents about chainsaws that must be known to workers, which should be present for the purposes of training and/or legislative requirements about chainsaw work. This paper only include a list but each of which has been clearly defined in the study, in order to facilitate their search in the materials analyzed later. The checklist had 46 items in total, 20 main categories and 26 subcategories. The subcategories include aspects that have been specially analyzed within each category, either by its special impact on safety or their importance. The main categories were: main features, maintenance, refuelling, displacement, start method, work planning, considerations prior to felling, working techniques, tractel or winches, chainsaw risks, special risk situations, ergonomics, Personal Protective Equipment (PPE), preventive measures, emergency actions, first aid, safety documentation, legislation, environmental safety and working at height.

3 RESULTS

A total of 61 documents were found. The analysis (analysis 1) showed that 69% of the documents found in spanish were published in Spain and 67% focused directly on the chainsaw or on activities where the chainsaw was the main work equipment (Felling, delimbing, crosscutting ...). The majority were focused on the forestry sector (69%) and the rest in the agricultural sector (8%), gardening (5%) or were not linked to any sector and focused on the chainsaw in general (13%), the prevention of forest fires (3%) or maintenance operations of the chainsaw (2%). In addition, more than 93% presented illustrations or photographs, since only 4 of the 61 documents did not include any illustrations or photographs. However, few more than 16% of the documents, ie 10 of the 61 documents, was focused on the work in height with chainsaws. The documents in spanish developed outside Spain come from countries such as the US, Uruguay, Argentina, Chile and other countries in Latin America.

In a second analysis (analysis 2), these documents were eliminated in order to study only the documents produced in Spain with the purpose of determining the situation in Spain, which was part of the collaboration of the University of Malaga in the mentioned European project. Thus, in analysis 2, a total of 42 documents were analyzed. It was obtained 62% focused on the chainsaw or on activities in which the chainsaw was the main work equipment. Also, 79% were focused on the forestry sector, the others were about agricultural sector (9%), gardening (5%) or were not linked to any sector and focused on the chainsaw in general (5%) or the prevention of forest fires (2%). In addition more than 90% presented illustrations or photographs. However, just over 16% of the documents focused on work in height with chainsaws, as in the previous analysis.

Once the documents were characterized and classified, the checklist was applied to each one individually, the results obtained in each analysis are shown in Table 1. In the first column, the categories and subcategories are included, that is, all the contents of the checklist that have been reviewed in the documents, assigning to each row a content. Thus, the others columns shows the percentage of documents, with respect to the total of documents analyzed in each analysis, which include the content assigned to that row.

It is observed that in analysis 1 and 2 the contents that less appear in the documents are the Techniques of rescue in height. Second would be the Good Environmental Practices and the Systems of subjection. In the analysis of the documents elaborated in Spain, two other less-mentioned contents would also appear in the second place: Legislation and Systems of Techniques of Climbing.

It is highlighted that if the less mentioned contents are selected in the documents, which have a frequency of occurrence of less than 12%, the same contents are obtained in both analyzes but with different order, except for the Health and Safety Documentation that only appears in the analysis of all documents (analysis 1). Thus, the category of Environmental Safety with all its subcategories (Oils and Waste, Good Environmental Practices, Legislation) is often not included in the documents available in spanish on the chainsaw. Neither do Pruning or Stacking Techniques usually appear. With respect to work in height the less mentioned are the contents referring to Rescue Techniques, Clamping Systems and Climbing Techniques.

If the contents that appear more frequently are observed, again very similar results are obtained in both analyzes for the contents with a frequency greater than 35%. The three contents that are most frequently present in the documents are the Preventive Measures, the PPE and the Risks of the chainsaw. Within this latter category the most mentioned are Rebounds and Noises and vibrations. In addition, there are frequently indications on the work positions to be adopted, ergonomics, and guidelines on chainsaw working techniques. Among the techniques the most mentioned are the techniques of felling or cutting. Also noteworthy are frequent references to the first aid kit in the documents, however, its main category, First Aid, appears less in the documents, since many do not include techniques

Table 1. Results of the all analysis of the contents that includes each group of document according to the designed checklist about minimum training requirements which must be included in documents.

Contents	Analysis 1 (61 documents)	Analysis 2 (42 documents)	Analysis 3 (32 documents)
	%	%	%
1. Main features	31	24	16
1.1. Security elements	34	33	31
2. Maintenance	39	33	25
2.1. Chainsaw	26	17	13
2.2. Guide bar	25	17	9
2.3. Lubrication	25	19	13
2.4. Sharpness	23	14	6
3. Refueling	36	33	31
4. Displacement	30	26	31
5. Start method	36	31	34
6. Work planning	18	16	3
7. Pre-use operational	36	29	28
7.1. Escape routes	36	24	19
8. Working techniques	49	45	47
8.1. Felling or cutting	45	41	34
8.2. Delimbing	31	26	22
8.3. Crosscutting	26	19	13
8.4. Stacked	7	10	6
8.5. Pruning	12	12	6
9. Tractel or winches	18	26	16
10. Chainsaw Risks	53	57	59
10.1. Rebounds	39	41	41
10.2. Recoil	34	33	38
10.3. Noise and vibration	38	38	38
10.4. Pulls	23	21	25
11. Special risk situations	31	24	22
12. Ergonomics	45	45	44
13. PPE	71	71	66
14. Preventive Measures	77	76	72
14.1. Rests	28	29	34
15. Emergency actions	20	24	19
16. First Aid	20	24	9
16.1. Botiquin Aid	38	45	28
17. Safety Documentation	12	17	3
18. Legislation	20	29	22
19. Environmental safety	10	12	0
19.1. Oils and waste	8	12	0
19.2. Good environmental practices	7	10	0
19.3. Environmental regulations	10	10	3
20. Works at height	25	26	19
20.1. Ladder	23	24	22
20.2. Elevated platforms	15	19	9
20.3. Clamping systems	8	10	3
20.4. Climbing Techniques	8	10	6
20.5. Pruning techniques in height	13	17	9
20.6. Rescue Techniques	3	5	0

Table 2. Contents that appear more and less frequently in all analyzes performed.

Contents that appear more frequently	Contents that appear less frequently
14. Preventive Measures	20.6. Rescue Techniques
13. PPE	19.2. Good environmental practices
10. Chainsaw Risks	19. Environmental safety
8. Working techniques	19.1. Oils and waste
12. Ergonomics	19.3. Environmental regulations
10.1. Rebounds	20.3. Clamping systems
10.3. Noise and vibration	20.4. Climbing Techniques
16.1. Botiquin Aid	8.4. Stacked
8.1. Felling or cutting	8.5. Pruning
2. Maintenance	17. Safety Documentation

of first aid. Finally, in the analysis of all the documents it was found that there are also frequently general contents about the Maintenance that the workers must realize to the chainsaw.

A significant fact to take into account in the analysis is that, in Spain, on February 28, 2012, the formative specialties of Chainsaw Operator, Cork Oak Operator and Restorer Pruner were replaced by the formative specialty of Forest Harnesses, linked to a professionalism certificate and their corresponding professional qualification. This change in the voluntary training related to the chainsaw is reflected in the textbooks analyzed, since the new professional qualification of forestry included six training modules and two of them was focused on the chainsaw. Thus, textbooks after February 2012, eight of the ten analyzed, was organized according to these new training contents.

For this reason, from analysis 2 a new analysis was developed without including any of the textbooks in the total of documents to be analyzed (analysis 3), resulting 32 documents.

Deleting the textbooks found that there were contents that were not included in any of the documents analyzed. These contents were within the category of Environmental Safety, specifically the Oils and wastes generated in the activity, which if not handled properly can be pollutants for the environment, and Good environmental practices, which should be perform during the activity in order to minimize the impact on the environment. Neither Techniques of rescue in height were mentioned, whose ignorance can have repercussions in the emergencies in works in height. All these contents already appeared in the less frequent contents, which were obtained in the initial analysis.

In analysis 3 can also be observed that the contents that appeared more frequently in the documents were the same ones that were obtained initially: Preventive Measures, Chainsaw risks, PPE, Working techniques and Ergonomics.

4 CONCLUSIONS

The only contents that appear in all analyzes performed with a frequency higher than 50% are Preventive Measures, PPE and Chainsaw Risks. This is a positive point, as they are basic elements of safety and health. However, it is necessary to use them as a starting point to continue improving on the contents that are usually included in the documents for the training of workers who use chainsaw. Since, on the other hand, there are contents included in the checklist of minimum requirements for chainsaw operators that often do not appear or appear infrequently. These contents focus on work in height and especially on techniques of rescue in height, clamping systems or climbing Techniques.

In the case of work techniques, it is emphasized that general concepts such as felling or cutting are often included, but there are other aspects such as stacking and pruning that are not usually included and which are important as well. In the case of pruning is one of the operations with chainsaws most common in sectors such as agriculture or gardening and it is essential that the workers be trained adequately on the work techniques that must follow in this operation.

Neither aspects pertaining to environmental safety appear normally and these should be included, as workers should know how to do their work minimizing environmental impact.

Another aspect that should be properly considered and included is knowledge about safety documentation such as prevention plan, safety and health plan or emergency plan. This is important because these documents contain all the information for proper safety and health management. For example, in the emergency plan, workers can know the evacuation routes that are essential in an emergency.

Finally, it emphasizes that new professional qualification and its content seem to have improved the voluntary training that existed previously. However, it is necessary to unify in a single document all the contents that a worker, who uses the chainsaw in his work, must know and apply. This document would serve as a reference for the use of that work equipment.

In conclusion, it is necessary to continue completing the contents of the documents available for the training and information of workers. To perform the work safely and minimize the risks and the possible impact on the health and safety of workers, it is essential that they know all the techniques and funda-mental aspects of the work, which they perform. Thus, the development of international standards and harmonization of skills qualifications worldwide would favor the improvement of the safety and health of the chainsaw workers, as well as the reduction of their accidents. Emphasizing that it is important to establish minimum mandatory training for chainsaw operators, filling the legal gap that currently exists in Spain and other countries.

ACKNOWLEDGEMENTS

Authors would like to acknowledge the contributions to this research from: Proyecto Erasmus+, "Vocational Education & Training Standards in Agriculture, Forestry & Environmental Safety at Heights: VET-SAFETY".

REFERENCES

Albizu Urionabarrenetxea, P. (2012). Diagnóstico de la seguridad en los aprovechamientos forestales a partir de registros empresariales, bases de datos oficiales y muestreos de campo: propuestas de actuación (Doctoral dissertation, ETSI Montes, UPM).

Axelsson SA, 1998. The mechanisation of logging in Sweden and its effect on occupational safety and health. J Forest Eng 9(2): 25–31.

Bell JL, 2002. Changes in logging injury rates associated with use of feller-bunchers in West Virginia. J Saf Res 33:463–471.

Cividino, S. R., Colantoni, A., Vello, M., Dell'Antonia, D., Malev, O., & Gubiani, R. (2015). Risk Analysis of Agricultural, Forestry and Green Maintenance Working Sites.

Enez, K., Topbas, M., & Acar, H. H. (2014). An evaluation of the occupational accidents among logging workers within the boundaries of Trabzon Forestry Directorate, Turkey. International Journal of Industrial Ergonomics, 44(5), 621–628.

European Chainsaw Certificate. (2014). Carnet Europeo de Motoserrista. ECS1: Mantenimiento de la motosierra y técnicas de tronzado.

European Chainsaw Certificate. (2014). Carnet Europeo de Motoserrista. ECS2: Técnicas básicas de tala de árboles (árboles pequeños).

European Chainsaw Certificate. (2014). Carnet Europeo de Motoserrista. ECS3: Técnicas avanzadas de tala de árboles (árboles medianos y grandes).

European Chainsaw Certificate. (2015). European Chainsaw Standars. ECS4: Windblow & Damaged Tree Techniques.

Lefort AJ, de Hoop CP, Pine JC, 2003. Characteristics of injuries in the logging industry of Louisiana, USA: 1986 to 1998. Int J For Eng 14: 75–89.

Neely G, Wilhelmson E, 2006.Self-reported incidents, accidents, and use of protective gear among small-scale forestry workers in Sweden. Safety Science 44: 723–732.

Nieuwenhuis M, Lyons M, 2002. Health and Safety Issues and Perception of Forest Harvesting Contractors in Ireland. Int J For Eng 13(2): 69–76.

Peters P, 1991. Chainsaw Felling Fatal Accidents. Transactions of the ASAE 6: 2600–2608.

Real Decreto 108/2008, de 1 de febrero, por el que se complementa el Catálogo nacional de cualificaciones profesionales, mediante el establecimiento de ocho cualificaciones profesionales de la Familia profesional agraria.

Resolución de baja de la Dirección General de Políticas Activas de Empleo por la que se dan de baja especialidades formativas en el Fichero de Empleo Público Estatal.

Robb, W., & Cocking, J. (2014). Review of European chainsaw fatalities, accidents and trends. Arboricultural Journal: The International Journal of Urban Forestry, 36(2), 103–126.

Shaffer R, Milburn J, 1999. Injuries on Feller-Buncher/ Grapple Skidder Logging Operations in the South-Eastern United States. For Prod J 49(7–8): 24–26.

Occupational Safety and Hygiene V – Arezes et al. (Eds)
© *2017 Taylor & Francis Group, London, ISBN 978-1-138-05761-6*

Risks and preventive measures in building demolitions

Cristina M. Reis & S. Paula
School of Sciences and Technology, UTAD, Vila Real, Portugal

J. Ferreira & C. Oliveira
Escola Superior de Tecnologia e Gestão, Instituto Politécnico de Viana do Castelo, Portugal

ABSTRACT: This research work is based on the act of demolishing since, with the increasing rehabilitation of buildings, this is has become an increasingly frequent situation. Law-Decree n. 273/2003, 29th October, which refers to construction security, has been taken into consideration. This work main purpose is to evaluate which risks and prevention measures should be taken in a building demolition work. A demolition work, situ in the city of Vila Real, Northern Portugal, was considered as a case study. The methodology followed in this research took into account the risks inherent to the demolition of the building under study. After characterization of the risks involved in this type of work, preventive measures were presented in order to avoid them. It was also found that there could be risks inherent in the inhalation of some type of chemical, present in the dust particles resulting from the demolition, which were also analyzed. From the laboratory analysis of the dust particles it was found that they were not very harmful. However this study warns against the fact that, often, no additional precaution measure is taken into account and materials such as asbestos and radon may be present after demolition. These require other preventive measures to avoid putting the lives of workers at risk.

1 INTRODUCTION

Demolition is the inverse act of building, or deliberately throwing down some construction in order to give another destination to it. In order to plan the demolition of a building, there must be some reasons for its occurrence. For instance, in old buildings: by regulatory impositions due to the existence of long-term deformations, anomalies and durability of materials, structural reinforcement, among others … and in recent constructions due to change in design, incompatibility between projects of different specialties, errors and deficiencies of design and/or construction, etc.

Demolitions are high-risk jobs, since it is the opposite of building and workers are more accustomed to building than demolishing. These should be carried out by companies specialized in this type of work because they are high risk operations. In the demolition process, initially a detailed study of the structure to be demolished, the existing infrastructures in place, of possible neighboring buildings and the state of degradation of these, the procedures of execution and inspection, possible existence of dangerous products/materials (as asbestos, pre-stressed concrete) should be made. It is important to to protect the public service elements that may be affected, so that none of these jobs endangers the safety of workers, neighboring

buildings and the public circulating in the vicinity of the area to be demolished. The plan should include the provision of collective and/or individual safeguards to be implemented during demolition.

In demolitions it is also necessary to assess the strength and stability of each part of the construction (in particular of the pavements), in order to be able to predict the type of demolition plan to be adopted, without jeopardizing the safety of workers and neighboring buildings. After this stage (site recognition), the choice of the demolition process is made.

This research work has as main objective to know the risks and preventive measures to be taken into account in a demolition work studied. It will also be analyzed if the materials from the demolition are harmful to the health of the workers, through the characterization of its chemical and morphological composition.

2 METHODOLOGY

In order to better portray the risks and preventive measures that must be considered in demolition works, it was decided to apply the study to a concrete case. The work under analysis is a building, ground floor and 1st floor, located in Vila Real, Portugal. This building was demolished, having only maintained the facades, as it may be seen in Figure 1. It is intended to build in it the headquarters of the Latin Club and the Artists House.

The demolition processes used in this work were manual and mechanical demolition. Being that the mechanics was made with a pneumatic-hammer and with rotating.

2.1 Main risks

The main risks to which workers or third parties may be affected in this work are:

- Particles or materials projections;
- Materials falls;
- Objects falls;
- Building collapse;
- Fall in height;
- Electrocution or electrification;
- Cutting or drilling;
- Crushing;
- Damage to neighboring structures;
- Uncontrolled destruction of all or part of the construction;
- Others.

2.2 Preventive measures

In this sub-chapter it will presented some of the preventive measures used to avoid the risks and, in turn, the work accidents. The demolition, in order to be carried out, must follow technical and safety standards to avoid the occurrence of undesired events, such as damage to the property, health or life risk of the persons who work in it or who come into contact. As preventive measures in demolition works, it is important to carry out a preliminary study to collect information on the nature of the structure and stability of its elements, as well as the survey of all existing infrastructures.

- Secure the perimeter of security to neighboring works. Elements of resistance scale out and their influence in the demolition process must be made.
- Avoid constraints on adjacent roads and networks
- Previous study of the hazardous materials and products that may exist in the work.
- Choose the most suitable demolition process for this purpose
- Use of appropriate personal protective equipment, such as a helmet, goggles, protective masks, steel toe boots, ear protectors and rubber or mechanical protection gloves.
- Make a careful analysis of the execution project or the final screens, if they exist, in order to study how it was built
- Analyze where the supporting and supported elements are.
- First demolishing the supported elements.
- Demolish from top to bottom.

3 CASE STUDY

3.1 Preventive measures of this work

The building under study is located in Vila Real, already in a state of considerable degradation and whose facade had to be preserved. So the first thing that was done before beginning to demolish was the containment of the facade, as may be seen in Figure 2.

Before the work began, a protective net was used around the building, in the zone of passage of the operators, to prevent the fall of debris as may be observed in Figure 3.

Another important preventive measure was the fence, so that people outside the work do not enter into it. In addition it is a place where there are several dwellings and a place of people passage. In order to reinforce security, opaque hard-wearing plates were used in the fence as shown in Figure 4.

In almost all the demolition, the process used was manual demolition, using the work of man and with manual tools, where first the removal of window frames and tiles was made, following the partition walls and only at the end the structural part was demolished.

Figure 1. Image of the facade with respective containment.

Figure 2. Facade of building with materials containment.

Figure 3. Use of safety net to prevent materials fall.

Figure 4. Fencing of the work.

Finally, mechanical demolition was performed by means of a mini-rotary with a pneumatic hammer connected to its arm, and manual pneumatic hammers for the dismantling of stone, in order to lower the elevation of the threshold. A crane truck was used to remove the removed materials.

During the demolition, a scaffold was also used (Figure 5) to carry out the work safely since workers would have a proper work platform and body guards to prevent the fall of people and materials.

3.2 Demolition materials characterization

When the workers were dismantling the stone to lower the quota threshold, with the aid of manual pneumatic drills, the material was being crushed and reduced to dust. Given that its particles were so fine, it was considered appropriate to ascertain whether they may be harmful to the workers' health and possibly to new users of the building, in case some of these materials would remain. This dust was collected on site, at several distinct points with the aid of a shovel. The sample was then observed using Scanning Electronic Microscope, SEM, in order to obtain morphological and chemical composition. The samples were mounted on aluminium supports for placing the microscope. In Figure 6 one of obtained SEM photomicrographs may be observed.

The sample appears to be more crumbled, with finer powders. Powders are practically all of angular shape. The sample varies greatly with regard to the degree of heterogeneity, since the powder shape and size are very dissimilar.

Figure 5. Use of scaffold to prevent people falling.

Figure 6. SEM photomicrograph of a collected sample.

605

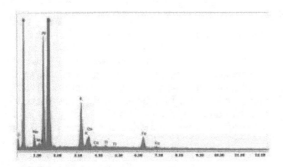

Figure 7. Samples EDS analysis.

Figure 7 shows the results of Energy Dispersive Spectrometry, EDS, analysis. Through the spectrum observations, and WDS analysis, it may be seen that:

– The sample main constituent is oxygen (O_2), silicon (Si) and aluminium (Al), although the aluminium may be misleading, because aluminium supports were used as a support for samples placing,
– It also comprises potassium (K), sodium (Na), iron (Fe) and Calcium (Ca), although in lower percentages;
– Magnesium (Mg) and Titanium (Ti) were also present, but in very low percentages, nearly zero.

4   CONCLUSIONS

As predicted the act of demolishing presents numerous risks that need to be minimized in order to avoid possible accidents. These works involve several preventive measures. The demolition materials must also be analyzed because it is often unknown what types of materials are being demolished, and the material may be harmful to the health of workers or futures users. The materials used were those used at the time of construction, partition walls in "tabique", wooden slabs and beams, and ceramic coatings. However in this work was only possible to analyze the dismantling of stone. This disassembles served to lower the quota threshold. The remaining materials had already been removed and disposed in landfill, according to the legislation indication.

Since one does not know the chemical composition of the stone (used to remove inert) prior to the construction work, it was not possible to correlate the chemical composition before the beginning of the work, with the inert current analysis results.

Apparently chemical elements that have emerged in stone demolished are not harmful and so they seem to have no negative interference in the workers' health. Often these hazard materials are present people involved with the demolition process may even Apparently chemical elements that have emerged in stone demolished this work does not have harmful materials that might interfere the health of workers. Because often these harmful materials exist and may appear in the demolition process it is important to make studies like this, or others of similar nature, to realize in what extent they can affect workers or residents of these rehabilitated buildings.

REFERENCES

Almeida, J.B. (1997). A Microscopia por Varrimento de Sensor (SPM). Boletim da Sociedade Portuguesa de Química, 64, pp. 50–53.
Araújo, J. (2002). Universidade de Évora no ano letivo 2002/2003. <http://materiais.dbio.uevora.pt/jaraujo/biocel/metecnicas.htm>. [Consultado em outubro de 2014).
Faria A. (2006). Reabilitação de coberturas em madeira de edifícios históricos. A intervenção no património; práticas de conservação e reabilitação, Université de Porto, École d'Ingénieur, 2006.
Law—Decree no. 273/2003, 29/10
Mecatrônica Fácil (2006). Microscópio eletrônico de varredura como ferramenta para micro e nanotecnologia, 31. [Em Linha]. Disponível em <http://www.mecatronicaatual.com.br/educacao/1632-microscpio-eletrnico-de-varredura-como-ferramenta-para-micro-e-nanotecnologia>. [Consultado em novembro de 2014].   <http://reabilitacaodeedificios.dashofer.pt/?s=modulos&v=capitulo&c=12136>[Consultado   em Dezembro de 2014].
Reis Christina, Oliveira Carlos, Mieiro Márcio, Marwen Bouasker And Muzahim Al-Mukhtar (2014). Restauration et réhabilitation d'un monument historique du XVIe siècle. 32èmes Rencontres Universitaires de Génie Civil. Université de Orléans, École Polytecnique de Orléans, França Junho de 2014.
Silva Paula, Madureira Reis Cristina, Oliveira Carlos (2013). Analysis of emerging risks: The nanoparticles. 39º IAHS The International Association for Housing Science. Prooceding of the thyrtis-ninth word congress on Housing science. Volume 2. ISBN 9788864930206. Pág. 245–250. Milão, Itália, 17 a 20 de Setembro de 2013.
Silva, P., Reis, C.M., Ferreira, J., Oliveira, C. 2015. "Nanoparticles characterization and potential hazard assessments", Occupational Safety and Hygiene III—Selected Extended and Revised Contributions from the International Symposium on Safety and Hygiene, III: 453–456. ISBN: 978-1-138-02765-7. London: CRC Press Taylor & Francis Group.

Occupational Safety and Hygiene V – Arezes et al. (Eds)
© 2017 Taylor & Francis Group, London, ISBN 978-1-138-05761-6

Author index